MARINE STRUCTURES ENGINEERING: SPECIALIZED APPLICATIONS

MARINE STRUCTURES ENGINEERING: SPECIALIZED APPLICATIONS

Gregory P. Tsinker, Ph.D., P.E.

SPRINGER-SCIENCE+BUSINESS MEDIA, B.V.

Cover photo courtesy of: Port Autonome de Nantes/St. Nazaire
(Photo: A. Bouquel)
Cover design: Edgar Blakeney

Originally published by Chapman & Hall in 1995
Softcover reprint of the hardcover 1st edition 1995

1 2 3 4 5 6 7 8 9 10 XXX 01 00 99 98 97 96 95

Library of Congress Cataloging-in-Publication Data

Tsinker, Gregory P.
Marine structures engineering : specialized applications / Gregory P. Tsinker.
p. cm.
Includes bibliographical references and index.
ISBN 978-1-4613-5865-7 ISBN 978-1-4615-2081-8 (eBook)
DOI 10.1007/978-1-4615-2081-8
1. Harbors—Design and construction. 2. Ocean engineering.
I. Title.
TC205.T75 1994
627'.2—dc20 94-30074
CIP

British Library Cataloguing in Publication Data available

**To dear mother
with all my love**

Contents

Preface

During my long career as a practicing waterfront consultant, considerable progress has occurred in the field of design and construction of port and navigation related marine structures. Progress in port design, and, in particular design of waterfront structures, has been strongly influenced by dramatic changes in vessel sizes and in modes of modern terminal operation. Multipurpose ports have been replaced by more specialized terminals, which result in dramatic effects on both the design of berth structures and layout of the terminal. Furthermore, marine structures for various purposes have been developed using new design and construction principles, and operation of these structures have been significantly improved by the introduction of new and better fendering systems, and efficient mooring accessories. New and better structural materials have also been introduced. For example, modern concrete technology now enables an engineer to use durable high-strength concrete, highly resistant to deterioration in harsh marine environments. New and better repair procedures and rehabilitation techniques for port structures have also been introduced.

Progress in development of new marine structures and modernization of existing structures was based on advances in analytical design methods as well as on results of numerous scale-model tests and field studies conducted all over the world. Today, marine structure design is a unique discipline in the field of civil engineering that is based on the use of highly advanced methods of soil foundation investigation and thorough understanding of the principles of soil interaction in the marine environment.

During recent years sophisticated computational procedures and mathematical models have been developed and used for design of various marine structures. It must be stressed, however, that in many cases the diverse and complex geology at various port locations results in a wide variety of geotechnical environments. Such conditions require a careful approach to the selection of structure type and use of the appropriate design method, which should not necessarily be highly sophisticated. It is a misconception that the sophisticated computer analyses, with their greater accuracy, will automatically lead to better design. Despite the highly sophisticated analytical methods available today, the marine structural designer must be aware that the design is not merely a stress analysis process. The use of computers has not diminished the value of some hand calculations. In fact, many questions about marine structure engineering are best answered with simple, often empirically based, but practical formulas.

Computers have revolutionized the process of structural engineering and greatly increased productivity of engineering consulting firms. Computer aided analyses are of great help when used in the proper context, for example when modeling of the structure is correct, the real boundary conditions are taken into account, and most of all when the output is examined and interpreted by an experienced engineer.

This work has been conceived as a two part treatise in which I have attempted to provide marine structure designers with state-of-the-art information and common sense guidelines to the design of basic types of marine structures associated with port activities.

The total material is presented in two separate volumes. This volume *Marine Structures Engineering: Specialized Applications* contains seven chapters. It covers important subjects such as evaluation of capacity of the in-service marine structures and methods of their remediation and maintenance (Chapter 1), construction and operation of the marine structures in cold regions with in-depth discussion on ice mechanical properties and ice loads acting on marine structures (Chapter 2), design and construction of marine structures used for construction and repair of vessels (Chapter 3), design of anchored offshore moorings and floating breakwaters (Chapters 4 and 5), design and construction of marinas (small craft harbors) (Chapter 6) and design and construction of the marine structures used in navigable waterways for protecting the bridge piers from ship impact (Chapter 7).

The second volume, which is tentatively titled "Design of Marine Structures," will be published in 1996. The material included in this volume will contain approximately ten chapters and provide the state-of-the-art information on design of miscellaneous port elements, i.e. breakwaters, port layout, access channel, port entrance and other; structural materials used in a port/harbor construction; and design construction and modernization of gravity type quay walls, sheet pile bulkheads, piled structures and dolphins of miscellaneous designs.

In both books each chapter includes a considerable list of relevant references intended to help the interested reader to study the subject in depth.

I have drawn from about 40 years of my own experience as a marine engineer involved with research and all practical aspects of structural design, construction, and project management. Also, worldwide experience has been examined and the best was included in this work. Subsequently, acknowledgements of material used in this book are given the appropriate places in the text and figures. I wish to extend my deepest gratitude to all the publishers, authors, and organizations from whom material for this work has been drawn.

This volume is not a one-man job. I am deeply indebted to many experienced individuals who have contributed materials and comments to this project. In attempting to make this most helpful and useful I have drawn from sources including the knowledge and experience of my former colleagues at Acres International Limited, who assisted in a variety of ways: Dr. A. Mee contributed information on redundant pier system design included in Chapter 6, Mr. T. Lavender made a number of useful comments on Chapter 2; special gratitude goes to Mr. R. Tanner for his review of a number of chapters and searching criticism and valuable recommendations; Messrs I. Shaw, D. Daw and D. Protulipac dedicated a good deal of their personal time to editing the text. These individuals, of course, are in no way responsible for faults that may remain.

I would also like to express my gratitude to Ms. M. Mitnick (Moffatt & Nicol, Baltimore office) who assisted with editing the final version of Chapters 4 and 5.

I wish to record my deep gratitude and to acknowledge enjoyable cooperation with Professor B. Mazurkiewicz (Gdansk Polytechnical Institute, Poland) who contributed Chapter 3 and Mr. J. Headland (vice president, Moffatt & Nichol Engineers) who contributed Chapters 4 and 5.

As usual my good friend Mr. R. Glusman has helped a lot with preparation of illustrations. My deep gratitude is extended to Ms. L. Dunn, who typed the manuscript and dealt ably with many difficulties in the process. Special thanks go to Sumitomo Rubber Industries, Ltd. for sponsorship of this project. I wish to extend my deepest gratitude to Messrs. M. Shiono and Ed Patrick of Sumitomo Canada for support given this project. I also wish to thank my publisher, Chapman & Hall, for cooperation and patience. I hope that a useful contribution to the profession has been made.

Any project of this magnitude requires many months and hundreds of hours of hard work in the evenings, during weekends, and on vacations. It cannot be successfully completed without tremendous understanding and support from many people, especially one's family. This is why I am especially grateful to my wife Nora for her commitment to leave me alone, undisturbed for many hundreds of hours and for her valuable assistance during preparation of this text. I also extend my gratitude to my grandson Daniel who helped with preparation of a subject index to this work.

GREGORY P. TSINKER

Introduction

In the past 30 to 50 years the worldwide seaborne tonnage has increased dramatically. This has created a strong demand for construction of new and modernization of existing ports and terminals. This process has been strongly influenced by dramatic increases in vessel sizes and in modes of modern terminal operation. During this period of time tankers for crude oil and ore carriers have reached 500 000 and 350 000 DWT respectively, and the largest container vessels now in use are 50 000 to 60 000 DWT. In most cases traditional multipurpose ports have been replaced by specialized terminals equipped with specialized technology to handle a certain specific kind of cargo, e.g., crude oil, bulk material, containers and other.

Introduction of new large vessels with side thrusters and bulbous flared bows created an almost unique condition for damaging and undermining of the traditional dock structures. Furthermore, propeller and side thruster induced scour can seriously compromise the integrity of structures constructed upon erodible foundations. In most cases the new cargo handling and hauling equipment that is required for servicing of larger vessels is much heavier than that previously used. Essentially, all aforementioned complicates the process of modernization of the existing ports or terminals.

To meet todays requirements the existing marine facilities must be carefully evaluated. Many port-related marine structures such as piers, wharves, and others presently in use were built in the early post-World War II years. Naturally, these older structures were designed for smaller vessels and less sophisicated and lighter cargo handling and hauling technology. Thus, evaluation of the real in-service structural capacity of older docks, as well as methods for their remediation and upgrading, as required to service the new seaborne traffic, is of a paramount importance. Not accidentally the latter has become almost a permanent topic of discussion of different specialty conferences. This subject is broadly discussed in Chapter 1 of this book.

In the past 20 to 30 years significant activities related to the offshore oil and gas exploration fields, in regions with cold climates, have led to a surge in research and engineering practices in the field of navigation in ice and construction of port facilities in the ice laden waters. Some of the common problems featured in the ice affected waterways and harbors are as follows: accelerated formation of ice in ship traffic lines, blockage of the open traffic lines by wind-driven ice features, damage of navigation aids by ice, ice growth and its adherence to marine structures, ice formation in berthing area, ice buildup on marine structures due to water spray, and others. The principal solution to the above problems is usually referred to as 'ice control' and/or 'ice management'. Naturally, the behavior and cost of marine structures operating in the ice-affected waters are greatly influenced by ice global or local loads. Hence, ice-structure interaction must be properly understood by the structure designer.

Additionally, in cold regions the marine structure designers usually have to deal with frozen soils, which are referred to as permafrost. Frozen grounds are highly complex in their interaction with structures contructed upon them. Frost/thaw related soil heave and settlement are the principal causes of unacceptable deformation of structures constructed in cold regions. Essentially, this phenomenon must be given proper attention during design process.

Last but not least the cold temperature can greatly affect performance of dock fender systems that are manufactured with rubber components. This must be carefully evaluated and treated with caution. Experience indicates that cold temperature effects upon rubber fenders, as well as effects of the ice buildup on and around fender units, in a great many cases is overlooked by the designers.

The aforementioned problems associated with navigation in ice affected waters and port design and operation in cold regions are discussed in Chapter 2 of this book.

Shipyards engaged with construction of a new and/or repair of vessels in-service include miscellaneous marine structures that allow complete dry access to a vessel for maintenance, overhaul, and repairs or for new construction and launching. These structures are shiplifts, marine railways, shipways and dry docks; they usually provide a means for transferring vessels to and from dry land as required. There are various types of these structures including those that lift the vessel from the water or launch them either by buoyancy force, vertical lift by means of miscellaneous vertical lift systems, or by use of marine railways.

Traditional and new approaches to design, construction and operation of shipyard related marine structures are discussed in Chapter 3 contributed to this book by Professor B. Mazurkiewicz (Gdansk Polytechnical Institute, Poland).

Offshore moorings play a very important role in port operation. They provide temporary or permanent berthing for vessels and for a wide range of port related floating marine structures such as piers, dry docks and other. Vessels are often moored temporarily at offshore moorings while waiting for their turn to be serviced at the berth. Tankers can be moored at offshore moorings during oil transfer operation.

Depending on their proposed use the offshore moorings can be of either single-point or multi-points designs. The offshore moorings design procedure is presented in Chapter 4. This chapter was written by Mr. J. R. Headland who is vice president for Moffat & Nichol and is current manager of this company office in Baltimore, Maryland. The material presented in Chapter 4 is based in part on the U.S. Navy's Design Manual "Offshore Moorings, Basic Criteria and Planning Guidelines" DM 26.5 which has been prepared by Mr. Headland, formerly the Technical Consultant for the U.S. Navy. It should be pointed out that the guidelines as are given in DM 26.5 are based on the static analyses approach to design of offshore moorings. In Chapter 4 Mr. Headland extended the standard design procedure further by developing methods for dynamic analysis of offshore moorings.

Mr. J. Headland is also responsible for contributing Chapter 5, "Floating Breakwaters". These breakwaters are gaining popularity for protecting water areas exposed to moderate waves, e.g., $H_w \leq 2\,m$, where H_w = wave height. They are particularly attractive where deep water makes construction of the conventional breakwaters cost prohibitive. In this chapter the guidelines for both static and dynamic analyses of floating breakwaters are presented.

Detailed discussions on design and construction of marinas that are sometimes referred to as small-craft marinas, or small-craft harbors is presented in Chapter 6. Construction of millions of small craft or pleasure boats worldwide during 1980s resulted in a strong demand for marinas. Many small-craft harbors have been built during the last two decades, and it is obvious that many more will be built in the future. In recent years, however, some marina developments have been curtailed by environmental groups and local residents concerned with the effects of large-scale marinas on the quality of environment, such as water pollution, visual pollution, noise, destruction of wildlife habitat and others. More stringent regulations concerning environmental impact of construction and operation of small craft marinas have been implemented. Chapter 6 emphasizes importance of the environmental design process for successful completion of marina design and construction. The material presented in this chapter enables the marina developer to design efficient, cost-effective facility. Part of the material included in this chpater, namely subsection 6.5.7.2 "Redundant Pier Systems" is contributed by Dr. A. Mee (Acres International Ltd.).

In the past three decades inland waterways navigation has been marked by catastrophic ship bridge pier collisions, resulting in a heavy structural damage and loss of human lives. In the period 1965 to 1994, an average of one catastrophic accident per year involving bridge collision by vessels have been recorded worldwide. More than half of these bridge collisions occurred in the United States. The bridge collision phenomenon has been subject of numerous publications and discussions which took place elsewhere in the world. The basic conclusions and recommendations drawn from the existing literature on a subject matter are presented in Chapter 7. This chapter covers state-of-the-art approach and provides basic guidelines to the design of marine structures installed in navigable channels for protection of bridges from collision with vessels.

Contributors

Mr. John R. Headland
Vice President
Moffatt & Nichol Engineers
2809 Boston Street
Suite 6
Baltimore, Maryland 21224
USA

Professor B. K. Mazurkiewicz
Technical University of Gdansk
ul Majakowskiego 11/12
PL 80-952 Gdansk
Poland

Dr. All. Mee
Project Engineer for Acres International Ltd.
5259 Dorchester Road
Niagara Falls, Ontario
Canada L2E 6W1

1

The Dock-in-Service: Evaluation of Load Carrying Capacity, Repair, Rehabilitation

1.1 INTRODUCTION

Provided that port marine structures are properly maintained, the design service life of such structures is usually considered to be 35 to 50 years.

During their lifetimes, however, these structures are susceptible to damage from time to time since original construction. In addition, constructed facilities deteriorate due to the effects of the natural environment, excessive use beyond that intended in the original design, aging of materials, and general obsolescence.

Different structural materials are affected in various ways by the marine environment. The most notable effects include corrosion of metals, degradation of concrete, attack on timber by marine organisms, fouling, and encrustation of virtually all materials.

Many port-related structures such as piers, wharves, etc., presently in service in North America and throughout the world were built before World War II or in the early post war years, and are still functional. Naturally, the older docks presently in service were designed for smaller vessels and less sophisticated cargo handling and hauling equipment than those presently in use.

To meet today's service requirements, the load-carrying capacity of these facilities must be carefully reviewed. It is generally accepted that dockside cargo handling and hauling equipment loadings are now much heavier, and when installed upon older dock structures may lead to a multitude of problems such as settlement, distortion, sliding, and local and general instability.

Dramatic changes in ship size and shape, particularly the introduction of bulbous and flared bows, created almost unique situations for piercing holes in sheet piling, damaging piles and even solid wall constructions at an inaccessible low level, and in some instances even causing ship–crane or ship–train collision. Furthermore, larger ships with increased draft, and therefore less keel and propeller clearance and with side thrusters, approaching the berth without tug assistance can cause considerable scouring effects, especially if the structure is built upon an erodible foundation. This may have serious consequences on the quay wall base, which may be undermined. Substantial scour in foundation material could also have a significant impact on sheet-pile bulkheads, because increase in sheet-pile wall height dramatically increases the bending moment and shear forces in sheet piles, and also increases pullout forces in the bulkhead's anchoring system.

It is worth noting that in the case of sheet-pile bulkheads, this process is irreversible, because as seen later, simple restoration of the sea floor to its original level alone cannot restore stresses in a bulkhead structure to its prescour condition.

In port operation, heavy static and dynamic loads imposed on dock structures by regular portal and container cranes, heavy mobile cranes, a variety of heavy miscellaneous unit loads such as loading arms, and a great variety of mobile cargo handling and hauling equipment are normally considered.

All of the above, coupled with natural wear and tear conditions, led to the problem of the assessment of safety of older dock operations. Thus evaluation of the real in-service capacity, of older dock rehabilitation and modernization, as well as preventive maintenance, and maintenance management of port-related structures, in the past two decades became almost a permanent topic of discussion at different specialty conferences.

In order to evaluate present capacity of a dock-in-service, or to select the most economical repair/rehabilitation scheme, the designer must have knowledge of the dock's present physical condition.

Basic information on inspection and maintenance of maritime structures exposed to material degradation is given in a report by the Permanent International Association of Navigation Congresses (1990) and Nucci and Connors (1990).

In the following sections of this chapter the reader will find discussions on general causes of deterioration of basic structural materials such as concrete, steel, and timber in the marine environment, and recommendations on cost-effective approaches to the inspection and evaluation of a structure's present capacity.

This chapter also deals with the problems associated with effects of propeller and side thruster jets on an in-service structure and with remedial work required for structure repair and rehabilitation.

1.2 DETERIORATION OF STRUCTURAL MATERIALS IN A MARINE ENVIRONMENT

Marine structures are subjected to various deteriorating agents throughout their service lives. The degree of deterioration depends on properties of ambient water, for example, whether it is sea- or freshwater, its seasonal fluctuation, tide range, climatic conditions, and chemical composition of construction materials.

Concrete, steel, and timber are the primary materials used in the construction of marine structures (as they are on land); therefore most of the discussion of marine environment effects is devoted to them.

1.2.1 The Marine Environment

There are five distinct zones that concern marine structure deterioration (Figure 1–1): atmospheric, splash zone, tidal zone, zone of continuous immersion, and seabed zone.

The atmospheric zone tends always to contain some amount of salt, which increases the rate of atmospheric corrosion of marine structure metal or deterioration of concrete over that of land structure materials. In timber structural components above the splash zone, freshwater may collect and stagnate, initiating rot. The splash zone constitutes an area from the high water

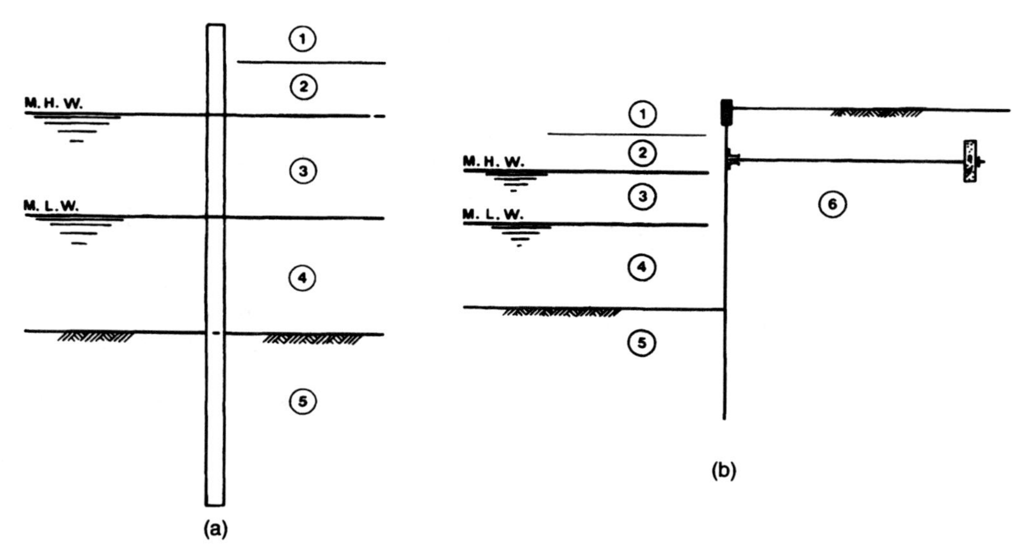

Figure 1–1. Zones of structure deterioration in the marine environment. (a) pile, (b) sheet pile and backfill buried parts of a structure. 1—Atmospheric zone, 2—splash zone, 3—tidal zone, 4—zone of continuous immersion, 5—seabed zone, 6—backfill zone.

level to the upper levels attained by spray. This zone also is subjected to intermittent wetting and drying as waves run up or break on the structure. The tidal zone is the usual range between high and low water levels which is periodically immersed. Below low tide level to the seabed the structure is continuously immersed and this is typically a zone of moderate to light attack on steel and concrete, but not wood.

Below the mudline the structure's elements are buried in the foundation material. There all materials are relatively well protected as the lack of oxygen prohibits oxidation and the existence of most organisms.

Generally seawater can be thought of as a solution containing a great number of elements in different proportions. Elements in solutions are found as ions, many of which combine and precipitate as salts on evaporation of the water. Concentrations of elements in solution are typically given in parts per million (ppm) by weight, for example, equivalent to milligrams per liter (mg/L). The concentration of elements in water or total amount of dissolved solids in water is defined as salinity. Salinity can also be considered as the amount of dissolved solids in parts per thousand (‰) by weight in a water sample. The mean salinity of seawater is approximately 35‰ and varies typically from 31‰ to 38‰. It can be considerably less for nearshore coastal waters because of the freshwater effect, especially in the vicinity of large river mouths.

Seawater contains 11 major elements, of which chlorine (1.9×10^7 μg/L) in the form of chloride ions (Cl^-) and sodium (1.08×10^7 μg/L) in the form of ions (Na^+) are by far the most abundant. The remaining nine elements in ionic forms are, in order of decreasing quantity, sulfur as sulfate (SO_4^{2-}), magnesium (Mg^{2+}), calcium (Ca^{2+}), potassium (K^+), carbon as bicarbonate (HCO_3^-), bromide (Br^-), strontium (Sr^+), boron (H_3BO_3), and fluoride (F^-). The ratios of these major ions do not vary significantly over the range of salinities normally observed in the seas. Therefore by measurement of concentrations of any particular ion the water salinity can be obtained by multiplying the value of ion concentration by its constant.

The chlorinity, or chloride ion concentration, is typically used for determination of water salinity (S)

$$S\ (‰) = 0.03 + 1.805\ \text{chlorinity}\ (‰). \qquad (1\text{–}1)$$

The density of seawater (approximately 1025 kg/m^3) is not very different from that of freshwater (1000 kg/m^3). Water density typically increases with decreasing temperature and with increasing salinity.

Generally, for most engineering purposes, water is considered an incompressible fluid. The error due to neglecting compressibility in most marine engineering calculations is negligible.

The heat capacity of water is among the highest of all liquids and solids. This property of water explains the absence of extreme temperatures in the oceans and helps to maintain uniform body temperature. Although the thermal conductivity of water is the highest of all liquids, it is still relatively low compared to that of many other substances.

The thermal conductivity of seawater is slightly less than that of freshwater. Chemical reactions in seawater are governed by such chemical factors as acidity and alkalinity as indicated by the pH, the oxidation–reduction potential, and solidifying ability.

$$\text{pH} = -\log_{10}(\text{H}^+) \qquad (1\text{–}2)$$

where H^+ = concentration of hydrogen ion.

The H^+ concentration of pure water is 10^{-7}, hence its pH = 7. The pH scale ranges from 0 to 14; a pH less than 7 is considered acidic and greater than 7 is basic. Seawater is slightly alkaline; the pH of seawater typically varies from 8.0 to 8.4. Knowledge of local pH is important to the marine structure designer because it helps him or her to assess the degree of biological activity and potential corrosive effects on structures.

Marine fouling is part of the marine environment. This is an accumulation of various marine growths and animal organisms on immersed and partly immersed surfaces of the marine structures. Fouling may have a dramatic effect on structure performance; it may increase drag due to increased surface roughness and projected area. It may also increase mass and gravity loads. In addition fouling may increase the rate of corrosion of metal parts of the structure due to destruction of protective coatings, and oxygen concentration cell effects in areas where certain organisms, for example, barnacles, exist.

The fouling creates difficulties in inspection and maintenance of the structure due to the presence at times of quite thick and tenacious growths. For example, in warm waters hard mussel fouling may be 300 mm or more thick. This will significantly increase gravity load and drag force upon a structure.

Experimental studies by Blumberg and Rigg (1961) showed that the drag coefficient may be increased due to fouling in the order of 50% to 70% for light to medium fouling on a smooth 900-mm diameter cylinder. Even in cold water the fouling effect may be substantial. Heaf (1979) presented data on fouling of a North Sea oil platform that indicated the possibility of a significant increase in both overall and local loads attributed to a marine growth about 50 mm thick.

Heaf correctly indicated that the consequences of fouling can be even more severe considering the effect on fatigue life compared to the design wave loading.

Fouling density varies from 1.0 t/m^3 (soft fouling) to 1.3 to 1.4 t/m^3 (t = metric tonnes). Generaly information on types and extent of fouling for many coastal areas of the world can be found in The US Navy World Atlas of Coastal Biological Fouling (1970).

The marine structure designer must recognize the fouling phenomenon when either designing a new structure or evaluating a load-carrying capacity of in-service structure.

Some metals are reported to exhibit a natural resistance to fouling (Efird, 1975) and some special paints are reported to be very effective in prevention of fouling (Offshore Engineering, 1984).

1.2.2 Concrete Deterioration in the Marine Environment

A marine environment is one of the most severe and aggressive on earth. Despite general evidence of long-term durability of concrete structures in a marine environment and their generally outstanding performance in the ocean, in some cases serious concrete deterioration has been reported (Feld, 1968; Gerwick, 1975; Gilbride et al., 1988; Mehta, 1980, 1988; Khanna et al., 1988; Toyama and Ishii, 1990; and many others).

The general cause of deterioration was noted to be cracking of concrete, resulting in corrosion of embedded reinforcing steel. Corrosion occurs in the presence of oxygen, moisture, and an electrolyte. Salt intensifies the electrolytic properties of concrete, thereby creating a corrosion cell, resulting in corrosion of steel.

Deterioration of concrete is usually limited to certain parts of the structure. From the above standpoint, the face of a structure exposed to a seawater environment can be divided into five characteristic zones, as previously stated (Figure 1–1):

- The uppermost zone exposed to the atmosphere.
- The splash zone, which is exposed to atmosphere wave impact, wind-driven spray, frost action, solar radiation, and drying winds which produce rapid evaporation.
- The tidal zone, which is exposed to repeated cycles of wetting and drying as well as to freezing and thawing in a saturated condition and to the impact of waves. These waves can contain pieces of floating ice, gravel, or sand, which cause an abrasive action.
- The zone that is continuously submerged in seawater.
- The seabed zone.

The upper zone is generally characterized by concrete cracking due to corrosion of reinforcement.

The splash and tidal zones are usually the most vulnerable and are characterized by cracking and spalling of concrete due to wetting and drying, frost action, corrosion of reinforcement, chemical decomposition of hydration products of cement, and dynamic effects of wave action.

The part that is constantly submerged in seawater is basically vulnerable to loss of concrete strength due to a chemical reaction between seawater and hydration products of cement. Corrosion of reinforcement steel bars is seldom a problem here.

Because of the symptoms of concrete distress and deterioration may be caused by more than one mechanism acting on the concrete, it is necessary to have an understanding of the basic underlying causes of damage and deterioration.

Deterioration of concrete is an extremely complex subject. It would be too simplistic to suggest that it is possible to identify a specific, single cause of deterioration for every symptom detected during an evaluation of a structure. In most cases, the damage detected will be the result of more than one mechanism. For example, corrosion of reinforcing steel may open cracks that allow moisture greater access to the interior of the concrete. This moisture could lead to additional damage by freezing and thawing. In spite of the complexity of several factors working simultaneously, given a basic understanding of the various damage-causing mechanisms, it should be possible, in most cases, to determine the primary cause or causes of the damage seen on a particular structure and to make intelligent choices concerning selection of repair materials and methods.

Generally, causes of distress and deterioration of concrete are as follows (US Army Corps of Engineers EM 1110-2-2002, 1986):

Chemical reactions
Construction errors
Corrosion of embedded metals
Design errors
Erosion
Freezing and thawing
Settlement and movement
Shrinkage
Temperature changes.

1.2.2.1 Chemical Reactions

This category includes several causes of deterioration that exhibit a wide variety of symptoms. In general, deleterious chemical reactions may be classified as those that occur as the result of external chemicals attacking the concrete (acid attack, agressive water attack, miscellaneous chemical attack, and sulfate attack) or those that occur as a result of internal chemical reactions between the constituents of the concrete (alkali–silica and alkali–carbonate rock reactions). Each of these chemical reactions is briefly described below.

Acid Attack

Portland-cement concrete is generally not very resistant to attack by acids. The deterioration of concrete by acids is primarily the result of a reaction between the acid and the products of the hydration of cement. In most cases the chemical reaction results in the formation of water-soluble calcium compounds that are then leached away. In the case of sulfuric acid attack, additional or accelerated deterioration results because the calcium sulfate formed may affect the concrete by the sulfate attack mechanism. If the acid is able to reach the reinforcing steel through cracks or pores in the concrete, corrosion of the reinforcing steel will result, which in turn will result in further deterioration of the concrete.

Visual examination will show disintegration of the concrete leading to the loss of cement paste and aggregate from the matrix. If reinforcing steel is reached by the acid, rust staining, cracking, and spalling may be seen.

Aggressive Water Attack

Some waters have been reported to have extremely low concentrations of dissolved minerals. These soft or aggressive waters will leach calcium from cement paste or aggregates. This phenomenon has been infrequently reported in the United States (Holland et al., 1980).

From the few cases that have been reported, there are indications that this attack takes place very slowly. For aggressive water attack to have a serious effect on marine structures, the attack must occur in flowing water, which contains a constant supply of aggressive water in contact with the concrete and which washes away aggregate particles that become loosened as a result of leaching of the paste.

Visual examination will show concrete surfaces that are very rough in areas where the paste has been leached. Sand grains may be present on the surface of the concrete which resembles a coarse sandpaper. If the aggregate is susceptible to leaching, holes where the coarse aggregate has been dissolved will be evident.

Alkali–Carbonate Rock Reaction

Certain aggregates of carbonate rock have been found to be reactive in concrete. The results of these reactions have been characterized as ranging from beneficial to destructive. The destructive category is apparently limited to reactions with impure dolomitic aggregates and are a result of either dedolomitization or rim-silicification reactions. The mechanism of alkali–carbonate rock reaction is covered in detail in US Army Corps of Engineers EM 1110-2-2000 (1985).

If not detected, alkali–aggregate reactions may impair concrete durability to such an extent that the load carrying capacity of a marine structure is dangerously jeopardized. Alkali–aggregate reactions produce cracks through which oxygen, water, and chlorides travel to attack reinforcing steel. In addition, weathering wears the concrete down.

Experts have found that chlorides can amplify the adverse effects of alkali-aggregate reactions and significantly increase concrete expansion. Chloride salts react with the products of cement hydration to generate additional or secondary alkalis. Oddly enough, alkali–aggregate reactions do not decrease compressive strength of concrete appreciably. Because structural adequacy of concrete in situ is often determined based on compressive core strength, the data obtained from cores can be misleading. Therefore, when evaluating the structural adequacy of concrete affected by alkali–aggregate reactions, it is advisable to determine compressive and tensile strength, as well as the modulus of concrete elasticity.

Visual examination of those reactions that are serious enough to disrupt the concrete in a structure will generally show map or pattern cracking and a general appearance that indicates that the concrete is swelling. A distinguishing feature that differentiates an alkali–carbonate rock reaction from an alkali–silica reaction is the lack of silica gel exudations at cracks (American Concrete Institute ACI 201.2R, 1985c).

Alkali–Silica Reaction

Some aggregates containing silica that is soluble in highly alkaline solutions may react to form either a solid nonexpansive calcium–alkali–silica complex or an alkali–silica complex that can imbibe considerable amounts of water and then expand, disrupting the concrete.

Visual examination of those concrete structures that are affected will generally show map or pattern cracking and a general appearance that indicates that the concrete is swelling. Petrographic examination may be used to confirm the presence of alkali–silica reaction. For more details on alkali–silica reaction the interested reader is referred to Hobbs (1988), Okada et al. (1989), and Acres International Ltd. (1989).

Miscellaneous Chemical Attack

Concrete will resist chemical attack to varying degrees depending on the exact nature of the chemical. American Concrete Institute ACI 515.1R (1985d) includes an extensive listing of the degrees of resistance of concrete to various chemicals. Most chemicals, in order to produce a significant attack on concrete, must be in solution form and must be above a certain minimum concentration. Concrete is seldom attacked by solid dry chemicals. Also, for maximum effect, the chemical solution needs to be circulated in contact with the concrete. Concrete subjected to aggressive solutions under positive differential pressure is particularly vulnerable. The pressure gradients tend to force the aggressive solutions into the matrix. If the low-pressure face of the concrete is exposed to evaporation, a concentration of salts tends to accumulate at that face, resulting in increased attack. In addition to the specific nature of the chemical involved, the degree to which concrete resists attack depends on: the temperature of the aggressive solution, the water/cement ratio of the concrete, the type of cement used (in some circumstances), the degree of consolidation of the concrete, the permeability of the concrete, the degree of wetting and drying of the chemical on the concrete, and the extent of chemically induced corrosion of the reinforcing steel (American Concrete Institute ACI 201.IR, 1985b).

Visual examination of concrete that has been subjected to chemical attack will usually show surface disintegration, spalling, and the opening of joints and cracks. There may also be swelling and general disruption of the concrete mass. Coarse aggregate particles are generally more inert than the cement paste matrix; therefore, aggregate particles may be seen as protruding from the matrix. Laboratory analysis may be required to identify the unknown chemicals that are causing the damage.

Sulfate Attack

Naturally occurring sulfates of sodium, potassium, calcium, or magnesium are sometimes found in soil or in solution in groundwater adjacent to marine structures. The sulfate ions in solution will attack the concrete. There are apparently two chemical reactions involved in sulfate attack on concrete. First, the sulfate reacts with free calcium hydroxide which is liberated during the hydration of the cement to form calcium sulfate (gypsum). Next, the gympsum combines with hydrated calcium aluminate to form calcium sulfoaluminate. Both of these reactions result in an increase in volume. The second reaction is responsible for most of the disruption due to volume increase of the concrete (American Concrete Institute, ACI 201.2R, 1985c). In addition to the two chemical reactions, there may also be a purely physical phenomenon in which the growth of crystals of sulfate salts disrupts the concrete.

Visual examination of concrete exposed to sulfate attack will show map and pattern cracking as well as a general disintegration of the concrete. Laboratory analysis can verify the occurrence of the reactions described.

1.2.2.2 Construction Errors

Failure to follow specified procedures and good practice or outright carelessness may lead to a number of conditions that may be grouped together as construction errors. Typically, most of these errors do not lead directly to failure or deterioration of concrete. Instead, they enhance the adverse impacts of other mechanisms previously identified. Each error is briefly described in the following paragraphs. It should be noted that errors of the type described in this section are equally likely to occur during repair or rehabilitation projects as during new construction.

Adding Water to Freshly Mixed Concrete

This practice will generally lead to concrete with lowered strength and reduced durability. As the water/cement ratio of the concrete increases, the strength and durability will decrease and shrinkage and permeability will increase.

Improper Consolidation

Improper consolidation of concrete may result in a variety of defects, the most common being bugholes, honeycombing, and cold joints.

Improper Curing

Curing is probably the most abused aspect of the concrete construction process. Unless concrete is given adequate time to cure at a proper humidity and temperature, it will not develop the characteristics that are expected and that are necessary to provide durability. Symptoms of improperly cured concrete can include various types of cracking and surface disintegration. In extreme cases where poor curing leads to failure to achieve anticipated concrete strengths, structural cracking may occur.

Improper Location of Reinforcing Steel

This may result in reinforcing steel that is either improperly located or is not adequately secured in the proper location. Either of these faults may lead to two general types of problems. First, the steel may not function structurally as intended, resulting in structural cracking or failure. Second, the concrete cover over steel is reduced which makes it much easier for corrosion to begin.

Movement of Formwork

Movement of formwork during the period while the concrete is going from a fluid to a rigid material may induce cracking and separation within the concrete.

Premature Removal of Shores or Reshores

If shores or reshores are removed too soon, the concrete affected may become overstressed and cracked. In extreme cases there may be major failures.

Settling of the Subgrade

If there is any settling of the subgrade during the period after the concrete begins to become rigid but before it gains enough strength to support its own weight, cracking may occur.

Vibration of Freshly Placed Concrete

Most construction sites are subjected to vibration from various sources, such as blasting, pile driving, and from the operation of construction equipment. Freshly placed concrete is vulnerable to weakening of its properties if subjected to forces that disrupt the concrete matrix during setting.

1.2.2.3 Corrosion of Embedded Metals

The major theory concerning the dangers of reinforcing steel corrosion and concrete deterioration, which still holds true today, was conceived as early as 1907 to 1909. Called the mechanical pressure theory, it postulated that a current passing from iron into concrete caused the iron to corrode. Ultimately, insoluble iron compounds—mainly oxides—form near the surface of the iron, taking up several times the volume of the original piece of iron.

Later, in 1910, experts studying the effects of chlorides on concrete noted a marked increase in corrosion when cement contained as little as 0.2% $CaCl_2$. However, $CaCl_2$ doses below that amount appeared to have no appreciable effect. Based on the results of this study, it was recommended that chlorides should not be added to concrete if the structure is likely to be subjected to electric currents (Novokshchenov, 1988).

Numerous subsequent studies have confirmed these early findings. As described by Uhlig (1971), Mehta (1986, 1988), Holmes and Brundle (1987), American Concrete Institute ACI 222R-85 (1985a), and many other experts, corrosion of metals in concrete is an electrochemical process during which the iron can corrode by chemical attack.

When carbon steel reinforcement is embedded in concrete the surface of the steel oxidizes to form a very thin surface film of ferric oxide (Fe_2O_3). This film is known as the passive film because it is extremely stable when embedded in the highly alkaline cement matrix (pH normally greater than 11). Provided the alkaline environment is sustained and the passive film remains intact the reinforcement will undergo virtually no further oxidation over an indefinite period and the reinforced concrete structure will therefore exhibit none of the problems associated with corrosion of the reinforcement.

However, if chloride ions are present in sufficiently high volumes at a reinforcing bar within the concrete they cause the passive film at that point to break down. This breakdown causes a difference in electrode potential between the exposed steel at the point at which depassivation has occurred and the oxide-coated steel on either side.

It should be noted that a commonly accepted chloride threshold value for depassivation of steel is about 0.2% total chloride ion by weight of cement or about 0.8 kg/m^3 for typical high-strength concrete.

Once the passivity of steel is destroyed, it is the cathode process and the electrical resistivity of the system that control the rate of corrosion. The cathode process, which is necessary to complete the corrosion cell, will not progress until a sufficient supply of both oxygen and moisture is available at the surface of steel at some distance away from the anode.

Reactions at the anodes and cathodes are broadly referred to as "half-cell reactions". At the anode, which is the negative pole, iron is oxidized to ferrous ions (Figure 1–2).

$$Fe = Fe^{2+} + 2e^-. \quad (1\text{–}3)$$

The Fe^{2+} in Eq. (1–3) is subsequently changed to oxides of iron by a number of complex reactions. As was mentioned earlier, the volume of the reaction products is several times the volume of original iron.

In an acid medium, the reaction taking place at the cathode is the reduction of hydrogen ions to hydrogen. As summarized by Holmes and Brundle (1987), the factors affecting the corrosion rate are as follows:

1. The chloride concentration at the surface of the reinforcing bar
2. The initial integrity of the passive layer on the surface of the reinforcement
3. The electrical resistance of the concrete

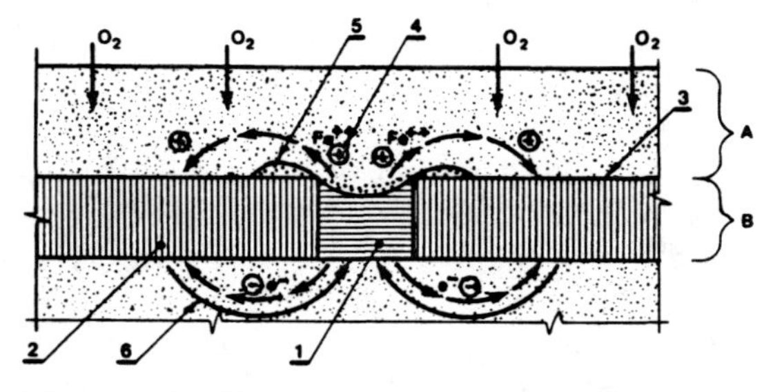

Figure 1–2. Diagrammatic presentation of electrochemical corrosion of steel in concrete. (A) Moist concrete as electrolyte, (B) steel. 1—Anode, 2—cathode, 3—surface film (Fe_2O_3), 4—iron ion, 5—product of corrosion ($Fe(OH)_2$), 6—current flaw.

4. The availability of oxygen to complete the cathodic reaction.

High concentrations of chloride ions within the concrete matrix are either derived from the concrete mix constituents themselves or enter through the surface of the hardened concrete from the external environment.

Deterioration of many marine structures over the past few years, to some extent, has been attributed to introduction of salts in the concrete mix used in the past. However, the external supply of chloride ions from the marine environment is mainly responsible for deterioration of concrete. The concentration of chlorides on concrete surfaces due to effects of the external (marine) environment can vary significantly from one part of the structure to another. For example, in the splash zone solar radiation and drying winds can produce rapid evaporation, which could lead to the deposition of salt crystals. Gradual buildup of these crystals can increase the chloride concentration in this zone.

In the tidal zone, salts are built up on the concrete surface during a falling tide and as the tide level increases the concrete surface is washed and the surface salt concentration is reduced.

In the submerged zone the chloride concentration at the surface of the concrete is relatively constant and the maximum concentration approximates the concentration of salts in seawater.

High concentrations of salts in contact with the concrete surface have only a limited effect on the integrity of the concrete matrix itself. Salt crystals can cause disruptive pressures within the surface layer of concrete which often results in a minor loss of cement paste or aggregate from the concrete surface; the expected long-term result can be loss of cover to the reinforcement.

Depassivation of the reinforcing steel is affected mainly by the rate of chloride penetration. Penetration of the salts into the concrete and concentration of chlorides at the level of reinforcing steel depend on concrete cover permeability, which in turn is dependent on concrete matrix density, and degree of cracking present in the concrete cover.

Concrete density is the most important factor affecting the rate of chloride penetration into concrete. Dense concrete with a relatively low porosity is less permeable to moisture-containing dissolved salts than a less dense concrete with a high porosity.

However, cracks within the concrete cover can have a significant bearing on the effective permeability of the concrete despite the presence of a highly dense and impervious concrete matrix. Cracks also play major role in permitting an abundant supply of oxygen to come into contact with the reinforcement. The availability of a supply of free oxygen is essential to the corrosion process and any limitation to this supply will restrict the corrosion rate. The fact that marine structures seldom exhibit signs of serious reinforcement corrosion in the submerged zone is believed to be largely due to the limited amount of free oxygen in the seawater.

Cracking in concrete can be caused by a large number of different factors and many of these factors are discussed in this chapter.

Once the passivity of the steel is destroyed, the electrical resistivity of the system controls the rate of corrosion. The electrical resistance of the concrete and the consequent rate of ionic transfer are governed to a large extent by the moisture content.

Moist concrete is a good conductor of electrons between the anode and the cathode. Because the corrosion process is a chemical reaction, the rate of reaction is affected by changes in temperature. This means that the higher ambient temperatures in countries with warm climates result in a much higher rate of corrosion of steel than in more temperate climates.

In the context of the above described fundamental principles underlying the process of corrosion of steel in concrete, it may be concluded that corrosion of the reinforcing steel can occur when the passive oxide film on the steel is destroyed by chloride ions, the excess of oxygen to this area is available, and an abundance of moisture is present there. Accordingly, different parts of marine structures within the maritime environment are at higher risk than others.

Reviews of the relevant case histories (Mehta and Gerwick, 1982; Buslov and Rojansky, 1983; Gilbride et al., 1988; Ingram and Morgans, 1986; and many others) clearly indicate that the parts of concrete structures most susceptible to steel corrosion are those that are most saturated and/or exposed to intermittent wetting and drying. The saturated tidal zone of marine structures has more potential than other parts of the structure for cracking due to cyclic wetting/drying conditions, freeze and thaw attack, and other factors described in this chapter. It is also the most vulnerable to corrosion of reinforcing steel. In many cases the splash zone has been reported as being vulnerable to corrosion of reinforcing steel.

In addition to the development of an electrolytic cell, corrosion may develop under several other conditions. The first of these is corrosion produced by the presence of a stray electrical current. In this case, the current necessary for the corrosion reaction is provided from an outside source. A second additional source of corrosion is chemicals that may be able to act directly on the reinforcing steel.

In the case of embedded metal corrosion visual examination of the existing marine structure will typically reveal rust staining of the concrete. This staining will be followed by cracking. Cracks produced by corrosion generally run in straight, parallel lines at uniform intervals corresponding to the spacing of the reinforcement. As deterioration continues, spalling of the concrete over the reinforcing steel will occur with the reinforcing bars becoming visible. One area where laboratory analysis may be beneficial is the determination of the chloride content in the concrete. This procedure may be used to determine the amount of concrete to be removed during a rehabilitation project.

1.2.2.4 Design Errors

Design errors may be divided into two general types: those resulting from inadequate structural design, and those resulting from lack of attention to relatively minor design details.

In the case of inadequate structure design the failure mechanism is simple—the concrete is exposed to greater stress than it is capable of carrying, or it sustains greater strain than its strain capacity.

Visual examinations of failures resulting from inadequate structural design will usually show spalling and/or cracking. To identify inadequate design as a cause of damage, the locations of damage should be compared to the types of stresses that should be present in the concrete. If the type and location of the damage and the probable stress are in agreement, a detailed stress analysis will be required to determine whether inadequate design is the cause.

Although a structure may be adequately designed to meet loadings and other overall requirements, poor detailing may result in localized concentrations of high stresses in otherwise satisfactory concrete. These high stresses may result in cracking that allows water or chemicals to get access to the concrete. In general, poor detailing does not lead directly to concrete failure; rather, it contributes to the action of one of the other causes of concrete deterioration described in this chapter.

Abrupt changes in section may cause stress concentrations that may result in cracking. A typical example is the use of relatively thin sections such as pier approach bridge decks rigidly tied into massive abutments.

Insufficient reinforcement at openings tends to cause stress concentrations that may cause cracking.

Poor attention to the details of draining a structure may result in the ponding of water. This ponding may lead to leakage or saturation of concrete and may result in severely damaged concrete if the area is subjected to freezing and thawing.

Inadequately designed expansion joints may result in spalling of concrete adjacent to the joints.

The use of materials with different properties (modulus of elasticity or coefficient of thermal expansion) adjacent to one another may result in cracking or spalling as the structure is loaded or as it is subject to daily or annual temperature variations.

1.2.2.5 Erosion

Erosion damage to a marine structure is usually caused by the action of ice, sediments, or miscellaneous floating debris that are rolling and grinding against a concrete surface. Ice is typically the principal source of concrete erosion. Moving ice, which can have compressive strength as great as 20 MPa, has been known to remove all the concrete cover and near-surface layers of reinforcement in marine structures (Hoff, 1988).

Ice abrasion at or near a waterline is a typical result of the combined effects of ice impact (or repeated impacts) and sliding of ice floes along the structure which create friction or drag on the concrete surface.

Wind- and/or current-driven ice floes can possess significant kinetic energy, much of which is dissipated into the concrete during collision with the structure. Some kinetic energy is lost in the crushing of ice. As driving forces continue to drag the ice floe against and along the structure a local failure occurs in both ice and concrete. The degree of failure of the ice/concrete system depends on ice and concrete characteristics, as well as on the dynamic response of the structure to the repetitive ice dynamic loading. With time this repetitive loading can affect the aggregate bond near the surface of the concrete, and cause or propagate microcracks in the concrete matrix. When eventually the integrity of the surface of the concrete has become impaired, the ice dynamic/abrasion action may cause particles of the surface to be removed. Oblique impact forces on exposed aggregates can be especially damaging. Some ice floes may contain grit which provides more abrasive impact on exposed concrete. Gjorv et al. (1987) suggested that wet concrete abrades more rapidly than dry concrete.

Environmental effects such as cyclic freezing and thawing and saturation of concrete in the splash zone and below the waterline can also be a contributing factor to weakening of the concrete matrix and aggregate bond weakening.

At any point in time, the degree to which each of the above factors has contributed to the overall deterioration and loss of concrete due to ice action is difficult to quantify. Recent field and laboratory studies of ice abrasion provided some useful information (Hoff, 1988). One useful (and quite obvious) conclusion of the above studies is that normal weight concrete is more abrasion resistant if hard and tough aggregates are used. The general perception of structural lightweight aggregates is that basically they do satisfy this requirement.

Visual examination of a concrete surface exposed to any kind of abrasion typically will reveal local scratches and a surface that looks worn and sometimes polished. The example of abrasion of concrete piles by moving sediments is illustrated in Figure 1–7.

1.2.2.6 Freezing and Thawing

As the temperature of critically saturated concrete is lowered during cold weather, the freezable water held in the capillary pores of the cement paste and aggregates expands on freezing. Is subsequent thawing is followed by refreezing, the concrete is further expanded, so that repeated cycles of freezing and thawing have a cumulative effect. By their very nature, concrete marine structures are particularly vulnerable to freezing and thawing simply because there is ample opportunity for portions of these structures to become critically saturated. Concrete is especially vulnerable in tidal and splash zones. Exposure in such areas as the face of walls, piers, deck structures, and bank protections enhances the vulnerability of concrete to the harmful effects of repeated cycles of freezing and thawing. Seawater accelerates damage caused by freezing and thawing.

American Concrete Institute report ACI 201.2R (1985c) describes the action of salts (deicing chemicals) on concrete surfaces as physical. It involves the development of osmotic and hydraulic pressures during freezing, principally in the paste, similar to ordinary frost action.

Visual examination of concrete damaged by freezing and thawing may reveal symptoms ranging from surface scaling to extensive disintegration. Laboratory examination of cores taken from structures that show surficial effects of freezing and thawing will often show a series of cracks parallel to the surface of the structure.

1.2.2.7 Settlement and Movement

Because concrete structures are typically very rigid, they can tolerate very little differential movement. As the differential movement increases, concrete components can be expected to be subjected to an overstressed condition. Ultimately, these components will crack or spall.

Situations in which an entire structure is moving or a single element of a structure such as a monolith is moving with respect to the remainder of the structure are basically caused by subsidence attributed to long-term consolidations, new loading conditions, or by a wide variety of other mechanisms. In these cases, the concerns are generally not cracking or spalling but rather stability against overturning or sliding. In situations in which overall structure movement is diagnosed as a cause of concrete deterioration, a thorough geotechnical investigation should be conducted.

Visual examination of structures undergoing settlement or movement will usually reveal cracking or spalling or misalignment of structural components. Because differential settlement of the foundation of a structure is usually a long-term phenomenon, review of instrumentation data, if any, will be helpful in determining of causes of the apparent movement.

1.2.2.8 Shrinkage

Shrinkage is caused by the loss of moisture from concrete. It may be divided into two general categories: that which occurs before setting (plastic shrinkage) and that which occurs after setting (drying shrinkage).

Plastic Shrinkage

During the period between placing and setting, most concrete will exhibit bleeding to some degree. Bleeding is the appearance of moisture on the surface of the concrete; it is caused by the settling of the heavier components of the mixture. Usually, the bleed water evaporates slowly from the concrete surface. If environmental conditions are such that evaporation is occurring faster than water is being supplied to the surface by bleeding, high tensile stresses can develop. These stresses can lead to the development of cracks on the concrete surface.

Cracking due to plastic shrinkage will be seen within a few hours of concrete placement. Typically, the cracks are isolated rather than patterned. These cracks are generally wide and shallow.

Drying Shrinkage

This is the long-term change in volume of concrete caused by the loss of moisture. If this shrinkage could take place without any restraint, there would be no damage to the concrete. However, the concrete in a structure is always subject to some degree of restraint by either the foundation, by another part of the structure, or by the difference in shrinkage between the concrete at the surface and that in the interior of a component. This restraint may also be attributed to purely physical conditions such as the placement of a footing on a rough foundation, or to chemical bonding of new concrete to earlier placements, or to both. The combination of shrinkage and restraints causes tensile stresses that can ultimately lead to cracking.

Visual examination will typically show cracks that are characterized by their fineness and absence of any indication of movement. They are usually shallow, a few centimeters in depth. The crack pattern is typically orthogonal or blocky. This type of surface cracking should not be confused with thermally induced deep cracking which occurs when dimensional change is restrained in newly placed concrete by rigid foundations or by old lifts or concrete.

1.2.2.9 Temperature Changes

Changes in temperature cause a corresponding change in the volume of concrete. As was true for moisture-induced volume change (drying shrinkage), temperature-induced volume changes must be combined with restraint before damage can occur. Basically, there are three temperature change phenomena that may cause damage to concrete. First, there are the temperature changes that are generated internally by the heat of hydration of cement in large placements. Second, there are the temperature changes generated by variations in climatic conditions. Finally there is a special case of externally generated temperature change, for example, fire damage. Because of the infrequent nature of its occurrence in marine structures, fire damage is not discussed in this work.

Internally Generated Temperature Differences

The hydration of Portland cement is an exothermic chemical reaction. In large volume placements, significant amounts of heat may be generated and the temperature of the concrete may be raised by more than 35°C over the concrete temperature at placement. Usually, this temperature rise is not uniform throughout the mass of the concrete, and steep temperature gradients may develop. These temperature gradients give rise to a situation known as internal restraint—the outer portions of the concrete may be losing heat while the inner portions are gaining (heat). If the differential is great, cracking may occur. Simultaneously with the development of this internal restraint condition, as the concrete mass begins to cool a reduction in volume takes place. If the reduction in volume is prevented by external conditions (such as by chemical bonding, by mechanical interlock, or by piles or dowels extending into the concrete), the concrete is said to be externally restrained. If the strains induced by the external restraint are great enough, cracking may occur. There is increasing evidence, particularly for rehabilitation work, that a relatively minor temperature difference in thin, highly restrained overlays can lead to cracking.

Visual examination will usually show relatively shallow isolated cracking resulting from conditions of internal restraint. Cracking resulting from external restraint will usually extend through the full section. Thermally induced cracking may be expected to be regularly spaced and perpendicular to the larger dimensions of the concrete.

Externally Generated Temperature Differences

The basic failure mechanism in this case is the same as that for internally generated temperature differences—the tensile strength of the concrete is exceeded. In this case the temperature change leading to the concrete volume change is caused by external factors, usually climatic conditions.

If the structure is not provided with adequately space expansion joints the externally generated temperature differences may result in concrete cracking and/or spalling at expansion joints.

1.2.2.10 Evaluation of Causes of Concrete Distress and Deterioration

Given a detailed report of the conditon of the concrete in a structure and a basic understanding of the various mechanisms that can cause concrete deterioration, the problem becomes one of relating the observation of symptoms to the underlying causes. When many of the different causes of deteroration produce the same symptoms, the task of relating symptoms to causes is more difficult than it first appears. One procedure to consider is based on that described by Johnson (1965). Johnson recommends the following steps to be taken for evaluation of the present condition of concrete.

1. Evaluate structure design to determine adequacy. First consider what types of stress could have caused the observed symptoms. For example, tension will cause cracking, whereas compression will cause spalling. Torsion or shear will usually result in both cracking and spalling. If the basic symptom is disintegration, then overstress may be eliminated as a cause. Second, attempt to relate the probable types of stress causing the damage noted to the locations of the damage. For example, if cracking resulting from excessive tensile stress is suspected, it would not be consistent to find that type of damage in an area that is under compression. Next, if the damage seems appropriate for the location, attempt to relate the specific orientation of the damage to the stress pattern. Tension cracks should be roughly perpendicular to the line of tension stress. Shear usually causes failure by diagonal tension, in which case the cracks will run diagonally in the web of a beam. Visualizing the basic stress patterns in the structure will aid in this phase of the evaluation. If no inconsistency is encountered during this evaluation, then overstress may be the cause of the observed damage. A thorough stress analysis is warranted in order to confirm this finding. If an inconsistency has been detected, such as cracking in a compression zone, the next step in the procedure should be followed.

2. *Relate the symptoms to potential causes.* For this step, Table 1–1 will be of benefit. Depending on the symptom, it may be possible to eliminate several possible causes. For example, if the symptom is distintegration or erosion, several potential causes may be eliminated by this procedure.

3. *Eliminate the readily identifiable causes.* From the list of possible causes remaining after relating symptoms to potential causes, it may be possible to eliminate two causes very quickly because they are relatively easy to identify. The first of these is corrosion of embedded metals. It will be easy to verify if the cracking and spalling noted are a result of corrosion. The second cause that is readily identified is accidental loading, as personnel at the structure should be able to relate the observed symptoms to a specific incident.

4. *Analyze the available clues.* If no solution has been reached at this stage, all of the evidence generated by field and laboratory investigations should be carefully reviewed. Attention should be paid to the following points:

(i) If the basic symptom is that of disintegration of the concrete surface then essentially three possible causes remain: chemical attack, erosion, and freezing and thawing. Attempts should be made to relate the nature and type of the damage to the location in the structure and to the environment of the concrete in determining which of the above possibilities is the most likely to be the cause of the damage.

(ii) If there is evidence of swelling of the concrete then there are two possibilities: chemical reactions and temperature changes. Destructive chemical reactions such as alkali–silica or alkali–carbonate attack which cause swelling will have been identified during the laboratory investigation. Temperature-induced swel-

Table 1–1 Relating Symptoms to Causes of Distress and Deterioration of Concrete

Symptoms	Construction Faults	Cracking	Disintegration	Distortion/ Movement	Erosion	Joint Failures	Seepage	Spalling
Accidental loadings		X						X
Chemical reactions		X	X				X	
Construction errors	X	X				X	X	X
Corrosion		X						X
Design errors		X				X	X	X
Erosion			X		X			
Freezing and thawing		X	X					X
Settlement and movement		X		X		X		
Shrinkage		X						
Temperature changes		X				X		X

fFrom US Army Corps of Enginers, EM 1110-2-2002, 1986.

ling should be ruled out unless there is additional evidence such as spalling at joints.

(iii) If the evidence is spalling, and corrosion and accidental loadings have been eliminated earlier, the major causes of spalling remaining are construction errors, poor detailing, freezing and thawing, and externally generated temperature changes. Examination of the structure should have provided evidence as to the location and general nature of the spalling that will allow identification of the exact cause.

(iv) If the evidence is cracking then construction errors, shrinkage, temperature changes, settlement and movement, chemical reactions, and poor design details remain as possible causes of distress and deterioration of concrete. Each of these possibilities will have to be reviewed in light of the available laboratory and field observations to establish which is responsible.

(v) If the evidence is seepage and it has not been related to a detrimental internal chemical reaction by this time, then it is probably the result of design errors or construction errors such as improper location or installation of a waterstop.

5. *Determine why the deterioration has occurred.* Once the basic cause or causes of the damage have been established, there remains one final requirement: to understand how the causal agent acted on the concrete. For example, if the symptons were cracking and spalling and the cause was corrosion of the reinforcing steel, what facilitated the corrosion? Was there chloride in the concrete? Was there inadequate cover over the reinforcing steel? Another example to consider is concrete damage due to freezing and thawing. Did the damage occur because the concrete did not contain an adequate air/void ratio, or did the damage occur because the concrete used was not expected to be saturated but for whatever reason was saturated? Only when the cause and its mode of action are completely understood should the next step of selecting a repair material be attempted.

1.2.3 Corrosion of Steel in the Marine Environment

The subject of metal corrosion in general and in the marine environment in particular is well covered in several fundamental works (Uhlig, 1948; Rogers, 1960; Dismuke et al., 1981; and others). Therefore, only a basic overview of corrosion problems peculiar to marine structures will be suggested to the reader, who is referred to the abundance of literature on the subject for further detailed information.

Corrosion occurs because of small physical and/or chemical differences present in metals such as minor impurities or local composition variations or environment, for example, changes in amounts of dissolved oxygen varying with the depth of immersion, nonuniform salt concentrations due to pollution, etc. Corrosion is an electrochemical process similar to that which take place in a common flashlight battery. Corrosion occurs at the anode and is accompanied by a flow of electrons through the external "wire" to the cathode.

Two types of corrosion are recognized: dry and aqueous. The former may be briefly described as the metal directly oxidizing, thereby returning to a lower chemical energy level. This type of corrosion is slow and relatively uniform. Its rate is determined by temperature and diffusion of oxygen through the oxide. Thus the thickness and physical stability of the rust layer are significant.

In practice, however, dry atmospheric conditions almost never exist in a marine environment.

A marine atmosphere where condensed moisture is very high (corresponding to 100% relative humidity) is a very aggressive environment for metals. Under such rather "wet" conditions, the corrosion process is analogous to that of continuous seawater immersion, except that the thin wet electrolytic film has a marked effect on the corrosion pattern, the corrosion products, and the ease with which oxygen is transferred to the metal surface, resulting in accelerated corrosion.

Aqueous corrosion may cause the metal to go into solution at one point, and allow oxygen to be taken up at a second point while depositing the corrosion product at a third point. In freshwater, which is not a good conductor, there is little corrosion because the corrosion rate is directly proportional to the current. The salts dissolved in seawater greatly increase the water conductivity and hence its corrosiveness. To initiate the corrosion process there must be a complete electrical circuit in both the structure and aquatic medium (electrolyte) such that negatively charged ions in the electrolyte flow from where they are produced, at the cathode toward the anode. In the structure itself, therefore, the ions flow from the anode to the cathode, unless an opposing voltage is applied with the aim of suppressing this current. The presence of negative ions near the anode encourages positively charged metallic ions to dissolve into the electrolyte, when they combine with any available negative ions to form a corrosion product. Thus the whole process continues unless the corrosion product itself forms a barrier to ionic movement. This so-called "passive" coating reforms and heals itself spontaneously provided oxygen is available but rapid corrosion can occur in crevices or under marine growth. The rate of corrosion in carbon steel is much higher than in other metals, including those that do not form a passive coating, because the magnetite produced in the process is a good conductor. The process of corrosion of metal immersed into seawater is shown in Figure 1–3. In addition, because a current can flow only in the presence of a potential difference any source of potential difference, for example, electrical; bimetallic [due to contact between different metals (Figure 1–4)]; physical, such as surface defects or stress concentration; chemical; or temperature difference, may cause corrosion.

The chemical reactions that take place on iron (the principal constituent of steel) corroding in seawater are as follows.

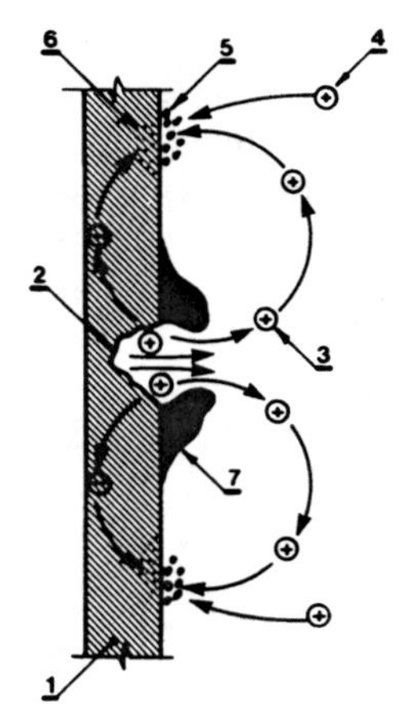

Figure 1–3. Corrosion of steel immersed in water. 1—Steel, 2—Pit, 3—iron ion, 4—hydrogen ion, 5—hydrogen film, 6—impurity, 7—product of corrosion $Fe(OH)_2$.

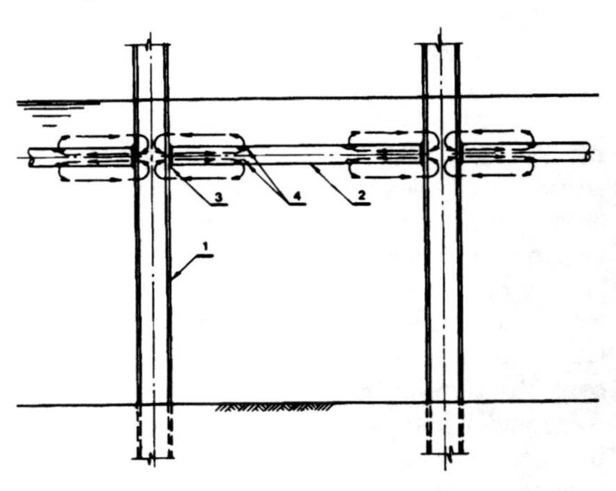

Figure 1–4. Example of galvanic corrosion couples (dissimilar-electrode cells). 1—A242 H-pile, low alloy steel (cathode), 2—mild steel pipe brace (anode), 3—weld, 4—pit. Note: Pitting occurs where current leaves the anode to enter the electrolyte.

At the anode iron goes into solution

$$Fe \rightarrow Fe^{2+} + 2e. \quad (1\text{–}4)$$

The electrons flow to the cathode through the metallic circuit. At the cathode oxygen converts hydrogen atoms into water

$$2H^{+} + 1/2\,O_2 + 2e \rightarrow H_2O \quad (1\text{–}5)$$

or converts water to hydroxyl ions.

$$H_2O + 1/2\,O_2 + 2e \rightarrow 2OH^{-}. \quad (1\text{–}6)$$

By adding the above Eq. (1–4) and (1–6),

$$Fe + H_2O + 1/2\,O_2 \rightarrow Fe(OH)_2. \quad (1\text{–}7)$$

Iron is converted to ferrous hydroxide. Other reactions can occur, such as conversion of ferrous hydroxide to ferric hydroxide ($Fe(OH)_3$) by further reaction with oxygen.

As previously stated the unique conditions of the marine environment create a series of environmental zones that affect metal corrosion rates. These zones are apparent when examining a corroded marine pile and result from different conditions of moisture, oxygen content, and other factors. Figure 1–5 is a graphic representation of the above noted five zones.

Average corrosion rates are normally calculated by total weight loss of metal or by metal thickness losses. In most cases, however, corrosion is not uniform over the surface of the structure. Severe pitting may form areas where structural stresses can concentrate. Pitting can be a more important concern than uniform corrosion. Pitting corrosion rates are generally greater than uniform rates, particularly for the first 10 years after installation. Average rates are those referred to by the terms "metal loss', rate or "corrosion rate".

The corrosion rate on steel piling typically varies considerably by zone. A representative corrosion rate profile of steel sheet piling is shown in Figure 1–6. There the varying cor-

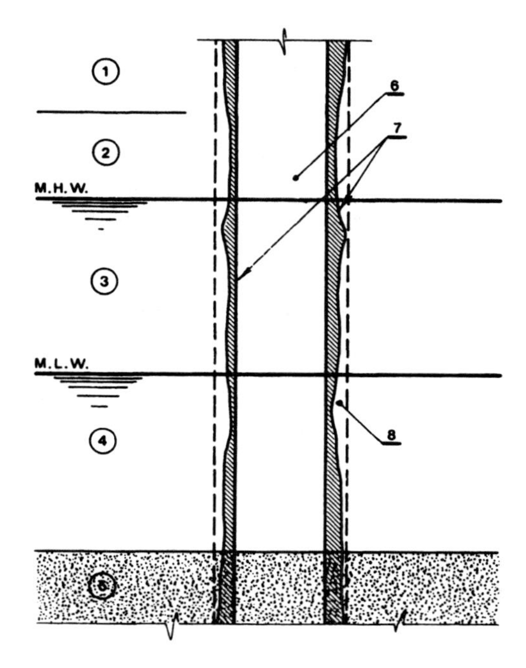

Figure 1–5. Typical corrosion of a steel H-pile in a marine environment. 1—Atmospheric zone, 2—splash zone, 3—tidal zone, 4—zone of continuous immersion, 5—seabed zone, 6—flange, 7—web, 8—corroded substrate.

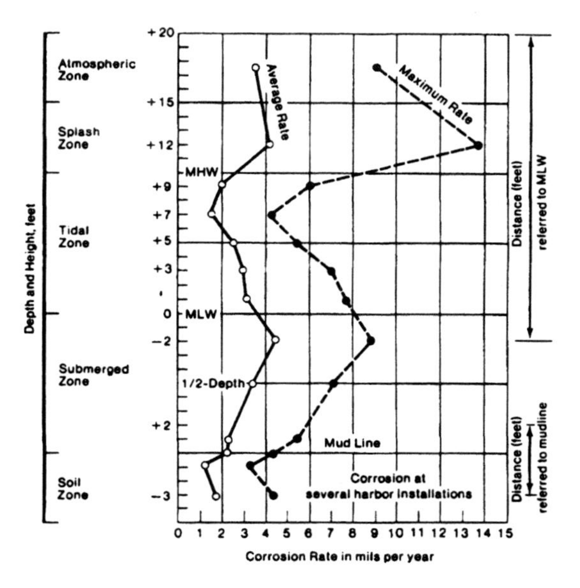

Figure 1–6. Corrosion rate profile of steel sheet piling. (From Edwards, 1963.)

rosion loss by zones is the average of eight harbor installations after 19 years of exposure. It is seen that the maximum loss of metal to corrosion occurs in the splash zone immediately above Minimum High Water Level (MHW). Marine structures usually suffer severe and rapid corrosion in a splash zone because of the high availability of both oxygen and electrolytes, and also because any protection that might be afforded by the corrosion product itself is rapidly washed away. A significant loss usually occurs in the immediate vicinity below Minimum Low Water Level (MLW).

The average rate of corrosion in the submerged zone is usually much smaller than in the splash zone, measuring about 0.15 mm/yr at first, and dropping to perhaps half this figure after about 15 years (Ractliffe, 1983). However, in warm climates the rate of steel corrosion in marine environments can be much higher (Buslov, 1979, 1983; Fukute et al., 1990).

Loss of metal due to corrosion is generally very low (sometimes even negligible) on areas of steel piling driven into undisturbed soils (Schwerdtfeger and Romanoff, 1972). The average corrosion rates there may be around 0.011 mm/yr (Ohsaki, 1982) at first, and decrease with time after driving. According to Ohsaki the corrosion rate is not affected by the soil type or permeability, even between such soils as clay, silt, sand, or gravel. However, highly acidic or highly alkaline soils with low pH value can be corrosive. Some accelerated corrosion may occur just below mudline as a result of an abrupt change in the environment of the structure as it passes from the water to the soil. This may be due to removal of corrosion product by water velocity, or sometimes due to water contamination. Average rates of atmospheric corrosion are similar to those in the submerged zone. A wet marine atmosphere is a very aggressive environment for steel and the corrosion process there is similar to that for continuous seawater immersion.

The rate of corrosion varies from location to location and depends on the following environmental factors:

Temperature of water
Oxygen concentration
pH value
Water salinity
Water velocity
Marine organisms
Pollution
Wind
Rain
Himidity
Sun
Ice.

Water Temperature

Chemical reactions, which include corrosion reactions, are in general accelerated in warmer water. However, the fouling rate is also accelerated in warmer water. The fouling provides a protective covering over the metal surface which produces a retarding effect on the corrosion rate due to decreased access of oxygen to the steel. Therefore, contrary to expectations, corrosion rates in warm seawater (immersion zone) have not been found to differ significantly from those measured in cold water.

Oxygen concentration

Oxygen concentration is a corrosive agent of steel in seawater. Rate of oxygen concentration on the surface of steel is a function of water depth. Areas of low concentrations of oxygen are anodic to areas of higher concentrations (US Army Corps of Engineers, TM-5-811-4, 1962).

The increased corrosion rate in the submerged zone just below MLW as compared to the tidal zone is attributed to the action of such a differential aeration cell.

pH Value

pH value or degree of acidity or alkalinity generally varies from 7.2 to 2.2 (Dismuke et al., 1981). A pH below 7 is acidic, and above 7 is alkaline. As alkalinity is increased above pH 9.5 iron tends to become passive and forms protective films that retard diffusion of oxygen to the surface. Below pH 4, alkaline protective films are dissolved and the acid acts directly on metal, accompanied by evolution of hydrogen. Hence, changes in pH values may affect the rate of corrosion.

Water Salinity

Water in the open sea has a salt content of about 3.5% consisting of various ions resulting from dissolved salts. The chloride ion is the most significant because of its high concentration. Chloride ions are able to penetrate the protective film formed by corrosion products to cause localized corrosion. The presence of chloride ions also affects the solubility of oxygen in water, hence increasing rates of corrosion.

Water Velocity

Water velocity increases the rate of steel corrosion. It is attributed to the increased availability and rate of diffusion of oxygen through the stagnant hydrodynamic sublayer of seawater on the surface of the metal (Schmitt and Phelps, 1969). With zero velocity the overall corrosion rate of steel in seawater is lower.

Marine Organisms

Organic matter in seawater has a marked effect on corrosion. As discussed in the previous section, the fouling community consists of a vast variety of marine plants and animals that attach themselves to marine structures. These marine organisms are generally found in fluidized mediums with pH values from 6 to 11 and can accelerate the corrosion rate in localized areas.

Although not strictly a type of corrosion this type of metal deterioration is a result of metabolic activity of living microorganisms.

The normal metabolic activities of these microorganisms affect corrosion in seawater by altering anodic and cathodic reactions, and creating a corrosion condition by producing differential aeration.

Organically induced corrosion is a result of activities of two classes of bacteria: aerobic, which require oxygen for their metabolic process; and anaerobic, which live and grow in an environment containing little or no oxygen (Cleary, 1969). The most troublesome kind of bacteria are sulfate-reducing anaerobic. These bacteria can affect deeply submerged steel components, such as piles. A discussion of microbial induced corrosion in steel is given in Scott and Davies (1992).

In a low-oxygen environment bacteria can utilize the hydrogen cathodically formed at the steel surface to reduce sulfate from the electrolyte and increase the local corrosivity of the environment. The corrosion product can be identified by its blackness. The corrosion typically results in the formation of pits on the steel surface.

Pollution

Pollution in harbors generally causes harm to the marine environment by its toxic effects or by depletion of the dissolved oxygen. The follow-up destruction of the oxygen-dependent fouling organisms in seawater may alter corrosion rates unfavorably.

Wind

Wind causes wave action, resulting in intermittent wetting in the splash zone. It also whips up the water surface and captures salt spray from breaking waves. The salt-laden spray evaporates and the remaining salt crystals are deposited on metal surfaces. These salt crystals accelerate the corrosion of steel to which they adhere by attracting and retaining moisture and forming aggressive local cells.

Rain

Water-dissolved salt deposits promote corrosion due to rapid diffusion of oxygen through the layer of electrolyte. Heavy rain can wash salt from a steel surface and thereby reduce the corrosion rate.

Humidity

Humidity forms a thin film of electrolyte on a steel surface in the atmospheric zone, and this promotes corrosion by diffusion of oxygen through this layer. In dry atmospheres corrosion is relatively mild and is the result of direct chemical attack, such as the oxidation of iron by air.

Sun

The sun affects relative humidity, the rate of evaporation, and the temperature of the structure.

Ice

Winter ice conditions cause removal of all corrosion products and effectively expose totally bare steel every spring. In addition to the ice forces, sediments are often carried in the ice and they act as an abrasive and essentially polish the steel surface. The loss of steel, which can reach above 25 mils per year in this case, usually refers to pitting and not to uniform or average corrosion loss (1 mil = 1/1000 inch).

All the above factors along with the damaging effects of ultraviolet exposure from the sun on pigmentation and composition of many coating systems affect the rate of steel corrosion in a complex way. Additional factors that can affect steel corrosion rate are

Steel composition
Use of dissimilar metals
Mill scale
Stray current.

Steel Composition

The addition of copper to carbon steel with some inclusion of nickel and phosphorus is reported to provide superior corrosion resistance in the splash zone (Dismuke et al., 1981). Tests of other steel composition such as ASTM A517 Grade F and ASTM A242 Type 1 (Cr-Si-Cu-Ni-P) have demonstrated even better performance (Schmitt and Phelps, 1969).

Use of Dissimilar Metals

Dissimilar metals coupled together can produce rapid corrosion in seawater because the galvanic current between different metals can be large unless steps are taken to insulate them from each other. As always, it is the anodic metal that corrodes. The most common anodic metals are magnesium, zinc, pure aluminum, aluminum alloy, clean steel, cast iron, chromium stainless steel, lead-tin solder, corroded steel, tin, mill-scale on steel, brasses, copper, bronzes, copper-metal alloys, and chromium stainless steel. The sequence of anodic metals listed above shows the relative position of metals and alloys in the so-called galvanic series. It must be remembered, however, that any of the series should be used with care, because the relative order can change under different conditions, for example, above 70°C steel can become anodic to zinc, whereas in slightly acidic solutions tin becomes anodic to steel (Ractliffe, 1983). Hence, if there is any doubt about a particular combination of metals and electrolyte it is best to conduct a test.

In marine pile construction, fasteners, welds, fittings, etc. are sometimes fabricated from metals other than plain carbon steel. If these metals are cathodic to the pile and are much smaller relative to the pile surface area, little corrosion should be expected on the adjacent anodic pile surface. Similarly cathodic metals also would be unaffected. In the reverse situation small anodic fitting attached to the cathodic pile may be rapidly corroded.

Mill Scale

Mill scale, a tenacious deposit of iron oxide on the steel surface (a result of the hot rolling process), is more cathodic than steel. Steel exposed with mill scale present can be pitted about three times as deeply as descaled steels (LaQue in handbook by Uhlig, 1948).

Stray Current

Stray current from external sources around marine structures can cause corrosion to the metal parts of the structure if such current is collected by the structure and leaves to enter the electrolyte.

One ampere of direct current passing from the structure to the electrolyte (seawater) can remove up to 10 mg of steel in one year. Typical sources of direct current are improperly grounded welding generators, electric railways systems, port electric service systems, etc. More information on effects of direct current is found in NAVDOCKS MO-306 (1964).

Corrosion deterioration limits the structure's service life. It may range from negligible to severe. An example of severe corrosion-affected steel sheet piling is seen in Figure 1–7a. Corrosion deterioration can develop into stress corrosion and/or corrosion fatigue.

Stress Corrosion essentially is the cracking of metal under the continued effects of tensile stress and a corrosive environment. A tensile stress creates an anodic area that pits and increases the stress concentration. Corrosion at the tip of the notch causes the stress corrosion crack to start there (Schmitt and Phelps, 1969). The cracks are very fine at the early stages of growth, but become visible before failure. The failure appears like a brittle fracture.

Corrosion fatigue is a term that defines the reduction of the metal's air fatigue strength in a corrosive environment.

In dry air, steel exhibits a fatigue limit defined as the stress below which the metal can sustain an unlimited number of load (stress) reversals. In the sea environment fatigue damage can actually occur at all levels of stress and fatigue strength in the high cycle range is reduced more than 40% of its value in dry air (Schmitt and Phelps, 1969). The rate of growth of fatigue cracks is about five times as rapid in sewater as in dry air, while cathodic protection can further accelerate crack propogation by encouraging the production of hydrogen at the crack root (Ractliffe, 1983).

1.2.4 Timber Degradation in the Marine Environment

Timber can and generally does give very satisfactory performance in a marine environment, particularly in the components continuously submerged below the lowest water level. Timber, however, is susceptible to degradation by a variety of marine organisisms, such as bacteria, fungi, mollusks, and crustaceans (Figure 1–8). Premature degradation of timber sheet piles, regular wooden piles, or other structural components made of wood is often attributed to the lack of special treatment or to the improper selection of preservative type for the particular marine organism present at the particular project location. It should be noted that usually in a freshwater environment wooden structural components submerged below the lowest water level do not require any treatment at all and with time may gain some additional strength.

Destructive and fouling organisms are present practically at every seaport location. However, tropical and subtropical areas provide an environment that supports more species and greater activity than the more temperate and colder regions.

Marine organisms throughout the world cause damage or destruction of timber and other construction materials that costs hundreds of millions of dollars to repair.

(a)

(b)

Figure 1–7. Deterioration of sheet pile bulkhead in a marine environment. (a) Corrosion of steel sheet piles in splash/tidal zones. (b) damaged concrete cope wall. (c) concrete erosion by moving sediments.

(c)

Figure 1–7. (*continued*)

(a) (b)

Figure 1–8. Degradation of timber pile in a marine environment.

Certain marine organisms in the seawater environment, primarily from the taxonomic groupings of *mollusks* and *crustaceans*, will bore and destroy timber. Marine borers are distributed worldwide. Fortunately no one specific location contains all borers and most areas have only a few different kinds (Menzies and Turner, 1957). The presence of a particular species at a particualr site is largely depedent on environmental factors, such as temperature and salinity of water. The most important mollusks found in

North American waters are the *teredinids* and the *pholads* (Atwood and Johnson, 1924). The teredinids, commonly known as shipworms, are perhaps the most treacherous of marine borers, because the extent of damage they produce is largely unknown until the structure becomes unserviceable or totally collapses. Teredinids begin their life as free-swimming larvae less than 0.3 mm long. During the early period of life, which may last a few days to a couple of weeks, the larvae may settle and begin to bore, if suitable, unpreserved wood is found. Because the larvae are very small, the entry holes are and remain almost invisible.

On settlement the mollusk begins to metamorphose into a worm by using the wood as a sustained source of food. Within a year or so the teredinid may grow up to 25 mm in diameter and 120 cm in length (Atwood and Johnson, 1924). Clearly if many such worms are present in a pile, or any timber member of the structure, it cannot retain much strength.

Teredinids have been known to destroy untreated pine piling in one month. Although teredinid penetration is common within the tidal zone, attack is generally the heaviest near the mudline (Chellis, 1961). Discussion of the complete anatomy and natural history of the teredinid can be found in the work by Turner (1966).

The pholads, known as rock borers, resemble ordinary clams somewhat, as their bodies are entirely inside their shells. Pholads burrow for shelter and may penetrate soft rock, poor grade concrete, and clay as well as wood. The genus *Martesia* has been known to penetrate the solid lead sheathing of underwater power cable and has attacked the concrete jackets on wooden piles in Los Angeles Harbor (Muraoka, 1962). Approximately 70 to 80 mollusks/m^2 were found inhabiting 16 of 18 piles.

Like teredinids, pholads are initially free-swimming and settle as larvae. Unlike teredinids, however, they do not burrow as deeply as teredinids, often no further than the length of their shell, about 50 to 65 mm. Their entrance hole (about 5 mm) is much bigger than that of teredinids (Richards, 1983).

Burrows from the crustacean group are related to crabs, shrimps, etc. They are quite small, but may be present in sufficient numbers to produce substantial damage in a relatively short time. The principal genera are *Limnoria* and *Sphaeroma*. The mode of attack of the crustacean borers is somewhat different from that of the mollusks.

The Limnorians, for example, first penetrate the wood, then create tunnels just below and parallel to the surface with occasional perforations (for respiration) to the surface. The thin layers of their burrows are easily washed away, leaving a surface with miriad fine tunnels and exposing new wood for the next attack. According to Richards (1983), under heavy *Limnoria* attack a pile radius in a tidal zone may diminish by 10 to 15 mm/yr, and probably more in some cases.

Members of the genus *Sphaeroma*, known as the pill bugs (so named because of their tendency to roll into a ball when disturbed), tend to be somewhat darker in color and considerably larger than *Limnoria*, often reaching 10 to 15 mm in length. Damage by these borers is easily recognized by the large (around 5 mm in diameter) round entry holes that are usually perpendicular to the wood surface and grain, and about 10 to 20 mm deep. They attack wood predominantly in the tidal zone but have also been reported at the mudline (Atwood and Johnson, 1924). All types of borers respond to differences in environmental conditions such as water temperature, salinity, and presence of a suitable substrate for boring. But other factors such as water depth, current, pollution, siltation, oxygen content, turbidity, and suspended organic matter may also be important (Turner, 1966). Changes in environmental conditions such as prolonged droughts, heavy rains, or seasonal changes in water temperature can also affect borers' activities mainly through regulation of reproduction. For example, the discharge of a power plant's warm water into a harbor can result in more severe teredinid damage (Bletchly, 1967). Borers are generally more active in the tropics and within normal ranges of oceanic salinity. Their activity is generally decreased by increased turbidity and siltation, by increased light, and by fast moving currents. For example, teredinids will not attack in currents in excess of about 1.4 knots and *Limnoria* in currents over about 1.8 knots (Chellis, 1961).

Generally no site should be considered to be free from borers. Site-specific records of marine borer activity can be found for the US coast and worldwide in NAVDOCKS, June 1950; works by W. F. Clapp Laboratories; bulletin of BC Research Ltd; the article by Johnson (1987); and many other sources.

1.3 DAMAGES ATTRIBUTED TO DOCK OPERATION

These can be attributed to the following factors: ship hard docking when ship impact forces exceed design values, effects of cargo handling and hauling systems, ship propeller induced scour which may threaten to undermine the dock structure, or a combination of all of these.

1.3.1 Physical Damage to the Structure by Vessel and/or Cargo Handling Systems

Physical damage to the dock structure caused by a vessel can be generally categorized as either accidental or attributed to the method of ship handling. Accidental loadings by their very nature are a short-duration, one-time event. These loadings can generate stresses higher than the strength of the structure, resulting in localized or general failure. Determination of whether accidental loading caused damage to a structure requires knowledge of the events preceding discovery of the damage. Usually, damage due to accidental loading is easy to diagnose.

Visual examination will usually show damaged steel or wooden stuctural components, and spalling or cracking of concrete that has been subjected to accidental loadings.

Accidental loadings by their very nature cannot be prevented. The accidental damage to the dock structure is usually attributed to ship mechanical problems such as power loss or poor weather conditions. At times poor coordination of berthing/unberthing operations creates an accidental situation when a ship approaches or leaves the berth at an unacceptable angle or with unacceptable velocity.

Ship–structure collision under these conditions sometimes results in damage to the structure or its elements such as the fender system, piling, and others, and/or to the ship.

Traditionally the problem of ship/dock structure damage due to collision impact existed since the time when the first primitive dock was constructed for Roman or Viking crafts to moor against. Increase in ship sizes, particularly in the category of containerships, ferries, and oil and bulk carriers, substantially increased damage to port structures. Large ships with bulbous bows have added a new factor to potential damage to port structures.

At times quite substantial damages to the berth structure caused by power failure, mechanical problems, or by errors in berthing of ships literally of all sizes, but mainly by large vessels with bulbous bows, have been reported. A bulbous bow is ideal for piercing holes in sheet pile bulkheads and in thin-walled concrete structures, damaging gravity type vertical walls and catching the piling under piled platforms and all types of piled dolphins. Experience indicates that the damage to the structure caused by bulbous bows usually occurs in the outer row of piles nearest the face of the wharf. Very often these piles were centered as close as 0.3 to 0.5 m from the face of the berth. In some incidents other support piling, and the second row in particular, were damaged but not to the extent or frequency of the outer row.

The damage to the structure usually occurred when the vessel had been approaching the wharf at an angle that exceeded 10 to 15° and/or with an unacceptable velocity from the fender system energy absorbing capacity point of view. Hard docking may be affected by weather conditions, for example, heavy wind or high waves. On occasion additional tugs may be required to combat the force of the winds to ensure safe docking of a ship with a large sailing area. Hard docking, absence of, or inadequate fendering may result in greater wear and tear on dock walls. Lack of a proper fender system and errors in berthing may also result in removal of mooring bollards and hand rails by flared bows. Flared bows have been responsible in a number of accidents involving cargo handling equipment, for example, container cranes. This author is aware of at least one case of vessel–train collision attributed to the ship's flared bow. In the latter case the train track was located at about 2.75 m from the berth's face. Hard docking very often results in damage or complete destruction of the fender system.

It must be noted that the fender system is usually the most vulnerable element of the berth. If not properly designed, or not properly used and maintained, it can be damaged or can be completely torn off the structure by the ship.

A number of case histories of physical damage to dock structure by ships are found in Thompson (1977) and Gorunov (1974).

Sometimes the effect of changing cargo handling systems at the dock results in increase in loadings from equipment and/or cargo and produces a detrimental effect on the structure. In general the structure in-service must cope with not only the deterioration produced by the heavy quasi-static loads produced by mobile, portal, and container cranes and by a variety of heavy unit loads but also with the deterioration produced by the high axle and wheel loads of carelessly driven mobile equipment. Lift trucks can be driven with their forks catching the deck material as can the goosenecks on tractors. Trailers and trucks can reverse into fences, handrails, lighting posts, mooring accessories, curbs, and other fixed objects.

In general the older quay walls built in the past 50 years have been designed for about $25 kN/m^2$, whereas today we are looking at basic structures to handle unit and other loads in the order of three and even more times that amount. Use of heavy cargo handling equipment which outstrips the quay's designed capacity to carry these loads creates material fatigue conditions resulting in heavy overstress and disintegration of structural components such as sheet piles, bearing piles, deck structure, etc.

Cargo handling equipment today is very mobile and can be used anywhere in a port, and this is not always appreciated by their operators. The use of a heavy outrigger or wheel load sometimes results in local damages to the structure.

1.3.2 Propeller-Induced Scour

1.3.2.1 Introduction

Dramatic increases in ship dimensions and installed engine power, introduction of new types of special purpose ships with bow thrusters, and use of roll-on/roll-off ships in recent years are frequently reported as a source of dangerous scour which in many cases threatens to undermine berth structures (Römisch, 1977; Blaauw and van de

Kaa, 1978; Bergh and Magnusson, 1987; Biswas and Bandyopadhyay, 1987; Chait, 1987; Clausner and Truitt, 1987; Fuehrer et al., 1987; Longe et al., 1987; Robakiewicz, 1987; Verhey et al., 1987; Hamil, 1988 and others). Velocities at propeller jets at the exits of propellers, as well as at side thrusters can easily be 11 to 12 m/s with resulting bed velocities from 3 to 4 m/s (Longe et al., 1987). Depending on circumstances the average depth of scour could reach as much as 0.5 m/month and more.

In some cases damage produced by propeller jets has extended to washing out grout seals between joints in gravity type walls and in concrete sheet-pile bulkheads (Chait, 1987).

Disintegration of the grout seals caused sand backfill to leak through the joints, producing sinkholes in the surfacing of the quay. If allowed to progress then crane rails, rail tracks, buildings, etc. can be undermined.

In general, scour is dangerous to any kind of marine structure, but particularly to those relying on a designed bottom level to support these structures against horizontal and vertical loads, for example, all kinds of piled platforms, sheet-pile bulkheads, gravity type structures built upon erodible foundation material, etc., and rip-rap-protected slopes under open platform type structures.

The main factors that affect the extent of damage done to a dock structure are:

Power of propulsion unit or side thruster
Draft of vessel when berthing or unberthing
Position of propeller or side thruster in relation to height above keel
Position of ship's rudder
Shape of the vessel hull and beam
Underkeel clearance
Distance of quay and direction of thrust.

Propeller-induced scour has proved to be a very serious factor to be considered in design of new port-related marine structures or in reevaluation of performance of structures-in-service.

The possible measures to consider are:

- Adjustment of the design in such a way that the effect of the scour is neutralized
- Installation of different kinds of bed protection systems
- Imposition of different kinds of operational constraints, for example, reducing the jet velocities by reducing the number of propeller revolutions or by increasing the underkeel clearance of the vessel.

A decision on cost effectiveness is usually made on the basis of a cost–benefit analysis, taking into account the initial capital cost and the cost of maintenance.

It must be noted that besides propeller jet induced scour, a variety of other factors can cause bed erosion, such as wave attack associated with ship movement, underkeel current associated with a ship being moved laterally by tugs against a quay wall, or ship motion, for example, surging or rolling caused by long waves in the harbor.

1.3.2.2 Theory of Jet Flows Induced by Propulsion Unit

A rotating ship propeller generates a turbulent jet, with axial, radial, and tangential velocity components. The term "jet" in this context means a continuous stream of a fast-moving fluid discharging from some type of nozzle into a body of the same fluid.

Römisch (1977), Blaauw and van de Kaa (1978), and Oebius and Schuster (1979) have shown that at a certain distance (2.8 to 10 propeller diameters) from the propeller the velocity field is rather similar to that caused by a momentum jet.

Propeller-induced seabed velocity directly under the propeller due to water inflow to the propeller is significant over a circular area of approximately twice the propeller diameter. This velocity along the seabed varies with blade type, clearance, and/or propeller diameter.

There are three phases of propeller thrust action on the surrounding fluid that have a bearing on the propeller-induced scour problem (Prosser, 1986):

1. The propeller draws fluid into itself from the surrounding area. This action produces considerable flow velocities on the harbor/basin bed immediately under the propeller, and scour takes place in this area.
2. For a distance of about twice the propeller diameter downstream, the flow is accelerating to form a jet. In this region the flow is nonuniform, with the fluid that passed close to the blades moving faster than the remainder of the flow. During this phase, the overall diameter of the jet decreases substantially, particularly at the low speeds of advance of typical vessels maneuvering near a berth. The fluid in the jet is rotating about the core as well as moving axially.
3. Finally the jet is spreading, gradually entraining more of the surrounding fluid. During this phase the jet becomes more uniform circumferentially and resembles more clearly the flow due to a simple axial jet issuing from a pipe. Due to the spreading process, the jet will eventually impinge on the harbor bed to form a second and larger scour area downstream from the propeller.

As stated previously, velocities in propeller jets and from bow thrusters can easily be up to 10 m/s and more. With a small bed clearance these velocities can impinge directly on the seabed and are sufficient to move very heavy stones. Recently, calculation methods have been developed to predict the velocities in the jet flow of main propellers (Blaauw and van de Kaa, 1978; Fuehrer et al., 1981; Robakiewicz, 1987; Verhey et al., 1987).

These methods distinguish between the velocities at the propeller plane and velocities in the slip stream of the propeller. The afflux velocity from the propulsion system is calculated with the momentum theory applied by simplifying the propeller into an actuator disc.

Actuator disc theory assumes the following axioms (Figure 1–9):

1. An actuator disc is a circular area of a diameter equal to the propeller diameter. Each particle of the fluid flowing through this area is subjected to an abrupt rise of the pressure.
2. A separate column of fluid in the form of a body of revolution around the actuator disc axis appears as a result of the actuator disc action. The diameter of this body of revolution is equal to the actuator disc diameter.
3. All changes of the momentum due to the actuator disc action are limited within the said column of fluid, being a beam of streamlines flowing through the actuator disc.
4. The velocity far upstream from the actuator disc is equal to the uniform, undisturbed flow velocity V_0. Far downstream from the disc as a result of the acceleration, the fluid velocity $V_2 > V_0$.
5. The velocity distribution in successive cross-sections of the screw race is uniform.
6. The screw race pressures far upstream and far downstream from the actuator disc are equal to the ambient pressure.
7. The fluid stream flow upstream and downstream from the actuator disc is defined by the flow continuity equation and by Bernoulli's theorem.
8. No viscous affects are considered.

The above assumptions are so severe that they can only be representative of average

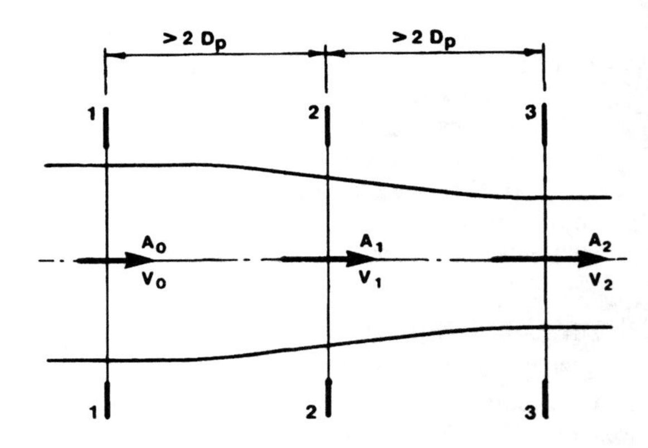

Figure 1–9. Actuator disc flow.

conditions at the propeller disc location. According to these assumptions it is obvious that

$$V_0A_0 = V_1A_1 = V_2A_2 \quad (1\text{–}8)$$

where A_0, A_1, A_2, and V_0, V_1 and V_2 are cross-sectional areas and velocities at cross-sections 1-1, 2-2, and 3-3 respectively.

The total thrust (T) on the propulsion unit can be derived from the momentum change in the jet between sections 1-1 and 3-3; therefore

$$T = \rho A_1V_1(V_2 - V_0) \quad (1\text{–}9)$$

where ρ = density of water.

The work done by the rotor (W) on the fluid at section 2-2 can be derived from the change in kinetic energy of the jet between sections 1-1 and 3-3.

$$W = 0.5\rho A_1V_1(V_2^2 - V_0^2) = T_rV_1 \quad (1\text{–}10)$$

where T_r = rotor thrust.

It must be noted that for a ducted propeller unit, part of the total thrust is carried by the duct so that T_r and T are not equal. For nonducted propellers $T_r = T$. Therefore, for a nonducted propeller unit, by combining Eqs. (1–9) and (1–10),

$$0.5\rho A_1(V_2^2 - V_0^2) = \rho A_1V_1(V_2 - V_0)$$

from which

$$V_1 = 0.5(V_2 + V_0). \quad (1\text{–}11)$$

Then from Eq. (1–8)

$$\frac{A_2}{A_1} = \frac{V_1}{V_2} = \frac{0.5(V_2 + V_0)}{V_2}. \quad (1\text{–}12)$$

From Eq. (1–12) the reduction of jet diameter behind the nonducted propeller can be obtained:

$$A_2V_2 = 0.5A_1(V_2 + V_0). \quad (1\text{–}13)$$

Hence, with $V_0 = 0$, $A_2 = \frac{\pi}{4}D_0^2$ and $A_1 = \frac{\pi}{4}D_p^2$

$$\frac{\pi}{4}D_0^2V_2 = 0.5\frac{\pi}{4}D_p^2V_2 \quad (1\text{–}14)$$

from which

$$D_0 = D_p/(2)^{0.5} = 0.7D_p. \quad (1\text{–}15)$$

The race contraction for a ducted propeller is much smaller and it is usually assumed that $A_1 = A_2$ and therefore $V_1 = V_2$ for such a propeller.

According to Verhey et al. (1987) using the momentum change caused by the propeller and the condition of mass conservation, the relationship between the total unit thrust (T), propeller diameter (D_p), and afflux velocity (U_0) can be expressed in the case of zero speed of advance ($V_s = 0$) as

$$\text{for an open propeller: } T = \frac{\pi}{8}\rho D_p^2U_0^2 \quad (1\text{–}16)$$

$$\text{for a ducted propeller: } T = \frac{\pi}{4}\rho D_p^2U_0^2. \quad (1\text{–}17)$$

For a definition sketch see Figure 1–10.

The afflux velocity can then be calculated using the thrust coefficient K_T which is defined as

$$K_T = T/(\rho n^2/D_p^4) \quad (1\text{–}18)$$

where n = number of revolutions of the propeller per second. This results in:

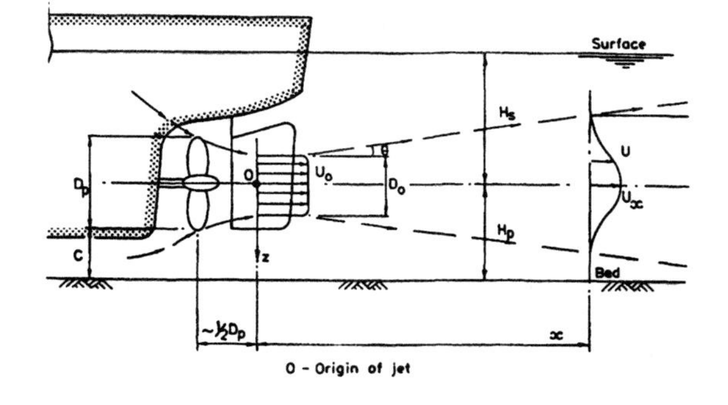

Figure 1–10. Propeller jet, definition sketch.

for an open propeller: $U_0 = 1.6nD_p(K_T)^{0.5}$ (1–19)

for a ducted propeller: $U_0 = 1.1nD_p(K_T)^{0.5}$. (1–20)

Actual values of K_T will depend on details of propeller and duct design but the primary parameter is the pitch/diameter ratio of the propeller. Prosser (1986) recommends that at zero speed of advance the values of $K_T = f(p/D_p)$ as given in Table (1–2). Verhey et al. (1987) suggested that if $K_T = K_{T_p}$ for a nonducted propeller, and $K_T = 2K_{T_p}$ for a ducted propeller, then this would result in one equation for U_0 for both propulsion systems:

$$U_0 = 1.6nD_p(K_{T_p})^{0.5}. \quad (1\text{–}21)$$

Generally side thrusters can be schematized as ducted propellers with the exception that the propeller sits in a tube that is substantially longer than a normal size duct would be. However, Verhey et al. (1987) indicated that the losses due to a propeller being mounted in a long tube may be ignored. Consequently, the same equation (1–21) can be used to calculate the afflux velocity (U_0) for bow thrusters as well as for the main ducted propellers. The flow behind a propeller can be considered as a three-dimensional jet flow from an orifice with homogeneous and constant afflux velocities. Then on the basis of jet flow theory, the axial velocities in propeller jets (U_x) can be calculated with Eq. (1–22) (Blaauw and van de Kaa, 1978; Blaauw et al., 1984; Fuehrer et al., 1981; Verhey et al., 1987)

$$U_x = U_{\max} \exp\left[-15.4\frac{r^2}{x^2}\right] \quad (1\text{–}22)$$

where

$$U_{\max} = AU_0(D_0/x)^a \quad (1\text{–}23)$$

The coefficient A and the exponent a are functions of vessel type, rudder configuration, and bed and lateral limitations.

Fuehrer et al. (1981) recommend the following values for exponent a:

$a = 6$ on limitation by bottom and water level

$a = 0.3$ on additional limitation by a lateral wall (quay).

Values of coefficient A are given in Tables 1–3 and 1–4.

Again it must be noted that Eqs. (1–22) and (1–23) are valid only for ships moving at very low speeds, when the effect of ship movement on the jet flow is small. An exception is a main propeller running astern, in which situation the jet flow is deflected to the bed.

For an undisturbed jet without bed and lateral limitations a maximum velocity near

Table 1–2 Thrust Coefficient (K_T) as Function of Propeller Pitch/Diameter Ratio

P/D_p	0.6	0.8	1.0	1.2	1.4
K_T (open)	0.26	0.37	0.48	0.57	0.54
K_T (ducted)	0.24	0.37	0.51	0.67	0.82

P = propeller pitch.

Table 1–3 Ship with Central Rudder $A = f(H_p/D_p)$

H_p/D_p	0.5	1.0	2.0	3.0	4.0	5.0
A	1.75	1.6	1.4	1.2	0.95	0.8

From Fuehrer et al. (1981)

Table 1–4 Ship Without Rudder $A = f[(H_s + H_p)/D_p]$

$(H_s = H_p)/D_p$	1.0	2.0	3.0	4.0	5.0	6.0	7.0	8.0	9.0	10
A	1.75	1.6	1.4	1.3	1.2	1.1	0.95	0.85	0.8	0.75

From Fuehrer et al. (1981)

the bed (U_b) may be expected at distances of about 5 to 10 times H_p behind the propeller. The value of U_b as recommended by Verhey et al. (1987) can be determined by Eq. (1–24):

$$U_b = 0.3U_0(D_0/H_p). \quad (1\text{–}24)$$

The coefficient 0.3 corresponds to average values of $A = 2.8$ and $a = 1.0$ but may vary with the ship type as well as propeller and rudder configuration.

It must be noted that estimates of U_0 and D_0 are approximate. Therefore one cannot expect to make an analysis of the jet scouring effect with mathematical precision and these calculations must be treated as approximate.

A direct impingement of a jet flow against a vertical wall occurs with the main propeller(s) of roll-on/roll-off ships with a stern ramp and also with bow or stern thrusters (Figure 1–11). The jet is deflected by the wall with a component turning downward to the bottom and then along the bottom. Because of the limited water area there will be very little flow entrainment, and consequently high bottom velocities will be generated.

Prosser (1986) suggests that at a first approximation the peak velocities experienced by the bed local to the ramp structure will be similar to the peak velocity in the propeller jet just before it strikes the wall. This velocity can be calculated by Eq. (1–23) or from the centerline decay equation given in Figure (1–12).

In addition it must be noted that the presence of the rudder has a significant effect on the propeller jet. With a normally positioned rudder mounted astern of the propeller in the vertical plane through the propeller shaft there is a significant effect on the propeller jet at zero or smaller advance ratios. In this case the rudder acts only as a splinter and the initial jet typically is divided into two jets, one of which is deflected sideways and upwards and the other sideways and downwards (Figure 1–13).

In the case of twin ruddered or twin screw ships, rudders are not mounted in the plane of the propeller shaft. In the latter the rudder effect on a screw jet at low or zero advance ratios may not be significant.

More information on rudder effects on propeller jets can be found in works by Verhey (1983), Fuehrer and Römish (1977), and Robakiewicz (1966, 1987).

The rudder effect on maximum jet velocities at the bed in a confined area is shown in Figure 1–14.

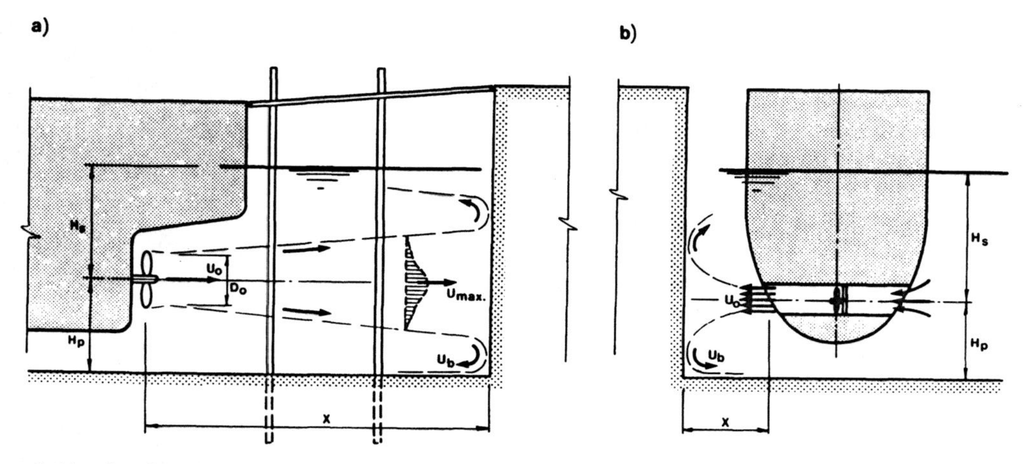

Figure 1–11. Jet flow deflection toward the bed. (a) Case of a Ro/Ro ship with astern ramp. (b) Case of a bow side thruster.

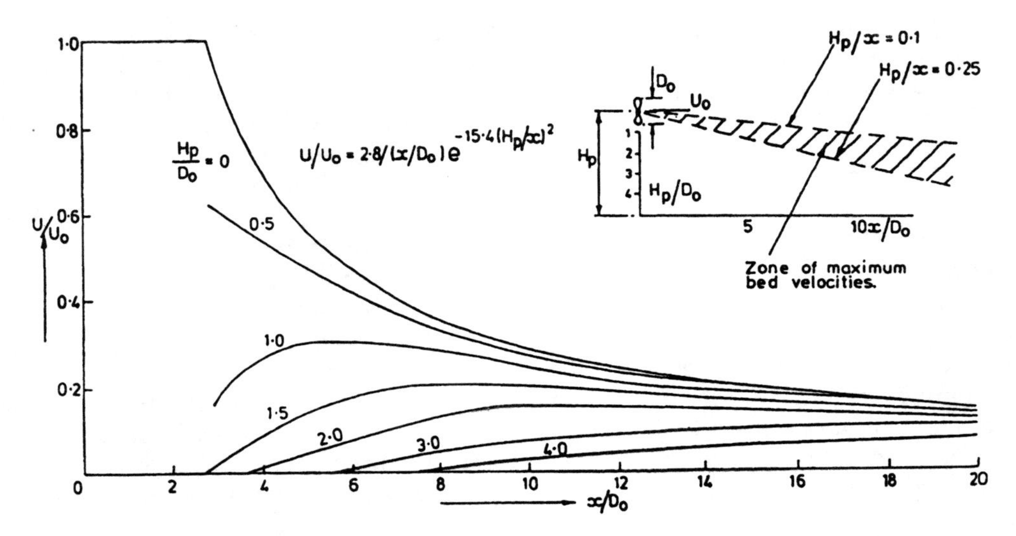

Figure 1–12. Velocity of seabed for various values of propeller clearance in absence of quay walls, berthing ramps, and other structures. (From Prosser, 1986.)

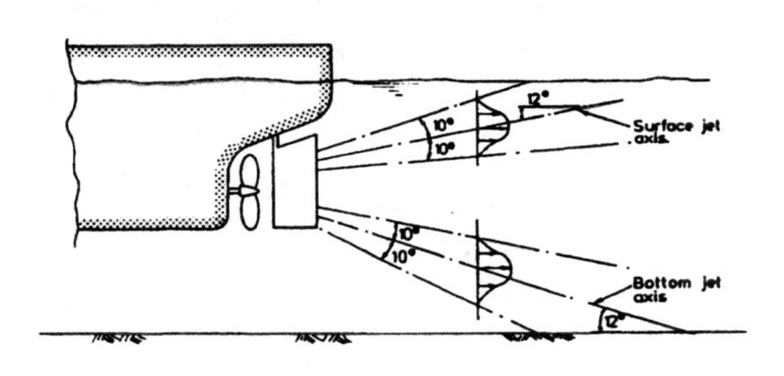

Figure 1–13. Effect of rudder on propeller induced jet. (From Fuehrer and Römish, 1977.)

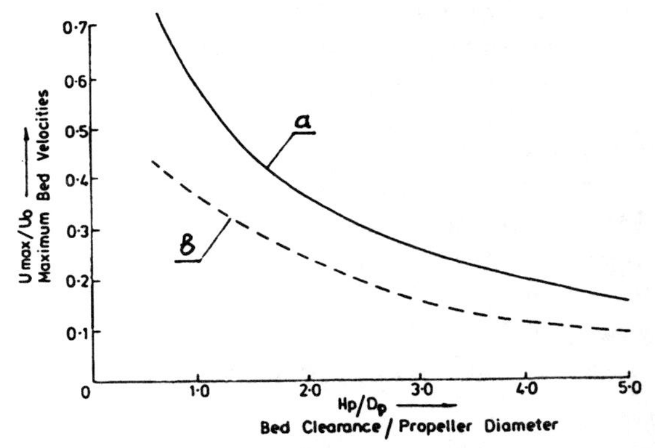

Figure 1–14. Correction of maximum bed velocities allowing for effects of rudder and quay wall. (a) Allowance for rudder and quay wall according to Fuehrer and Römish, 1977. (b) Data from Figure 1-12—no rudder or quay wall effect. (From Prosser, 1986.)

It is worth noting that due to geometric constraints of a single hull in a typical multi-screw ship design the space between propeller discs is unlikely to exceed one, or at the most two, propeller diameters. With such close spacing, the jets from each propeller will merge together fairly rapidly, so that by about 10 jet diameters downstream, the velocity distribution is similar to that produced by a single, larger jet. In practice, however, propellers of multiscrew ships are likely to be used individually for maneuvering purposes, for example, short periods of running one propeller ahead and one astern. This will set up very complicated flow patterns, thus making prediction of bottom velocities very difficult.

1.3.2.3 Bottom Erosion and Scour

Flow patterns and impingement velocities resulting from ship hull and propulsion unit characteristics, as well as berth geometry, exert a load upon the bed material. The resistance of this material to erosion depends primarily on material characteristics.

Bed erosion does not present a real threat to the structure if it occurs at such a distance from the berth structure that its stability is not endangered.

It can, however, create a problem if sedimentation of the displaced material will substantially reduce underkeel clearance in the access channel, turning basin, or near the berth structure.

Generally, significant bed scour in the immediate vicinity of a gravity type quay wall, sheet-pile bulkhead, piled platform, or other structure whose stability is based on a particular design bed level can be extremely detrimental to short- or long-term stability of these structures.

Piled structures provide obstruction to the jet flow. In general, the effect of such an obstruction in a water jet is to produce higher velocities at the bed within an area of one or two pile diameters from the pile. In this area a maximum scour depth of two or three times the pile diameter can be expected.

The search for accurate and physically correct relationships describing bed erosion and sediment movement has been conducted all over the world for well over a century by hundreds of researchers and literally hundreds of technical articles have been published on the topic. The behavior of sediments in an erodible bottom is a complex phenomenon and only the basic approach is outlined here; further details can be found in Chow (1959), ASCE (1966), Jansen et al. (1979), Lane (1979), Herbich et al. (1984), and many other works.

Generally speaking, sediment is transported by flowing water as a bed load, a suspended load, or as a combination of both. The bed load comprises particles that roll or slide in continuous contact with the rest of the bed plus particles that move by a saltation, which is the process when particles "bounce" along the bed in a series of very low trajectories. The suspended load consists of small particles that are kept in suspension by the upward components of the turbulence in the flow. In 1936 Shields produced a diagram that defines a threshold of sediment movement (Figure 1–15). This diagram is based on experimental data collected from flume experiments, and a curve $F_s = f(R_e)$ defines the onset of bed material movement.

$$F_s = \frac{\tau_0}{(w_s - w)d_s} \quad \text{and} \quad R_e = Vd_s/\nu$$

where F_s = dimensionless shear stress;
τ_0 = bed material shear stress.

Blaauw and van de Kaa (1978), with reference to experiments at the Delft Hydraulics Laboratory, recommend using

$$\tau_0 = 0.5c_f\rho V^2.$$

In the above equations
c_f = 0.06 to 0.11;
w_s = specific weight of bed material;
w = specific weight of water;
d_s = diameter of bed material when of uniform size
R_e = particle Reynolds number
$V = (\tau_0/\rho)^{0.5}$ = shear velocity
ν = kinematic velocity
ρ = bed material density $\rho = w/g$, where g = gravitational acceleration.

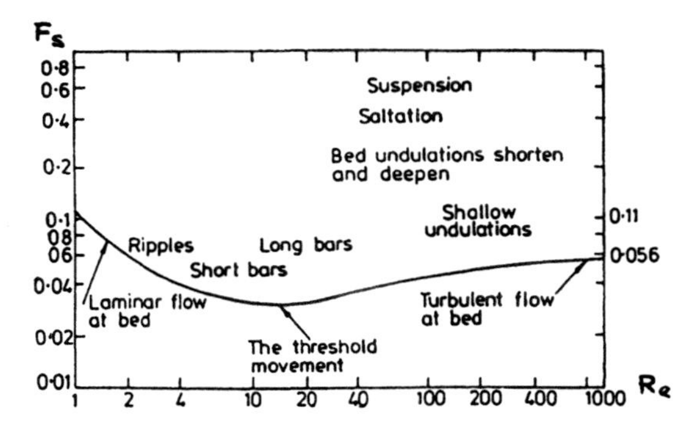

Figure 1–15. Shields diagram. Threshold of bed material movement. (From Shields, 1936.)

With reference to Figure 1–15, if for a given condition the value of F_s falls below the line, then there will be no sediment transport.

For conditions that are normally encountered with beds composed of coarse grain sands, gravels, and stones the Reynolds number R_e is relatively large ($R_e > 100$). Therefore under $F_s = 0.004$–0.006 one cannot expect any significant bed movement. The resistance of noncohesive bed material to erosion can be verified with the following Isbash formula. $D_{50} > KU_b^2/g\Delta$, (Verhey, 1983; Blaauw et al., 1984; Verhey et al., 1987), where D_{50} = mean diameter of bed material and γ = relative density of bed material ($\Delta = \rho/\rho_w - 1$, where p_w = density of water). According to Verhey (1983) and Fuehrer et al. (1981) the coefficient K varies between 1.0 and 2.0 (the average value usually used is $K = 1.3$).

Generally, once the local bed shear stress and hence local velocity increases above the threshold for sediment movement, scour will develop until the shear stress within the scour hole falls to below the threshold value.

For noncohesive, coarse sediment the Permanent International Association of Navigation Congresses (1987) recommended for most practical cases the formula (1–25) developed by Pilarczyk (1985, 1986) as a safe estimate of the critical velocity causing bed erosion.

$$\frac{D_{50}}{h} = \left(\frac{U_{b(cr)}}{B_1(k^1\psi_{cr}g\Delta h)^{0.5}}\right)^{2.5} \tag{1–25}$$

where

h = water depth

$U_{b(cr)}$ = critical bottom velocity

B_1 = coefficient estimated as follows:
major turbulent flow $B_1 = 5$–6
normal turbulence $B_1 = 7$–8
minor turbulence $B_1 = 8$–10

k^1 = slope reduction factor $k^1 = (1 - \sin^2\alpha/\sin^2\phi)^{0.5}$ where α = slope angle and ϕ = angle of internal friction of bottom material

ψ_{cr} = Shields parameter:
for absolute rest $\psi_{cr} = 0.03$
for start of instability $\psi_{cr} = 0.04$
for movement $\psi_{cr} = 0.06$.

Pilarczyk's formula can also be used to estimate the critical velocity causing erosion of granular cover layers such as small size riprap.

However, whatever method is adopted to estimate the critical velocity that can cause bed erosion it is stressed that experience and sound engineering judgement must play a large role at this stage of the design.

Verhey (1983) proposed the empirically derived formula to predict the maximum scour produced by the propeller jet (Z_{max}) as a function of the bed material (w_s and d_s) and the propeller jet's initial characteristics (D_0, U_0, and H_p):

$$Z_{max} = 0.004H_p\left(\frac{F_0}{H_p/D_0}\right)^{2.9} \tag{1–26}$$

where F_0 = Froude number:

$$F_0 = U_0/\left[gd_s\frac{(w_s - w)}{w}\right]^{0.5} \tag{1–27}$$

The formula (1–26) is limited to $0.1\,\text{m} < d_s < 0.3\,\text{m}$.

The extent of the scour for a given fixed position of a ship can be estimated from Figure 1–12, by determining the length over which the jet velocity is greater than the threshold velocity for the bed material. The possible width of scour can be estimated from the fact that the jet diffusion angle cannot be greater than ±12° from the jet centerline.

Cohesive materials such as clay generally have higher resistance to erosion than noncohesive materials. At present the approach to determination of critical velocity for cohesive material still relies heavily on empirical data based on various experiments and in situ observations (Chow 1959; ASCE, 1966; Delft University Press, 1984).

For the first approximation to the erosion resistance of various subsoil material the following value may be used:

Fairly compacted clay (void ratio = 0.5)	0.80 m/s
Stiff clay (void ratio = 0.25)	1.50 m/s
Grassed clay	up to 2.00m/s.

1.3.2.4 Effect of Astern Rotation of the Propeller

Quite different conditions of propeller operation occur during astern motion (Robakiewicz, 1987). The propeller characteristics for astern motion are worse than for forward motion because of the unfavorable pitch distribution and disadvantageous flow conditions. An average loss of propulsive efficiency by 50% should be taken into account. It has been found during the laboratory tests that the depth of the bottom erosion cavity for astern motion of the propeller is about three times less than for forward motion provided that the rudder position is the same and the greatest cavity depth occurs in the longitudinal axis of the ship.

1.3.2.5 Duration Effect

Not very many studies produced to date have addressed the effect of duration of propeller jet induced velocities on bottom scour. It is reasonable to expect that the longer the propeller jet flow acts on bottom material, the larger and deeper the scour cavity that will be produced, up to some limiting time, after which little additional scour will take place. Studies done by Oebius (1984) indicated that in a model test conducted for 10 h, over 50% of the final scour depth occurred in less than 30 min. Hamil (1988) has found that the depth of maximum scour from a propeller jet (ε_{max}) depends on jet velocity, propeller tip clearance (C), and sediment size (D_{50}), and is proportional to the natural log of time (t).

$$\varepsilon_{max} = 45.04^{\Gamma-6.98}(\mathrm{Ln}t)^{\Gamma} \tag{1–28}$$

where indices Γ can be obtained by the following formula (1–29):

$$\Gamma = 4.1135(C/D_{50})^{0.742}(D_p/D_{50})^{-0.522}F_0^{-0.682} \tag{1–29}$$

Equation (1–28) is recommended for use on any sand within the medium to coarse range. Its use on gravels and very fine sands would require further verification of the equation, though it would provide an estimate of the scour to be expected.

1.3.2.6 Recommendations

In view of a continuing increase in ship sizes and installed engine power, the introduction of side thrusters, and use of roll-on/roll-off ships, it is very important to consider potential propeller jet induced erosion at quay structures, either new or in-service.

Possible measures to protect the quay wall are as follows:

1. Protect the bed at locations exposed to the propeller jet induced scour.
2. Increase the depth of water, which means increasing the underkeel clearance of the ship and results in an increase in the height of the quay wall.
3. Imposition of operation constraints such as reducing the jet velocities by reducing the number of propeller revolutions.

A decision has to be made on the basis of a cost–benefit analysis, taking into account the initial capital cost and the maintenance cost.

In some cases it may be more cost-effective to install a cheaper bed protection which requires regular maintenance work instead of applying an expensive but maintenance-free protection (Verhey et al., 1987).

1.4 COST-EFFECTIVE APPROACH TO EVALUATION OF THE DOCK IN-SERVICE

In spite of some deterioration and quite obvious overloading conditions (compared with original design criteria), some older and newer dock structures function without visible deformation and deflection and generally require just regular maintenance. Others bear obvious signs of deterioration, overstress, and fatigue, and if reduced structural capacity is not properly addressed in time, this may eventually result in increased risk of consequential damage to the facility and to ships. A recently completed review of maintenance methods in use at conventional steel and concrete dock facilities (Bakun, 1986) indicated that, in some cases, contractors had conducted unfounded, expensive, and wide-ranging inspections of deteriorated docks. Structures were often restored to their original condition without a prior evaluation of consequences of the deterioration. It has also been noticed by many investigators that, in a great many instances, in-service dock structures (even those substantially deteriorated) still possess adequate service capacity and safety standards. There are many good reasons for this, but the most common include the following.

1. The original design was based on conservative assumptions. Usually, this included assumed geotechnical soil parameters, combination of design loads, ignoring some specifics of soil–structure interaction, etc.
2. The design may have included substantial allowance for corrosion.
3. Concrete had gained additional strength over the original assumed value.
4. Original timber or rubber tire fenders were replaced by a new sophisticated fender system which substantially reduced ship impact forces.

The above factors may contribute to the fact that at times even substantially deteriorated or damaged structures do not compromise a dock's safe operation. In such cases a so-called "judicious neglect" method of repair may be applicable. (The term was suggested by US Army Corps of Engineers, EM 1110-2-2002, 1986.)

As the name implies, judicious neglect is the repair method of taking no action. This method does not suggest ignoring situations in which damage to the structure has been detected. Instead, after a careful review of the circumstances (i.e., "judicious"), the most appropriate action may be to take no action at all.

Judicious neglect would be suitable for those cases of deterioration in which the damage to the structure material is not causing any current operational problems for the structure and that will not contribute to future deterioration of the structure. For example, cracks in concrete such as those due to shrinkage or some other one-time occurrence are frequently self-sealing. This does not imply an autogenous healing and gain of strength, but merely that the cracks clog with dirt, grease, or oil, or perhaps a little recrystallization occurs, and so on. The result is that the cracks are plugged and problems that may have been encountered with leakage, particularly if leakage is due to some intermittent cause rather than to a continuing pressure head, will disappear without any repair having been done. Hence, before decisions on costly maintenance work, or use of existing dock facilities for servicing larger ships, or use of much heavier cargo handling and hauling equipment are made, it is prudent to establish the actual capacity of the facility and possible load limitations associated with the facility's present and expected use for the design period of time.

A similar philosophy was advocated by Buslov (1992), who suggested that a cycle of about 5 years, to repair only well-developed deteriorated concrete could be appropriate for most marine structures, operated in a saltwater environment.

The successful cost-effective evaluation of the dock structure actual capacity usually comprises the following basic stages:

Inspection
Engineering evaluation.

A repair/rehabilitation method is then established. If required, a long-term maintenance plan and procedures are developed.

1.4.1 Inspection

The inspection is essentially the process of data gathering. The final objective of inspection is to determine the structure's present condition and to identify only those repairs that are required to ensure safe operation of the dock facility for the design life of the structure.

The success of the inspection is largely dependent on the *inspection criteria* (specification) and inspection execution.

Inspection criteria typically include:

- Data required for inspection, that is, general description of the structure, construction, and service history; operation requirements; and original design and "as-built" drawings
- Purpose of the particular inspection task, and parts of the structure where the inspection is to be performed
- Inspection method and required types of equipment to be used
- Sets of logs and forms for manual records
- Scope of inspection work, which includes hydrographic, geotechnical, diving, and material survey programs
- Work schedule.

Inspection execution is significantly affected by the inspection method selected, as well as by the objectives to be achieved. For example, searching for cracking or material deterioration subjected to marine growth requires thorough surface cleaning, or searching for propeller-induced bottom scour requires involvement of divers equipped with specialized equipment to enable them to determine the depth and extent of scour. During inspection, it is imperative that the particular spot being inspected be properly identified. The most common method is to relate these spots to established profiles, or pile bents along the structure, and at given elevations.

The inspection procedure usually comprises two phases: preliminary and detailed inspection.

1.4.1.1 Preliminary Inspection (Phase I)

This consists of a visual scan (both above water and underwater) of the structure. It is conducted for preliminary assessment of the general condition of the structure and to identify potential problem areas requiring follow-up action.

The inspecting diver must be skilled as an engineer and as a diver. All diving operations must conform with standards established by the appropriate authority.

It should be noted that preliminary underwater inspection is essentially limited to visual observation by divers (often obstructed by suspended sediments). If visibility is less than 0.25 m structures should be inspected using large sweeping motions of the diver's hand.

The main objective of the preliminary inspection is to ascertain whether any damage has occurred to the structure. The basic qualitative data obtained from visual inspection are generally inadequate to accurately assess the condition of the structure. Therefore, if any significant damage to the structure has been found in the process of visual inspection, then another inspection phase is usually ordered.

The most important goal of the Phase I stage is to define future required action. Properly conducted, inexpensive preliminary inspection can prevent costly, wide-ranging, detailed inspection and limit it to that reasonably required. Actually, a prudent owner would conduct a preliminary inspection every 1 to 3 years, depending on structure deterioration rates. Concise documentation of each such inspection is essential to the success of cost-effective decisions on detailed inspection, and to the evaluation of the actual capacity of the structure.

A visual inspection of the structure is aimed at identifying and defining potential areas of distress. It would normally include a mapping of the current state of various types of structural deficiencies such as construction faults, distortion or movement, and material deterioration.

Construction faults typically found during a visual inspection include use of structural elements and materials that deviate from the original design and specifications. They may also include local deficiencies such as honeycomb evidence of cold joints, exposed reinforcing steel, irregular surfaces caused by improperly aligned forms, and a wide variety of surface blemishes and irregularities in concrete, and poorly welded or insufficiently bolted connections in steel structures, and in improperly treated timber elements. These faults are typically the result of negligence, poor workmanship, or failure to follow accepted good practice.

Distortion and movement, as the terms imply, are simply changes in alignment of the structural components, which include buckling, settling, tilting, warping, or differential settlement of adjacent sections of gravity type structures. Review of historical data such as periodic inspection reports may be helpful in determining when movement first occurred and the apparent cause and rate of movement.

Material deterioration typically includes corrosion of steel, decay of timber, cracking, disintegration (blistering, delamination, weathering, dusting), and spalling of concrete. During visual underwater inspection the presence and extent of the propeller jet induced erosion, for example, washout of sealing grout between concrete blocks or concrete sheet piles, as well as presence and extent of bottom scour shall be established. Visual survey of the structure is normally formalized by a clear, concise report supplemented by surface mapping.

Surface mapping is a procedure to survey a structure in which deterioration of the structural material is located and described. Surface mapping may be accomplished using detailed drawings, photographs, or video tapes.

Typically mapping begins at one end of the structure and proceeds in a systematic manner until all surfaces are mapped. Use of three-dimensonal isometric sketches showing offsets of distortion of structural features is occasionally desirable. Areas of significant distress should be photographed for later reference. A familiar object or scale should be placed in the area to show the relative size of the feature being photographed. It is important to describe each condition mapped in clear, concise detail and to avoid generalizations unless reference is being made to conditions previously detailed in other areas.

During a visual survey particular attention should be paid to the condition of joints. Opened or displaced joints (surface offsets) should be checked for movement if appropriate; various loading conditions should be considered when measurements of joints are taken. All joints should be checked for potential defects such as spalling or cracking, chemical attack, evidence of emission of solids, etc. Conditions of joint filler if present should also be examined. More useful information on visual inspection of marine structures is found in US Army Corps of Engineers EM 1110-2002 (1986) and in the article by Milwee and Aichele (1986).

1.4.1.2 Detailed Inspection (Phase II)

The objective of a detailed inspection is to gather quantitative data so that an engineering evaluation of dock capacity or actual strength of its particular elements can be performed. Specialized inspection equipment and techniques are required to accurately gather quantiative data to assess the condition of underwater parts of the structure.

New tools and techniques for above water and underwater inspection are discussed by Smith (1987), De Lange et al. (1985), Milwee and Aichele (1986), and in American Concrete Institute ACI 228.1R (1988).

It must be noted that over the years many new inspection techniques have been proposed for estimating the in-place properties of materials. The rapid progress in development of new techniques still continues and no doubt will continue to progress in the future. Hence, a detailed discussion on new system capabilities is likely to be quickly outdated. Accordingly, no attempt is made in this work to review all of these techniques; only those methods and instruments that are customarily used for the inspection of structures are briefly discussed.

In-place tests, which have also been called *nondestructive tests*, are normally used in detailed inspection to obtain information about the properties of materials as they exist in a structure.

These tests are usually performed for two basic reasons: the evaluation of the existing structure, and monitoring strength development during new construction. Inspection methods commonly used for detail structure survey, both above water and underwater, are: ultrasonic material testing, concrete testing by rebound hammer or the probe penetration method, material coring, and sampling. Some recently developed techniques include photogrammetry and alternating current potential, surface potential testing. Traditional tools for concrete testing such as a rebound hammer and probe penetration are still indispensable for material investigation.

All inspection tehniques require some cleaning of the surface to obtain accurate observations and measurements. The degree of cleaning required is dependent on the inspection technique to be used and the structural material. The surface cleaning usually involves removal of marine fouling and corrosion. It is usually a time-consuming operation; therefore, improved methods of surface cleaning are needed to decrease surface preparation time.

Among the presently used techniques for surface cleaning are powered and hand brushes, scrapers and grinders, and high-pressure waterjet cleaning systems. The latter is the most efficient tool. It can deliver a waterjet to the cleaning surface at 82.5 MPa (Smith, 1987). It is usually available in the form of a handheld waterjet pistol with various nozzle assemblies that can be easily changed at the work site to match requirements. Because of the high pressure, however, these devices must be used with great caution as the high pressure jets tend to remove sound material on wooden and concrete surfaces.

Rebound hammer, a traditional tool for concrete testing, is used for both above water and underwater material testing. The rebound hammer consists of a steel mass and a tension spring in a tubular frame (Figure 1–16). When the plunger of the hammer is pushed against the surface of the concrete, the steel mass is retracted and the spring is compressed. When the mass is completely retracted, the spring is automatically released and the mass is driven against the plunger, which impacts the concrete and rebounds. The rebound distance is indicated by a pointer on a scale that is usually graduated from zero to 100. The rebound readings are termed R-values. Determination of R-values is usually outlined in the manual supplied by the hammer manufacturer. R-values indicate the coefficient of restitution of the concrete; the values increase with the "strength" of the concrete.

Most hammers come with a calibration chart, showing a purported relationship between compressive strength of concrete

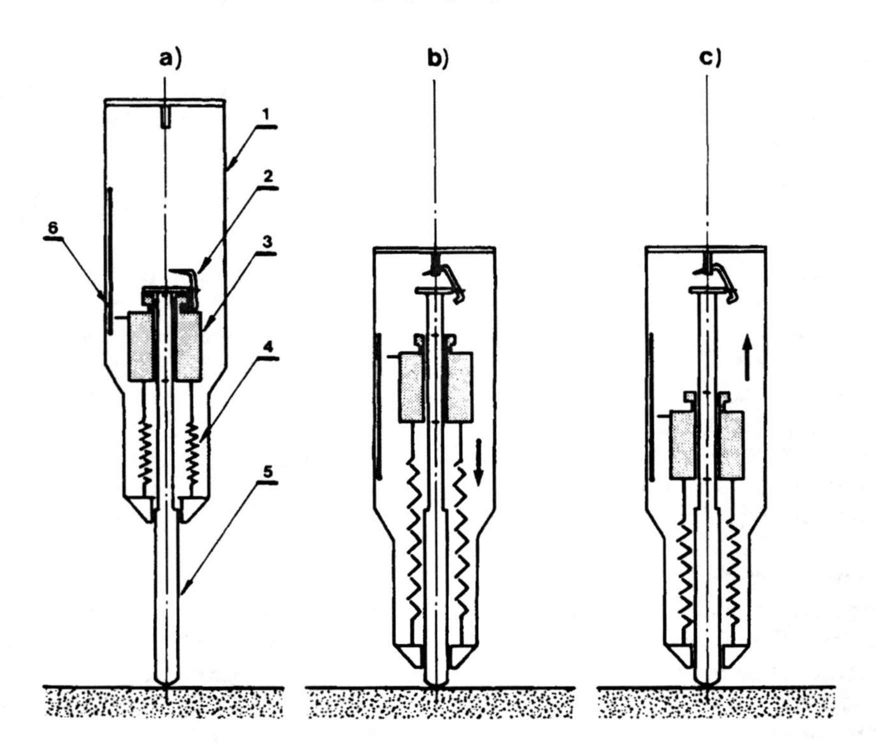

Figure 1–16. Schematic illustration of the rebound hammer and its operation. (a) Rebound hammer: typical assembly. (b) The body of instrument is pushed toward the concrete component up to its limit. The action causes an extension of the spring. The latch releases and the spring pulls the hammer toward the concrete member. (c) The hammer impacts the shoulder area of the plunger and rebounds. The indicator records the rebound distance. 1—Body, 2–latch, 3—hammer, 4—spring, 5—plunger.

and rebound readings. However, rather than placing confidence in such a chart, users should develop their own relations for each concrete mixture and each rebound hammer.

For underwater testing the hammer is equipped with a special auger, connected to a topside recording instrument that collects and stores data (Smith, 1987). Structurally the rebound hammer for underwater testing is about the same as for testing in air.

It must be pointed out that although the rebound hammer test is very simple to perform, there are many factors other than concrete strength that may affect the test result. In rebound hammer testing, only the concrete in the immediate vicinity of the hammer's plunger affects the hammer rebound value. Hence, the test is sensitive to the local conditions where the test is performed. For example, the location of plunger over a hard or soft aggregate particle or over a large void will result in increased or lower rebound readings. The presence of a layer of carbonization can result in higher readings than are indicative of the interior concrete. Likewise, a dry surface will result in higher rebound numbers than for the moist interior concrete, and so on. To account for different possibilities, ASTM C 805 requires that a minimum of 10 rebound readings be taken for a test. If one of these readings differs by more than seven units from the average, then this particular reading should be discarded and a new average should be computed based on remaining readings. If more than two readings differ from the average by seven units, then the entire set of readings is to be discarded.

Despite some disadvantages the rebound hammer is a simple and quick method for nondestructive testing of concrete in place. The equipment is inexpensive and can be operated by field personnel with a limited amount of instruction. The rebound hammer is very useful in assessing the general quality of concrete and for locating areas of poor quality concrete. A large number of measurements can be rapidly taken so that large exposed areas of concrete can be mapped within a few hours.

The probe penetration technique from a fundamental point of view is similar to the rebound hammer test, except that the probe impacts the concrete with much higher energy than the plunger of the rebound hammer.

The apparatus most often used for penetration resistance is a special gun that uses a 0.32 caliber blank with a precise quantity of powder to fire a high-strength steel probe into the concrete. The probe penetrates into the concrete to the distance required for the absorption of its initial kinetic energy which is governed by the size of the powder charge to fire the probe. An essential requirement of this test is that the probe has a consistent value of initial kinetic energy. ASTM C803 requires that the exit velocities of probes should not have a coefficient of variation >3% based on 10 tests by approved ballistic methods.

The probe tip travels through the mortar and aggregate, creating a cone-shaped region (fractured zone) (Figure 1–17). The important characteristic of the probe penetration test is that the type of coarse aggregate has a strong effect on the correlation, between concrete strength and probe penetration. For example, for equal compressive strength, concrete with soft aggregate will result in greater probe penetration than concrete with hard aggregate. Therefore, the penetration probe should be calibrated in each individual case of material investigation.

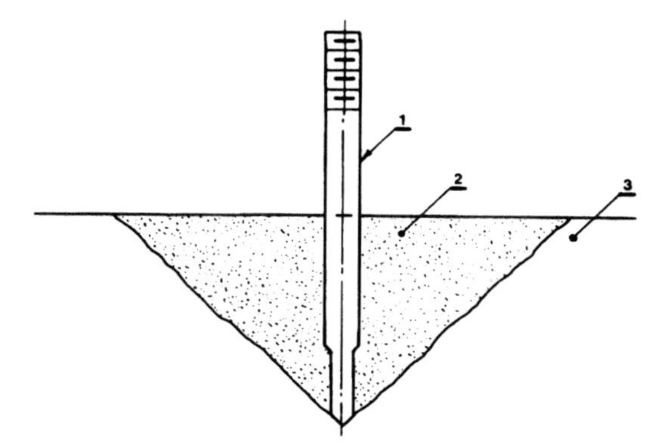

Figure 1–17. Penetration test. Approximate shape of failure zone in concrete during probe. 1—probe, 2—heavy fractured concrete, 3—tested concrete.

Detailed information on the effect of aggregate type on correlation relationships can be found in Malhotra (1976) and Bungey (1982). The advantage of this technique is that the probe equipment is simple, durable, requires little maintenance, and can be used in the field by inspectors with little training. The probe test is very useful in assessing the general quality and relative strength of concrete in different parts of a structure.

However, care must be exercised whenever this device is used because a projectile is being fired. The probe primarily measures surface and subsurface hardness; it does not yield precise measurements of the in situ strength of concrete. However, useful estimates of the compressive strength of concrete may be obtained if the probe is properly calibrated. The probe test does damage the concrete, leaving a hole of about 8 to 10 mm in diameter for the depth of the probe, and may cause minor cracking and some surface spalling.

Ultrasonic determination of material thickness and location of subsurface discontinuities and flaws (both in dry condition and underwater) is probably the most commonly used technique. As described in ASTM C597 the ultrasonic technique is used to determine the propagation velocity of a pulse of vibrational energy through a structural member.

The method is used routinely on steel and concrete structures, usually with good results.

Concurrently, thickness measurements on underwater steel structures are obtained with conventional digital ultrasonic thickness gauges. These gauges have been reported as having some limitations when used to measure the thickness of corrosion pitted steel (Naval Civil Engineering Laboratory, 1983; Scola, 1989; Russell, 1989). This occurs due to multiple front surface reflections from corroded surfaces which can be misinterpreted as real thickness readings by these gauges.

An improved ultrasonic system has recently been developed by the Naval Civil Engineering Laboratory (1985). The new system is capable of measuring the thickness of steel in the presence of severe corrosion with greatly reduced effects by the multiple reflections.

It must be realized that while taking measurements on concrete structural elements some factors such as moisture content or presence of rebars can affect pulse velocity, thus producing erroneous readings. It is reported that as the moisture content of concrete increases from the air-dry to saturated condition, pulse velocity may increase by up to 5% (Bungey, 1982). Because the pulse velocity through steel is about 40% greater than through concrete, the pulse velocity through a reinforced concrete component may be greater than through plane concrete. This may turn troublesome when reinforcing bars are oriented parallel to the pulse propagation direction, because the pulse may be retracted into the bars and transmitted to the receiver at the pulse velocity in steel. Malhotra (1976) and Bungey (1982) proposed correction factors to account for the presence and orientation of reinforcement. However, American Concrete Institute ACI 228.1R (1988) questioned the reliability of the proposed corrections.

Although ultrasonic method does not always provide a precise estimate of the material's present condition, it has sufficient power to penetrate deeply (15 to 20 m) good continuous concrete; the test on any material can be performed quickly and the required equipment is reliable and portable. Skilled personnel are usually required for analysis of test results.

The hole drilling strain gauge method can also be qualified as nondestructive and can be used for measuring actual residual stresses near the surface of steel structural components, for example, sheet piles, regular piles, deck components, etc. This method is covered by ASTM E837-85 and discussed by Flaman and Manning (1985), Flaman et al. (1985), Schajer (1981), and others. In principle the stress measurement procedure consists of the following.

1. Strain gauges, in the form of a three-element rosette, are placed in the area under consideration (Figure 1–18).
2. A hole is drilled in the vicinity of the strain gauges to a depth slightly greater than its diameter.
3. The residual stresses in the area surrounding the drilled hole relax. The relaxation strains are measured with a suitable strain-recording

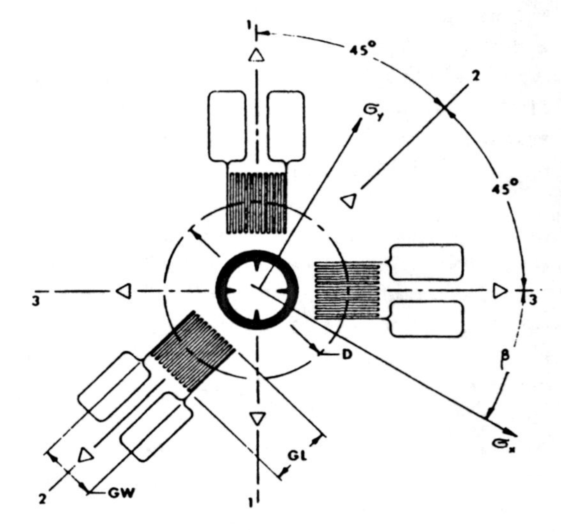

Figure 1–18. The hole drilling strain gauge method. Typical three-element strain gauge rosette.

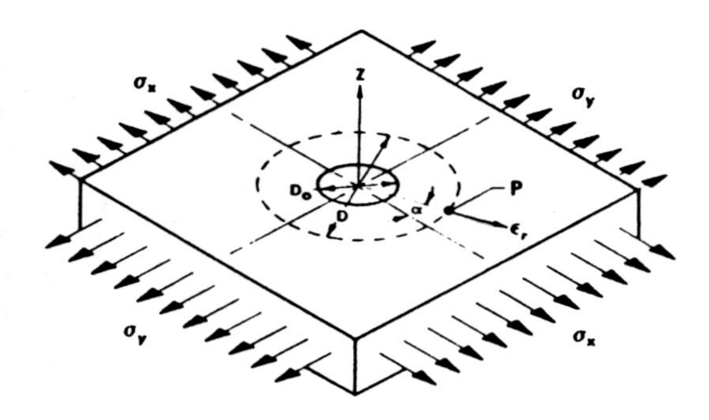

Figure 1–19. The hole drilling strain gauge method. Definition of symbols in eq. 1–30.

instrument. Within the close vicinity of the hole, the relaxation is nearly complete when the depth of the drilled hole approaches 1.2 times the diameter.

4. The surface strains relieved are related to the relieved principal stresses by the following relationship. (For definition symbols see Figure 1–19.)

$$\epsilon_r = (A + B\cos 2\alpha)\sigma_x + (A - B\cos 2\alpha)\sigma_y \quad (1\text{–}30)$$

where

ϵ_r = radial strain relived at point P

$A = \dfrac{1+\mu}{2Er^2}$

$B = -\dfrac{1+\mu}{2E}\left[\dfrac{4}{(1+\mu)r^2} - \dfrac{3}{r^4}\right]$

σ_x and σ_y = principal stresses present in the structure before drilling

α = angle between the directions of ϵ and σ_x

E = Young's modulus

μ = Poisson's ratio

$r = D/D_0$ where D = diameter of gauge circle, and D_0 = diameter of drilled hole.

5. Measuring the relieved radial strains ϵ_1, ϵ_2, and ϵ_3 at points P_1, P_2, and P_3 respectively provides sufficient information to calculate principal stresses σ_x and σ_y and their orientation β, with respect to an arbitrary selected reference.

It is assumed that the variation of the stresses in the x and y directions is small and the variation with depth is negligible.

The nondestructive methods of material inspection can be supplemented by cutting samples of steel material or core drilling for further laboratory analysis.

Core drilling to recover concrete for laboratory analysis or testing is the best method of obtaining information on the condition of concrete within a structure. However, because core drilling is expensive, it should be considered only when sampling and testing of interior concrete is deemed necessary.

The presence of abnormal conditions of concrete at exposed surfaces may suggest questionable quality or a change in the physical or chemical properties of the concrete. These conditions may include scaling, leaching, and pattern cracking. When such observations are made, core drilling to examine and sample the hardened concrete may be necessary.

Depth of cores will vary depending on intended use and type of structure. The minimum depth of sampling concrete in massive structures should be about 50 cm. The core samples should be sufficient in number and size to permit appropriate laboratory examination and testing. For compressive strength, and static or dynamic modulus of elasticity, the diameter of the core should not be less than three times the nominal maximum size of aggregate.

Core samples must be properly identified and oriented with permanent markings on the material itself when feasible. Locations of borings must be accurately described and marked on photographs or drawings. Cores should be logged by methods similar to those used for geological subsurface exploration. Logs should show, in addition to general information on the hole, conditions at the surface; depth of obvious deterioration; fractures and conditions of fractured surfaces; unusual deposits, coloring, or straining; distribution and size of voids; locations of observed construction joints; the size of voids; and contact with the foundation or other surface. The concrete should be wrapped and sealed as may be appropriate to preserve the moisture content representative of the structure at the time of sampling, and should be packed so as to be properly protected from freezing or damage in transit or storage, especially if the concrete is very weak.

When drill hole coring is not practical or core recovery is poor, a viewing system such as a borehole camera, borehole television, or borehole televiewer may be used for evaluating the interior concrete conditions.

A description and information on the availability of these borehole viewing systems can be found in publication EP 1110-1-10 by US Army Corps of Engineers.

Once samples of concrete, steel, or other materials have been obtained, whether by coring or other means, they should be examined in a qualified laboratory. In general, the examination should include petrographic, chemical, or physical tests.

Evaluation of concrete compressive strength is done by conducting tests on samples prepared from the test specimens in the laboratory.

The same specimens could be used to determine the chloride content. Experience indicates that concrete with evidence of extensive deterioration usually shows higher values of chloride content than concrete with minor or no deterioration at all.

Concrete strength in structural components inaccessible for coring, for example, anchor piles, piled relieving platforms, etc., could be estimated on the basis of concrete age. With age (usually within a few years after construction) concrete gains strength. Strength increase is dependent on numerous factors such as type of cement, curing temperatures, and water/cement ratio, and can reach 100% and more of the concrete compressive strength at 28 days.

The usual procedures for relating concrete strength at different ages involve the use of the maturity computation for various ages. For example, see the articles by Plowman (1956) and Kee (1971), which discussed the state-of-the-art and provide a reference on the subject. Approximate strength values for matured concrete can be obtained by the following equation:

$$f'_{c(n)} = f'_{c(28)} \frac{\log(n)}{\log(28)} \qquad (1\text{–}31)$$

where

$f'_{c(n)}$ = matured concrete strength in n days after placing

$f'_{c(28)}$ = 28 days strength of concrete.

The above equation (1–31) is invalid for concrete that is too old, say older than 5 years, because the concrete cannot gain much strength once all the available cement has been hydrated.

Shroff (1988) reported results of evaluation of a 50-year old concrete bridge built in 1934. Original concrete used for the cast-in-place bridge's two abutments had specified 28-day strength of 17 MPa (mix proportion of 1 : 2 : 4 for cement, coarse aggregate, fine aggregate). Eight core samples (four from each abutment) were taken and tested. All cores tested had an average compressive strength of 46 MPa. The chloride content tests on the cores from abutments indicated an area with excessive chloride content to a depth of over 125 mm with an average chloride content of 0.9 kg/m^3. In addition 22 core speciments, including two from rehabilitated concrete, were taken from 19 pile bents. All piles used on this project are of precast concrete with a minimum specified strength of 24 MPa (mix proportion 1 : 1.5 : 3). All cores tested had an average compressive strength of about 54 MPa.

Nine cores were tested for chloride content at various depths from the pile face. These test results indicated an average chloride content of 7.8, 4.9, and 2.5 kg/m^3 at depths of 25, 75, and 100 to 150 mm respectively. Samples were tested in accordance with recommendations by the American Concrete Institute, ACI 214 (1965).

Material survey, such as noticeable loss of steel due to corrosion, spalled and crumbling concrete, shear and diagonal cracks in concrete elements and its remaining strength, as suggested by Hassani (1986), is best recorded as a percentage of the minimum remaining section of the component. In some cases of underwater concrete structures and timber piles and cribs below minimum water level, one may expect some gain in material structural strengths.

Special attention must be paid to material conditions in tidal and splash zones, in previously repaired areas of relatively thin concrete elements such as pier deck, piles, and sheet piles. In addition, the condition of existing coating systems, cathodic protection, fendering systems, and different kinds of mooring accessories must be surveyed as a part of a detailed inspection.

Typically, a *hydrographic survey* and/or *geotechnical investigation* are essential parts of the detailed investigation.

The objective of a hydrographic survey is to reveal erosion or deposition of material in the vicinity of the structure, and to identify areas where propeller-induced scour has taken place and to detect if substantial undermining of the piling or solid wall foundation has occurred.

Different types of acoustic mapping systems are used for evaluation of the sea floor. These systems use the sonar principle; that is, transmitting acoustic waves and receiving reflections from underwater surfaces (Garlich and Chrzastowski, 1989).

They can be used to perform rapid, accurate surveys of submerged horizontal surfaces in water depths of 1.5 to 10 m and produce survey results with accuracies of ±50 mm vertically and ±300 mm laterally, and are very useful in fixing and investigating large scour holes and sediment buildup. It should be mentioned, however, that accuracy of these systems will decrease at depths > 10 m.

Geotechnical investigation is usually performed to determine if the foundation and fill materials have undergone changes that might affect the strength of the structure. At this stage, previously collected geotechnical data should be creatively reevaluated.

A cost-effective, detailed inspection would concentrate on the areas that may control the overall structural strength or stability, that is, points of maximum deterioration, damaged elements, points of maximum stress, structural joints, foundation, or condition of the drainage system behind all kinds of soil-retaining structures, sheet-pile bulkhead anchor systems, etc.

For more information on a subject of dock inspection the reader is referred to Collins (1987), Buslow (1987) and Buslov and Skola (1991).

1.4.2 Engineering Evaluation

At this stage, two basic tasks are to be performed:

1. Evaluation and interpretation of inspection survey data
2. Structural analysis to evaluate the structure's actual capacity.

Evaluation and interpretation of inspection survey data is an important, sometimes controversial, and often a very difficult phase of the assessment process. In a great many cases, this is more art than science, and essentially requires involvement of highly experienced personnel.

In evaluation of concrete components, it must be recognized that the function of concrete in a reinforced concrete structure is to provide adequate compressive strength and to protect reinforcing steel against corrosion.

The latter ensures the durability of reinforced concrete components. To prevent corrosion of reinforcing steel, and to be durable, concrete must be dense and nonporous, have low capillarity and low permeability, and have aggregates and cement that are non-reactive to each other and to the environment. Many of these properties are related to compressive strength, but usually are achieved by the controlled amounts of cement, entrained air, slump, water/cement ratio, and by controlled procedures for mixing, placing, and curing. Therefore, the following criteria should be considered in evaluation of existing concrete:

- Actual compressive stress
- Ability to prevent a corrosive environment forming at the level of reinforcing steel due to the presence of chloride, for example, chloride content threshold for active corrosion of steel is approximately 0.8 kg/m^3 of free chloride at the steel reinforcement level.
- Ability of concrete to prevent exposure of the reinforcing steel to water and to the atmosphere.

In addition, an overall evaluation of concrete and other structural materials in terms of past performance and repairs should be conducted.

For example, on occasion contractors pull misplaced piles into design position, thus inducing in them initial bending stress. This may be the cause of premature pile deterioration or overstress, and should be taken into account in structural analysis.

Johnson (1989) pointed out that spalling concrete at the soffit of a beam or slab may produce a completely different effect on the remaining portion of beam or slab depending on the kind of reinforcement used in these structures. In the case of regular reinforcement the spalling concrete will result in increased tension in the bottom reinforcement; on the other hand in the case of bottom located prestressed reinforcement the spalling concrete will produce additional tension in the beam/slab upper zone due to arch action.

Conversion of data provided by the inspection report into design parameters, which are used in structural analyses, requires a good knowledge of inspection methods and capabilities, as well as structural engineering, and understanding of the mechanism of loads–structure interaction.

Structural analysis is intended to define the actual strength of an in-service structure, and to compare it to the required capacities. The required capacities are based on the present, or new short- or long-term conditions and suitable factors of safety.

In the case of a damaged or badly deteriorated structure, the purpose of structural analysis is to determine the degree of redundancy present, that is, structural evaluation of reserve strength. The accuracy of such analysis depends on the engineer's ability to establish the actual strength of each subcomponent of the structure, realistic load criteria, and properly modeled load–structure interaction.

An example of strength assessment of corrosion-damaged reinforced concrete slabs and beams is given by Eyre and Nokhasteh (1992).

An important part of these analyses is also the correct assessment of energy absorption capacity of the fender system, that is, performance of fender with rubber components under hot and low ambient temperature, and capacity of different kinds of mooring accessories, such as bollards, cleats, etc. Actual or expected long-term capacity of the structure must be determined on the basis of projected rates of concrete or timber deterioration, and corrosion of steel components of the structure.

The initial analysis phase provides the engineer with an answer to the question of whether or not the in-service structure (new, deteriorated, or damaged) can satisfactorily cope with the required (present or projected) service loads for the projected period of time. If the answer is no, then a second phase of analysis associated with design of rehabilitation, repair, modernization, preventive maintenance, or corresponding downgrading shall be performed.

All port-related marine structures can roughly be classified as

Open pile constructions (i.e., piers, marginal platforms, dolphins, etc.)
Soil-retaining structure (i.e., sheet-pile bulkheads, relieving platforms, and gravity type quay walls).

Provided that design parameters are properly established, analysis of piled structures is generally straightforward and the actual capacity of the structure in terms of allowable maximum displacement tonnage of the berthing ship, allowable surcharge and point loads, mooring forces, etc., for the existing site features such as depth of water, tide, and prevailing wind and current, are fairly easy to determine.

This kind of analysis typically involves the following phases:

1. On the basis of existing as-built or original drawings and calculations, and the results of the material survey an engineer will evaluate the actual strength of each subcomponent of the structure under consideration. This will also include evaluation of the capacity of fender system and mooring accessories, and then the average strength/capacity of each structural component will be established.
2. On the basis of results obtained at Phase I an engineer can easily determine the value of horizontal or vertical loads which can be sustained by the strucutre with required factors of safety. If these loads are consistent with the design loads, then the structure is satisfactory without any modifications.

   If obtained loads are below required values, then an engineer has the following options:

   (*i*) In order to reduce impact force from berthing ship install new, or improve existing fender system.

   (*ii*) Increase capacity of structural components such as piles and deck, install additional new piles, and/or install additional bracing system to reduce free height of piles.

The size (displacement tonnage) of the theoretical ship that can use the facility under consideration regardless of actual strength of the structure shall be determined on the basis of specific site features such as length of the wharf, depth of water, geotechnical conditions of the sea floor, and berthing and mooring conditions.

Again, if the structure under consideration cannot sustain impact from the theoretical ship then a new fender system can be installed and/or the structure must be reinforced accordingly. If the latter is considered to be an uneconomical solution then the use of the structure shall be limited to receiving smaller ships.

Establishing the actual capacity of gravity-type retaining structures, generally of different types, also does not present significant problems.

Gravity-type retaining structures are usually constructed over a bedrock foundation, and therefore little deflection of the structure is expected. In this case, the soil pressure against wall normally exceeds values obtained by the classical theories of soil pressure and is classified as 'pressure at rest.' Again, accuracy of these analyses would depend heavily on the quality of investigation of soil design parameters.

A much more complicated task is establishing the actual capacity of the sheet-pile bulkheads.

Most of the troubles reported in anchored bulkheads are caused by wrong design or poorly constructed anchorages. (Some untypical sheet-pile failures are described by Sowers and Sowers, 1967; and Broms and Stille, 1976.) Less frequently, bulkhead failures are reported attributed to highly overstressed sheet piles, or undermining by overdredging or by propeller-induced scour.

Sheet-pile bulkheads are usually considered as "flexible" structures because of their ability to deform slightly. The problems involved in the design of "flexible" retaining structures are somewhat more complicated than those applying to regular gravity-retaining walls. The soil pressure is the main force acting against the sheet-pile wall. The magnitude of this pressure depends on the physical properties of the soil and the character of the interaction of the soil–structure system.

Movement (deflection) of the structure is a primary factor in the development of earth pressures. However, this problem is highly indeterminate. Earth pressures are affected by the time-dependent nature of soil strength, such as consolidation due to vibration, ground water movement, soil creeping, and chemical changes in the soil.

Despite various theories that have been suggested to derive analytical solutions for the bending moments, reaction force, and the sheet-pile submerged depth, most designers in the field still rely heavily on past practice, good judgment, and experience.

1.4.3 Structure Repair/Rehabilitation

After the actual capacity of the structure is established cost-effective decisions on follow-up actions can be made.

If the analyses indicate that in its present condition the structure can safely carry design service loads then just regular preventive maintenance/repair to mitigate deterioration is necessary to maintain its operability for the design period of time; thus costly rehabilitation is avoided. At this stage, provided that all major structural elements are in good shape, repairs could be performed to restore only deteriorated components of the structure.

If structural analysis or stress/strain/deflection surveys indicate that the structure is overstressed, then structural restoration (rehabilitation) or downgrading of the structure's capacity through operational limitations is required. The rehabilitation could include restoration of all or just major structural components to their original or required strength, installation of a more efficient fender system, or major reconstruction of the structure.

In the case of piled structures, the actual solution to the repair, or rehabilitation, is quite straightforward. This generally includes repair of piles (seriously damaged piles are usually replaced by new piles), and deck structure. In some cases, addition of new piles, replacement of the existing fender system, and repair of under-platform rip-rap is required. Repair/rehabilitation of the soil-retaining structures usually presents more serious problems. As stated previously, these structures typically are gravity-type retaining walls or anchored sheet-pile bulkheads.

Before any repair/rehabilitation method is selected it is prudent to establish not only the existing and required capacity of the facility, but also the length of time for which the desired capacity will be valid. It is also necessary to establish the limitations, such as ship allowable approach speed and approach mode (e.g., assisted by tugs, or direct approach) and acceptable load combinations.

1.4.3.1 Planning

To be successful the repair/rehabilitation of the marine structure should be carefully planned. The main stages of planning procedure include the following steps:

1. Determination of all available basic relevant repair procedures. This involves a review of previous experience of similar repair/rehabilitation work, and development of a new idea, relevant for the case under consideration.
2. Examination of technical feasibility of the selected repair/rehabilitation procedures. This can be done either on the basis of previous experience, or through specific analysis relevant for the case under consideration.
3. Economic evaluation of selected repair/rehabilitation methods and selecting the most economical alternative. This must take into

account cost of repair/rehabilitation and cost of the possible disruption of dock operation.

4. Preparation of the repair/rehabilitation specifications which will typically cover detailed procedures, sequence of operation, materials, quality control, tolerances, and other relevant subjects.

The following is a discussion on the basic repair/rehabilitation methods used for different types of marine structures.

1.4.3.2 Repair/Rehabilitation of Gravity-Type Walls

Erosion of structural material (basically superficial) and erosion (scour) of the wall's mattress or foundation material are basic problems associated with gravity-type quay wall operations. Deterioration of structural material very seldom presents problems to the integrity of the structure.

Repair of Scour Cavities at the Foot of the Wall

Depending on the size of the cavity at the wall's foot, the cavity could be filled by using one of the following methods or a combination (Figure 1–20):

- Tremie concrete
- Grout injecting concreting (preplaced aggregate with grout injection)
- Installation of concrete-filled bags, subsequently filling gaps between the base of the wall and installed concrete filled bags by using cementitious pressure grout.

For a good repair/rehabilitation before any of the above concrete installation methods are employed the exposed surface of the existing concrete has to be thoroughly prepared; damaged and unsound concrete, as well as marine growth, have to be removed to ensure good bond with repair concrete.

Practically all dry land techniques such as sand and grit blasting, water jetting, or jack hammering can be applied underwater. Sand or grit blasting are also useful for removal of corrosion product from rebars and other metal parts. It must be noted that marine fouling which prevents bonding between old and new concrete can present serious problems to structure repair. It is good practice to have any marine growth removed just before placing of new concrete.

According to Steijaert and De Kreuk (1985), during the growing season (June to September) concrete can be covered by tiny barnacles or mussels in 24 to 36 hours. Therefore, in areas where marine growth can present a problem the new concrete has

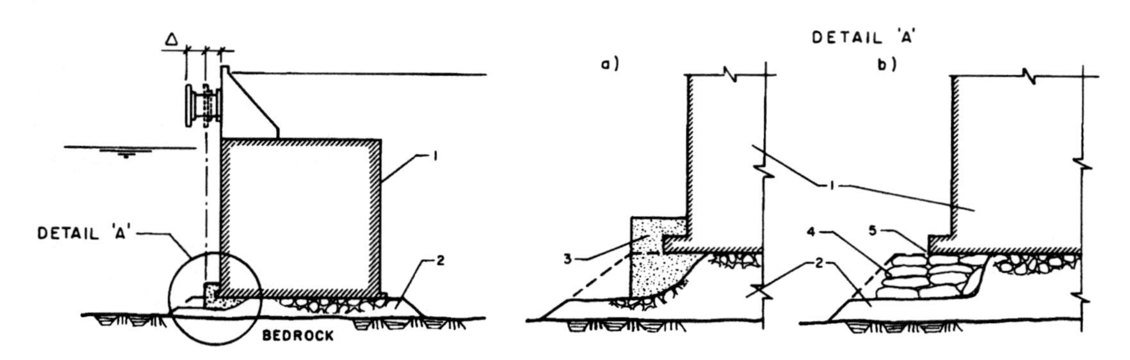

Figure 1–20. Repair of propeller/current or other induced scour. Δ—design compression of fender system. 1—Gravity type quay wall, 2—rockfill mattress, 3—tremie concrete, prepacked concrete, or special admixture, 4—concrete in bags, 5—pressure grout. (From Tsinker, 1989.)

to be installed within 1 to 2 days after cleaning the repair area.

TREMIE CONCRETE. The basis of the tremie method is to pour concrete down a flexible or rigid tube 150 to 300 mm in diameter complete with a hopper into which fresh concrete is fed.

At the end of the operation, the water in the void is replaced with concrete.

For better results a chamfer should be provided at the high end at the feed side of the cavity. Some extra height should be provided for the outside wall to ensure complete filling. The best results in filling voids are obtained with stable concrete mixes of high workability (slump 15 to 20 cm), rich in cement (minimum 350 kg/m^3), free of bleeding, and resistant to washing out action.

During the concrete installation operation, free fall through water should be avoided.

The tremie concrete can also be placed by a technique developed in the Netherlands and called "The Hop–Dobber" (Kohne, 1983). In this technique instead of a solid steel pipe a collapsible fabric tube is used. This is kept closed by water pressure until the weight of concrete in the hopper overcomes the hydrostatic pressure and skin friction. A slug of concrete then descends slowly and the tube is again sealed behind it by the pressure of the water. The advantage of this method over tremie is that the concrete is fed into the hopper intermittently, and thus the problem of ensuring continuity of the concrete supply is avoided. However, the need to ensure the relative position of the bottom of the tube and the surface of the concrete remains. During the Hop–Dobber operation the hopper level remains unchanged, and the lower section of the tube moves up and down to suit the level of the concrete. Tremie concrete can also be delivered by employing a concrete pump.

GROUT INJECTING CONCRETING. Under this method, initially the void is filled with coarse aggregates, then the free space between these aggregates is filled by grout injection.

In recent years, new efficient antiwash admixtures have been developed that effectively prevent segregation of concrete or grout when placed underwater (Fotinos, 1986; Sancier and Neeley, 1987; McDonald et al., 1988 and Staynes and Corbeft, 1988). A diver can simply distribute this type of concrete into the void using an underwater hose.

CONCRETE-FILLED BAGS. Under this method the diver installs 50 to 70% concrete-filled bags into the cavity. Because the cement paste is normally squeezed out through the porous bags a certain cementation within the bag structure takes place. Subsequently for better interaction the space between the bags and the wall under question is filled by pressure grout, or tremie concrete.

Surface Repair/Rehabilitation

The depth of concrete deterioration can range from surface scaling to a meter and more, which at times can reach several cubic meters in volume. In the latter case, to prevent further material deterioration that eventually may compromise the integrity of the wall the cavity must be filled with concrete. The installation of new concrete would not present any problem in an underwater zone of the wall. It must be noted, however, that substitution of the lost concrete with conventional concrete should not be used in situations where an agressive factor that has caused the deterioration of the concrete being replaced still exists. For example, if the deterioration noted has been caused by acid attack, aggressive water attack, or even abrasion–erosion, it is doubtful that repair by conventional concrete placement will be successful unless the cause of deterioration is removed.

Before placement of new concrete the existing concrete in question must be removed until sound concrete has been reached. To ensure a good bond between new and existing concrete the minimum depth of the cavity should be at least 150 mm. For better repair the cavity should have vertical sides normal to the formed surface or keyed as necessary to lock the repair into the structure. The top inside face should be sloped up toward the front. Slope 1 : 3 is usually sufficient to prevent air pocket formation inside the repaired cavity.

Surfaces of existing concrete must be thoroughly cleaned by sandblasting, or by another equally satisfactory method, followed by final cleaning with compressed air, water, or other satisfactory methods of cleaning. Sandblasting effects should be confined to the surface that is to receive the new concrete. If required dowels and reinforcement can be installed to make repair self-sustaining and to anchor it to the underlying concrete, thus providing an additional safety factor.

Before receiving the new concrete the surface is first carefully coated with a thin layer of epoxy modified slurry not exceeding 3 to 5 mm and having the same water/cement ratio as the concrete to be used in the replacement.

To minimize strains due to temperature, moisture change, shrinkage, etc., concrete for the repair should generally be similar to the old concrete in maximum size of aggregate and water/cement ratio, and should be thoroughly vibrated; internal vibration should be used if accessibility permits. A tighter patch results if the concrete is placed through a chimney at the top of the form (Figure 1–21).

It is good practice that immediately after the cavity has been filled, a pressure cap is placed inside the chimney. Pressure is applied while the form is vibrated. This operation should be repeated at 30-min intervals until the concrete hardens and no longer responds to vibration. The projection left by the chimney should normally be removed the second day.

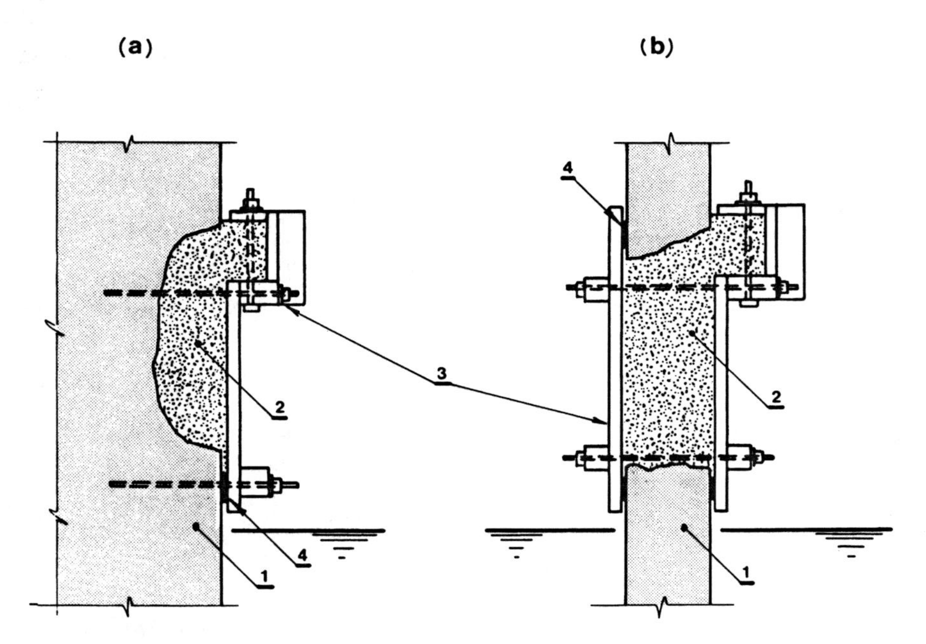

Figure 1–21. Details of formwork for concrete repair above water level. (a) Local cavity repair in mass concrete wall. (b) Repair of thin concrete wall. 1—Existing concrete, 2—new concrete, 3—formwork, 4—soft seal (foam plastic).

An underwater concrete repair, although similar in technique to repair "in dry," is, however, more complex in execution (Figure 1–22).

All requirements and precautions normally applied to surface preparation for the above water repair should be considered for underwater concrete installation.

As previously pointed out, underwater concrete (grout) must be proportioned for workability, and not for a given strength level. Normally strength achieved will be more than adequate. (US Army Corps of Engineers, 1986.)

The concrete (grout) placement equipment must be adequate to handle the designed concrete mixture at a designed rate of placement.

Concrete (grout) is normally injected into the cavity from the bottom up (Figure 1–22). In the process of grouting the water escapes from the cavity through the upper port. To ensure good quality of repair the concrete (grout) must continue to be injected some time after it starts escaping from the upper port.

Installation of some reinforcement anchored inside the cavity can substantially enhance cohesiveness between new and existing concretes. The formwork is attached to the wall surface by means of anchor bolts. The size and number of these bolts must be proportioned in a way to take pressure from fresh concrete and self weight. In water colder than 10°C formwork should be insulated to ensure proper curing of concrete. If the size of the cavity is substantial then several concrete injection ports can be installed in formwork, and a diver will orderly move concrete carrying hose from port to port as required. To prevent concrete escape the space between formwork and the face of the wall, as well as potential voids between separate concrete blocks inside the cavity, must be thoroughly sealed.

A great many marine structures in North America were built 40 and more years ago. The concrete in most of these structures presently in-service (a great deal of them are nonreinforced mass concrete quay walls, and walls in navigation locks) does not contain entrained air and is therefore susceptible to freeze–thaw deterioration. The

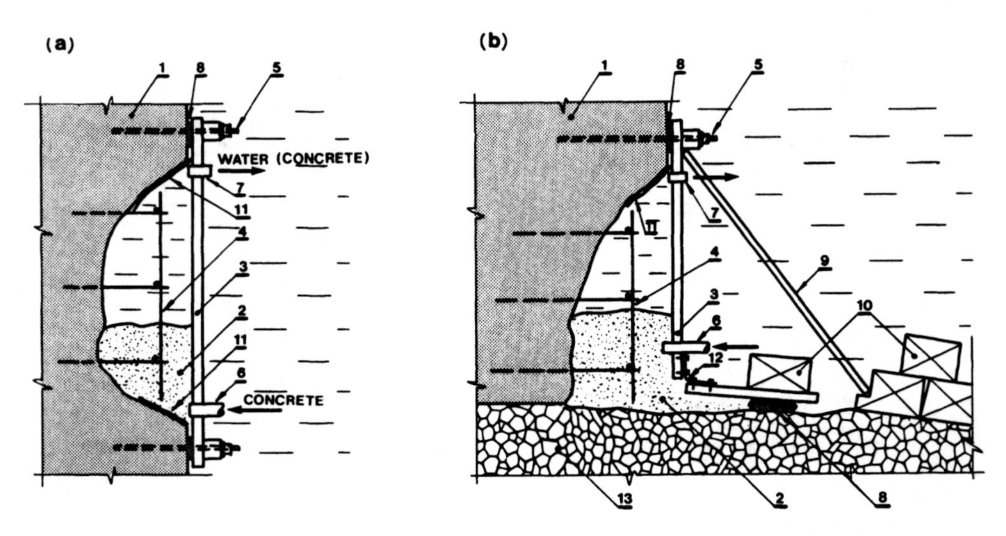

Figure 1–22. Underwater concrete repair. (a) Cavity at the face of the wall. (b) Cavity at the foot of the wall. 1—Existing concrete, 2—new concrete, 3—formwork, 4—new reinforcement, 5—anchor bolt, 6—concrete injection port, 7—water (concrete) excape port, 8—soft seal (foam), 9—prop, 10—weight, 11—epoxy bonder, 12—sheet metal (hinge), 13—mattress.

general approach in wall repair/rehabilitation in such a case is to remove 0.2 to 1.0 m of concrete as required from the face of the wall and replace it with new concrete using conventional concrete forming and placing techniques.

The deteriorated concrete is usually removed by blasting techniques. This is accomplished by drilling a series of small-diameter holes along the top of the wall filled with charges. After the deteriorated material is blasted off, any remaining loose or deteriorated concrete materials ought to be removed by chipping, grinding, or water blasting to the extent that the bond surface is clean and free of materials that could inhibit bond.

Dowels are normally used to anchor the new concrete facing to the existing wall. They usually are 20 to 25 mm in diameter and are typically spaced 1.2 to 1.5 m center-to-center in both directions. Dowel diameter, embedment, and spacing are proportioned to take the load of fresh concrete or a potentially large hydrostatic pressure which could develop behind the reinstalled concrete.

It is recommended that at least 0.25% of all dowels installed be field tested. The embedment length is considered to be adequate when applied pull-out load is equal to or greater than the calculated yield strength of the dowel. Mats of reinforcing steel usually 15 to 20 mm in diameter (300 to 500 mm center to center) are hung vertically on dowels and fixed into position with provision for 75 to 100 mm cover. Once the reinforcement and formwork are in position, replacement concrete is placed by using any conventional technique.

If the thickness of the replaced concrete exceeds 25 to 30 cm use of conventional formwork and conventional concrete replacement techniques are in general the more economical solution to wall refacing. The advantages of the latter techniques include: the new (replacement) concrete mixtures can be proportioned to simulate the existing concrete substrate, thus minimizing strains due to material incompatibility; proper air entrainment in the new concrete can be obtained by use of admixtures to ensure resistance to cycles of freezing or thawing; materials, equipment, and personnel experienced in conventional concrete application are readily available in most areas.

One of the most persistent problems in this type of wall rehabilitation is cracking in the new concrete. These cracks, which generally extend completely through the new concrete, are attributed primarily to restraint of volume changes resulting from thermal gradients and drying shrinkage. To reduce or minimize deterioration effect concrete materials, mixture proportions, and construction procedures that will reduce concrete temperature differentials or minimize shrinkage of volume change should be considered.

For better quality (and appearance) a permanent formwork system of precast concrete panels can be used (Figure 1–23). Use of high-quality, durable concrete panels attached to the prepared concrete substrate has significant potential in resolving the cracking problem encountered in refacing deteriorated walls. Also use of precast formwork panels should shorten the time a quay wall must be out of service for repair/rehabilitation. Permanent concrete formwork panels have been successfully used for refacing concrete wharves, navigation locks, and dams. The practical example of complete refacing of the deteriorated blockwork wall is depicted in Figure 1–23b. There the precast concrete panels were installed on the steel dowels driven into the shore bedding. The top of the panel was hooked on the upper tiebacks installed into existing concrete superstructure. The space between panels and the wall was filled with concrete by using tremie technique.

New capping structure was constructed and jointed with the existing superstructure be steel dowels. Finally, the existing wall was solidified by the reinforcing bars installed

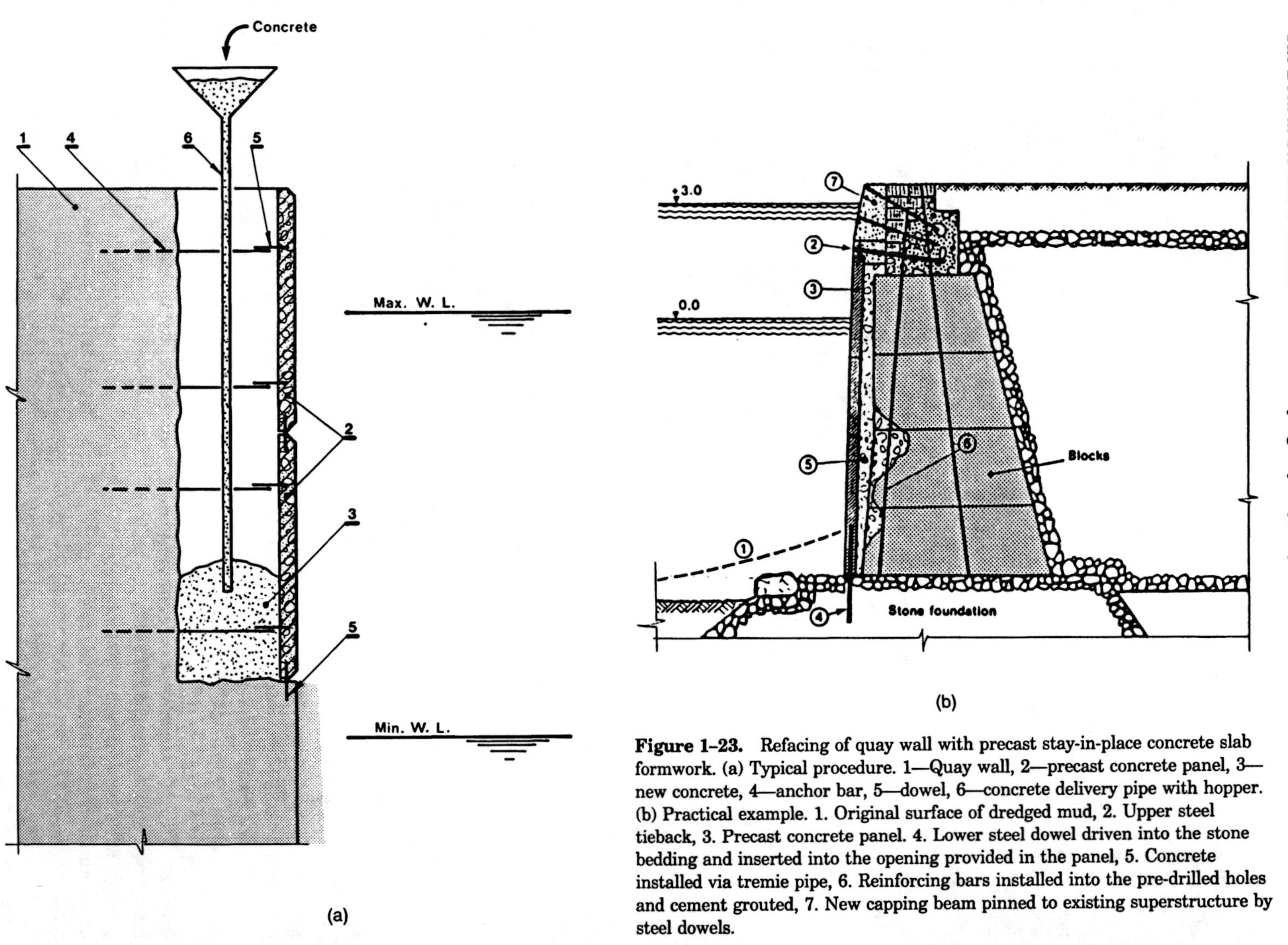

Figure 1–23. Refacing of quay wall with precast stay-in-place concrete slab formwork. (a) Typical procedure. 1—Quay wall, 2—precast concrete panel, 3—new concrete, 4—anchor bar, 5—dowel, 6—concrete delivery pipe with hopper. (b) Practical example. 1. Original surface of dredged mud, 2. Upper steel tieback, 3. Precast concrete panel. 4. Lower steel dowel driven into the stone bedding and inserted into the opening provided in the panel, 5. Concrete installed via tremie pipe, 6. Reinforcing bars installed into the pre-drilled holes and cement grouted, 7. New capping beam pinned to existing superstructure by steel dowels.

into predrilled holes and grouted there by cement grout. More relevant examples are described by Davis et al. (1948), Lundin (1983), and others. Precast concrete formwork panels are typical in hydropower project constructions in the former USSR.

For concrete replacement up to 150 to 250 mm deep both conventional and shotcrete techniques are used.

SHOTCRETING is the process of sand/cement mortar application at high velocity onto a concrete surface. It is done with the help of special equipment. As a material, shotcrete is a fine aggregate concrete with various sand/cement ratios, depending on the application. Shotcrete repair is typically used for large areas of deteriorated concrete. It is very effective for variable depth repairs of vertical and horizontal surfaces.

Shotcrete repair typically involves surface preparation, reinforcing, and concreting. A distintegrated area of concrete to be shotcrete treated is prepared in the same way as discussed earlier in this section. Galvanized wire mesh is then attached to the exposed reinforcing bars. A bonding agent is usually used to ensure a good bond between the old and new concrete. Finally, well-graded concrete (sometimes with epoxy additives) with the proper moisture content is applied to the repaired surface at a required pressure. If properly, done the end result is a dense, high strength, and durable shotcrete repair.

For repair of deteriorated concrete components thinner than 250 mm in dry conditions, shotcrete is generally more economical than conventional concrete because of the saving in cost of formwork. Shotcrete is a strong and durable material and is capable of very good bond with concrete and other construction materials. The resistance of shotcrete to cycles of freezing and thawing is generally good despite lack of entrained air. This is attributed in part to the low permeability of properly proportioned and properly applied shotcrete which minimizes the ingress of moisture, thus preventing the shotcrete from becoming critically saturated. Use of steel fiber shotcrete can greatly enhance the quality of concrete repair (Hoff, 1987; Gilbride et al., 1988; Morgan, 1988). Hoff found that in dense, low-permeability, steel fiber reinforced concretes the depth of surface carbonization is no more than a few millimeters over a period of 10 years or more. Although any fibers in a carbonized surface layer can be expected to corrode and disappear, the internal fibers beyond this zone will remain unaffected.

If for some reason the existing non-air-entrained wall behind a shotcrete repair can become critically saturated by water and migration from beneath or behind and the shotcrete is unable to permit the passage of water through it to the exposed surface, it is likely that the existing concrete will be more fully saturated during future cycles of freezing and thawing. If frost penetration exceeds the thickness of the shotcrete section under these conditions, freeze-thaw deterioration of the existing wall should be expected. Numerous examples of the above phenomenon are found in US Army Corps of Engineers EM 1110-2-2202, 1986.

Rehabilitation of Distressed Foundation Material

A distressed foundation causing deformation to the gravity wall can be improved by one of the following methods:

Injection grouting, more commonly known as grouting
Jet grouting
Soil mixing technique.

GROUTING. Grouting is the technique by which foundation materials can be impregnated under pressure with grout that is allowed to set. The type and gradation of the soil dictates the type of grout to be used.

General information on soil grouting as well as a comprehensive list of references is found in Caron et al., published in a

Handbook edited by Winterkorn and Fang (1975).

JET GROUTING. Jet grouting is a rather remarkable technique now in use for well over 15 years. Although several techniques of jet grouting exist, the process of greatest commercial application to date derives from a 1971 Japanese patent granted to Nakanishi.

The method is used for modifying relatively soft soils to achieve general material improvement, or to construct subgrade structural or load-bearing members without prior excavation. Chief uses include underpinning existing structures threatened by subsidence. It is also used for providing seepage control, making cofferdams, limiting subsidence over tunnel excavation sites, etc. Jet grouting is radically different from traditional pressure grouting with cement or chemicals.

It is essentially a partial soil replacement technique whereby a column is formed composed of a mixture of injected grout and in situ soil. The final product, called "soilcrete," is made by complete, hydraulically induced mixing of cement slurry with the native soil. It is usually done in the shape of a cylindrical column, and the column size and shape depend on the brand of equipment, and can be varied. The properties of the soilcrete depend on the native soil and the way the process is applied. In a typical jet grouting technique initially a pilot drill hole about 100 to 150 mm is installed to the required depth using standard drilling equipment. Then a grout injecting stem capable of withstanding extremely high pressure is inserted. At the base of the stem is a "monitor" that typically has two nozzles: the lower for injecting grout and the upper, typically 2 mm in diameter, for a high pressure water jet.

The water (sometimes water–air) jetting is usually operated at around a water delivery rate of 70(±) L/m and at a pressure of about 400 bar. Grout is force fed via the lower nozzle at a pressure of about 40 bar. Portland cement and water mixed in 1 : 1 ratio is the grout mix that is most commonly used. However, any pumpable material that solidifies when mixed with soil can be used, for example, bentonite slurry, which is typically used for cut-off walls.

Once at full depth, the stem is slowly extracted and rotated, and grout expelled radially from one or more base nozzles of the jetting monitor. The result of the high pressure jetting is that the in situ soil structure is destroyed and the soil particles thoroughly mixed with the grout. A continuous,

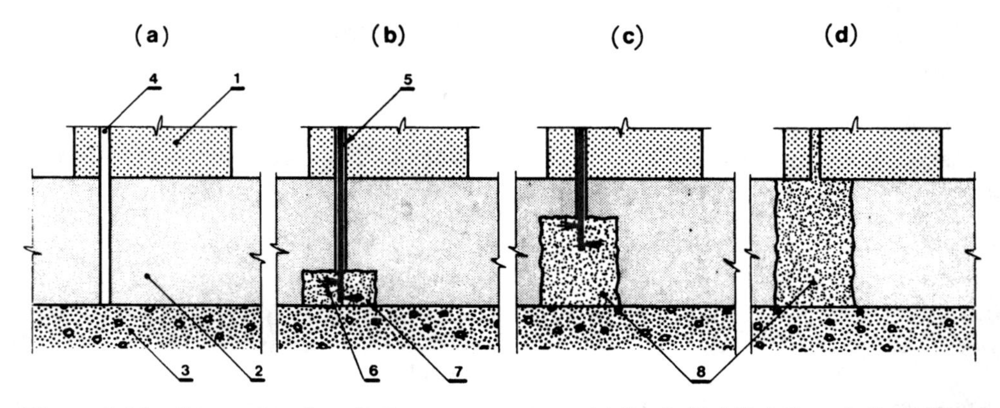

Figure 1-24. Jet grouting. Installation sequence stages. (a) Borehold drilled through the wall and distressed foundation material. (b, c, d) Initial, intermediate and final stages of jet grouting. 1—Quay wall, 2—distressed foundation material, 3—firm foundation material, 4—borehole, 5—jet grouting stem, 6—water jet, 7—grout, 8—soilcrete.

full height mortar column from the bottom up is formed at a single pass of the jetting head (Figure 1–24). When cement and water are mixed 1 : 1 the grout's typical compressive strength values can be about 7 MPa for clays and about 20 MPa for clean sand.

The diameter of the column formed can vary from 0.3 to 2 m depending on the in situ soil conditions, the jetting pressures, the speed of rotation, and the rate of extraction.

The column spacing and alignment need to be controlled, though some tolerance is available when forming a wall by overlapping columns; also, some correction can be achieved by locally increasing the diameter of the column to give a connection to a misaligned column.

The primary concern in the case of the cement grouts used with jet grouting techniques is the ability of the grout to set, cure, and develop appreciable strength at relatively low temperatures (say below 5°C).

Neat cement grouts consisting of normal Portland cement and water will not set at about −1.8°C at all. Therefore, it will be necessary to use high early cement and/or accelerators.

During the jetting process, a mixture of the grout and in situ soil may be expelled from the hole. The amount expelled depends on the void ratio of the in situ ground mass. In the case of sandy materials, little, if any, material is expelled whereas with consolidated clays, more can be expected.

When the equipment is suitably modified, jet-grouted panels rather than columns can be formed. In this way considerable material savings can be obtained as a relatively thin (say 300 to 500 mm) wall can be formed.

The jet grouting can be stopped at any level below ground level, hence the technique can be used to underpin an existing structure or connect a grout wall to the base of a structure.

Jet grouting was developed using cementitious grouts, but the ambient conditions, for example, the retardation effects of the extreme low temperatures, may preclude the use of cement grouts. However, the technique can be adaptable to chemical grouts.

The individual columns of soilcrete are the basic building unit produced by jet grouting. They can be combined in various shapes and sizes to solve special problems related to new construction, or encountered in repair/rehabilitation projects.

Groups of interconnected columns can be so arranged as to make the group support required for vertical and horizontal forces. Typical layouts include: interconnected rows, which can form cut-offs; or retaining walls, staggered grid pattern, which is used to improve the bearing capacity of foundation soil for a proposed structure, or for underpinning an existing structure or circular walls which are used to provide a watertight barrier (cell) for the local soil excavation.

Jet grouting offers several advantages. First of all it is the only technique for the relatively predictable treatment of clay, which is typically resistant to any kind of pressure grouting. Second, the process creates minimum vibration and therefore can be safely used in sensitive builtup areas. Finally, it requires no excavation prior to installation. The method's main disadvantage is that it can be applied only in relatively soft soils.

Since the jet grouting method has been introduced thousands of linear meters of jet grouting have been completed in Japan, Europe, and South America. However, in North America until the late 1980s it was relatively unknown. Part of the reason why jet grouting has been little used in North America compared with other parts of the world lies in the fact that North American designers seem to be more conservative than their European and Japanese counterparts and less receptive to new techniques until they have been proven. For example, several other geotechnical innovations such as slurry walls, tie-back anchors, reinforced earth, and soil pressure relieving systems were pioneered and used elsewhere before they were commonly accepted in North America. Now the jet grouting method is gaining popularity among North American contractors and it seems that the pace of its use is getting more frequent. Pettit and Wooden (1988) describe a variety of examples of jet grouting use in North American civil engineering practice in the late 1980s, for example, use of jet grouting in tunneling, construction of cut-off walls, underpinning, and miscellaneous repairs. For the purpose of this work it is interesting to note the case of the anchored sheet-pile bulkhead repair in Galveston Bay described by Pettit and Wooden. The wharf comprises 61 cm wide precast concrete tongue-and-groove sheet piles driven into the bay mud and anchored with steel tieback rods secured to deadmen. Probably because of absence or poor quality of the filter material behind the wall the backfill material was washing out through sheet pile joints. Large gaps have been formed, which led to extensive pavement distortion. The repair was done by sealing joints between adjacent sheet piles by the jet grouting method. This procedure permitted safely carrying out repair work between tieback rods without disturbing them. A special latex/cement slurry bonded the bulkhead sheet piles, and in 5 days, over 150 seams were sealed.

Coomber (1987) reported an interesting case of successful restoration of a quay wall by using (as part of rehabilitation work) the jet-grouting technique (Figure 1–25). There were two rows of columns 1.5 m in diameter (total of 81 columns) which were installed to underpin the existing quay wall. This was achieved by drilling a 150 mm diameter pilot hole to the required depth of about 2.8 m below the base of the wall, and then inserting the 90-mm diameter rotating stem through which a mix of three parts pulverized fuel ash to one part cement was discharged into the soil, which had been previously cut by water–air jet (water delivery rate was 70 L/m at 400 bar pressure).

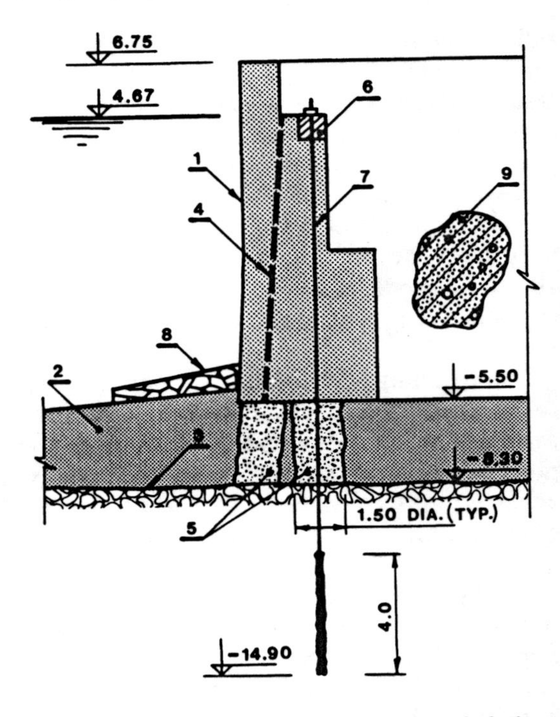

Figure 1–25. Restoration of the Brocklebank dock West quay wall. (From Coomber, 1987.) 1—Quay wall, 2—stiff brown fine gravelly clay becoming very sandy and silty locally (glacial till), 3—sandstone bedrock, 4—jet grout borehold 150 mm diameter, 5—jet grout column (soilcrete), 6—cast-in-situ new concrete beam, 7—new post-tensioned rock anchor, 8—antiscour mattress, 9—miscellaneous backfill.

The foundation soil was stiff brown fine gravelly clay that became very sandy and silty locally (glacial till). When the rear row of columns was installed, they were redrilled, to take about 20 m long rock anchors secured into the sandstone bedrock. Initially these anchors were stressed to 1.25 times their design load of 630 kN, and subsequently distressed to the design load. It was reported that the whole job was done with a minimum disruption of dock operations.

SOIL MIXING TECHNIQUE. In this technique the soil structure is destroyed by mechanical means, using an expanding bit type device attached to the head of a drill string. As the drill is slowly rotated and extracted a cement slurry is injected and a lean cement/soil column is formed.

The technique was developed for improving ground condtions using cement stabilization techniques to increase various strength parameters.

GROUTS. To date, cement grouts are the only grouts that have been used with different grouting techniques. An advantage of using cementitious grouts is that seawater can be used for the mix, whereas freshwater is generally mandatory for chemical grouts. Cement grouts vary in terms of the types and/or proportions of Portland cement, pozzolan admixtures, and water/cement ratio.

Chemical grouts are typically used in a pressure (injection) grouting technique. They vary considerably in terms of their chemical composition and end products (solids and gels). Grouts in each category, chemical or cement, have been used to grout different kinds of soils.

The category of grout and the type of grout within the category, which can be used in a specific type of soil, will depend primarily on its rheological behavior (the manner in which it deforms and flows). The physical characteristics that will determine the rheological behavior of a grout are viscosity, rigidity, and granular content. Some grouts are nongranular and liquid (no rigidity) and will maintain the same viscosity with time until mass polymerization takes place. Because of the water-like nature of these grouts, they are well suited for grouting sand and silts.

1.4.3.3 Repair/Rehabilitation of Open-Pile Structures

Open-pile structures are normally built in the form of piers or marginal wharves. The latter are usually built with an underdeck slope, protected with rip-rap, or with sheet piling placed behind the platform.

Typical damages that structures of the above construction usually experience are:

Broken piles
Pile and deck material deterioration
Physical damage to the deck elements inflicted by the docking ship or cargo handling/hauling machineries
Scour of the seafloor in front of the dock, and/or underdeck slope protection inflicted by propeller induced jet, or wave actions
Damage or deterioration of elements of the fender system
Damage or deterioration of the dock mooring accessories
Bed or underdeck rip-rap erosion (scour).

Pile Restoration

As stated earlier, broken piles are usually removed and substituted with new piles. This is usually done by providing access to the broken pile at the deck level. After the broken pile is removed, a new pile is driven through the window in the deck structure, after which the deck structure is integrated with the new pile by restoring the deck's original reinforcement, or installing new as required with subsequent filling of the window with relevant material (most often with concrete).

Pile material deterioration usually occurs in the splash and/or tidal zone. For pile restoration, a technique called jacketing is typically used. It is especially useful where all or a portion of a pile repair section is underwater. Essentially jacketing is a method of restoring or increasing the strength of an existing pile by encasing it into a new concrete. The original pile can be concrete, steel, or wood.

When properly applied, jacketing will strengthen the repaired component as well as provide some degree of protection against further deterioration. However, if for example a concrete pile is deteriorating because of exposure to acidic water, jacketing with conventional Portland cement concrete will not ensure against future disintegration.

The removal of the existing damaged or other material such as marine growth, dirt, etc., is usually necessary to ensure that the repair material bonds well to the original material that is left in place. Corrosion products on the steel should likewise be removed down to bare metal. If a significant amount of removal is necessary, it may be necessary to provide temporary support to the strucutre during the jacketing process. Material of any suitable form may be used for jackets. A variety of proprietary form systems are available specifically for jacketing. These systems employ fabric, steel, or fiberglass forms (Johnson, 1965). Use of a preformed fiberglass jacket for repair of a concrete pile is shown in Figure 1–26. Reinforced concrete jackets for repair of steel pile piles are widely used in Japan (Fukute et al., 1990). Once the form is in place, it may be filled using any suitable material. Choice of the filling material

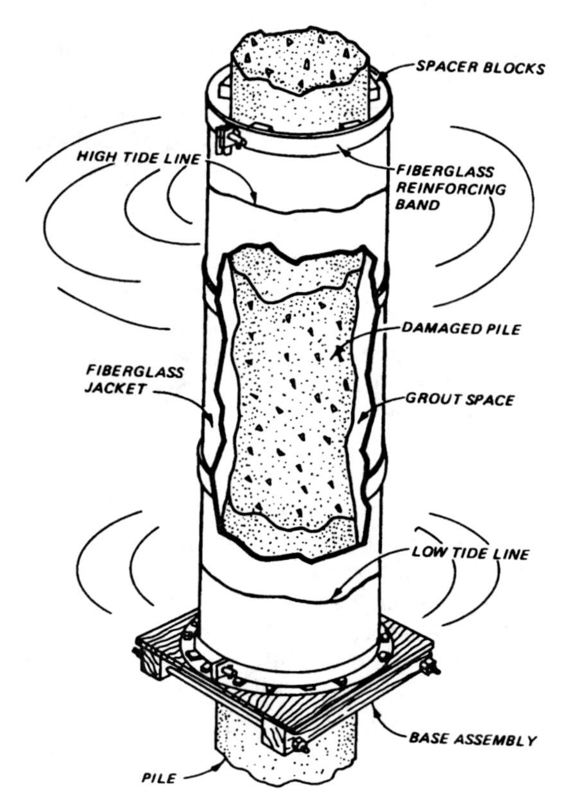

Figure 1–26. Typical repair of deteriorated concrete pile. (From US Army Corps of Engineers, 1986.)

should be based on the environment in which it will serve as well as a knowledge of what caused the original material to fail. Filling may be accomplished by pumping, by tremie placement, or by prepacked aggregate techniques. The new concrete must adhere well to the previously cleaned surface of the existing pile and should attain required level of strength and durability. This can be achieved by including in concrete mix moderate amounts of silica fume and antiwashout admixtures (Khayat, 1992); silica fume increases the strength, adhesion, and durability of the concrete, while the antiwashout admixtures, that are water-soluble polymers, are able to absorb some of the mixing water in concrete, thus reducing its migration from the cement past, along with some suspended cement and fines, as it comes in contact with moving water. According to Khayat such concrete can provide in-place compressive strength of 62 MPa and low chloride ion permeability values at upper sections of repair lifts.

For better bond between a steel pipe pile and the concrete jacket steel studs are welded to the repaired pile. This is done in dry conditions and underwater by using a specially designed welding gun (Fukute et al., 1990).

If required, steel reinforcement placed in the annulus formed by the jacket around existing pile is used. Khana et al. (1988a) reported successful use of a steel-fiber reinforced concrete for repair of distressed hollow core concrete piles at Rodney Terminal, Canada. In the tidal zone these piles have experienced miscellaneous kinds of marked deteriorations, for example, deep longitudinal cracks, multiple fine hair cracks, circumferential cracks and severe surface degradation, basically due to freezing and thawing, spalling, and delimination (Figure 1–27).

In-depth investigation of the problem has been carried out. It was concluded that severe winter thermal stresses, exposure to intermittent saturation, and a large number

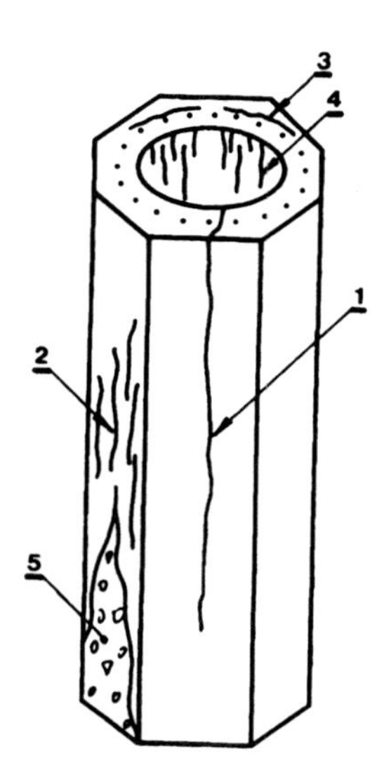

Figure 1-27. Typical type of distress, experienced by concrete piles at Rodney Terminal, St. John's New Brunswick, Canada. (From Khana et al., 1988a.)

of freezing and thawing cycles acting on the piles, combined with relatively low manufacturing quality, inadequate air entrainment, and inadequate circumferential steel were responsible for the above-noted problems (Khana et al., 1988b).

Initially, 90 piles at Rodney Terminal were repaired using a method by which insulated jackets made from fiberglass and polyurethane insulation formed an annulus around deteriorated piles into which steel reinforcement had been placed and that was subsequently pumped full of grout. This type of repair, however, proved to be difficult in terms of obtaining satisfactorily fabricated jackets as well as in transporting, handling, and installing them without damage, so as to provide a leakproof form for the grout. Bearing in mind very difficult working conditions under the dock's deck (diurnal tidal cycle 6 to 8m and a 2 m/h maximum change in water level), a simpler repair method has been proposed and implemented. The distressed piles were repaired by simply pumping a steel-fiber reinforced concrete mix into a reusable heavy-duty steel leakproof form installed around the pile to construct a 150-mm thick jacket of strong and freeze and thaw resistant concrete which was intended to preserve (restore) the strength of the existing piles and act as a wearing coat (Figure 1-28). The steel fiber replaced the conventional reinforcement in the concrete and provided restraint against shrinkage and thermal cracking right out to the surface of the concrete. A typical concrete mix per cubic meter

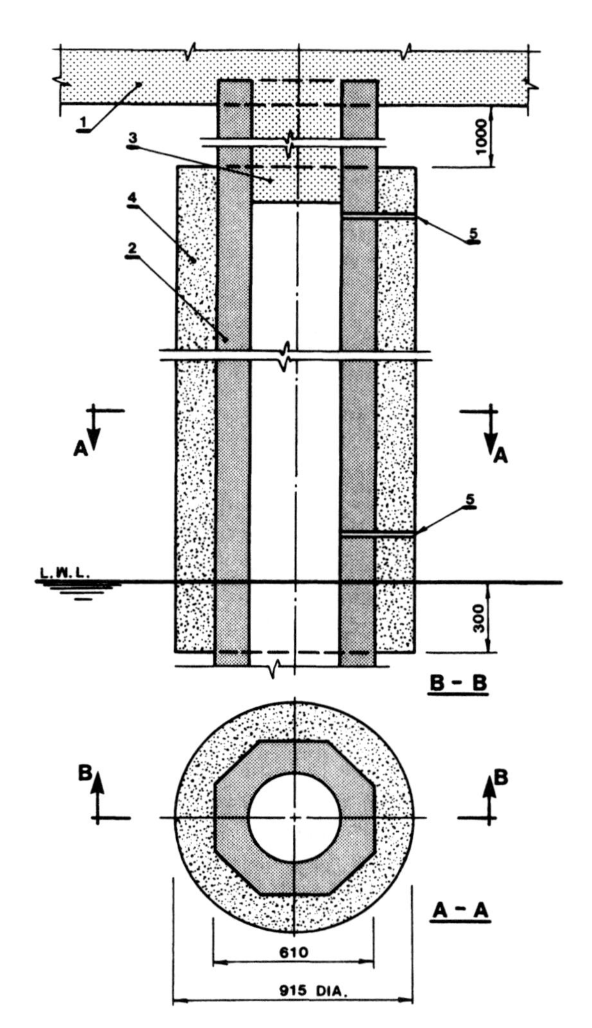

Figure 1-28. Rodney Terminal, New Brunswick, Canada. Restoration of distressed piles. 1—Deck structure, 2—hollow core octagonal concrete pile, 3—concrete plug, 4—steel fiber reinforced concrete jacket, 5—25 mm diameter drain hole.

with specified strength at 28 days of about 40 MPa included 415 kg of cement type 50, 105 kg of fly ash type F, 80 to 215 kg of sand, collated and uncollated $\ell = 50$, d = 0.5 steel fiber. The typical fresh concrete test results showed the following average properties: slump 158 mm, air content 8.5%, temperature 24°C, density 2336 kg/m^3, and excellent workability.

In a period from 1984 to 1987 a total of 362 piles were repaired using steel-fiber reinforced concrete (SFRC). The piles were cleaned prior to installation of the form (usually a day or two ahead of concreting) by a high-pressure freshwater blaster, operated in the range of 28 to 35 MPa. The underwater part below the low level was washed by a diver who used the same as for the above water wash nozzle gun.

Following pile cleaning the clean formwork was assembled around the pile to be repaired. The form was completely sealed around all components of the system and the pile. The concrete was pumped from the bottom up via a steel pipe line 125 mm in diameter.

Three vibrators operated continously until 4 to 5 min after completion of pumping. One of those vibrators was attached to the form just above the inlet nipple to help in avoiding blockage at this point. Two other vibrators were attached to the upper third of the form. The experience gained at Rodney Terminal indicated the following:

1. The SFRC pile repair method is a relatively simple method for jacketing marine piles.
2. When air entrainment is required for freezing and thawing resistance, the specified concrete strength and air void parameters must be met.
3. The SFRC method of marine pile jacketing is particularly attractive in tidal zones with a large range and strong currents because it minimizes work with marine plant over water.
4. The quality of the SFRC jackets may vary over the height of the jacket due to a nonuniform steel fiber distribution and a migration of air to the upper portion of the jacket. This, however, was not found to be a serious problem because the lower part of the jacket, where the SFRC quality is better, protects the pile zone where the freezing and thawing distress is maximum.
5. Collated steel fibers used on the project demonstrated a better toughness index than the uncollated fibers for the same loading, partly because of the much larger number of collated fibers for the same weight.

The success of the SFRC pile jacketing at Rodney Terminal was significantly affected by the pumping equipment and set-up, as well as a good maintenance program.

Deteriorated bearing piles of any kind under waterfront structures can be repaired by fabric-formed jackets filled with concrete. This type of repair has been applied to timber, concrete, and steel pile in diameters ranging up to 3.0 m and in lengths ranging up to 17.0 m (Lamberton, 1989). After cleaning the pile to be jacketed and placing a reinforcing sheet if necessary, the fabric forms were suspended from the pier structure, closed around the pile at the bottom of the jacket, and the sleeve closed vertically, usually with a zipper. The sleeve was filled with concrete through vertical injection pipes placed symmetrically at opposite ends at a jacket down through the annulus between the pile and the jacket. Injection on one side only can cause the jacket to take a banana shape that is very difficult to straighten once formed. A factor of safety of 1.5 to 2.0 applied to 80% of the break load is usually considered in fabric jacket components design. This specifically applies to selection of zippers, which are available today with strengths up to 280 g/mm. Fabric sewing is also critical. Fabric at seams should be folded before stitching and typical specification usually requires development of 80% of fabric break load at the seam.

When longer jackets are required, the fabric jackets may be reinforced with circumferential bands of steel strapping, which should sustain the full hydrostatic pressure of concrete. Piles of a moderate batter (say 1 : 10) are filled from the high side, while jackets located in moderate current are filled from the upstream side. Fabric pile jackets are not appropriate for use in wave action such as that occurs in breaking wave zone, or in flowing water with a current velocity over 1.0 m/s.

Piles also can be repaired by miscellaneous polymers. Polymers in various forms have been used in marine environments for pile repair and protection for over 20 years. From simple coatings to plastic wraps, to membranes and composites, they have been used with widely varying degree of success. Snow (1992) attempted to identify basic reasons for distress and failure of polymer systems that occurred in the past in order to present a viable solution for pile encapsulation.

Deck Repair/Rehabilitation

As noted in previous sections detailed inspection followed by engineering evaluation will produce data concerned with the average remaining live load capacity of the deck structure and its individual members for cargo handling and hauling equipment, for storage of cargo if required, and for remaining strength to sustain the reactive berthing and mooring forces. Then the decision will be made as to whether the structure should be repaired, strengthened, or just left as it is with no action. The basic elements of concrete deck structure deterioration are shown in Figure 1–29.

Concrete restoration or rehabilitation can be accomplished by one of the following methods: patching, grouting, shotcreting, or strengthening, or with combinations of the three.

Patching is the process of replacing loose, spalled, or crumbling concrete with new material (Figure 1–30). A good patch should restore structural integrity, be compatible with the surrounding concrete, and last as long as the structure does. Concrete deterioration is symptomatic of many kinds of structural damage, so patching is part of almost every major concrete repair project. Suitable materials for patching deteriorated concrete pier structures typically include

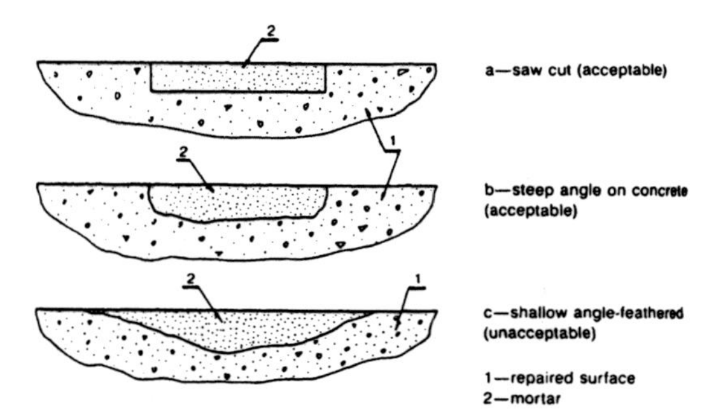

Figure 1–30. Surface repair (patching).

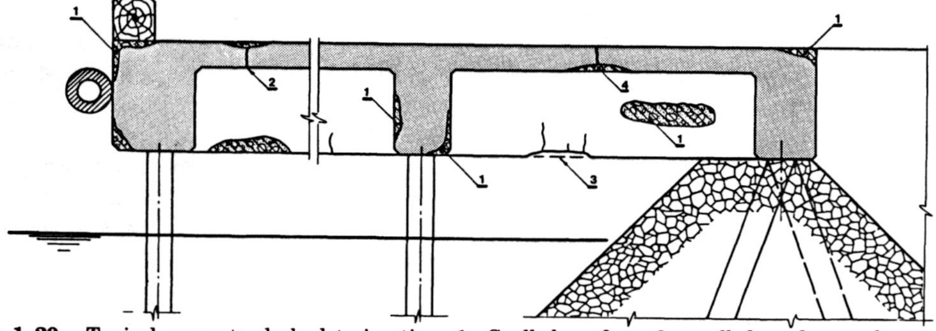

Figure 1–29. Typical concrete deck deterioration. 1—Spalled surface, 2—spalled surface and leaking crack, 3—underside spalled concrete and cracks, 4—underside spalled concrete and leaking crack.

epoxy-modified Portland cement mortar/concrete for repair in dry conditions, and epoxy mortar for underwater repair. Sometimes Portland cement mortar/concrete or special cements are used. The choice of material depends on such factors as patch thickness, condition of placement (dry, wet, underwater), shrinkage, and cost. Selecting the right material is essential. For example, epoxy mortar has high adhesive qualities but is expensive; it is thus a good choice for those applications where high adhesion is important, such as a freeze-thaw-resistant pier elements or a wear-resistant deck. For a deep patch, a better choice might be epoxy-modified Portland cement mortar/concrete because it costs less and is more compatible with original concrete. Successful patching involves:

1. Good surface preparation (Gaul, 1984). All deteriorated concrete must be removed, and substrata must be brought down to a clean, sound finish. All foreign particles and materials such as dust, laitance, grease, curing compounds, impregnations, and waxes must be removed by mechanical abrasion methods such as sandblasting. Exposed reinforcement must be cleaned by similar methods. Reinforcing that has lost a substantial sectional area must be replaced.
2. Installing epoxy-modified Portland cement mortar/concrete or epoxy mortar. To achieve best results, the concrete should be moistened, allowed to dry until damp, and then coated with epoxy-modified slurry. The latter is especially important when regular Portland cement mortar/concrete is used for repair.

Mortar installation should be followed by tamping or ramming of the mortar in place to produce close contact between the mortar and the existing concrete.

If Portland cement mortar is used, in order to minimize its shrinkage in place, the mortar should stand for half an hour after mixing and then be remixed prior to use (American Concrete Institute, Committee 224, 1984). It is good practice to place it in thin layers (about 10 mm thick). There is no need to wait between layers. The above-water repair is cured by using either water or a curing compound.

Grouting is the process of placing materials into a surface fissure, and/or interior crack. Typically, grouting is used to restore the monolithic nature of a structure, and to restore the bond of reinforcing steel to concrete. Grouts are liquids that solidify after application. They are pumped into place, usually under substantial pressure, to ensure that the void is filled completely. The major grouting materials are chemical, epoxy, cement-based, and polymer grouts.

Chemical grouts typically consist of solutions of two or more chemical components that combine to form a gel. They are usually flexible materials that never solidify completely. Their low viscosity makes them the best choice to stop water flow. Chemical grouts can fill cracks in concrete as narrow as 0.05 mm and can be applied in a moist environment. The disadvantage of chemical grouts is their lack of strength.

Because of the excellent tensile quality of *epoxy grouts*, they are often used to seal cracks as narrow as 0.05 mm. However, the presence of contaminants in cracks, including water, reduces the effectiveness of the epoxy to repair structural cracks. To achieve best results in repair work, epoxies should be mixed strictly according to the manufacturer's instructions.

Portland cement-based grouts consist of suspensions of solid cement particles in fluid. These grouts are versatile materials used to fill large voids and wide cracks. Depending on crack width, the cement-based grout may be composed of cement and water, or of cement, sand, and water. To minimize shrinkage of this grout, the water/cement ratio should be kept as low as practical.

Polymer grouts are liquids that consist of small organic molecules capable of forming a solid, strong, and durable plastic that greatly enhances concrete properties.

Polymer grouts are effective only if applied to a dry surface. If the cracks contain moisture, the repair will be unsatisfactory.

The typical procedure involved in grout injection consists of:

1. Cleaning the cracks of oil, grease, dirt, and fine particles.
2. Sealing the surface cracks to keep grout from leaking before it has hardened. This is usually done by brushing an epoxy or similar material along the surface of the crack, then allowing it to harden. A strippable plastic may also be used to seal the crack if high injection pressure is not required. In the latter case, the crack should be cut out 15 mm deep and 20 mm wide, in a V-shape, then filled with epoxy.
3. Installing entry ports, which could take the form of special nipples or fittings placed flush with the concrete face over the crack.
4. Mixing the grout in accordance with specifications provided by grout manufacturers.
5. Injecting grout, which is usually done by hydraulic pumps, paint pressure pots, or air-actuated guns. To seal vertical cracks, grout should be injected first at the lowest injection port until the grout level reaches the entry port above. The process is successively repeated. For horizontal cracks, the process is similar: the grout injection proceeds from one end of the crack to the other. Typical examples of concrete repair by patching and grouting are shown in Figures 1–31 and 1–32. More detailed information on crack grouting is found in US Army Corps of Engineers EM1110-2-2002, 1986.

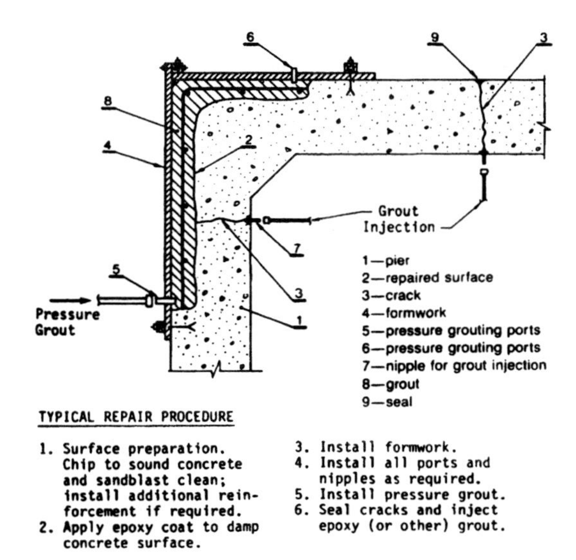

Figure 1–31. Typical examples of concrete deterioration and repair.

The *shotcreting technique* has been described in a previous section. As noted before, shotcreting is typically used for repair of large areas of deteriorated concrete.

Concrete repair practice such as patching, grouting, and shotcreting varies among locales. The character of concrete deterioration or damage, climatic conditions, materials, and economics dictate procedure. Successful concrete repair depends on an understanding of the effect of the variables just discussed.

Deck Strengthening

When deck structure loadings become too great to bear, the deck may crack or even fail. The failure problems typically result from design or construction errors, structural overloads, outside factors such as weather or operational conditions that alter or weaken structural materials over time or material deterioration. Solutions must be specific to the problem and sometimes can be quite complex. Typical solutions to failure problems include:

1. *Post-tensioning*, which is essentially strengthening of the deck structure with prestressing bars or strands, to which tension is applied so that the existing steel will not fail under the tension of service conditions (Figure 1–33). The length of the prestressed portion of the deck should be carefully evaluated so that the problem does not simply migrate to another part of the structure. Secondary bending moment and corresponding compression result from post-tensioning should also be carefully considered (Lin and Burns, 1981).

Problem		Solution	
Description	Sketch	Description	Sketch
Leaking crack	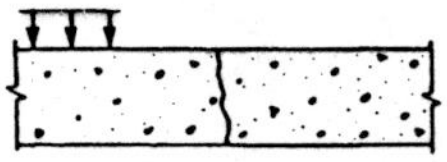	Route and clean crack then install sealant	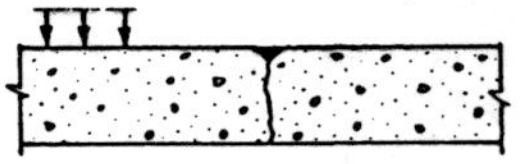
Spalled surface	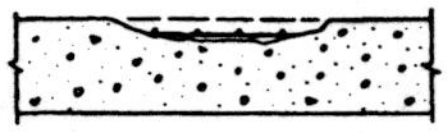	Remove unsound concrete, clean surface of concrete and reinforcement, install grout	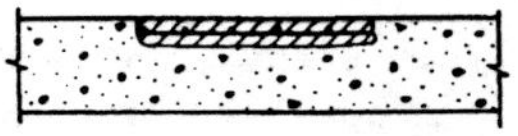
Spalled surface and leaking crack	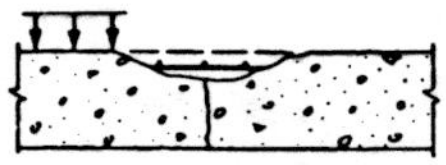	Remove unsound concrete and rout crack, clean surface of concrete and reinforcement, install sealant and grout	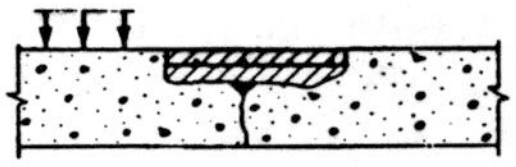
Underside spalled surface and leaking crack	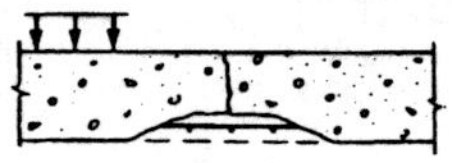	Rout and clean crack, remove unsound concrete and clean surface of concrete and reinforcement, install sealant and grout	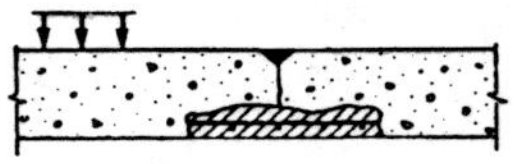

Figure 1–32. Typical examples of concrete deterioration and repair.

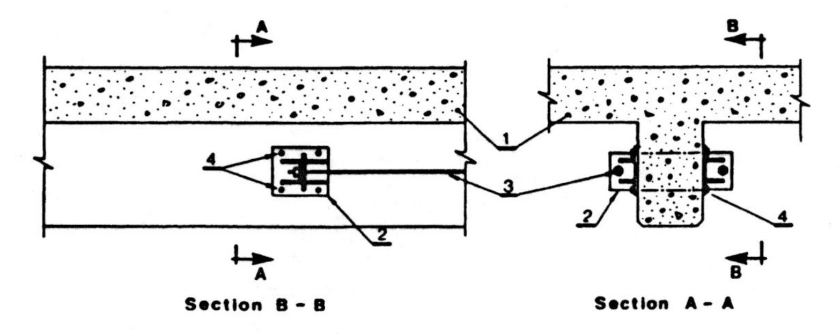

Figure 1–33. Strengthening the deck structure by post-tensioning.

Several practical examples of strengthening of existing structures with post-tensioning technique is given in Vejvoda (1992).

2. *Strapping*, which is local reinforcing of the pier structure with exterior steel stitching dogs (Figure 1–34). This method could be used when tensile strength across major cracks must be reestablished. Stitching dogs should be properly anchored into holes predrilled on both sides of the crack. Nonshrink grouts or epoxy-based systems are used to anchor the stitching dogs into the holes.

Strapping prevents further crack propagation and typically is combined with simultaneous crack repair. However, local stiffening of the pier structure by means of stitching dogs may cause the concrete to crack at

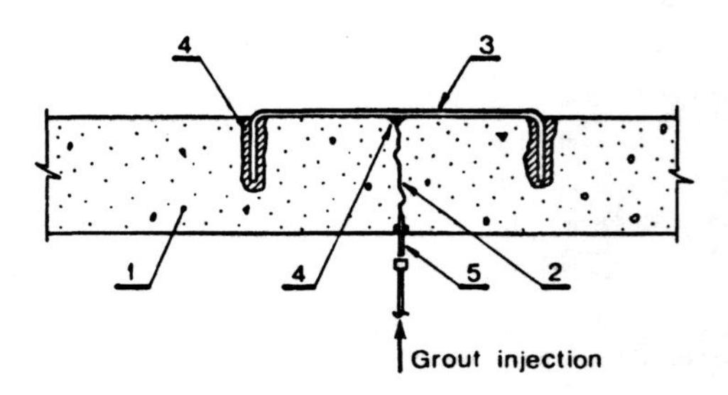

Figure 1–34. Crack repair by strapping.

other locations. Therefore, it may be necessary to combine several techniques—for example, stitching and post-tensioning—to increase deck structural strength.

3. Installation of additional conventional reinforcement. This technique consists of sealing the crack, drilling holes (20 mm in diameter) at 90° (or closer) to the crack plane, filling the hole and crack plane with epoxy grout pumped under low pressure (0.4 to 0.6 MPa), and placing a reinforcing bar into the drilled hole (Figure 1–35). Typically bars 15 to 20 mm in diameter are used, extending at least 500 mm on each side of the crack. The epoxy grout bonds the bar to the walls of the hole, fills the crack plane, bonds the cracked concrete surfaces together in one monolithic form, and thus reinforces the section.

A temporary elastic rack sealant is required for successful repair. Gel-type epoxy crack sealants work very well within their elastic limits. Silicone or elastomeric sealants work well and are especially attractive in cold weather or when time is limited. The sealant should be applied in a uniform layer approximately 2 mm thick and should span the crack by at least 20 mm on each side.

The reinforcing bars can be spaced to suit the needs of the repair. They can be placed in any desired pattern, depending on the design criteria and the location of the in-place reinforcement.

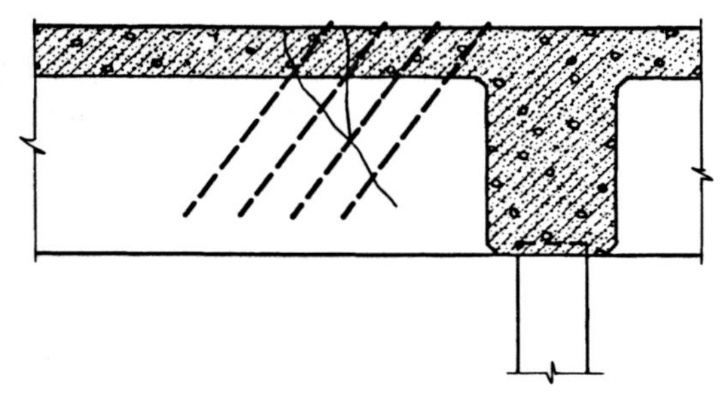

Figure 1–35. Crack repair using conventional reinforcement installed into predrilled holes filled with epoxy grout.

The added steel is proportioned to carry the tensile (shear) forces that have caused cracking in the concrete.

Stratton et al. (1982) described successful repair of cracked reinforced concrete bridge girders by using the above technique.

More information on epoxies used for cracked reinforced concrete structures can be found in American Concrete Institute, Committee 224, 1984.

Additional information on concrete repair is found in Johnson (1965) and Shilstone and Kenney (1979). A comprehensive list of recommended codes, regulations, and other helpful references on this subject can be found in American Concrete Institute, Committee 224 (1984) and American Concrete Institute (Concrete International, 1984).

1.4.3.4 Repair/Rehabilitation of Sheet-Pile Bulkheads

Repair or rehabilitation of sheet-pile bulkheads usually presents quite a complicated task. This is because stress and/or deflection conditions inherited by a sheet-pile bulkhead as a result of overloading, material deterioration, scour, or overdredging, or due to other reasons is practically irreversible.

The latter is best illustrated in Figure 1–36. As can be seen in Figure 1–36, scouring or overdredging reduced the value of the original passive pressure in front of the anchored sheet-pile bulkhead (E_p to $E_{p(1)}$), increased bulkhead active span ($l_1 > l$), and produced additional deflection (Δ) and higher values of stress in sheet piles and anchor force ($R_{A(1)} > R_A$). Restoration of the foundation soil up to its original level produces insignificant active pressure (E_A) in addition to remaining passive pressure ($E_{p(1)}$) and therefore cannot help much to restore the bulkhead's passive pressure to its original value (E_p). It also cannot bring stress in sheet piles and anchor forces to their prescour level, because it is practically

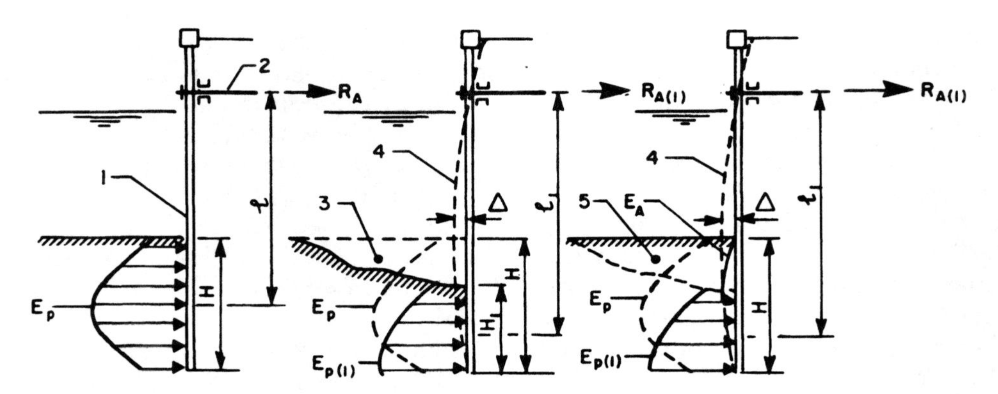

Figure 1–36. Change in passive soil pressure in front of sheet-pile bulkhead due to scour effect. 1—Sheet pile, 2—anchor system, 3—scour, 4—deflection line, 5—restored foundation material.

E_p Pre scour passive pressure
$E_{p(1)}$ Post scour passive pressure $(E_{p(1)} + E_A) < E_p$
E_A Active pressure due to restoration of eroded foundation material
R_A Pre scour anchor force
$R_{A(1)}$ Post scour anchor force $(R_{A(1)} > R_A)$
Δ Scour induced deflection
l Pre scour effective span
l_1 Post scour effective span $(l_1 > l)$

impossible to change sheet-pile deflection, that is, to bring it to the prescour position.

Practically, the situation as described could be improved by installation of an additional anchor system located below the existing anchors. New anchors must be tensioned in order to prevent or limit sheet-pile deflection under design live load. This would minimize additional stress in a sheet-pile structure. New ground or rock anchors are typically drilled through sheet piles into the backfill.

The problem with such types of repair/rehabilitation, however, is that in most cases this additional anchorage must be placed underwater because the original tie rods are usually placed as close as possible to the mean low water level if the body of water is tidal. Consequently, this type of construction may involve use of an underwater technique or installation of a special portable cofferdam around the working area. A conceptual scheme of such cofferdam is shown in Figure 1–37. Cofferdams of this type are used for isolation of a certain working area under question to conduct repair, rehabilitation, or other specific work in dryness without constructing of conventional cofferdams. The most recent examples of installation of the portable cofferdam are a cofferdam used for repair of the heavily corroded sheet-pile bulkhead at Harwich U.K. (The Dock and Harbour Authority, October, 1989), and a "blister" cofferdam used for the rehabilitation work on water intake at the hydro project in Canada (Figure 1–38).

Ishiguro and Miyata (1988) published a report on new technology developed in Japan for installation of additional anchor tie rods. This technology allows for installation of new tie rods through existing or new deadman and backfill toward sheet piles, and then at desired locations through sheet piles (Figure 1–39).

The new rod installation proceeds in the following sequence:

1. The existing backfill is excavated just behind the existing anchor block. The width of the excavation should be large enough to accommodate a boring machine.

2. Precise survey of the sheet piling is done and the boring machine is set up in

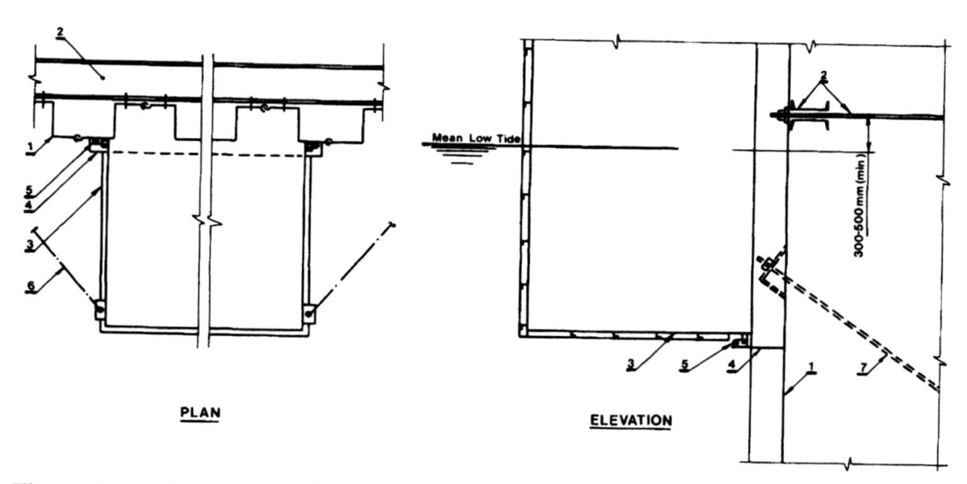

Figure 1–37. Portable cofferdam. 1—Sheeting, 2—anchor system, 3—cofferdam, 4—bottom plate welded to sheet piles, 5—rubber seal, 6—bracings, 7—new ground (rock) anchor.

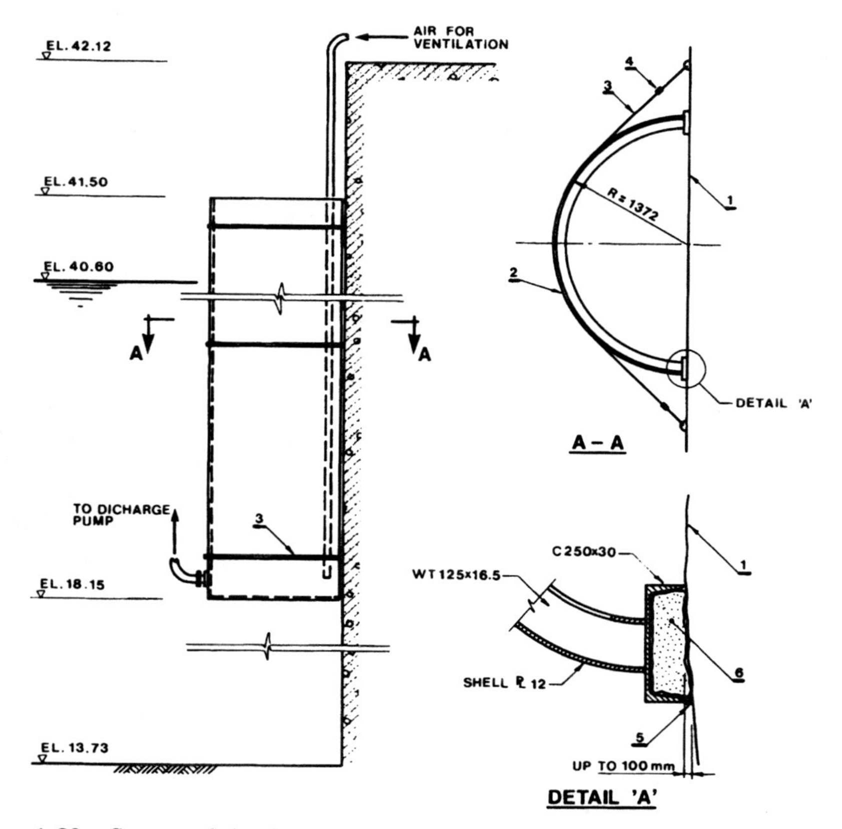

Figure 1–38. Conceptual sketch of the blister cofferdam used for rehabilitation work on water intake structure at the hydro project in Canada. 1—Concrete face, 2—blister cofferdam, 3—anchor cable 20 mm diameter, 4—turnbuckle, 5—oversized geotextile tube filled with cement grout (seal), 6—grout.

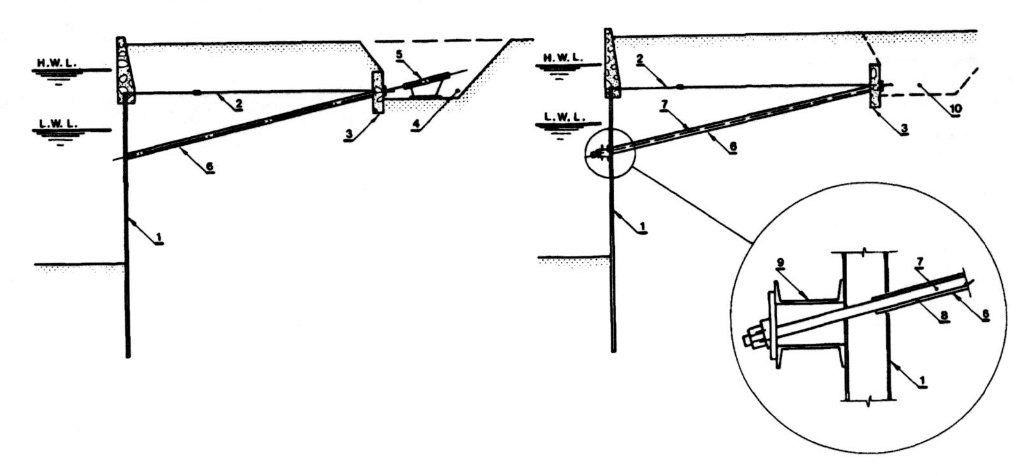

Figure 1–39. Installation of a secondary row of tie rods. (From Ishiguro and Miyata, 1988.) 1—Sheet piling, 2—original tie rod, 3—anchor block, 4—excavation behind anchor block, 5—boring machine, 6—casing, 7—new anchor rod, 8—grout, 9—new wale, 10—fill.
Construction stages:
1—fill behind anchor block excavated, 2—borehole drilled through fill and sheeting, 3—new anchor rod installed and tensioned, 4—drill hole grouted and previously excavated fill put back.

the exact position for the drilling operation. To avoid deviation of the drill from both vertical and horizontal directions, the boring machine has to be firmly fixed in the predetermined position.

3. The borehole is drilled through the fill by employing a casing pipe (diameter 100 to 150 mm depending on diameter of the tie rod).

The boring process is aided by a special electronic sensor system that ensures an extremely accurately guided boring process. Cutting a hole through the sheet piling is allowed only if the survey indicates that the deviation of the borehole from a designed position is within an acceptable range.

After cutting the hole through the sheeting the boring machine is shifted to a new position.

4. The new tie road is installed, and secured by divers at the new wale, and the rod is tensioned at the anchor block location.

The magnitude of the tension load applied is predetermined by design computation. Tensioning of the lower tie rod system alone cannot significantly affect existing tension load in the upper anchor rods. However, even the slightest backward movement of the sheeting induced by tensioning of the lower anchors can change the tension load at the upper tie rods. Therefore, it is imperative to keep tension loads at both levels under control.

It must be noted that in most cases the existing anchor block or anchor systems cannot safely take the combined load from both existing and new anchor rods, and therefore it must be reinforced. This can be done in a variety of ways, for example, piles can be driven in front of anchor blocks, regular soil in front of anchors can be replaced with well compacted rock fill, etc. If necessary, new tie rods may be secured to a new anchor system.

5. Finally, the space between the new anchor and the casing pipe is grouted from the water side upwards, and the previously excavated material is backfilled to its original elevation. Grouting is usually used to provide tie rods with corrosion protection and to prevent possible loosening of backfill material after corrosion of the casing pipe.

Typical damage to sheet-pile bulkheads involves corrosion of steel sheet piles, deterioration of concrete sheet piles, and distress or failure of the anchorage systems. Flexural failure of the sheeting is generally acknowledged to be quite a rare event. If corrosion of the sheeting is a problem at a given site, it is likely to progress most rapidly in the tidal/splash zone, where alternate wetting and drying occur. Hence, a corrosion problem will likely be more severe in an area where bending moments are relatively small. However, the existence of corrosion-induced holes may create some dock operational problems. For example, local damages to the sheet piling may result in backfill loss, settlement of apron, and damage to the crane truck, but almost never result in total structural failure. An example of badly deteriorated steel sheet piling is shown in Figure 1–7. It should be noted that in the case presented in this figure sheet piling was so badly corroded, particularly in the splash/tidal zone, that the dock operator was forced to take this particular berth out of service.

Reinforced concrete sheet piles are typically deteriorated due to corrosion of reinforcing steel and concrete deterioration. The problem of sheet piling deterioration is particularly acute for bulkhead structures in a warm saltwater environment. Ultimately, sheet pile deterioration which generally origiantes in the splash/tidal zone area provides access to the water into the bulkhead's interior. This in turn may trigger corrosion of the wale, tie rod, and connections.

It must be specifically noted that the wale is a very important part of a sheet-pile bulkhead and is very often the most vulnerable component of this structure. It is typically located just above mean high tide level, where in most cases the rate of steel sheet-pile corrosion or concrete deterioration reaches its maximum value.

In most cases of sheet-pile bulkhead construction the wale (usually composed of two channel sections placed back to back) is installed behind sheet piles. This is done in order to protect it from being damaged by ship impact, and for better corrosion protection. Therefore, the wale is practically inaccessible for inspection and regular maintenance and repair. It has been observed, in a great many cases, that seawater that has penetrated through holes in corroded steel sheet piles has been capable, in a very short period of time, of severely corroding the wale and adjacent portions of anchor rods.

Wales, as part of the sheet-pile bulkhead anchor system, is vitally important to the integrity of the structure (failure of a wale is practically failure of an anchor system and therefore its strength should never be compromised).

If any damage has been done to the wale structure, then its remaining strength must be carefully evaluated and a decision must be made on whether its present strength is sufficient to provide for safe service, or whether it has to be repaired, or reinstalled. If the remaining strength of the wale is sufficient to transfer service load to the tie rods then in order to preserve its status quo and eliminate sources of corrosion, which in most cases would be seawater penetration into the bulkhead interior, the voids around wale and in adjacent backfill material should be grouted. Large voids can be filled simply by placing tremie concrete through holes in sheetpiling; backfill can be grouted by any means previously discussed. A typical example of possible repair of steel sheet-pile bulkhead is shown in Figure 1–40. For additional corrosion protection of existing steel, installation of sacrifical anodes is highly desirable. If the wale's remaining strength is insufficient for safe operation then it must be repaired, or a badly corroded part of it must be reinstalled. In both cases, the wale has to be exposed, by excavation of backfill material. The repair procedure typically may involve installation of additional steel to the wale structure, or encasement of the damaged portion of the wale into the reinforced concrete jacket. In the latter case

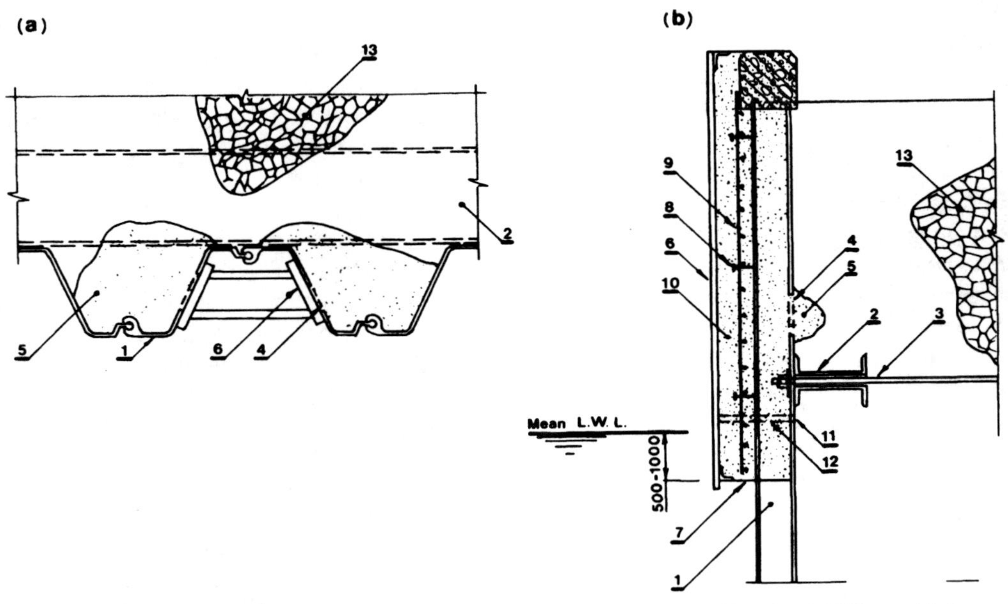

Figure 1–40. Repair of corroded in tidal/splash zone steel sheet pile bulkhead. (a) Filling voids in bulkhead interior. (b) Bulkhead refacing. 1—Sheet piling, 2—wale, 3—achor rod, 4—corrosion hole, 5—interior grout, 6—formwork, 7—steel plate welded to sheet piles, 8—stud, 9—steel reinforcement, 10—concrete face, 11—existing weeping hole, 12—new weeping hole, 13—backfill.

to prevent a stress concentration at the tie rod the latter must not be restrained from free rotation as much as possible. In the case of replacement of a damaged wale it should be disconnected from the sheeting, then the new wale would be installed, and the tie rods retensioned to the required value.

Experience indicates that most sheet-pile bulkhead failures are attributed to damage of the anchorage system. This can be due to either corrosion of the anchorage structural components or heavy local overload. The latter may be caused by placement of the unusually heavy point load in the immediate vicinity of the sheet pile wall, for example, unspecified mobile crane load, or unspecified heavy surcharge load, or by substantial settlement of the backfill material surrounding tie rods. In addition, heavy overstressing of both sheet piling and anchor system can be attributed to scour of the sea floor. Failure of one or two anchor ties, however, would not necessarily be followed by the failure of the bulkhead.

Because a wale is usually designed with substantial redundancy it may be capable to transfer a load from the failed anchor rod to the adjacent rods, thus overloading them. In case of failure of one or a few anchor rods the adjacent dock zone must be restricted to smaller loads than designed load or completely prohibited for operation. In special cases some of the backfill has to be excavated in order to reduce load and prevent local or progressive failure of the structure.

The wale/anchor tie system can be analyzed as a beam on elastic supports. Elastic properties of each individual tie rod should include elastic properties of both the tie rod and anchor system, for example, anchor block, anchor sheet pile wall, piles, etc. The latter is best established by a field or model test.

In order to bring stresses in a wale system to an acceptable level repaired or new tie rods must be tensioned. It is not practical to expect that installation and tensioning of new or repaired rod(s) will significantly reduce load in adjacent overstressed rods. This cannot happen because tensioning of these rods cannot bring the backfill behind the sheeting into its original boundaries. As a matter of fact an attempt to do this may fail the cantilever portion of bulkhead due to development of passive pressure behind it.

Therefore, to prevent further overstressing of the existing rods immediately adjacent to a distressed or failed rod, new tie rods should be installed as close as required to the existing rods (Figure 1–41).

If material deterioration has not caused significant overstressing in bulkhead components, then in most cases just a cosmetic repair might be sufficient. It would basically include plugging of holes in sheeting to prevent backfill material from washing away.

This is usually done by grouting voids behind sheeting, or by constructing a new face in front of deteriorated sheet piles. Typical details of this type of repair are shown in Figure 1–40. The latter repair technique would not only patch the existing holes but also provide additional strength to the deteriorated portion of bulkhead and extend the useful life of the structure.

Again, it is very important to realize that just mere replacement of steel lost to corrosion by new plates welded to the sheet piles, or installing new concrete to replace deteriorated concrete in concrete sheet piles, cannot bring the stress level within allowable limits and therefore cannot increase the capacity of these sheet piles or relieve stress in the bulkhead's anchor system.

Some more practical examples of sheet-pile bulkhead repair can be found in papers by Horvath and Dette (1983), Kray (1983), Porter (1986), and in many other works.

A properly engineered sheet-pile bulkhead repair must consider maximum use of the existing material that is still in good condition.

This is best illustrated by a relatively unusual design for steel sheet-pile bulkhead rehabilitation shown in Figure 1–42.

The steel sheet-pile bulkhead, 12.7 m high, anchored with tie rods and secured at the concrete anchor block, was badly deteriorated basically in the splash zone. A great deal of steel was lost from all components of the structure. Through holes in the steel sheet piles, one could see quite badly corroded wale and adjacent ends of steel tie rods. This situation forced a dock operator to take the dock off service, and consultants were commissioned to design rehabilitation works.

Inspection of the structure was conducted. It was found that steel sheet piles below minimum mean water level did not lose much steel to corrosion, and remained in

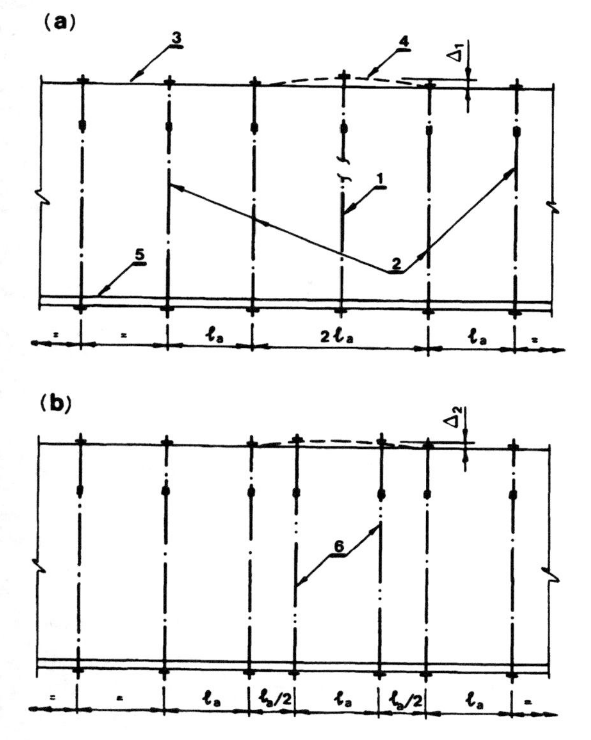

Figure 1–41. Example of replacement of failed or distressed anchor tie rod with new rods. (a) Plan of sheet-pile bulkhead with damaged or distressed tie rod. (b) Possible rehabilitation of bulkhead's anchor system. 1—Failed (distressed) anchor tie rod, 2—heavy overloaded anchor rods, 3—sheeting, 4—bulge in sheeting and wale, 5—tie rod anchorage, 6—new tensioned anchor rods.

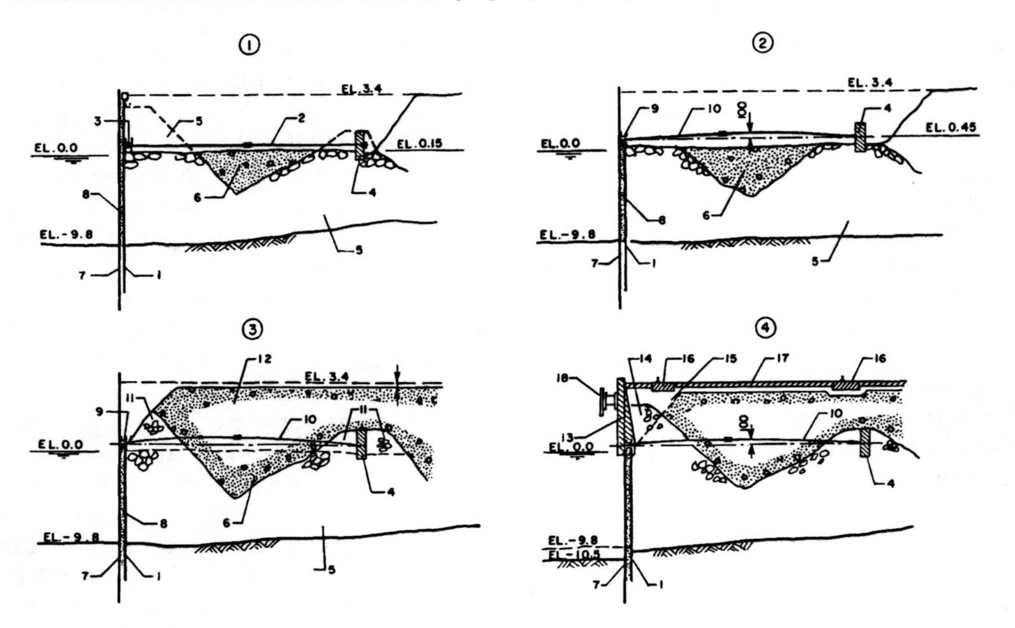

Figure 1–42. Example of steel sheet-pile bulkhead rehabilitation. 1—Badly deteriorated sheet piles, 2—existing tie rod, 3—existing badly deteriorated wale, 4—existing solid concrete anchor block, 5—existing rubble, 6—existing fill, 7—new steel sheet piles, 8—gravel fill between new and existing sheeting, 9—new wale, 10—new and/or reconditioned tie rods, 11—partly reinstalled rubble, 12—new fill, 13—concrete cope, 14—complete rubble, 15—complete fill, 16—crane way, 17—new apron, 18—new fender system.

rather good shape. It was also concluded that tie rods with the exception of 1 to 1.5 m immediately adjacent to the sheet piles and anchor concrete block could be reused. The terms of reference for dock rehabilitation called for increase in the berth height by 1.2 m.

The design rehabilitation work comprised the following stages:

Stage 1

New steel sheet piles (7) are driven as close as practicable to the existing sheet-pile bulkhead (1), and the space between the old and new sheet piles is filled with gravel (8). Next, the existing fill (6) and rubble (5) are to be excavated just below the existing tie rods (2). The excavation was designed to progress from the face of the wall toward the concrete anchor block (4). The structural analysis indicated that, in combined action, new and existing sheet piles could take lateral soil (rubble) pressure and live load imposed against them by construction equipment with acceptable deflection, while performing as a cantilever wall. It has been suggested that the initial excavation (6) should expose no more than five existing tie rods (2), which brought the total initial excavation width to about 7.5 m.

Stage 2

Existing tie rods (2) should be dismantled, cleaned, extended (in some cases tested), properly recoated, and certified for the purpose of reinstallation to support the new sheet-pile bulkhead. Because used rods' preparation and certification take time, initially five new tie rods (10) should be installed.

At this stage, existing concrete anchor block (4) should be examined, and if found to be in unsatisfactory condition, it should be dealt with accordingly, that is, repaired or demolished, and a new anchor block of similar construction should be built. Then, existing sheet piles can be cut right below the existing wales (3), and new wales (9) and refurbished existing tie rods (10) should be installed.

Stage 3

Backfilling, which would include reinstallation of previously partly excavated rubble (11) and new selected fill (12), should be conducted. The backfill should proceed from the anchor block toward the face of the berth. Space in the immediate vicinity of the new sheet-pile bulkhead should remain unfilled. This would allow for uniform initial tension in all anchor tie rods before a concrete cope (13) is cast in situ.

Stage 4

A concrete cope (13) extended from its lowest level located 60 cm below minimum mean water level up to 40 cm above the apron is installed and completed with a fender system (18). Then the space between the cope and fill is filled with rubble (14) and granular fill (15), apron (17) and craneway (16) built, and mooring accessories installed.

The above example demonstrates a rational approach to bulkhead rehabilitation in which all useful structural components are reused with benefit to the owner.

1.5 REHABILITATION OF DISTRESSED SOIL-RETAINING STRUCTURES

In soil-retaining structures the horizontal component of soil pressure is the main force that tends to destabilize the wall. Soil pressure and lateral components of live load transmitted to the wall through the fill depend on soil properties, such as density, cohesion, and angle of internal friction. Therefore, if the wall performance, in terms of sliding and/or overturning stability or bearing pressure needs to be improved, this can be done by improving soil performance. Primarily, this could be achieved by replacement of poor soil behind the wall with a good granular material, and/or reinforcing the backfill.

The other methods of improving wall stability are installation of horizontal, vertical, or inclined post-tensioned ground anchors, or use of different kinds of pressure relieving systems (Figure 1–43). Some of these methods are discussed in the following paragraphs.

1.5.1 Soil Replacement

Replacement of poor soil with a good granular material such as selected rockfill, gravel, or coarse grained sand is the easiest and in a way the most popular (however not the most efficient) method of reduction of soil horizontal pressure. Porter (1986) reported a successful sheetpile bulkhead repair by using lightweight expanded shale for partial replacement of existing fill.

The reduction in earth pressure is typically achieved because of a smaller value for the coefficient of lateral pressure of the new fill.

The angle of internal friction of good quality granular material is usually taken as equal to 40 to 45°. Therefore replacement of say silty sand with a good quality rockfill can reduce soil lateral pressure by about 2 times (see data presented in Table 1–5).

To be effective new fill has to be extended far enough beyond the slip line. If left within a potentially unstable slip wedge new fill may promote even heavier later soil thrust against wall than the original. Examples of use of good quality granular fill as replacement for poor soil are shown in Figures 1–44a and 1–45b.

As can be seen from Figures 1–44a and 1–45b the slip line behind the wall crosses the body of new fill. Use of granular material for

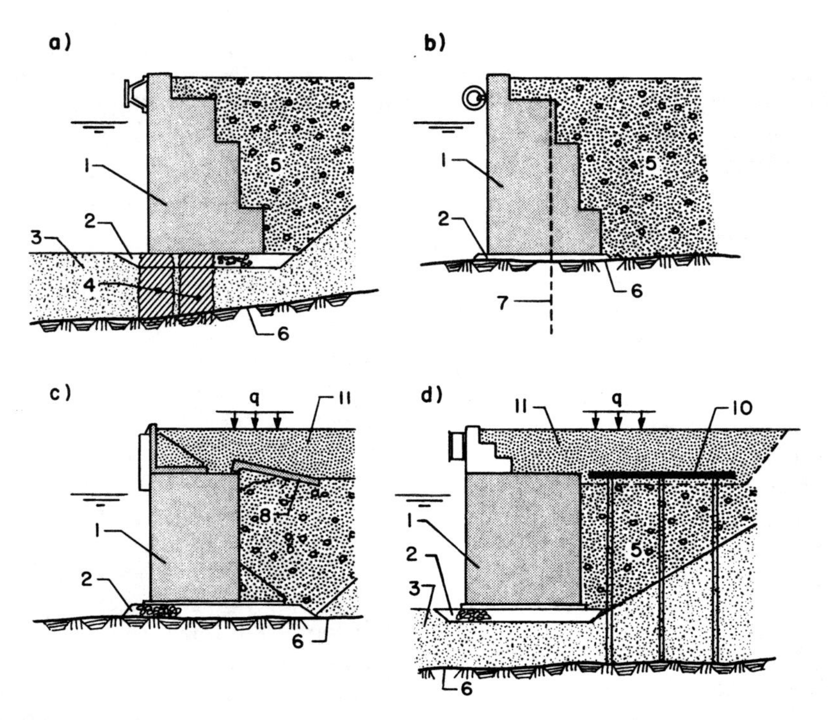

Figure 1–43. Improvement of performance of a soil retaining quay wall. 1—Quay wall, 2—rockfill matress, 3—distressed foundation material, 4—jet grout column, 5—backfill material, 6—outline of bedrock, 7—ground/rock anchor, 8—lateral soil pressure relieving slab, 9—excavation outline, 10—pressure relieving piled platform, 11—new backfill.

Table 1–5 Effect of Replacement of the Silty Sand with Quality Rockfill

Characteristics	Conditions	Silty Sand	Quality Rockfill
Density (γ)	Dry	1.75	1.9
(t/m^3)	Wet	1.0	1.1
Angle of internal friction	Dry	28	45
(degree)	Wet	26	45
Coefficient of lateral pressure	Dry	0.36	0.17
K_A	Wet	0.39	0.17
Reduction in soil pressure	Dry	$\frac{1.75 \times 0.36}{1.9 \times 0.17} = 1.95$	
$\frac{(\gamma K_A) \text{ silty sand}}{(\gamma K_A) \text{ quality rock}}$ (times)	Wet	$\frac{1 \times 0.39}{1.1 \times 0.17} = 2.09$	

soil pressure reduction is most efficient where it is available locally.

Generally, for better performance new fill should be placed from the excavation edge toward the wall. This will allow for maximum mobilization of shear forces within body of a fill.

Excavation of poor soil from behind the gravity type retaining wall almost always will result in reduction in soil-bearing stres-

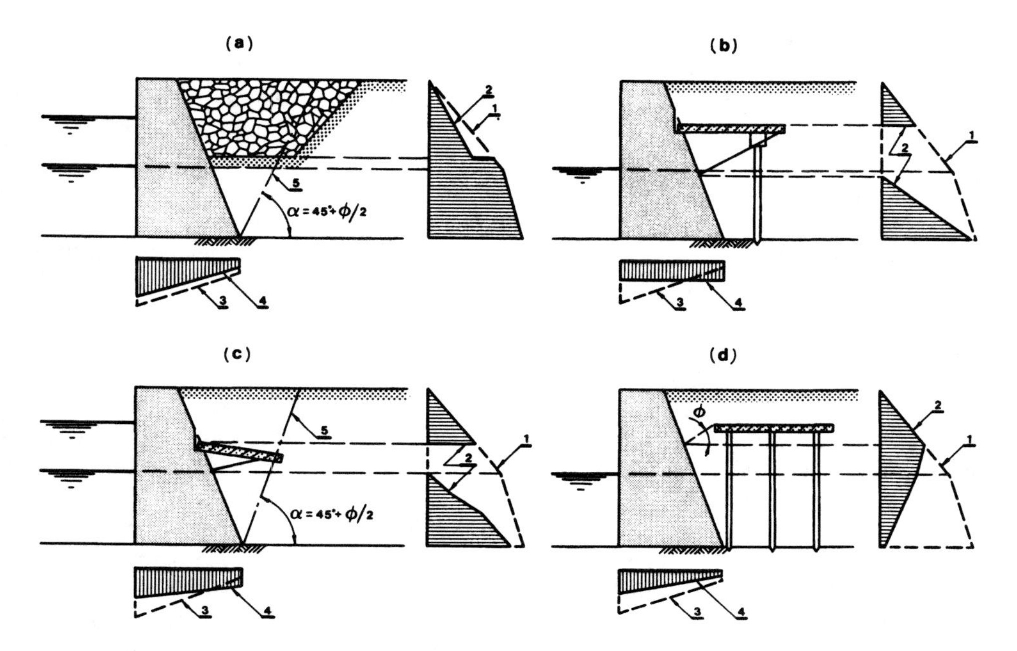

Figure 1–44. Methods of reducing of lateral soil pressure upon retaining wall. (a) Replacement of poor backfill soil with good quality granular material. (b) Pressure relieving slab supported on wall interior and piles. (c) Pressure relieving slab supported on wall interior and backfill. (d) Pressure relieving piled platform placed behind wall. 1—Original soil lateral pressure diagram, 2—new soil lateral pressure diagram, 3—original bearing pressure diagram, 4—new bearing pressure diagram, 5—slip line, 6—excavation.

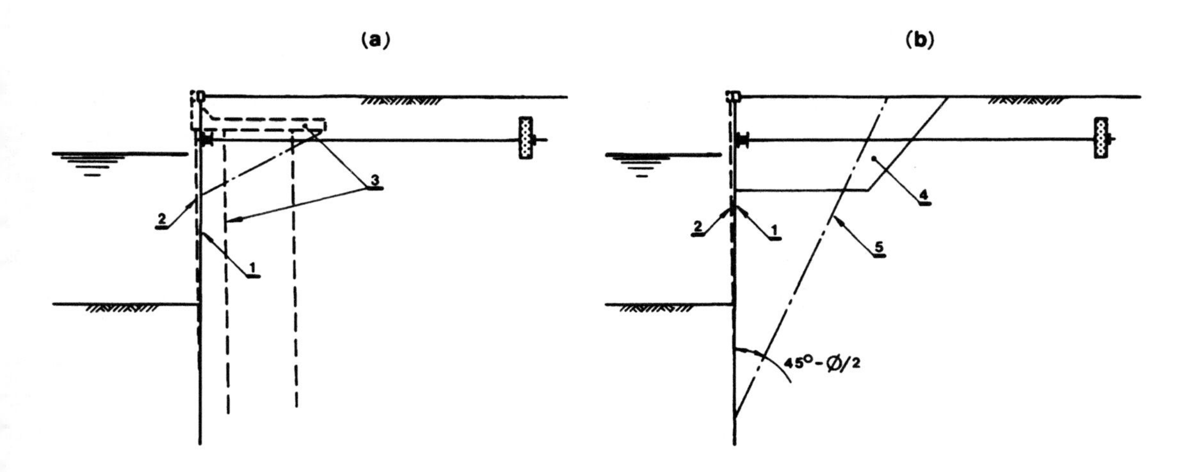

Figure 1–45. Rehabilitation of distressed sheet-pile bulkhead. (a) By incorporating of piled relieving platform. (b) By partial replacement of poor backfill material with good granular fill. 1—Existing (distressed) sheeting, 2—new position of existing sheeting after loosened tie rod nut or turnbuckle, 3—new pile relieving platform, 4—new granular fill, 5—slip line, 6—excavation.

ses due to reduction in both vertical and horizontal soil pressure. However, one should not expect significant reduction in stresses in a sheet-pile bulkhead just due to excavation of poor soil. As stated before, the sheeting cannot rebound in its middle section after excavation is done. Therefore only the upper portion of the bulkhead will change its curvature after excavation of part of the existing fill.

The stress reduction in the most stressed middle portion of the sheeting can be achieved by distressing the anchor rod to allow for just sligh sheet pile movement away from the fill. Practically this may be achieved by loosening the nut at the anchor block.

The value of new soil pressure against the retaining wall as shown in Figure 1–45 may be determined by the conventional methods.

1.5.2 Use of Slabs and Piled Platforms for Reduction of the Soil Pressure

Pressure relieving slabs, placed behind soil-retaining structures, are used to reduce bearing stresses and increase a wall's sliding and overturning stability. If water level allows then for better performance these slabs are typically placed at a depth of about one-fourth the wall height. They should be of sufficient length to provide for the required shileding of the wall from the effects of the soil above the slab. The minimum length of such a slab is usually taken as equal to $0.6H$ for gravity walls and $0.7H$ for sheet-pile bulkheads (H = height of the wall from the dredge line to the top). In the case of a gravity type retaining wall the slab may be supported by one end on the wall while the other end is supported upon backfill or piles (Figures 1–44b and c). In this case, a combination of reduced horizontal thrust and effect of the vertical load transmitted from slab to the interior face of the wall will result in a much smaller, as well as a better distributed bearing pressure, for example, rectangular (or close to it) shape of bearing pressure diagram vs. a trapezoidal diagram.

If extended far enough beyond the slip circle the pressure relieving slab will also act as an anchorage to the wall. In the latter case, to reduce the value of bending moment in the slab due to increased span it may be broken into two parts jointed together through a hinge(s) (Figure 1–46).

In some cases piled platforms independent from the wall structure or incorporated into it are used for the same purpose (Figures 1–

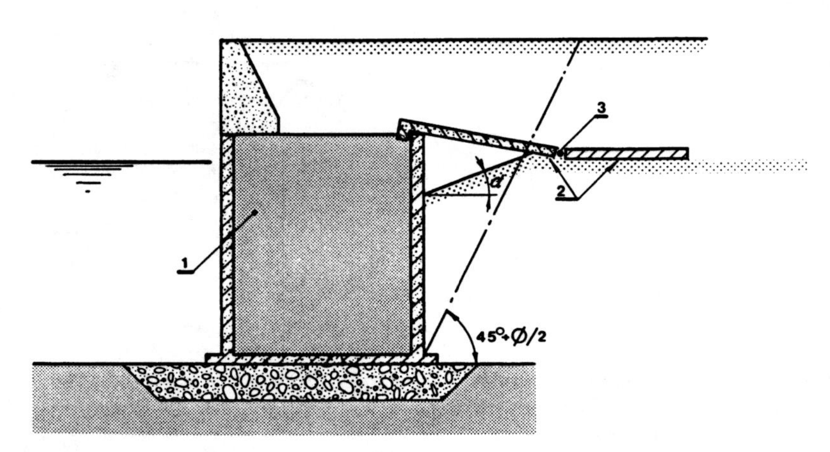

Figure 1–46. Pressure relieving/anchor slab. 1—Quay wall, 2—slab, 3—hinge.

44d and 1–45a). The effect of reduction in soil horizontal thrust attributed to piled platform is shown in Figure 1–44d).

In the case of a sheet-pile bulkhead the pressure relieving slab may be supported wholly on piles or partly on bearing piles and partly on the sheet piling (Figure 1–45a). Anchorage of the new system is still provided by the existing tie rods, secured to any kind of anchors. Again, to relieve existing stresses in sheet piling nuts an anchor tie rods should be slightly loosened.

It should be noted that soil pressure relieving slabs (platforms) are probably one of the oldest techniques that have been used by engineers for the relief of soil pressure. For example, in Russia they were used more than 130 years ago (Dubrova, 1959). In the past they were often used in other European countries. Some examples of the typical use of pressure relieving platforms or slabs are discussed below.

Figure 1–47 demonstrates an example of use of the pressure relieving concrete slab, which sits upon the interior of quay wall and a row of piles. This was a solution used to prevent progressive distress of a quay's foundation material. This wall was constructed upon weak foundation soil. The concrete casissons composing the quay structure were placed directly upon the existing foundation without prior installation of a rockfill mattress. In the process of placing backfill material the wall began to lean toward the water side.

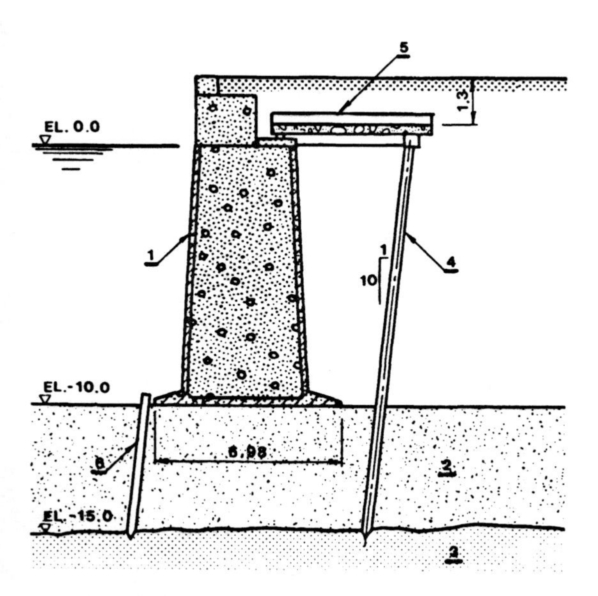

Figure 1–47. Rehabilitation of quay wall Mollo DiPonento, Venice, Italy. 1—Caisson type quay wall, 2—weak foundation soil, 3—firm soil stratum, 4—concrete piles 400 × 400 mm, $L = 15$ m, 5— T-shaped concrete slab, 6—concrete sheet-pile wall, $L = 5.3$ m.

Rehabilitation work included installation of continous sheet piling in front of the wall and construction of a pressure relieving system, comprising concrete piles 16 m long (cross-section 0.4 × 0.4 m) and a T-shaped concrete slab. Sheet piles and pressure relief system piles were driven through the layer of weak soil to the firm soil stratum.

A quay wall rehabilitation project implemented in the Port-of-Hamburg is shown in Figure 1–48. There, a pressure relieving slab was incorporated into a distressed quay wall structure. The slab also was supported on piles and joined with the quay wall superstructure by means of steel anchors.

Figure 1–49 demonstrates an example of use of an independent pressure relieving

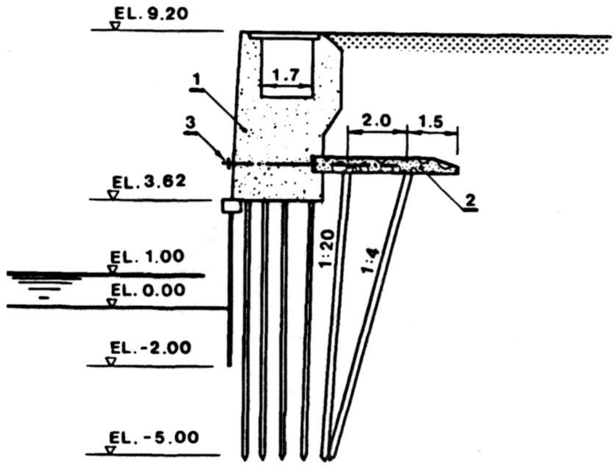

Figure 1–48. Rehabilitation of quay wall in Hamburg, Germany. 1—Existing quay wall, 2—added pressure relieving piled concrete slab, 3—steel anchor.

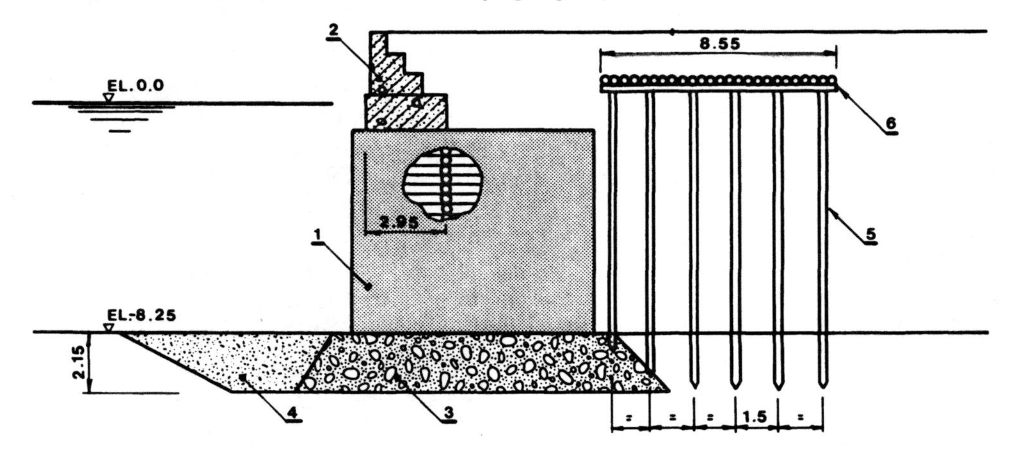

Figure 1–49. Rehabilitation of timber crib quay wall in Russia. 1—Rock-filled timber crib, 2—concrete superstructure, 3—rockfill mattress, 4—gravel, 5—new timber piles, 6—new timber pressure relieving platform.

platform in the USSR. The platform was built from wooden elements, because wood was a cheap, locally available material. Six rows of wooden piles 0.25 to 0.3 m in diameter were driven in partly completed backfill material just behind the quay wall structure. This type of construction reduced horizontal soil pressure to about 50%, which permitted completion and safe operation of the quay wall under question. To design a soil pressure relieving system(s) the designer has to establish the following:

1. The actual horizontal and vertical soil pressures acting upon structure.
2. Safe load the structure can take
3. The balance of load that has to be taken by the pressure relieving system.

This is actually a trial and error process, during which the size of the pressure relieving system and location are determined.

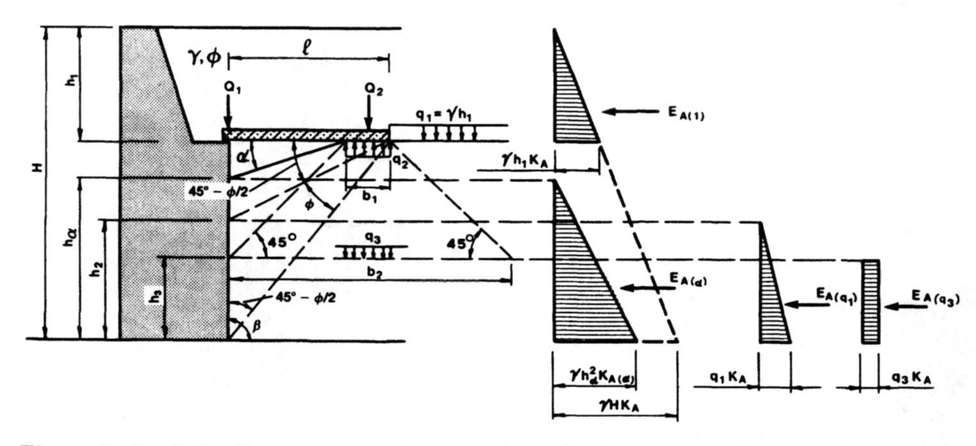

Figure 1–50. Soil active pressure against retaining wall with pressure relieving slab. Practical example.

An example of determination of lateral soil pressure against a wall is shown in Figure 1–50. The soil pressure is calculated by the conventional method found in standard texts on soil, mechanics and foundation engineering. As seen in Figure 1–50 the pressure relieving slab is resting upon a gravity type wall interior and backfill. The total vertical load on slab $Q = \gamma h_1 \ell$ is transmitted to wall structure (Q_1) and backfill (Q_2). The resulting soil bearing pressure at fill side $q_2 = Q_2/b_1$ shall not exceed the value of allowable bearing pressure for fill material. Total lateral active soil pressure against wall structure will include the following:

1. Soil pressure at the upper portion of the wall $E_{A(1)} = 0.5\gamma h_1^2 K_A$.
2. Soil pressure under the pressure relieving part of the wall:
   (i) Soil pressure due to underslab soil action $E_{A(\alpha)}$:

$$E_{A(\alpha)} = \gamma h_\alpha^2 K_{A(\alpha)}$$

   (ii) Soil pressure due to load Q_2. In granular soil this load may be assumed as acting at a 45° angle in all directions. Hence $q_3 = Q_2/b_2$ where $b_2 = 2l - b_1$ and $E_{A(q_3)} = q_3 h_3 K_A$.
   (iii) Effect of soil surcharge, $q = \gamma h_1$, located beyond pressure relieving slab: $E_{A(q_1)} = 0.5 h_2 q_1 K_A$.

In the above formulations

γ = soil unit weight

K_A = coefficient of soil active pressure $K_A = tq^2(45° - \phi/2)$

$K_{A(\alpha)}$ = coefficient of soil active pressure for constant slope angle, α; it is obtained from Coulomb's theory as transformed by Kezdi (1975) as follows

$$K_{A(\alpha)} = \left[\frac{\sin(\beta - \phi)}{[\sin(\beta + \delta)]^{0.5} + \left(\frac{\sin(\phi + \delta)\sin(\phi - \alpha)}{\sin(\beta - \alpha)}\right)^{0.5}}\right]^2$$

where

α and β = as defined in Figure 1–50

ϕ = angle of internal friction

δ = angle of friction between the wall and the soil, commonly referred to as wall friction; for practical purposes $\delta = 0$ is commonly used.

From the above calculations it is quite obvious that $\Sigma E_A = E_{A(\alpha)} + E_{A(q_3)} + E_{A(q_1)}$ is significantly smaller than total active soil pressure without a relieving slab $E_A = \gamma H^2 K_A$.

If a pressure relieving slab used in a sheet-pile bulkhead structure transmits vertical load directly to the sheeting (Figure 1–51) then the stress level in sheet piles that are normally designed to take horizontal load only must be checked for the combined effect of both horizontal and vertical loads.

It is also important to check the adequacy of sheet pile embedment into the soil.

Sheet pile resistance (Q_R) to vertical loads ($Q_1 + Q_s$) should confirm to equilibrium $Q_R = 1.5(Q_1 + Q_s)$, where Q_s = weight of sheet piling, Q_1 = part of weight of soil plus any kind of live load transmitted through slab to sheet piling, and 1.5 = safety factor.

Sheet piling resistance to vertical load in general will be $Q_R = Af$, where A = area of embeded part of sheet piling, and f = soil friction resistance.

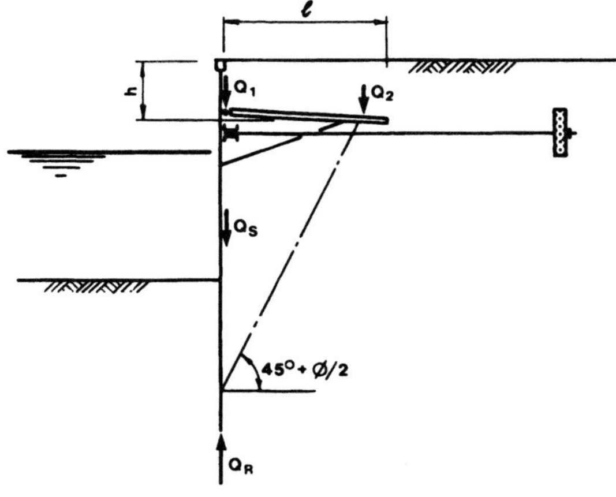

Figure 1–51. Sheet-pile bulkhead with pressure relieving slab. Design diagram.

Reaction $Q_1 \simeq 0.5l(\gamma h + q)$, where γ = soil unit weight, and q = live surcharge load. Hence $Af = 1.5(0.5l(\gamma h + q) + Q_s)$. From where the required minimum length of th soil pressure relieving slab, $l_{\min}$, is formulated as follows:

$$l_{\min} \leq \frac{1.33Af - 2Q_s}{\gamma h + q} \quad (1\text{–}32)$$

1.6 SCOUR PROTECTION

For any important situation where there is a possibility of bed scour resulting in undermining of a new or existing berth structure, a physical model test should be conducted to aid and confirm the design of the required bed protection.

The most typical scour protection structures are (Figure 1–52):

Rip-rap
Gabions (alternatively concrete in bags)
Prefabricated concrete slabs
Different types of flexible structures, for example, cable-linked prefabricated concrete blocks of various constructions and different kinds of fabric or bituminous mattresses
Flow deflectors. (Figure 1–60)

As will be seen from the following discussion geotextiles are almost indispensable components used in modern marine engineering practice, including bed protection to control erosion and scour in front of waterfront structures.

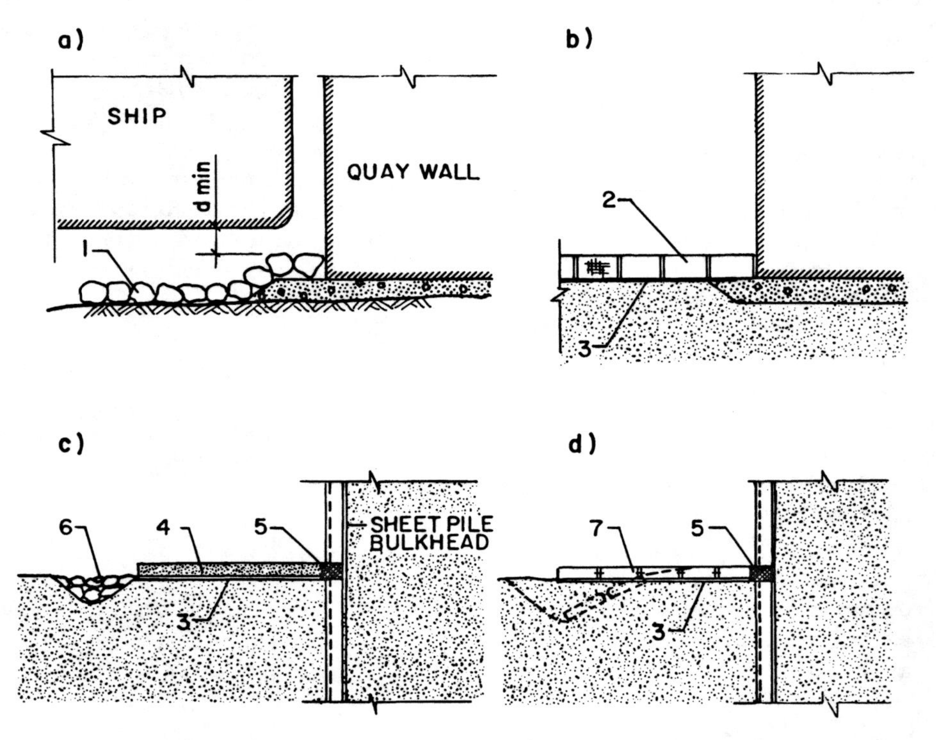

Figure 1–52. Typical scour protection systems. 1—Rockfill, 2—gabions (alteratively concrete blocks, concrete in bags, etc, 3—geotextile filter, 4—prefabricated concrete slab, 5—tremie concrete, 6—rip-rap, 7—cable linked concrete blocks of various constructions (alternatively different types of fabric or bituminous mattresses). (From Tsinker, 1989.)

1.6.1 Geotextiles

Basically geotextiles are a family of synthetic materials such as nylon, polyester, polypropylene, or similar substances employed in geotechnical engineering. In their common form, geotextiles are flexible, permeable, durable sheet fabrics, resistant to tension and tear and able to retain soils.

In modern marine engineering practice, in their most common form geotextiles are used to provide functions of reinforcement, filtration, and containment.

Geotextiles are relatively new materials in marine engineering practice; therefore when considering the use of geotextiles in bottom protection, or in any other applications, for example, in bank revetments, filter or drainage system behind quay walls or backfill reinforcement, the designer must be aware of exactly what geotextiles can and cannot do and how to select the best product available. It must be noted that misuse of geotextiles may have a detrimental impact upon structure performance.

There are two ways of dealing with the problem of material selection: analytical and empirical. The first requires a detailed knowledge of technical properties of selected material such as permeability, filtration characteristics, and strength. However, often the analytical approach does not cover various potential secondary aspects of design and therefore needs to be supplemented by empirical observations or by reference to relevant practical experience.

The basic data needed to describe a geotextile as a construction material are the type, size, and shape of its constituent elements; manufacturing technique; permeability; and soil retention properties. Typically all these data is found in the manufacturer's literature.

Geotextiles are materials made essentially from fibers or filaments. There is a great variety of geotextiles, but the most common ones are woven or nonwoven fabrics made from polymer fibers or filaments. Woven geotextiles are relatively homogenous. Strain at failure of woven geotextiles is usually between 15 and 40%,

Nonwoven are fabrics laid in a nonregular manner and bonded together by thermal, mechanical, or chemical processes. Nonwoven fabrics are relatively thick with high porosity value (close to 0.9). Compressibility of nonwoven fabrics under load is noticeable but the entanglement of fibers is such that the porosity usually remains as high as 0.8.

Porosity of a geotextile fabric (n_g) can be found from the mass per unit area (m_g) and the thickness (t_g) and density (ρ_g) of the polymer:

$$n_g = 1 - m_g/\rho_g t_g. \tag{1–33}$$

The usual values of geotextile porosity are between 0.7 and 0.9.

A geotextile's permeability (k_g) can be determined by the following equation:

$$k_g = c_1 n_g^3/s_g \tag{1–34}$$

where

c_1 = coefficient dependent on the geometry of the geotextile filter and shape of fibers and their rugosity.

s_g = specific surface. s_g depends on fiber diameter as well as production technique. Values of s_g may vary between 500 and 5000 m^2/m^3.

It must be noted that to obtain reliable values of k_g standard laboratory tests must be carried out.

These properties allow geotextiles to be used in foundation engineering to function as soil reinforcement and to be used for filtration, drainage, separation, containment, or protection purposes. When incorporated in bed protection system, geotextiles are primarily used as filters.

The durability of selected material and its long-term performance must be of the designer's concern. The geotextile's properties and its performance may deteriorate with time. First, characteristics of the physical or chemical (or both) environments can lead to degradation of the fabric material. Second, performance of the geotextile filter may be reduced due to clogging, fatigue, and creep, which inevitably occur to any kind of filter structure. Therefore these must be considered at the design stage.

It is advisable that at the specification stage recommendations by RILAM Technical Committee on Geotextiles (1985), Ingold and Miller (1988), and other relevant recommendations be considered. Hydraulic properties of geotextiles fabric depend on its permeability and filtration characteristics.

The geotextile filter primary function within a bed protection system is to prevent migration of underlying soil particles out of underlying soil.

To avoid soil erosion due to transportation of soil particles through the geotextile, and/or clogging of the geotextile by soil particles thus reducing the free flow of water through the geotextile, the geotextile filter must be designed to satisfy two important criteria: soil retention and fabric permeability. The above assumes that the geotextile must have an opening small enough to prevent soil particle migration, and must be permeable enough to allow free flow of water without inducing uplift load upon bed protection structure.

Geotextile permeability is the rate of flow per unit area per unit hydraulic gradient, where hydraulic gradient is the ratio of head to the thickness of the geotextile.

Because is can be difficult to measure the fabric thickness during the test Permanent International Association of Navigation Congresses (1987) proposed a simplified approach to geotextile filter design for permeability as follows. The geotextile filter must at all design periods of time maintain a permeability equal to or greater than that of the underlying soil. In the process of performance there will be a reduction in the permeability of the virgin fabric due to clogging. This will depend on fabric thickness and pore structure, as well as on the grain structure of the underlaying soil.

In general the permeability criterion requires that

$$\eta k_g = k_s \tag{1–35}$$

where

η = reduction in factor
k_s = soil permeability.

The value of the η depends to a large extent on the type of fabric and may be used as follows:

- For needle-punched nonwovens and other nonwoven fabrics thicker than 2 mm (measured at a normal stress of 2 kN/m^2) $\eta = 1/50$.
- For woven fabrics the reduction factor is dependent on the permeability of the geotextile and the d_{10} of the soil, and can be obtained from Figure 1–53. Here d_{10} = grain size diameter corresponding to 10% by weight of finer particles in millimeters.

The soil retention criteria for the filter are dependent on the grain-size distribution of the soil. For this reason it is proposed to

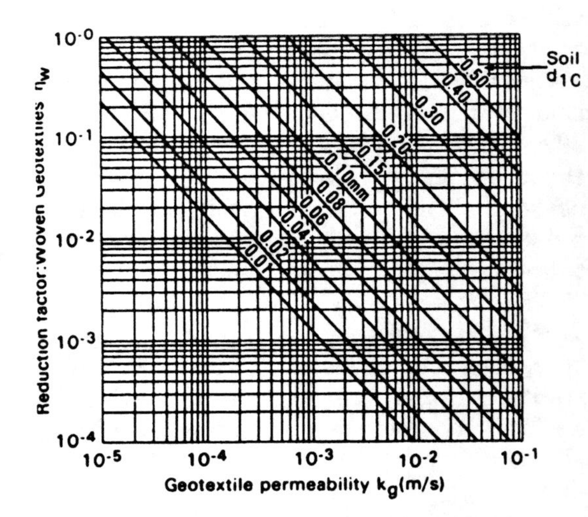

Figure 1–53. Permeability reduction factor. (Adopted from Permanent International Association of Navigation Congresses, 1987, supplement to Bulletin No. 57.)

classify soil grains in the following three basic ranges.

1. Forty percent or more of the soil particles will be smaller than or equal to 0.06 mm.
2. Fifteen percent or less of the soil particles will be smaller than or equal to 0.06 mm.
3. Between 15% and 40% of the soil particles will be smaller than or equal to 0.06 mm.

Once the range has been established the filter can be designed for soil retention. The criteria for soil retention is based on the value of the effective opening size O_{90}, which is vital for filter design. Generally O_{90} is the opening size of the geotextile which corresponds to d_{90}—the largest soil particle able to pass through the geotextile.

$$O_{90} < \lambda d_{90}, \tag{1–36}$$

where

λ = coefficient that depends on the type of fabric, type of soil, soil density and grading, and magnitude of hydraulic gradient and is usually provided by the material fabricator. It may be determined by the testing method established in Swiss Standard SN640550 or equivalent.

To protect a geotextile from being damaged during installation of a coverlayer, for example, when dumping rip-rap, a granular sublayer 0.2 to 0.3 m thick is used. Granular sublayers can greatly relieve downward hydraulic gradients in the subsoil. It is important, however, to realize that a sublayer can promote uplift load beneath coverlayers of low porosity, for example, tightly placed concrete blocks. In such instances it is better to omit the sublayer and take all precautions necessary to protect the geotextile during installation of a coverlayer. Otherwise the coverlayer must be designed to resist the consequential uplift load. Once the required geotextile properties are established they must be properly specified to ensure that an appropriate geotextile is selected. Guidance for the specification preparation may be found in Federal Highway Administration "Geotextile Engineering Manual," in Velduijzen Van Zanten (1986) and Ingold and Miller (1988).

The various construction techniques and methods involved in handling and placing of geotextiles are described in the US Army Corps of Engineers (1977) and in the report by Keown and Dardeau (1980).

1.6.2 Rip-Rap

Rip-rap is a flexible bottom protection and is probably the simplest one; it can usually be constructed using locally available materials. It generally comprises randomly placed quarried rock. Rip-rap is made up of durable stone of sizes ranging typically from 100 to 500 mm.

Depending on the hydraulic conditions much larger stones may be required. Stones used in rip-rap are specified by weight and typically range from 10 to 500 kg. They are normally placed in one or two layers and a sublayer is often incorporated. The rip-rap stability is dependent on shape, size, and weight of stones as well as on their gradation. For greater stability the stones used in rip-rap should have a blocky shape and uniform size and lie within the layer thickness. An oversized stone protruding from the layer can cause a weakness that may lead to a progressive failure. It must be noted that if not properly proportioned rocks included in the rip-rap cover could be swirled by the propeller jet and heavily damage the propeller's blades.

Because of its rugged nature rip-rap can cause damage to the geotextile filter layer; hence it must be placed with great caution. The best way to protect a geotextile layer from a concentrated load from stones is to use a granular sublayer of smaller size stones, or use a geotextile resistant to concentrated load. Where rocks are locally not available precast concrete blocks can be used instead. These blocks should be designed for a specific minimum crushing strength with a minimum cement content and durable heavy aggregates. When used with an appropriate granular sublayer material and/or geotextile filter concrete blocks provide a stable armor system. The smooth faces, however, reduce interlocking capabilities of adjacent cubical concrete blocks.

For information on granular filters for impervious soil the reader is referred to Sherard and Dunnigan (1989).

The stability of rip-rap composed of large stones can be enhanced by grouting the space between adjacent rocks or by granular material.

With a small bed clearance the effect of propeller jet induced bottom velocities may be profound. It may be sufficient to move stones of several tonnes in weight. The propeller jet induced bottom velocities may be considered analogous to highly turbulent flow situations that exist in stilling basins of hydraulic structures such as spillways. Under such flow condition Prosser (1986) recommends the following equations for dimensioning of rip-rap stones:

$$D_{50} = U_b^2/23 \qquad (1\text{–}37)$$

$$W_{50} = \frac{\pi W_s}{6} D_{50}^3 \qquad (1\text{–}38)$$

where

D_{50} = mean size of stones
W_{50} = mean weight of stones
W_s = density of stone $W_s \simeq 2050\,\text{kg/m}^3$.

Note: Equation (1–38) assumes that stones are perfectly spherical.

Examples of stone size and weight as a function of bottom velocities as determined by Eqs. (1–37) and (1–38) are presented in Table (1–6). Typical rip-rap protection is shown in Figure 1–54.

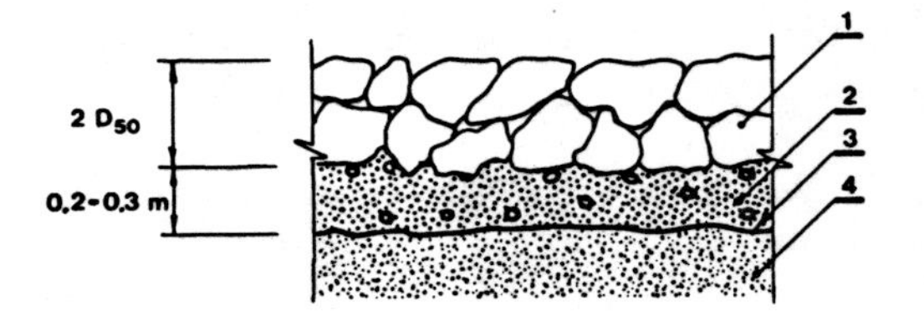

Figure 1–54. Typical rip-rap protection. 1–Rip-rap, 2—fine stones, (gravel), 3—synthetic fabric filter, layer, 4—natural bed.

Table 1–6 Stone Size to Give No Scour for Given Bed Velocity

Velocity (m/s)	Mean Size of Stone D_{50} (cm)	Mean Weight of Stone W_{50} (kg)
1	4.5	0.10–0.15
2	17.5	7.5
3	40	90
4	70	475
5	110	1850
6	160	5700

1.6.3 Concrete Blocks

There are many different designs that use precast concrete blocks to protect the seabed in front of a quay wall from propller-induced scour. To provide a stable armor system these blocks are typically used with an appropriate sublayer and/or geotextile filter.

Concrete blocks can be classified as follows: loose noninterlocking blocks, interlocking blocks, cable-connected blocks, and blocks connected to geotextile.

Loose noninterlocking blocks of different geometric forms are placed directly on sublayers or geotextiles with no connection to each other. Similar to the rip-rap stability of the cover layer formed by these bocks is dependent on the stability of the individual blocks. Therefore if an unanticipated force would cause displacement of the one block, this may be a start of a larger failure. Placement of loose blocks is a straightforward operation and is generally carried out by divers. Similar to rip-rap protection the stability of these blocks can be enhanced by grouting the space between adjacent blocks with a granular material or with tremie concrete.

This develops an interlocking effect between adjacent blocks which provides additional stability by mobilizing the weight of adjacent blocks. When granular material is used, it is essential to have it remain in place and it is therefore important to select blocks with inclined interfaces.

A loose noninterlocking block system requires a regular inspection and maintenance program.

Interlocking blocks of different designs have been used mainly for bank protection. They are placed by hand or by mechanical means over a sublayer or a geotextile. The interlocking block systems are generally very sensitive to settlements during which connections between adjacent block can be separated. Replacement of individual blocks can be difficult if not impossible.

Cable-connected block systems are much more stable than just interlocking block covers. In these systems blocks (either simple or interlocked) are connected by cables running through the blocks in one or more direction, which provides for greater stability against block displacement from either unanticipated forces or bed settlement. An additional advantage of cabling is that it reduces the risk of localized failure.

Connecting cables can be either marine chains, galvanized wire ropes, or synthetic fibers. In all cases splicing of cables should be avoided. However, when a splice is needed the connector should be able to take up tendon slackness. The connector also should be easily installed and released in the event of a block panel requiring replacement. Permanent International Association of Navigation Congresses (1987) suggests that cabling should generally not be taken into account in the design for stability of cable-connected blocks systems against hydraulic loading.

However, if a system were to rely upon the cabling for its stability under normal working loads then the risk of the blocks being repeatedly lifted off the sublayer or textile filter material and thereby causing a pumping failure of a subsoil material or abrasion of cables and/or geotextile must be considered. The flexibility of the cabled system must also be considered. The flexural strength derived from the cabling means that blocks can bridge undesirable deformations in the subsoil, thus allowing for some erosion in those areas. For this purpose the normal strength of a cable may be reduced by up to 40% when used as a sling and/or is subject to acute bending and poor splicing techniques.

The cabling facilitates construction work as the blocks are made up into panels and placed under water using a crane and spreader frame. Also, if repair is required the full panel can be lifted out and replaced.

Heavy armor rocks (2 to 4 tonnes) chained together, thus creating a cable-connected blocks system, has been used for bottom protection at the Frederikshavn, Denmark ferry terminal (Kristensen, 1987). The rocks were fitted with eye bolts which were used for placing and for chaining of these rocks together. A chain 20 mm in diameter was used to join every rock with each other (Figure 1–55). The rocks were placed upon a protection mat, which consists of woven sand pipes 5 cm in diameter. The mat was used as a sealing layer to protect subsoil material against waves and water flow effects.

A mat made from propylene fiber fabric replaced a layer of a conventional granular sealer. It possesses great strength against mechanical stresses which makes it suitable for use as a sublayer for rock armors. It weighs about 50 kg/m^2 and was made in the form of 5 × 5 m to 5 × 15 m covers. The total average thickness of this type of bed protection is about 1.2 m. The voids between rocks and sheet piles of the quay wall were filled with concrete. This type of bed protection

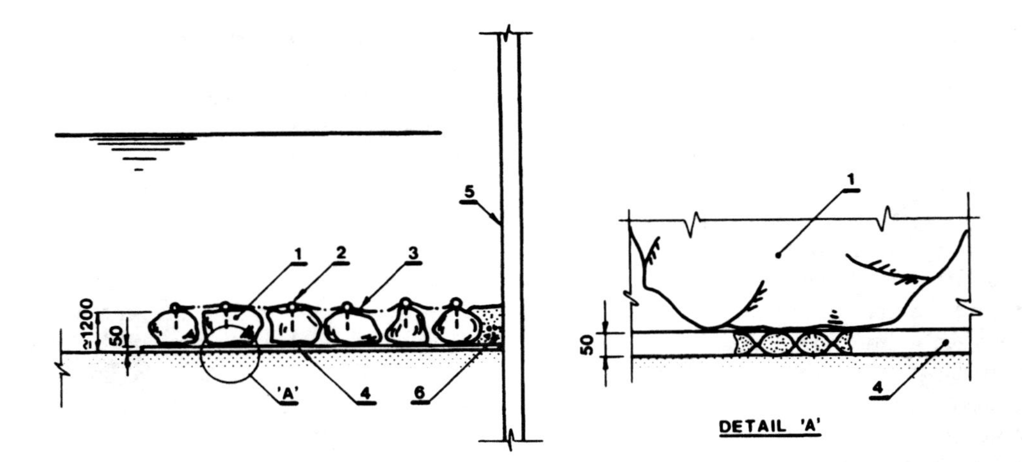

Figure 1–55. Port of Frederikshavn, Denmark; scour protection system. 1—Rock (2–4 tons), 2—eye bolt, 3—chain 20 mm diameter, 4—sealing woven sand pipe mat, 5—steel sheet-pile bulkhead, 6—sealing tremie concrete.

proved to be reliable against propeller action when ship engine power was up to 30 000 hp and more.

Blocks connected to geotextile systems are a relatively new type of bed protection. Several concrete block systems of this type have been developed, where the blocks are connected directly to the geotextile (Kirstensen, 1987).

Blocks may be either glued to the geotextile by a special type of adhesive (Figure 1–56) or joined by mechanical means (Figure 1–57). Such systems have advantages and disadvantages similar to those of cabled systems, although as there is normally a clear space between each block no inter-block friction can be developed. In addition, the geotextile fabric must possess adequate strength and durability as a connector, as well as fulfill filtering functions. The nature of this system also precludes in most cases the incorporation of a sublayer between blocks and fabric which limits application of this type of system to a hydraulically stable subsoil materials.

The panels of preassembled covers are placed underwater with the help of a crane and divers. Separate panels are joined underwater by divers. The detail of a joint is seen in Figure 1–57.

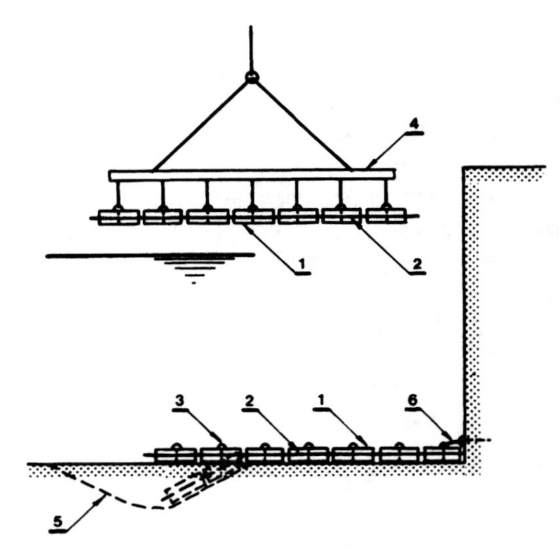

Figure 1–56. Bed protection system comprises concrete blocks connected to geotextile. 1—Concrete block, 2—geotextile, 3—eye bolt, 4—lifting frame, 5—contour of potential bed erosion, 6—connection to the wall.

Although this system is very flexible, the best result is obtained if the bed is reasonably flat. Therefore bed leveling before placing of the cover is necessary.

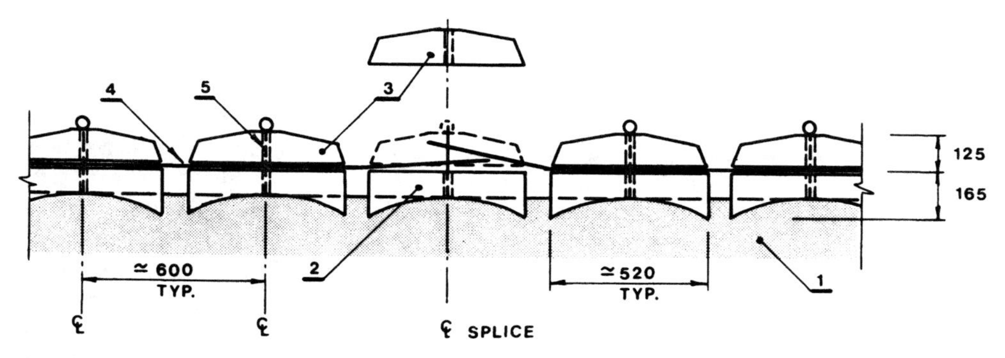

Figure 1–57. Detail of bed protection system comprises concrete blocks connected to geotextile as used for bed protection at Frederikshavn ferry terinal, Denmark. 1—Seabed, 2—bottom concrete block, 3—upper concrete block, 4—geotextile, 5—eye blt.

Blocks connected to geotextile have been used at the port of Frederikshavn ferry terminal, Denmark (Kristensen, 1987). Concrete blocks were composed of double hexagonal concrete units ($f_c' = 40\,\text{MPa}$) bolted together by means of eyebolts through the geotextile (polypropylene fiber) covering (Figure 1–57).

The individual blocks were composed out of two parts; the upper part (50 kg per unit) is formed in a way to reduce resistance to flow of water, and the bottom of the lower part is concave, thus forming a vacuum which binds a block to the bed. The weight of lower part is 83 kg. The average weight of the cover is 420 kg/m^2. The weight of cover along outside edges is approximately 500 kg/m^2.

The separate sheets of cover about 3.0 × 6.45 m were assembled on a leveled surface. The weight of one sheet was about 8 tonnes. Each cover unit was lifted by crane equipped with a special lifting device able to pick up the cover at each block. Finally, the adjacent sheet of cover were joined together as is shown in Figure 1–57. The cover was secured to the sheet pile wall by chains or hooks fixed to brackets welded to the sheet piles.

1.6.4 Gabions

Gabions used for bed protection are stone filled, box-shaped compartmentalized containers made of either polymer mesh or galvanized and/or polyvinyl chloride coated steel wire mesh. The gabion container can also be manufactured as a basket or as a mattress.

Boxes are commonly about 1.0 m high and 0.5 to 1.0 m wide. Gabion mattresses are typically 0.3 m thick and are usually used in a single layer to form a flexible free-draining protective cover to the bed surface. Typical gabion sizes are given in Table 1–7.

Table 1–7 Typical Gabion Sizes

Length, Width, Depth (m)	Number of Diaphragms	Volume (m^3)	Weight (tonnes)
2 × 1 × 1	1	2	4.2
3 × 1 × 1	2	3	6.3
4 × 1 × 1	3	4	8.4
2 × 1 × 0.5	1	1	2.1
3 × 1 × 0.5	2	1.5	3.2
4 × 1 × 0.5	3	2	4.2
2 × 1 × 0.3	1	0.6	1.3
3 × 1 × 0.3	2	0.9	1.9
4 × 1 × 0.3	3	1.2	2.5

The mattress is subdivided into equally sized compartments that are assembled by lacing the edges together with steel wire or polymer rope. The individual units are then tied together and filled with stone. The lids are finally closed to form a large heavy mattress. To ensure best performance, properly sized filler rock should be specified. Flat stones and interior liners of any kind in most cases are not recommended.

For easy handling and shipping the gabions are supplied folded into a flat position. They are readily assembled by unfolding and by simply wiring the edges together and the diaphragms to the side. Then the gabions are filled to a depth of one connecting about 25 to 30 cm and wire is placed in each direction and looped around two meshes of the gabion wall. This operation is repeated until the gabion is filled. The containers should be filled tightly to prevent any significant filler rock movement. They should be refilled as necessary to maintain tight packing. After the gabion is filled the top is folded shut and wired to the ends, sides, and diaphragms. The land prefabricated gabions are then installed underwater by divers. An advantage of a gabion structure is that it can be built without heavy lifting equipment and using locally available stone.

The combination of the following features gives gabion structures some technical advantages over conventional rigid structures, particularly on sites where settlement or undermining is anticipated. Gabions are relatively flexible in the presence of unstable ground and/or moving water. This allows a gabion structure to settle and deform without failure and loss of efficiency.

Gabions are strong enough to withstand substantial flow velocity. They also are permeable, and therefore usually do not require any sublayer drainage systems. Gabions have a quite high resistance to corrosion due to the well-bonded zinc coating on the wire (approximately 260 g/m^2). Also, gabions with polyvinyl chloride plastic coating are available for structures in locations where severe corrosion problems exist.

Maccaferri Gabions Construction Company developed a regular type of gabion mattress and gabion type linings consolidated or sealed with a sand asphalt mastic to be used for underwater installation. (Maccaferri Gabions. Flexible Reno Mattress and Gabion Linings. Commercial Booklet.)

Where the bottom is soft, the toe of the apron may sink, making frequent and expensive maintenance operations necessary. Therefore flexible gabion type aprons are designed to settle without fracture and to adhere to the ground as scour occurs. To offer effective protection the apron must have sufficient length to reach the bottom of the expected scour. The projection of the apron in front of the structure is generally one and a half to two times the estimated scour depth. Normally a thickness of 0.3 to 0.5 m is required so that the apron can have enough flexibility and weight to remain adhered to the ground even during the most severe propeller jet induced velocities. Similarly to regular box-shaped gabions, gabion mattresses are fabricated on site near the site of installation and then are placed underwater by crane or by other means.

Typical sizes of gabion mattresses of regular construction and consolidated with sand asphalt mastic are as follows: width 2 to 3 m, length 3 to 6 m. The thickness of sand asphalt mastic consolidated mattresses can be reduced to 0.15 m. The asphalt grouted mattress retains its flexibility while the density of the fill is increased with subsequent increase in efficiency of the mattress performance. The sand asphalt mastic also offers additional corrosion and abrasion protection to the wire mesh.

Some technical data and minimum application rates of bituminous mastic for partial and full penetration of gabion mattress and recommended thicknesses of grouted gabions in relation to water velocities can be obtained from Tables 1–8 and 1–9.

Table 1–8 Some Technical Data of Gabion Mattresses Consolidated with Sand Asphalt

Thickness (m)	Filling Stones		Design Velocity (m/s)	Limit Velocity (m/s)
	Stone Size (mm)	d_{50} (mm)		
0.15–0.17	70–100	85	3.5	4.2
	70–150	110	4.2	4.5
0.23–0.25	70–100	85	3.6	5.5
	70–150	120	4.5	6.1
0.30	70–120	100	4.2	5.5
	100–150	125	5.0	6.4
0.50	100–200	150	5.8	7.6
	120–250	190	6.4	8.0

From Maccaferri Gabions.

Table 1–9 Minimum Application Rates of Bituminous Mastic for Partial and Full Penetration of Gabion Mattress

Thickness (m)	Sand Asphalt Mastic Quantity	
	Partial Penetration (kg/m^2)	Total Penetration (kg/m^2)
0.15	60– 90	120–140
0.17	80–100	130–150
0.23	90–120	190–220
0.25	100–130	200–240
0.30	120–150	240–280
0.50	150–200	400–450
1.00	200–350	700–800

From Maccaferri Gabions.

The minimum thickness of the gabion mattress (t_m) can be related to the stone size (D_{50}) and can be determined by the following formula, but should not be less than 0.3 m:

$$t_m(\min) \simeq 1.8D_{50}. \qquad (1\text{–}39)$$

As recommended by Permanent International Association of Navigation Congresses (1987) the size of stone (D_{50}) can be determined by the following formula:

$$\frac{D_{50}}{h} = \left(\frac{U_b}{B_1(k^1\psi_{cr}g\Delta_m h)^{0.5}}\right)^{2.5} \qquad (1\text{–}40)$$

where

h = depth of water
U_b = propeller-induced velocity at bed level
B_1 = factor dependent on flow condition $B_1 = 5$–6
k_1 = slope (α) reduction factor
k^1 = $(1 - \sin^2 / \sin^2\phi)^{0.5}$. Here ϕ = angle of internal friction $\phi = 45°$. For a horizontal bed $k_1 = 1$.
ψ_{cr} = Shields paramter. For stone filled gabions use $\psi_{cr} = 0.1$.
Δ_m = relative density of fill material.

It is obvious that in no case must the size of the stone not be less than the size of gabion mesh opening.

If $U_b > 3.0$ m/s then a granular sublayer (0.2 to 0.3 m) should be incorporated into the gabion type bed protection system. In all other cases it is sufficient to place the gabion directly upon the geotextile filter only.

Recently gabion mattresses have been used for the bottom protection works at port of Stockholm ferry terminal (Silja basin) (Bergh and Magnusson, 1987). the design bottom velocity generated by the ferry (maximum machinery installed power from 31 000 to 36 000 hp) was estimated as being in order of 10 to 12 m/s at some local areas.

Theoretically these velocities would require stone diameters on the order of 4 to 6 m to protect the bottom from severe erosion. It is quite obvious that use of rockfill with stones of the above size is not practical. In fact a rockfill protection is not likely to be a suitable solution if the bottom velocities exceed 5 m/s.

For the scour protection at Silja basin several types of structures have been tested. One of them was a gabion mattress 0.3 m thick (Figure 1–58a).

Two years after gabion installation it was found that at the locations exposed to the direct propeller jet, for example, free edges at short ends, and in front of the loading ramp, the mattress had been rolled up.

At these locations the outer mattresses were completed with new mattresses placed underneath the previous ones and fixed to them in order to increase the weight at the free edges. The follow-up observation conducted 3.5 years after the repair works were completed found the gabions in generally good condition. However, near the outer end of the mattress there was a slight tendnency of rolling up over a length of some 15 m.

The main disadvantage of gabions are their relatively high cost and durability. They are labor-intensive to construct. Although some gabion manufacturers claim that some of their gabions have been in service for decades, nevertheless the long-term durability of the wire in severe marine environment is always in question. Even polyvinyl chloride-coated wire is subject to deterioration at gaps in the coating, which can be created by the placement of rough, angular stones. A damaged basket will eventually develop a hole and start losing stone.

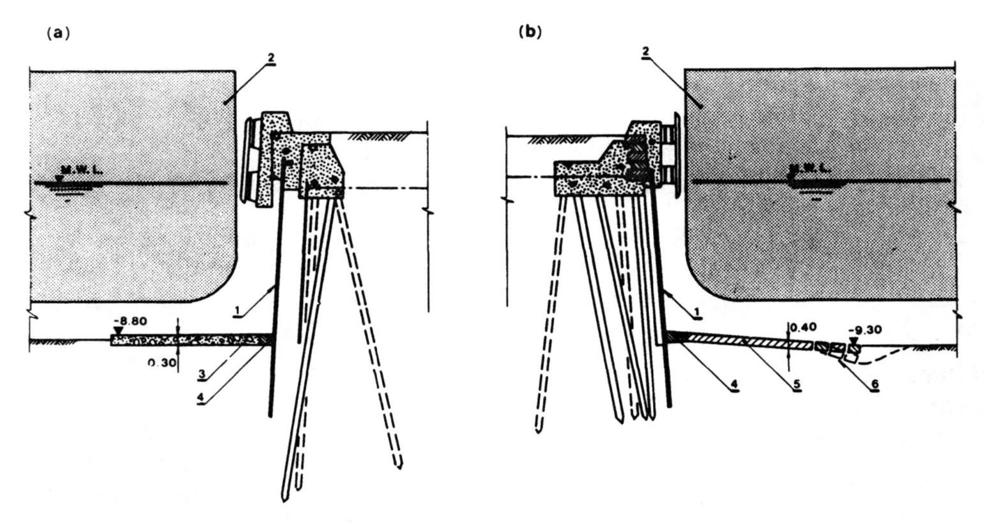

Figure 1–58. Port of Stockholm ferry terminal (Silja basin). (a) Section through the north quay. (b) Section through the south quay. 1—Quay wall, 2—ferry, 3—gabion mattress 2 × 2 × 4 m, 4—tremie concrete, 5—precast concrete slabs 9 × 9.5 × 0.4 m, 6—concrete flaps 3 × 2.5 × 0.4 m. (From Bergh and Magnusson, 1987.)

1.6.5 Precast Concrete Slabs

This type of bottom protection is very reliable although not the most economical solution. The size of precast slabs basically depends on the heavy lift equipment available. The slab reinforcement is usually designed to take stresses associated with handling during transportation and installation. The slab outer edge is typically protected against undermining by rip-rap, or by different kinds of flexible covers attached to this edge. The space between slab and the quay wall structure is usually sealed with tremie concrete. A typical example of the use of prefabricated concrete slabs for bed protection is a recent modification of the south ferry terminal (Silja basin) at Port of Stockholm, Sweden as shown in figure 1–58b.

There the precast concrete slabs 9.5 m wide, 9.0 m long, and 0.4 m thick with the outer end protected against propeller induced scour by three concrete flaps 3.0 m × 2.5 m have been installed there. To reduce uplift forces, flaps were equipped with drainage holes. Before placing the slabs, the bottom was levelled by trailing a heavy steel beam. The slabs, weighing 85 tonnes each were lifted and placed in position about 1 m from the sheet piling and about half a meter from each other by means of a floating crane. Subsequently the space between sheet piling and slabs and space between adjacent slabs were sealed with tremie concrete.

This type of construction proved to be satisfactory even in areas exposed to very high velocities. As expected, erosion had occurred at some of the flaps which then have moved down as anticipated and stopped further erosion.

1.6.6 Fabric Containers Filled with Concrete

High-strength, water-permeable fabrics made of synthetic fibers are being used as forming material for concrete in a wide variety of applications, particularly those that involve placement under water. An early use of fabric as form for concrete was proposed in a patent issued to J. Störe of Norway in 1922. In contemporary marine and coastal engineering practice fabric containers filled with concrete, for example, bags, tubes, or mattresses of different constructions, made of porous woven geotextiles, are effectively used for bottom protection and underwater repairs of different kinds of marine structures. In bottom protection they are typically placed in a single or a double layer upon filter material.

Although, theoretically the choice of fabric is independent of the concrete mix proportions or placement methods, in practice fabric used for concrete forming is usually limited to specialized applications and the choice of mix and placement method should be appropriate to these applications (Lamberton, 1989). When concrete is injected in fabric forms underwater the pressure within the delivery hose must be sufficient to expel water through the water-permeable fabric and substituted it with concrete. At the same time the external water must not penetrate the system to contaminate the fresh concrete. Obviously, the above conditions apply only where the form is fully collapsed prior to injection of concrete; when the form is suspended in an open manner, a normal tremie concrete technique can be employed.

For reasons of accessibility nearly all concrete is placed in fabric form by pump; the pumping concrete should be plastic enough to accommodate the demands imposed by the lack of coarse aggregate.

In North American practice only woven fabric is regularly used as a concrete forming material.

If fully filled with concrete the smooth rounded contours of the bags may present an interlocking problem, and therefore they can slide easily. For this reason bags are usually filled between 50 and 70% to permit the diver to place them with good contact with each other. Since the cement paste is normally squeezed out through the porous bags a certain cementation within the bag structure takes place. The bags are normally placed in bond similar to block walling.

To provide for better and stronger performance the diver can drive reinforcing bars through bags. Bags also can be stabilized by cabling. An example of bed protection by concrete-filled tubes is shown in Figure 1–59.

Woven textile bags and tubes are available in various diameters and lengths. Tubes can be placed on the bed, and then grouted with concrete. They are usually placed on a woven filter cloth. A small tube, factory-stitched to the outer edge of the filter cloth, can provide toe protection.

Mattresses are usually designed for placement directly on a levelled bed. They are laid in place when empty, joined together, and then pumped full of concrete. This results in a mass of pillow-like rather flexible concrete units with regularly spaced filter meshes for hydrostatic pressure relief. The filter is usually a normal geotextile material as previously described.

In general polyester fabric must not be used, for any kind of fabric containers that will be filled with concrete, as this is susceptible to attack by chemical reactions generated from hardened concrete.

With respect to use of concrete-filled geotextile containers in marine applications where they must withstand the impact of different kinds of hydraulic loads their filling in situ causes them to mould closely together and hence they do not rock. For this reason they need not have the strength of precast concrete units. In most practical applications a compressive strength of concrete or grout of 10 MPa is sufficient, and certainly more than adequate for the bed protection units exposed to propeller action. The specific gravity of in situ concrete may approximate 2.0.

In most instances a fluid concrete or mortar mix is pumped in as would be a grout, via a 50 mm in diameter injection hose or greater for ease of flow. The typical mix would be at a 0.55 to 0.6 water/cement ratio. This kind of mix would generally include aggregates 1.5 to 10 mm diameter with corresponding aggregate/cement ratio between 2 : 1 and 3 : 1.

It is important that concrete or mortar "fluid" run to all points in the container readily without the aid of vibrators. For this reason the tubes are placed parallel to the face of the quay wall, continuously

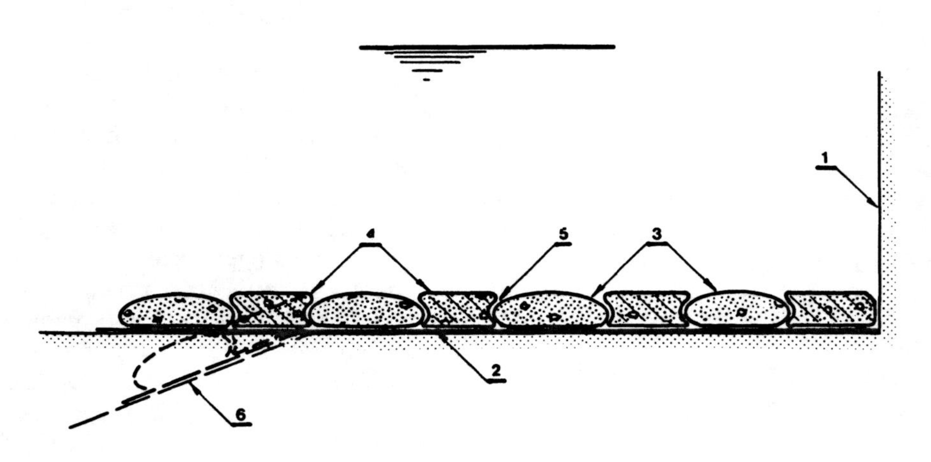

Figure 1–59. Bed protection by concrete-filled geotextile tubes. 1—Quay wall, 2—geotextile filter, 3—tube laid first, 4—subsequent unit, 5—tongue and groove joint, 6—contour of potential scour.

being spaced approximately one tube apart. Then space between these tubes is filled with similar units. This makes a tongue-and groove-like connection with those already in place (Figure 1–59). It should be noted that tubes at the edge could slump into a scoured trench. It is necessary then to continue the required protection far enough. For this purpose it is better to have edge tubes relatively shorter in length in order to cope with different scouring dimensions along the length of high velocity jets. For better performance edge tubes can be cabled; super-plastisicers should be included in the mix which reduces the water content required and hence maintains the strength (Silvester, 1983, 1986). The use of ground blast furnace slag to replace a large proportion of the cement can make in situ concrete filled tubes very economical.

For short-term application nonsetting materials such as sand, bentonite combined with 10% of cement, and others can be used as in-fill materials.

The equipment used for concreting is very simple; it usually comprises a mobile hopper or concrete pump and a boat or barge for feeding out the fabric sheath. The concrete mix can be supplied from ready-mix trucks that pour into the hopper continually.

1.6.7 Deflectors

Longe et al. (1987) found that the deflector constructed at the foot of the quay can mitigate the effect of perpendicular jet against wall generated by propeller or side thruster. The deflectors shown in figure 1–60 according to Longe et al. have offered the best protection against bow thruster induced scour.

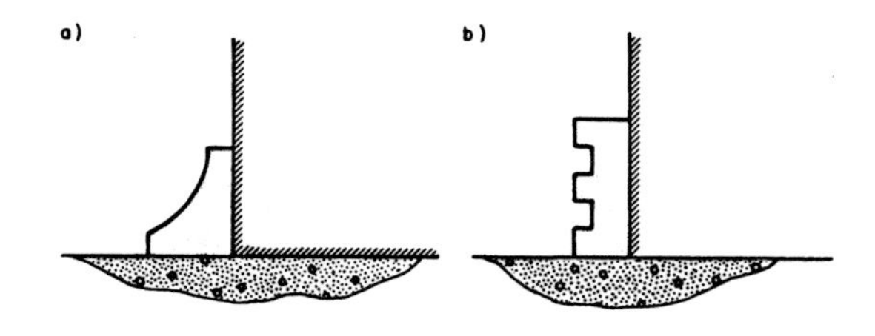

Figure 1–60. Jet deflectors at foot of the quay.

REFERENCES

ACRES INTERNATIONAL LIMITED, 1989. "Alkali Aggregate Reaction in Hydraulic Structures." Proceedings of 20th Annual Seminar. April 14 and 15, Niagara Falls, Ontario.

AMERICAN CONCRETE INSTITUTE, 1965. Recommended Practice for Evaluation of Compression Test Results of Field Concrete (ACI 214–65), Detroit.

AMERICAN CONCRETE INSTITUTE, 1984. Concrete Repair and Rehabilitation, Current ACI Guides. Concrete International, Vol. 6, No. 7, July.

AMERICAN CONCRETE INSTITUTE, Committee 224, 1984. Causes, Evaluation and Repair of Cracks in Concrete Structures. ACI Journal, May–June.

AMERICAN CONCRETE IINSTITUTE, 1985a. Corrosion of Metals in Concrete, ACI 222R-85, Detroit.

AMERICAN CCONCRETE INSTITUTE, 1985b. Guide for Making a Condition Survey of Concrete in Service ACI 201.1R, Part 1, Detroit.

AMERICAN CONCRETE INSTITUTE, 1985c. Guide to Durable Concrete. Manual of Concrete Practice, ACI 201.2R, Part 1, Detroit.

AMERICAN CONCRETE INSTITUTE, 1985d. Guide to the Use of Waterproofing, Dampproofing, Protective, and Decorative Barrier System for Concrete, ACI 515.1R, Part 5, Detroit.

AMERICAN CONCRETE INSTITUTE, 1988. In-Place Method of Determination of Strength of Concrete, ACI 228.1R. Detroit.

AMERICAN SOCIETY of CIVIL ENGINEERS, March, 1966. "Sediment Transportation Mechanics: Initiation of Motion." Progress Report. Task Committee on Preparation of Sedimentation Manual. Journal of the Hydraulic Division.

ATWOOD, W. G. and JOHNSON, A. A., 1924. "Marine Structures, Their Deterioration and Preservation." National Research Council Washington, DC.

B. C. RESEARCH, LTD. "Tidelines—A Current Review of Pacific Coast Marine Borers." Bulletin, Issued monthly.

BAKUN, G. B., 1986. "Cost-effective Approach to Pier Inspection and Maintenance. In Maritime and Offshore Structure Maintenance." Proceedings of the Institution of Civil Engineers Conference, Thomas Telford, London.

BERGH, H. and MAGNUSSON, N., 1987. "Propeller Erosion and Protection Methods Used in Ferry Terminals in the Port of Stockholm." Permanent International Association of Navigation Congresses, Bulletin No. 58, Brussels, Belgium.

BISWUS, A. N. and BANDYOPADHYAY, K. K., 1987. "Scour of Haldia Oil Jetty on the Hugli Estuary." Permanent International Association of Navigation Congresses Bulletin No. 58, Brussels, Belgium.

BLAAUW, H. G. and VAN DE KAA, E. J., 1978. "Erosion of Bottom and Sloping Banks Caused by the Screw Race of Maneuvering Ships." Paper presented at the 7th International Harbor Congress, Antwerp, May 22–26.

BLAAUW, H. G., VAN DEN KNAAP, F. C. M., DE GROOT, M. T., and PILARCZYK, K. W., 1984. "Design of Inland Navigation Fairways." Publication 320, Delft Hydraulic Laboratory.

BLETCHLY, J. D., 1967. "Insect and Marine Borer Damage to Timber and Woodwork." Ministry of Technology, Her Majesty's Stationery Office, London.

BLUMBERG, R. and RIGG, A. M., 1961. "Hydrodynamic Drag at Supercritical Reynolds Numbers." American Society of Mechanical Engineers. Petroleum Session.

BROMS, B. and STILLE, H., 1976. "Failure of Anchored Sheet-Pile Walls." Journal of the Geotechnical Division, ASCE, Vol. 102, No. GT3, pp 225–251.

BUDIN, A. Y. and DEMINA, G. A., 1979. Quays. Handbook. Stroyisdat, Moscow (In Russian).

BUNGEY, J. H., 1982. "Testing of Concrete in Structures." Surrey University Press, Glasgow.

BUSLOV, V. M., 1979. "Durability of Various Types of Wharves." ASCE, Proceedings of Specialty Conference Coastal Structures '79.

BUSLOV, V. M., 1983. "Corrosion of Steel Sheet Piles in Port Structures," ASCE Journal of Waterway, Port, Coastal and Ocean Engineering, Vol. 109, No. 3 August.

BUSLOV, V. M., 1987. "Inspection, Maintenance, and Repair of Offshore Platforms: A System Approach. paper OTC 5386. Proceedings of Annual Offshore Technology Conference, Houston, Texas.

BUSLOV, V. M., 1992. "Marine Concrete—When to Repair, What to Repair." Concrete International, May.

BUSLOV, V. M. and ROJANSKY, 1988. "Assessment of Corrosion Damage to Concrete Marine Structures," Proceedings of FIP Symposium in Israel, September.

BUSLOV, V. M. and SCOLA, P. T., 1991. "Inspection and Structural Evaluation of Timber Pier: Case Study." ASCE Journal of Structural Engineering Vol. 117, No. 9 September.

CARON, C., HERBST, T. F., and CATTIN, P., 1975. "Injections." Published in Foundation Engineering Handbook, edited by H. F. Winterkorn and H-Y. Fang. Van Nostrand Reinhold, New York.

CHAIT, S., 1987. "Undermining of Quay-Walls at South African Ports due to the use of Bow Thrusters and Other Propeller Units." Permanent International Association of Navigation Congresses, Bulletin No. 58, Brussels, Belgium.

CHELLIS, R. D., 1961. "Pile Foundations." McGraw-Hill, New York.

CHOW, VEN TE, 1959. Open Channel Hydraulics. McGraw-Hill, New York.

CLAPP, W. F. LABORATORIES, Under B.U.Y.D. Conf No. NBy-17810. Progress Reports on Marine Borer Activity in Test Boards. Thirteen issues, annually, 1946 through 1959.

CLAUSNER, J. E. and TRUITT, C. L., 1987. "Propeller Wash Effects on Protective Armour Layer Design for Container Aquatic Disposal Sites." Permanent International Association of Navigation Congresses. Bulletin No. 58, Brussels, Belgium.

CLEARY, H. J., 1969. "On the Mechanism of Corrosion of Steel Immersed in Saline Water." Proceedings of Offshore Technology Conference, Houston, Texas.

COOMBER, D., 1987. "Restoration of Quay Wall." Civil Engineering, January/February, London.

DAVIS, R. E., et al., 1948. "Restoration of Barker Dam." American Concrete Institute Journal, Vol. 19, No. 8 (Apr) pp 633–688.

DE LANGE, J. I., VERHULST, K., and BRUECUWER, R., 1985. "Underwater Inspection of Concrete Structures." Behavior of Offshore Structuring. Elsevier Science Publishers B.V., Amsterdam.

DELFT UNIVERSITY PRESS, 1984. "The Closure of Tidal Basins." Netherlands.

DISMUKE, T. D., COBURN, S. F., and HIRSH, C. M., 1981. Handbook of Corrosion Protection for Steel Pile Structures in Marine Environments. American Iron and Steel Institute, Washington, D.C.

DUBROVA, G. A., 1959. "Design of the Economically Efficient Hydrotechnical Structures." River Transport Publishing House, Moscow (In Russian).

EDWARDS, W. E., 1963. "Marine Corrosion: Its Cause and Cure." Proceedings of the Eighth Annual Appalachian Underground Short Course, Technical Bulletin. No. 69. West Virginia University Bulletin, October.

EFIRD, K. D., 1975. "The Interrelation of Corrosion and Fouling of Metals in Seawater." National Association of Corrosion. Engineers Proceedings Corrosion Conference, Toronto, Ontario.

EYRE, J. R. and NOKHASTEH, M. A., 1992. "Strength Assessment of Corrosion Damaged Reinforced Concrete Slabs and Beams, Paper 9851." Proceedings Institute Civil Engineers Structures and Buildings, No. 94, May.

FEDERAL HIGHWAY ADMINISTRATION. Geotextile Engineering Manual. Washington, DC.

FELD, J., 1968. "Construction Failures." John Wiley & Sons, New York.

FLAMAN, M. T. and MANNING, B. H., 1985. "Determination of Residual-Stress Variation with Depth by the Hole-Drilling Method." Experimental Mechanics, September.

FLAMAN, M. T., MILLS, B. E., and BOAG, J. M., 1985. "Analysis of Stress-Variation-with-Depth Measurment Procedures for the Center-Hole Method of Residual Stress Measurement." Experimental Technique, Vol. 11, No. 6.

FOTINOS, G. C., 1986. "New Structural Developments for Concrete Marine Structures." ASCE Proceedings Specialty Conference Ports '86. Oakland, California, May 19–21.

FUKUTE, T., KIYOMIYA, O. and MINAMI, K., 1990. "Steel Structures in Port and Harbour Facilities: Actual Conditions of Corrosion and Counter-Measures. PIANC, Bulletin No. 68, Brussels, Belgium.

FUEHRER, M. and RÖMISCH, K., 1977. "Effects of Modern Ship Traffic on Island and Ocean Waterways and Their Structures." Proceedings PIANC, 24th Congress Leningrad, USSR.

FUEHRER, M., RÖMISCH, K., and ENGELKE, G., 1981. "Criteria for Dimensioning the Bottom and Slop Protections and for Applying the New Methods of Protecting Navigation Canals." Proceedings XXVth Congress of Permanent International Association of Navigation Congresses, S.1-1 Edinburgh, UK.

FUEHRER, M., POHL, M., and RÖMISCH, K., 1987. "Propeller Jet Erosion and Stability Criteria for Bottom Protections of Various Constructions." Permanent International Association of Navigation Congresses. Bulletin No. 58, Brussels, Belgium.

GARLICH, M. J. and CHRZASTOWSKI, 1989. "An Example of Sidescan Sonar in Waterfront Facility Evaluation." American Society of Civil Engineers, Proceedings of Specialty Conference Ports '89, Boston, May.

GAUL, R. W., 1984. "Preparing Concrete Surfaces for Coating." Concrete International, Vol. 6, No. 7.

GERWICK, B. C., 1975. "Durability of Structures Under Water." Proceedings AIPC-PIP-CEB-RILEM-IASS Symposium on Behavior of Concrete in Service. Liege, Belgium.

GILBRIDE, P., MORGAN, D. R., and BREMNER, T. W., 1988. "Deterioration and Rehabilitation of Berth Faces in Tidal Zones at the Port of Saint John." Proceedings of the Second International Conference on Concrete in Marine Environment—SP-109. St. Andrews-by-Sea, Canada.

GJORV, O. E., BAERLAND, T., and RONNING, H. R., 1987. "High Strength Concrete for Highway Pavements and Bridge Decks." Proceedings,

Utilization of High Strength Concrete Symposium Stavanger, Norway.

GORONOV, B. F., 1974. "Operation and Maintenance of Port Structures." 'Transport' Publishing House, Moscow (In Russian).

HAMIL, G. A., 1988. "The Scouring Action of the Propeller Jet Produced by a Slowly Maneuvering Ship." Permanent International Association of Navigation Congresses Bulletin No. 62.

HASSANI, J. J., 1986. "Capacity Rating of Existing Piers and Wharves." ASCE Proceedings of Specialty Conference Ports '86, May 19–21. Oakland, California.

HEAF, H. J., 1979. "The Effects of Marine Growth on the performance of Fixed Offshore Platforms in the North Sea." Offshore Technology Conference paper No. 3386, Houston, Texas.

HERBICH, J. B. (ed.), 1989. Handbook on Coastal and Ocean Engineering. Gulf, Houston, Texas.

HERBICH, J. B., SCHILLER, E., WATANABE, R. K., and DUNLAP, W. A., 1984. "Seafloor Scour." Marcel Dekker, New York.

HOBBS, D. W., 1988. "Alkali–Silica Reaction in Concrete." Thomas Telford, London.

HOFF, G. C., 1987. "Durability of Fiber Reinforced Concrete in a Severe Marine Environment." Catharine and Bryant Mather International Conference on Concrete Durability. Atlanta, Georgia, April.

HOFF, G. C., 1988. "Resistance of Concrete to Ice Abrasion—A Review." Proceedings of the Second International Conference on Concrete in Marine Environment SD-109. St. Andrews-by-Sea, Canada.

HOLLAND, T. C., HUSBUANDS, T. B., BUCK, A. D., and WONG, G. S., 1980. "Concrete Deterioration in Spillway Warm-water Chute," Raystown Dam, Pennsylvania. MP SL-80-19. U.S. Army Engineer Waterways Experiment Station. Vicksburg, Mississippi.

HOLMES, C. W. and BRUNDLE, S. G., 1987. "The Effect of an Arid Climate on Reinforced Concrete in Marine Environment." Proceedings of Conference on Coastal and Port Engineering in Developing Countries. Vol. 1. China Ocean Press, Beijing, China.

HOPE, B. B., IP, A. K. and MANNING, D. G., 1985. "Corrosion and Electrical Impedance Concrete." Cement and Concrete Research, Vol. 15.

HORVATH, J. S. and DETTE, J. T., 1983. "Rehabilitation of Failed Steel Sheet-Pile Bulkheads." ASCE Proceedings Specialty Conference Ports '83, New Orleans, Louisiana, March 21–23.

INGOLD, T. S. and MILLER, K. S., 1988. Geotextiles Handbook. Thomas Telford, London.

INGRAM, C. J. and MORGANS, G. W., 1986. "Repair of Marine Structures." Proceedings of the Second International Conference on the Maritime and Offshore Structure Maintenance. Thomas Telford, London.

ISHIGURO, K. and MIYATA, Y., 1988. "Reinforcement of an Anchored Sheet Pile Wall with Additional Lower tie-Rods." PIANC, Bulletin No. 61.

JANSEN, E. W., et al., 1979. "Principles of River Engineering: The Nontidal Alluival River." Pitman Publishers, Aulender, North Carolina.

JOHNSON, B. R., 1987. "Protection of Timber Bulkheads from Marine Borers." In American Society of Civil Engineering, Geotechnical Special Publication, No. 7, april 29.

JOHNSON, S. M., 1965. "Deterioration, Maintenance and Repair of Structures." McGraw-Hill, New York.

JOHNSON, S. M., 1989. "Design of Marine Structures: Life Cycle Cost Factor." American Society of Civil Engineers, Proceedings of Specialty Conference Ports '89, Boston, May.

KEE, CHIN FUNG, 1971. "Relation Between Strength and Maturity of Concrete." ACI Journal, Proceedings, Vol. 68, No. 3 March.

KEOWN, M. P. and DARDEAU, E. A., 1980. "Utilization of Filter Fabric for Streambank Protection Applications." Technical Report HL-80-12, US Army Engineer Waterways Experiment Station. Vicksburg, Mississippi.

KEZDI, A., 1975. "Lateral Earth Pressure," Chapter 5 in Foundation Engineering Handbook, editors Winterkorn, H. F. and H.-Y. Fang. Van Nostand Reinhold Company, New York, N.Y.

KHAYAT, K. H., 1992. "In-situ Properties of Concrete Piles Repaired Under Water," Concrete International, March.

KHANNA, J., GILBRIDE, P., and WHITCOMB, C. R., 1988a. "Steel Fiber Reinforced Concrete Jackets for Repairing Concrete Piles." Proceedings of

Second International Conference on Concrete in Marine Environment SP-109. St. Andrews-by-Sea, Canada.

KHANNA, J., SEABROOK, P., GERWICK, B., and BICKLEY, J., 1988b. Invesgigation of Distress in Precast Concrete Piles at Rodney Terminal, Saint John, New Brunswick. Proceedings of the Second International Conference on Concrete in Marine Environment SP-109. St. Andrews-by-Sea, Canada.

KOHNE, J. K., 1983. Storten van onderwaterbeton met de Hop-dobber. Vol. XXXV.12.

KRAY, C. J., 1983. "Rehabilitation of Steel Sheet Pile Bulkhead Walls." ASCE Proceedings of Specialty Conference Ports '83. March 21–23, new Orleans, Louisiana.

KRISTENSEN, M., 1987. "Seabed Erosion in Ferry Berths." PIANC bulletin No. 58.

LAMBERTON, B. A., 1989. "Fabric Forms for Concrete." Concrete International, December.

LANE, E. W., 1979. Progress report on Studies on the Design of Stable Channels of the Bureau of Reclamation. Proceedings of the American Society of Civil Engineers, No. 280.

LIN, T. Y. and BURNS, N. H., 1981. "Design of Prestressed Concrete Structures," Third Edition John Wiley & Sons, New York.

LIU, T. C. and HOLLAND, T. C., 1981. "Design of Dowels for Anchoring Replacement Concrete to Vertical Lock Walls." TR C-78-4, Report, US Army Engineer Waterway Experiment Station. April, Vicksburg, Mississipi.

LONGE, M. J. P., HERGERT, M. P. and BYLK, M. R., 1987. "Problems d'erosion aux ouvrages de quai existants causes par les propulseurs d'etrave et les helices principles des novires lors de leurs accostages aou appareillages." PIANC, Bulletin No. 58.

LUNDIN, R. P., 1983. "Rehabilitation of Cyclopian Concrete Wharf." ASCE Proceedings of Specialty Conference Ports '83. New Orleans, Louisiana.

MACCAFERI GABIONS. Flexible Reno Mattress and Gabion Linings. Commercial booklet.

MALHOTRA, V. M., 1976. "Testing Hardened Concrete: Nondestructive Methods." ACI Monograph No. 9, American Concrete Institute/Iowa State University Press, Detroit.

MCDONALD, J. E. and LIU, T. C., 1988. "Evaluation and Repair of Concrete Navigation Structures." Proceedings of the Second International Conference on Concrete in Marine Environment SP-109. St. Andrews-by-Sea, Canada.

MEHTA, P. K., 1980. "Durability of concrete in Marine Environments." ACI SP-65, American Concrete Institute, Detroit.

MEHTA, P. K., 1986. "Concrete: Structure, Properties, and Materials." Prentice Hall, Englewood Cliffs, NJ.

MEHTA, P. K., 1988. "Durability of Concrete Exposed to Marine Environment—A Fresh Look." Proceedings of the Second International Conference on Concrete in Marine Environment SP-109, St. Andrews-by-Sea, Canada.

MEHTA, P. K. and GERWICK, B. C., 1982. "Crack–Corrosion Interaction in Concrete Exposed to Marine Environment." Concrete International, Vol. 4, No. 10, Detroit.

MENZIES, R. J. and TURNER, R., 1957. "The Distribution and Importance of Marine Wood Borers in the United States." In ASTM Special Technical Publication No. 200, Proceedings of a Symposium on Wood for Marine Use and its Protection from Marine Organisms.

MILWEE, W. I. and AICHELE, W. F., 1986. "Modern Inspection Techniques in Port Maintenance." ASCE Proceedings Specialty Conference Ports '86, Oakland, California, May 19–21.

MORGAN, D. R., 1988. "Recent Developments in Shotcrete Technology—A Materials Engineering Perspective." World of Concrete '88 Las Vegas, Nevada, February.

MURAOKA, J., 1962. "The Effects of Marine Organisms on Engineering Materials for Deep Ocean Use." U.S. Navy, Civil Engineering Laboratory, Port Heuneme, California.

NAVAL CIVIL ENGINEERING LABORATORY, 1983. Contract Report "Evaluation of Commercial Digital Ultrasonic Metal Thiskness Measurement Instruments." Southwest Research Institute, San Antonio, Texas.

NAVAL CIVIL ENGINEERING LABORATORY, 1985. Contract Report "Design of the Underwater Steel Inspection System." Southwest Research Institute, San Antonio, Texas.

NAVDOKS, 1950. "Harbor Reports on Marine Borer Activity." U.S. navy P-43, Washington DC, June.

NAVDOKS, Mo-306, 1964. "Corrosion Prevention and Control." Washington DC.

NOVOKSHCHENOV, V., 1988. "Stopping the Cracks." Civil Engineering, ASCE Vol. 58, No. 11.

NUCCI, L. R. and CONNORS, C. C., 1990. "Inspection and Maintenance of Port Facilities for Improved Service Life." Proceedings PIANC 27th Congress s. II-3. Osaka, Japan.

OEBIUS, H. U., 1984. "Loads on Beds and Banks Caused by Ship Propulsion Systems." In Flexible Armored Revetments. Thomas Telford, London.

OEBIUS, H. and SCHUSTER, S., 1979. Analitische und Experimentelle Untersuchungen uber dan Eiufluss vou Schrauben propellern ant lewegliche Gewassersohlen Teil 1: Der ungestorte Propellerstrhl. Versuchsanstalt fur Wasserbau und Schifflau, Berlin.

OFFSHORE ENGINEER, 1984. "Anti-Fouling Paint Makes Barnacles Lose their Grip." July, pp 73–74.

OHSAKI, Y., 1982. "Corrosion of Steel Piles Driven in Soil Deposits." Soils and Foundation. Japanese Society of Soil Mechanics and Foundation Engineering, Vol. 22, No. 3, September.

OKADA, K., NISTHBAYASHI, S. and NAKAMURA, M., 1989. Proceedings 8th International Conference on Alkali–Aggregate Reaction. Tokyo, Japan.

PERMANENT INTERNATIONAL ASSOCIATION OF NAVIGATION CONGRESSES, 1987. "Guidelines for the Design and Construction of Flexible Revetments Incorporating Geotextiles for Inland Waterways." Report on Working Group 4 of the Permanent Technical Committe 1 Suplement to Bulletin No. 57, Brussels, Belgium.

PERMANENT INTERNATIONAL ASSOCIATION OF NAVIGATION CONGRESSES, 1990. Inspection, Maintenance and Repair of Maritime Structures Exposed to Material Degradation Caused by a Salt Water Environment. Report of Working Group No. 17 of PTC II. Brussels, Belgium.

PETTIT, P. and WOODEN, C., 1988. "Jet Grouting: The Pace Quickens." American Society of Civil Engineers. Civil Engineering, Vol. 58, No. 8.

PILARCZYK, K. W., 1985. "Stability of Revetments Under Wave and Current Attack." the Twenty-First International Association for Hydraulics Research Congress Melbourne.

PILARCZYK, K. W., 1986. "Design Aspects of Block Revetments." Post Graduate Course on Bank and Dike Protection. Delft University of Technology Civil Engineering Department, PATO, Delft.

PLOWMAN, J. M., 1956. "Maturity and the Strength of Concrete." Magazine of Concrete Research, Vol. 8, No. 22, London, March.

PORTER, D. L., 1986. "Innovative Repairs to Steel Sheet Pile Structures." ASCE Specialty Conference, Ports '86, Oakland, California, May 19–21.

PROSSER, M. J., 1986. "Propeller induced Scour." The British Port Association, RR2570, London.

RACTLIFFE, A. T., 1983. "The Basis and Essentials of Marine Corrosion in Steel Structures." Proceedings of the Institute of Civil Engineers, Part 1, No. 74, London, November.

RICHARDS, B. R., 1983. "Marine Borers." Proceedings of the American Wood-Preservers' Association.

RILAM, 1985. "Synthetic Membranes." Draft Recommendations 47—SM Technical Committee.

ROBAKIEWICZ, W., 1966. "The Influence of the Action of the Jet Behind the Screw on the bottom. Examples of model and field measurements with trawler B-20." Rozprawy Hydrotechiczene, 19 (In Polish).

ROBAKIEWICZ, W., 1987. "Bottom Erosion as an Effect of Ship Propeller Action Near the Harbor Quays." Permanent International Association of Navigation Congresses. Bulletin No. 58, Brussels, Belgium.

ROGERS, T. H., 1960. The Marine Corrosion Handbook. McGraw-Hill, New York.

RÖMISCH, K., 1977. "Damage of Waterways and Hydraulic Structures Caused by the Attack of Propeller Jet." Proceedings of 24th Congress of Permanent Association of Navigation Congresses. Section 1, Subject 3, Leningrad, USSR.

RUSSELL, S., 1989. "Inspection and Construction of Underwater Repairs." American Society of Civil Engineers, Proceedings of Specialty Conference Ports '89, Boston, May.

SANCIER, K. L. and NEELEY, B. D., 1987. "Antiwashout Admixtures in Underwater Concrete." Concrete International, May.

SCHAJER, G. S., 1981. "Application of Finite Element Calculations to Residual Stress Management." Journal of Engineering Materials and Technology, Vol. 103/157, April.

SCHMITT, R. J. and PHELPS, E. M., 1969. "Corrosion Performance of Constructional Steels in Marine Applications." Talk prepared for presentation at the First Annual Offshore Technology Conference, Houston, Texas.

SCHWERDTFEGER, W. J. and ROMANOFF, M., 1972. "NBS Papers on Underground Corrosion of Steel Piling 1962–1971." National Bureau of Standards Monograph 127, March.

SCOLA, P. T., 1989. "US Navy Underwater Inspection of Waterfront Facilities." American Society of Civil Engineers, Proceedings of Specialty Conference Ports '89, Boston, May.

SCOTT, P. J. B. and DAVIES, M., 1992. Microbiologically Induced Corrosion. Civil Engineering, May.

SHERARD, J. L. and DUNNIGAN, L. P., "Critical Filters for Impervious Soils," ASCE Journal of Geotechnical Engineering, Vol. 115, No. 7, July.

SHIELDS, A., 1936. "Auwendung der Aenlich keitsmechanic und des Turbulenz forsehung anf die Gesehiebebeweguug." Mitteilungeu der Preussischeu Versuchsantalt fur Wasserbau und Schiffbau. W. P. Oft and J. C. van Uchelen (tr.), California Institute of Technology, Pasadena.

SHILSTONE, J. M. and KENNEY, A. R., 1979. "Evaluation and Retrofit of Concrete Structures." Paper presented at the ACI Annual Convention, Milwaukee, Wisconsin.

SHROFF, A. C., 1988. "Evaluating a 50-year Old Concrete Bride." Concrete International, Design and Construction, No. 5.

SILVESTER, R., 1983. "Design of In Situ Cast Mortar Filled Armor Units of Marine Structures." Proceedings of the 6th Australian Conference on Coastal and Ocean Engineering.

SILVESTER, R., 1986. "Use of Grout Filled Sausages in Coastal Structures." American Society of Civil Engineers, Journal Waterway, Port, Coastal and Ocean Engineering, Vol. 112(1).

SMITH, A. P., 1987. "New Tools and Techniques for the Underwater Inspection of Waterfront Structures." Proceedings of the Offshore Technology Conference, Houston, Texas, pp. 291–298, April 27–30.

SNOW, R. K., 1992. Polymer Pile Encapsulation: Factors Influencing Performance. Concrete International, May.

SOWERS, G. B. and SOWERS, G. F., 1967. "Failures of Bulkhead and Excavation Bracing." Civil Engineering, ASCE, Vol. 37, No. 1, pp 72–77.

STAYNES, B. and CORBEFT, B., 1988. "Underwater Concreting with Polymers." Civil Engineering, London, March.

STEIJAERT, P. D. and DE KREUK, J. F., 1985. "Underwater Repairs with Cement-Based Concretes." Behavior of Offshore Structures, Elsevier Science Publishers B.V., Amsterdam.

STRATTON, F. W., ALEXANDER, R. and NOLTING, W., 1982. "development and Implementation of Concrete Girder Repair by Post-Reinforcement." Report No. FHWA-KS-82-1. Kansas Department of Transportation, Topeka, Kansas.

THE DOCK and HARBOUR AUTHORITY, vol. LXX, No. 812, October, 1989. Limpet Dam used for quay upgrading at Harwich.

THOMPSON, W. J., 1977. "Damage to Port Structure by Ships with Bulbous Bows." American Society of Civil Engineers, Proceedings of the Specialty Conference, PORTS '77.

TOYAMA, S. and ISHII, Y., 1990. "The Treatment of the Deterioration of Port and Harbour Concrete Structures in Japan." PIANC Bulletin No. 68, Brussels, Belgium.

TSINKER, G. P., 1983. "Anchored Sheet Pile Bulkheads: Design Practice." ASCE Journal of Geotechnical Engineering, Vol. 109, No. 8, August.

TSINKER, G. P., 1986. "Floating Ports." Gulf Publishing Company, Houston, Texas.

TSINKER, G. P., 1989. "The Dock-in-Service," Evaluation of Load Carrying Capacity, Repair, Rehabilitation. American Society of Civil Engineers, Proceedings of Specialty Conference Ports '89, Boston, May.

TURNER, R. D., 1966. "A Survey and Illustrated Catalogue of the Teredinidal." Harvard University.

UHLIG, H. H., 1971. "Corrosion and Corrosion Control," Second Edition. John Wiley & Sons, New York.

UHLIG, H. H. (ed.), 1948. "The Corrsion Handbook." John Wiley & sons, New York.

US ARMY CORPS of ENGINEERS TM-5-811-4, 1962. "Corrosion Control." Washington, DC.

US ARMY CORPS of ENGINEERS, 1977. Plastic Filter Fabric: Civil Works Construction Guide Specification.

US ARMY CORPS of ENGINEERS, 1985. Engineering Manual EM 1110-2-2000, Standard Practice for Concrete, Washington, DC.

US ARMY CORPS of ENGINEERS, 1986. Evaluation and Repair of Concrete Structures. Engineering Manual EM 1110-2-2002. Washington, DC.

US ARMY CORPS of ENGINEERS, 1982. Publication EP 1110-1-10., May

US NAVY, 1970. "World Atlas of Coastal Biological Fouling." USNOO, I.R. No. 70–51, September.

VEJVODA, M. F., 1990. "Strengthening of Existing Structures with Post-Tensioning," concrete International, September.

VELDHUIJZEN VAN ZANTEN, R. (ed.), 1986. "Geotextiles and Geomembranes in Civil Engineering." A. A. Balkema, Rotterdam, Boston.

VERHEY, B., 1983. "The Stability of bottom and Banks Subjected to the Velocities in the Propeller Jet Behind Ships." Proceedings of the 8th International Harbour Congress. Antwerpan, Belgium.

VERHEY, H. J., BLOKLAND, T., BOGAERTS, M. P., VOLGER, D., and WEYDE, R. W., 1987. "Experiences in the Netherlands with Quay Structures Subjected to Velocities Created by Bow Thrusters and Main Propellers of Mooring and Unmooring Ships." Permanent International Association of Navigation Congresses, Bulletin No. 58, Brussels, Belgium.

WINTERKORN, H. F. and FANG, H-Y. (eds.), 1975. Foundation Engineering Handbook. Van Nostrand Reinhold, New York.

2

Marine Structures in Cold Regions

2.1 INTRODUCTION

Areas of the earth with seasonally frozen grounds as well as such regions as the arctic and the subarctic are generally referred to as cold regions.

The navigation and port and harbor operations in cold regions are generally associated with the ice cover which hinders or sometimes prohibits navigation.

In North America during the winter season and early spring the ice cover extends as far south as Chesapeake Bay, through all the Great Lakes, the Upper Mississippi River and its principal tributaries—the Ohio, Illinois, and Missouri rivers, and the St. Lawrence Seaway—and hinders navigation along the east coast of the United States and Canada, in the Cook Inlet and the Baring Sea of Alaska, as well as throughout the U.S. and Canadian arctic regions.

Ice problems and ice effects on navigation in waterways, ports, and harbors vary greatly, depending on geographical location, harbor and waterway geometry and size, and vessel traffic.

As pointed out by PIANC (1984), some of the common problems featured in the ice-affected waterways and harbors are as follows:

1. Ship traffic is restricted to certain lanes, both offshore and in the channel. The lanes are typically kept open by icebreakers. However, the ice breaking itself accelerates ice formation in that large amounts of broken and refrozen ice may accumulate after repeated passages.

2. Occasionally during the winter season, wind-driven landfast or offshore ice outside the open lanes, or ice from the river upstream, may break up and enter the ship lanes.

3. Navigation aids must be able to withstand heavy ice forces. Furthermore, buoy icing may start when freezing spray builds into a heavy ice accumulation on the buoy cage. This accumulation may eventually capsize the buoy, or at least cause it to tilt excessively, thereby reducing its visibility. Furthermore, once a buoy and its anchor chain become submerged, they become hazards to navigation.

4. Ice growth and its adherence to structures such as navigation locks and docks may hamper and disrupt the port/waterway operation. Usually it is necessary to remove this ice. In addition, ships may transport broken ice from the channel or harbor area into the lock or in dock area.

5. In-harbor large areas have to be kept open to provide ships with sufficient maneuvering room into and away from the docks.

6. Ice forming in berthing areas must be removed in order to permit the ship safe mooring operation.

The principal solutions to the above problems include a set of methods usually referred to as "ice control" and/or "ice management."

"Ice control" and ice management by definition comprise a set of methods for reducing or eliminating the growth of ice in navigational areas and on berthing structures. They may include mechanical, thermal, chemical or other means and contain a wide variety of engineering approaches aimed at improving existing ice conditions, to manipulate the action of ice, and to reduce or eliminate the ice forces on marine structures.

The behavior of marine structures in ice-affected waters also is influenced by ice action. Depending on the geographical area, ice can be present in many forms, ranging from sheet ice to icebergs. The motion of ice against a structure leads to generation of usually substantial forces. These forces may depend on structure geometry, ice parameters, and other environmental forces such as wind, waves, and currents.

Effects of global and local ice loads must be of interest to the structural marine engineer; global ice forces in some cases may control general stability of a structure, and the local loads may determine the required strength of some structural components. Naturally, both type of loads have a significant effect on cost of construction.

In order to properly understand and specify ice loads upon marine structures an engineer has to understand how ice as a material acts under stress. This is why the topic of ice mechanics is closely linked to that of ice loads.

Although formal studies in ice engineering date back to the turn of the century the field of ice mechanics has not yet arrived at a mature stage and a good deal of judgment and experience is still required to achieve satisfactory results in determination of ice loads upon marine structures.

In recent years, significant activity relating to the arctic offshore oil and gas exploration has led to a surge in research and engineering practices relating to ice mechanics and ice loads exerted upon marine structures. As a result some improved theories on ice action against marine structures have been introduced (Kry, 1978, 1980; Croasdale, 1980b). The recent state-of-the-art has been described in works by Cammaert and Muggeridge (1988), Croasdale (1985), Hallam and Sanderson (1987), Määttänen (1987), Metge et al. (1981), Sinha et al. (1987), Sodhi and Cox (1987), and Watt (1982). Hager and Klein (1990) summarized a European experience and discussed basic principles of ice formation in inland waterways and dimensioning of marine structures exposed to ice loading included in the Recommendations of the Committee for Waterfront Structures (EAU, 1990). Most recently Richter-Menge (1992) compiled the major investigations in ice mechanics conducted in the United States in the period 1987 to 1990 with emphasis on lateral movement of the ice against a vertical structure.

In spite of extensive research efforts practicing engineers still rely heavily on either rough or semi-empirical methods of analysis when dealing with ice problems. The latter are basically attributed to lack of proper understanding of mechanical properties of ice, which vary greatly depending on ice conditions, and the fact that it is extremely difficult to correlate results of small-scale model tests to a prototype condition.

Therefore, to improve basic understanding of ice mechanics and to develop more practical models of ice–structure interaction, there still exists a great need for full-scale field measurements.

In a great many cases marine structure designers have to deal with frozen soils which exist in nature in large geographical areas of the world and include about 22 million square kilometers in Russia, Canada, Nordic Countries, and Alaska. A situation where the grounds remain frozen for several years is referred to as permafrost.

In cold regions the ground temperature usually ranges from 0° to −15°C, and the depth of freezing varies from a few meters to hundreds of meters. The frozen grounds are highly complex multiphase compositions of low temperature material comprising mineral soil particles, ice, unfrozen water, and gases. The quantity and quality of the three latter components vary with temperature and pressure.

The freezing of the ground is associated with a volume increase called frost heaving phenomenon, and its thawing is associated with a volume decrease, called the thaw settlement phenomenon. Frost heaving and thaw settlement are the principal causes of unacceptable deformations of structures constructed in cold regions, if frozen soil characteristics are not taken into account in the design.

The foundation problems associated with construction in cold regions are discussed in depth by Legget (1966), Brown (1970), Tsitovich (1975), Andersland and Anderson (1978), Johnston (1981), Hutcheon and Handegorg (1983), Phukan (1985), and Youssef (1987).

The major port components that can be affected by the cold region environment as discussed earlier are: harbor area, breakwaters, shore protection, dock structures and their fender systems and mooring accessories, all waterside and dockside equipment, and aids to navigation.

In navigable waterways it may also include navigation locks and bridge protection structures. For efficient winter operation (ice navigation) all the above components must be designed to function reliably under severe winter conditions.

2.2 ICE COVERS

The purpose of this section is to provide a concise discussion of the formation of ice and its various forms, conditions, and properties to facilitate better understanding of the issues of ice mechanics and ice loads upon marine structures.

For detailed information on each particular subject the reader will be referred to an appropriate source of information. In addition, a comprehensive review of state-of-the-art of ice properties is given in recent publications by Chakraburtty et al. (1984), Sinha et al. (1987), and Cammaert and Muggeridge (1988).

2.2.1 Ice Microstructure and Morphology

The many processes that can occur during the formation and growth of sea or fresh-water ice cause it to assume a variety of crystalline structures with a hexagonal crystallographic symmetry.

The typical size and shape of an ice grain which actually is an individual crystal vary greatly. The ice grain may range from 1 mm to several centimeters, and its shape can be tabular, granular, or columnar. The axis of molecular symmetry of an ice crystal is parallel to the basal plane. Different types of ice are conveniently classified and described by Pounder (1965), Michel and Ramsie (1971), Hobbs (1974), Michel (1978), and Doronin and Khesin (1977).

The number of bonds to be broken for deformation parallel to the basal plane are less than in other directions. Therefore, the deformation and strength of ice, at least on a microscale, are significantly affected by the direction of stress in relation to crystal orientation.

Snow ice is generally composed of granular crystals of random orientation and can therefore be considered isotropic. In contrast, ice that grows on the surface of water will be mostly composed of long continuous columnar crystals. The basal planes will generally be vertically oriented but not necessarily parallel. Hence, in the latter case the ice can be considered to be anisotropic in its material properties.

Broken and then refrozen ice will generally contain randomly arranged columnar ice pieces so that on a larger scale for practical purpose such ice is usually considered as isotropic.

The structure of ice may be complicated further by the presence of salts and air. During ice growth these impurities are expelled to the plate boundaries. At the usual temperature of sea ice in nature these brine pockets may remain unfrozen and hence have an effect on the strength of ice. The latter in fact results in sea ice being weaker than freshwater ice at the same temperature. Furthermore, as noted by Croasdale (1985), "... on a larger scale than the crystal structure, ice in nature contains flaws and cracks. It is now generally believed that these play a more significant role than microstructure in determining large scale deformation and strength."

2.2.2 Ice Formation

Under calm water surface conditions and relatively low temperature gradients, a very thin, supercooled water layer may appear at the water surface. The temperature of this layer will actually be several degrees lower than the freezing temperature of the water. In such calm, supercooled conditions, ice crystals will grow in the form of needles with vertical axes at random orientation to the water surface. At small temperature gradients, crystallization will proceed at a slow pace, and at large temperature gradients the solidification of the ice cover will occur more rapidly with the initial needles interlocking with each other.

In most cases, sea ice will begin to form when wind and waves agitate the surface layer; supercooling will then extend to deeper levels. Natural nucleation will then cause the formation of frazil particles in the form of small discoids. As these grow, they will adhere to each other first to form slush, only to refreeze quickly and form the first layer of ice (primary ice). The crystal orientation in this layer will usually be random.

If relatively undisturbed the smooth sheet ice grows continuously throughout the winter season. It can vary from a few centimeters in the southern regions of North America to 1.0 to 2.5 m in the Bering Sea area and Canadian Arctic Islands, respectively.

As previously noted a layer of primary sheet ice first forms when ice crystals begin to concentrate on the surface. Because sea surfaces are almost always in motion, these crystals form a soupy layer. Such ice can coagulate and form a continuous sheet, or, if there is strong wind and wave action, it can agglomerate into "pancakes" of ice that can reach substantial dimensions. Such ice is usually called *frazil ice*.

Any ice that remains frozen to the shoreline for the remainder of the winter is labeled *fast ice*. In most areas, however, ice is affected by movements resulting from winds, waves, tides, currents, or thermal changes. Continuous, large-scale movements of ice can be caused by major current systems, such as those in Cook Inlet, Alaska. Movements of a shorter duration can occur from wind stresses or wave action.

The growth of a sea ice cover is difficult to predict, as it requires accurate information on climatological factors, such as solar radiation and snowfall, and the thermal properties of ice and snow. However, when field measurements are not available, initial estimates of ice growth can be obtained by approximate analysis.

Nakawo and Sinha (1981) developed a numerical method for calculation of ice thickness that accommodates variations in snow conditions and the thermal properties of ice and snow.

A more complex mode of ice growth has been formulated by Maykut and Untersteiner (1971). This is a one-dimensional thermodynamic model that includes the effects of snow cover, ice salinity, and various forms of heat flux through ice and snow.

The air temperature is the principal factor that governs ice melt rate, although snow-ice formation, snow-cover depth, solar radiation, currents, and wind also play a role (Bilello, 1961, 1980).

Sea ice is generally classified as either first-year or multiyear. The characteristics of each type of ice change significantly in time and are also highly variable within a given region.

Much of the sheet ice in the Arctic Ocean survives more than one season. During the cycles of warming and cooling the brine is gradually expelled so that this multiyear ice is fresher and stronger than annual or first-year sea ice. Furthermore, first-year sea ice can be overrridden by the ice sheets of the same or varying thicknesses which tend to consolidate quickly into a single thicker sheet. In addition primary ice can frequently be superimposed by the ice caused by flooding through ice cracks, refreezing of melted ice, or formation of snow-ice. All the above factors may result in a consolidated multiyear ice sheet of about 4 to 5 m thick. The structure of multiyear ice is however greatly affected by seasonal surface melting, and a layered structure often results (Weeks, 1982).

Much of the sea ice is present in the form of ridges and rubble fields. Ice pressure ridges are formed by ice pressure and appear as ridges on the surface of the ice. The ice pressure that forms ridges is largely the result of wind drag acting over the ice which causes the ice to move.

The ridge building process is discussed in detail by Parmerter and Coon (1983).

Because the wind is variable and because ice usually cannot move freely, the wind drag creates ice pressure which leads to ridge deformation.

The cross-sections and properties of ridges can vary greatly. An example of cross-section of the grounded first-year ridge is depicted in Figure 2–1.

First-year ridges are initially a loose accumulation of ice blocks held together by gravity and buoyancy forces.

As the winter progresses, the "core" of the ridge consolidates as the water between the ice blocks near the waterline freezes. Typically, the ridge keel (below waterline portion) is 4.5 to 6 times greater than the sail height (above waterline portion). The total thickness of a ridge can reach about 20 to 30 m.

First-year ridges gradually consolidate and freeze into multiyear ridges, or ridges that have survived a number of melt seasons.

In the process the pore water between ice blocks continues to freeze, and melt water produced during the summer months drains down into the ridge core and refreezes quickly. The accumulated ice will then become a solid, resistant mass.

Multiyear ridge geometry and some properties obtained from field measurements are discussed in detail by Kovacs and Mellor (1971), Kovacs et al. (1973, 1975), Kovacs and Gow (1976), Dickins and Wetzel (1981), Voelker et al. (1981), Kovacs (1983), and Wright et al. (1978).

A typical multiyear ridge model is depicted in Figure 2–2.

Ridges are generally considered to be linear features. When they become areal in extent because of pressure acting from several directions they are often referred to as ice rubble fields. Ice rubble fields also form

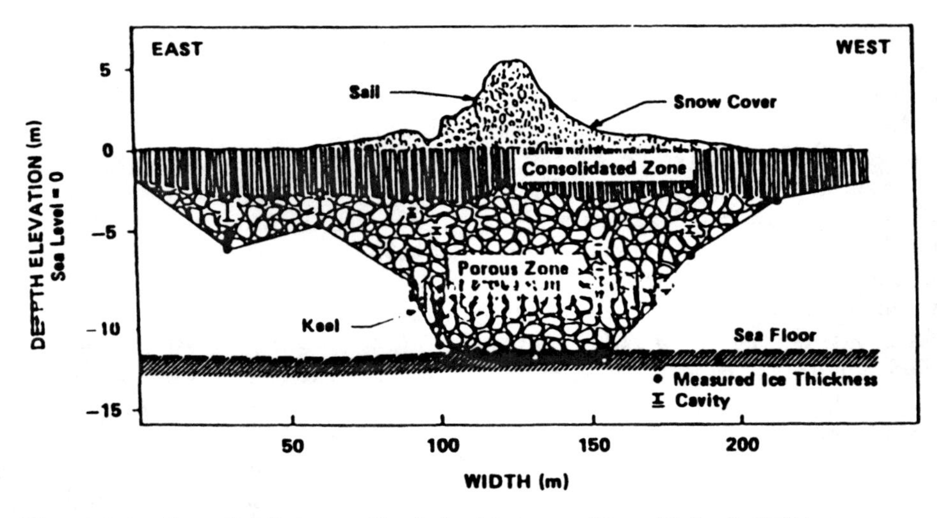

Figure 2–1. Grounded first-year ridge in land-fast zone. (From Gladwell, 1976.)

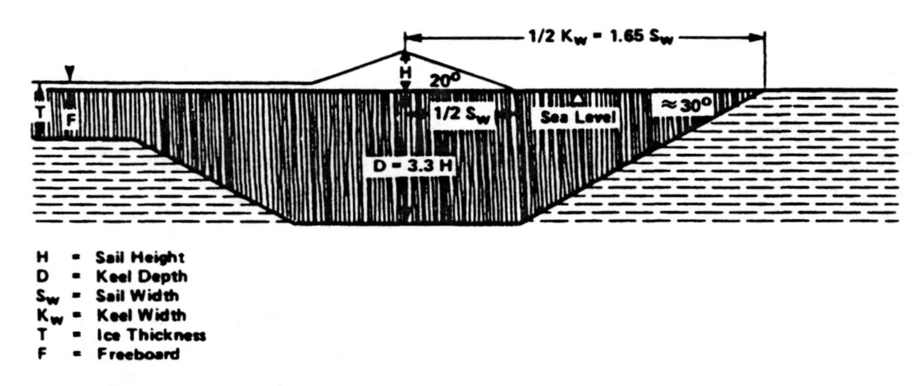

Figure 2–2. Multyear ridge mode. (From Wright et al., 1978.)

around grounded objects. If ice rubble fields survive and become multiyear features they are usually referred to as *hummock fields*.

Such extreme ice features may control the design of offshore structures of any kind in the Arctic Ocean.

The internal structure of rubble fields is usually composed of a consolidated ice layer at the waterline; under this is a randomly mixed distribution of solid angular blocks, voids, and soft or slush ice. Grounded rubble in the Beaufort Sea has been observed to have sail heights up to 20 m.

Since ports and marine terminals are usually located in protected harbors, then ice ridges of any kind are seldom considered in port-related marine structures design.

2.2.3 Sea Ice Characteristics (Parameters of Importance)

2.2.3.1 Grain Size and Crystal Orientation

Parameters of grain size and orientation are the characteristics defining homogeneity (or inhomogenity) and isotropy (or anisotropy) of ice. Most available test results for salt water ice are for speciments of columnar ice which represents best the average properties of ice sheets as far as horizontal strength is concerned since it usually forms the thickest zone. Wang (1979, 1981) and Wang and Poplin (1986) presented compressive strength results for granular and columnar salt water ice; generally, finer grained ice offers a higher strength.

2.2.3.2 Brine Volume

Due to the columnar structure of most ice sheets, uniaxial strength properties vary widely if specimens are loaded parallel or perpendicular to the plane of the ice sheet. The grain boundaries and the basal planes of individual crystals, which are the planes where brine pockets are located, represent weak zones in the ice where failure occurs more easily when loaded perpendicular to the growth direction. Typically, the strength of the vertical direction is about three times that in the horizontal direction.

The relationship between brine volume and strength is explained by a model developed by Weeks and Assur (1967). Schwarz and Weeks (1977) suggest that strength in a direction perpendicular to the ice columns can be computed from the following

$$\sigma = \sigma_0[1 - (\nu_b/\nu_0)^{0.5}] \qquad (2\text{–}1)$$

where σ_0 and ν_0 are constants and ν_b = brine volume in ‰. Large amounts of data are presented in terms of the brine volume or square root of the brine volume. However, Timco and Frederking (1986) suggest that a better correlation exists between strength and square root of the total porosity (brine and air).

Cox and Weeks (1983) have derived an expression that takes into account the presence of both air and solid salts in the ice.

2.2.3.3 Strain and Stress Rate

The deformational behavior, failure mechanism, and strength of ice are highly affected by the strain rate. Ranges of strain rate may be classified as ductile, transitional, and brittle. At very high rates of loading, stresses are affected by strain-rate effects, the impact velocity, and drainage effects.

When subjected to low rates of loading or sustained loads, ice creeps. As strains are slowly increased, a ductile failure is reached: the maximum stress that is attained is referred to as the yield strength. The behavior in this zone is controlled by the number of mobile dislocations occurring along the basal planes in the ice crystals. Figure 2–3 shows the effect of strain rate on the yield strength of sea ice.

For low rates of compression strains in constant strain rate tests, the dependence of the yield stress on the strain is in good agreement with the dependence of secondary creep rate on the stress in a constant load test.

Rapidly applied loads cause a sudden, brittle failure. At high strain rates, dislocation velocities are too slow to allow ductile behavior so that the mechanisms of fracture initiation and propagation control the strength in this range of strain rates.

At strain rates intermediate between the ductile and brittle zones lie a so-called transition zone where the mode of failure is unstable and either ductile or brittle failures may be encountered. The range of strain rates corresponding to the transition zone is temperature dependent for both compressive and tensile strengths.

In addition to the prevalent increase, a decrease at high rates in compressive strength with increasing strain rate has been observed by some investigators (Peyton, 1966; Wu et al., 1976; Michel, 1978). Although this has been reported to occur in both fresh and salt types of ice, in the case of saltwater ice, conditions are often not as well controlled and the additional effect of salinity usually produces a larger dispersion of strength values at high strain rates. Sinha and Frederking (1979), in an investigation of the effect of test system stiff-

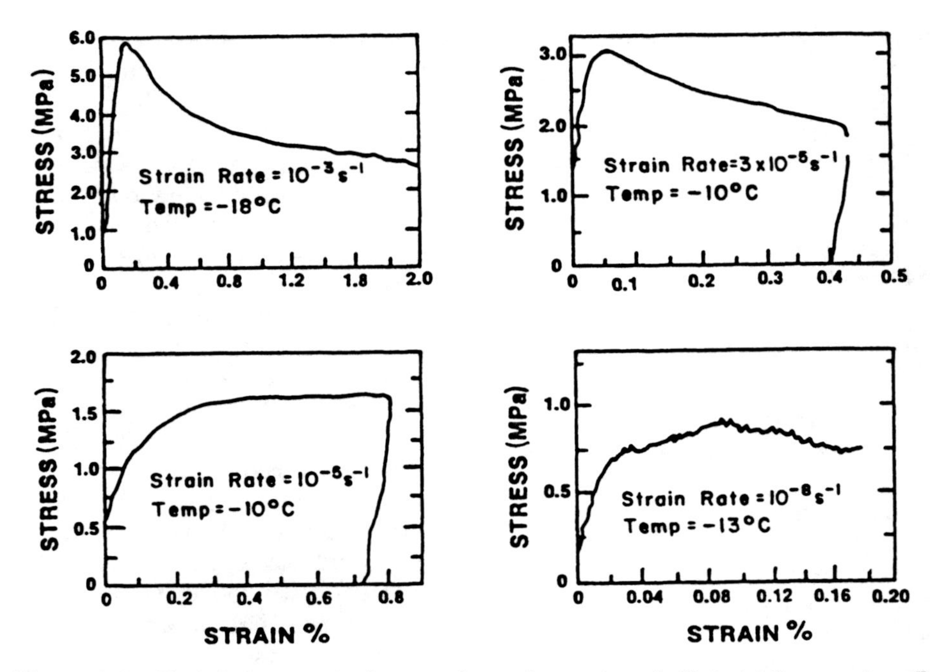

Figure 2–3. Typical stress-strain diagrams for sea ice specimen in Uniaxial Compression. (From Wang, 1981.)

ness on strength, noted that a constant rate of cross head movement of the testing machine produces an essentially constant stress rate through most of the loading but that actual strain rates, measured with strain gauges, approach the nominal strain rate only at the time of failure. At a given nominal strain rate, defined as the ratio of cross head velocity to specimen length, higher stress rates are produced by a stiff testing machine compared to a flexible system. A stiffer system thus tends to produce a more brittle failure than a flexible test apparatus.

Strain or stress ratio is affected by the temperature effects and confinement.

The effect of temperature on the mechanical properties of sea ice is usually accounted for by the effect of brine volume (Michel, 1978). Temperature, however, has a negligible effect on tensile strength, grain size being the most significant parameter. Temperature as well as grain size are important parameters in the prediction of brittle compressive strength.

Compressive strength variations with temperature may be illustrated for various salinities of sea ice by computing the compressive strength at the temperature of interest and dividing it by the compressive strength at −10°C, a temperature at which a large number of tests have been conducted.

Materials in a confined state being crushed in front of a structure exhibit strength properties that are quite different from that of an unconfined test specimen. Confinement is typically expressed in terms of a structure diameter to ice thickness aspect ratio. Analysis of the problem therefore requires determining an assumed failure criterion for the material which allows a prediction of the forces exerted against the structure. The failure criterion is based on one or several index strength parameters such as the unconfined compressive and tensile strength. Some failure criteria also required multiaxial data for their definition (Frederking and Timco, 1983).

Triaxial and compressive strength tests were completed by Nawwar et al. (1983) on laboratory-grown saline ice. They found that the mean strength ratio of samples taken parallel to the ice growth direction to samples taken perpendicular to the ice growth direction was 3.0 at a strain rate of $3 \times 10^{-4}\,s^{-1}$. Similar behavior was reported by Blanchet and Hamza (1983).

Timco and Frederking (1984, 1986) measured the full failure envelope using both vertical and horizontal loading arrangements for granular/discontinuous columnar sea ice. They presented results from five confinement arrangements, fitting these data to a modified n-type yield function for temperatures of $-2°$ and $-10°C$ at a nominal strain rate of $2 \times 10^{-4}\,s^{-1}$.

Relatively recent work has related scale effects to fracture characteristics of sea ice. Important investigations were conducted by Urabe and Yoshitake (1981) and sample analyses have been carried out by Hamza and Muggeridge (1984) on ice forces related to fracture toughness.

Scale effects may be defined as the effect of the size of a model test specimen on the actual behavior at full scale in nature. The importance of scale on the strength of brittle materials is well established. As a brittle fracture is usually initiated by a stress concentration at the tip of a crack, and if it is assumed that these imperfections are Poisson-distributed, a larger volume of material is more likely to offer a flaw at an unfavorable orientation than a smaller volume and therefore is likely to fracture at a lower stress value. Weeks and Assur (1969) proposed the following relationship:

$$\frac{f_s}{f_L} = 1 + \frac{L}{L_0} \qquad (2\text{–}2)$$

where f_s and f_L are the measured strengths of the small and large scale specimens respectively, L is the square root of the area of the failure surface, and L_0 is a constant determined from test results that was found to vary with temperature.

2.2.4 Mechanical Properties

These usually include compressive, tensile, flexural, shear, and adhesion strengths of ice, and also its deformation characteristics such as elastic modulus and Poisson's ratio. All of the above properties heavily depend on a brine volume, or in other words whether the ice is formed from salt- or freshwater.

By definition strength of ice is the maximum stress that an ice specimen can support. It is of great importance when dealing with determination of the ice forces on a structure. To date numerous investigators have conducted research on ice strength and its deformation characteristics.

The following is a brief review of the state-of-the-art.

2.2.4.1 Sea Ice

Compressive Strength

The strength of ice in compression is of fundamental importance in almost all aspects of ice mechanics. Studies of ice strength are usually performed by conducting *uniaxial* compression in which a uniform uniaxial stress is applied to a specimen, *multiaxial* compression in which a more complex stress state is applied to a specimen, and *in situ* test, where the ice is tested in its natural state in the ice cover. Three commonly applied loading conditions for determining strength are: constant displacement rate, controlled constant strain rate, and controlled load or stress rate (Sinha et al., 1987).

As noted by Timco and Frederking (1983), in general there is no substitute for measuring both the load and ice specimen deformation. With known stress and strain histories analysis can be made simpler, more straightforward, and devoid of misinterpretation.

The compressive strength of sea ice, which varies with brine volume, strain rate, temperature, porosity, and grain orientation, is affected by the ratio of grain size to specimen size as well as the scale effect between specimen size and prototype (Iyer, 1983).

The uniaxial unconfined compressive strength test is the most common test made on ice to determine its properties. It is usually easily performed on cylindrical or prismatic specimens.

A representative set of uniaxial unconfined compressive test results for −2° to −29°C has been reported by Butkovich (1959), Peyton (1966), Schwarz (1970), Saeki et al. (1978a), Wang (1979), Frederking and Timco (1984a), Timco and Frederking (1986), and Sinha (1983). Butkovich obtained compressive strength values ranging from 7.6 MPa at −5°C to roughly 12.0 MPa at −16°C from vertical cores of sea ice. He found average values of 2.1 MPa and 4.2 MPa for horizontal cores having the same two temperatures.

In tests performed by Imperial Oil Ltd. (Croasdale, 1974) ice crashing strength values from 4 to 6 MPa were measured.

Weeks and Assur (1967) proposed the following relation between compressive strength, σ_c, and brine volume, ν_b, of sea ice:

$$\sigma_c = 1.65[1 - (\nu_b/275)^{0.5}] \tag{2–3}$$

where σ_c is in MPa and ν_b is in ‰.

For a given creep strain rate Wang (1979) suggested the following equation for determination of the compressive strength of granular ice σ_c in MPa:

$$\sigma_c = 30\dot{\epsilon}^{0.22} \tag{2–4}$$

where $\dot{\epsilon}$ = strain rate in units of s^{-1}.

The curve (Figure 2–4) drawn through the data points from Frederking and Timco

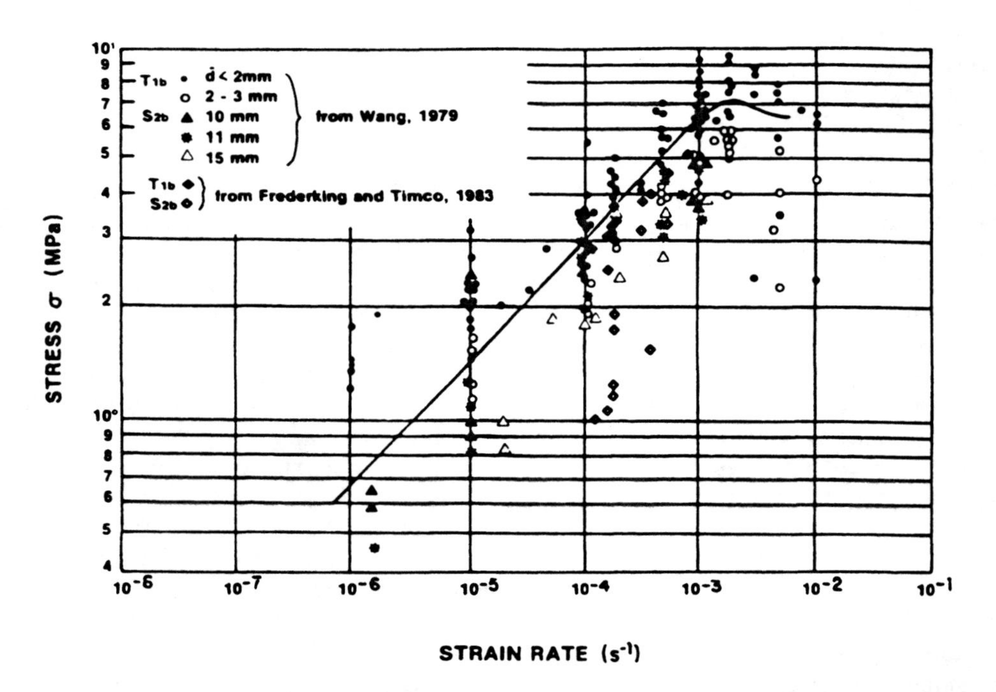

Figure 2–4. Crushing strength of sea ice for different strain rates. (From Nadreau and Michel, 1984.)

(1983) and Wang (1979) indicates the general dependence of strength on strain rate.

More results of sea ice compressive tests are found in Tsutae et al. (1983), and Det Norske Veritas (1984).

Comparison of miscellaneous unconfined tests indicated that the strength of the sea ice loaded vertically is approximately four times higher than strength of the same ice but loaded horizontally.

Some information on strength of a multi-year sea ice derived from over 200 unconfined compression tests on vertical ice samples is found in Mellor et al. (1984) and Cox et al. (1984a,b).

Relatively few confined tests on sea ice have been reported to date (Hausler, 1981; Nawwar et al., 1983; Timco and Frederking, 1984, 1986; Blanchet and Hamza, 1983). The major conclusion drawn was that although confinement conditions do not appreciably affect the strength of the ice, the stress level can be over four times higher than the strength of the ice with no confinement.

The following empirical values of an "ice-pressure resistance" for the German coastal region were proposed for inclusion in the EAU-90 (Hager, 1990):

North Sea ice:	1.5 MPa
Baltic Sea ice:	1.8 MPa
Freshwater ice:	2.5 MPa.

Tensile Strength

Uniaxial tensile strength is determined through a direct tension test.

Conducting a uniaxial tensile stress test is much more difficult than a compression test because of the difficulty in maintaining perfect specimen axiality and in avoiding perturbations of the stress field.

Direct-tension test results have been reported by Peyton (1966). Dykins (1970, 1971) obtained tensile strength results for both vertical and horizontal directions as a function of the brine content. Dykins' results showed that the ice was two to three times stronger when the tension was applied in the vertical direction rather than in the horizontal.

Schwarz and Weeks (1977) analyzed the results of Dykins and proposed the following relations for ice tensile strength:

$$\sigma_1 \text{ (vertical)} = 1.54[1 - (\nu_b/311)^{0.5}] \qquad (2\text{–}5)$$

$$\sigma_1 \text{ (horizontal)} = 0.82[1 - (\nu_b/142)^{0.5}] \qquad (2\text{–}6)$$

where the tensile strength, σ_1, is in MPa and the brine volume, ν_b, in ‰.

Saeki et al. (1978a) recommended the following equation for tensile strength as a function of temperature when the load is applied perpendicular to the grain structure:

$$\sigma_t = 95 + 50(-T) \qquad (2\text{–}7)$$

where σ_t is in kPa and T is in degrees Celsius.

Flexural Strength

It is not a basic material property but should be considered as an index value (Schwarz et al., 1981). It may be calculated as follows:

$$\sigma_f = 6PL/bh^2 \qquad (2\text{–}8)$$

where σ_f is the flexural strength, P the failure load, L the span, b the width, and h the thickness of the ice beam.

Ice flexural strength has been measured in several test programs involving both laboratory and field tests. The two tests principally used during such testing programs include simple beam and cantilever beam tests. In situ testing is considered to better preserve natural ice conditions; however, cantilever beam tests have been the subject of much controversy, especially concerning the problem of beam-root stress concentrations, variations in grain structure, steep temperature gradients, and scale effects.

Results on ice flexural strength tests have been reported by Määtänen (1976),

Butkovich (1956, 1959); Weeks and Anderson (1958), Weeks and Assur (1967), Dykins (1968, 1971), Frederking and Hausler (1978), Saeki et al. (1981b), Tabata (1966), Katona and Vaudrey (1973), Murat (1978), Saeki et al. (1978a), Lainey and Tinawi (1981), and Timco and Frederking (1982).

Nadreau and Michel (1984) have combined the work of Butkovich (1956), Brown (1963), Weeks and Anderson (1958), and Tabata et al. (1975) on flexural strength of sea ice as a function of brine volume in Figure 2–5. The data are represented by the following equation for $\nu_b^{0.5} < 0.33$:

$$\sigma_f = 0.75[1 - (\nu_b/0.202)^{0.5}] \qquad (2\text{–}9)$$

where σ_f is in MPa and ν_b is the brine volume (a number divided by 1000).

A similar expression is given by Dykins (1971), who tested large in situ beams with ice thickness of up to 2.4 m, as follows:

$$\sigma_f = 1.03[1 - (\nu_b/0.209)^{0.5}]. \qquad (2\text{–}10)$$

where σ_f is in MPa and the brine volume ν_b with values up to 0.012. Brine porosity, which can be derived from a knowledge of temperature and salinity profiles, can therefore be used to predict flexural strength.

Elastic (Young's) Modulus (E)

This is usually obtained by direct test of ice samples. The results of some tests, summarized by Voitkovskii (1960), show considerable scatter in the value of Young's modulus which ranges from 0.3 GPa to 10 GPa for static measurements and 6 GPa to 10 GPa for dynamic measurements.

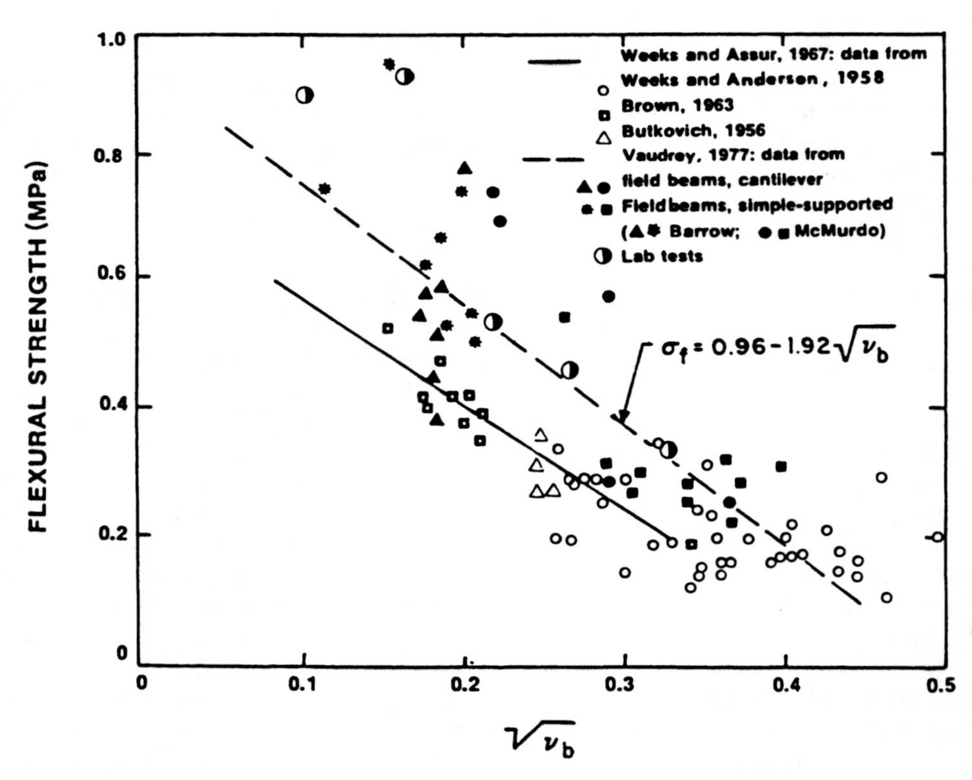

Figure 2–5. Flexural strength of sea ice. (Nadreau and Michel, 1984.)

As stated by Cammaert and Muggeridge (1988), the reason dynamic measurements of Young's modulus yield a larger value than do static measurements is that static tests measure strain after the application of a load and thus the deformation can be measured only after a finite time interval. Due to the viscoelastic behavior of ice, this interval is invariably sufficient to permit viscous, as well as the desired elastic, deformation. Therefore, a static modulus determination specified by the measurement of total deformation does not characterize the resistance of ice to an instantaneous elastic deformation. The greater the stress, the more important the viscous and creep components become. Hence, the value of the modulus calculated on the basis of strain measurements decreases as stress increases.

On the contrary, the value of modulus determined by dynamic methods is based on the rate of propogation of vibrations in the ice or by exciting resonant frequencies in small-scale specimens. The displacements in such measurements are extremely small, and elastic effects can be neglected.

Test results on dynamic measurements of ice Young's modulus have been reported by Weeks and Assur (1967), Kohnen (1972), Pounder and Stalinsky (1961), Abele and Frankenstein (1967), and Langleben and Pounder (1963). Butovich (1959), Tabata (1966), Dykins (1971), Schwarz (1975), Lainey and Tinawi (1984), and Cox et al. (1984a,b) reported static test result conducted on sea ice.

Traetteberg et al. (1975) have indicated that Young's modulus does not depend explicitly on the rate of strain or stress but only on the time of application of the constantly increasing load.

The comprehensive theoretical model of Young's modulus of ice is yet to be developed. It is however, very difficult, because as indicated by current experimental observations the realistic model of modulus of sea ice has to consider a variety of variable factors such as brine volume, crystal orientation, and porosity which in addition includes a random distribution of pores.

Porosity and brine volume specifically have a substantial effect on the variation of the modulus of sea ice.

Poisson's Ratio (μ)

This is defined as the ratio between the strain in two directions perpendicular to each other in simple tension. Murat and Lainey (1982) measured Poisson's ratio under various stress and test speed rates and found values of 0.5 and 0.35 to 0.40 for low stress and high test speed rates. They discovered that high ratio values decreased with temperature from 0.40 at $-5°C$ to 0.37 at $-30°C$ and $-40°C$.

The value of μ also appears to vary with the structural orientation of columnar ice. Wang (1981) found that saltwater ice was much stiffer in a direction parallel to the basal planes of columns (vertical) than in the perpendicular direction (horizontal) and resulted in a range from 0 to 0.2 vertically and 0.8 to 1.2 horizontally.

As long as ice deformation remains elastic, Poisson's ratio can be considered constant with a value of 0.3. With lateral confinement and plastic deformation Poisson's ratio increases to values of 0.8.

Weeks and Assur (1967) proposed the following equation based on Lin'kov's (1958) data:

$$\mu_D = 0.333 + 6.105 \times 10^{-2} \exp(-T/5.48) \quad (2\text{–}11)$$

where

μ_D = the dynamic Poisson's ratio
T = ice temperature in °C.

At very low stress rates, the effective values of Poisson's ratio approach the limit of 0.5, as predicted by Mellor (1983) from the equation

$$\mu = 0.5 - 0.167(E/E_0) \quad (2\text{–}12)$$

where

μ = the effective Poisson's ratio
E = the effective Young's modulus
E_0 = the true Young's modulus for zero porosity.

Additional information on Poisson's ratio is found in Saeki et al. (1981a) and Wang (1981).

Shear Strength

In many engineering problems there is a need to know the shear strength of ice. Shear is an ice property characterized by lateral movement within a material, that is, angular distortion or change in shape. Test results on sea ice shear strength have been reported by Butkovich (1956), Serikov (1961), Paige and Lee (1967), Lavrov (1969), Dykins (1971), and Frederking and Timco (1984b). The results obtained suggested that sea ice shear strength may vary in a range from 0.3 to about 3.4 MPa. It basically depends on ice salinity and ambient temperature.

The test results obtained by different investigators vary substantially.

Frederking and Timco (1984b) reviewed the previous work on shear strength of sea ice and noted that no apparent explanation was available for the discrepancy between the results previously reported.

Adhesion Strength

Adhesion forces primarily occur when an ice cover adfreezes to a structure and by horizontal and vertical movement of the ice field causes considerable forces to be transferred to the structure. It thus becomes an important parameter in the design of structures operating in cold regions.

Oksanen (1983) found that brine cells weaken the contact surface between ice and other materials as a result of imperfections at the boundary that act as centers of stress concentrations. This causes a dramatic breakdown in the adhesion strength even at low brine volume which seems to remain at a constant level with increased concentrations.

Static and dynamic friction coefficients for different materials in contact with seawater ice have been reported by Saeki et al. (1979) and are illustrated in Figure 2–6.

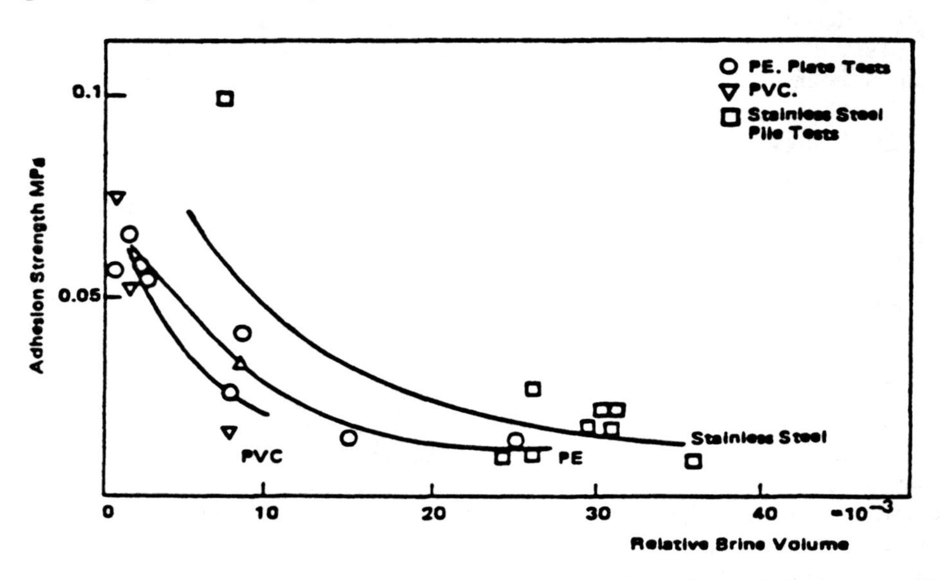

Figure 2–6. Adhesion strength as a function of the brine volume. (From Chakraburtty et al., 1984.)

Saeki et al. (1981a) found that the adhesion strength of saltwater ice to various piles was observed to have peak values at 0.1 to 10 mm/s and with 0.5 to 2.0 kg.cm^2·s with changes of pushout speed and stress rate and increased with ice thickness and decreasing pile diameter and ice temperature. The latter authors showed that the ahdesion strength of saltwater ice to painted steel piles was one-tenth of that to steel piles and much lower than the adhesion strength for other materials primarily due to the roughness of the pile materials.

Sackinger and Sackinger (1977) experimented with uncoated, untreated steel cylinders. The ice temperature in their experiments ranged from −1.5°C to −23°C, and the salinity varied between 0.4 and 20‰. Sackingers' results indicated a variation in bond strength between 250 kPa and 1.7 MPa; strengths generally decreased with an increase in salinity and a decrease in temperature.

Cammaert et al (1986) have reviewed the published information on adhesive strength and developed an analytical approach for calculating adfreeze loads on conical structures. Parameswaran (1987) reported results on adfreezing strength of freshwater ice to model piles made of various materials such as wood, steel, and aluminium.

Most adfreeze strength values are quoted (and measured) at high strain rates. In landfast ice areas, however, ice movement events start at slow strain rates, even if eventually quite high moment rates occur. This movement pattern is usually recognized when calculating forces against a vertical structure by associating low strain rate ice strength with a contact factor of 1.0 (see Section 2.4.3 for a discussion of the contact factor). Subsequent higher strain rates have associated with them lower contact factors, the implication being that breakout has occurred at the slow strain rates (with associated lower ice strength).

This approach can be used in considering adfreeze forces. For example, it might be assumed that the adfreeze bond is broken at strain rates as low as 10^{-5} or $10^{-6}\,s^{-1}$. However, no actual test data are available to support this assumption. According to Croasdale (1984) a conservative adfreeze strength of 700 kPa at high strain rates corresponds to a value of 100 kPa at low strain rates.

Fracture Toughness

Fracture toughness or critical stress intensity factor is a critical stress at which a crack in a brittle solid becomes unstable and propagates.

Irwin (1957) related the critical value of the stress intensity factor in the opening mode of failure, K_{1c}, to the strain energy release rate, G, as follows:

For plane stress,

$$K_{1c} = (GE)^{0.5}. \qquad (2\text{–}13)$$

For plane strain,

$$K_{1c} = [GE/(1-\mu^2)]^{0.5} \qquad (2\text{–}14)$$

where

E = Young's modulus

μ = Poisson's ratio

G for a purely brittle solid has a value twice that of the surface energy.

Vaudrey (1977), using a four-point bending test, obtained values of K_{1c} between 28 and 126 kPa m$^{0.5}$ for saline ice with a brine volume between 18 and 40‰ at test temperatures of −10° and −20°C. His results indicated that K_{1c} increased with decreasing test temperature.

Urabe et al. (1980) reported an average value of 70 kPa m$^{0.5}$ for sea ice of grain size between 3 and 30 mm.

Urabe and Yoshitake (1981) have summarized their experimental results as well as those of other investigators for pure and saline ice. They indicated that K_{1c} increases with increasing grain size. Plane strain fracturing tests in freshwater ice have been reported by Hamza and Muggeridge (1979).

Timco and Frederking (1983) used a four-point loading configuration to test samples obtained from the Beaufort Sea. The results indicated that fracture toughness was relatively constant at about 110 kPa m$^{0.5}$ in the top granular region of the ice cover, where the salinity ranged between 4 and 7‰. With increasing depth in the columnar layer, where salinity ranged between 3 and 5‰, the toughness of the ice increased to a value of about 145 kPa m$^{0.5}$. These tests were conducted at a loading rate of 10 kPa m$^{0.5}$/s and at a temperature of −20°C.

Urabe et al. (1980) and Urabe and Yoshitake (1981) reported values of K_{1c} for sea ice at −2°C from the Sea of Okhotsk.

Friction Coefficients

Static and dynamic friction coefficients for different materials in contact with sea ice have been reported by Saeki et al. (1979). Table 2–1 summarizes some of the available data. From this table it can be seen that the static-friction coefficients are appreciably larger than the dynamic coefficients.

Static coefficients are relatively independent of surface pressure, whereas dynamic coefficients decrease rapidly with increases in surface pressure up to 0.5kPa and then usually remain constant, with values between 0.04 and 0.11.

Thermal Expansion

The coefficient of thermal expansion of freshwater ice decreases slightly with decreasing temperature, the average value being 5×10^{-5} (°C^{-5}). The termal expansion coefficient of saltwater ice is similar to that of freshwater ice although much more variable and complex as a result of the variability of crystalographic brine entrapment at variable freezing rates combined with brine crystallization at various temperatures.

2.2.4.2 Freshwater and Low-Salinity Ice Characteristics

Compressive Strength

Many tests have been conducted to measure the uniaxial compressive strength of freshwater ice. These studies include those of Parameswaran and Jones (1975), Ramseir (1976), Frederking (1977), Michel

Table 2–1. Friction Coefficients

Test Material	Static	Kinetic	Temperature
Brackish ice (0.9‰)			
Enkvist (1972)			
smooth steel	—	0.025–0.045	−5°C
rough steel	—	0.11	
Snow			
smooth steel	—	0.09–0.19	
rough steel	—	0.31	
West snow			
smooth steel	—	0.03–0.10	
rough steel	—	0.14	
Sea ice (Finke, 1974)			
steel	—	0.045–0.065	−4.5°
Sea ice (Airaksinnen, 1974)			
steel	0.4–0.7	0.07–0.25	
Sea ice (Grothnes-Spork, 1974)			
steel	0.3–0.5	0.12–0.23	−7°C

From Frederking, 1983.

(1978), Sinha (1982), and Jones and Chew (1983).

The uniaxial compressive strength of freshwater and of low salinity ice types varies with strain rate temperature and porosity. It varies for nonsaline ice at $-5°$ to $-10°C$ by three orders of magnitude (0.01 to 10 MPa) as strain rate varies from about 10^{-11} to $10^{-2}\,s^{-1}$. Gammon et al. (1983a) found iceberg ice at $-5°C$ had a uniaxial compressive strength at failure that averaged 5.3 MPa at a strain rate averaging $10^{-3}\,s^{-1}$. Sinha (1984) found that the compressive strength of multiyear ice averaged 2 to 4 MPa at strain rates between 10^{-5} and $10^{-4}\,s^{-1}$, when loaded in the horizontal plane.

At high strain rates compressive strength is not highly sensitive to temperature whereas at very low strain rates, the variation with temperature can be deduced from the dependence of minimum creep rate on temperature.

According to Weeks and Mellor (1983) at high strain rates compressive strength decreases as porosity increases (density decreases). For low strain rates the effect of porosity is more indirectly related.

Grain size does not appear to affect compressive strength systematically at strain rates lower than $10^{-4}\,s^{-1}$. At high strain rates compressive strength is expected to decrease as grain size increases (Chakráburtty et al., 1984).

Tensile Strength

At low strain rates ($<10^{-7}\,s^{-1}$) there is little difference between tensile and compressive strength for isotropic ice. In both cases the ice yields by shearing, and the difference of normal stress does not seem to have much effect. According to Hawkes and Mellor (1972) critical strain rate ($10^{-6}\,s^{-1}$ at $-7°C$) is a bifurcation in the stress–strain rate curves for tension and compression, presumably because internal microcracks can form and affect the failure at high rates. At high rates ($\geq 10^{-5}\,s^{-1}$), tensile strength tends to a limiting value, while compressive strength continues to increase. For fine-grained ice at $-7°C$, the high rate limit of tensile strength is about 2 MPa, with compressive strength around 10 MPa.

The effect of temperature on tensile strength is the same as its effect on compressive strength at low strain rates. By contrast, the lack of sensitivity to strain rate at very high rates leads to the expectation that there will be a corresponding insensitivity to temperature at that range.

Similar to compressive strength, tensile strength decreases with decreasing density (increased air porosity) at high strain rates.

The grain size d has a considerable effect on tensile stress σ_t at moderate strain rates ($10^{-6}\,s^{-1}$). The effect can be described by the Hall–Petch relation:

$$\sigma_t = a + bd^{-0.5} \qquad (2\text{–}15)$$

where a and b are constants (Currier and Schulson, 1982). This type of behavior is expected to prevail at strain rates higher than $10^{-6}\,s^{-1}$, but at very low strain rates d may not have much effect. For practical purposes the value $\sigma_t = 2\,\text{MPa}$ for fine-grained ice can be regarded an upper limit. More coarse-grained nonsaline material that is encountered in glacier ice and lake ice will usually have tensile strength much lower than 2 MPa (typically 1 MPa or less).

Flexural Strength

Gow (1977) conducted flexural strength tests on freshwater ice and found that lower temperature structurally unmodified ice under simple supported, center loaded modes averaging from 0.5 to 1.5 MPa.

Butkovich (1959) conducted tests on glacial ice. He reported values of flexural strength at $-5°C$ as ranging from 1.1 to 3.2 MPa, with mean values of 2.1 to 2.4 MPa.

Shear Strength

Roggensack (1975) carried out direct shear tests on columnar grained ice with shear plane surface perpendicular to and parallel to growth direction. He found that shear stress (f_v) in MPa can be described by the following expression:

$$f_v = 0.7 + 0.47 f_n \qquad (2\text{–}16)$$

where f_n = normal stress in the range 0.5 to 1.4 MPa.

Extrapolating Eq. (2–16) to the case of zero normal stress, a shear stress of 0.7 MPa is obtained.

Elastic Modulus

For polycrystalline low and nonsaline ice of low porosity (density approaching 0.917 mg/m^3), high-frequency dynamic measurement of Young's modulus E gave values of approximately 9.0 to 9.5 GPa in the temperature range −5 to −10°C (Hobbs, 1974; Gammon et al., 1983b). As temperature decreases, E increases nonlinearly (Chakraburtty et al., 1984) but the effect is small for "true" Young's modulus (as opposed to "effective" values of E which inclue creep effects). Porosity, n, which can be expressed alternatively as bulk density ρ, has a significant effect on E and it is interesting to note that E drops sharply below the density which represents close-packing grains ($\rho \approx 0.55$ mg/m^3).

Poisson's Ratio

Poisson's ratio μ, as measured by dynamic tests, has values close to 0.3 for nonsaline ice of low porosity (Mellor, 1974), and there is not much variation with porosity over the range where the material is regarded as "ice" rather than "snow" ($\rho > 0.8$ mg/m^3).

Adhesion and Friction

The work of Barnes et al. (1971), Evans et al. (1976), and Oksanen (1980, 1981, 1983) presents results of friction of freshwater ice and theories explaining the results. Rate, temperature, material, and contact load can affect the results. The coefficient of friction for freshwater ice can be in the range of 0.01 to 0.1, dependent on the above-mentioned factors. Some results on friction of sea ice for several materials are summarized by Frederking (1983) in Table 2–1. They show that generally the friction coefficient of sea ice is higher than that for freshwater ice.

Measurements of adhesion of freshwater ice to a number of different materials is summarized in Table 2–2. It can be seen that ice adhesion to low contact angle materials (wood, metals, concrete) is higher than to high contact angle materials such as polymers. In general, adhesion is greater for ice

Table 2–2. Adhesion Strength of Some Materials to Ice

Material	Temperature (°C)	Strength (kPa)	Adhesion Source
PVC (1981)	−2.5 to −3	65	Frederking and Karri (1981, 1983)
PE	−2 to −3	59	
Concrete	−1 to −3.5	440	
Wood	−2 to −4	470	
Steel	−1 to −3.5	480	
Wood	−6	1380	Parameswaran (1981)
Concrete	−6	840	

From Frederking, 1983.

with minimal or no salinity than for ice with a high brine porosity. Adhesion also increases with loading rates.

2.3 ICE–STRUCTURE INTERACTION: TYPICAL PROBLEMS AND PRACTICAL EXAMPLES

2.3.1 General

The bottom-mounted marine structures in ice-covered ports and harbors are classified as either rigid or flexible. For example, gravity-type structures of various constructions are usually classified as rigid, and piled structures fall into the category of flexible structures. The combination of both types is also used. The type of the structure has significant impact on the character of ice–structure interaction and the approach for calculating of ice global and local forces, which depend on and are affected by the ice mode of failure.

Ice can fail in a variety of ways, including crushing, buckling, bending, shearing, and splitting. Even in vertical face structures, where crushing is normally expected, ice can fail in the buckling mode if it is relatively thin, or in the bending mode if ice rubble has formed in front of the structure.

Relatively recent experience gained in the Beaufort Sea suggested that even large-diameter gravity-type structures can experience vibration problems and that ice dynamic effects on marine structures should always be kept in mind.

The ice forces on rigid structures generated by continuous ice covers are commonly considered as pseudostatic.

The dynamic interaction between flexible structure and ice is usually more pronounced; it has been observed that structures having natural frequencies between 0.5 and 15 Hz may exhibit problems with ice-induced vibrations (Määtänen, 1980).

To analyze ice forces acting on flexible structures, it is necessary to obtain additional information on ice properties, such as the dependence of crushing strength on loading rate. It is also necessary to know the mass, stiffness, and damping characteristics of the structure, as well as damping effects due to water, foundation, and friction.

It should be noted that for vertical structures, dynamic effects due to ice bending or buckling failures may also be significant. In general, however, because ice force is dependent on the dynamic characteristics of both the ice and the structure, the analytical determination of ice forces on flexible structures is quite complex.

Therefore if the mode of ice–structure interaction is complex or the method of calculating ice loads is unreliable then a model test may have to be performed. As might be expected, the most reliable way to determine ice forces is to conduct a large-scale model test with verification against a suitable full-scale prototype.

The ice forces acting on any type of structures are affected by miscellaneous environmental parameters such as winds, currents, tides, and temperatures. These environmental parameters in some cases may represent the upper limit to ice loading when they are concentrated on a structure by large ice sheets.

Last but not least, the ice–structure interaction is affected by size and configuration of a structure, as well as ice movements around it.

From the viewpoint of the durability of the structure, the abrasion of structural materials due to ice action is an important parameter of the ice–structure interaction process to be considered.

As suggested by Itoh et al. (1988), the abrasion process of concrete without surface treatment can be divided into the following three stages, irrespective of the type of aggregate or the concrete strength. These are: *the surface region*, where only the cement paste on the surface of concrete is abraded; *the transition region*, where coarse aggregate is gradually exposed on the concrete surface; and *the stable region*, where coarse aggregate is also abraded.

Itoh et al. concluded that the concrete's wear rate can be represented by that in the stable region, because the wear depth in the stable region causes severe damage to structures.

Generally, the concrete wear rate depends mainly on ice temperature and contact pressure. As suggested by Itoh et al. relative velocity, concrete aggregate type, and compressive strength of concrete are not as critical for abrasion as ice temperature and contact pressure.

Surface treatment such as polyurethane resin lining, or resin mortar lining, which reduce the friction between structures and sea ice, is very effective for wear protection.

As a result of ice–structure interaction the structure may be exposed to various types of ice forces such as static and dynamic horizontal thrust, and/or downgrade load from weight of ice built up on a structure, or an uplift force due to effect of water fluctuation.

Depending on type of structure and the character of its exposure to ice either static or dynamic effects could be critical.

Ice static load is likely to occur in bays, large harbors, and shallow waters due to thermal stresses, and/or wind and current actions on ice sheets adhering to the structure.

On the other hand, wind- or current-driven large ice sheets, or bergy pieces of ice could deliver a heavy, dynamic thrust on a structure. As pointed out earlier, the water fluctuation under certain conditions may be responsible for the heavy ice built up on the face of a vertical wall structure, or growth of the ice bustles on piles.

This may create dangerously destabilizing downgrade or uplift forces, threaten to fail piles in compression or tension, or add an additional overturning load to the vertical face of a gravity-type structure.

Furthermore, an exaggerated ice growth on a vertical face structure may be sufficient to hamper ship berthing operations. Probably the best documented case of such an increase in ice accumulation due to tidal action is an experience of operation of the Nanisivik dock built at Strathcona Sound, Canada (Girgrah and Shah, 1978). It is discussed later in this section.

It should be noted that the review of the existing practice of design and construction of the marine structures in cold climate regions suggested that sometimes this practice suffers from a lack of knowledge and proper understanding of the mechanism of ice–structure interaction.

The problem is compounded because often for similar ice conditions different types of structures exhibit quite different modes of ice–structure interaction.

All the above, as well as practically a very limited body of experience, sometimes make it difficult to arrive at a sound engineering judgment.

The following are several typical examples of marine structure constructions in ice-affected water that can help provide a better understanding of the ice–structure interaction phenomenon.

2.3.2 Port of Anchorage, Alaska

The port was built in an area with an extreme tidal range of 12.8 m and in a body of water where ice floes 1.2 m thick carried by strong currents tend to impinge against the marine structures.

The port was built primarily to handle bulk petroleum products, containers, and Ro/Ro trailers. The terminal's berthing structures were built in the form of a concrete platform, supported upon the long unbraced steel pipe piles driven into rather poor foundation soils.

A typical cross-section of this structure is depicted in Figure 2–7.

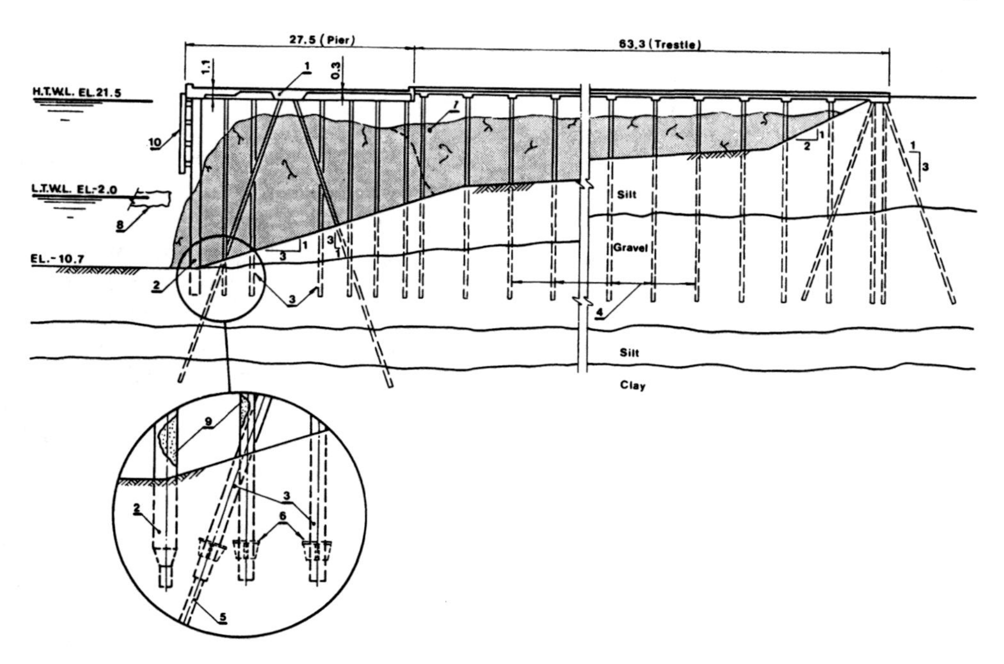

Figure 2–7. Port-of-Anchorage, Alaska. Typical cross-section through pier and trestle. 1—Concrete deck, 2—concrete-filled steel caissons 1067 mm in diameter, 3—concrete-filled steel piles 610 mm in diameter, 4—concrete-filled steel piles 406 mm in diameter, 5—H-piles 14 BP 73, l = 12.5 m, 6—steel collar, 7—ice buildup due to tide effects, 8—ice floe, 9—concrete fill, 10—fender system.

The design loads supported by the wharf foundation included a deck live load of about 3.0 tonnes/m^2; about 200 kg/m^2 snow loads; railroad, trailer truck, and crane loadings; vessel mooring and docking forces; the dynamic force of ice floes impinging on the dock structure; and the weight of about 6.1 m of ice surrounding the piles. Furthermore, the structure was designed to sustain earthquake forces.

To absorb the heavy impact forces from the current- and wind-driven ice floes and to withstand the docking impact forces of the vessels the outboard row of piles was made of 1067-mm diameter concrete-filled caissons, spaced apart in bents at about 4.06 m.

All other piles were 610 mm and 406 mm diameter steel pipes filled with sand. The pile heads were embedded in 1.2 m wide and 1.07 m deep concrete cap beams. The platform's deck slab was 305 mm thick.

The harbor bottom was dredged to 10.7 m below mean low water level. Therefore, many of the piles have an unsupported length exceeding 20 m to accommodate the dredge depth and extreme high tides. Cross-bracings have not been used because of the pressures they would sustain after ice buildup on the wharf structure due to ebb and flow of the tide which causes extensive ice buildup during the winter season.

Support of the piles, many of which sustain a load of about 110 tonnes in soils at the site posed unusual problems. The port site is a tidal flat consisting mainly of dark estuarine silt 60 to 20 m thick underlain by a very thick light gray stiff to very stiff clay. In order to develop a pile design load of 110 tonnes without having them excessively long they were driven into the dense material (mainly gravel) overlying the clay structures.

However, the layer of this material was reduced to as little as about 6.0 m in thickness as a result of dredging. Thus, to give the piles the required capacity in dense material, closed end piles enhanced with load bearing collars 0.9 m in diameter attached near the bottom of these piles have been used. The latter piles, designed to resist tensile forces, were driven deeper into the silty materials. Pile driving was controlled such that the tips of the vertical bearing piles stayed in the gravel high enough to reduce bearing load upon underlying compressible materials by spreading it upon a larger area. The bearing collars, placed about 1 to 1.5 m above the pile tip, were intended to enhance the pile load spreading effect.

Pile driving was carefully monitored to achieve desirable results. Subsequent pile load tests proved the actual bearing capacity at three times the design load.

The long (flexible) unsupported length of piles may have helped the structure to survive the 1964 earthquake which measured 8.4 on the Richter scale by giving it a longer natural period. However, the cracks formed in the deck at the tops of the batter piles may indicate that the main horizontal thrust due to the earthquake had been absorbed by these piles.

Extra-thick metal was used for piles to allow for corrosion losses, which was estimated at 1.6 mm over a period of 20 years. (The rate of corrosion in cold waters is quite low.)

As pointed out in Chapter 1, there are many reasons for corrosion of steel in the marine environment. First is the normal surface oxidation, rust. This is often caused by differential aeration, a condition created by variation in oxygen concentrations. Second is electrolysis, which requires an electrolyte or moist conductor, and water and particularly salt.

Corrosion of metals in the sea environment typically shows two peaks: one in the splash zone and the other just below low tide.

In Port-of-Anchorage the corrosion in a splash zone due to tide variation extended to nearly 15.0 m. To date the metal loss due to corrosion in this zone, although it did not seriously weaken the structure, did however constitute a substantial amount of the original corrosion allowance of 1.6 m. (The port began year-round operation in 1964 to 1965.) The relatively low rate of corrosion in the splash zone was attributed to the piles' encasement into the ice which in this case during about 6 months each year acts as a protective coating by eliminating factors contributing to the corrosion (Perdichizzi and Yasuda, 1978).

The extreme tidal range (12.8 m) is also beneficial because a widely distributed splash zone results.

Generally metal losses occur on all piles below mean lower low water with the highest rate of corrosion occurring near the mean lower low waterline. The higher corrosion loss near the low waterline would be expected because of the anodic area resulting from differential aeration.

Exclusive of the selective corrosion of the welds, this general corrosion is progressing at a rate of approximately 0.08 to 0.10 mm (3 to 4 mils) per year in the vicinity of mean low water (MLW) which is equivalent to a loss of approximately 1.5 mm (60 mils) of metal in 20 years.

The deepest penetration of the selective corrosion occurs at each edge of the weld metal, forming doubled grooves around the pile.

Where this type of corrosion of the weld metal has occurred the mean rate of penetration into the pile was approximately 0.4 mm (15 mils) per year.

It should be noted that the large splash zone at Port-of-Anchorage has militated against use of the cathodic protection. However, cathodic protection can be the most viable approach to preventing corrosion attack against welded joints below or in vicinity of MLLW.

The experience gained at the Port-of-Anchorage operation suggested that every attempt should be made during the construction of new piling to minimize field welds or splices in the zone between MLLW and the mud line.

At Port-of-Anchorage pile coatings was not considered because the ice forces and ice abrasion would peel or chip them off.

The major factors in design of the marine structures in the area were the tidal range, ice buildup, current- and/or wind-driven ice floes, and earthquake load.

Among other things the large tidal range affected terminal operations which include dock fendering and operation of a mooring system.

During the first few years of seasonal operation a camel fender system, stayed by cables, was used. This floating system had to be removed each fall so the ice flows would not destroy it. A design using truncated timbers has been used since 1965. This system is considered by port operatives as a compromise between the fendering that the ship owners desired and the rigidity that ice flows dictate.

The tide is responsible for heavy ice accumulation first around the piles and eventually it forms a body of ice as indicated in Figure 2–7.

The first icing occurs on the piles as the tide recedes, leaving a film of water. Once the ice lasts through a complete tidal cycle, each wave deposits another layer of ice in the splash zone as the tide levels change.

As explained by Pounder (1961), for most saltwater, a density thermocline will not form, so the body of saltwater will tend to be of uniform temperature and, therefore, unable to melt the ice accumulation when the tides submerge it.

Hence, gradually the ice layers grow outward from each pile along the full length of the tidal zone. Eventually the ice cylinders form around piles join forming a latticework of ice ribs between piles. Finally, the holes in the lattice are plugged (Figure 2–8). Growth rates for the ice cylinders around piles have been measured at over 25 mm per day.

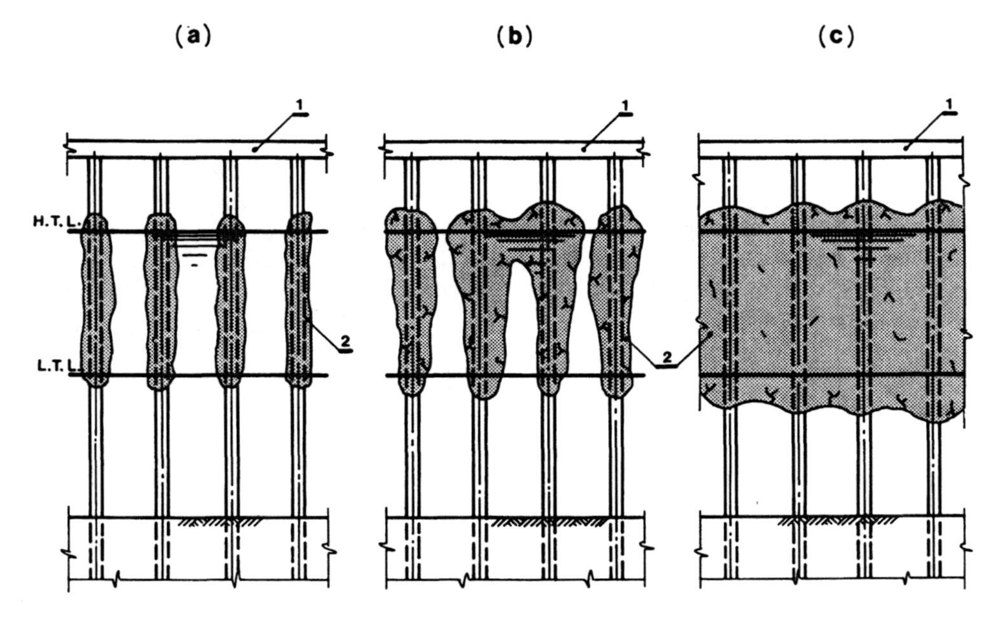

Figure 2–8. Stages of ice formation on piles due to tide effects. 1—Piled structure, 2—Sequence of ice formation at different stages.

Density is low due to air entrainment so the ice weighs about 560 to 610 kg/m^3.

Ice buildup on the piles was a major design consideration. Because the body of ice between piles is honeycombed, a design weight of 640 kg/m^3 was assumed, so the design ice load was about 4.0 tonnes/m^2. This was added to a dead load and a live load of 2.9 tonnes/m^2 for a normal design load condition. In fact the ice nearly doubled the load placed on each pile and ultimately the subsurface strata.

The uplift load on piles with ice cylinders adhered to them is typically limited by the adhesive strength of ice on pile material. In Port-of-Anchorage the design shear strength of the bond for ice adhering to piles was used as equal to about 830 kPa. From this assumption the most critical piles have been designed for the maximum uplift load of about 21.0 tonnes. This load was designed to be resisted by structure dead load and piles in tension.

Ice buildup may significantly affect the structure's response to earthquake. In Port-of-Anchorage the horizontal earthquake force of 10% of the dead and live loads was assumed to act at deck level, and 10% of the ice loads was assumed as a horizontal force acting at elevations between 7.6 m and 1.5 m.

The presence of ice, however, may help the structure to resist miscellaneous horizontal loads, including earthquakes. In Port-of-Anchorage a 6.1 m thick ice block provided encasement for all piles. The ice stiffened the structure considerably by providing lateral support to individual piles against lateral deflection. It was found that with this added stiffness vertical piles can absorb about 60% more and battered piles about 40% more of horizontal load. As noted before, the marine structures at Port-of-anchorage withstood the March 1964 catastrophic earthquake with just minor damages.

For more information on Port-of-Anchorage design and construction the reader is referred to Erkizian (1976) and Perdichizzi and Yasuda (1978).

2.3.3 Wharf at Godthab, Greenland

A pattern of ice buildup upon wharf structure similar to that at Port-of-Anchorage has been reported by Hulgaard (1985). The fishing dock 210 m long at Godthab, Greenland, has been built as a cantilevered timber truss bolted at 3.5 m to the face of a steep bedrock slope (Figure 2–9). The front piles, 300 mm in diameter, although they have some bearing capacity, nevertheless have not been considered as vertical load-carrying structural components. The tidal range at the wharf location was about 4.4 m. The 4 m thick ice buildup resulted from the tide fluctuation, nearly totally engulfing the structure (Figure 2–9a). Typically, the top and the bottom of the ice correspond to the high and low water levels, respectively, and the front face of ice is almost vertical. The seward ice cover moves vertically with tide along the vertical face of the ice confined to the wharf structure.

During the cold temperature season the ice adheres so well to the bedrock slope that despite substantial downward and uplift forces due to tidal fluctuation only a small fraction of these forces is transmitted to the structure.

During the spring the ice thickness is reduced. Simultaneously, however, the bond between the ice and the bedrock slope is severed when the ice is melted by water running down the slope. At a certain period of time during the spring season the ice loosens its bond to the bedrock completely and its still nonmelted portion (about 1.0 to 1.5 m thick) transmits vertical loads to the truss (Figure 2–9b).

In some cases this resulted in damages to the truss components and required some repair/maintenance work. The latter

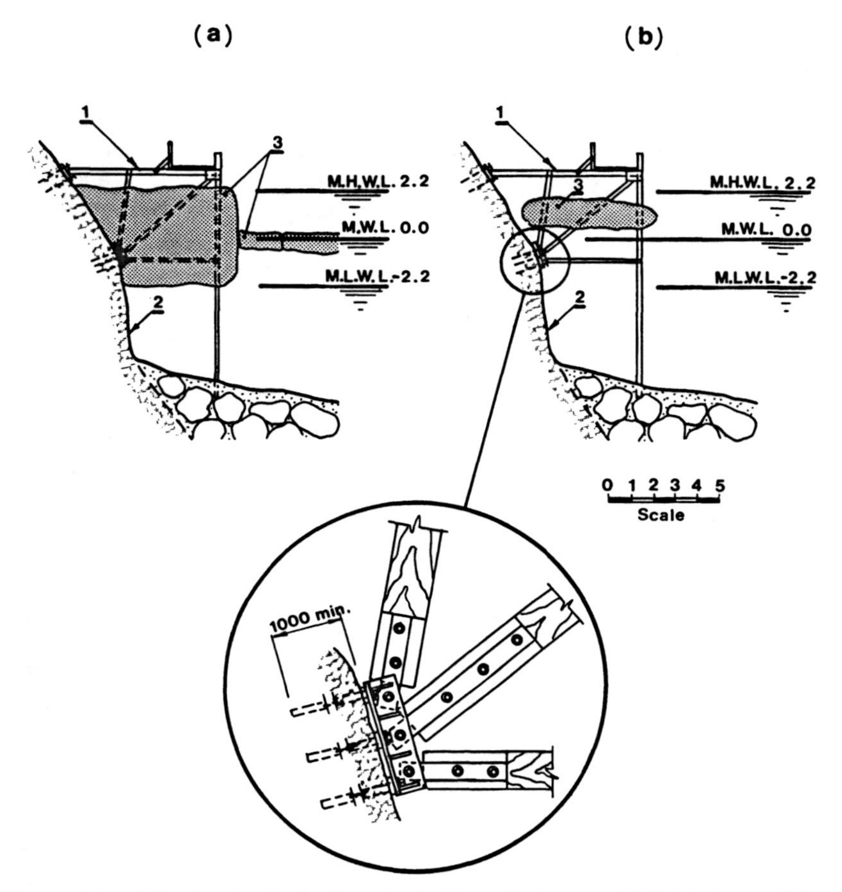

Figure 2–9. Wharf at Godthab, Greenland. (From Hulgaard, 1985.) (a) Ice conditions on February 1983 and 1984. (b) Ice condition on May 1983 and 1984. 1—Wharf structure, 2—Bedrock, 3—Ice.

included installation of additional anchor bolts and steel brackets to reinforce timber structural components. However, despite the effort to reinforce the structure about 40 m of its length sustained heavy damages and finally collapsed.

2.3.4 Wharf at Nanisivik, Baffin Island

In some cases, the tides cause the ice formed at the dock to effectively form an "active zone," which responds to the tides and acts as a hinge between the outer, larger ice sheet and the dock structure. For very low vessel frequencies, this active zone may build outward to a distance of several meters from the dock, as well as increase in thickness to several times the ice thickness due to repeated flooding. This ice growth at the dock may severely hamper the docking operations.

Probably the best documented case of such increased ice accumulation due to tidal action is a description by Frederking (1980a) and Frederking and Nakawo (1984) of ice behavior at the Nanisivik dock at Baffin Island, Strathcona Sound.

Strathcona Sound (73°N to 85°W), on the northern coast of Baffin Island, is one of several fjords branching off Admiralty Inlet. The wharf site is located on south shore of the sound, about 20 km from its outlet. The dock structure was designed to accommodate a 50 000 DWT bulk carrier.

Nanisivik is located in the zone of past and active glaciation and continuous permafrost with extreme cold conditions and relatively little precipitation (less than 150 mm per year). The annual mean temperature is −14°C with a high of 10°C in summer and low of −40°C in winter. The maximum tidal range is about 2.5 to 2.8 m, and the currents occurring in the sound are basically of tidal nature confined mainly to the upper 7.0 and 7.5 m layer of water. The magnitudes of current velocities are small, rarely exceeding 0.5 knots. The ice sheet when formed (typically between September to July) rides up and down on the water with tides. There is no apparent horizontal movement of the ice cover.

In shallow water along the shore the hinge is formed in the ice sheet at either nonfloating or partly floating ice. The maximum thickness of ice (measured in May) is around 2.0 m.

The berth structure was built in the form of three steel sheet-pile cells 21.3 m in diameter aligned in a straight line and spaced at about 38.1 m center to center. The cells were filled with coarse granualr material and completed with a 0.46 m thick reinforced concrete slab (Figure 2–10). The fender system at each cell was composed of three sets of large-diameter used rubber tires. Each set was made up of two types bolted together and suspended from the guardrail.

Minimum water depth at the outer edge of the cell was about 13.5 m.

The sheet-pile cells were installed from the ice which served as a construction platform. Details on cell construction are given in Girgrah and Shah (1978).

Observations made at the wharf by Frederking (1980a) and Frederking and Nakawa (1984) identified a typical pattern of ice conditions. Shortly after freeze-up an ice "bustle" begins to form on the vertical face of the cells. Seawater freezes and adheres in layers to the face of cells between the high and low tide levels, forming a parabolic shaped protuberance, extended seaward by about 1.0 m, by mid-November. Between the first-year ice sheet and the bustle there is an active zone 2 to 2.4 m wide, which is typically hinged at the junction with the natural ice and is moving up and down with the tide. It is occasionally flooded when the edge at the cells is tilted downward (against the bustle) during high tide.

By mid-winter (February) this flooding occurs during nearly every high tide and new layers of ice are grown not only on the horizontal surface, but also at the vertical face of the cells. The ice buildup at this face averages about 2 cm per tidal cycle. An active zone in general expands in width by about 3.0 m in the 90-day period from December to March, for an average rate of 1.8 cm per tidal cycle. The average density of the ice was found to be around 865 kg/m^3.

By the end of June the width of the active zone is usually over 11 m and its thickness is about three times that of the natural ice. At this point the active zone may float freely between the natural ice cover and the wharf (Figure 2–11).

As a result this very thick ice active zone presents problems for docking vessels early in the navigation season and may be a great obstacle for the extended season operation required in the future. Furthermore, in a year-round operation it may combine with ice buildup from repeated passages through the winter to result in a compounded problem, although it is quite likely that in the latter case the reduction in ice consolidation due to vessel traffic could lessen the problem due to tidal effects. The major problem would then be that due to accelerated buildup of ice from frequent vessel passages.

According to Frederking and Nakawo (1984) from considerations of buoyancy

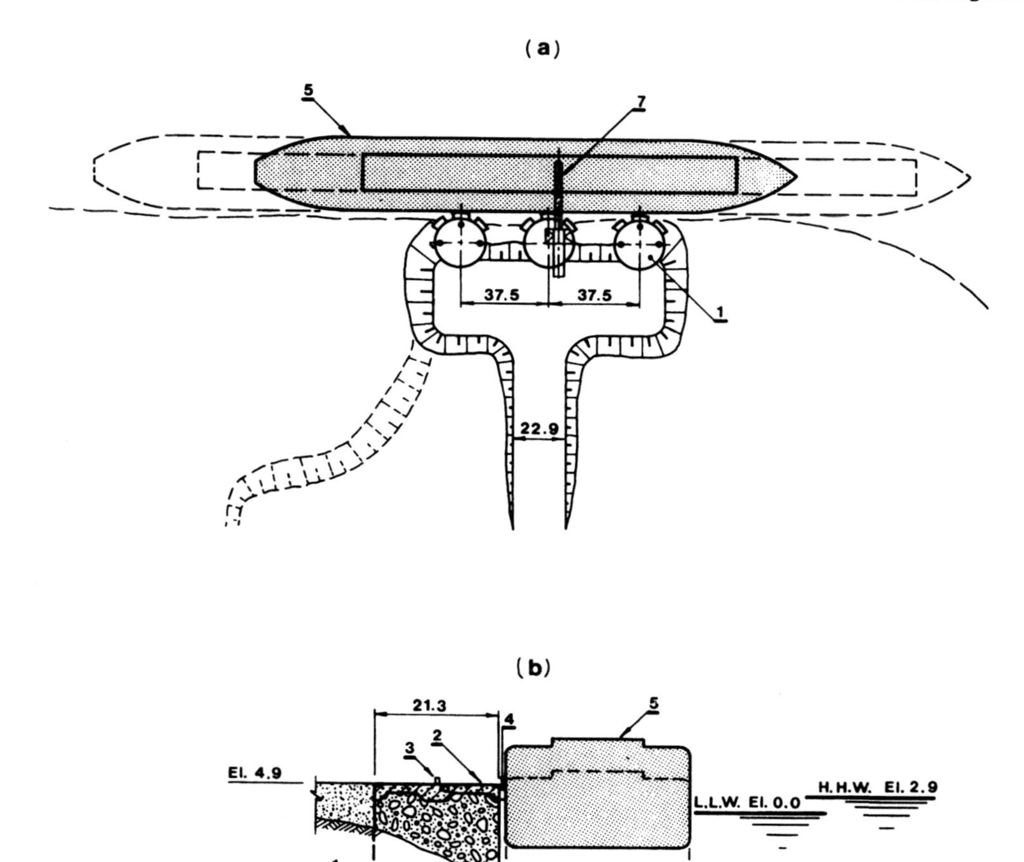

Figure 2–10. Wharf at Nanicivik. (From Girgrah, 1987.) 1—Steel sheet pile cell, 2—concrete deck 455 mm thick, 3—bollard, 4—fender, 5—ship, 6—original slope, 7—ore loader.

and the movement of the active zone during a tidal cycle, the ice load generated upon sheet-pile cell can reach values of about 6.9 and 27 kPa at high and low tide, respectively.

2.3.5 Offshore Oil Loading Terminal in Cook Inlet, Alaska

In the late 1960s several oil companies joined together to build offshore oil loading terminal at about 3 km from the mouth of the Drift River, Alaska. The terminal was designed to accommodate 30 000 DWT tankers in the winter and up to 60 000 DWT during ice-free periods. Environmental conditions at the terminal site are almost identical to those at Port-of-Anchorage.

Spring tides plus wind setup or draw down there range from −1.5 m to about 9.45 m referenced to a zero datum of MLW. Due to the large tidal prism contained by the long inlet, combined with a standing wave effect, the maximum tidal currents in inlet may attain velocities of up to 7.5 knots, with a

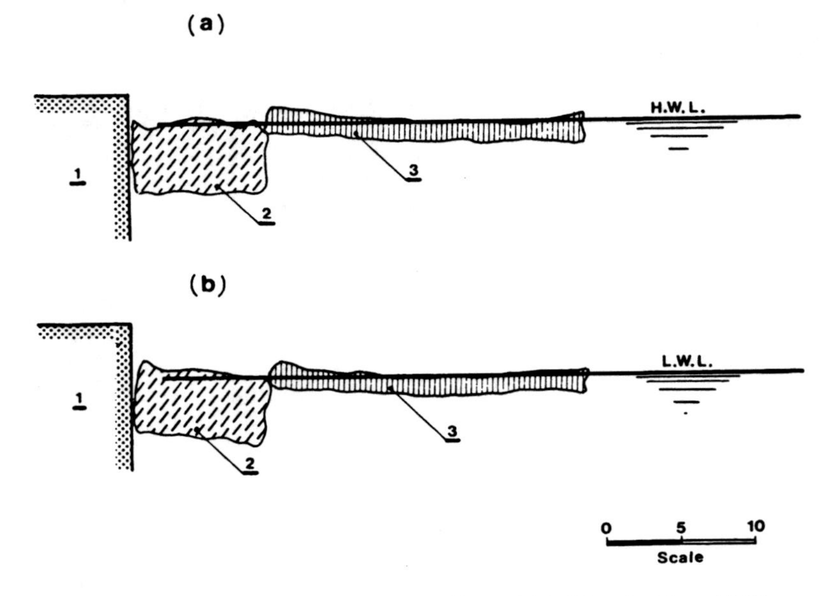

Figure 2–11. Wharf at Nanisivik. Ice profiles at high (a) and at low (b) tides. (From Frederking, 1980.) 1—Central cell, 2—Ice active zone, 3—first-year ice.

maximum value of about 6 knots at berth location.

While the currents in the inlet are primarily caused by the tides, the ebb may exceed the flood by 1 knot due to the addition of runoff water from tributary river and streams.

Furthermore, the terminal location is susceptible to sudden intense storms. The design minimum temperature is about −45.5°C. Due to the extreme range of the diurnal tides, the inlet does not freeze into a continuous ice sheet. Instead, large ice floes of about 800 m or more in diameter and about 1.1 m thick travel with tides at the velocity of currents. While tidal currents are the primary cause of ice movement, wind drag force on the ice surface could drive an ice floe at right angles to the direction of tidal flow. Thus dynamic ice forces could be expected on the berth facility from any direction. An additional feature of ice–structure interaction is vibrations caused by the brittle failure of the ice. The latter results in erratic ice cracking which imparts random vibrations to the structure.

To prevent ice buildup on structures similar to that at Port-of-Anchorage the structural elements are not so closely spaced as to permit the bridging of ice between adjacent legs.

The approach to design of the docking structures took into consideration breaking the ice floes in a crushing mode of failure. For this all structural components exposed to ice impact were designed in the form of steel tubular elements.

The berth included the following basic structures: loading platform and two breasting and two mooring dolphins, all linked with walkways (Figure 2–12).

To avoid ice pack formation between the ship and the fender system which would hamper the dock operation, a movable fender system was installed at breasting dolphins. The system remains a couple of meters above the water surface at all times and thus avoids direct engagement with ice, while remaining in a position to fender effectively an approaching or moored ship.

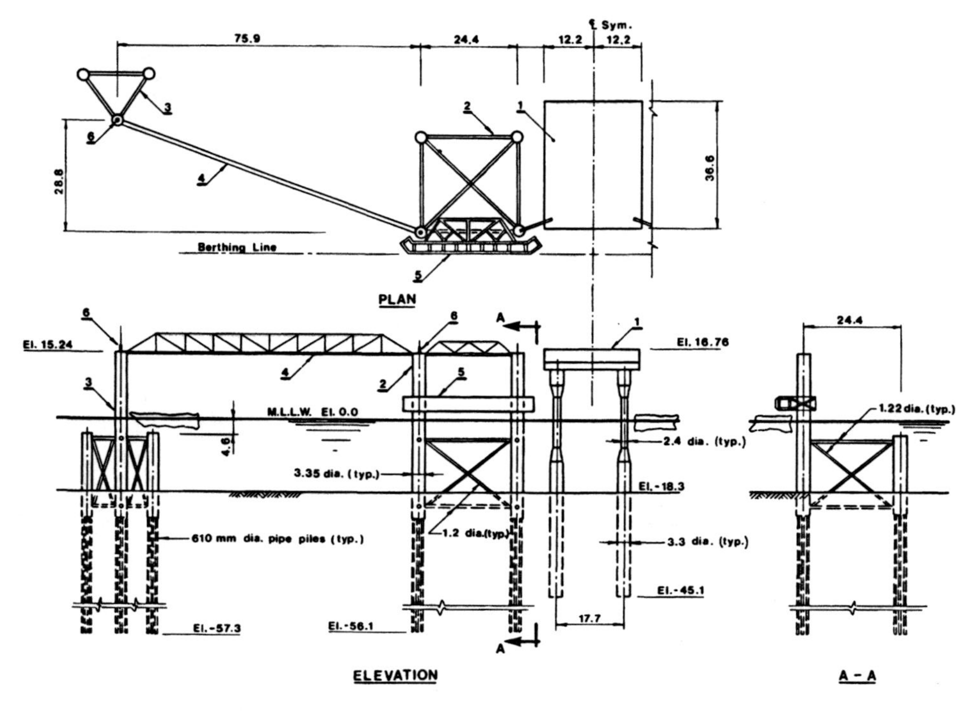

Figure 2–12. Tanker Terminal at Cook Inlet, Alaska. 1—Loading platform, 2—breasting dolphin, 3—mooring dolphin, 4—gangway, 5—movable fender, 6—quick release hook.

Typical views and cross-sections of the dock structures are depicted in Figure 2–13. It should be noted that the berth facility can be expended to accommodate 100 000 DWT tanker. The latter may be accomplished by adding two new breasting and two additional mooring dolphins.

The fender system is a crucial element in dock operation. It was designed to absorb docking energies of 30 000, 60 000, and 100 000 DWT tankers loaded with up to 30% with ballast water approaching the berth at the speed, normal to the berth's face about 30, 25, and 20 cm/s, respectively.

These relatively high velocities were selected because no tug assistance was considered.

Buckling type fender units with a special elastomer which maintains nearly constant elastic characteristics from the normal to the minimum design temperatures were used.

The fender units are equipped with Teflon shoes that slide on continuous Teflon surfaces attached to the dolphin caissons. Both the fixed and movable teflon surfaces are bonded to replaceable metal bars. Sliding contact surfaces are designed to be spaced sufficiently apart with adjacent structural components so that excessive ice buildup would not hamper their operation. In the event that ice would form on these surfaces, fender weight or fender lifting motor power are adequate to permit the sliding sections to act as scrapers and shear the ice off the surface.

The ice design criteria were based on a 100-year occurrence, which is considered to be an ultimate load situation. Therefore the allowable design stresses were based on the yield stress of the steel.

The design ice load which exceeds all other loads is assumed to result from a solid ice floe impact. The loading platform

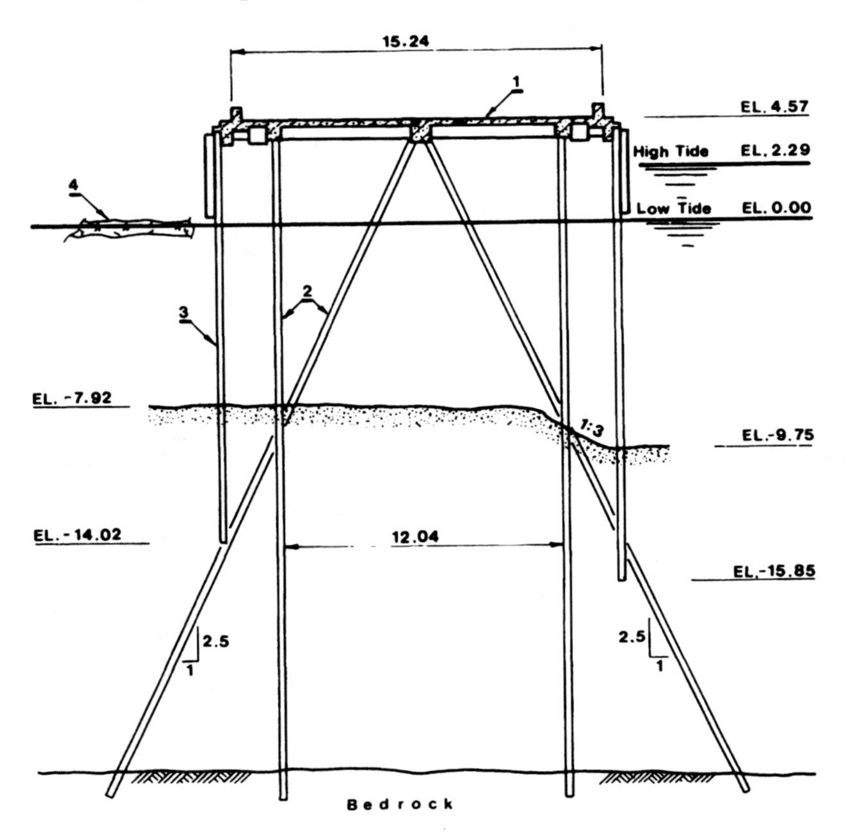

Figure 2–13. Pier at Caps Noirs, Quebec. (From Girgrah, 1987.) Typical cross-section. 1—Concrete deck, 2—concrete filled steel piles 356 mm in diameter, 3—Greenheart fender pile, 4—ice sheet.

is designed as a rigid three-dimensional frame able to resist ice loads impinging from any horizontal direction. The platform's legs are composed of steel caissons 3.3 m diameter driven to about 27.0 m below the sea floor, and steel pipes 2.44 in diameter extend through the ice shear zone to the platform's deck. The legs are placed 17.7 m apart in a direction along the berthing line and at 16.6 m perpendicular to the direction of this line.

The 760 mm diameter oil-carrying submarine pipe line was attached to the rear leg of the platform and then extended from within the leg to the top of the platform's deck.

The breasting dolphin consisted of four 3.35 m diameter steel legs braced below MLW and the ice level by 1.22 m diameter steel components. The upper parts of two front legs at the dock face above the water surface served as the guides for the movable fender system, housing for the fender hoisting equipment and as mooring points. The horizontal shear and overturning forces were resisted by means of 0.61 m diameter steel piles at the bottom of each leg.

These piles were driven through the temporary bulkheads placed inside of legs with the help of a pile follower. Once all piles were in position a tremie concrete seal was installed, tubes were dewatered, and all piles were joined together by means of shear plates subsequently welded to the tube inner wall. Finally, piles were totally encased in concrete. Mooring dolphins were very similar in construction to the breast dolphins. Further details on the Cook Inlet offshore oil terminal is given in Gaither and Dalton (1969).

2.3.6 Caps Noirs Wharf, Quebec

According to Girgrah (1987) this wharf was built in the form of a finger pier 15 m wide and 122 m long, and linked to the shore by an approach trestle approximately 213 m long. The depth of water alongside the pier was 9.8 m below MLW on one side and 7.9 m on the other.

The wharf and the trestle consisted of a reinforced concrete deck supported on 350-mm diameter concrete-filled steel pipe piles.

The typical pile bent is shown in Figure 2–13. Bents were spaced 5.3 m apart. The dock fender system consisted of greenheart piles with cylindrical rubber fenders sandwiched between the deck face and the fender piles. Although tide fluctuation in the area is about 2.3 m, no substantial ice formation on piles in the tidal zone similar to that observed at Port-of-Anchorage has occurred.

The pier was designed to resist ship impact forces, to support design surcharge deck load, and to resist all relevant environmental forces. However, the ice impact load and related ice–structure interaction apparently were underestimated.

An ice floe about 0.45 m thick and with an average diameter of 150 m and an estimated weight of about 16 400 tonnes broke off at low tide from an offshore ice cover and, driven by a strong wind, crashed against the pier structure at a speed of about 1.2 m/s, causing extensive damage to the structure. The ice crashing strength at the time of the accident was estimated as 2.6 MPa.

The post crash observation suggested that the piles actually failed without causing failure to the ice sheet in a crashing or fracture mode. The follow-up investigation of an ice–structure interaction led to the conclusion that the repair/rehabilitation procedure should include an installation of concrete-filled steel piles 0.75 m in diameter to replace damaged piles.

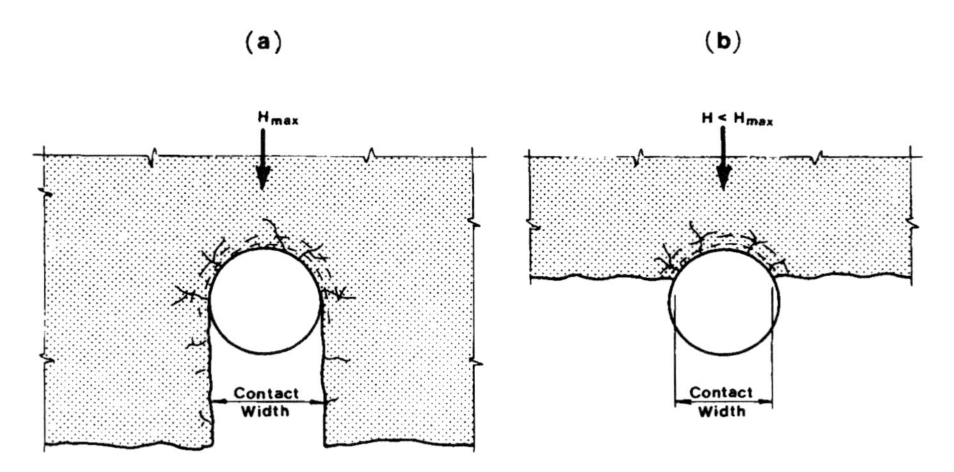

Figure 2–14. Illustration of steady-state and transitional state of ice loading. (a) Steady-state case of maximum constant contact width. (b) Transitional case of increasing contact width.

2.4 ICE FORCES ON STRUCTURES

2.4.1 General

As discussed earlier, ice acts as a medium to transmit such environmental forces as winds, currents, and tides to the structure.

If the driving forces are sufficiently high the structure may be fully enveloped in the ice feature (Figure 2–14a). In this case, the contact width is at its maximum value and subsequently the ice load exerted upon the structure reaches its maximum value (H_{max}). The ice–structure interaction may end at an earlier stage, however, if the driving forces are insufficient to overcome forces exerted by the structure upon the ice feature. In the latter case the contact width does not reach its maximum value (Figure 2–14b) and the ice floe may be stopped or change direction of motion without exerting a maximum load upon the structure.

The applicability of either of the aforementioned conditions depends on the ice–structure interaction scenario and the magnitude of the driving forces in comparison with the potential local forces between the ice and the structure. Examples of the above conditions may be movement of the first-year ice sheet against the structure, or interaction between structure and a large ice feature such as multiyear floe, ice ridge, or similar.

In the former case the ice sheet is usually large enough so that the structure can be fully embedded in the ice, and the ice load can be calculated from the local loads along the length (width) of the structure. In the latter case, if the driving forces are relatively small in comparison with the potential local forces the maximum ice force may not be reached. This would mean that use of

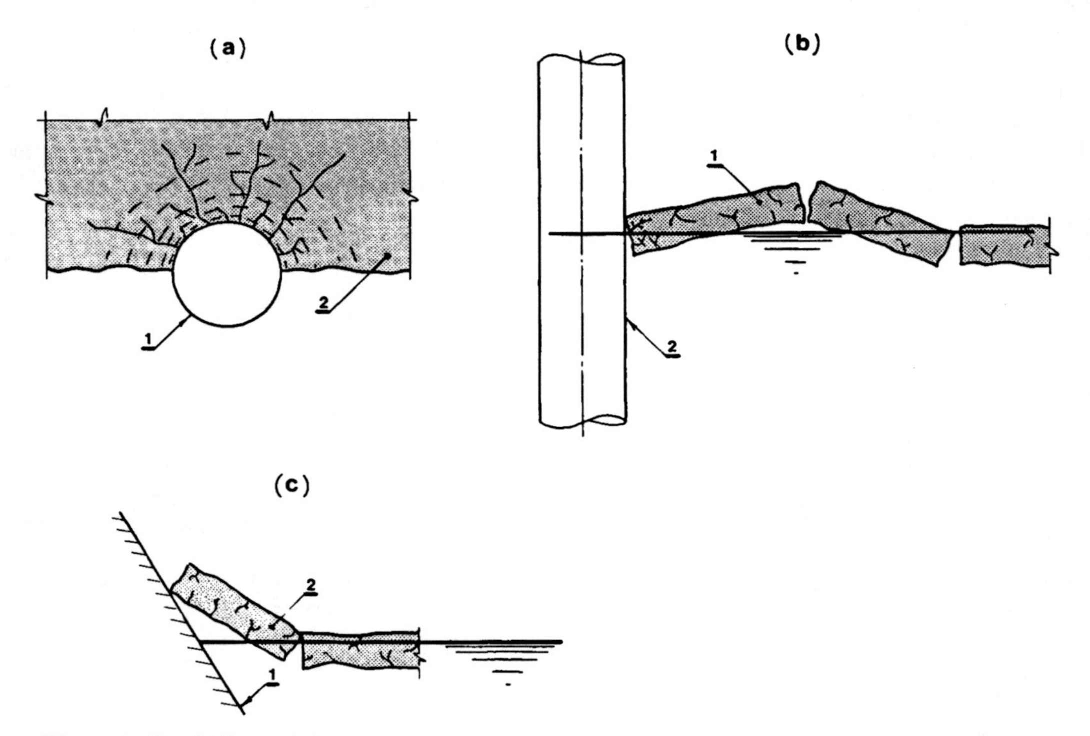

Figure 2–15. Different failure modes from ice sheet. (a) Crashing, (b) buckling, (c) bending. 1—structure, 2—ice sheet.

potential ice forces rather than those that practically could be reached could lead to the overestimation of the loads.

Furthermore, it should be noted that ice loads depend on the deformational behavior of ice (e.g., bending, buckling, compression, shear) which in turn depends on the geometry of the structure (e.g., vertical or sloping).

Vertical structures would typically include narrow vertical structures such as column (pile) supported platforms, as well as solid vertical walls. Sloping structures are those with sloping sides, intended to fail ice in bending.

Ice sheets or ice floes are common ice features to be considered in port marine structure engineering. When engaged with a strucutre they can crash or creep in the indentation mode (Figure 2–15a), buckle (2–15b), or fail in bending (Figure 2–15c).

Ice movement against a vertical structure is one of the most common ice–structure interaction scenarios in marine engineering.

On the basis of field observations and test results Sodhi and Nevel (1980) concluded that for ice sheets that collided with vertical structures the occurrence of the crushing or buckling mode of failure depends on the aspect ratio (D/t), where $D =$ width of the structure and $t =$ thickness of ice sheet. Buckling was observed to occur for higher aspect ratios (generally greater than 6) whereas a crashing failure is characteristic for aspect ratio smaller than 6.

Ice floes, depending on size, geometry, and strength may fail in crashing mode of failure in an area of contact with a vertical structure, bending in horizontal plane (Figure 2–16a), and shear in vertical planes (Figure 2–16b).

As noted by Nessim et al. (1987) " . . . under the assumption that failure due to global fracture does not occur, the main question is to estimate the stresses in the ice during indentation." The latter can be achieved by proper modeling of the ice–structure interaction as well as of the constitutive behavior and size effect in ice.

In the general case the irregularity of the properties and morphology of ice make the assumption of a single failure mode difficult, particularly for the case of wide structures, where several failure modes may be in effect at the same time or separately (Kry, 1978).

For example, the initial crashing mode of ice failure from a wide structure may result

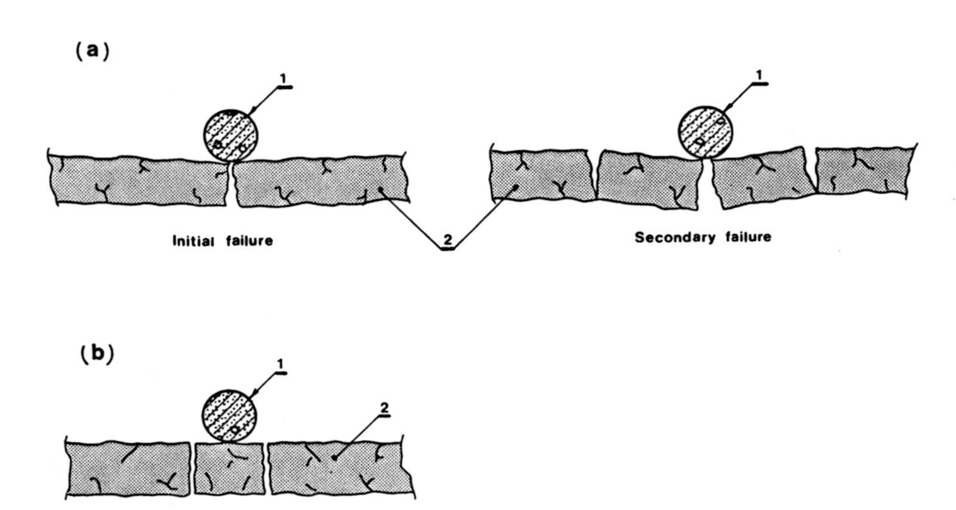

Figure 2–16. Ice floe failure modes on narrow vertical structure. (a) Bending in horizontal plane, (b) shear on vertical planes.

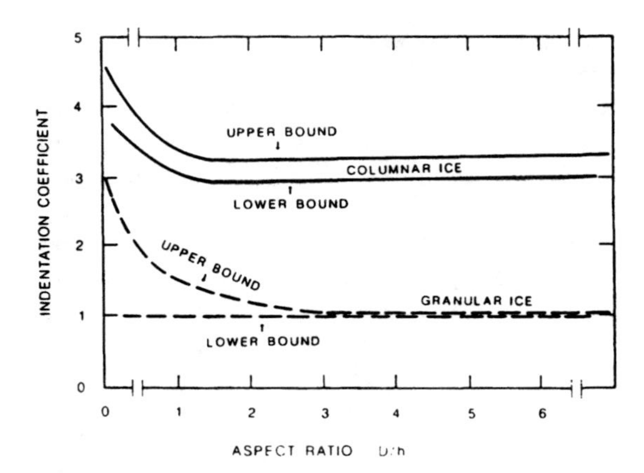

Figure 2–17. Indentation coefficient versus aspect ratio for columnar and granular ice. (From Ralston, 1978.)

in rubble pile buildup. This may follow with an ice sheet failing in the bending mode.

Comprehensive reviews of the computation of ice forces on vertical structures have been published by Neill (1976), Michel (1978), Croasdale (1980b), Machemehl (1983), ASCE (1983), Fenco Engineers (1984), Det Norske Veritas Canada (1984), Sakhuja (1985), and Cammaert and Muggeridge (1988).

Jordaan (1987) reviewed numerical and finite element techniques used in calculation of ice loads exerted upon marine structures.

Empirical, analytical, and numerical approaches are typically used to determine ice forces against marine structures.

2.4.2 Environmental Driving Forces

Ice motion in rivers and natural harbors which is typically associated with spring breakup begins with local melting and weakening of the ice cover. The following massive ice movement is typical in many rivers and large sea harbors.

Ice motion in large bodies of water (e.g., natural harbors or lakes) is caused mainly by wind friction. The air friction force exerted on an ice cover (T_w) can be calculated as follows:

$$T_w = \rho C V_w^2 A \qquad (2\text{–}17)$$

where

ρ = the air density; $\rho = 1.34\,\text{kg/m}^3$ at $-10°\text{C}$

C = the drag coefficient at the 10-m level. An average value of $C = 0.0022$ for a rough ice cover could be used (Croasdale, 1980b). According to Tryde (1989) $C = 0.006$.

V_w = the wind speed at the 10-m level

A = the fetch area.

In rivers and navigational channels, water friction forces on underside of the ice sheet are the principal cause of ice motion. Michel (1971) developed the following equation to determine the water friction force (T_c) acting on a floating ice sheet.

$$T_c = 1.26\,\frac{\gamma_w V_0^2 n_1^{1.5} n^{0.5}}{H_0^{1/3}} \qquad (2\text{–}18)$$

where γ_w = density of water, and

$$n = \left(\frac{n_1^{1.5} + n_2^{1.5}}{2}\right)^{2/3} \qquad (2\text{–}19)$$

The notations are shown in Figure 2–18.

In Eq. (2–18) and (2–19), n_1 and n_2 are Manning's roughness coefficients, which vary considerably. For example, $0.005 < n_1 < 0.04$ is used, depending on conditions of ice formation and accumulation. More details on the values of n_1 can be found in Michel (1971), Larsen (1973), and Pariset et al. (1966). Extensive discussion on the values of the bottom roughness coefficient n_2 is given in Van Te Chow (1959). Note that in Eq. (2–18) and (2–19) SI units are used.

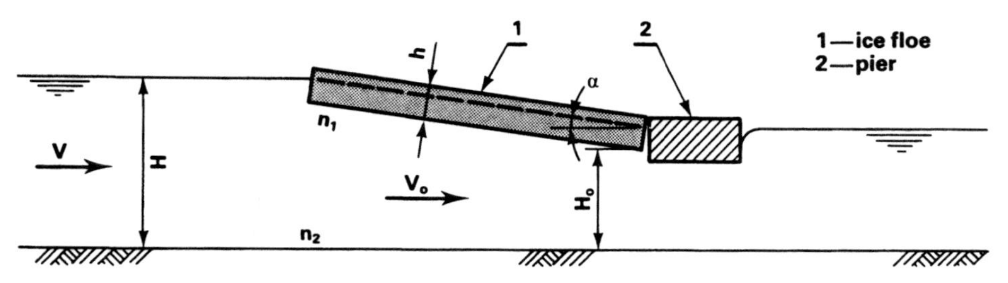

Figure 2–18. Driving forces on ice cover.

According to Tryde (1989) the current shear on underside of the ice sheet can be expressed as follows:

$$\tau_w = 0.006(0.5\rho_w)V_{wt}^2 \quad \text{kPa} \qquad (2\text{–}19)$$

where

ρ_w = density of water

V_{wt} = velocity 1 m below the underside of the ice sheet

The maximum value of environmental driving forces is reached when wind- and current-driven forces coincide. In real designs, the environmental external forces on the ice floe depend on many factors, such as ice accumulation, shape, and curvature of the river or navigational channel; geotechnical characteristics of the riverbanks; and other objects restricting ice movement. More information on this very special subject is given in Michel (1971) Eranty and Lee (1981), Fenco (1984), and in more recent work by Wake et al. (1987).

2.4.3 Ice Crushing Load

By definition the ice crushing process is the complete failure of granularization of the solid ice sheet into particles of grain or crystal dimensions; no cracking, flaking, or any other failure mode occurs during pure crushing. The immediate clearing of the broken ice by extrusion along the structure wall or indenter face follows the failure (Blanchet et al., 1989).

When the ice is sufficiently thick then crushing is one of the most common modes of failure, particularly for ice interacting with narrow vertical structures. Kry (1980) noted that the ice crashing mode of interaction follows the indentation problem, where a triaxial state of stress develops in the ice due to the confinement effect of the ice sheet in the ice–structure interaction area.

As summarized by Blanchet et al. (1989) an observation of indentation tests showed the following.

- First, microcracks are formed in the vicinity of the indenter.
- Immediately in front of the indenter, ice then separates into fine grains of dimensions less than 1% of the indenter width (Timco, 1986).
- The failed ice is ejected out of the contact zone and accumulates in front or on the sides of the indenter (Timco and Jordaan, 1987).
- In actural tests measured periods of crashing loads decrease from 1 s to 0.06 s with increase in penetration rate from 0.01 m/s to 0.21 m/s (Sodhi and Morris, 1986).
- The sizes of the ice pieces decrease with increase in the penetration rate but they do not vary with the aspect ratio (Kato and Sodhi, 1983).
- The recurrence period of crushing varies from 0.02 s to 1 s. (Sodhi and Morris, 1986; Timco, 1986; Timco and Jordaan, 1987; Michel and Blanchet, 1988).

On the premise that ice failure due to fracture does not occur, then the load exerted by the ice sheet due to crushing mode of failure (F_c) can be expressed as follows:

$$F_c = p\,D\,h \tag{2–20}$$

where

p = effective ice stress
D = width of indentor
h = ice thickness.

The essence of the ice force problem on narrow vertical structures is determination of the appropriate value of the ice effective stress p for the particular ice–structure interaction conditions.

Both empirical and theoretical methods are used to determine values of p.

In empirical methods, the effective ice pressure is usually related to the ice uniaxial compressive strengths by introduction of a number of coefficients to account for contact condition, indentation, and shape of the structure.

Korzhavin (1962) proposed the following empirical relationship (2–21) for p which according to Neill (1976) is valid for strain rates of 10^{-3} to $10^{-4}\,s^{-1}$ and gives values of the effective pressure in the range of $0.9\sigma_c < p < 1.6\sigma_c$.

$$p = If_c m\sigma_c \tag{2–21}$$

where

I = indentation coefficient which accounts for the confining (scale) effects on ice compressive strength.
f_c = contact coefficient which accounts for nonsimultaneous contact between the indentor and the ice feature. Essentially for perfect contact $f_c = 1.0$. Generally f_c is taken as $0.4 < f_c < 0.7$.
m = shape factor which takes into account shape of the indentor. It is usually taken as equal to 1.0 for flat and 0.9 for circular and semicircular indentors, and $0.85\,(\sin\alpha)^{0.5}$ for indentors with wedge angles of 2α between 60° and 120°.
σ_s = uniaxial unconfined compressive strength of ice.

I is a dimensionless number, and is the ratio of the observed failure pressure to the measured unconfined compressive strength for the same ice sample (e.g., for a sample the same crystal size, temperature, brine content or salinity, and so on). There are several different ways to calculate the value of I.

Assur (1975) recommended I to be determined as equal to $I = 1 + 2^{(1-D/h)}$, which gives a value of $I = 3$ at $D/h = 0$.

As the indentor size increases, the nature of the stress field in the ice more closely matches that of plane stress and the effective contact crushing stress is reduced (Ralston, 1977).

An alternative approach is found in Kry (1978) and Iyer (1983).

To some extent settling of f_c to less than 1 accounts for the brittle nature of the ice and its nonsimultaneous failure. Korzhavin (1962) recommended values of f_c depending on the velocity of ice as shown in Table 2–3.

Michel and Toussaint (1977) suggested that Korzhavin's formula be modified to account for strain rate. The effective strain

Table 2–3. Values of Contact Coefficient f_c

	Velocity of Ice	Floe Movement	(m/s)
Width of Indentor (m)	0.5	1.0	2.0
3–5	0.7	0.6	0.5
6–8	0.6	0.5	0.4

After Korzhavin, 1962.

rate was estimated as $\dot{\epsilon} = V/4D$ (which is a matter of convenience, as the strain rate varies behind the indentor) and used to adjust the maximum compressive strength in the ductile range, $\sigma_{c(\max)}$, according to empirical relations similar to that shown in Figure 2–4, as follows.

For the ductile zone:

$$p = If_c m \sigma_{c(\max)} (\dot{\epsilon}/\dot{\epsilon})^{0.32}$$
$$\text{for} \quad 10^{-8} < \dot{\epsilon} < 5 \times 10^{-4}\,\text{s}^{-1} \qquad (2\text{–}22)$$

where $I = 2.97$; $f_c = 1.0$ for full contact and 0.6 for continuous crushing or initial incomplete contact; $m = 1.0$ for a rectangular indentor; $\dot{\epsilon}_0 = 5 \times 10^{-4}\,\text{s}^{-1}$.

For the transition zone:

$$p = If_c m \sigma_{c(\max)} (\dot{\epsilon}/\dot{\epsilon}_0)^{-0.126}$$
$$\text{for} \quad 5 \times 10^{-4} < \dot{\epsilon} < 10^{-2} \qquad (2\text{–}23)$$

where $f_c = 0.25$; I, M, and $\dot{\epsilon}_0$ are as in Eq. (2–22).

For the brittle zone:

$$p = If_c m \sigma_{cb} \quad \text{for} \quad \dot{\epsilon} > 10^{-2} \qquad (2\text{–}24)$$

where

$I = 3$

$f_c = 0.3$

$m = 1.0$

σ_{cb} is the uniaxial crushing strength under brittle conditions.

More information on effects of such factors as temperature and strain rate on effective strength of ice is given in Vivatrat and Slomski (1983) and Sinha (1981).

Results of relatively recent significant field experiements reported by Danielewicz and Blanchet (1987), and Johnson and Benoit (1987) suggested that the highest real ice loads can be significantly lower than those measured at smaller scale or predicted by theoretical models. Furthermore, the existence of a pressure-area relationship has been confirmed in which nominal contact pressure between the ice and indentor decreases with increasing contact area.

Croasdale et al. (1977) applied the theory of plasticity to the solution of indentation problem. The latter authors suggested that the value of the effective stress can be determined by the following expression:

$$p = I\sigma_c \qquad (2\text{–}25)$$

where I = indentation factor which can be determined from upper and lower bound solutions.

The expression (2–25) assumes perfect contact between the structure and the indentor. Assuming a mouth flat indentor, the problem reduces to the classical Prandtl indentor, for which $I \approx 2.57$. For homogeneous and isotropic ice, provided that D is much larger than h, the lower bound solution becomes equal to 1.0. Between these limits, the problem is three-dimensional where I depends on the aspect ratio, D/h, as follows:

Rough contact:

$$I = 1.45 + \frac{0.35}{D/h} \leq 2.57. \qquad (2\text{–}26)$$

Smooth contact:

$$I = 1.15 + \frac{0.37}{D/h}. \qquad (2\text{–}27)$$

A number of other theoretical solutions have been proposed. These included use of a four-parameter generalized von Mises criterion to analyze randomly oriented columnar ice (Ralston, 1978; Figure 2–17), the reference stress method (Ponter et al., 1983) for low strain rate problems, and fracture analysis for high strain rate problems (Miller, 1980; Palmer et al., 1983; Hamza and Muggeridge, 1984).

The variation of contact pressure on the local scale of ice–structure interaction may be determined by nonlinear finite element

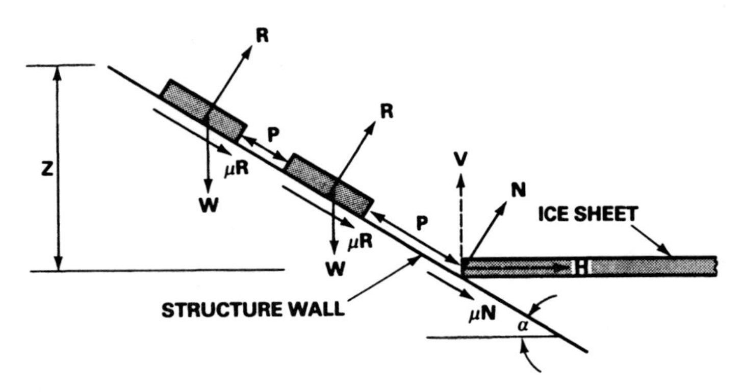

Figure 2–19. Interaction between ice and a sloping structure. (From Croasdale, 1978.)

formulations (Bercha, 1986; Walden et al., 1987).

The ice crushing failure at a wide vertical structure ($D/h \geq 5$) is somewhat different from crushing against a narrow vertical indentor. In wide structures because of the rugged surface of the ice sheet, the failure of the ice sheet edge does not take place simultaneously along its entire length, but only in a few areas along the edge.

The latter results in a dramatic decrease in a global ice pressure upon structure. Bercha (1987) provides the most recent review of the state-of-the-art on the effect of the structure size on ice pressure. A comprehensive discussion on mechanism of wide structure–ice interaction is given by Fenco (1984).

It should be noted that at present the use of empirical data has become the norm when it is desired to estimate ice loads on wide vertical sided structure. As a rule of thumb Bercha (1987) recommends computation of the first-year ice sheet load on a wide structure by multiplying ice contact area (Dh) by 1.75 MPa. In the latter case D is the width of a structure if it is smaller than the ice sheet; on the other hand it may be considered as effective length of the ice in contact with a structure in case an ice sheet is smaller than the structure.

Ayoub and Brown (1991) suggested that the magnitude of the ice force acting on a structure during an ice–structure interaction is very sensitive to the crack orientation. The latter investigators stated that their proposed fracture-mechanic based model is able to predict stresses similar to the ranges of the actual values measured on Hans Island (Danielewicz and Cornett, 1984) and also produces results that are in a good agreement with Timco's (1988) experiemental tests.

As stated earlier, ice crushing load is basically characteristic for the vertical face structure–ice interaction. However, this kind of load can also exist under condition where ice sheets interact with inclined structure with angles to the horizontal plane of greater than 75° (Michel, 1978). For structures with inclined surfaces the vertical component of the ice crushing load should be considered. Danys and Bercha (1975) provide formulas for calculating horizontal and vertical ice forces on conical light piers. In general, if the docking facilities are unshielded, ice pressures on the order of 2100 kPa may result. On the other hand, walls with faces sloped greater than 25° from the vertical will reduce moving ice loads by a factor of up to 4 (PIANC, 1984). The most recent review of the state-of-the-art on local ice loads (pressure) are given by Iyer (1989).

2.4.4 Loads Due to Ice Buckling Mode of Failure

As indicated by Blanchet et al. (1989), buckling results from the elastic instability of the ice sheet under a compressive horizontal load when the penetration rate is above the creep range. For elastic buckling to occur, the yield strength of the ice must be greater than the critical buckling strength.

Ice buckling load typically is not critical for the design of port marine structures.

Ice buckling may occur in early season when ice is rather thin. It may buckle, even under the full mobilization of compressive strength. The buckling load, F_b, when distributed over structure width D, is given by Sodhi and Hamza (1977) as

$$F_b = k\ell^3\{(D/\ell) + 3.32[1 + (D/4\ell)]\} \qquad (2\text{–}28)$$

where

k = foundation modulus (equal to the weight density of water)

ℓ = characteristic length, equal to $[Eh^3/12(1-\mu^2)k]^{0.25}$; where E = elastic modulus of ice

D = width of structure

μ = Poisson's ratio

Equation (2–28) essentially assumes that thickness of the ice sheet is uniform through the characteristic length ℓ and the elastic modulus is constant through the ice thickness.

It also assumes a floating ice sheet as a beam or plate resting upon an elastic foundation, which is conditionally valid as long as the top surface of an ice sheet is not submerged below the water, or the bottom surface does not emerge out of the water. However, these assumptions may be considered as adequate because the deflections of an ice sheet in general are assumed to be small.

Sodhi (1987) provided a comprehensive discussion on theoretical and experimental work on ice buckling loads on structures.

The major contributions in this area consist mainly of a few experimental studies (Michel and Blanchet, 1983; Sodhi et al., 1983; Sodhi, 1983; Sodhi and Adley, 1984) and theoretical analyses (Sodhi and Hamza, 1977; Kerr, 1978b, 1980; Sjölind, 1984, 1985) of the sheet ice buckling phenomenon.

It should be noted that most of the experimental work related to ice forces due to the buckling mode of failure of the ice sheet has consisted of small-scale studies to verify miscellaneous theories. In general, experimental studies confirmed an overall agreement between buckling loads from experiments and linear buckling analysis. Sodhi (1987) concluded that estimation of ice forces as a result of the buckling mode of failure of an ice sheet can be made with a fair degree of confidence if ice–structure interaction leads to this mode of failure.

2.4.5 Horizontal and Vertical Loads Due to Ice Bending Mode of Failure

These loads are characteristic for the sloping or cone-shaped structures which tend to induce bending failure in an ice sheet (Figure 2-19). The resulted ice load is substantially lower as compared to the case of the ice sheet failing in the crushing mode.

Actually, two basic loading conditions can result in ice bending failure. The first one may occur because of ice compressive loading in combination with ice sheet natural asymmetry and eccentricity of the load which results in tensile stresses in the ice sheet. The second process occurs when a vertical load is applied to the edge of an ice sheet as a result of riding up (or down) the side of a sloping (conical) structure. The low tensile strength and limited ductility of ice result in formation of cracks that break the sheet into blocks typically four to five thicknesses in diameter (Blanchet et al., 1989).

As an ice sheet approaches a sloping structure, at initial contact it begins to crush at the interface (on the underside of the ice sheet for an upward breaking slope). The resulting interaction force, acting normal to the face of the structure, has a vertical and a horizontal component. There will also be a frictional force along the slope. The vertical and frictional components produce bending in the ice sheet and the ice will fail in bending when these components are increased to a certain critical level. Once the ice sheet has failed the smaller pieces of ice are pushed by the advancing ice sheet and begin to ride up the face of the structure. This causes a larger interaction force to be generated, as additional force is required to push the broken pieces of ice up the structure.

On a wide structure the ice may get turned back on itself, creating additional ice on the slope of the structure, which may lead to creation of the ice rubble in front of the strucutre. The latter in turn may inhibit simple bending failure of the advancing ice sheet against the sloped structure.

Croasdale (1978) presented a two-dimensional analysis model of ice interaction with a sloping structure in which the horizontal H and vertical V components of the normal load N (Figure 2–19) are determined as follows:

$$H = N \sin\alpha + fN \cos\alpha \qquad (2\text{–}29)$$

and

$$V = N \cos\alpha - fN \sin\alpha \qquad (2\text{–}30)$$

where f is the friction coefficient. When the moment capacity of the ice sheet is related to corresponding vertical force required to initiate failure, the horizontal force per unit width of the structure is computed by

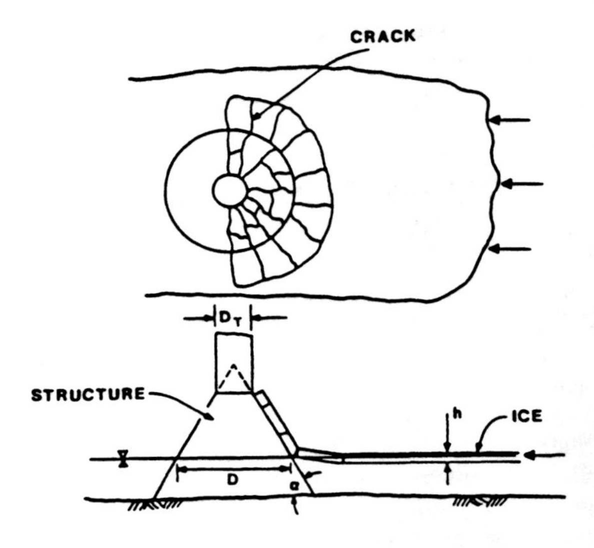

Figure 2–20. Three-dimensional representation of ice–structure interaction. (From Ralston, 1977.)

$$H/D = 0.68\sigma_f(\rho_w g h^5/E)^{0.25} C \qquad (2\text{–}31)$$

where

D = width of the structure
σ_f = ice bending strength
$\rho_w g$ = weight density of water
h = ice thickness
E = elastic modulus of ice
C = coefficient that relates H and V in relationship to α; i.e., $H = VC$ where $C = (\sin\alpha + f\cos\alpha)/(\cos\alpha - f\sin\alpha)$.

Once the ice has failed, the broken pieces start to ride up the face of the structure, and an additional force is experienced by the structure; the latter is computed from the following

$$H/D = C_1\sigma_f(\rho_w g h^5/E)^{0.25} + zh\rho_i g C_2 \qquad (2\text{–}32)$$

where

C_1 = 0.68 C
C_2 = $C(\sin\alpha + \mathrm{f}\cos\alpha) + (\sin\alpha + \mathrm{f}\cos\alpha)/\tan\alpha$
z = the maximum ride-up height
$\rho_i g$ = the weight density of ice.

As follows from the two-dimensional model of the ice forces acting upon the sloping structure the effects of friction and slope angle become significant above an angle of 45°. [Some examples are provided by Marcellus et al. (1987)]. The ice strength affects the ice breaking component but not the ride-up component and in two-dimensional elastic analysis, the ride-up force in a typical interaction is larger than the breaking force.

Ice thickness is the most significant parameter affecting ice loads on sloping structures. As follows from Eq. (2–32) in two-dimensional analysis the ice breaking component is proportional to about h^5 and the ride-up force is proportional to ice thickness.

For inclined narrow or conical structures the failure zone typically extends around the structure (Figure 2–20) and usually not all broken pieces of ice are riding up the slope. Typically they clear around the structure.

Edwards and Croasdale (1976), on the basis of model tests conducted on 45° cones with $f = 0.05$, suggested the following formula to predict a maximum horizontal force exerted upon conical structures:

$$H = 1.6\sigma_f h^2 + 6.0\rho_w g D h^2. \qquad (2\text{–}33)$$

In this equation the first term represents the ice breaking component, while the second is the ice clearing component.

Other empirical formulas were derived by Afanas'ev et al. (1971) and Tryde (1977).

Ralston (1977) used three-dimensional plate theory and plastic limit analysis to predict the horizontal H and vertical V forces exerted by ice failing in the bending mode upon conical structure. He suggested the following formulations for predicting of H and V:

$$H = A_4[A_1\sigma_f h^2 + A_2\rho_w g h D^2 + A_3\rho_w g h (D^2 - D_T^2)] \qquad (2\text{–}34)$$

and

$$V = B_1 H + B_2\rho_w g h_R (D^2 - D_T^2) \qquad (2\text{–}35)$$

where

A_1, A_2, A_3, A_4, B_1, B_2 = dimensionless coefficients; A_1 and A_2 are given in Figure 2–21 and A_3, A_4, B_1, and B_2 are given in Figure 2–22.

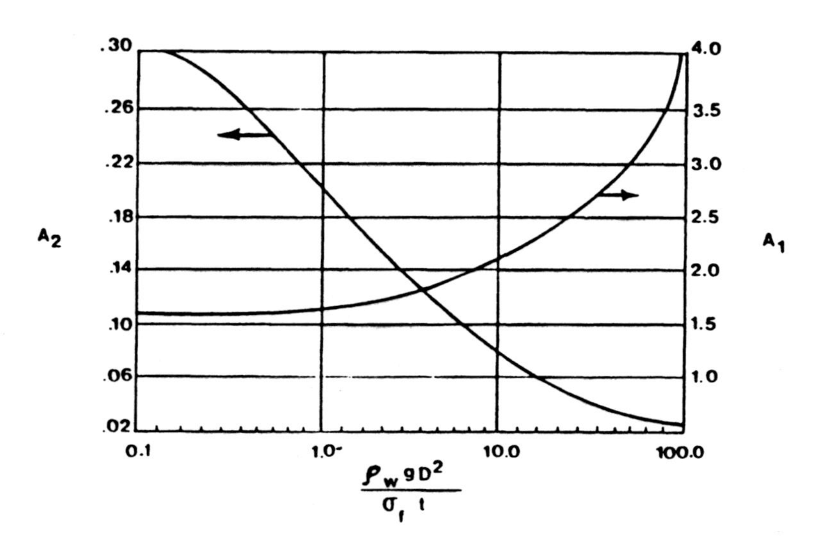

Figure 2–21. Ice force coefficients for forces on conical structures; A_1 and A_2. (From Ralston, 1977.)

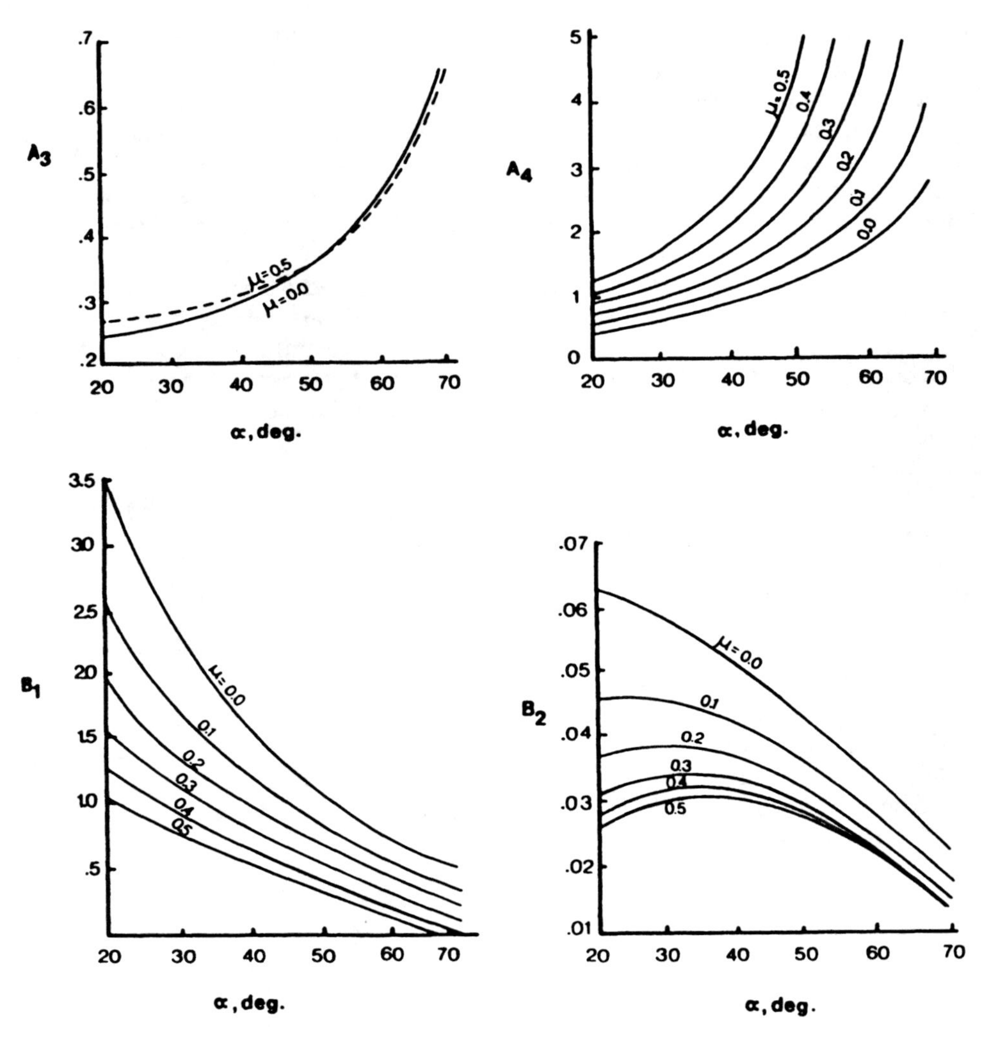

Figure 2–22. Ice force coefficients for forces on conical structures; A_3, A_4, B_1, and B_2. (From Ralston, 1977.)

$\rho_w g$ = weight density of water
σ_f = the flexural strength of ice
D = waterline diameter of the cone
D_T = top diameter of the cone
h = ice thickness.

The first term within brackets of Eq. (2–34) is due to ice breaking; the second is due to buoyancy and the third is due to clearing.

Although the above load formulations have been developed for an upward breaking cone, the same formulas and coefficients may be used to determine loads on a downward breaking cone. In the latter case ρ_w should be replaced by $\rho_w/9$ in the appropriate terms of Eq. (2–34) and (2–35). The factor of 9 accounts for the ratio between the forces to lift an ice floe out of the water and to submerge it completely into the water.

Ralston (1977a,b) reported reasonably good agreement between his prediction of load and the results of model tests of upward breaking cones, conducted by

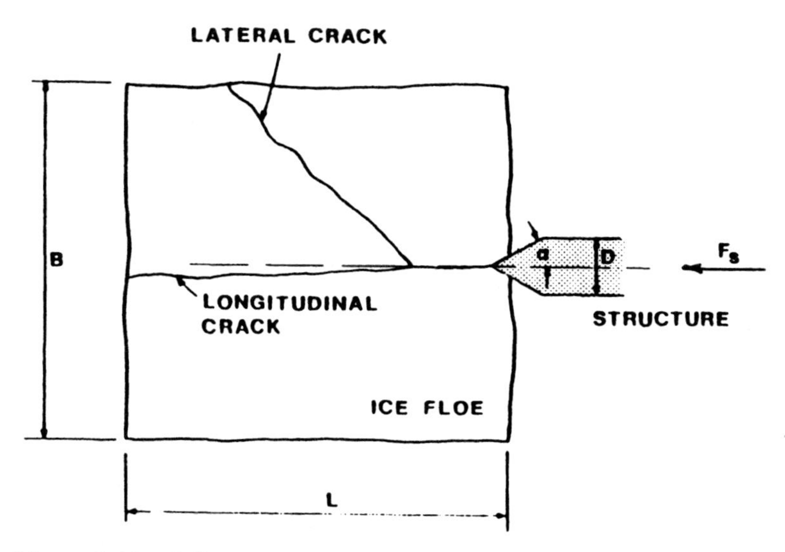

Figure 2–23. Splitting of an ice floe by a wedge-shaped structure. (From Michel, 1978.)

Afanas'ev et al. (1971) and Edwards and Croasdale (1976).

Ralston's (1979) analysis indicates that ice failing against a downward breaking slope does so at lower ice loads than upward breaking. He pointed out that this is primarily because the effective weight of the ice pieces below water is much smaller than the weight above water, and therefore clearing forces are smaller.

Other differences between ice failing against upward and downward slopes that might be accounted for include the following:

- The difference in flexural strength of the ice between upward and downward breaking
- The difference in ice-to-structure friction, recognizing that the upper surface of the ice usually has a snow layer.

Very little experimental data exist on downward breaking structures. Some preliminary expressions were reported by Abdelnour (1981). His experiments indicated reductions in forces by a factor of about 2, compared with upward breaking. Results of model tests on a downward icebreaking cone carried out in an ice tank are reported by Lau and Williams (1991). Effects of some parameters, for example, ice thickness, approach speed, and cone waterplane diameter, on ice load are discussed.

Most recently Nixon et al. (1993) published results of a total 57 tests carried out on model of a cable-moored floating platform with inverted conical geometry.

The basic findings from this study were that the forces on the structure increased monotonically with ice layer thickness; resonance effects and the process of ice collar formation around the platform complicated the variation of forces with ice speed such that a clear trend of platform reaction (mooring) forces with ice speed is not readily apparent; and that stiffness of moorings has a marked effect on a platform resisting forces.

More information on numerical estimation of ice forces acting on inclined structure is given in Lindström (1990).

When considering the appropriate value to use for ice flexural strength in ice loading equations, it is necessary to recognize that ice flexural strength can vary as a function of:

Salinity and temperature distribution
Strain rate

Crystal structure
The size of the ice beam (or thickness of the ice sheet).

Whether the same flexural strength should be used for upward breaking and downward breaking will depend on the salinity and temperature profiles through the ice. In this context it is also necessary to recognize that moment capacity is perhaps more relevant than flexural strength, as the position of the neutral axis within the ice will also vary with salinity and temperature profiles.

The issue of size effects in flexural failure is important because most real situations involve ice features that are much larger than have been tested in the measurement of flexural strength. It is not clear, however, whether a size effect exists in flexural failure. The ice flexural strength values discussed in the literature show a range from nearly zero to about 3.5 MPa, but it is clear that the high values have been obtained from small scale tests.

This trend is confirmed by Marcellus et al. (1987), who indicated that for large beams a flexural strength value of 0.7 MPa or less would be appropriate. The results also show that multilyear sea ice appears to be weaker than low-salinity columnar sea ice. The appropriate value of flexural strength to use in ice load calculations for typical multiyear ice features, with cross-sections 10 to 500 times greater than the largest tests to date, is still an issue for speculation.

For a comprehensive review on the state-of-the-art on ice sheet–conical structure interaction the reader is referred to Marcellus et al. (1987). Some additional information on the subject is found in Izumiyama et al. (1991).

Specific information on ice loads on sloping or cone-shaped structures is given in Lane et al. (1988), Frederking (1980b), Nevel (1972), Morrison et al. (1988), and Wessels and Kato (1989).

2.4.6 Forces Due to Adfreeze Mode of Failure

In addition to the predominant mode of bending failure, it is also necessary to consider adfreeze forces that may arise if an ice sheet has developed a bond with the sloping structure during a period of no ice movement.

The latter is characteristic for the near-shore arctic environment and inland waterways where the ice surrounding a structure can remain stationary long enough to adfreeze to the structure. There, the vertical motions of the ice due to tidal action or water level fluctuation can be so small that an adfreeze bond can develop between the ice sheet and the structure. Once the ice sheet starts to move again, sliding motion between ice and a structure first requires breaking the adfreeze bond. In the process ice adhering to marine structures may cause substantial horizontal and/or vertical forces on these structures. According to Cammaert et al. (1986) the load required to fail an adfreeze bond on a conical structure can be much larger than the load associated with bending failure.

As suggested by Cammaert et al. (1986), the load required to break the bond between the ice sheet and a conical structure over an angle 2Θ can be obtained from the following formulation.

$$H_a = C_a C_s D h \tau_a I / \sin\alpha \qquad (2\text{–}36)$$

where

C_a = 0.3 to 1.0 = adfreeze factor to account for incomplete bonding
C_s = 0.7 to 1.0 = stress factor to account for nonuniform stress distributions
D = structure diameter at the water
α = cone angle to the horizontal
I = elliptic integral which varies with α and Θ = failure zone angle. Some representative values of I are given in Table 2–4.

Table 2–4. Values of Factor I in Eq. (2–36)

Failure Zone Angle θ	Cone Angle				
	40°	50°	60°	70°	80°
75°	1.524	1.672	1.892	2.240	2.889
80°	1.612	1.760	1.981	2.329	2.987
85°	1.700	1.848	2.069	2.417	3.066
90°	1.787	1.936	2.157	2.505	3.152

From Cammaert et al., 1986.

Other models that consider effects of adfreeze forces on ice horizontal forces on marine structures have been proposed by Croasdale (1980b) and Gershunov (1985). It should be noted that all three approaches produce substantially different force values.

With regard to the adfreeze shear strength, several studies have been performed, usually on a very small scale, and typically at high strain rates. The range of values obtained from these studies is quite wide. For example, Sackinger and Sackinger (1977) measured about 0.5 MPa with sea ice and uncoated steel at −15°C. At higher temperatures, the adfreeze strength was somewhat less.

Saeki et al. (1981a) measured adhesion strengths for steel, concrete, painted steel, and corroded steel for sea ice at −2.0°C. Typical values for clean steel were in the range 0.1 to 0.3 MPa, for painted steel much lower, and for corroded steel in the range of 0.4 to 0.56 MPa.

Oksanen (1983) performed adhesion strength tests with fresh water ice on a variety of materials at temperatures of −5, −10, and −15°C. For uncoated steel he measured 0.46 MPa at all temperatures. With epoxy-coated steel the strength was 0.13 MPa.

Alliston (1985) reported results of comprehensive laboratory tests of a low friction and adfreeze coatings to reduce frictional resistance and adfreeze.

As stated earlier, ice adhering to the marine structure may cause substantial vertical uplift or downgrade load. For example, on large linear structures the uplift load may range from 15 to 30 kN/m around the periphery.

Downward loads may result when water levels fall. Hanging ice may span distances of 5 to 10 m between vertical supports.

An upper limit to these loads is provided by the adhesion bond between the ice and the structural material, shear strength of ice, or by the bending strength of ice sheet. The limit force of two former conditions is calculated from

$$F_a = \tau S \qquad (2\text{–}37)$$

where

τ = ice adfreeze (shear) strength

S = adfreeze (shear) area

2.4.7 Load Due to Ice Splitting Mode of Failure

Often a crack will propagate away from the immediate locality. This is identified as splitting. Splitting of an oncoming ice sheet or ice floe occurs most commonly when an intermediate-sized floe that is neither big enough to be stopped by a structure nor to be crushed is split along a line of minimum resistance (Michel, 1978).

For shear cracks, consider a floe of size $L \times B \times h$ as depicted in Figure 2–23. Neglecting friction forces and assuming the crack to be a straight line, Michel (1978) obtained the following splitting force for a crack reaching the side of a floe:

$$F_s = 2\tau_0 Bhtg(\alpha/2) \qquad (2\text{–}38)$$

where

τ_0 = the average shearing strength of the ice

2α = the angle of a wedge shaped structure.

If the floe is not long enough, the force causing a longitudinal crack in the axis of the structure is

$$F_s = 2\tau_0 Lh \sin\alpha \qquad (2\text{–}39)$$

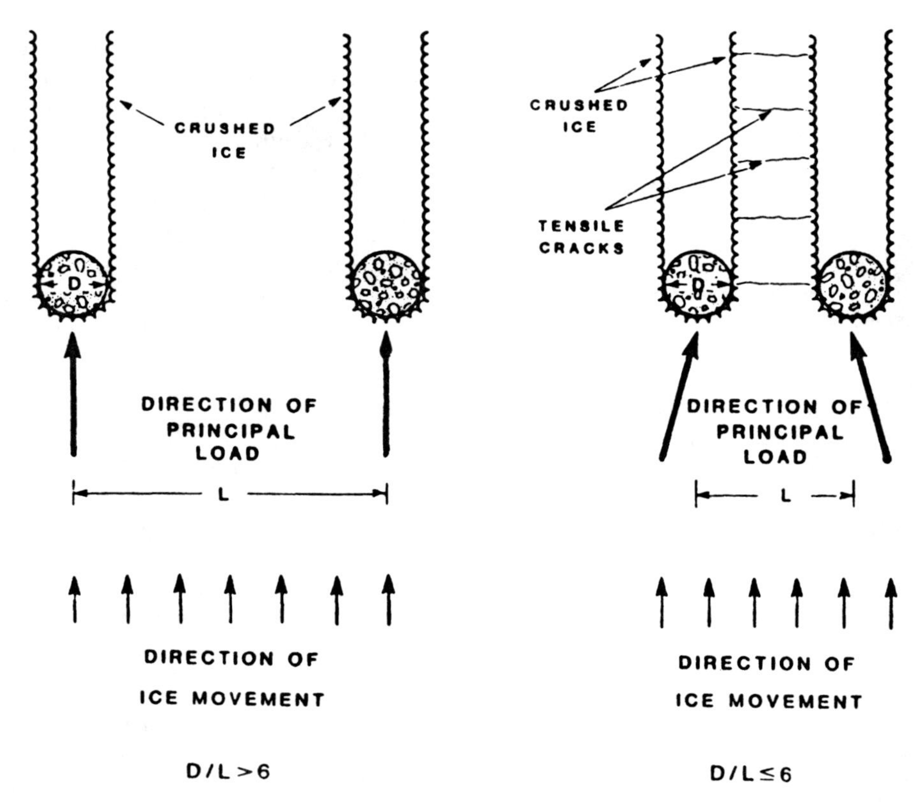

Figure 2–24. Ice failure modes and direction of principal loads on a multilegged structure determined using physical modeling techniques. (From Timco, 1987.)

2.4.8 Ice Load on Multilegged Structures

2.4.8.1 Horizontal Load

The interaction of ice with marine structure in general is a very complex phenomenon. This complexity is more pronounced when determining the ice load for a multilegged structure, such as, for example, piled piers.

For this type of structure it is necessary to know both the total load on the whole structure, and the loads (both in magnitude and direction) on each leg of the structure. If the legs are spaced widely apart, each may be considered as a single isolated pile and treated accordingly. However, if the legs are close together, interference effects may occur that would affect both the load on the structure and the failure mode of the ice. As such, in trying to design the optimum spacing of the legs of a multilegged structure for strength and economy, information on both the magnitude and direction of the loads, ice failure modes, and ice feature sizes is required. To date, there has been very little reported on the subject in question.

Schwarz (1970) reported on the ice loads on a marine pier in the North Sea. Jizu and Leira (1981) and Wang (1983) reported on the problems experienced on the Bo-hai Gulf in China by two platforms that were designed without consideration of ice effects. The investigators outlined the forces on the platforms and the problems associated with ice jamming.

Peyton (1968) and Blenkarn (1970) reported on the ice conditions and ice loads on a platform in Cook Inlet, Alaska over a number of years. The work includes reports on measured prototype values and a discussion of problems associated with the dynamic response of the structure to ice action.

The comprehensive review by Neill (1976) assesses various analytical, full-scale data, and small-scale laboratory tests as they pertain to the estimate of forces due to impact of moving ice on piles, piers, and towers.

The most recent review of the state-of-the-art on multilegged structure–ice interaction is offered by Marcellus et al. (1987) and Timco (1987).

To date, very little information has been published on the ice loads (either measured or calculated) for multilegged structures. In general, the approach in calculating the loads on a multilegged structure has been to treat it as if it were composed of a number of individual piles subjected to ice loading. The loads on each leg are calculated in turn using the formulations developed for the situation of a single isolated pile in ice. Then, the total load on the whole structure is determined by summing the loads on each leg. At the present time, the validity or correctness of this approach is not known. As noted by Timco (1987) one major uncertainty in this area is whether the legs of the platform can be treated individually with summing of the individual loads, or whether there is interference between them that will affect the failure mode of the ice.

The ice forces on multilegged structures have been investigated using physical modeling techniques by several investigators. Test results have been reported by Saeki et al. (1978b), Noble and Singh (1982), Kato and Sodhi (1983), Wessels (1983), Timco and Pratte (1985), and Evers and Wessels (1986). As summarized by Timco (1987), all test results reported are in a reasonable agreement with the following.

1. The ice action on two legs has no interference effects if the spacing between the legs L is greater than six times the diameter D of one of the legs.

2. As the leg spacing decreases (for $L/D < 6$), the horizontal force on each leg in the direction of ice movement is less than that for an isolated leg under the same conditions.

Over this same range, however, the magnitude of the force on each leg in the direction perpendicular to the ice movement increases in such a way that the total horizontal force on the leg is the same as that on an isolated leg.

3. This mechanism causes the resultant direction of the ice horizontal loading on each single leg to change if the legs are close together. The model tests indicate that the direction of loading may deviate by as much as 12° from the direction of motion of the moving ice. Thus, when the legs are wide apart, the failure of the ice is symmetrical about each leg, whereas when the legs are close together the failure tends to occur on one side. The proximity of the other leg causes the legs to be pushed together (Figure 2–24). The resultant force on the entire structure, however, decreases with decreasing leg spacing, as the lateral components of the single leg forces compensate each other. This was confirmed by model test results on three- and four-legged platforms (Evers and Wessels, 1986).

4. When $L/D > 6$, the failure of the ice is pure crushing, sometimes with radial cracks. For $L/D < 6$, tensile cracks usually develop in the ice such that the ice moving between the legs is broken into smaller pieces (Figure 2–24). In general, the length of each ice piece is a function of the leg spacing such that it increases with increasing spacing.

5. For broken ice moving against the front legs of a multilegged structure, there is a definite advantage in ice clearing and total load for a structure that has a few large-diameter and widely spaced legs as compared to more numerous thinner legs.

6. The magnitude of the load depends on the angle of the ice movement relative to the structure's legs. If, for example, ice movement is normal to the structure, then front legs will simultaneously penetrate into the ice; the back legs are shielded by front legs, and therefore, will not be exposed to any appreciable load. If, however, the direction of ice movement is changed then under certain conditions all legs can be exposed to ice loads. In the latter case, the back legs will experience a broken ice field and the ice may fail by bending or shearing rather than crushing. Consequently, the peak loads on each individual leg do not occur simultaneously, and the maximum ice load on the total structure is less than the sum of the maximum loads on each leg if treated in isolation. The amount of reduction will depend on the structure design and angle of ice attack. It should be realized, however, that for frozen-in conditions with complete 100% contact, simultaneous occurrence of peak loads may occur. This may increase the total force on the structure by a factor of up to 4 compared to the force during level ice penetration.

In order to calculate ice loads, representative values of the mechanical properties of the ice are necessary. Furthermore, in the design of a structure for a specific site, it would be imperative to have reliable information on the type, extent, and potential involvement of the ice cover at the site.

The ice load on one single leg (pile) can be estimated from the Korzhavin (1962) equation as discussed in Section 2.4.3. Again, it should be remembered, that under $L/D < 6$ condition piles may be "pushed together." For the highest overall load on the structure, the individual load components should be simply summed.

2.4.8.2 Vertical Loads on Piles or Piers Due to Changes in Water Level

As water level changes through, for example, tidal action or seasonal fluctuation, the ice sheet adhering to a pile or pier can deflect and exert a vertical force on it.

The problem of computation of vertical forces on piles is usually solved using the theory of a floating elastic ice plate subjected to static loads with the following governing differential equation (Kerr, 1976, 1978a, 1986; Gold, 1984):

$$D_i \nabla^4 \Delta = q - k\Delta \tag{2–40}$$

where

D_i = flexural rigidity of the ice plate. $D_i = Eh^3/12(1-\nu^2)$ where E = Young's modulus, h = ice cover thickness, and ν = average Poisson's ratio across the ice cover thickness

∇^4 = biharmonic operator

Δ = vertical deflection of point at the plate reference plane

k = specific weight of water ($k = \rho_w g$)

q = vertical load distribution.

The Eq. (2–40) may be valid even when E varies across the thickness of ice cover, provided that flexural stiffness of the ice plate is determined from the following expression:

$$D_i = 1/(1-\nu^2) \int_{-Z_0}^{h-Z_0} z^2 E(z)\, dz \tag{2–41}$$

where

z_0 = coordinate of ice plate reference plan which is determined from the condition

$$\int_{-Z_0}^{h-Z_0} z^2 E(z)\, dz = 0.$$

Solutions for a variety of ice cover problems that are based on Eq. (2–41) were reviewed by Kerr (1976). However, as stated by Timco (1987), Kerr's approach can be used reliably under the condition that the water level changes are small, so that no cracks occur in ice cover.

In most cases, however, ice cracking will occur. For this situation the complete analytical model to predict the uplift forces on a multilegged structure is yet to be developed. Therefore in the latter case, it is necessary to treat each leg independently. One approach to the problem of an isolated pile is to treat the failure of the ice in terms of radial cracking and subsequent circumferential failure of the wedges. Based on the work of Nevel (1972), an approximate estimate of the uplifting forces F_v on an isolated circular pile or pier is given as follows:

$$F_v = 1.54\sigma_f h^2(1.05 + 2\alpha + 0.5\alpha^3) \qquad (2\text{–}42)$$

where

σ_f = the flexural strength of ice
h = the ice thickness
α = nondimensional pile radius (r/ℓ), where r is the radius of the pile (pier) and ℓ is the characteristic length [this is the same definition as in Eq. (2–28)].

Calculated by Eq. (2–28) an uplift load would be appropriate in several situations. However, in the case of multilegged structure this approach can be used as a guide only because the deformation and failure of the ice will be affected greatly by the proximity of the other legs in the structure. In the latter case just a simple summing of the loads on each individual pile may underestimate the actual total vertical load on the structure. In those cases where interference of the ice deformation due to other structures is of concern, perhaps the best approach may be to use a finite element analysis to determine the failure mechanism (Eranti and Lee, 1981). This approach, however, still involves many simplifying assumptions.

The results of the most recent study on design approach for piles subject to ice-induced vertical loads were reported by Edil et al. (1988). In this study the investigators used existing design experience to arrive at means by which an engineer could predict the level of pile uplift damage for a given pile design due to ice jacking.

As mentioned earlier, there have been very few reports on the ice loads or icebreaking behavior around multilegged structures. Doud (1978) reported results of measurements of the uplift forces on a series of piles that were ice-jacked by the freshwater ice. Tryde (1983) reported result of observations on ice-lifting forces in a number of Danish marinas. He also provided a description of the pile-lifting mechanism (Tryde, 1989). Wortley (1984, 1987) reported many examples of severe uplift of multilegged structures.

A comprehensive review of experimental studies of uplifting forces exerted by adfrozen ice on marine piles is given in the work by Christensen and Zabilansky (1985).

In general terms, depending on intact ice sheet thickness and ambient temperature, water level fluctuations may produce uplift and downdrag forces on dolphins, single piles, or small groups of piles ranging from 100 to 400 kN.

In practice these loads can be minimized by reducing ice adhesion values to piles. As discussed earlier this can be achieved by using low friction materials, miscellaneous spray-on coatings, and jackets and wrappings around piers and piles. In this way the ice forces can be reduced but not eliminated. Additionally bubble systems can be used to protect the piers and piles from ice adhesion.

2.4.9 Ice Load of Thermal Origin

When an ice cover is subjected to a temperature increase it will expand. This will result in forces exerted on the surroundings, which may include miscellaneous marine structures such as piled piers and gravity-type vertical walls. The magnitude of these forces will depend on a number of parameters such as:

- Temperature variation as a function of time
- Material properties of the ice cover
- Ice thickness, presence of cracks and other irregularities, and geometry of ice cover
- Restrictions to expansion along the boundaries of the ice cover.

A number of theories on ice thermal loadings have been suggested and several laboratory experiments and full-scale observations have been conducted to verify these theories. A comprehensive review of the state-of-the-art ice forces is given in Kjeldgaard and Carstens (1980). The latter investigators concluded that in spite of the considerable amount of experimental work conducted in laboratories, there is a great deal of divergence concerning the difficult question of which stress–strain relationships should be considered appropriate as the basis for the method of computation of thermal ice forces. Drouin (1970) made a comparison between results obtained by some of the thermal ice pressure theories using specific conditions: the initial temperature of ice −40°C is raised at a rate of 2.8°C/h; the ice cover is uniaxially restricted and without a snow cover; no solar energy is absorbed. The pressure values were computed for two thicknesses of ice cover: 0.45 m and 0.9 m.

Later the list of results on ice thermal loads tabulated by Drouin was extended by Kjeldgaard (1977) and Bergdahl (1978).

The data collected by the above investigators are shown in Table 2–5. As pointed out by Kjeldgaard in conjunction with data presented in Table 2–5 the approach used in the former USSR Standard SN76-66 which replaced SN76-59 does not distinguish between different types of ice, treats the ice pressure values as a function of wind velocity, and temperature data included in SN76-66 formulation are air temperature data. Hence, the ice loads related to SN76-66 assume that the initial air temperature is −40°C and that the increase rate of air temperature is 2.8°C/h. Furthermore, the values of ice forces computed by means of SN76-66 are dependent on the speed of wind, for example, on the thermal boundary layer in air above the ice surface.

It should also be noted that SN76-66 assumes that the extent of the ice cover does not exceed 50 m. Otherwise the values of ice load should be reduced by an appropriate factor as stipulated by SN76-66. Furthermore, this standard made no distinction between uniaxial and biaxial restraints of an ice cover.

Subsequently, the values of the thermal ice pressure as stipulated in SN76-66 and given in Table 2–5 have been computed by Kjeldgaard for the wind speeds of 0, 5, and 20 m/s. It is worth noting that Drouin and Michel (1971) do not consider the wind speed as an important factor in development of ice thermal pressure.

The first approximation of thermal ice loads at different locations and climatic zones was traditionally done on the basis of available empirical values. For example, in contemporary Canadian practice empirical values of ice loads of thermal origin for rigid structures such as dams or gravity-type marine structures vary from 150 to 220 kN/m, and for the design of relatively flexible structures such as sluice gates values of 70 to 75 kN/m are commonly used.

These values are suggested by Michel (1970). Drouin (1970) recommended even higher values of ice thermal pressure (up to 300 kN/m) to be used in the design of gravity-

Table 2–5. Thermal Ice Force Computed by Different Theories

Source		Ice Force in kPa for Ice Thickness	
		0.45 m	0.9 m
Rose (1947)[a]		47	86
Monfore (1954)[a]		222	232
SN76-59 (1959)[a]		128	255
Drouin and Michel:[b]	S1 Ice	330	390
	Snow ice	220	270
SN76-66 (1966):[b]	0 m/s	30	60
	5 m/s	310	440
	20 m/s	410	580
Bergdahl (1978):	0 m/s	459	752
	5 m/s	502	830
	20 m/s	531	829

[a] Computed by Drouin (1970).

[b] Computed by Kjeldgaard (1977).

Kjeldgaard and Carstens, 1980.

type structures, and PIANC (1984) suggested that intact ice sheets may exert thermal thrusts in the range of 75 to 300 kN/m regardless of ice thickness. Rigid structures and structures in ice sheets confined by harbor (basin) geometry will experience the larger loadings. More flexible structures and structures in more open or sloping-sided harbors (basins) will have smaller loads.

It has to be noted that smaller values of ice pressure exerted upon flexible structures must be treated with great caution because in some practical cases (e.g., sluice gates) load relaxation at the middle of the structure may result in a heavy load concentration at the bearing points.

In recommendations by the Canadian Department of Environment (1971) it is assumed that the ice thrust varies linearly with ice thickness, and therefore recommended ice loads of 150 and 220 kN/m subsequently correspond to ice thicknesses of 0.3 and 0.6 m, respectively.

In cold Siberian regions of Russia an ice pressure of 300 kPa is commonly used whereas for somewhat less severe conditions ice pressures ranging from 150 to 200 kPa are more common (Starosolsky, 1970).

In Norway an ice load of 100 kN/m is typically used for an average ice condition and under especially unfavorable conditions the value of an ice load may be as high as 150 to 200 kN/m (Kjeldgaard and Carstens, 1980).

2.4.10 Icing

When droplets generated from seawater fly in cold air, cool, and hit an object, spray ice will form. Problems caused by spray ice for vessels have been known as long as man has navigated cold sea areas. A more recent problem is accumulation of ice on marine structures in cold sea areas.

Icing can cause considerable loads on horizontal and vertical surfaces of a marine structure and inconvenience to berth operation. Asymmetrical ice accretion on a structure above the waterline can affect the center of gravity of the structure and increase exposure to wind area.

The icing condition is typically the result of freezing rain and/or freezing spray. In some regions the ice fog also can contribute substantially to structure icing. In general, however, frequencies of freezing precipitation are quite low, and therefore, icing due to ice fog and precipitation is almost never considered in practical marine structure design. Typically icing due to a wind-generated spray that is the result of a direct whipping of wave crests by the wind, and icing due to spray generated by waves hitting a structure is taken into consideration.

In strong wind and low temperature the spray freezes on a structure if the air temperature is lower than the freezing point of water.

Shehtman (1968) found that rapid ice accumulation may occur from sea sprays with air temperatures as low as −2°C and wind speeds of 10 m/s.

Unfortunately the exact mechanism of spray cloud generation on wave impact with a structure is unknown and there is at present no theory that could be used to estimate the spray liquid water content and its variations with height under various wave impact conditions. Quantitatively, it is obvious that the amount of spray water caused by splash depends mainly on the wave energy and the geometry of the impact surface. Hence, the mean liquid water content at any height above sea level should be a function of the significant wave height, group velocity, steepness and period of the waves, and of the shape of the structure. Unfortunately there are very limited data available to derive the relationship between the aforementioned parameters. Various parameterizations for the liquid water content of wave-generated spray have been proposed, basically, based on scarce data from ship observations.

At present, a modified Stallabrass (1980) icing model is usually used for computing accretion of ice on a structure due to splashing a structure with wave impact generated spray.

The input in this model requires information on air temperatures, sea-surface temperature, wind speed, wave height and period, and sea-surface salinity.

The model computes icing rates on a surface by considering the primary heat transfer terms, and assuming that icing is a continuous, quasi-steady-state process.

The equilibrium surface temperature T_s may be solved by the Tabata et al. (1968) empirical relationship.

$$T_s = (1 + n)T_f \qquad (2\text{–}43)$$

where

T_f = freezing temperature of seawater
n = freezing fraction (can be determined from the heat transfer terms).

The following modifications are typically made in application of the Stallabrass model to vertical face structures.

1. Liquid water content (LWC) expression is replaced with an expression that allows LWC to be computed as a function of elevation.
2. LWCs from wave-generated and wind-generated spray are included.
3. The model is modified to compute icing rates over 0.5-m intervals.
4. Relative humidity changed to about 85% from 90%.
5. Droplet flight time was made a function of elevation.
6. A "correction" factor is added to allow a smooth transition from wet to dry growth. The poor transition is a result of the simple physics employed by the model, and the empirical relationship from Tabata et al., which does not hold as n approaches unity.

Unfortunately, very little information exists on the characteristics of spray generated by offshore structures such as piers. This information is crucial to properly assess the vertical distribution of accreted ice. A number of LWC expressions are available in the literature for wave-generated and wind-generated spray. Wave-generated spray is the dominant source of LWC (Zakrzewski, 1987). However, most of the available expressions are based on spray measurements taken on fishing vessels. These may not be appropriate for fixed marine structures which are likely to generate less spray than small fishing vessels. In the absence of the required LWC information, three expressions for the wave-generated spray, summarized in Table 2–6, can be used for calculations.

A number of other expressions are available. However, these represent minor modifications of those given in Table 2–6.

Wind-generated spray (LWC_w) is considered as secondary to LWC in the icing process and usually is ignored. Information on LWC_w and method of computation is given in the work of Horjen and Vefsnmo (1984).

The most recent numerical icing model on wave-generating icing was proposed by Makkonen (1989). According to Makkonen comparison results obtained by his theory and icing wind tunnel experiments show good agreement. Furthermore, new theories show good agreement with results obtained by Stallabrass (1980) and with Lozowski et al. (1983) numerical models.

According to Makkonen, for estimation of the icing ice load, it can be assumed that spray ice density in wet growth is close to the density of pure ice, especially at high water salinities (Laforte and Lavigne, 1986).

On the other hand when the growth process is dry, it is conceivable that the density of the ice is lower, because the individual droplets may freeze on impact, resulting in a porous material (Macklin and Payne, 1968). Empirical models have been developed to predict the density of accreted ice in this situation (Makkonen and Stallabrass, 1984). A theoretical solution was suggested by Makkonen (1989).

A comprehensive list of references on the icing phenomenon is given in Zakrzewski (1987) and Makkonen (1989) and useful information is found in publications by the US Army Corps of Engineers CRREL edited by Minsk (1983).

Miscellaneous deicing and anti-icing techniques that include use of steam, hot water, ice-phobic coatings, chemical freezing point depressants, electric heat tracing, infrared radiation, and diteren mechanical methods are discussed by Lousdale and Norrby (1985).

Table 2–6. Expressions Used for Wave-Generated Spray (kg/m^3)

Source	Expression	Comments
Zakrzewski (1986)	$LWC_s = 6.1457e^{-5} H_s V_r^2 \exp(-0.55z)$ where z = elevation above deck of Soviet fishing trawler	From Soviet trawlers. Found to overpredict ice loads on drill rigs, height distribution good for severe events.
Horjen and Vefsnmo (1984)	$LWC_s = 0.1H_s \exp(H_s - 2z)$, $z > H_s/2$	From Japanese trawlers. Found to give good predictions of ice loads on drill rigs but underestimated the height of ice under severe conditions.
Brown and Roebber (1985)	$LWC_s = 4.6 \exp(2z/H_s)^2$	Based on Rayleigh distribution for waves and Soviet trawler data. Found to under predict ice loads and vertical ice distribution on drill rigs.

H_s = significant wave height (m);
V_r = wave period (s);
z = elevation (m)

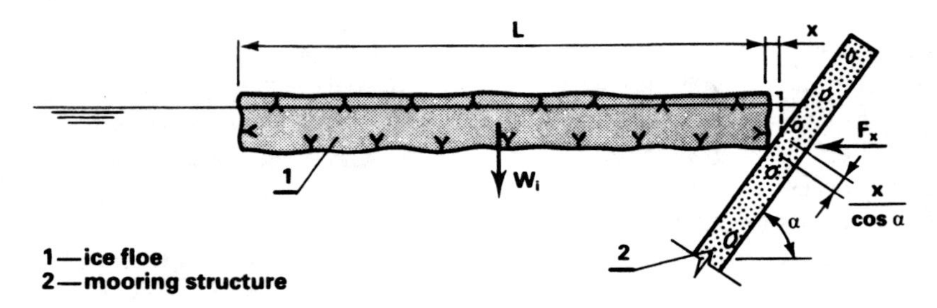

Figure 2–25. Impact of ice floe on a flat-surfaced structure. (From Cammaert and Tsinker, 1981.)

2.4.11 Dynamic Ice Forces

Drifting ice floes can move with considerable speed under the action of environmental driving forces, especially in spring during ice breakup. In this case, the velocity of the ice is such that the failure of the ice occurs in the brittle or transition zone. The failure modes that may be involved are crushing, spalling, ice cracking, and ice wedging.

Generally the crushing mode is the most reasonable to expect. A fundamental procedure for treating the case of the ice sheet impacting a structure is based on an energy balance equation. In this, the initial kinetic energy of the ice feature is equated to the energy dissipated by crushing the ice.

A simple analytical model based on the energy balance equation to compute ice dynamic forces has been proposed by Cammeart and Tsinker (1981) for the following set of assumptions:

1. The leading edge of the ice floe fails in progressive crushing.
2. The ice floe is of a constant thickness.
3. The energy absorbed in the elastic deformation of the ice is minimal.
4. The deformation of the structure and foundation is neglected because it is normally insignificant in the overall displacement.
5. Ice dimensions and strengths are such that buckling and bending will not occur.
6. For large pieces of ice, the volume of water in motion with the ice will affect impact energy. The virtual displacement of the ice floe is obtained by adding this volume to the ice displacement.
7. The impact scenario may be approximately modeled by neglecting the time-varying effects and selecting constant values of ice strength for the duration of impact.

The following analyses represent three different cases of an ice collision with differently shaped marine structures.

2.4.11.1 Collision of an Ice Floe with a Flat-Surface Marine Structure

Assume that the ice floe, with mass W_i, moving with velocity V, crashed at a flat-faced marine structure. This would create an impact load F. The collision force can be expected to reach its maximum when the steadily increasing contact area due to ice crushing reaches a certain critical value A_c. This will happen when kinetic energy of the floe E_c is completely dissipated. The contact area (Figure 2–25) is defined by

$$A_x = Bx/\cos\alpha \qquad (2\text{–}44)$$

where

B = width of the contact area
x = ice floe penetration during collision
α = slope angle of the structure.

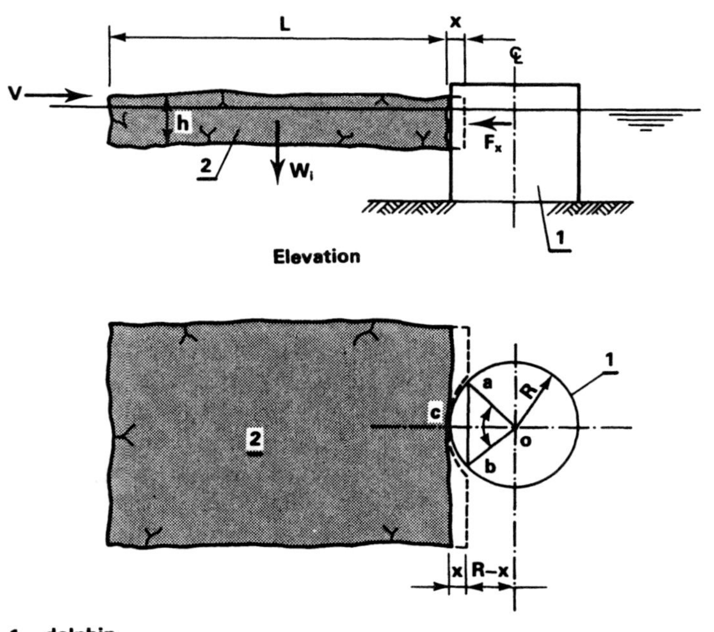

Figure 2–26. Impact of a large ice feature on a cylindrical dolphin (From Cammaert and Tsinker, 1981.)

Hence, the impact load as a function of penetration depth x can be expressed as

$$F_x = A_x p = Bxp/\cos\alpha \qquad (2\text{–}45)$$

where

p = average ice crushing (effective) pressure (see Section 2.4.3).

The virtual mass of the ice flow W can be expressed as

$$W = (1 + C_m)W_i \qquad (2\text{–}46)$$

where

W_i = mass of ice feature

C_m = added mass factor.

The expression $(1 + C_m)$ is defined as an added mass coefficient.

An added mass factor C_m takes into account the mass of ambient water that decelerates together with the ice while impacting upon the structure. During an ice–structure collision the surrounding water originally moving with the ice feature decelerates simultaneously and induces an inertia force on the structure. This inertia force is proportional to the deceleration of the ice mass during impact and is considdred as the added hydrodynamic mass. The latter depends on water depth, roughness of ice surfaces in contact with ambient water, speed of movement, and angle between direction of the ice movement and the current's flow.

Because very little information is available on the hydrodynamic coefficient of the floating ice features with "realistic" shapes practicing engineers normally use approximate values of C_m.

In most practical cases $0.2 \leq C_m \leq 0.5$ is assumed, where the lower limit is used for the ice underkeel clearance larger than

$0.5\,h_u$ and the upper limit is used for the underkeel limit less than $0.1\,h_u$, where h_u = draft of the floating ice feature. For the underkeel clearance ranging from $0.1\,h_u$ to $0.5\,h_u$, a linear distribution of the added mass coefficient may be assumed. For more information on C_m values the reader is referred to works by Bass and Sen (1986), Isaccson and Cheung (1988), and Isaacson and McTaggart (1990).

The maximum value of ice penetration, x_m, can be determined by equating the kinetic energy of the ice floe, E_k, to the energy absorbed by the crushing of the ice, E_c.

$$E_k = \frac{WV^2}{2g} = \frac{(1+c_m)W_iV^2}{2g} \tag{2–47}$$

$$E_c = \int_0^{x_m} F_x\,dx = \frac{Bp}{\cos\alpha}\int_0^{x_m} x\,dx = \frac{Bpx_m^2}{2\cos\alpha} \tag{2–48}$$

Setting $E_k = E_c$ will give

$$x_m = V\left[\frac{(1+C_m)W_i\cos\alpha}{Bpg}\right] \tag{2–49}$$

where

g = gravitational acceleration.

Subsequently the maximum impact force F_m is expressed as follows:

$$F_m = Bpx_m/\cos\alpha. \tag{2–50}$$

2.4.11.2 Impact of Large Ice Features on Cylindrical Structures

Driven by environmental forces large ice features, such as ridges and large ice floes, are commonly observed to rotate and it is very probable that they will rotate on collision with the structure. The latter, of course, will lessen to some extent the effect of ice momentum. Cammaert and Tsinker (1981) suggested that for the purpose of preliminary calculations the kinetic energy to be dissipated during collision is one-half of its total value. Hence kinetic energy (E_k) may be determined as follows:

$$E_k = 0.5\,\frac{(1+C_m)W_iV^2}{2g} = \frac{0.25(1+C_m)W_iV^2}{g}. \tag{2–51}$$

As discussed in the previous case the critical impact force upon a cylindrical structure will occur when the ice feature while rotating about the structure penetrates it by x_m (Figure 2–26).

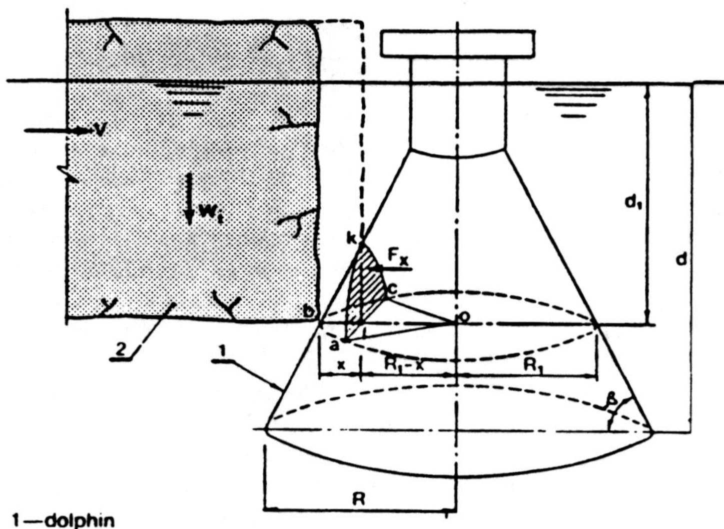

Figure 2–27. Impact of a large "blocky" ice feature on a conical dolphin. (From Cammaert and Tsinker, 1981.)

For a cylindrical structure and an idealized shape of the ice feature of a constant thickness the maximum impact force F_m is determined as follows: it is assumed that the arc length (acb) is approximately equal to the chord length (ab), if the total penetration is small. Hence, for a penetration x, the contact area (A_x) is expressed as follows:

$$A_x = 2h(2Rx - x^2)^{0.5} \qquad (2\text{–}52)$$

where
h = depth of ice feature
R = structure radius.
The impact load is then defined as

$$F_x = 2hp(2Rx - x^2)^{0.5} \qquad (2\text{–}53)$$

and the energy dissipated during crushing is

$$E_c = 2hp \int_0^{x_m} (2Rx - x^2)^{0.5}\, dx. \qquad (2\text{–}54)$$

Because for large cylindrical structures such as dolphins ice penetration is much smaller than radius ($x_m \ll R$). Then by dropping x^2 in Eq. (2–54) we obtain

$$E_c \simeq 1.89hpR^{0.5}x_m^{1.5} \qquad (2\text{–}55)$$

Setting $E_k = E_c$ will give

$$x_m = [0.132(1 + C_m)W_iV^2/hpR^{0.5}g]^{2/3}. \qquad (2\text{–}56)$$

Finally we find the maximum impact load is equal to

$$F_m = 2hp(2Rx_m - x_m^2)^{0.5}. \qquad (2\text{–}57)$$

2.4.11.3 Impact of Large Ice Features on Conical Structures

Conical structures are primarily designed to fail ice in the bending mode. However, in collisions involving an ice feature of substantial thickness and length a dynamic force may result. This force can be determined similarly to that described previously for a cylindrical structure. To calculate the value of x_m assume the idealized "blocky" shape of the ice feature.

Again, for small penetrations, the area ($abck$) in Figure 2–27 is assumed equal to the projected parabolic area (akc), so that

$$A_x = 2/3(ac)(kf) = 4/3 \tan\beta x(2R_1x - x^2)^{0.5} \qquad (2\text{–}58)$$

where
R_1 = radius of the structure at the point of impact, which is

$$R_1 = R - (d - d_1)\tan\beta \qquad (2\text{–}59)$$

where
R = radius of the structure base
β = base angle
d = water depth
d_1 = draft of the ice feature.
The parameters F_x, E_c, x_m, and F_m are defined as

$$F_x = 4/3 \tan\beta px(2R_1x - x^2)^{0.5} \qquad (2\text{–}60)$$

$$E_c = \frac{4}{3}\tan\beta p \int_0^{x_m} x(2R_1 - x^2)^{0.5}\, dx \qquad (2\text{–}61)$$

$$\simeq 0.75tg\beta pR_1^{0.5}x_m^{2.5}$$

Again, by equating E_k obtained from Eq. (2–51) to E_c we obtain

$$x_m = [0.333(1 + C_m)W_iV^2/tg\beta pR_1^{0.5}g]^{0.4} \qquad (2\text{–}62)$$

and

$$F_m = 4/3 \tan\beta px_m(2R_1x_m - x_m^2)^{0.5}. \qquad (2\text{–}63)$$

As mentioned earlier a certain arbitrary assumption such as "blocky" shape of the ice feature and percentage of ice energy dissipation due to ice rotation has been proposed by Cammaert and Tsinker (1981) for the above calculations.

In reality ice is of more random geometry, and the probability of having contact with a structure over the full thickness is virtually nil. In real design of important structures, the assumed shape of floating ice features should be based on an ice profile generated statistically from field measurements. A more rational approach for determining the design portion of ice kinetic energy would be based on probabilistic estimates of ice trajectories for each particular location.

Furthermore, in more precise calculations in addition to ice feature deformation and rotation the deformation/deflection of the foundation material may be considered.

The most recent studies on ice impact upon marine structures include the following: Cammaert et al. (1983) presented an idealized numerical model based on the laws of motion for a two-mass, three degrees-of-freedom system, where the large ice feature is assumed to translate, while the structure (floating platform) can both translate and rotate. Curtis et al. (1984) developed several computer models that solved the equations of motion for a two-mass system to predict the collision force as a function of time.

A number of methods for the solution of the impact loads have been presented by Arockiasamy et al. (1983). Among them are the energy approach, which equates the kinetic energy of the oncoming ice feature to the total absorbed by the structure; the initial velocity condition approach which idealizes the system by certain initial velocity conditions; and the additional degrees-of-freedom for local deformation, which reformulates the equations of motion for the floating platform with the only approximation being that, during collision, the zone of impact provides stiffness coupling only—not mass coupling.

Croteau et al. (1984) developed an impact model that considers the effect of inertia, damping, foundation compliance, and slippage at the base, in addition to a crushing failure mechanism in the ice at the contact zone.

Cox (1985) proposed a similar model, except that it does not take foundation response into account. The model accomodates strain rate effects, nonsimultaneous failure zones, and changes in ice–structure geometry. Cox's method relates the ice impact problem to ice crushing and takes into account nonsimultaneous failure zones.

Guttman et al. (1984) have developed a finite element model in which nonlinear elastic elements characterize the ice–structure interaction zone, and nonlinear elastoplastic elements comprise the foundation response.

Kreider (1984) presented a numerical time-domain solution for ice-floe impacts in which the predominant ice-failure mode is crushing. In respect to this mode, the force exerted on a structure during impact is determined by solving in one dimension the equation of motion for the floe. Ice failure is assumed to occur at the ice–structure contact.

According to Kreider the maximum impact force (F_m) for head-on collision can be found by the following equation (2–64) which was obtained by equating the ice feature kinetic energy and work done.

$$F_m = 2.66h(\rho_i R_s)^{1/3}(pR_iV_i)^{2/3} \tag{2–64}$$

where

h = ice thickness
ρ_i = ice mass density
R_s = structure radius
p = average ice crushing pressure
R_i = ice floe radius
V_i = floe velocity.

Evans and Parmerter (1985) have addressed the loads on sloping structures that result from the impact of large floes.

Energy-based solutions have been developed by Johnson and Nevel (1985), Isaacson (1985), and Gershunov (1986). Tunik (1987) proposed a numerical model for predicting ice impact on structures of

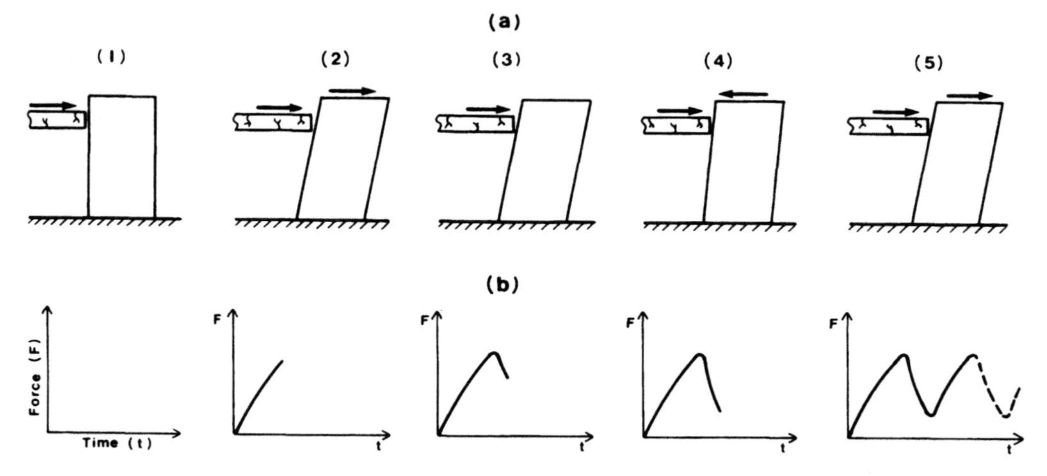

Figure 2–28. Ice–structure interaction that induces vibration $[F = f(t)]$.

different shapes which considers ice properties and mode of failure and shape, mass, and velocity of the ice feature.

It should be pointed out that during full-scale multiyear ice impact load measurements taken at Hans Island off the northwestern coast of Greenland (Danielewicz et al., 1983) and at Adams Island in Lancaster Sound (Frederking et al. 1986) surprisingly low failure stresses have been recorded. These stresses contrast sharply with predictions made from models based on small-scale strength data for ice. The discrepancy is basically due to the presence of a failure mechanism other than just crushing in large-scale processes and the inherently lower strength associated with natural flawed ice rather than a perfect laboratory specimen.

Walden et al. (1987) proposed a model of ice impact phenomena that is calibrated with Hans Island data. The model is based on the assumptions of dissipation of the kinetic energy of the ice feature through failure of the ice as it penetrates the structure, and that the strength of ice in the brittle zone is independent of floe kinetic energy, but varies with scale in a progressive indentation process according to an inverse square root of contact area relationship.

2.4.12 Ice-Induced Vibration of Structures

The moving ice–structure interaction results in loading and deformation of both the ice and the structure. Eventual brittle failure of ice at some finite stress resulted in sudden unloading of the structure. The transient response of the structure (spring-back) may execute a few cycles of damped vibration at its natural frequency. When contact between the structure and the intact ice is reestablished a new cycle of ice brittle failure resulting in structure vibration is initiated.

Depending on the severity of ice action, even a rigid massive structure may vibrate at a frequency close to its natural frequency, as was observed during crashing failure of multiyear ice floes against Molikpac, a 110 m wide structure deployed at Amanligak in the southern Beaufort Sea (Wright et al., 1986; Jefferies and Wright, 1988).

For a wide structure Jefferies and Wright (1988) proposed the following conceptual framework for the dynamic ice–structure interaction during ice crushing in the following steps (Figure 2–28):

1. The ice feature approaches the structure at a constant velocity.
2. After initial contact the structure deflects.
3. The ice crushes when failure stress is reached.
4. Because the crushed ice cannot support the interface stresses, there is a spring-back of both the structure and the ice due to release of stored elastic energy.
5. The process repeats through steps 1, 2, 3, and 4 as an ice feature advances toward the structure.

During the spring-back of the structure, the crushed ice is squeezed out of the zone in front of the structure and offers added damping beyond that already present in the structural support.

Very little has been written on the flow of crushed ice. For a comprehensive discussion on a progressive damage of ice and factors that affect damage initiation and evolution the reader is referred to Jordaan and McKenna (1989).

For design and analysis of marine structures operated in ice-covered waters, it is essential to know not only the magnitudes but also the variations of the ice forces generated through a continuous ice–structure interaction. These forces vary with time and depend on speed of ice movement and flexibility of the structure.

In the case of a slow moving ice feature against a massive rigid structure the ice crushing may not cause an extreme structural vibration. In the latter case, the typical pattern of ice–structure interaction may be simple, similar to that depicted in Figure 2–28. In the case of slender flexible structures, however, the variation in the ice force may cause the structure to vibrate or oscillate at its natural frequencies, resulting in a complicated ice–structure interaction.

The vibration of a flexible structure exposed to the continuous ice force is characterized as "self-excited vibrations" (Määtänen, 1987, 1989).

The state-of-the-art of the theoretical and experimental studies on structure vibration during ice–structure interaction is quite advanced. A comprehensive list of the most recent relevant works is found in Jordaan and McKenna (1989), Määttänen (1987, 1989), and Sodhi (1989).

Sodhi (1989) postulated that to simplify analysis, one needs only to consider those modes of structural vibration that will be excited by ice–structure interaction. He also suggested that in some instances a good theoretical model based on the finite element method can be developed if the structure is modeled for its fundamental mode of vibration by a single-degree-of-freedom system composed of a mass M attached to a spring of stiffness K and a linear damping element of coefficient C, such that the differential equation governing motion of the strucutre may be expressed as follows:

$$M\ddot{x} + C\dot{x} + Kx = F(t) \tag{2–65}$$

where

x = displacement
$\dot{x}$ = velocity
$\ddot{x}$ = acceleration of the mass
$F(t)$ = interaction force generated as the result of ice–structure interaction.

Most theoretical models proposed to date for ice–structure interaction treat the structure as a spring-mass-damper body. Some of the recent theoretical models proposed to date are discussed by Sodhi (1989) and Määtänen (1987, 1989).

2.5 HARBOR OPERATION: BASIC DESIGN CONSIDERATIONS

2.5.1 General

2.5.1.1 Ice Problems

Extension of port operation into the winter season or year-round operation is today's reality. Hence, the requirements for rational design of ports and marine terminals located in cold regions are becoming more demanding.

The restraints imposed on port operation by the ice conditions during the winter season dictate more stringent and demanding design considerations than would normally be required for just seasonal operation. Depending on the local condition of the harbor and the ice regimen, the problems experienced could be due to one or all of the following.

1. The accelerated buildup of ice on the dock structure due to tidal action
2. The interaction of the dock structure with ice driven by environmental forces
3. The buildup of ice "active zone" in front of the dock due to tidal action
4. The accelerated buildup of ice in the berthing area from repeated passages of vessels due to the freezing of water that is repeatedly exposed to atmosphere
5. The ice buildup on the dock vertical and horizontal faces due to wind or wave generated icings.

As repeatedly emphasized earlier the ports and marine terminals in most cases are built at sheltered locations, either natural or man-made protected harbors. A harbor by definition offers better protection against waves and currents. However, sheltering usually increases the length of the ice season by cutting down effects of waves and currents which otherwise may break ice formed in calm water and/or retard freezing by replacing the cool surface water with warmer water from below. At some locations harbors have a tendency to accumulate drifting ice. The latter is well demonstrated at the port of Montreal, where the harbor basin sometimes acts as a storage for ice produced upstream. Shore harbors are filled less regularly than harbors located in inland waterways. The unpleasant fact is that nature tends to bring ice more readily into the harbor than out of it.

Ice problems in harbors can be reduced considerably if advantage is taken of the local currents and thermal energy sources. For example, in straits and river mouths the incoming flow may reduce the accelerated growth of brash ice because it will be mixed and contain some thermal energy.

Furthermore, winter time dock operators are quite familiar with icing of the dock wall faces, which hampers the berthing and mooring operations. Dock walls covered with ice bulges may also be a hazard because the ice may fall off and hit people.

Effects of each of the above problems can be mitigated by proper sitting and layout design of the port (terminal) as well as by employing miscellaneous ice control techniques.

2.5.1.2 Geotechnical Conditions

Site geotechnical condition is a factor that affects selection of type of docking structure.

In northern areas, a permafrost on land or subsea permafrost may affect a marine structure design.

Permafrost is defined solely by the thermal condition of the ground. For a soil or rock to be classified as permafrost its temperature must remain below 0°C for at least two consecutive winters and the intervening summer (Brown and Kupsch, 1974).

Climate is the principal factor affecting the formation and existence of permafrost. Severe frigid climate has resulted in freezing of soil and rock formations to depths of hundreds of meters. For example, in the Arctic Archipelago, the maximum thickness of permafrost is about 1000 m at inland locations. In the land areas adjacent to the Beaufort Sea, permafrost thickness is typically about 100 m. The extent of permafrost varies with climatic conditions and there is a broad correlation between mean annual air temperature and the distribution of permafrost. The thickness of permafrost varies generally with the amount that mean annual air temperature is below freezing and is limited by heat flow from the interior of the earth. Permafrost distribution and occurrence on land are also affected by other factors such as local terrain conditions, vegetation, drainage, presence of lakes and rivers, snow cover, and type of subsurface conditions.

Permafrost occurring below the sea bottom is defined as subsea permafrost. The formation and distribution of subsea permafrost in the cold regions is a complex phenomenon, the details of which have not been fully determined at present.

The land underlined by permafrost can be divided into two main regions: the continuous and the discontinuous zones (Brown, 1970). The latter may range from sporadic occurrence of permafrost to widespread permafrost.

In offshore areas, the conventional association of ice with soils at subzero temperatures does not apply. The salt in seawater affects the physical characteristics of the near seabed permafrost soils. The salt concentration in water prevents nucleation of ice down to temperatures of about −1.8°C, depending on concentration.

Salt in seawater also increases its density so as to minimize thermal reversals and thus subzero temperature water remains at the seabed year round. The subzero temperatures at seabed either preserve submerged land-formed permafrost or aggrade permafrost into recently deposited sediments. Hence, depending on the geological history of an area, subsea permafrost can occur in various thermal conditions. Where equilibrium conditions are achieved with respect to seawater temperature, subsea permafrost is warm with temperatures between about −1.5°C and 0°C.

Subsea permafrost can occur in an ice-free condition because the salt concentration prevents formation of ice by depressing the pore fluid's freezing temperature. The behavior of saline, ice-free permafrost is not well understood but some soils have been found to behave the same as their saline, ice-bonded counterparts (Domaschuk et al., 1983). Saline, ice-free permafrost includes recent seabed sediments that have settled through saltwater.

Subsea permafrost may also contain ice (and be ice-poor or ice-rich) depending on whether or not it contains salt and to what degree it is salt free.

In areas where a major river flows into the sea, the brackish water temperature would be warmer than that of saltwater and thus the thermal regimen in subsea permafrost would be affected. Also, in these areas, fresh or brackish water would occur in the pores of the soil. Where such conditions occur in near shore areas, in conjunction with freezing to the bottom, there would be potential for the formation of ice lenses in frost susceptible soil deposits. Freshwater ice lenses may also be present in permafrost that is submerged by sea rise.

For more information on cold regions geotechnique the reader is referred to Fenco (1984), Osterkamp et al. (1987), and Clark and Guigne (1988).

2.5.2 Site Selection

It is evident from previous discussions of ice problems that in the process of port site selection, in addition to the typical site parameters, such as winds, waves, tides, currents, and bottom conditions, consideration should be given to a number of ice effects.

Typically, location or siting of a port or marine terminal is determined by the location of an industrial demand or settlement. In considering the effects of ice on port siting, it is clear that maximum protection from moving ice is desirable, while at the same time, maximum accessibility to the harbor must be provided. Protection is best provided by locating the harbor in landfast ice and with the entrance to the harbor oriented away from the direction of the prevailing wind. This will also encourage ice floes to move out of the harbor during summer breakup.

If the harbor is exposed to moving ice, particularly multiyear ice, the ice loads on the dock would be more severe and port operations could be significantly hindered. Where harbors must be located in moving ice, as for example in the case of offshore loading terminals, provision should be made in the design to minimize ice interaction with the dock structure and interference with vessel operations.

Accessibility to the harbor will be determined by ice conditions outside the harbor. If large areas of high concentration multiyear ice exist outside the harbor, accessibility will be more difficult than if only first-year fast or pack ice exists. Likewise, the occurrence of ridges will hinder access to the harbor.

Because of the ice problems that can result from tidal effects, the harbor should be located where possible in an area of low tidal range. Otherwise, considerable amounts of ice could accumulate on the dock structure which could lead to berthing problems as well as result in increased loads on the dock structure and the foundation. This problem would be manifested more in a seasonal rather than a year-round operation.

Where possible, the harbor should be located in an area that has a natural source of ice control. This could be in an area that shows a thermal stratification in the water column with warmer water at depth, or is near a freshwater lake whose bottom water could be transported to the harbor and used as a means of ice control. If the harbor is to serve an industrial complex, it should be located as close as possible to the plant to utilize any waste heat that may be available from the plant.

In general, ice control is made easier when there is good circulation in the harbor. In this regard it is not favorable to surround a harbor with breakwaters or to locate it in a well-protected bay as far as ice control is concerned.

From an environment aspect the selected location should require minimum dredging. The increasing public desire to obtain a high-quality environment requires the designer to strictly follow requirements of the existing guidelines and regulations. Last but not least, the site geotechnical condition must be suitable for construction of economically sound docking facilities.

The selected site is usually the subject of a thorough investigation.

In general, the scope of site investigation for port or marine terminals located in cold regions typically does not differ much from that for port facilities located elsewhere. This would typically include data collection on the marine environment (e.g., bathymetry, meteorology, hydrology, and oceanography), ice condition, subsurface condition, and availability of construction materials.

The depth of investigation usually depends on stage of investigation, for example, feasibility study, preliminary site evaluation, or final design stage, and on size and importance of the project.

2.5.3 Subsurface Investigation

2.5.3.1 General Scope

In general, the scope of a subsurface exploration program is to determine the lithology of soils and rocks, existence of permafrost, the form and distribution of ground ice, groundwater conditions, depth of active layer, presence of taliks (unfrozen layers), and ground thermal regimen. Site exploration should also include bathymetric surveys to define sea bottom configuration.

The extent of subsurface exploration depends on the type and size of the proposed structure and its interaction with the surrounding soils. The minimum depth of borings and soundings is generally taken as the width of the proposed structure unless a competent bearing stratum or sound rock is encountered at a shallower depth. Borings in thawing permafrost should be carried out to a depth of at least 3 m below the maximum estimated depth of thaw penetration. When piles or sheet piles are contemplated, the exploration depth should extend at least 3 m below the pile tips.

Comprehensive exploration programs are conducted to obtain undisturbed samples of frozen and unfrozen soils (including saline soils) to determine the laboratory frost heave characteristics, thaw-settlement properties, strength and creep properties, and thermal properties (unfrozen water content, conductivity, and heat capacity). In situ testing may be carried out in test holes and special instrumentation may be installed and left in place. Large-scale tests such as trenching, blasting, pile load tests, and trial embankments may be carried out to obtain practical data for foundation design and to establish optimum construction methods.

2.5.3.2 Drilling and Sampling Permafrost

Onshore permafrost samples can be secured from natural exposures, hand borings, test pits, and core drilling (Johnston, 1981).

In cold permafrost at greater depth and in offshore permafrost core drilling is required.

Test holes in shallow (less than 30 m of water) offshore areas can be drilled either from a vessel or from the surface of the ice during the winter or spring. Because of the warm nature of subsea permafrost, the permafrost can readily be thawed during drilling and therefore great care is required in the drilling procedure. Rotary drilling with a drilling fluid that can be maintained below 0°C is commonly used. Drilling fluids that have been used are diesel fuel, sodium chloride and calcium chloride brine solutions, and antifreeze (alcohol or glycol). To maintain the drilling fluid below 0°C chillers may be needed. In selection of drilling fluids, due consideration should be given to their effect on the environment. Vibratory (resonant) drills have also been used to obtain mechanically disturbed but stratigraphically intact seabed cores (Huck and Hull, 1971; Hayley, 1979). A variety of lightweight drilling equipment has been used in areas of landfast ice and very shallow offshore areas (Osterkamp and Harrison, 1976, 1981) with different degrees of success. These tools are usually suited for preliminary site investigations rather than for final detailed drilling.

The principles of sampling frozen materials offshore are similar to those of diamond drilling in rock.

When samples are brought back to the surface, they are visually inspected and classified according to the present of ice (Pihlainen and Johnston, 1963). Permafrost samples are normally photographed and carefully protected and stored to preserve their undisturbed frozen state for laboratory testing. The storage and shipping temperature must be as close as possible to the in situ temperature. Routine testing to establish index properties and ice content is usually carried out in a portable site laboratory.

2.5.3.3 In Situ Soil Testing

In situ testing is often used to obtain engineering soil parameters and minimize problems associated with scale effect and sample disturbances.

Pressuremeter tests have been used on land to study the strength and deformation properties of frozen soils (Ladanyi and Johnston, 1973; Rowley et al., 1975).

In the unfrozen soil strata the common type of in situ tests are vane shear strength tets in cohesive soils and static cone penetration test (CPT).

Static penetrometers as well as electric and acoustic piezo-penetrometers have been used in the Beaufort Sea to determine engineering properties and distribution of material types, including distribution of ice-bonded sediments (Blouin et al., 1979).

2.5.3.4 In Situ Ground Temperature Measurements

Various types of temperature sensors, inclduing mercury thermometers, thermocouples, thermistors, and diodes, are available for the measurement of ground temperatures. Thermocouples and thermistors are the most widely used for engineering field investigations. In situ temperature measurement techniques have been reviewed by Johnston (1981) and Judge (1973).

2.5.3.5 Laboratory Testing of Frozen, Freezing, and Thawing Soils

Laboratory tests routinely carried out on permafrost soils include density and water content tests. These index properties are useful in assessing the thaw settlement and thermal properties of frozen soils. Other index tests carried out on frozen and unfrozen samples include particle size distribution, Atterberg limits, and specific gravity.

Unfrozen moisture content in permafrost samples can be measured using a number of techniques (Anderson and Morgenstern, 1973), but recently a new technique, time domain reflectometry, was used with satisfactory results in frozen nonsaline soils (Patterson and Smith, 1981). Fenco (1984), however, stated there are doubts that in saline permafrost the latter method is effective because of the dielectric properties of saltwater.

For the determination of engineering properties needed in design, test methods routinely used in soil and rock mechanics must be modified to include provisions for accurate control of test temperature and long-term testing of frozen materials.

Strength and deformation properties of frozen soils and rocks vary greatly with temperature, moisture (ice) content, and duration of loading. Laboratory testing programs must realistically model site loading conditions to produce the most relevant geotechnical design parameters.

Tests easily carried out on frozen soils include simple consolidation (without temperature or pore pressure dissipation monitoring) and unconfined compression. The first is to assess total potential settlement on thawing and the second gives a qualitative strength measurement. These tests are used only for simple design problems. When designing major structures in sensitive warm permafrost more specialized tests with deformation rate and temperature control using specially designed equipment and facilities are carried out.

The testing of freezing soils is carried out in "frost cells." Although no uniformity yet exists in testing equipment and methods, certain basic features are common to all laboratory equipment of this type. Mageau and Sherman (1983) have reviewed all equipment used by other researchers and proposed a frost cell design and operating procedure.

Testing of thawing soils is carried out in specially designed odometers with a heating plate and pore pressure measurement sensor on top and temperature sensors on the side. Such an apparatus allows study of the entire thaw-consolidation process (Morgenstern and Smith, 1973; Nixon and Morgenstern, 1974).

2.5.4 Layout

The layout of the facility plays a major role in the ability of ships to reach and be berthed at a dock.

In port layout design it is important that ice navigation aspects be taken into account as early as possible in planning a new harbor. Experiences from existing harbors with ice navigation can provide a valuable contribution to the planning work.

In a well-planned harbor the use of different ice-control methods should be optimized to minimize future ice problems.

The siting and layout of harbor components should be carried out so that the ice problems are minimized. The dock orientation should be such that movement of the ice cover under the effects of environmental forces such as wind, current, and waves would result in minimum loads on the dock and minimum interference with harbor operations.

The connection between the open sea and the harbor should be as short, wide, and straight as possible. The entry tract to a harbor must not be the same as the departure tract. Separate tracts eliminate large turning areas where large amounts of broken ice can hinder maneuvering vessels.

The layout would be particularly important in a river port where moving ice could have serious effects on dock operations if the dock orientation is such that the ice is blocked during its downstream movement.

The possibility of loads being exerted on the dock structure by moving ice during breakup and freeze-up should be accommodated in the design of the structure. However, in some cases, particularly from an operational point of view, it might be desirable to design the layout in a way to prevent the moving ice from impacting the dock or entering the berthing area where it could interfere with vessel berthing. In these cases it is necessary to contain or divert the ice floes, and the appropriate ice control techniques will be required to do this.

The proper layout design is also important from the point of view of vessel maneuvering in ice, particularly if large vessels such as very large crude carriers or liquid natural gas (LNG) carriers are being considered. If the vessel is to make a turning circle in the harbor, a considerable area will be required. This area will depend on the vessel design and the degree of tug assistance provided. Without tug assistance, it has been estimated that the turning diameter of a 300 000 DWT vessel in arctic ice is approximately 12 km (Griesbach and Kremer, 1973). The turning diameter of the ice breaking type LNG carrier of 234 000 displacement tonnes designed for the Arctic Pilot Project was estimated to be 4 km (Gill et al., 1983). In such cases, the dock and other harbor facilities should be sited so as to allow the maximum area for vessel maneuvering.

In general, the minimum turning radius of a large ship traveling under its own power in track broken in a continuous ice cover by the vessel itself will depend on the width of the track in ice which has attained its maximum natural growth thickness and the shape of the vessel. For preliminary design, assuming that the track is about 3 to 4 m wider than the vessel's beam and that the critical points of contact are at the quarter points of the vessel length, the maximum turning radius in meters (R_{max}) may be determined as $R_{max} = L^2/29.5$, where $L =$ vessel length in meters.

2.5.5 Effects of Vessel Operation on Ice Growth in the Ship Track

In layout design consideration must be given to the possibility of ice buildup in the ship track.

The passage of vessels along the same track in a continuous ice sheet results in an accelerated formation of ice through the repeated exposure of water to the atmosphere. After a certain number of passages, the accumulation of ice may reach the point where further passages will be precluded. At this time, the track may have to be abandoned in favor of an alternative route. The same ice buildup will occur in the docking area and, if the ice accumulation proves to be excessive, ice control techniques will be required to permit proper berthing of vessels.

Two mathematical models for this type of ice buildup have been proposed by Michel and Berenger (1975) and Ashton (1974b).

(a)

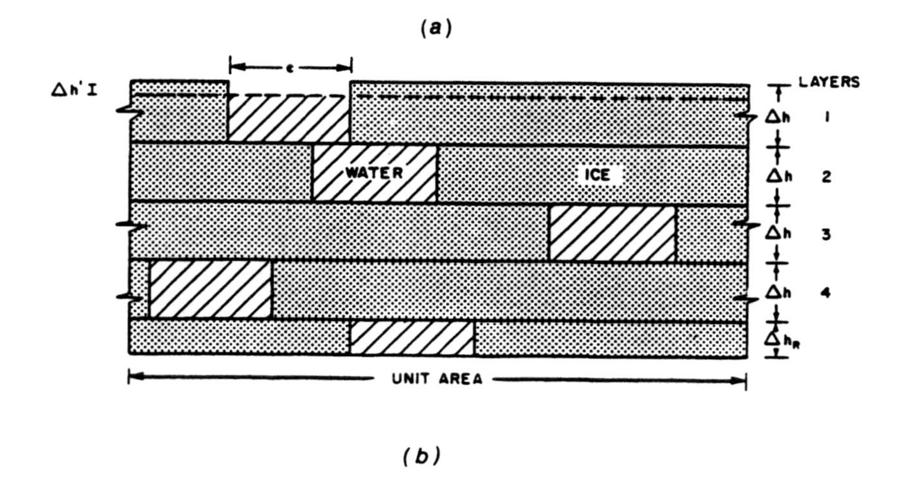

(b)

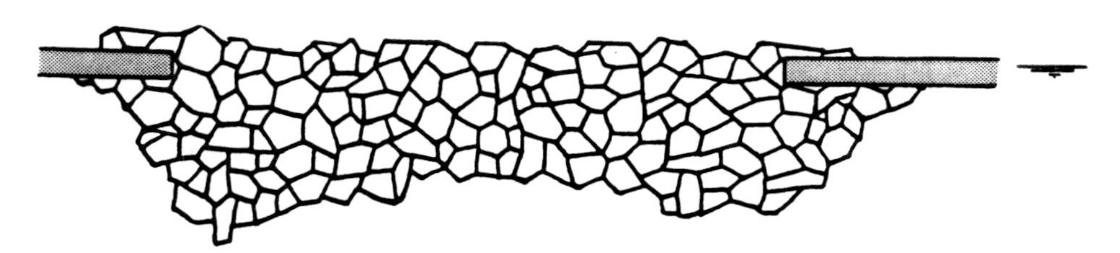

Figure 2–29. Model of ice growth in ship track. (From Michel and Berenger, 1975.) (a) Physical model. (b) ice profile in ship track after repeated passages.

After observing the modes of accumulation of broken ice and ice growth in ship tracks, Michel and Berenger (1975) proposed the simple physical model shown in Figure 2–30a. It is made up of a matrix of superimposed ice pieces of porosity ϵ and of basic thickness Δh with a residual ice thickness Δh_r at the bottom. A typical ship track profile is shown in Figure 2–29b.

The additional ice produced in the surface layer is written as follows:

$$\Delta_T = \Delta_1 + \Delta_2 + \Delta_3 \qquad (2\text{–}66)$$

where

Δ_1 = additional ice produced in the surface layer

Δ_2 = layer of ice inside the matrix contribute growth

Δ_3 = increase in residual ice formed at the bottom.

All the above parameters can be described by the following algorithms:

$$\Delta_1 = \epsilon/(1-\epsilon)\{(\Delta h - \Delta h^1) + \xi[(\alpha^2 \Delta D)^{0.5} - (\Delta h - \Delta h^1)]\} \qquad (2\text{–}67)$$

$$\Delta_2 = \epsilon \sum_{i=2}^{k} [(\{(i-1)\Delta h^2\} + \alpha^2 \Delta D)^{0.5} - (i-1)\Delta h] \qquad (2\text{–}68)$$

$$\Delta_3 = \nu(1-\epsilon)^{n-1}\{([n\Delta h + h_r]^2 + \alpha^2 \Delta D)^{0.5} - (n\Delta h + h_r)\} \qquad (2\text{–}69)$$

where

k = the critical number of layers required to stop the formation of residual ice, $k \leq 1/(1-\epsilon)\epsilon$

ξ = 1 for $\alpha(\Delta D)^{0.5} \leq (\Delta h - \Delta h^1)$

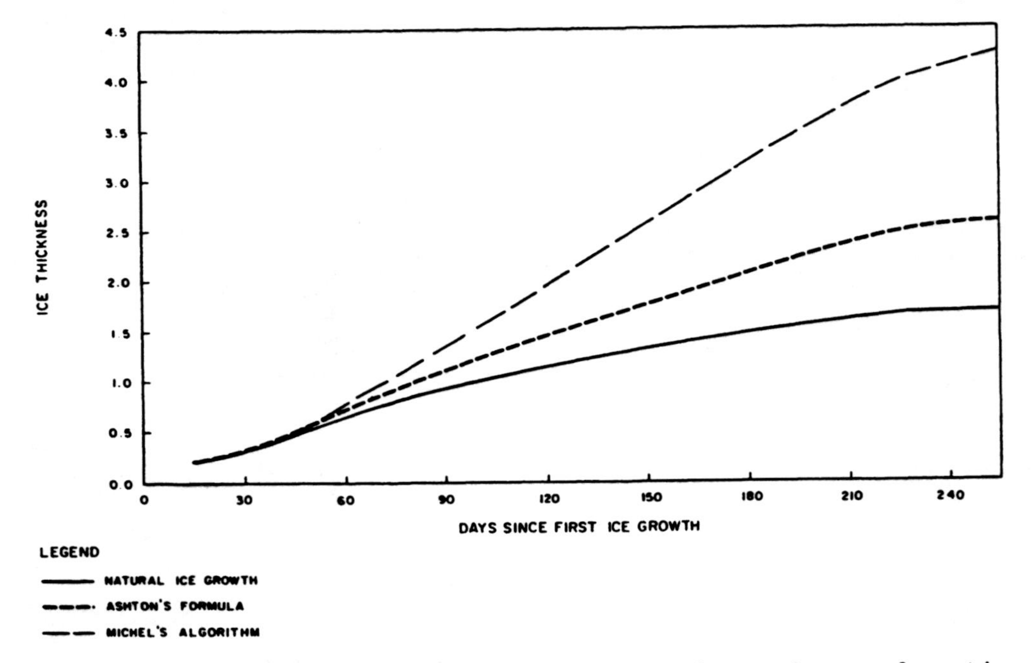

Figure 2–30. Example of theoretical ice growth in a ship track with repeated passages from a trip once every 15 days. (From Acres Consulting Services Ltd., 1983.)

ξ $= \epsilon$ for $\alpha(\Delta D)^{0.5} > (\Delta h - \Delta h^1)$
ν $= 1$ for $n \leq k$, and $\nu = 0$ for $n > k$
Δh^1 = distance between the top of the ice and the water surface
Δh = thickness of ice layer i, as calculated from $\alpha(\Delta D)^{0.5}$
α = local heat exchange coefficient
ΔD = number of degree-days of frost between successive ship passages
n = number of layers (passages) $1 < n < k$.

Sandkvist (1981) stated that the average field data observed at Lulea Harbor in the Gulf of Bothnia corresponded reasonably well to the results obtained by the above formulation.

The algorithm developed by Ashton (1974b) is based on a somewhat different assumption of ice growth in a ship's track. According to Ashton's observations, several passages of a vessel usually result in a more or less uniform coverage of the channel area. The percentage of open water surface area is in the order of 20% to 30%. As a result of this open water area, the production of ice tends to increase.

It is assumed that on each day of traffic, a fraction (ϵ) of open area is exposed in the channel and this fraction freezes and thickens as an initial ice cover. The other fraction $(1 - \epsilon)$ continues to thicken, but the effective initial thickness at the start of the freezing interval is assumed to be $h_{i-1}/(1 - \epsilon)$ to account for the fact that the broken ice is contained in the channel track and, hence, presents an effective thickness greater than before. Thus, the average thickness is given by

$$h_i = (1 - \epsilon)\left[\frac{h_{i-1}^2}{1 - \epsilon} + \alpha \Delta D^{0.5} + \epsilon \alpha \Delta D\right]^{0.5} \quad (2\text{–}70)$$

where

h_i = the average thickness at the end of the period of ice growth
h_{i-1} = the effective thickness at the beginning of the period.

Computer programs were developed for both of the above algorithms and these were used to generate example accelerated ice buildup curves based on an assumed number of freezing degree-days and vessel frequency (Acres, 1983). The results are illustrated in Figure 2–30. The normal undisturbed ice growth is also shown for comparison purposes.

Ettema and Hung-Pin Huang (1990) and Ettema (1991) produce the most recent state-of-the-art review on ice formation in frequently transited navigation channels and also reported results of extensive laboratory experiments conducted in ice tanks using model vessels.

A conclusion based on some field observations and ice-tank experiments tentatively drawn by the above investigators is that the ice formation in navigation channels is significantly affected by the hull form and length of transiting vessels, the channel geometry, and the presence of significant water currents. For example, quite different patterns of ice formation may occur in river channels that are plied by bluff-bowed, flat-bottomed tows or have significant water currents than occur in harbor and coastal channels, which are plied by ship-form hulls and are subjected to comparatively weak currents. The essential difference in the patterns is that transiting tows and water currents may cause broken and brush ice to be conveyed along channels and may consequently lead to the markedly nonuniform ice formation.

The latter investigators also concluded that frequent transit of ice-covered navigation channels resulted in increased volume of ice growth, but the increase may not be as dramatic as might be expected based on existing numerical models.

As fewer ice-breaking transits result in less ice growth, then convoying appears attractive because it reduced the number of ice-breaking transits. In their studies Ettema and Huang-Pin Huang found that a problem peculiar to tow transit of ice-covered channels is the accumulation of broken and brush ice beneath and along the flat bottom of tows. Ettema and Huang-Pin Huang also produced an extensive list of references on the subject of effects of vessel operation on ice growth in ship track.

2.5.6 Effects of Ice Buildup Due to Tidal Action

In tidal harbors, marine structures can accumulate exaggerated ice growth due to the freezing of water on vertical surfaces following a high tide. Such a growth creates a buildup of ice that may be sufficient to hamper ship berthing. In addition, this ice accumulation must be considered during design to ensure that the dock and the foundation can accommodate the added loads.

The initial layer of ice on the piles or columns forms during falling tide when a film of water is left on the pile. This layer gradually becomes built up with each tidal cycle and forms an ice "bustle". Depending on the dock structure, these "bustles" may eventually join and form a lattice of ice ribs between the piles or they may become incorporated into the larger ice sheet as it grows adjacent to the dock. The exact process depends on the dock configuration and the tidal extent.

In some cases, the tides cause the ice formed at the dock to effectively form an active zone that responds to the tides and acts as a hinge between the outer, larger ice sheet and the dock structure. For very low vessel frequencies, this active zone may build outward to a distance of several meters from the dock as well as increase in thickness to several times the regular ice thickness due to repeated floodings. This ice growth at the dock may severely hamper docking.

Some characteristic examples of ice buildup due to tidal action are discussed in Section 2.3 of this chapter.

2.5.7 Effect of Harbor Operation

From an operational point of view, minimizing the ice problems in harbors would expediate the normal procedures relating to ship and harbor operations. The intended operation of the harbor is obviously a basic input to the overall harbor and dock design, and in that sense, the harbor/dock system should be designed to mitigate ice interference with harbor and vessel operations. The degree of toleance of ice interference will depend to some extent on the use of the harbor. For example, if a vessel needs to berth immediately adjacent to the dock for loading and/or unloading, it would be essential not to have ice jammed between vessel and dock. On the other hand, if the vessel uses loading arms, it might be able to tolerate some ice in that area, and the ice control requirements might be lessened.

The degree of the ice problem will also depend on whether the operation is supported by a dedicated ice-breaking tug in the harbor. If ice-breaking tugs were available to aid in berthing, the ice problems facing the berthing vessel would be reduced. On the other hand, consideration would have to be given to the effect of the extra volumes of ice generated.

2.6 ICE CONTROL

2.6.1 General

Most of the ice problems in port can be mitigated by introduction of effective ice control techniques suitable for a particular location.

By definition, the ice control techniques are methods for reducing the growth of ice in navigational areas and on structures in navigation channels and in harbors, and methods for breaking ice and keeping the broken ice away from navigation channels. In general these methods can be classified as mechanical, thermal, or chemical.

Typical solutions for ice control fall within the following categories:

1. Ice breaking, which involves any possible means for ice fragmentation to ease and facilitate berthing maneuvers.
2. Ice suppression, which inhibits the formation of ice
3. Ice diversion, where the moving ice is deflected in order to prevent collision with a vessel or mooring structure. The latter is practiced primarily in dynamic ice situations involving movement of large ice masses
4. Ice removal and disposal, which involves the physical displacement of ice to minimize vessel interference.

2.6.2 Icebreaking

This method involves bending, breaking, crushing, and shearing of ice and includes use of equipment such as conventional icebreakers; air cushion vehicles; explosives; and mechanical, thermal, and chemical ice cutters that allow the passage of a vessel.

Icebreaking is commonly accomplished by various classes of icebreakers. The icebreakers can be classified according to their working environments, as this often determines the main characteristics of the ships. Accordingly, there are harbor, lake, river, and coastal arctic and subarctic icebreakers.

The existing harbor icebreaker fleet ranges from ordinary ice-strengthened tugboats to specially designed icebreakers for operation in harbor areas. Their tasks include towing and assisting ships, breaking ice, and clearing ice away from docks. These ships have to have good maneuverability, good towing capabilities, icebreaking ability suited to local conditions, firefighting capabilities, and pollution control capabilities.

The characteristics of a lake icebreaker depend on the size of the lake. On large lakes the icebreakers are similar to those designed for sea or coastal operations. On smaller lakes the vessels are more like harbor or river icebreakers.

River icebreakers must satisfy some extra demands compared to harbor and lake icebreakers. They must be able to operate in shallow water, be able to pass through canal locks, and be able to pass under bridges.

Coastal arctic and subarctic icebreakers operate between harbors and the open sea. Their characteristics are a mixture of those of the harbor, lake, river, and open ocean icebreakers.

Icebreakers provide by far the most popular and proven technique for ice control in areas where year-round shipping is conducted. However, this technique alone is often not adequate for year-round port/harbor operation where accelerated buildup of ice in the berthing area is taking place as a result of repeated passages of vessels or the icebreakers themselves.

In the latter case, the use of icebreakers in combination with ice removal techniques could provide practical solutions for conducting year-round berthing operations.

It should be noted that icebreakers are constantly undergoing engineering changes, making work in ice-affected waters more promising. Water jets, air bubblers, and cutting devices have been tried, adding new features to the methods of breaking only with the hull.

During the last decade, icebreaking using air cushion technology has been carried out successfully in Canada (Acres, 1983). Two modes of icebreaking are possible when the air cushion technology is used. At low speeds, icebreaking is achieved by the introduction of an air cavity under the ice cover, which causes the cover to fail under its own weight. The second mode of icebreaking, carried out at high speed, generates a wave in the ice cover behind the vehicle, which leads to cracking of the ice. Several passes may be required to break the ice, depending on its thickness. This method is suitable only for rivers and channels where currents are sufficiently strong to move out the broken ice. In calm water the cracks will refreeze, resulting in no benefit.

Application of the low-speed icebreaking technique in northern harbors has proven successful. It has also been found that the use of a low-speed air cushion icebreaking bow greatly enhances the ability of a ship to break ice in the confines of a harbor. For speeds up to about 8 km/h the thickness of ice that can be broken is directly related to the cushion pressure and is about 0.9 the cushion pressure head.

Experience of ice cutting with mechanical cutters of various types suggested that this method is practical and can provide ice-free channels through the solid ice (Mellor, 1980). Furthermore, it has been suggested that a series of low-pressure stream jets or high-pressure water jets can cut grooves in the top surface of an ice cover, allowing easier fracture by an icebreaker (Coveney, 1981; Dery, 1981).

Ice can be effectively blasted by use of different kinds of explosives. This is probably the oldest method for icebreaking and is still in use to effectively remove ice jams in inland waterways and to protect bridges placed across river from impact by large ice floes.

Detailed guidelines for blasting floating ice sheets are given by Mellor (1986).

Miscellaneous ice-cutting techniques such as use of a hot wire, lasers, microwave ablation, and fluid injection are discussed by Gammon et a. (1988).

2.6.3 Ice Suppression

Ice suppression is used to prevent ice formation and includes such techniques as thermal discharge, bubblers, use of chemicals, and ice dusting.

2.6.3.1 Thermal Discharge

Where waste heat from the cooling system of industrial plants such as nuclear power stations, thermopower stations, gas liquefaction plants, or other industrial processes is available, it may be a valuable source of heat used for and to inhibit growth of ice.

This may be accomplished by discharging warm water directly into the ice control area through a system of diffusers or nozzles. The effectiveness of a warm water outlet depends very much on the local hydrography, the ice conditions, the outlet water properties, and the outlet method.

The density difference between the recipient and the outlet water is a function of temperature and salinity. Warm water with no salinity and a temperature of +10°C that is let out on the surface of cold, fresh water becomes heavier while losing heat. With a surface temperature below +8°C the outlet water sinks and its ice-reducing effect is lost. In more saline water the outlet water will stay at the surface with all its heat available for deicing purposes.

To avoid heat concentrations in small areas, with heat losses and ice fog problems, the outlet water must be spread over large areas and have an initial temperature of just a few degrees above the freezing point. Many outlet points may be used. The heat losses are reduced considerably when a thin ice cover is kept on the water surface.

If warm water is discharged on the bottom of a harbor entrance area or in a ship channel, means have to be provided to transfer this water to the surface. This can be accomplished with an air bubbler system. The air bubbler line should be attached or placed close to the warm water discharge pipe.

Probably the most efficient way to use heat is to eject the warm water vertically and into the immediate vicinity of the ice cover (Gill et al., 1983). By keeping the warm water discharge arrangement in close proximity to the ice cover, the heat loss to the ambient water would be minimized and the temperature of the water contacting the ice surface would be maintained at a higher level. Although numerous examples exist of warm industrial waste water discharge into ice-covered areas there is little documentation on the design of such systems for the purpose of ice control and ice management to mitigate navigation or ship maneuverability during docking operations. Useful information on this subject is found in Carstens (1978), Carey (1979), Cammaert et al. (1979), and Acres (1983).

2.6.3.2 Bubbler System

Air bubbler systems suppress ice formation and thereby allow navigation in harbors, ports, and waterways during periods when thick ice would otherwise halt navigation. They are also used to suppress ice in and around dock structures and recreation vessels. They suppress ice growth when the bubble plume entrains the warmer bottom water and brings it to the surface.

The most important requirement for the successful performance of an air bubbler system is a supply of warm water. In most harbors winter water temperatures are not as high as 4°C but are generally more than 0.2°C above the freezing point, which is usually adequate to supply the necessary thermal energy.

The release of air bubbles at depth below the water level from a system of pipes creates an upward circulation of the warmer water layer to replace cooler water at the surface. This heat, in addition to turbulence caused by the surfacing bubbles, retards the formation of an ice cover in the immediate area.

In addition to this application, which has been used in bodies of water with a natural thermal stratification, bubblers may also be used as a means of

- Increasing heat transfer to an ice cover in areas of an articially generated heat reserve, that is, in conjunction with a thermal discharge

- Providing a barrier to heat loss through water flow in thermal discharge systems by generating a countercurrent (a pneumatic barrier)
- Providing a barrier to the movement of ice pieces.

In designing a low-flow bubbler system the output pressure must be sufficient to overcome the hydrostatic pressure and the frictional losses in the supply and diffuser lines, and yet provide a pressure differential at the orifice to drive the air out at the desired rate. The required air discharge rate is controlled by the system geometry (length, pipe diameter, etc). Typical air discharge rates used in field installations have been on the order of $2.1 \times 10^{-3}\,m^3/s$ per 30 m (PIANC, 1984). The supply and diffuser line diameters should be large enough that the pressure drop due to friction along the line is small so that a uniform air discharge rate can be maintained along the line. Typical line diameters at field installations are between 40 mm and 80 mm.

Submergence depth is generally governed by operational limitations such as depth of water body or required clearance for vessel drafts. The deeper the submergence, the water will be moved by a given discharge rate, and hence the more suppression effected, but a larger compressor pressure is needed as the depth increases. In some cases, very great depths require pressures that make it desirable to suspend the line above the bottom.

Typical orifice diameters are in the order of 1.2 mm, and typical spacing is about one-third of submergence depth. Orifice diameters that are too large can result in all the air leaving the diffuser line at one end. The pressure, diameter, and discharge are interrelated and cannot be separated easily from the total system design.

Several types of bubbler arrangements have been used for the retarding of ice growths, but typically a system consists simply of perforated pipes that are placed on the bed or at predetermined depths of water and connected to an air compressor. The application of air bubblers to create ice-free conditions in freshwater is common but installations in seawater generally have been unsuccessful, although there are isolated examples of limited success.

In the latter case, the ice suppression apparently resulted from the movement of water with higher salinity and thus a lower freezing point.

More and detailed information on bubbler systems and their effects on ice suppression is given in Ashton (1974a,b and 1979), Carey (1979), Cook (1982), and Acres (1983).

High-flow air screens based on use of a bubbler system are being used to keep floating ice away from docking facilities and out of navigation locks. In the latter case, the rising air flow entrains and accelerates a large column of water, which must be deflected outwards when it reaches the surface. This resulting surface current is sufficient to prevent ice from being carried across the air screen line.

Air screen systems have been placed across navigation channel entrances, with subsurface emplacement depth varying from 5 to 10 m. As production of high water velocity at and near the surface is the goal, a high air flow through bottom nozzles is required. High pressures are also required to overcome the hydrostatic pressure and pressure losses in the supply line, while still leaving an excess for sufficient pressure to provide the needed orifice (nozzle) flow. The design parameters for an air screen system would include the submerged depth, supply line size, orifice size and spacing, air supply flow rate and pressure, and air screen manifold size and length.

An air bubble curtain is in effect a mechanism for transporting water to the surface, thus creating a horizontal current. The same effect can be obtained by using mechanical pumps (flow developers) in restricted areas. The choice between the two pumping methods will be dependent on local conditions.

2.6.3.3 Ice Dusting

Both laboratory and field experiments have shown that dusting of an ice surface with coal dust or the dark granular material can increase the absorption of solar radiation and so increase the rate of melting of the ice.

A typical value for the required change in albedo, A_d/A, is 0.3, where A_d equals the albedo of the covered ice and A equals the albedo of the original ice. The ratio of that part of the surface to be covered by particles to the total melting surface is taken as 0.7 (Acres, 1983).

For these requirements, dusting rates for two different particle diameters have been established. For a diameter of 0.5 mm, the dusting rate is 0.20 kg/m^2; and for a diameter of 0.2 mm, the rate is 0.07 kg/m^2.

Air temperature and the amount of sunshine during the early stages of breakup are critical factors in ice dusting operations. Ice dusting is not very effective if the daily minimum air temperature falls much below 0°C. If the air temperature is too low, heat losses from the surface by convection, evaporation, and long-wave radiation will offset the increased solar energy abosrbed by the darkened ice surface. Approximate calculations, based on the amount of solar radiation during the breakup period, indicate that dusting can increase melting rates up to about 1 cm/d in latitudes of 45°N and up to 2 to 3 cm of ice a day in latitudes of 70° to 80°N (Williams, 1970). Several days of weather conditions conducive to increasing melt rates are needed to decrease significantly the length of the breakup period or weaken thick, stable ice covers. The effectiveness of dusting is decreased by snow accumulation and melt puddles, the latter actually decreasing the albedo by essentially the same amount as the dust.

Successful dusting operations require detailed studies of past weather records, past ice conditions, and dusting techniques. Information is needed on the frequency of new snowfall during potential dusting periods, the availability of suitable dusting material, and the cost of transporting and applying the dust. If potential benefits exceed anticipated expenditures, large-scale trials may have to be carried out before the most effective technique is developed for a site.

2.6.3.4 Chemical Methods

Reduced ice strength at the time of ice breaking is one benefit achieved when chemicals are used. The chemicals, for example salt, can be spread on an existing ice cover or can be mixed with the water before freeze-up.

Due to negative environmental effects the use of chemicals must be carefully localized; for example, piles can be protected from ice-lifting forces by salting the adhesive ice around them.

The many outlets that discharge more or less polluted waste water increase the general level of chemicals in the water mass, resulting in reduced ice strength.

Chemicals along with electric heating can be used for preventing ice from adhering to wharf or lock walls.

2.6.4 Ice Diversion

Ice diversion is the deflection of moving ice floes to prevent collisions with harbor structures. Movement of ice in harbors and waterways can be controlled by numerous methods. Among them ice booms, small islands, breakwaters, large cells, dolphins, clusters of piles, and flow developers of miscellaneous constructions are only a few. The most common of these structures is the ice boom.

2.6.4.1 Ice Booms

Booms have been used to form an ice cover primarily for the control of river ice and mainly for the purpose of preventing slush ice intake to power plants rather than the prevention of ice interference with port and harbor structures. They are now being used as navigation aids to hold in place ice broken by ship passage, to prevent it from flowing downstream and blocking channels, or to protect docking facilities and keep them operational. Booms are also used to help form and retain stable ice covers, thus keeping navigation tracks free from drifting ice. Booms have been built in many configurations; some cross the entire channel, some have a gap to allow for ship passage, and some restrain ice on only one side of a channel. There is considerable discussion in the literature on the installation of a number of booms on the St. Lawrence Seaway system. A study by Lewis and Johnson (1980) focused on the use of booms on a portion of the Seaway to allow the safe transit of vessels while minimizing the movement of ice downstream and maintaining a stable ice cover. Navigation channels were maintained by using booms with gaps to allow ship movement. Such an installation on the St. Mary's River connecting Lake Superior and Lake Huron has proved to be successful (Perham, 1977).

The successful Lake Erie–Niagara ice boom installation is discussed by Bryce and Berry (1968).

The tension T in the boom cables connecting the boom elements is given by Perham (1974, 1985) as

$$T = \frac{xF_D}{\cos\theta[4(s/x)^2 - 1]^{0.5}} \qquad (2\text{–}71)$$

where

x = one half the cable span
s = boom cable length
F_D = total drag force per unit length on the boom due to water and ice
θ = boom cable angle.

The total force on the boom is the sum of the following forces:

The hydrodynamic force at the upstream edge of the ice cover
The frictional drag of the water flowing under the ice and the air above
Downstream component of the weight of the ice cover
Impact force of collecting ice floes
Frictional force from the shore line.

Parfenov et al. (1973) described the use of floating booms to prevent broken ice from entering berthing areas. One such design formed an adjustable barrier consisting of a series of floats pressed together, forming a barrier when broken ice was driven against it. As a ship passed through the barrier, the floats located directly in front of the hull were parted and formed compact accumulations of ice to the sides. After the ship has passed through, the floats returned to their initial positions. However, there is no indication in the literature as to whether there were any installations based on this design.

Tsang (1981) described a finned boom that was developed for the purpose of recovering oil in ice-affected rivers. Model studies in synethic ice indicated that such a boom could be used in harbors to allow passage by vessels while still retaining the broken ice pieces. Consideration might be given to the use of such booms in a harbor to prevent ice broken by an exiting vessel from floating back into the sea in front of the dock. This would minimize the ice buildup in front of the dock because most of the ice in that area is pushed by the vessels as they depart the dock.

As reported by PIANC (1984) the most common boom section in use today is made from a number of $0.3 \times 0.6 \times 6.0$ m Douglas Fir timbers. Some use a double pontoon which has two parallel rows of timbers. These timbers are chained to and support a wire rope connected to bottom and shore anchors at 30 to 130 m intervals, depending on the load.

Ice booms are designed to allow ice to override them when the ice load becomes excessive. A typical single boom has a load capacity before override of approximately 7.0 kN/m (70.0 kN/m for a double pontoon boom).

The cost of building ice booms varies substantially according to their construction and the type of anchors used. The cost of the anchors can amount to over 40% of the total.

2.6.4.2 Structures

Harbors that are subject to large, massive ice runs will require a structure other than a typical ice boom. In some areas, it might be feasible to consider the use of structures such as artificial islands, breakwaters, and berms to modify ice movement in an ice control zone. Such structures would be strategically located to prevent the impact of ice on the dock facilities by immobilizing the ice or by diverting it away from the control area. Such structures might also be used in conjunction with other ice control techniques to provide an overall control system. An example would be the use of a berm serving as an enclosure for a thermal discharge system while at the same time preventing the ice beyond the control zone from impacting on the activities or strucutres in the control area.

To date, the application of such ice control methods has been almost totally limited to river ice. Artificial islands have been constructed in the St. Lawrence River to stabilize the ice cover and prevent jams and flooding (Danys, 1979). At the Golden Eagle petroleum terminal opposite Quebec city, three sheet-pile cells are located at the upstream end of the wharf in a line at right angles to the berth to stop the ice and deflect it toward the center of the river (Dery,1981). An ice deflector is also used at a ferry landing wharf at Levis, again near Quebec City. This deflector consists of a concrete box filled with rock and causes ice moving downstream to be deflected past the berth area with minimum restriction.

A recent application of a protective structure in the Arctic is the use of by Dome Petroleum Limited of a dredged island to protect a fleet of drill ships in their winter mooring basin in McKinley Bay. The berm is meant to prevent landfast ice movements due to west and northwest winds from impacting the fleet. Such a structure was deemed to be necessary following substantial ice movement in McKinley Bay in December 1979.

In recent years, designs for offshore arctic petroleum production and loading terminals have focused on the use of artificial islands or underwater berms supporting concrete or steel caissons to provide a protected "harbor" enclosure. These structures would provide protection from the moving pack ice but would still require some form of ice management inside the enclosure area in a manner similar to other dock facilities that are operated in landfast ice conditions.

River harbors or docking facilities are often subjected to damage from floating ice. Piles, single or in clusters, placed upstream of these facilities have been used to deflect the ice. The piles are driven deep enough in the soil to withstand the force of the moving ice, and are banded together if placed in a cluster. More than one cluster would be required for larger facilities. Breakwaters and artificial ice islands are also used to stabilize the ice cover.

2.6.4.3 Flow Developers

In a broken ice regimen, flow developers might be used to move or direct ice away from dock structures. Flow developers could take the form of a standard air bubbler, an air cannon mixer ("burper"), wave generator, or propeller system.

Air bubbling units are used as pneumatic barriers by the St. Lawrence Seaway Authority to prevent ice from entering the locks where it gets crushed against the lock walls by ships passing through (Poe, 1982). An extension of this application would be as a barrier to prevent broken ice from moving into the berthing area in front of a dock as a vessel exits. This would minimize the buildup of ice in the berthing area and would facilitate berthing.

Flow developers are commonly used in the Baltic ports of Finland and Sweden to maintain ice-free areas in special berthing places such as Ro/Ro ramps. The successful functioning of flow developers is based on three criteria: thermal reserve of water, surface currents that prevent freezing, and removal of ice blocks. Engine power should be 30 kW in order to produce significant ice-free areas when the water temperature is less than 0.25°C above the freezing point (PIANC, 1984). Flow developers can successfully be used together with warm water discharge.

Propellers could be used behind docks, where the design allows, to push ice to the rear thereby creating a low resistance zone into which ice could be pushed by a berthing ship. This could be quite feasible with an open-face dock that is a sufficient distance from shore to allow for displacement of the ice.

2.6.5 Ice Removal and Disposal

A number of possible techniques may be addressed for the removal and disposal of ice in critical ice control areas. It may include ice lifting, towing, or pushing, and use of conveyors and slurry pipelines for ice disposal. However, feasibility and practicality of the above techniques are yet to be established. This would require a substantial amount of research and development.

2.6.6 Ice Management in a Berthing Zone

Previous studies for the year-round navigation in cold climate regions and particularly in the Arctic have indicated that an effective ice management system is a key component to the dock operation. An important function of an ice management system is to alleviate the problems associated with ship movement into the dock berthing area and the reduction of any kind of forces (static or dynamic) imposed by ice on the dock structure.

In addition to previously discussed ice control techniques, ice management in a berthing zone may include use of floats and pontoons of miscellaneous designs placed in front of the dock structure, and utilization of special "hold tanks" of warm water, where previously crushed ice can be effectively melted.

2.6.6.1 Floats and Pontoons

Depending on the ice control problem, techniques that incorporate structures in the berthing area could minimize ice buildup. For example, if the growth of the ice due to tidal effects is the primary concern, than a sliding face on the dock support structure could move with the tides and thereby prevent the exposure of a layer of water to the atmosphere. The common ice collar or "bustle" would then be prevented from forming.

If the primary requirement is to minimize the ice buildup due to vessel traffic, a removable structure could be placed in the berthing area. A conceptual design of such a structure is the submersible pontoon shown in Figure 2–31 (Acres, 1983). On approach of a vessel, the pontoon would be filled with water and sunk to the bottom; when the vessel departs, the water would be pumped out and the pontoon refloated. The pontoon would be hinged in order to guide its up-and-down movements. The top surface could be

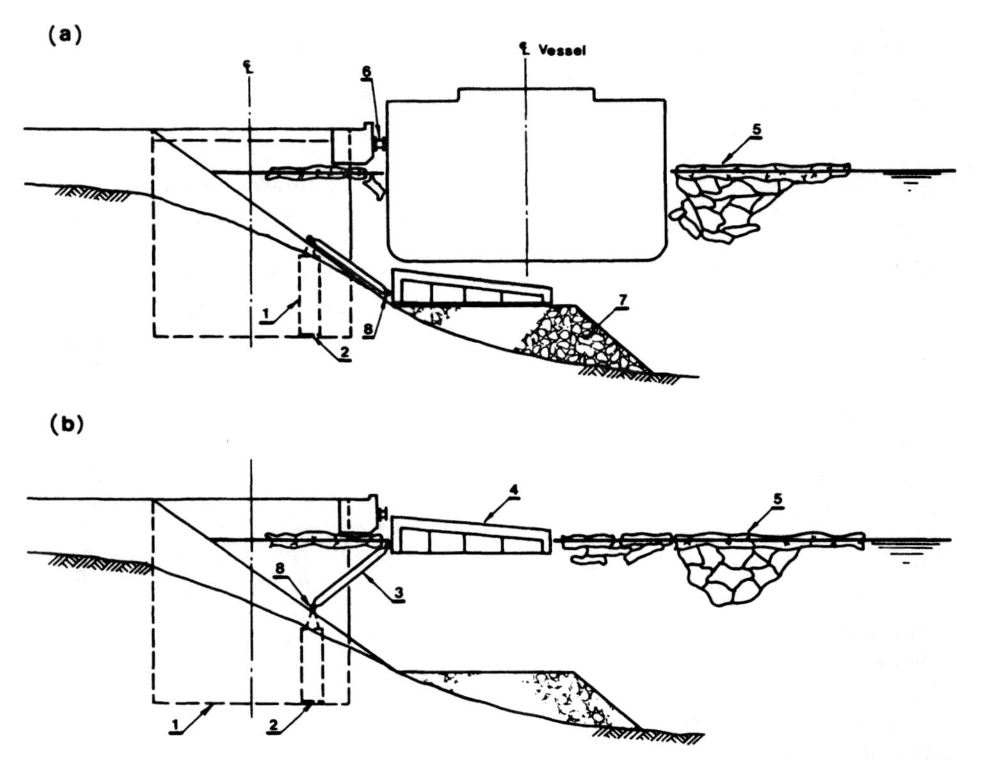

Figure 2–31. Conceptual design of ice management system. (a) Ship is moored at the dock and insulated submersible pontoon sits underwater on a rockfill berm. (b) Ship left the dock area; the pontoon is refloated to keep the berthing area ice free. 1—berthing structure, 2—foundation for the articulated boom, 3—articulated boom, 4—insulated submersible pontoon, 5—ice cover, 6—fender system, 7—rockfill berm, 8—universal hinge.

sloped so that any ice in the berthing area would be lifted by the pontoon and would slide to the outer edge of the berth. Insulation on the inside of the pontoon would prevent the water from freezing.

To avoid the problem of ice fog, which would hamper navigation during berthing, the pontoon would be submerged before the vessel reaches port so that a thin layer of ice will have had an opportunity to form in the berthing area.

An important consideration arises in coping with the ice that would be pushed back into the berthing area by the departing vessel. It would probably be necessary to prevent most of this ice from entering the berthing area; otherwise, large amounts of ice would be eventually built up on the outer edge of the area. Additional systems could be implemented to prevent this. Such systems could consist of ice booms that allow the vessels to pass through, or surface propeller units to repel the ice. Alternatively, it might be possible to stop the vessel immediately beyond the berthing area until the pontoon is raised.

There is no precedence for such an application to ice management and further investigation would be required to establish its feasibility.

2.6.6.2 Insulating Covers

Antifreeze films or foaming compounds placed on the surface of the water can decrease the heat loss from the water to the atmosphere, thus preventing formation of fog and ice (Grove et al., 1963; Stahle,

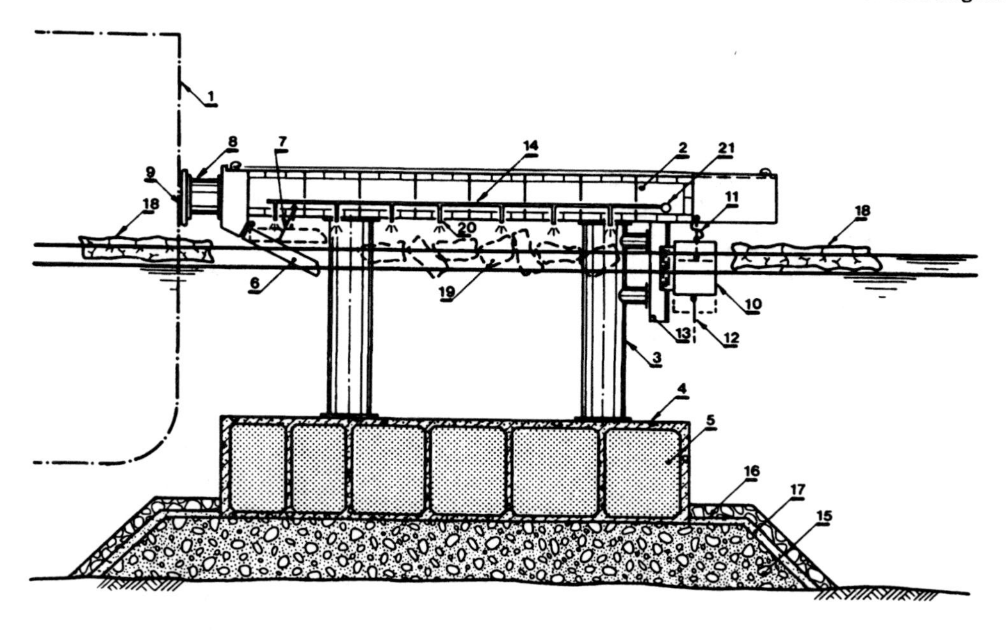

Figure 2–32. Arctic pilot project. Year-round operating LNG terminal 1—LNG icebreaking tanker, 2—insulated steel deck, 3—steel column, 4—concrete caisson, 5—granular fill, 6—articulated ice deflector, 7—ice deflector servomotor, 8—pneumatic rubber fender, 9—continuous rubbing panel, 10—steel pontoon, 11–flexible upper curtain, 12—bottom curtain, 13—pontoon guiding system, 14—warm water distributor, 15—rubble mattress, 16—protective armor, 17—gravel, 18—ice sheet, 19—broken ice, 20—water jet, 21—manifold.

1964; Parfenov et al., 1973). The insulation also could be placed on a relatively thin ice cover to retard ice formation.

The insulating material must be light enough to stay afloat and be substantial in thickness to provide the required degree of protection over the insulated area.

As the method of insulation is highly susceptible to wind, waves, and current action as well as to ship movements, it cannot be considered reliable or practical to maintain an ice-free area in front of a dock.

2.6.6.3 Ice Crushing and Melting

Instead of being transported away by pipeline or other methods from the berthing zone, the crushed or chipped ice can be dumped (pushed) into a "holding tank" of warm water. This method could be well illustrated by conceptual design of the Arctic Pilot Project (APP) ice management system (Cammaert et al., 1983) (Figure 2-32).

The APP is located at Bridgeport Inlet which is a natural harbor opening into Viscount Melville Sound. The major problem with ice management there is caused by the potential volume of ice produced due to the exposure of open water after each ship passage. Natural ice growth there averages about 2.2 m throughout the winter. An analysis of the accelerated ice growth caused by ship arrivals and departures shows that an ice thickness of over 7 m may be generated during winter season.

Consequently, it was necessary to implement an ice management system that would control the volume and hence the thickness of the ice to a value that would allow the ship to maneuver into appropriate position at the berth.

The ice management system for the APP liquid natural gas (LNG) terminal has been designed to allow the ice to be pushed underneath the deck structure by the vessel while docking. Once displaced under the deck, the ice would be melted there by employing a special network of jets of warm water from the condensate discharge of the gas liquefaction plant (Figure 2–32).

It is obvious that under continuous and thick ice cover condition the ship must approach the dock without the use of tugs. The unusual concept of using the LNG carrier to move the ice and a need to operate the very large ship (136 000 DWT) without the tug assistance means that the ship must approach the dock at an angle, make contact, and slide along the dock face until it is brought to a complete stop. The latter of course means that the dock structure requires an exceptionally powerful fender system, with a continuous rubbing face to berth the vessel.

At APP the total rate of heat output available from the cooling water was estimated as equal to 111 MW, and it was determined that only a fraction of this heat would be required to melt the ice cleared from the berthing zone.

The design criteria for the ice management system was that it should be able to melt the ice under the dock in a period of 12 days, which was considered an average interval between ship arrivals (Acres, 1981).

Heat distribution to the ice management zone was by spray jet network, which distributes warm water above and below the ice. To prevent heat dissipation from the ice melting zone, two barriers in front of the dock and on the rear have been designed.

A heat barrier in front of the dock is designed in the form of an articulated ice deflector operated by the hydraulic system; the deflector is raised while the vessel is docking to permit the ice fragmented by the ship to move inside the ice melting chamber. After the ship is safely secured alongside the dock, the deflector is lowered beneath the water level to a minimum depth of about 2 m. Warm water would then be discharged through nozzles placed strategically at the bottom of the deck and piers.

The rear barrier for heat containment was designed in the form of an ice-resistant floating pontoon with two curtains—an upper flexible curtain connected to the deck and pontoon, and the bottom curtain suspended from the pontoon.

There also was a system of nozzles placed at the outer faces of each pier to form the secondary spray jet network. The purpose of this system was to inhibit the ice cover in front of the dock in a period between ship arrivals. A large-scale model study with the objective of investigating the maneuvering capability of the very large LNG tanker during the docking procedure and the ability of the dock ice management system to absorb broken ice under the deck has been conducted by Acres (Cammaert et al., 1983a). The tests have been conducted in a test basin 8 m by 20 m. A special synthetic ice was prepared, consisting of polypropylene pellets sprayed with a mixture of paraffin and oil. The ice properties were adjusted to give required scaled-down values of density and breaking length. A 1 : 100 scale model of the LNG carrier was constructed, and fitted with a dual electric propulsion system, with separate radio control for rudders and each motor. The terminal was constructed from acrylic plastic to allow observation of ice movement.

Individual tests were carried out with different docking maneuvers and various configurations of the ice deflector at the face of the dock. These were followed by a simulation of the ice buildup over a complete winter, and for a typical ship arrival schedule.

One of the early conclusions of the test program was that it is extremely difficult to push ice under a rigid deflector. The model ice tended to break up by the action of the ship and contact with the dock. It typically formed a pile of rubble between the ship and the dock during berthing. When the rigid deflector was replaced by a hinged or a flexible deflector, ice easily passed into the dock enclosure. The degree of berthing difficulty was observed to increase with ice thickness, as expected. Nevertheless, with the anticipated ice thicknesses in front of the dock, berthing should be possible throughout the winter. The approach angle, however, is a critical element to docking capability; berthing will not likely be possible at angles greater than 15°. Similarly, a parallel docking maneuver, where the carrier is pulled into the dock by mooring line winches, will also be very difficult because of the force required to break ice in crushing at each of the piers. Berthing at higher approach angles and parallel berthing also pushes an excessive amount of ice under the dock, increasing the volume to be melted before the next ship arrival.

The unique capability of the dock to clear ice in front of the dock at each ship arrival is an important component of the total ice management system. It was observed in the model tests that an area approximately equivalent to 45% of the ship track width and the overall length of the ship is pushed under the dock. When the ship leaves the dock, the ice is redistributed by the action of the ship propellers, over at least three ship lengths. This means that on average, the net volume cleared in from the dock is about 15% of the original volume, for each ship arrival.

The conceptual design of an LNG dock for APP was in many aspects a milestone, an unprecedented departure from the conventional practices of arctic dock design.

APP ice management system design, however, is not free from some of the following deficiencies:

- Limited protection for unwarranted heat dissipation into the ambient water at the front of the dock (due to limited depth of ice deflector when closed)
- Potentially high heat conductivity of an upper rear flexible curtain and associated heat loss into the ambient atmosphere. Furthermore, because of its large sailing area this curtain is also highly susceptible to wind forces and could be in danger of being torn off the structure.

Tsinker (1986) proposed a novel concept of an ice management system relatively free from the aforementioned deficiencies (Figures 2–33 through 2–35).

The dock structure for which this new ice management system is applied is basically the same as was proposed for the APP.

The basic elements of the dock are concrete piers, steel deck, and the floating ice melting system. The dock structure concept has been developed in a way to make maximum use of prefabricated components, thus minimizing the work on site to the limited construction window available in the summer season. The deck is designed to carry live loads and to transmit berthing loads to the piers.

With reference to Figure 2–33 and 2–34, the ice melting system is designed in the form of a floating frame-like structure places under the deck along the dock. The structure is split into individual sections installed between adjacent piers. This arrangement makes the system more flexible in its response to environmental forces (ice and waves) and mitigates potential problems associated with construction, operation, and maintenance. The floating frame consists of front (3) and rear (4) floats, joined together by transverse floats (buoyancy tanks) (8) and an insulated deck (5). Low friction bearings (10) are provided on both the floating frame and the piers to allow for relatively unrestrained, vertical movements of the frame in response to the systems's operational requirements and tidal fluctuations.

The face of the front longitudinal float is sloped to break sheets of ice in the bending mode of failure. All surfaces exposed to ice action are covered with a low friction ma-

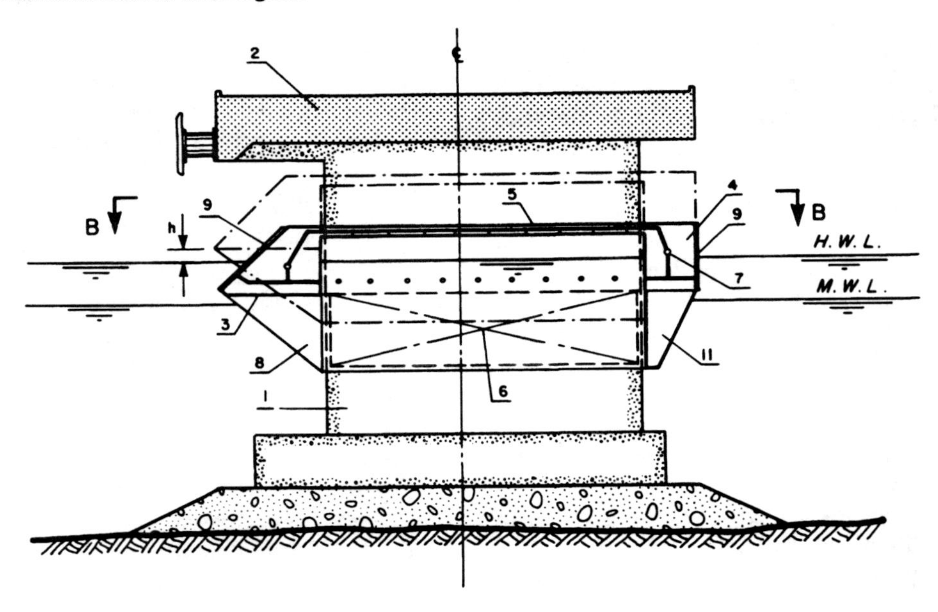

Figure 2–33. Floating concept of ice management system. (From Tsinker, 1986.) Typical cross-section (A-A). Plan (B-B). 1—Pier, 2—deck structure, 3—front float, 4—rear float, 5—insulated cover, 6—ballasting system, 7—warm water (air) discharge system, 8—buoyancy tank, 9—low friction cover, 10—low friction guides, 11—ice containment, 11—ice containment structure, 12—floating insulation (e.g., closed cell foam).

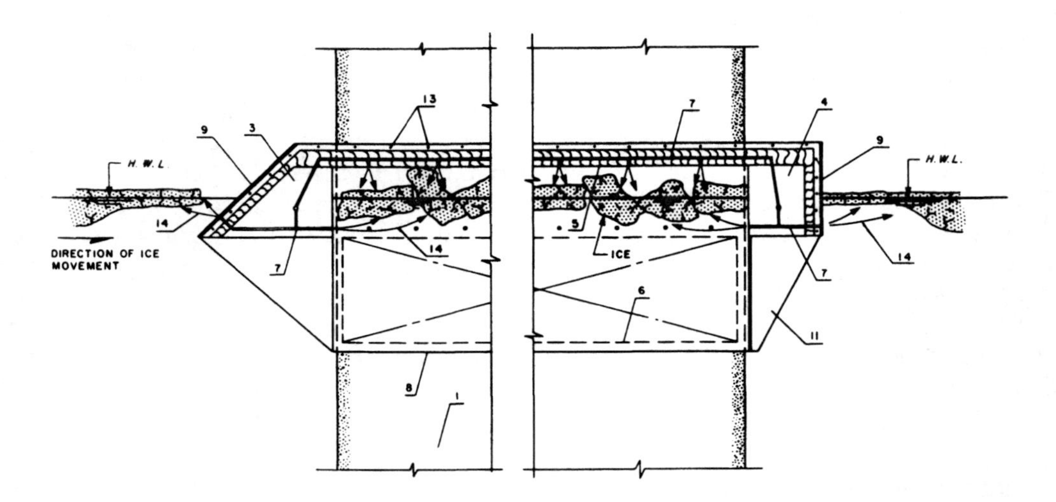

Figure 2–34. Floating concept of ice management system, details. (From Tsinker, 1986.) Note: For legend see Figure 2–33. Additionally: 13—heater, 14—warm water jet.

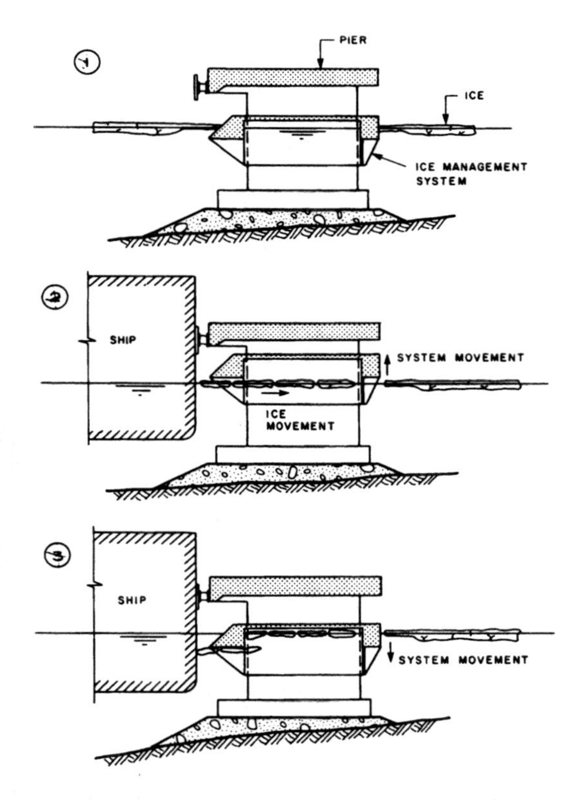

Figure 2–35. Floating concept of ice management system. Sequence of operation. (From Tsinker, 1986.) 1—System in submerged position between scheduled arrivals of a ship, 2—system is raised to provide an open space for broken ice pushed by the arrived ship, 3—system is brought down to confine ice sheets within melting enclosure.

terial (9) to reduce ice friction forces and alleviate possible ice adhesion problems. In addition to low friction cover the latter surfaces, if required, may be kept heated (13). The space between the transverse floats is used to accommodate piles of ice pushed toward the dock structure by the berthing ship.

In principle the new ice management system would operate in a manner similar to the original APP facility design.

While a ship is docking, the floating frame is raised up with help of the ballasting system (6) to provide a gap (h) between the bottom of the front float and the water level (Figure 2–35, Phase 2). The gap should be sufficient to permit free movement of pieces of ice pushed by the docking ship into the ice melting enclosure. After the ice is trapped inside the chamber provided by the rear and transverse floats, the system is ballasted. This brings the system down, thus effectively isolating the space between all floats (Figure 2–35, Phase 3).

The ice trapped between floats could be effectively melted by utilization of bubbler systems to convey heat from the deeper water (if available) to the surface. Alternatively, as it may be necessary in the Arctic, the warm water or warm air can be discharged through a special piping network placed inside of a floating system (7). Naturally, where appropriate, both the bubbler system and warm water and/or warm air application can be combined. The warm water could also be discharged to the outside of the floats in order to control the ice growth outside of the ice management system. The heat provided to melt the ice would be used in the most efficient way. This is because the surface of water inside of the ice melting space is completely and effectively insulated by the floating frame structure.

The front float (3) has an inclined face exposed to the ice. This would help to break the wind-driven large ice sheets in a bending mode of failure. However, at a great many locations, specifically at ports located in inland waterways, or at artificial atoll type harbors, protected against ice movement, the face of the front float could be vertical. Floats for the ice management system could be prefabricated from concrete or steel, and subsequently installed in situ.

The proposed new concept of ice management system has the following advantages.

1. The system's performance could not be hampered by tide. It would fluctuate with the tide, thus preventing ice active zone buildup in front of the dock.
2. The system offers an efficient use of the heat supplied for the ice melting process.

3. Structurally the system is simple. It could be manufactured off site and then barged to the site and mounted into place.
4. The system does not require sophisticated equipment for its operation and is relatively simple for on-site maintenance.

Additional discussion on some conventional and unconventional methods for ice control at vertical face structure is offered by Nakabo et al. (1989).

2.6.7 Environmental Aspects of Ice Control Management

The environmental aspects of ice control/ management system implemented at the harbor that have to be taken into account depend on the type of a port activity and the site environmental conditions. For example, ice navigation itself changes the ice conditions; several ship passages, with breakups and refreezing periods, cause an accelerated growth of brash ice. Ice boom arrangements forming a stable ice cover in a river may change the risk of ice jamming and flooding at breakup.

Erosion problems in narrow, shallow sections in dredged channels and under docks may be caused by currents generated by propellers of powerful vessels. Furthermore, the extra propeller thrust often used for maneuvering ships in ice may cause severe erosion and damage the foundations of piers and other structures.

Thermal systems have an effect on the environment. The effects of warm water outlets depend on the local hydrographical conditions. Subsequently, the expected spreading and mixing of outlet water have to be examined to avoid "thermal ponds".

Ineffective mixing of warm water may generate an open, ice-free area near the outlet point. The heat losses from such areas will be large, and during cold days fog will be generated, reducing visibility for navigation. The fog may also cause severe icing on equipment and structures.

In harbors at river mouths a warm water outlet may hinder migrating fish on their way upstream. Environmental aspects have to balance the ice-reducing aspects. On the other hand, some ice control measures can increase the oxygen content in the water and in this respect improve conditions for fish. For example, an air bubbler system will reduce stratified conditions in a water body by mixing.

2.6.8 Ice Control (Management) Selection Criteria

The effectiveness of different ice management techniques could be evaluated on the basis of the following criteria.

1. *Environmental Effects*
   The negative and positive effects on the local environment must be identified.
2. *Effectiveness*
   This should be evaluated on the basis of a system performance and its cost; the cost of construction and cost of operation must be considered.
3. *Reliability*
   Because a high degree of reliability is required, "passive systems" that do not require mechanical equipment for operation purpose are usually more preferable. Such a system must be operational under all ice and weather conditions, and when necessary to be able to be replaced or repaired as soon as possible. For the latter, replacement parts should be available for quick delivery.
4. *Operational Safety*
   The system must not have adverse effects on the safe operation of the dock, and should be acceptable from a personnel safety point of view.
5. *Power Requirements*
   Power requirements for system operation should be kept to a possible minimum.

Table 2–7. Summary of Ice Control Measures

		Effectiveness		Annual Cost						
	Alternatives	Docks	Harbor	Docks	Harbor	Reliability	Safety	Environmental Impact	Energy Consumption	Experience
Thermal, chemical methods	Air bubbler	Yes	Low	Low	Low	High	High	Low	Low	Proven
	Thermal discharge where available	Yes	Yes	Low	Medium	High	High	Low	Low	Proven after few studies conducted
	Insulation	Yes	No	Medium	—	—	High	Some	Low	Unproven
Mechanical methods	Icebreaker	Yes	Yes	High	High	High	High	Low	High	Proven
	Air cushion vehicle	No	Yes	—	High	Medium	Limited	Noise	High	Proven
	Soviet LTS device	No	Yes	—	High	High	High	Low	High	Proven
	Mechanical ice cutter	Yes	No	Low	—	Low	Low	Low	Low	Little
Ice-stabilizing structures	Navigation boom, artificial island	—	Yes	—	Low	High	High	Low	—	Proven

From PIANC, 1984.

6. *Construction*
   The ice management system should, if possible, be capable of being manufactured and tested off the site prior to installation. Systems that require long construction periods, unusual techniques, or extensive on-site labor should be avoided.
7. *Precedents*
   Limited precedents exist for ice control techniques, particularly under Arctic conditions. Because each particular location requires special, often unique ice management techniques, there is unlikely to be a precedent available where relative success can be demonstrated. However, performance of any proposed system should be predicted with a high level of confidence, either by theory or by model tests.

Preliminary evaluation of ice management techniques for a particular application could be carried out by establishing a set of "weighing" factors for all the criteria. From this evaluation a number of possible solutions can be identified and these should be subjected to a more detailed evaluation.

A summary of ice control measures is given in Table 2–7.

2.7 DOCK STRUCTURE: DESIGN CONSIDERATIONS

2.7.1 Loading

Design of docking facilities in general and in cold regions in particular is site-specific. While basic inputs to the design of any dock will be the intended use of the dock and the characteristics of the associated vessels, docks designed for operations in ice should exhibit features that minimize resistance to the berthing vessel and ice forces on the structure, and that mitigate ice problems in the berthing area.

As discussed earlier, in cold regions many types of ice can be present in and around docking facilities. The predominant type will be broken-up pieces that, if not controlled, can cause extensive damage to structures and hinder a dock operation. If there is infrequent vessel traffic, sheet ice will form, which can cause damage from uplift forces. Pancake ice can be blown into harbor areas and make docking maneuvers difficult which in turn may result in heavy-impact berthing forces. The problems caused by ice as well as by cold ambient temperature can be minimized by proper structure design and by the use of appropriate ice control methods.

In general, the ice loads that are typically considered in dock design come from the following sources; these are:

1. Wind- and current-driven ice either in the form of individual floes or large sections of the ice cover
2. Ice thermal expansion
3. Small ice pieces jammed between the vessel and dock during berthing maneuvers
4. Ice buildup on vertical surfaces due to tidal effects which increase foundation loads and may cause horizontal and vertical loads due to hinging at these surfaces. In earthquake zones, the "added mass" of this ice would have to be considered in the design process to evaluate the effect on the structure response to earthquake loading
5. Ice buildup on vertical and horizontal surfaces due to icing effects, which may substantially increase foundation load and also hamper a dock operation.

Extra bearing loads resulting from (4) may be obtained easily once the ice accumulation resulting from tidal effects is determined; the effect on bearing surfaces may then be estimated. The loads resulting from the hinging effect created by the ice "active zone" described in Section 2.3 have not been well quantified to date. However, for substantial dock support structures, it is likely that these forces would govern design. The most favorable design would, of course, eliminate such loadings by preventing ice from accumulating on and around the dock structure. Such designs would incorporate ice control techniques to achieve this.

Ice forces upon dock structures are usually recognized as either global or local. Global forces are typically generated by massive solid ice formations such as large wind-driven large ice floes or by thermally induced ice pressures.

These forces are usually considered in computation of general stability of the structure, for example, sliding and overturning. In some cases heavy ice forces may exist locally. This usually depends on the nature of the ice feature and the geometry of the structure.

The local pressure is related to the confined crushing strength of the ice and can be one order of magnitude higher than the simple compressive strength.

The local pressures are oscillating loadings, varying much more rapidly than the global pressure. The local pressure variations are mainly due to the ice crushing mechanism itself.

In dock design, potential ice pile-up in front of the structure and/or ice ride-up and associated forces should be considered.

The heights of ice pile-up and ride-up usually depend on the geometry of the structure and depth of water. In some cases ice pile-up may reduce direct ice thrust on a structure. In some instances ice loads may be mitigated or even completely eliminated by means of miscellaneous ice control (management) systems.

Furthermore, the ice forces on marine structures can be reduced if the structure design considers ice failure in the bending rather than in the crushing mode of failure. In this case the structure's components must be configured accordingly.

2.7.2 Foundation Design

In general the geotechnical considerations used for the design of marine structures in southern regions are also applicable in regions with a cold climate, the arctic zone included.

In addition, however, the potential presence of permafrost and changes of thermal regimen must be taken into consideration. If soil exploration indicates that unfrozen soils only are present and future construction will not produce foundation soil freezing, then established conventional approaches to soil–structure interaction should be used.

In some practical cases construction of the marine structure on frozen soils can produce either the thawing of permafrost with resultant settlement and strength reduction, or the freezing of unfrozen soil with resultant frost heave. Therefore the heat transfer and the potential of ground thermal regimen changes and their effects on the proposed structure have to be considered.

Typically thaw-stable or non-ice-bonded permafrost provides a competent foundation for marine structures and earthworks as long as the cold regimen is maintained. In most cases adequate foundation support is maintained when this type of permafrost thaws.

On the other hand the ice-rich permafrost is the most adverse kind of foundation. In the short term, the ice-rich soils provide excellent support. In the long term, however, ice-rich soils and massive ground ice could be a problem due to creep movement under sustained loading. Such creep movements can adversely affect a marine structure even when thawing of the permafrost does not occur. Therefore, structures on ice-rich soil or massive ground ice should be designed for a limiting creep occurring during the life of the structure. Creep is a function of temperature, stress level, and to a lesser extent soil type and ice content. Thawing of ice-rich permafrost and massive ground ice may result in unacceptable settlement.

An example of a large settlement due to thawing of ice at the Nanisivik mine in the high Arctic is described by DeRuiter (1984).

At this site the mill building, founded on bedrock, showed signs of subsidence several years after startup of the mill. It was established that settlement of the structure resulted from partial melting of ice lenses some 1 to 1.5 m in thickness and located about 10 m below the bedrock surface.

Naturally, the design of a marine structure in cold regions may involve various combinations of soil conditions that have to be treated with great care.

In the case of permafrost and particularly in the case of thaw-unstable permafrost conditions the design approaches to marine structures are as follows.

1. The ground thermal regimen is maintained by artificial cooling methods and ground insulation. This approach is usually used where the permafrost is well established and is not in a marginal condition.
2. Unfavorable foundation materials are modified prior to construction. This approach is usually difficult and may be uneconomical.
3. The changes in ground-thermal regimen conditions caused by the construction and operation or by natural permafrost degradation are accepted and accounted for in the design.

For comprehensive discussion on specifics of foundation engineering in permafrost zone the reader is referred to work by Tsitovich (1973).

2.7.3 Earthworks

2.7.3.1 General

Similarly to conventional marine structures, construction earthworks associated with construction in cold regions involve nearshore and offshore excavations and fill for onshore facilities and backfill for soil-retaining marine structures.

One major distinction between onshore and offshore excavation is, besides the presence of water, the warmer permafrost temperatures in the offshore which makes excavation much easier.

On land the temperature of soils near the ground surface varies seasonally with air temperature. At a depth of about 10 to 15 m ground temperature does not vary seasonally and is generally several degrees warmer than mean annual air temperature. For example, in the high Arctic the steady ground temperatures at about 10 to 15 m depth may in land areas be −10 to −17°C.

In arctic offshore areas the seabed soils experience warmer temperatures, generally being warmer than −1.5°C where equilibrium conditions have been reached. The surface of the seabed, except in shallow water where the ice freezes to the bottom during the winter, is unfrozen because of the presence of saline water. This unfrozen layer may be underlain by either unfrozen material or zones of subsea permafrost.

The warm subsea permafrost is more easily penetrated and excavated, but can also be more easily disturbed by new construction in comparison to colder permafrost. Therefore, the effect of earth fills or excavation on the thermal stability of subsea permafrost requires careful consideration in design.

2.7.3.2 Backfill

Similarly to conventional marine structures fill material used in cold region marine structures construction is normally restricted to granular materials such as sands, gravels, and crushed rock from quarries. These materials can be readily excavated, placed, and compacted if not frozen.

Granular deposits that were previously submerged in seawater or that were formed in saltwater marine environments may have significant salt content. The presence of saline borrow materials should therefore be checked and the implications of their use for concrete aggregate and fill should be evaluated.

Ideally, the structure backfilling should be performed during the summer season when the backfill material is not frozen and ice free. The use of frozen backfill material during the winter season may result in water in the backfill zone instantly turning into ice. Subsequent thaw of such a backfill will result in a substantial settlement.

2.7.3.3 Excavation

Excavation can be classified as onshore and offshore. The extent to which excavation soil onshore can be carried out is greatly dependent on time of year, depth of deposit, and ice/water content. A review of various methods of excavating frozen ground is given by Johnston (1981). The general conclusions derived from this review are as follows.

1. Excavation of frozen fine-grained soil using mechanical equipment is difficult to impossible.
2. Frozen fine-grained soil can be broken by blasting; this has been done for small excavations.
3. Frozen fine-grained soil and ice-bonded sands and gravels can be excavated by following a method of thawing and farming during the summer months. The type of soil and the amount of water in the soil after thawing governs the "mud" condition within the excavation. Silty sands, sands, and gravel can be dozed into ridges and drained while drying of clayey soils is more difficult.
4. Dry frozen clean sands and gravels can be excavated and lumps broken even during the coldest part of winter. However, densities normally specified in temperature climates cannot generally be achieved during freezing temperatures.

Offshore excavation may be required for several following reasons: use of excavated material for backfilling and onshore fill; increasing the water depth at the dock and in the harbor; and in the approach/departing channels. Presently there is a very limited information on the excavation of subsea permafrost.

Seabed soils in the high Arctic are submerged by seawater colder than 0°C and normally not below −1.8°C. In this temperature range, soil conditions and excavation of soils can be summarized as follows (Fenco, 1984):

1. Clays, silts, sands, and gravels are normally ice free and their properties and behavior are similar to that occurring in temperate climates; therefore experience gained there is applicable.
2. Fine silts and clays from glacial periods may or may not contain excess ice. Generally speaking these materials are stiffer than thawed soils but excavation may be possible with heavy equipment.
3. Added cohesion to subsea bedrock due to ice will marginally increase the strength of the rock.

2.7.3.4 Compaction

Similarly to conventional construction practice, in temperate climate marine structures built in cold climate regions usually involve fills that require compaction. Compaction of nonfrozen fills in above freezing temperatures is similar to procedures used in temperate zones. However, the use of frozen fill materials and/or fill placement at temperatures below freezing requires special consideration.

A comprehensive study of cold region engineered earthworks has been carried out by Clark et al. (1983). This study documents current state-of-the-art information on placement and compaction of frozen fill materials and other aspects of cold region earthworks. Some of the more relevant information from this study regarding compaction and performance of frozen fills, as it applies to marine structures, is summarized as follows.

1. The most suitable type of frozen fill consists of sands and gravels with a low ice content. Although these materials, in a frozen condition, cannot be compacted to densities nor-

mally specified in temperature zones, they are generally stable on thawing.

2. Frozen clay and silt soils are not commonly used as fills. These soils on thawing are unstable and large deformations should be expected. Use of these soils for marine structure application should be avoided.
3. Frozen fills are used only to a limited extent because of potential problems on thawing of the fill, except where permafrost is expected to rise into the fill.
4. Considerable information is available regarding compaction characteristics and limitations to use of various frozen soils.
5. Frozen fills can be readily used in applications where only the weight of the soil is required, such as surcharging and stabilizing berms.

2.7.3.5 Soil Improvements

In some practical cases when confronted with poor soil conditions improvement or replacement of a poor underwater foundation material is unavoidable. For this a variety of appropriate techniques such as vibroflotation, blasting, grouting, fiber reinforcement, and others employed for soil improvement in temperate climate areas may also be used in cold regions.

- *Vibrofloatation*, which is used to densify loose granular soils by lowering a vibrating tool into the deposit and backfilling with sand. The method has merit in improving loose cohesionless soils.
- *Blasting*, which is also used to densify loose granular waterbearing soils by blasting explosives buried below sea level.
- *Grouting* is used to stabilize poor soils by injecting Portland cement, clay, or chemical grouts. Grouting trials are usually required to confirm an adequate grout penetration and resulting soil improvement before starting a full-scale grouting.
- *Grouting of poor soil with a good granular material*. This is usually combined with densification of a new soil by the vibrofloatation method. It must be noted that replacement material must be of temperate temperature. As stated earlier the very cold (frozen) granular material may instantly turn an ambient water into the ice, thus precluding the fill densification (Tsaliuk and Semerenko, 1981).

The other unconventional methods of soil improvement are use of *fabric reinforcement* and *artificial refrigeration*.

Relatively recent developments in the use of fabrics to support embankments on poor soils and to reinforce earth embankments may find use in marine construction in regions with cold climates. The advantages of fabrics are the low cost, easy installation, and versatility (Proceedings 2nd International Conference on Geotextiles, Las Vegas, USA, 1982).

Freezing of unfrozen seabed soils in regions with cold climates and particularly in the Arctic has some merit for consideration because of low seabed and air temperatures. The types of refrigeration that can be considered are mechanical, thermosyphons, or ventilation (natural or forced), although thermosyphons should be used with caution. Thermosyphons may be considered as a type of thermal pile that uses natural convectional systems to remove heat from the ground. It has no moving parts, requires no external power for operation, and functions only when air temperatures are lower than the ground temperatures. Various aspects of thermosyphons and thermal pits are reviewed by Johnston (1981).

2.7.3.6 Instrumentation

The interaction of a structure with permafrost may lead to warming of the ground and eventually to complete thaw, thus resulting in severe foundation deformation. The presence of ice-rich permafrost and massive ice lenses can result in large deformations as they are weaker and creep faster at higher temperature. These materials involve a significant reduction in foundation support on thaw due to the accumulation of excess pore water unable to excape through the soil matrix. Such conditions, which frequently occur in fine-grained soil and sedimentary rock deposits, are called thaw-unstable and require special design approaches.

Therefore, where the integrity of a structure depends on the stability of the ground thermal regimen, it is of utmost importance to monitor ground temperature beneath and around the foundations.

Hence, an appropriate ground instrumentation must be installed to monitor existing ground conditions, temperature included. A well defined data collection and analysis program must be developed. Data analysis should be done on a regular basis which will allow monitoring of ground conditions and detection of an unusual or unexpected foundation soil behavior. The type and number of instruments to be installed on a structure depends on site conditions, the type of structure, and its expected behavior. For detailed information on geotechnical instrumentation for monitoring field performance the reader is referred to Dunnicliff (1988).

2.7.4 Piles in Permafrost

Different types of piles have been extensively used in permafrost (Andersland and Anderson, 1978).

In a permafrost zone piles are installed in ways that minimize the heat that is added to frozen soil so that adequate pile adhesion is obtained. Because it is not always possible to drive piles into permafrost, different pile installation techniques are usually used.

There have been reports on successful steel H-piles and open-end piles driving in permafrost with impact, vibratory, and sonic hammers (Ladanyi, 1983; Nottingham and Christopherson, 1983; Phukan, 1985). Optimum conditions for pile driving occur where soils are fine grained, the ice content of the soil is low, and the temperature of the permafrost is high. As noted by Heydinger (1987), these, however, are not optimum conditions for adfreeze strengths. It is not possible to drive piles into frozen gravel and soils containing cobbles without the use of pilot holes of somewhat smaller diameter than the piles. Piles can be installed by placing them into oversized bore holes with follow-up backfilling of granular material around them. Cement grout can also be used. Sufficient space must be provided between the pile and the hole to ensure proper placement of granular backfill material. This material can be compacted by vibration.

To shorten the freeze-back time the temperature of backfill material, just before installation, should be kept near 0°C. In some instances artificial freeze-back of a backfill material is required. This can be achieved in a short period of time by circulating a refrigerant in a borehole.

Artificial freeze-back is usually used where it is necessary to apply loads to the piles before freeze-back could occur naturally.

In general the state-of-the-art practice in designing piles in permafrost is to design for the capacity based on pile adfreeze, or allowable settlement. The pile adfreeze used for design is the minimum that occurs in late summer or early fall when the temperature of the permafrost is the highest. In the case of cohesive soils pile design should also consider a check for the potential uplift forces due to frost heave. Designs for pile settlements typically use creep theory for ice and frozen soils.

When heat transfer from the structure to the foundation is undesirable then the insulation material, for example, neoprene, can be placed between the pile cap and the superstructure. For the same reason the piles above sea level could be painted white to reduce heat absorption from the sun.

A discussion of the state-of-the-art of pile in frozen soils design and a comprehensive list of references on the subject is given by Heydinger (1987).

The detailed design procedure on piles installed in permafrost is discussed by Markin (1977) and Foriero and Ladanyi

(1991); the dynamic response of piles embedded in partly frozen soils is discussed by Vaziri and Han (1991).

2.7.5 Structural Materials

Wood, steel, and concrete are typically used for construction of marine structures in cold regions.

All basic specifications applicable for the above materials used for construction of marine structures in temperate climate zones are equally applicable for use in cold climate regions.

2.7.5.1 Wood

Wood is usually used when available locally. Structures built from wood, for example, timber cribs, bulkheads, or piled piers and platforms proved to be economical and quite adaptable to the static ice conditions. The useful life of the underwater portion of wooden marine structures built in cold regions may be quite extended because of the absence of marine organisms characteristic for warm waters.

On the other hand resistance of wooden structures to ice dynamic loads as well as to ice abrasion is rather low.

2.7.5.2 Steel

When steel elongates, the lateral dimensifons contract, owing to Poisson's ratio effect. If the lateral dimensions are restrained (fully, or even partly), the steel will pull apart without fully developing the yield potential; this type of failure is termed a *brittle fracture*. Essentially, a brittle fracture is a failure that takes place without material yielding.

In common practice when a large impact load is placed on a steel structure or steel structural elements the designer relies on the steel elastic properties, yielding plasticity at stress concentration zones and ability to absorb energy. All steels behave in a plastic manner above a certain temperature, generally referred to as the *transition temperature*. Below this temperature the steel behaves in a brittle manner, can absorb little energy, and if exposed to overstressing loads may fail in a brittle mode of failure.

Steel brittle characteristics are characterized by *notch toughness*. The notch toughness is evaluated by the Charpy V-notch test in which the specimen is broken by a falling pendulum over a range of temperatures. Then energy is plotted against temperatures (Figure 2–36). In a significant number of investigations, a reasonable correlation has been obtained that indicated that structural steel is unlikely to fail above the transition zone. The 20 J energy level at the test temperature of 17 to 22°C usually has been selected as a transition temperature for most structural steel.

For more details on minimum Charpy V-notch requirements for structures operating in miscellaneous temperature zones the reader is referred to ASTM Standard A709. If should be realized, however, that selection of a notch tough steel alone will not prevent a brittle fracture if poor design details or inferior fabrication practices are used.

For this reason, for example, the seamless steel pipe piles are usually specified where

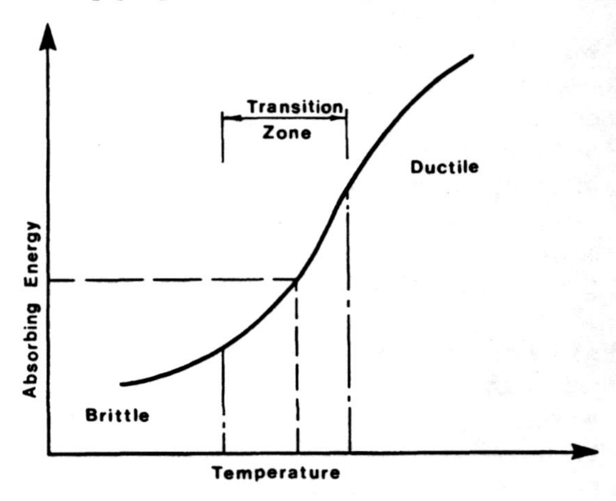

Figure 2–36. Charpy transition temperatures.

designed for driving under low temperatures. Commonly the Charpy V-notch levels of 20 to 35 J at the lowest expected ambient temperature is specified for driven steel piles used in cold climate regions.

In cases of exceptionally harsh cold conditions pipe piles such as A333 grade 6 have been specified.

When piles are placed in a pre-augured hole and therefore are not driven, they are subjected to little or no impact loading and carry mainly static loads. In the latter case the Charpy V-notch level of 20 J at −30°C for pile material may be sufficient.

If steel as a structural material is selected then the heavy ice load usually demands design of steel structures with heavily stiffened thick plate elements. This may result in rather severe fabrication problems usually associated with weld shrinkage, residual stresses, and also with the possibility of lamellar tearing in thick parent metal in the presence of large size welds. These problems can be controlled by judicious selection of weld procedures and by keeping plate thicknesses down to a reasonable size (say no more than 50 mm). Preheat treatment may be difficult for some specific site conditions.

Because welding on site under cold conditions can be very costly and harmful for steel it should be done either during moderate temperature windows, or be avoided altogether; structural bolting should be used whenever possible. In North America, the design and detailing of welded connections is done in accordance with the latest version of AWS D.1 or CSA W59.

It should be noted that conventional bolts ASTM A325 and A490 do not have certified low-temperature notch toughness. Therefore designers of structures in cold climate regions in each specific case of construction must consult steel and fastener producers. Other fasteners that may be required in structures for cold regions are anchor bolts. They are usually made from normal high-carbon brittle rebar material. When better quality bolts are required than the designer have to specify mechanical and chemical properties of these bolts. Stelco (Canadian Steel Mill) developed special quality deformed bar anchor bolts ($F_y = 485$ MPa) with the trade name HIPACT for its high impact resistance of 20 J at −29°C. Other mills also produce high-quality specialty anchor bolts and fasteners with good notch toughness properties and the designer should investigate their availability and economy before specifying regular bolts or paying a high premium for custom-made items.

Typically steel brittle fractures can be controlled in several ways:

1. Steel components should be detailed to avoid or minimize concentration of stress. Thus abrupt changes in cross-sectional areas should be avoided, and notches should be machined into a smooth transition.
2. The potential rate of tensile strain should be reduced; this can be achieved by application of a pretension stress at elevated or normal temperatures.
3. Steels that are specially alloyed for low-temperature environments (ASTM A633 or CSA G.40.21) should be used.

There is no single design code (guideline) that addresses all potential problems of designing of steel structures in cold climate zones (particularly in arctic and similar zones). In North American practice, steel structures in the cold region areas are usually designed in accordance with provisions of American Petroleum Institute (API) Bulletin 2N (1981) and Bulletin RP 2A (1984). However, these bulletins address some of the basic problems of designing cold region marine (offshore) structures in a rather preliminary fashion. In European practice marine structures for cold regions are usually designed according to Det Norske Veritas (DnV) rules.

Other design rules used are American Bureau of Shipping Rules and Canadian

Arctic Shipping Pollution Prevention Regulations (CASPPR). These rules and regulations, strictly speaking, are applicable to the design and construction of ships operating in ice environments but may be lifted from this context to apply to any relevant steel structure.

2.7.5.2 Concrete

Concrete structures have been built for many years in locations exposed to severe ice and cold temperature attacks. Numerous examples of these structures now exist in northern latitudes elsewhere in the world. Harbors located within the Arctic Circle have also utilized concrete piers and breakwaters.

The most recent state-of-the-art on use of concrete for construction marine structures in cold environments is given in American Concrete Institute (1985), report ACI 357.1R-85; a comprehensive list of marine structures constructed in North American regions with cold climates is given in Fenco (1984).

In general, recommendations for concrete in cold environment marine structures include limitations on the following:

Maximum water/cement ratio
Minimum compressive strength
Minimum cement content
Minimum air content
Maximum chloride content.

The above characteristics are specified to obtain durable concrete of low permeability. The requirements for concrete used for construction in cold environments are usually more stringent than for those used for ordinary marine structures. Owing to recent developments in superplasticizing admixtures modern concrete technology allows for the use of a water/cement ratio of 0.35 and even lower. Use of supplementary cementation material such as fly ash, slag, and silica fume has proven to be very beneficial in concrete for marine structures built for marine environments.

Silica fume in combination with superplasticizers has proven to be an extremely attractive means of producing high-strength concrete of low permeability. Low-permeability concrete can also be obtained with use of pozzolans.

Resistance of concrete to freezing and thawing is achieved with a proper amount of entrained air, adequate pore size distribution, and spacing. Limitations on the hardened air void system have to be included in concrete specifications. Approximate limits for concrete used in cold climate marine environments are as follows:

Calculated space factor: less than 0.2 mm
Surface area of air voids: more than 0.1 cm^2/cm^3
Number of voids per centimeter of traverse: significantly greater than percentage of air in harden concrete.

These recommendations are based on information provided in American Concrete Institute (1986), report ACI 357.IR-85.

With respect to concrete properties in low-temperature environments, tests indicate increases in strength and elastic modulus (Marshall, 1985; Marshall and Drake, 1987; Lee et al., 1988). Durability of concrete in cold climate environments is related to its ability to resist weathering, freezing and thawing, chemical attack, abrasion, or other deterioration processes. In marine environments the aforementioned types of deterioration are usually related to the zone of exposure, such as atmospheric, splash, and submerged zones.

Items to be considered in these zones are:

1. Atmospheric zone: freeze–thaw durability and corrosion of reinforcement and hardware.
2. Splash zone: freeze–thaw durability, corrosion of the reinforcement and hardware, chemical

deterioration of concrete, and abrasion due to ice effects.

3. Submerged zone: chemical deterioration of concrete, corrosion deterioration, and abrasion due to ice effects.

The splash zone is considered the most critical with regard to concrete durability. The ice movements around marine structures sometimes cause extensive concrete abrasion, which is a combination of direct wear, as by grinding, and a plucking out of discrete particles of coarse aggregates weakened by freezing and thawing effects.

Resistance of concrete to abrasion is a function of its compressive strength, aggregate properties, curing and finishing methods, and the ability to resist freezing and thawing deterioration.

For more information on concrete deterioration the reader is referred to Chapter 1.

Cold-region (including arctic) concrete structures generally demand the use of high-strength (30 to 45 MPa) concrete.

The latter is usually achieved by use of relatively large quantities of cement (360 to 415 kg per cubic meter) along with supplementary cementation materials (fly ash, slag, silica fume, pozzolans).

The recommended minimum of water/cement ratios for basic exposure zones and 28-day compressive strength of concrete designed for use in cold regions are given in Table 2–8.

hen high cement content (more than 415 kg per cubic meter) is used in relatively thick sections or mass concrete, precautions must be taken to prevent cracking from severe temperature stress gradients that may result as concrete temperature rises due to cement hydration. This may require the selection of cements that have a low heat of hydration, use of set of heat-retarding admixtures, use of reduced rate of placement, precooling of aggregates, use of water with added ice, use of pozzolans, and circulating of cold water through the network of pipes installed within concrete elements. Combination of all the above techniques may also be used as appropriate.

For additional information on concrete and its components in cold regions the reader is referred to publications by Carino (1988), Lane (1988), Pekar (1988), Maage and Helland (1988), Brook et al. (1988), and McFadden (1990).

Reinforcing and prestressing steel used for marine structures in cold environments is essentially the same as that used in general marine structure construction. However, ductility of reinforcing steel at low service temperature must be ensured.

With regard to effects of cold temperature on reinforcing and prestressing steels the available data indicate that the most significant effect of cold temperature is to reduce maximum elongation, and impact capacity. The latter is a function of ambient temperature and steel chemistry. Steel conforming to ASTM A706 is considered suitable for cold temperature environments. In general the design should be such as to minimize impact loads.

Concrete covers over reinforcement should be similar to those recommended for marine concrete structures operated in a general marine environment. However, where severe scouring action is expected to cause concrete abrasion, the minimum concrete strength should be increased to

Table 2-8. Water/Cement Ratios and Compressive Strengths for Three Exposure Zones

Zone	Maximum W/c Ratio	Minimum 28-Day Cylinder Compressive Strength (MPa)
Submerged	0.45	35
Splash	0.40	35
Atmospheric	0.40	35

45 MPa. Additional protection can be achieved by using concrete aggregates having equal or higher hardness than the scouring material. Where appropriate the epoxy resins may be utilized for waterproofing, sealing construction joints, repairing cracks, and other similar usages.

Lightweight aggregate concrete has a proven record in the marine environment (Holm, 1980). According to ABAM (1983) air-entrained lightweight concretes with fresh unit weights of about 1800 kg/m^3 can be produced to achieve design strengths in excess of 48 MPa. This can be achieved with low-absorption, high-strength/weight ratio lightweight aggregates. Guidelines on the use of lightweight aggregate concrete are given in publications by American Concrete Institute (1979) and FIP (1978, 1983).

Placement and curing of in situ concrete under cold weather conditions is usually done in accordance with recommendations by American Concrete Institute (1978, 1987) reports ACI 306.1-87 and ACI 306R-78. This typically involves use of miscellaneous admixtures, heating systems, and heat loss preventing insulated formworks. In some cases a special protective system was built to protect the cast-in-situ concrete structures (Mangus, 1988; Moore et al., 1988).

2.8 DESIGN ASPECTS

2.8.1. General

As stated earlier the primary purpose of port marine elements is to provide basic facilities and services to safely berth vessels, service them, and load and/or unload the cargo.

Major port elements that may be affected by the cold environment are harbor water area, breakwaters and ice booms, shore protection, wharf structures, miscellaneous mooring (floating and bottom fixed) facilities, and aids to navigation.

As previously stated, in general ports built in cold regions were located in harbors protected from ice action. This was available through natural geographical features or by man-made structures such as breakwaters or ice booms. If a protected area is formed by man-made structures then it is essential to assess the impact of these structures on the natural environmental conditions of the site. Of particular interest should be impacts on future ice regimen in the area where site conditions require construction of breakwaters or installation of ice booms. Furthermore, the effects of these structures on currents, littoral drifts, and wave climate must also be evaluated. Naturally, in the design of breakwaters and ice booms miscellaneous effects of ice action on these and associated structures must be considered.

In some cases construction of port marine facilities will substantially modify existing environmental conditions at port location, ice regimen included. In some cases this may require heavy shore protection works (Shah, 1978).

The wharf structure is the prime physical element providing interface between the vessel and terminal. In general and especially under ice conditions it must provide safe approach, berthing, and departure of the ship under site-specific environmental conditions. Therefore, when siting the port or marine terminal in cold regions, the key elements that must be considered carefully are navigational aspects, structure design, constructability, and operational aspects. The latter must consider problems associated with ice control in the harbor and ice management in the berth area. Furthermore, particular attention should be paid to the selection, operation, and maintenance of the fendering systems which may be highly susceptible to cold temperature effects, icing, and ice damage.

If offshore mooring facilities are required then they have to be designed to provide for safe mooring. In the latter case effects of wind, current, waves, and moving ice must be considered.

The very important navigational and operational aspects in cold regions (or say in ice-affected waters) is design, construction, installation, timely removal when required, and maintenance of permanent bottom-founded navigation aids, for example, lighthouses and floating navigation aids, such as marker buoys.

The provision of navigation aids is usually the responsibility of the Coast Guard, which must be consulted for any harbor developments.

2.8.2 Structures

Because ports and terminals in the cold regions are typically located in remote areas most, if not all, construction materials must be brought to the site and often from far away. This requirement speaks to the advantage of use of simple prefabricated components easy for assembly at the site.

Construction also can be facilitated greatly by using modular technology. In the latter case a large preassembled modulus, either crane installed or floated-in can be used for construction of port related structures. The advantages and disadvantages of this type of construction are discussed by Buslov (1985).

Marine structures built in cold regions are typically designed to withstand all operational and environmental forces, ice forces included. In cold regions, depending on the location, the ice forces in most cases are the governing factor. They may exceed other environmental and operational loads by several magnitudes.

This is basically why in most practical cases of marine structure constructions in ice-affected waters the bottom-fixed structures represent the most practical solution.

Depending on the character of ice loading and geotechnical conditions these structures are basically designed and built in the form of ice-resistant gravity-type structures, such as steel sheet pile cells, floating-in concrete caissons, or timber cribs.

Sheet-pile bulkheads, steel jacket type structures, and piled structures have also been used. In the latter case piles usually have been protected from direct impact by ice features.

This usually was accomplished by installation of ice-resistant large-diameter steel piles in front of the conventional piling, or by driving sheet piles in front of conventional piling with subsequent filling of the space between sheet piles and conventional piles with granular material.

Piled and jacket type structures have been used in moderately heavy ice conditions such as exist, for example, at Cook Inlet, Alaska. However, it is doubtful that piled structures, even those with piles well protected from direct contact with ice features, can survive heavy ice loads such as those that exist in arctic or subarctic regions.

As stated earlier, the presence of permafrost may sometimes preclude use of piles on sheet piles. However, in some cases it may be possible to drive heavy section H-piles into permafrost.

To prevent excessive ice accumulation on piled or jacket type structures nonvertical elements such as brackets and braces that can be affected by the tidal range should be avoided.

The structure itself should be designed for adequate buoyant forces and weight of builtup ice on structural members.

In some case platforms placed upon large diameter steel or concrete caissons, designed to take heavy ice static and dynamic loads, can perform well under heavy ice conditions.

To date circular steel sheet-pile cells have been the most popular type of marine structure constructions in ice-affected waters. Large-diameter steel sheet-pile cells were built in arctic waters at Deception Bay, Nanisivik, Little Cornwallis, and many other locations (Fenco, 1984). Recently steel sheet-pile cells were used for construction of a 91.5 m long × 30.5 m wide dock (depth of water about 6.5 m) at Nome, Alaska (Anonymous, 1991). This type of construction can resist large lateral loads by its large mass and/or by embedment into the sea floor. It may be installed as a prefabricated unit upon a prepared mattress, or can be driven into the seabed. The shear strength of granular fill within the cell provides significant resistance to lateral forces and prevents excessive global deformation. Natural freezing may greatly strengthen the cell. It is difficult however to predict the extent to which freezing may take place. In this regard, freezing of a granular fill below water level and seasonal ice levels would not be expected. However, freezing of a granular fill above water level and to some extent within the tidal zone, if the latter is significant, would be expected in areas where the average annual air temperature is below −1°C. Growth of permafrost into steel sheet-pile cells from land sidefills and from underlying permafrost is also a consideration. Thermal monitoring of actual steel sheet-pile cells is considered to be important for adequate comprehension of the mechanisms involved.

As mentioned earlier, sometimes use of frozen granule material for backfill may turn water inside the cell into the ice. This would hamper the fill consolidation with the possibility of a substantial settlement in the future due to thaw.

In general steel sheet-pile cells with well compacted fill could effectively resist local buckling of sheet piles in collision with ice features with crashing strength up to 7 MPa.

Guillermo and Lawrence (1987) investigated effects of ice pressure on steel sheet-pile cells. In the case under question was design of the steel sheet-pile cells 22.6 m in diameter, proposed for installation at AIDA Seaport, the Chuckcha Sea, Alaska. These cells are exposed to sea landfast ice pressure. The design ice thickness is 1.6 m, and design late winter ice movement rate is about 1.83 m/h. It was assumed that the ice fails in crushing against a single cell. Thus it was assumed that the ice is made up of a top layer of "cold" granular ice and bottom layer of "warm," columnar ice, which is much weaker than the top layer.

Therefore, the design global ice load is estimated as equal to about 1.43 MN/m. It was estimated that local ice pressures up to 3.8 MPa may be attained over small areas.

Finite element computer programs, developed at Virginia Politechnic Institute and State University (Clough et al., 1987), were used for ice–cell interaction analyses. The calculations show that the ice loading applied in these analyses is much larger than that used to simulate flood effects on river cofferdams. Furthermore, the ice loading is more concentrated than the loads occurring during a flood type event. The following specific observations and conclusions have been drawn.

1. The cell deflection profiles indicated that the maximum lateral movement occurs at the elevation of the ice load in the form of a localized bulge.
2. The lateral movements of the cell walls diminish below the point of ice loading.
3. Stiffer fill reduces the size of the yielding zones in the natural soils, but has little impact on the degree of the yielding in the cell fill.
4. The performance of the cell is better for drained conditions in the foundation material, and for a stiffer fill.

In some cases marine structures in ice-affected waters have been built in the form of circular or box-like floated-in concrete caissons. These structures are typically built at shipyards or other protected locations, then launched and towed to the site of deployment and finally sunk on a prepared pad and filled with granular material. The pad is usually constructed by placing a gravel or crushed stone mat upon bedrock and then thoroughly levelling it. In some cases to prevent large settlement the compaction of a pad is required.

Figure 2–37. Fender system plugged by ice.

This type of construction is intended to resist heavy ice forces by virtue of its own weight.

Timber cribs have been used at remote locations where wood is locally available construction material. The crib is usually completed with a concrete superstructure extended from the lowest water level to the final grade. The wooden portion of such a structure is usually placed below the lowest water level to protect wood from damage by ice and from petrification. In cold climate regions timber cribs typically are constructed in relatively shallow water. Timber cribs can sustain large vertical and horizontal movements and can withstand large horizontal forces from the sea direction because of resistance of granular fill behind the cribs. Cribs derive their support in part through shear resistance of the mass of granular fill within them.

Permanently installed floating mooring systems such as navigation aids or floating docks are basically used at protected locations with no substantial ice movements. At unprotected locations where substantial ice movement is expected a retractable mooring buoy system is usually considered, and floating docks are usually not practical.

2.8.3 Dock Fendering

The ship impact force during berthing operation is perhaps the second (after ice load) largest horizontal force imposed upon a dock structure. To protect both the ship and the structure from being damaged an appropriate fender system should be installed at the dock face.

In the process of selection of the fender system for installation on structures operating in cold regions consideration should be given to effects of cold temperature and ice action on fender performance.

The cold temperature stiffens the rubber which results in much larger reaction force during ship–fender interaction.

In addition to cold temperature the ice buildup on individual fender units may severely hamper their effectiveness. An example illustrated in Figure 2–37 demonstrates an ice and frozen snow buildup around a generally very effective cell-type fender.

In this case the ice collar and frozen snow around the fender's body were formed basically due to water spray generated by waves, and atmospheric precipitation. This type of icing in combination with low temperature effects on solid rubber may severely hamper fender performance.

If incorrectly designed and/or installed the fenders may be torn off the face of the dock by moving ice. In this respect the pile-supported fenders, or units mounted within tidal or ice-affected zones should be avoided, as such fenders are subject to ice damage and would likely be costly to maintain.

In areas where tides are small it may be possible to mount fender units above the ice-affected zone high enough to prevent severe icing. In areas with large tides it may be practical to maintain removable or retractable fender systems that can be readily placed in position for the duration of the navigation season, or in extreme conditions can be positioned for each specific berthing. The most effective systems for operation under cold temperatures are pneumatic fenders, both fixed to the dock face and floating. Floating pneumatic fenders have been successfully used in areas with large tides at Deception Bay and St. David de Levis (Golden Eagle Terminal), Quebec.

2.8.4 Basic Design Principles

Similarly to any conventional marine structures those built in cold climate regions have to be evaluated and analyzed for the following conditions:

- Bearing capacity of foundation
- Overturning and foundation edge pressure under lateral loading
- Sliding at structure base or at mattress–soil interface
- Global stability or shear through foundation soil
- Lateral and vertical capacity of piles
- Settlement due to compression/consolidation of unfrozen or partially frozen soils
- Settlement due to thaw consolidation of permafrost or frost heave due to ice formation in nonfrozen seabed soils
- Creep of frozen soil or ice
- Seismic effects on structure stability, lateral loads, and bearing capacity
- Liquefaction potential of foundation soils.

Geotechnical design methods used in southern regions for marine structures are also applicable in regions with cold climates with consideration given to the potential presence of permafrost, cold temperature effects on earthworks, changes in thermal regimen and large horizontal and vertical ice loads. Furthermore, implicit with all geotechnical design methods is appropriate consideration of seismicity in respect to potential liquefaction of soils and seismic load conditions.

Naturally, the marine structure design would be greatly dependent on the presence or lack of the permafrost. If soil exploration indicates that unfrozen soils only are present under the seabed and the future construction will not produce freezing, then conventional design methods can be used. Marine construction in cold regions may have potential for thawing of permafrost if present, or freezing of unfrozen soil.

In the former case the interaction of a structure with permafrost and especially with ice-rich permafrost may lead to severe deformation of foundation due to buildup of an excess of pore water unable to escape through the soil matrix and reduction in shear strength of soil.

Such conditions, which frequently occur in fine-grained soil and sedimentary rock deposits, are called thaw-unstable and require special design approaches. Generally, two design methods are used: the passive and the active methods.

In the passive method, the ground thermal regimen is maintained by artificial cooling methods and ground insulation. This approach should be used only where the permafrost is well established and is not in a marginal condition.

In the active method, two avenues are possible. First, the changes in ground thermal regimen conditions caused by construction and operation or by natural permafrost degradation are accepted and accounted for in the design. This is the most common design approach in offshore foundation design and is valid for any type of structure as long as it can accept the deformation imposed on it. Second, if economically justified unfavorable foundation materials can be modified prior to construction. On the other hand freezing of an unfrozen soil may result in frost heave.

Therefore, the potential of ground thermal regimen changes and their effects on the structure have to be thoroughly evaluated. For more information on the subject the reader is referred to Fenco (1984).

2.9 MARINE STRUCTURES IN COLD REGIONS: SOME CHARACTERISTIC CASE HISTORIES

Although a considerable amount of research and conceptual designs of port-related structures for construction in cold regions have been conducted in recent years, not very many innovative designs have been practically implemented. A comprehensive list of port-related marine structures built in North American cold regions is given in Fenco (1984) and some relevant information on the rubble mound structures, such as breakwaters and causeways, is given in Bruun and Sackinger (1985).

Very useful information on state-of-the-art of offshore concrete structures for the Arctic is provided in American Concrete Institute (1986) ACI 357.1R-85 report.

In this section, in addition to some information on marine structures in cold regions discussed in preceding sections of this chapter, several more characteristic case histories are discussed.

The latter include discussion of gravity-type structures, built all over the world; piled structures; and single-point moorings, and give examples of offshore terminals for construction in moving ice environments.

2.9.1 Gravity-Type Structures

The present experience indicates that where heavy ice forces on marine structures are expected in most cases the port designers prefer to use bottom-founded gravity-type structures, such as large-diameter steel sheet-pile cells or floated-in concrete caissons.

As noted earlier gravity-type marine structures constructed in cold regions are usually represented by steel sheet-pile cells and concrete caissons. These structures have been built at many locations and proven to be reliable from the structural standpoint. However, as in case of the Nanisivik Wharf discussed earlier in this chapter formation of an ice "active zone" in low-tide areas may severely hamper a dock operation. In general this phenomenon may apply to any type of vertical face structure operating in low-tide cold-climate areas.

However, this pattern may be changed by introduction of a suitable ice management system, or by covering the dock face by low-friction material in order to break ice adfrozen to the structure.

The following are some typical examples of gravity-type marine structures designed and constructed in cold-climate regions.

2.9.1.1 Deep Water Coal Loading Pier in Norway (Instanes, 1979)

The project is located at the end of Van Mijen Fjord (Svea Bay), which is almost completely closed at the mouth by the island. The fjord is sufficient for passage of vessels up to 60 000 DWT. The fjord ice is normally formed in early November and breaks up in mid-July. Therefore the normal navigation period is practically limited to 16 weeks.

The maximum ice thickness in the fjord is about 1.55 m. The ice load on a pier has been estimated as equal to 2.0 MN/m. The foundation soil at the dock site is represented by the recent marine moraine containing some permafrost at shoreline. The bedrock lies approximately 40 m below the water level (Figure 2–38). The 120 m long pier with water depth at low tide equal to 16 m is composed of five steel sheet-pile cells 22 m dia-

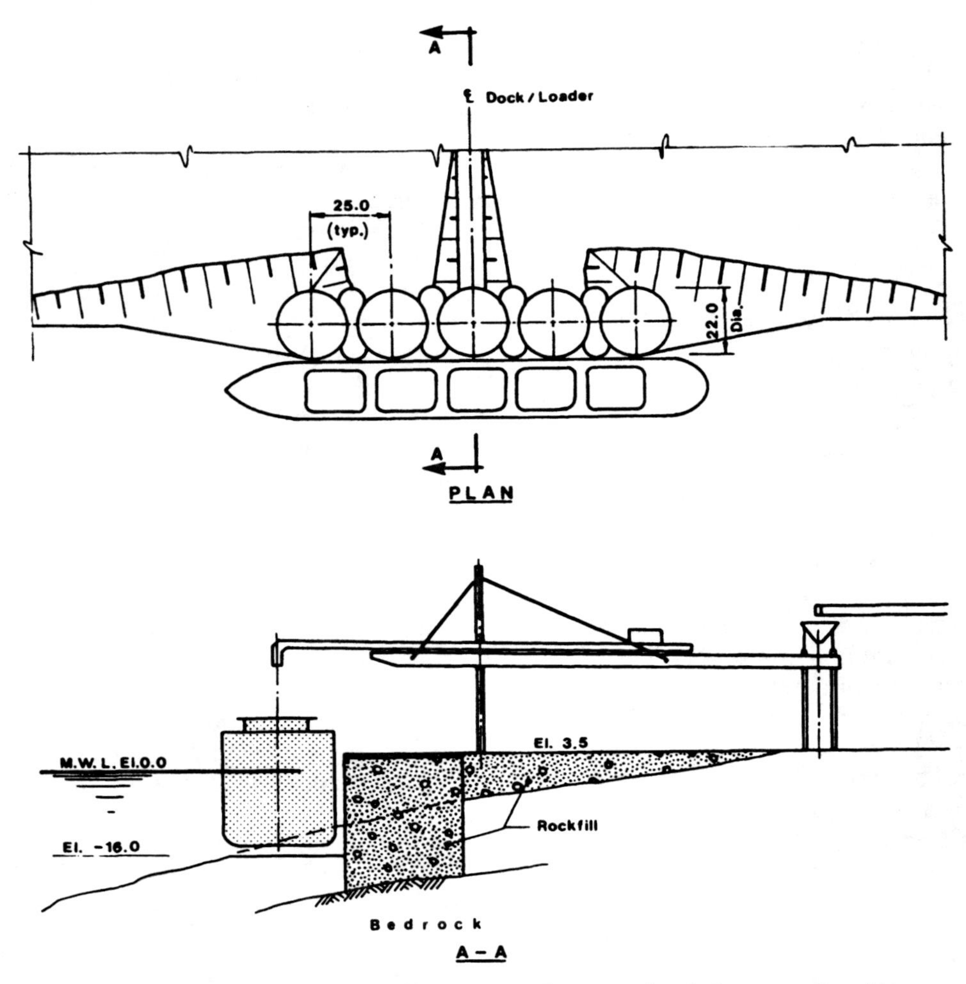

Figure 2–38. Deepwater pier at Svea Bay, Norway. Layout and typical cross-section. (From Instanes, 1979.)

meter placed at 25-m intervals. Cells are joined by sheet-pile arcs and filled with rock and gravel from a local quarry. At the top face of the pier there is a continuous concrete beam that spans the entire pier and is used for installation of fenders and mooring accessories. Unfortunately information on operational condition at this dock as well as hydraulic and ice–structure interaction conditions are not available.

2.9.1.2 Deep Water Pier in the former USSR (Tsaliuk and Semerenko, 1981)

This 245-m long pier was built in the Russian arctic region (Figure 2–39). It is composed of 14 steel sheet pile cells 16.3 m in diameter placed at 17.5-m intervals. Cells are joined by sheet-pile arcs and filled with

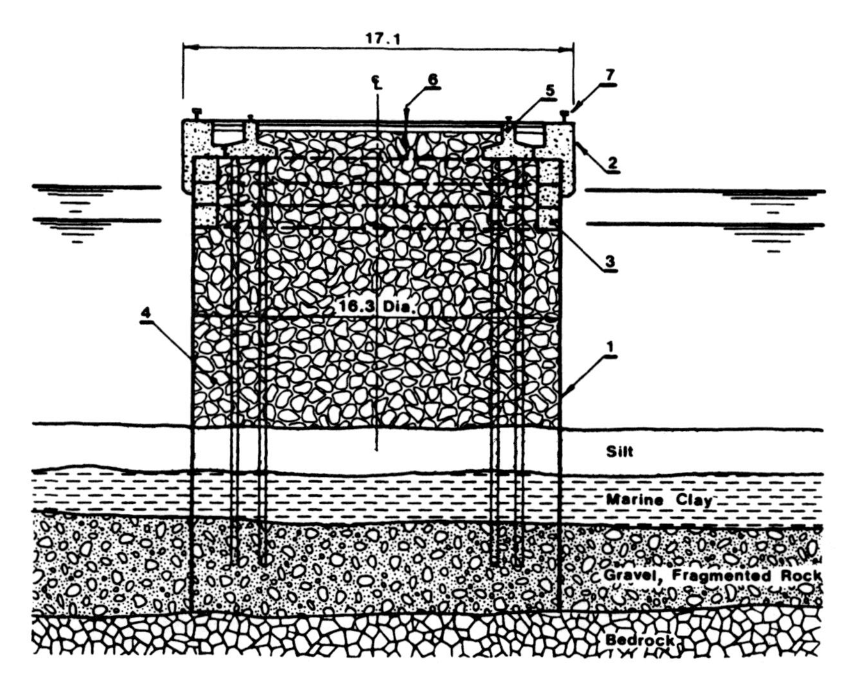

Figure 2–39. Deepwater pier in Russian Arctic. Typical cross-section. (From Tsaliuk and Semereenko, 1981.) 1—Steel sheet pile cell at 17.5 m, 2—concrete cope with face protected by steel plate, 3—prefabricated concrete rings, 4—rockfill, 5—craneway, 6—prefrabricated concrete slabs placed on a layer of gravel, 7—bollard.

random rocks. The pier was constructed during winter and the ice cover, after reaching a minimum thickness of about 1.0 m, was used as a working platform.

In the course of the pier construction the craneway piles where driven first. Then they were used to support a template for driving of sheet piles. The template was installed at about 60 to 70 cm above ice. The latter provided sufficient space for the staged ice cutting during sheet pile driving. Subsequently the ice inside of a cell was cut into pieces by steam pressure and removed. To strengthen the pier at the ice action level the pier design called for installation of a cast-in-situ concrete wall inside the cell directly behind the sheet piles. However, heavy water filtration into the cell during a dewatering operation made it virtually impossible. The latter occurred owing to the very low temperature of fill material (about −30°C to −40°C), which after placement inside the cell instantly turned the water there into ice. The latter dramatically reduced the expected fill lateral pressure inside the cell which in turn resulted in a substantial reduction in sheet-pile interlocking forces. These interlocking forces appeared to be insufficient to seal the cell against the inflow of water.

In final analysis the cast-in-situ concrete wall was replaced by the prefabricated concrete elements installed underwater.

To further reinforce the pier structure against ice actions and to diminish the ice abrasion effects a monolithic concrete cope wall, installed at the pier outside perimeter, was protected by a 10 mm thick steel plate.

After construction of the concrete craneway beams the structure was completed by installation of the prefabricated concrete slabs on top of the pier.

2.9.1.3 Deep Water Oil Terminal at St. David de Levis *(Saunders and Timascheff, 1973)*

This terminal, also known as "Golden Eagle Development", was built on the south shore of the St. Lawrence River at St. David de Levis, near Quebec City, Quebec. The terminal was designed to receive super tankers of up to 100 000 DWT on a year-round basis. The terminal site has a history of severe ice conditions. Normally, ice at that location reportedly built up to a thickness of about 75 cm. However, in certain winter seasons the observed ice thickness was about 110 cm. The ice data also

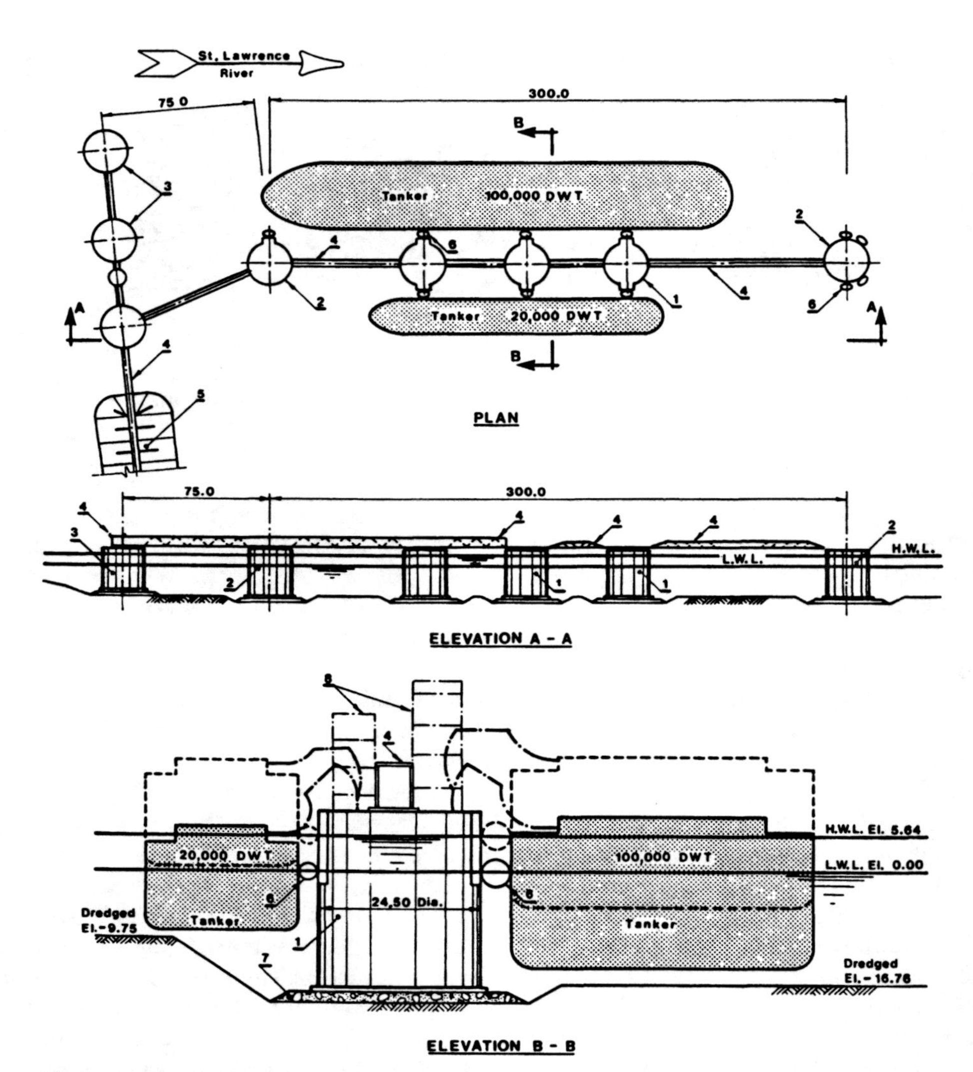

Figure 2–40. Golden Eagle Oil Terminal, Quebec. (a) Layout, (b) typical elevations. 1—Breasting dolphin, 2—Mooring dolphin, 3—Ice-breaking cell, 4—bridge, 5—causeway, 6—pneumatic rubber fender, 7—rockfill mattress, 8—hose handling towers.

indicated possible presence of medium-size ice floes much thicker than 110 cm. There also existed the possibility of ice ridges, hummocks, and ice hanging dams with a maximum thickness up to 18.0 m.

The river at the terminal site has a pattern of reversing flow and the maximum reversing current at terminal location in the spring may reach about 2.15 m/s with only a minor reduction during the other seasons. The tide at some locations may exceed 6.0 m.

The most dangerous masses of drifted ice in the area were a direct result of strong winds and current, and tidal actions.

The thorough evaluation of the site environmental conditions led the terminal designers to the conclusion that for safe operation during the winter season in general but basically for the protection of the ship moored at the outer berth from impact by moving ice the terminal must be protected by ice deflectors composed of three ice-breaking dolphins, placed at about 75 m upstream of the dock structure.

The number of cells in ice, the ice deflector, and the open space between them were chosen to provide the desired hydraulic characteristics at the dock location on the one hand, and to limit the size of ice floes that could pass through the system on the other hand.

It was established that the dock structure and upstream dolphins must be designed to absorb the impact of a very large ice floe (estimated volume of about 740 000 m^3 of ice). Accordingly, the structure was designed in the form of individual floated-in concrete caissons, seated on prepared rock mattresses.

The dock is composed of three breasting and two mooring dolphins with identical dolphin construction. The concrete caissons forming the substructure of the wharf are interconnected with each other by the steel truss bridges carrying a vehicular roadway along the bottom cords and pipelines on the top. A 300-m long rockfill causeway links the wharf with the river bank. The slopes of the causeway are suitably armored to protect them against ice- and wave-induced damages.

The wharf structure provides two berths. The offshore berth is designed primarily for receiving crude oil; it provides a minimum water depth of about 17.0 m. The inshore berth which has a minimum water depth of about 10.0 m has been designed for use by smaller tankers distributing refinery products.

At the terminal location the river bed is underlined by approximately 4.6 m of silty sand and gravel with boulders, overlying approximately 18.0 m of hard clay on shale bedrock.

The poor upper layer of foundation soil was excavated and thoroughly levelled rockfill mattresses were placed on a prepared bottom, to spread the bearing stresses from cells over a wider area of the underlying soil strata.

Concrete cylindrical caissons about 24.0 m in diameter, and 27.4 m high stiffened from inside by vertical concrete diaphragms in two directions have been constructed in a tidal dry dock, then floated in and installed on the mattresses. Caissons were designed to withstand an impact load equal to about 4365 tonnes and those portions of caissons subjected to the most intense ice forces were designed to withstand a static ice pressure of approximately 2.75 MPa.

Flat faces at breasting dolphins were cast for fenders as required, and all caissons were thickened from the inside at locations where an ice impact load was expected to act. The cells were filled with granular material. Under high ice loads of short duration, a base pressure of 100 kPa was allowed.

The steel truss bridges were fabricated onshore, then barged to the site at high tide and set upon bearings during the falling tide. To suit high tidal range floating pneumatic fenders 3.5 m in diameter at the outer berth were installed. Similar fenders but of smaller diameter have been installed at the inner berth.

To avoid damage by floating ice floes and for protection from being carried away during a natural clearing of ice the lifting mechanism was provided to retract fenders from water whenever no vessels were at the dock.

The fenders were designed to absorb energy of a docking ship and also to absorb energy of the ice floes hitting the ship moored at the dock. For overall safety and efficiency of operations quick release hooks were provided throughout the installation. To date the terminal survived more than 20 winter seasons with no serious problems reported.

2.9.1.4 *Waste Product Receiving Terminal at Glatved, Denmark* *(Lisby and Hartelius, 1991)*

The site is located on the east/west coast of Jutland (South of Grenaa), Denmark. The dock was designed to receive self-unloading ships of 1000 to 3000 DWT.

The dock was completed in 1989. The dock site is exposed to heavy ice forces, generated by ice sheets 0.6 m thick, and to loads produced by waves up to 3.0 m high.

The dock was designed and built in a traditional manner with a centrally located unloading platform supported on two cells, and two breasting and two mooring dolphins, all linked with each other by footbridges. The pier itself is linked with the shore by a 150 m long foot/conveyer bridge, which is supported by two cells. All cells used for the dock construction are of identical design.

The solution to the heavy ice forces was found in construction of gravity-type steel sheet-pile cells 7.64 m in diameter with a conical superstructure made of concrete (Figure 2–41). The conical shape of a cell superstructure reduces the design ice forces to less than 10% of the horizontal ice load exerted on a corresponding cell but of a constant diameter.

Thus the governing forces are basically those attributed to ship impact. The structure was constructed in the following sequence. First, sheet piles were driven to the required depth and to about 1.0 m above minimum water level. Then the cell was filled with gravel up to the level located at about 1.6 m below minimum water level. Next water within the upper portion of a cell was pumped out and the precast heavily reinforced concrete members of a conical portion of a cell and steel anchors between sheet piles and the conical superstructure were installed and the interior of the conical structure was filled with cast-in-situ concrete. Finally, sheet piles originally extending above the bottom of the conical portion of a cell were cut off as indicated in Figure 2–41. The super-arch type fenders were fixed to the steel frame-like support structure, extended up from the bottom of the conical superstructure, and fixed at the top of the dolphin. The distance between adjacent fenders was 17.0 m.

To ensure stability of the dolphin against ice forces the fender supporting structure which may interfere with ice was designed to collapse in case of exposure to ice sheets thicker than 0.4 m. It has been determined that the probability of this event occurring is once in 10 years.

It should be noted that in the course of the dock operation the fender supporting structures have been damaged on several occasions by small ships rammed into them and at least in one case even a foot-bridge between mooring and breasting dolphins, although located 4 m behind berthing line, has been damaged by the errant ship.

2.9.2 Piled Structures

In the preceding discussion on the types of structures used for the construction of the

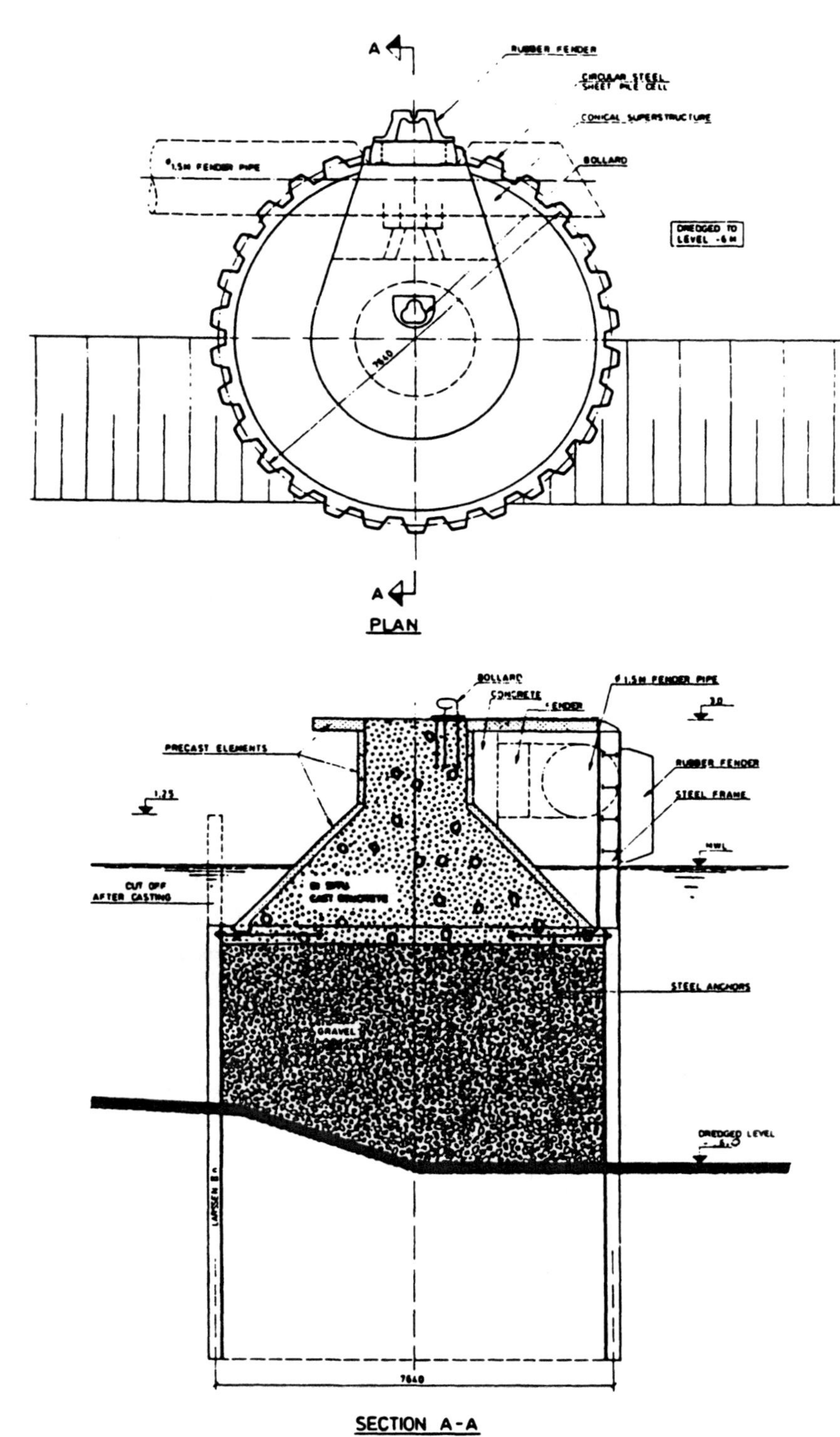

Figure 2–41. Dolphin structure for the Glatved Pier, Denmark. (From Lisby and Hartelius, 1991.)

dock facilities in ice-affected waters, it has been noted that conventional piles, protected in one way or another from the direct impact of the large ice features, can be used for marine construction in moderately heavy ice conditions. Under the latter conditions the conventional piling can be protected by the large-diameter steel pipe piles or by installation of sheet piles in front of the conventional piling. The following two recent case histories demonstrate the potential solution to the problem.

2.9.2.1 Bulk Material Unloading Terminal at Asnaes, Denmark *(Lisby and Hartelius, 1991)*

The terminal facility is located in Kalundborg Fjord on the west coast of Sealand and is designed to receive a fully loaded 120 000 DWT coal carrier or a partly loaded ship up to 175 000 DWT.

The design ice thickness at location is 0.4 m based on the new Danish Code of Practice EDS410. The design strength of ice was considered equal to 1200 kPa to 1600 kPa which is translated into a horizontal ice load of about 0.5 MN/ln.m to 0.65 MN/ln.m.

Considering all local design conditions such as waves 1.4 m high, geotechnical conditions, loads from two grab coal unloaders and conveyers, and ship impact load the structure was built in the form of a piled pier 210 m long, 20 m wide, and 21.0 m high, linked to the shore by a 100 m long conveyer trestle.

The pier's typical cross-section is illustrated in Figure 2–42. The pier pile arrangement includes vertical front steel pipe piles 762 mm in diameter and rear vertical and battered steel pipe piles 660 mm in diameter, embedded into foundation material, then complemented by injected ground anchors protected by 300 mm diameter steel

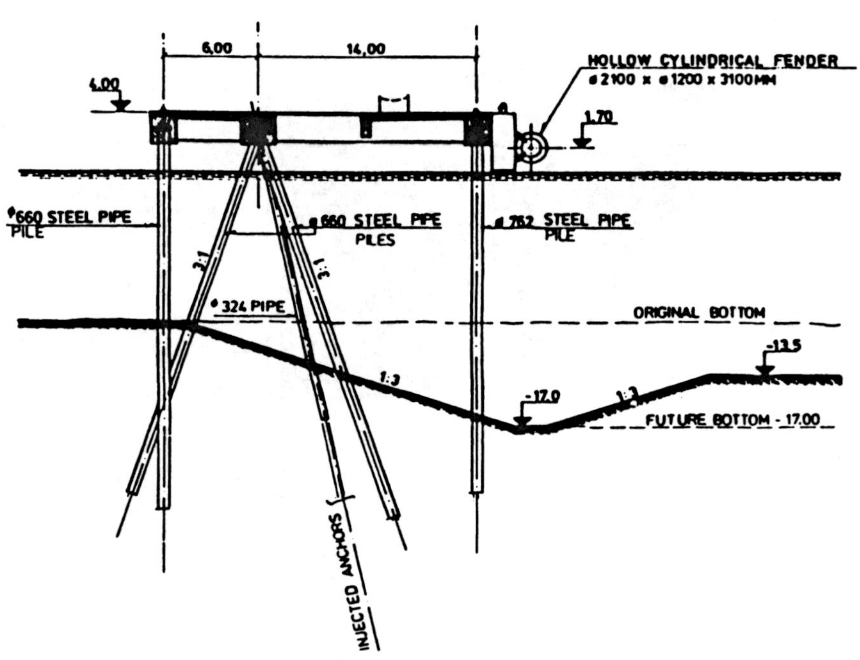

Figure 2–42. Coal unloading terminal at Asuaes, Denmark. Typical cross-section through the pier. (From Lisby and Hartelius, 1991.)

pipe piles. Unfortunately Lisby and Hartelius do not provide information on the geotechnical condition of the site and details on ground anchors. The pier deck is of a conventional cast-in-situ reinforced concrete construction.

The pier was commissioned in 1980. Since then it has been reported that the lower portion of the deck structure has experienced some local corrosion of reinforcing bars, basically attributed to the chloride attack.

2.9.2.2 Wharf at Stigsnaes, Denmark (Lisby and Hartelius, 1991)

This wharf was completed in 1979. It is located South of Kalundborg, at the Stigsnaes thermo power station. It is designed to receive fully coal-laden carriers of up to 150 000 DWT. The design ice thickness at the wharf location is 0.4 m, and the design wave height is 1.4 m. The site has a history of a piled structures (oil pier) being severely damaged by floating ice sheets. Therefore, during the design phase the decision was made to use conventional pile supporting and anchoring systems, protected, however, by a sheet-pile bulkhead (Figure 2–43). The water depth in front of the wharf is 18.0 m. To enable the use of regular steel sheet-pile profiles under such conditions, the steel anchor rods were placed below water level in order to reduce the free-standing portion of these sheet piles. This, however, produced high anchor load, which needed to be supported by an additional anchor system. The latter included a combination of A-type concrete piles 300 × 300 mm in cross-section and anchor rods 75 mm in diameter secured to the concrete anchor slabs.

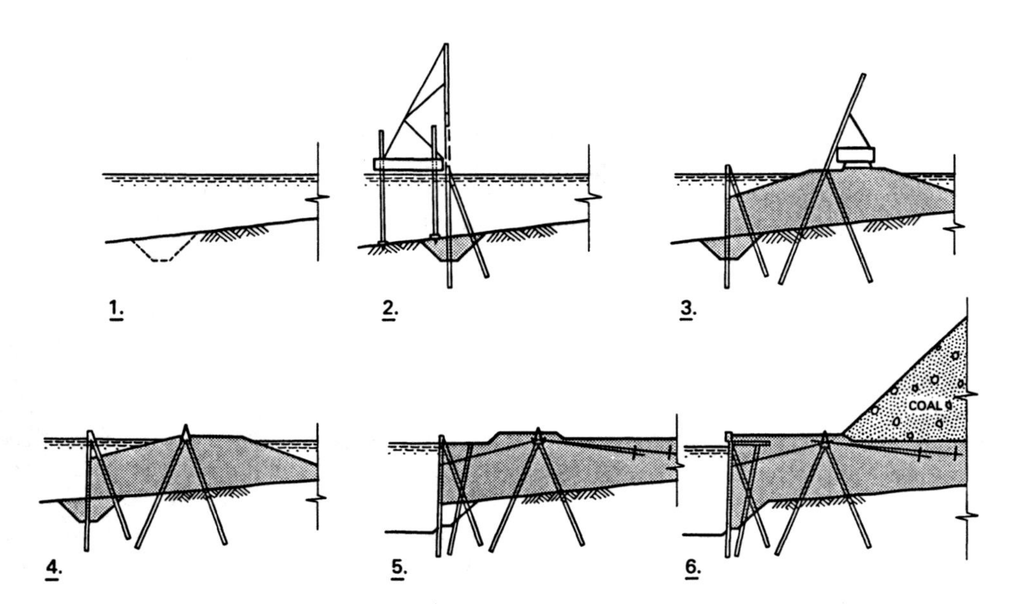

Figure 2–43. Coal unloading terminal at Stigsnaes, Denmark. Construction sequences. (From Lisby and Hartelius, 1991.) 1—Trench excavation for sheet piles driving, 2—driving sheet pile and anchor piles from jack-up platform, 3—driving concrete anchor piles from partly completed fill. Note that at this stage the sheeting is partly anchored and can take some of soil lateral load, 4—additional (underwater) anchor rods are installed, 5—bulkhead pile arrangement, additional anchorages, and backfill are completed, 6—concrete deck is completed and dock is in operational condition.

Besides sheet piles the dock bulkhead system includes two rows of battered piles. Eventually the whole pile arrangement is surmounted by the heavily reinforced concrete slab.

The sequence of the dock construction is shown in Figure 2–43. The regular granular fill was placed behind the bulkhead structure, in the course of construction, in the sequence illustrated in Figure 2–43.

The dock construction was extended to the winter season, and in order to mitigate the ice problem the water from a closely located thermo power station cooling system was discharged to the rear of the bulkhead. After completion the dock was equipped with two grab unloaders and conveyers for unloading and transporting of the coal to the storage facility and/or to the power station.

2.9.3 Single-Point Moorings

At some southern locations single-point moorings (SPMs) linked to the shore-based storage facilities by an underwater pipeline and used for loading/unloading of crude oil Liquid Natural Gas (LNG) or miscellaneous petroleum products represent economical and viable alternatives to the harbor located floating or bottom fixed terminals.

In ice-affected waters, however, conventional SPM systems are not suitable and cannot be reasonably adopted for use in the moving ice environment. A useful discussion on the subject is given in Pollack (1985).

In recent years a number of SPM concepts for year-round operation in ice-affected waters have been proposed. Actually, all proposed solutions can be grouped into three basic categories:

1. Bottom-fixed gravity-type structures designed to resist the ice and mooring forces and to provide a lee area free of moving ice.
2. Collapsible (articulated) systems, designed to yield under heavy ice loads, then rebound back to the design position when the attacking ice feature is broken up or moved away.
3. Submersible systems sitting under the ice cover and raised to the surface of water when required.

2.9.3.1 Bottom-Fixed Gravity-Type SPM

This concept of a SPM may be exemplified by a proposal made by Michel (1970). He suggested that the SPM operating in an area of moving ice be designed in the form of a large-diameter cylindrical dolphin seated on the sea floor and designed to crush and deflect the moving ice and form an ice-free lee area. Also, when there is no ice it would reduce the amplitude of storm waves to acceptable smaller values in the lee (Figure 2–44). The structure was designed to operate offshore in deep water, away from landfast ice. The bow of a ship berthed in the lee area would be secured at the dolphin such that the longitudinal axis of the ship would extend radially from the dolphin.

The dolphin itself can be constructed in the form of a floated-in concrete caisson, seated on a rockfill mattress. The caisson should be filled with granular material to provide sufficient weight needed to resist ice, mooring, wave, and current forces. In an ice moving zone the dolphin may have an ice-breaking mechanism, for example, conical capital which will help to reduce ice loads by failing ice in a bending rather than in a crushing or splitting mode of failure. The conical capital built on the top of a dolphin may also have an inclined outer surface to break the ice downward, which is more efficient not only to break, but also to clear the ice underneath on both sides of a dolphin.

The key to successful SPM operation is its proposed mooring system which is designed to keep the ship in the lee area free of moving ice radially to the dolphin at all times during loading/unloading operation.

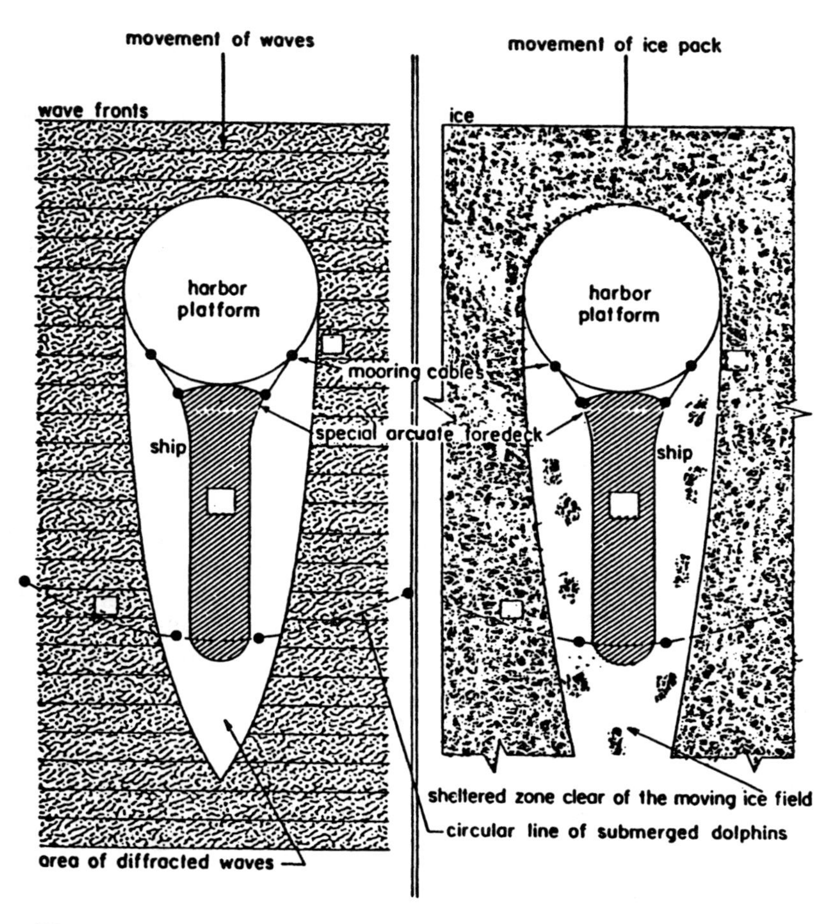

Figure 2–44. Arctic "single-point" mooring structure (SPM); principle of operation (proposed by Michel, 1970).

To obtain a high resisting moment at ship bow and high resistance to lateral forces in general, it was proposed that a special fore deck be installed on the ship. It should be shaped in the form of arcuate front and be as large as practical. Subsequently, the ship will be secured to the dolphin at mooring accessories placed at the ship and on the periphery of the dolphin as shown in Figure 2–44.

In the stern area the ship will be secured to underwater moorings (dolphins) distributed on a circle circumscribing the mooring head at a distance such as to allow mooring of the stern of the ship on both sides. The moorings of the cleat type are unaffected by moving ice and provide for holding the ship at any position within the lee area. According to Michel the mooring lines can be wound around the cleats by different means, but basically with the help of an icebreaking tugboat.

It must be noted that some critics of Michel's proposal suggested that the mooring dolphin may not produce ice-free lee area at all, or its shelter effect may be very minimal (Kovacs, Edvards, Dick, and Tsang, in the article by Michel, 1970).

2.9.3.2 Collapsible (Articulated) SPM System

Quite a few modifications of such a system have been proposed in recent years. A concept developed by Wang Quin-Jian (1987) may perhaps exemplify efforts in this direction (Figure 2–45). The system as proposed by Wang Quin-Jian consists of a buoy, connected to a deep-set foundation by means of a universal hinge. The buoy consists of three sections: a lower section, which is semispheroid; a middle section, which is a cone on which the floating ice feature will act; and an upper section, which is a column. The concept involves the following basic principles:

1. The structure provides resistance to ice forces not by its stiffness but by its buoyancy and its mooring to the sea floor.
2. Sea ice acts on the surface of the buoy's cone section. When the ice load will exceed the cone's ice-breaking capacity the whole buoy will rotate about the hinge. This will reduce angle between the ice feature moving against SPM and the contact surface. The latter in turn will increase the cone's icebreaking capability.
3. The system will be prefabricated and assembled in dry dock, then towed to the site of deployment, ballasted down, and fixed to the sea floor after which the buoy will be brought into a vertical position by pumping out the ballasting water. As a result of this operation the system will obtain the tension similar to that in tension leg platforms. In extreme conditions the ice can ride over the buoy. The buoy will rebound to the original position as soon as the extreme ice condition is over.

2.9.3.3 Submersible SPM

The system is designed for year-round operating installations in landfast ice areas, or in the harbors with no significant ice movements during the winter season (Figure 2–46) (Tsinker, 1984).

The proposed concept assumes that the SPM system will be kept under the ice cover between scheduled ship arrivals. It will be raised to the surface through the ice cover just before arrival of a tanker. This will be accomplished by discharging the required

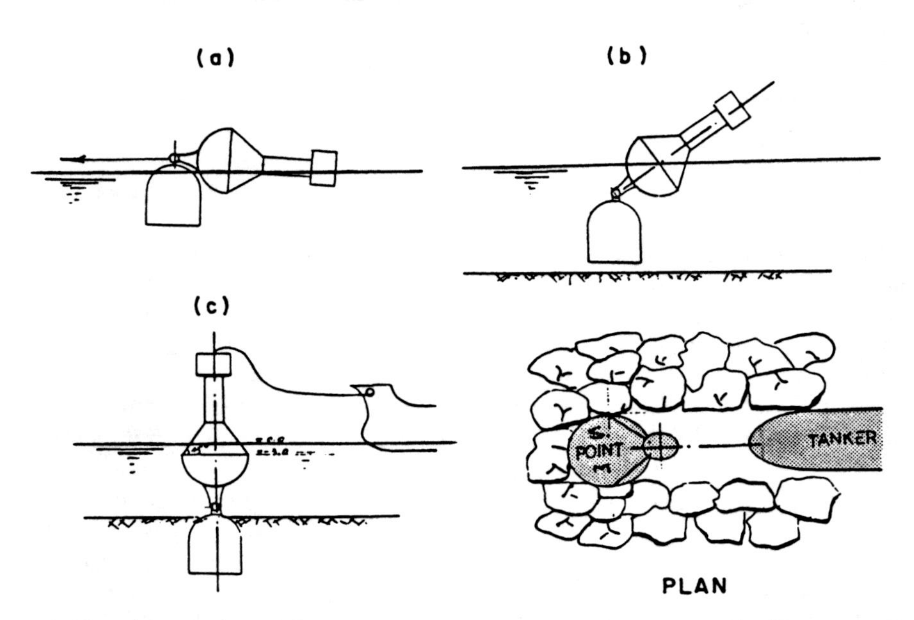

Figure 2–45. Yielding to ice pressure single-point mooring (proposed by Wang Quin-Jian, 1987). (a) In towing, (b) during installation or raising, (c) working state.

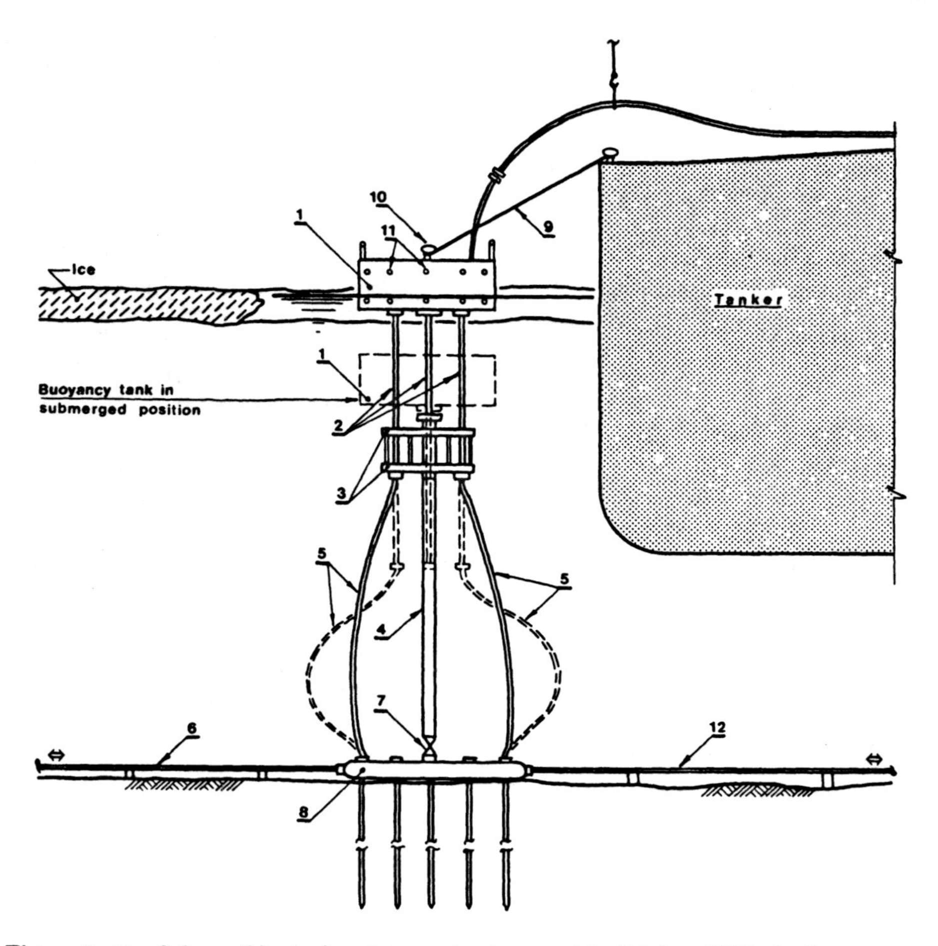

Figure 2–46. Submersible single-point mooring (proposed by Tsinker, 1984). 1—Buoyancy tank, 2— and 3—guilding system, 4—articulated buoyant column, 5—flexible hoses, 6—warm water (air) supply line, 7—universal hinge, 8—foundation base, 9—mooring line, 10—mooring accessories, 11—warm water (air) discharge outlet, 12—oil (LNG) supply line.

amount of warm water or air, or a combination of both bia a buoyancy tank placed just underneath the ice cover.

The buoyancy tank with its guiding mechanism, which constitutes the movable upper portion of the system, will pop up to the surface of the water as soon as the ice is melted. Then the ice-free zone around the SPM can be maintained by continuing discharge of a certain amount of warm water/air or by operating a bubbler system.

The foundation structure (8) of the system is fixed to the sea floor by suitable means which, basically, depends on sea floor geotechnical conditions. The articulated buoyant column (4) with a guiding mechanism (3) at the top is linked to the foundation structure by the universal hinge (7). The buoyancy tank (1) whose vertical movements are controlled by the guiding system (2) has the nozzels (11) for discharge of a warm water and/or warm compressed air. Flexible hoses (5) which link together the loading/unloading lines (6 and 9) and warm water/air supply lines (6) are attached to the buoy guiding system which also serves as a stiff loading/unloading line.

The mooring accessories (10) are placed on the top of a buoy. Depending on local environmental conditions the complete installation may include one loading/unloading SPM only, or for more secure mooring arrangements an additional four mooring systems structurally identical to loading/unloading SPM may be considered (Figure 2–47).

2.9.4 Offshore Terminals in Moving Ice

The choice of the type of an offshore terminal design in moving ice is dictated largely by ice management considerations. The ship(s) that will be using terminal must be able to approach the berth under a wide range of ice conditions, to pull alongside the dock structure, and to remain stationary there to allow for safe loading/unloading operations. Also, the "on-land" facilities such as oil storages, power generating facilities, heliport, living quarters, and others must all be protected from a moving ice pack.

This basically can be achieved by constructing an offshore facility protected by an island designed to resist or deflect the moving ice field. In shallow waters it can be accomplished by construction of an earth-filled island with protected slopes. In relatively deep waters, the island can be formed with the help of floating-in caissons, placed on a prepared pad. The latter solution will drastically reduce the amount of granular material required for construction of an island. It must be pointed out that in deep waters (more than 20 m) construction of SPMs (platforms) is usually more economical than construction of conventional islands.

Typically, large movements of ice cause extensive rubble to form around the island (atoll), which stabilizes later in the season.

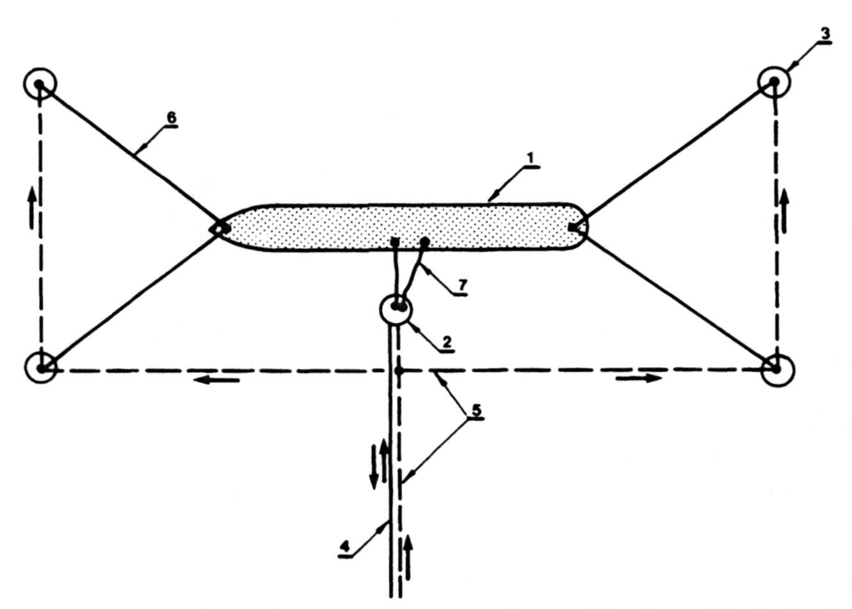

Figure 2–47. Submersible loading/unloading installation with submersible single-point mooring layout (proposed by Tsinker, 1984). 1—Tanker, 2—submersible loading/unloading system, 3—submersible mooring system, 4—material line, 5—warm water (air) supply line, 6—mooring line, 7—flexible hose.

A typical sequence of ice action on an exposed island is explained by Croasdale and Marcellus (1978). According to these investigators the initial movements of thin ice fail the ice sheet at low loads in bending. At this stage the ice is too thin to ride up a slope so that rubble piles are formed. In some cases the rubble forms a ramp upon which the oncoming ice sheets advance and fail in the bending mode. At other times the ice penetrates the rubble, but is again failed in bending. Once the rubble grounds it grows seaward. The active ice zone remains on the outside of the rubble, which consolidates and which has been observed to freeze into rigid ice annuals around an island. As the degree of consolidation increases through the winter the rubble becomes more competent and then transmits steadily, increasing loads to the structure.

Kry (1977) suggested that a rubble field around an island would increase the influence area for load transfer, thus leading to higher ice forces. However, Wong et al. (1991) with reference to field observations and field measurements of ice pressures reported by Gulf Canada Resources Inc. (1983), and Marshall et al. (1989) stated that in fact a grounded ice rubble field may reduce the ice loads transmitted to an offshore structure by 25 to 50%. The former authors suggested that this load depends on a number of factors including the rubble morphology (its geometry and composition, void ratio), the extent of grounding, the degree of grounding (the ice-berm interface property), and the mechanical properties of ice rubble.

As discussed earlier in this chapter there is no agreement among observers as to what may happen in the lee of a man-made structure placed within moving ice. Some maintain that the lee is ice free, and others state that the lee closes up very rapidly behind an island because of the pressure existing in the moving field which is perpendicular to the ice motion. Therefore, to protect the dock area from moving ice a second island strategically placed to prevent a lee close-up situation is sometimes required. This island will also protect the berth from waves and wind actions during an ice-free season.

Two basic types of layout design can be used to mitigate ice problems in the lee. They are islands with a double-entry basin (DEB) and islands with a single-entry basin (SEB). The former (Figure 2–48a) allows the ship to approach the berth from either direction normal to the principal ice movement. Under this scheme ships are able to leave in the same direction as they enter, and no maneuvers or turns within the protected areas of the island are involved. One major concern is the safe maneuverability of the ship while entering the protected area under variable dynamic ice conditions outside the atoll. When the ship enters the protected basin, ice conditions change abruptly from large floes and ridges to a fairly uniform, although relatively unconsolidated, ice cover. This may cause the ship to surge ahead if a maximum thrust has been exerted. When a ship enters the protected basin its initial speed may be dangerously high and therefore its stopping distance within this area must be carefully evaluated.

IN SEB design (Figure 2–48b) the ship approaches the island parallel to and against the principal ice movement. This maneuver may be easier than in DEB because the ship meets less resistance due to ice cross movements. Again, the basin must be long enough to prevent the ship from hitting the island at the end of the protected basin due to potential surge in speed.

In both aforementioned design alternatives the ice management system must be in place to ensure a ship safe entry into a protected basin and to minimize adverse effects of ice on a ship during berthing operation.

Although the pack ice and large ice features are unlikely to intrude into the protected basin, the natural ice growth over the basin by the end of a winter season, as

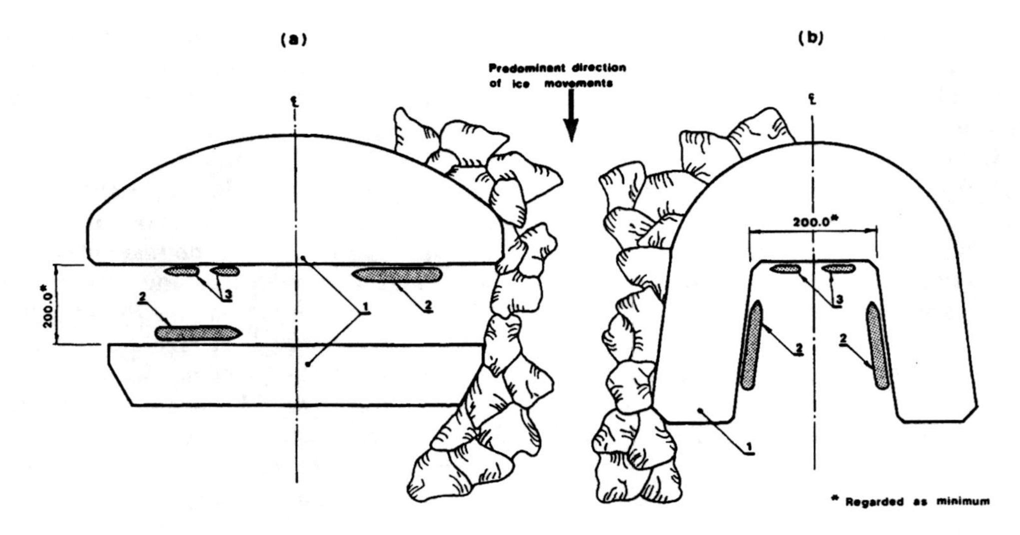

Figure 2–48. Typical layouts of an offshore terminal in moving ice. (a) Double entry basin (DEB). (b) single entry basin (SEB). 1—Island (atoll), 2—icebreaking tanker, 3—icebreaking tug.

well as an ice buildup in a ship tract, may be substantial. The ice control within the basin and ice management in the berth area may include different means as discussed earlier in this chapter.

The size of the main (front) island is governed primarily by the area required for the on-land facilities. Its crescent shape is designed to resist ice loading and to encourage formation of a rubble field to help guide the moving ice past the island and to control ice forces on the island perimeter structures.

In view of the limited open water season it is necessary to minimize the amount of on-site construction works by maximizing the use of prefabrication and modular construction techniques with on-site work carried out by the ice-strengthened marine equipment. This may involve the use of heavy armor blocks or rocks designed to be stable under ice movements on and around the island, and larger floated-in steel or concrete fill retaining caissons to be placed at the island perimeter and designed to withstand ice and wave actions.

An example of the armor used for the causeway protection at Nome, Alaska built in 1985 to 1986 is illustrated in Figure 2–

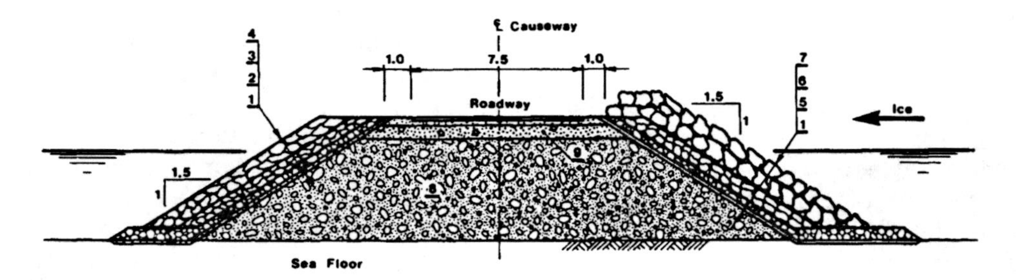

Figure 2–49. Port of Nome, Alaska Causeway. Typical cross-section. 1—Crushed stone, 2—rocks 40 kg, 3—two layers of 800-kg rocks, 4—two layers of 8-ton tocks, 5—two layers of 200-kg rocks, 6—two layers of 2-tone rocks, 7—two layers of 22-ton rocks, 8—quarry run/dredge tailings, 9—nonfrost suceptible material top 2 m (compacted).

49 (Horton, 1982; Bruun and Sackinger, 1985; Campion et al., 1987).

In recent years the offshore structures in moving ice have become numerous in the US and Canadian Arctic waters and in the St. Lawrence Seaway. They range from the monopod built in Cook Inlet, Alaska, to many caisson, cone-shaped islands and fills combined with caissons. Cammaert and Muggeridge (1988), Wortley (1990), and Bruun (1985) give numerous examples of such structures and an observation of their performance.

The most recent examples of caisson-retained islands are construction by Dome Petroleum Limited of Tarsiut Island, and Esso Resources Canada Ltd. built Kadluk Island.

The former is the first concrete caisson-retained island constructed in a 22-m water depth in the Beaufort Sea. The sea floor at the construction site was excavated to a depth of 2 to 3 m below the mudline, then carefully controlled sand berm with an average slope of 1 : 6.5 (compared to 1 : 15 for previous islands) was constructed. The berm was completed with a layer of thoroughly leveled layer of gravel. The concrete caissons were built on the west coast of Canada, then barged to the site of deployment. Each caisson was 69 m long, 11.5 m high, 15 m wide, weighed 5300 tonnes, and had 17 cells to hold the fill material. Each caisson was reinforced with 1100 tonnes of reinforcing steel, so it can withstand the pressure of winter ice in the Beaufort Sea. This amount of reinforcing steel is about four times that of similar structures built in southern regions. A total of 2200 m^3 of high-strength concrete (40 MPa) was used to build one caisson. To accommodate the high density of reinforcing bars, the use of superplasticizer was required. The caisson was designed to transmit global forces directly through diaphragms into the rear wall behind the point of application of the load and from there to the retained soil (Fitzpatrick and Stenning, 1983). The designed local pressure of 700 kPa was distributed through the structure with help of the T-beams placed at the caisson's front wall. A typical cross-section of Tarsuit caisson-retained island is depicted in Figure 2–50.

More information of Tarsuit Island design and construction is given in Hnatiuk (1983)

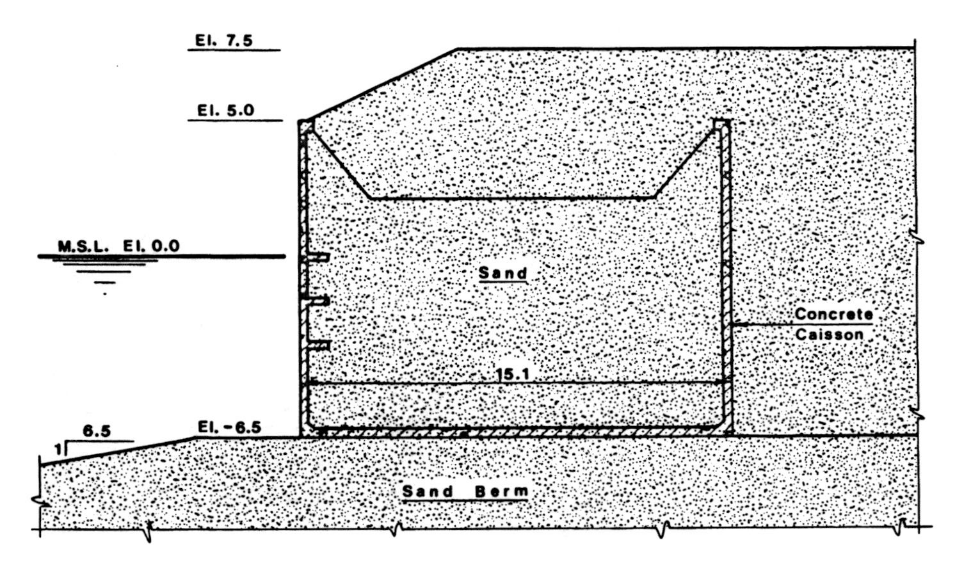

Figure 2–50. Tarsuit concrete caisson-retained island (Beaufort Sea). Typical cross-section.

and in Concrete International, Vol. 4, No. 3, March 1982.

Esso's Kadluk Island was constructed in a 15-m depth of water. In this case the island perimeter was designed and built in the form of a steel caisson (Figure 2–51). The caisson was assembled in an octagonal ring of eight separate units held in place by a system of stressed cables that encircled the caisson through interior compartments. Each caisson was 49 m long, 73.1 m wide, and 12.2 m high. About 1000 tones of steel were used to build all eight caissons. Together they form an octagon 117 m across. The system was built in Japan, then towed to the site of deployment, where it was ballasted down and finally filled with granular material. It should be noted that after a single winter season at designated locations, the system was raised, split in two halves, and towed to a new site.

The structure was designed for global (1700 kPa) and local (4800 kPa) ice forces and avoidance office ride-up onto the island working surface. First-year ice and multiyear floes were assumed to fail in flexure or in continuous crushing, depending on structure slope. First-year ridges were assumed to fail by plug-type shear across the caisson and multiyear ridges by bending across the berm. Clearing forces caused by pack ice and/or ridges were also considered in island design.

Although in island design the presence of rubble was ignored the latter experience has indicated that when rubble occurs it would likely ground and absorb some of the forces exerted by moving ice. The rubble diameter could be greater than island diameter. The particulars of Kadluk island design are discussed in detail in de Jong and Bruce (1978), Hawkins et al. (1983), and Comyn (1984.)

REFERENCES

ABAM ENGINEERS, INC., 1983. "Development, Design, and Testing of High Strength, Lightweight Concrete for Marine Arctic Structures." Phase I. AOGA Project No. 198. Tacoma, Washington.

ABDELNOUR, R., 1981. "Model Tests of Multiyear Pressure Ridges Moving onto Conical

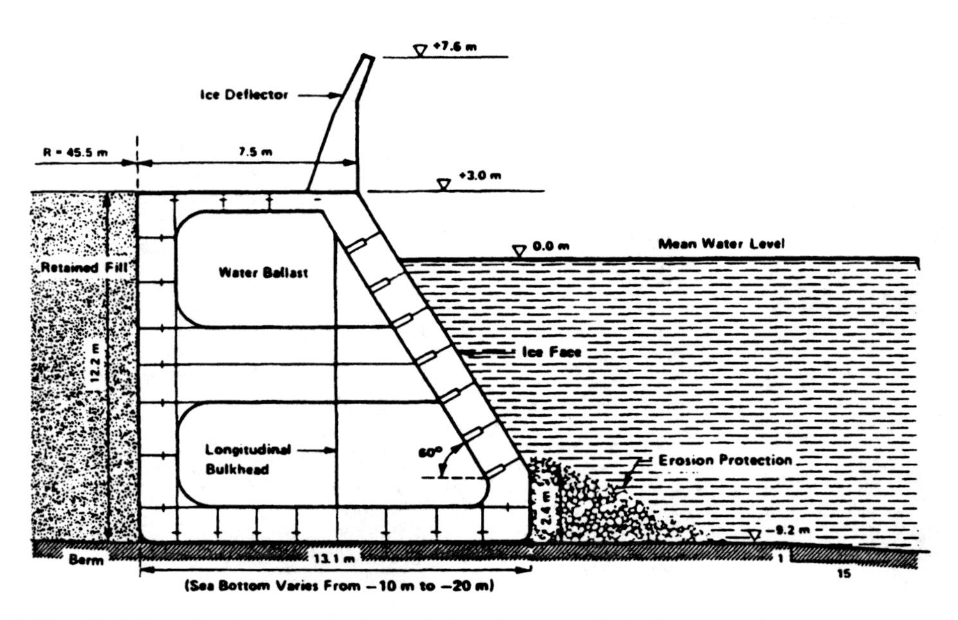

Figure 2–51. Kadluk steel caisson-retained island (Beaufort Sea). Typical cross-section.

Structures." Proceedings IAHR Syposium on Ice. Quebec City, Quebec.

ABELE, G. and FRANKENSTEIN, G., 1967. "Snow and Ice Properties as Related to Roads and Runways in Antarctica." CRREL Technical Report 176. Hanover, NY.

ACRES CONSULTING SERVICES LTD., 1981. "Arctic Pilot Project." Ice Management for Revised Terminal Layout. Petro-Canada. Calgary, Alberta.

ACRES CONSULTING SERVICES LTD., 1983. "Ice Control for Arctic Ports and Harbours." Final Report, Vol. 1. Transportation Development Centre, Transport Canada. Montreal and Canadian Marine Transportation Administration, Transport Canada Ottawa.

AFANAS'EV, V. P., DOLGOPOLOV, Y. V., and SHRAISHTEIN, Z. I., 1971. "Ice Pressure on Individual Marine Structures." In Ice Physics and Ice Engineering. Israel Program for Scientific Translation, 1973.

ALLISTON, G. R., 1985. "Low Friction and Adfreeze Coatings." Proceedings Conference Arctic 1985.

AMERICAN BUREAU OF SHIPPING, 1981. "Rules for Building and Classing Steel Vessels." New York, New York.

AMERICAN BUREAU OF SHIPPING, 1983. "Rules for Building and Classing Offshore Installations." Part 1, Structures. New York, New York.

AMERICAN CONCRETE INSTITUTE, 1978. Cold Weather Concreting ACI 306R-78. Detroit, Michigan.

AMERICAN CONCRETE INSTITUTE, 1979. "Guide for Structural Lightweight Aggregate Concrete." Published in Concrete International, Vol. 1, No. 2.

AMERICAN CONCRETE INSTITUTE, 1986. "State-of-the-Art Report on Offshore Concrete." Structures for the Arctic ACI 357.1R-85. Detroit, Michigan.

AMERICAN CONCRETE INSTITUTE, 1987. Standard Specification for Cold Weather Concreting ACI 306.1-87. Detroit, Michigan.

AMERICAN PETROLEUM INSTITUTE, Bulletin 2N, 1981. Bulletin on Planning, Designing, and Constructing Fixed Offshore Structures in Ice Environments. Dallas, Texas.

AMERICAN PETROLEUM INSTITUTE, Bulletin RP 2A, 1984. API Recommended Practice for Planning, Designing and Constructing Fixed Offshore Platforms. Dallas, Texas.

ANDERSLAND, O. B. and ANDERSON, D. M., 1978. "Geotechnical Engineering for Cold Regions." McGraw-Hill, New York.

ANDERSON, D. M. and MORGENSTERN, N. R., 1973. "Physics, Chemistry, and Mechanics of Frozen Ground: A Review." Proceedings 2nd International Conference on Permafrost. Yakntsk, USSR, North American Contribution. National Academy of Science, Washington, D.C.

ANONYMOUS, 1991. "Nome Dock Withstands Climate, Helps Commerce." Civil Engineering, Vol. 61, No. 10, October.

AROCKIASAMY, M., REDDY, D. V., and MUGGERIDGE, D. B., 1983. "Dynamic Responses of Moored Semisubmersibles in an Ice Environment." ASCE Structures Congress, Preprint S-12.

ASCE, 1983. "Design for Ice Forces," edited by S. R. Caldwell and R. D. Crissman. ASCE State of the Practice Report, New York.

ASHTON, G. D., 1974a. "Use of Air Bubbler Systems to Suppress Ice." U.S. Army CRREL, Special Report 210. Hanover, New Hampshire.

ASHTON, G. D., 1974b. "Evaluation of Ice Management Problems." U.S. Army CRREL Report. Hanover, New Hampshire.

ASHTON, G. D., 1979. "Suppression of River Ice by Thermal Effluents." U.S. Army CRREL Report 79-30. Hanover, New Hampshire.

ASSUR, A., 1975. "Problems in Ice Engineering." Proceedings Third International Symposium on Ice Problems. Hanover, New Hampshire.

AYOUB, A. S. and BROWN, T. G., 1991. "Fracture Analysis of Ice Forces." ASCE Journal of Cold Regions Engineering, Vol. 5, No. 4, December.

BARNES, P., et al., 1971. "The Friction and Creep of Polycrystalline Ice." Proceedings of the Royal Society of London, A., Vol. 324.

BASS, D. W. and SEN, D., 1986. "Added Mass and Damping Coefficient for Certain 'Realistic' Icebergs." Cold Regions Science and Technology No. 12. Elsevier Science Publishers B.V., Amsterdam, Netherlands.

BERCHA, F. G., 1986. "Ice–Structure Interaction: Engineering Design and Construction Criteria." Vol. 1 formal Report AES/SAG 1-2:86-17V1. For Public Works Canada.

BERGDAHL, L., 1978. "Thermal Ice Pressure in Lake Ice Covers." Department of Hydraulics Chalmers University of Technology. Report Series A:2.

BILELLO, M. A., 1961. "Formation, Growth and Decay of Sea Ice in the Canadian Archipelago." Arctic, Vol. 14, No. 1.

BILELLO, M. A., 1980. "Maximum Thickness and Subsequent Decay of Lake, River, and Fast Sea Ice in Canada and Alaska. CRREL Report 80-6, Hanover.

BLANCHET, D. and HAMZA, H., 1983. "Plane-Strain Compressive Strength of First Year Beaufort Sea Ice." Proceedings of the International Conference POAC, Vol. 3, Helsinki.

BLANCHET, D., CHURCHER, A., FITZPATRICK, J., and BARDA-BLANCHET, 1989. "An Analysis of Observed Failure Mechanisms for Laboratory, First-Year and Multi-Year Ice." In Special Report 89-5 by IAHR, edited by G. W. Timco.

BLENKARN, D. A., 1970. "Measurement and Analysis of Ice Forces on Cook Inlet Structures." Proceedings OTC, Paper No. OTC-126. Houston, Texas.

BLOUIN, S. E., CHAMBERLAIN, E. J., SELLMAN, P. V., and GARFIELD, D. E., 1979. "Penetration Tests in Subsea Permafrost Prudhoe Bay, Alaska." U.S. Army CRREL, Hanover, New Hampshire.

BROOK, J. W., FACTOR, D. F., KINNEY, F. D., and SARKER, A. K., 1988. "Cold Weather Admixture." Concrete International, October, No. 10.

BROWN, J. H., 1963. "Elasticity and Strength of Sea Ice." In Ice and Snow Properties, Processing and Applications, edited by W. D. Kingery. MIT Press, Cambridge, MA.

BROWN, R. D. and ROEBBER, P., 1985. "The Ice Aeration Problem in Canadian Waters Related to Offshore Energy and Transportation." Canadian Climate Centre, Report N. 85-13. Atmospheric Environment Service, Downsview, Ontario.

BROWN, R. J. E. and KUPSCH, W. O., 1974. "Permafrost Technology." National Research Council, Canada Committee Geotechnical Research Technical Memo 111.

BROWN, R. R. J., 1970. "Permafrost in Canada; Its Influence on Northern Development." edited by R. F. Legget, University of Toronto Press, Toronto, Canada.

BRUUN, E. and BRUUN, P., 1985. Index, POAC 1971–1985. Danish Hydraulic Institute, Greenland.

BRUUN, P. and SACKINGER, W., 1985. Brief Presentation on Port and Coast Structures in Ice—Some American and Canadian Experiences. Proceedings International Conference POAC 1985. Narssarssuaq, Greenland.

BRYCE, J. B. and BERRY, G. T., 1968. "Lake Erie–Niagara Ice Boom." Canadian Geotechnical Journal, Vol. 51/2, February.

BUSLOV, V. M., 1985. "Modular Construction Technology for Arctic Islands," Proceedings ASCE Conference Archive '85, March. San Francisco, California.

BUTKOVICH, T. R., 1956. "Strength Studies of Sea Ice." U.S.A. SIPRE Research Report 20.

BUTKOVICH, T. R., 1959. "On the Mechanical Properties of Sea Ice." Thule, Greenland, 1957. U.S. Army Snow, Ice and Permafrost Research Establishment (U.S.A. SIPRE), Research Report 54.

CAMMAERT, A. B., KIMURA, T., NOMA, N., YASHIMA, N., YANO, S., and MATSUSHIMA, Y., 1986. "Adfreeze Forces on Offshore Platforms." Proceedings International OMAE Symposium, Vol. 4, Tokyo.

CAMMAERT, A. B. and MUGGERIDGE, D. B., 1988. "Ice Interaction with Offshore Structures." Van Nostrand Reinhold, New York.

CAMMAERT, A. B. and TSINKER, G. P., 1981. "Impact of Large Ice Floes on Icebergs on Marine Structures." Proceedings International Conference POAC., Vol. 2. Quebec City, Quebec.

CAMMAERT, A. B., MILLER, D. R., and GILL, R. J., 1979. "Concepts for Ice Management at an Arctic LNG Terminal." Proceedings International Conference on Port and Ocean Engineering Under Arctic Conditions. Trondheim, Norway.

CAMMAERT, A. B., TANNER, R. G., and TSINKER, G. P., 1983a. "Design of Ice Management Systems for Arctic LNG Dock." Proceedings International Conference on Port and Ocean Engineering Under Arctic Conditions POAC. Helsinki, Finland.

CAMMAERT, A. B., WONG, T. T., and CURTIS, D. D., 1983b. "Impact of Icebergs on Offshore Gravity and Floating Platforms." Proceedings

International Conference POAC., Vol. 4. Helsinki, Finland.

CAMPION, D., SACKINGER, W. M., and WIDDIS, J., 1987. Construction of the Port of Nome, Alaska. Proceedings International Conference POAC '87. Fairbanks, Alaska.

CANADIAN DEPARTMENT OF ENVIRONMENT, 1971. "Review of Current Ice Technology and Evaluation of Research Priorities." Report Series No. 17, Inland Water Branch. Ottawa, Canada.

CAREY, K. L., 1979. Data Report. Siginaw Bay Ice Suppression Test. CRREL Draft Report.

CARINO, N. J., 1988. Specifications for Cold Weather Concreting. Concrete International, October, H10.

CARSTENS, T., 1978. "Ice in Estuaries and Harbours." Proceedings JAHR Symposium on Ice Problems. Lulea, Sweden.

CASPPR, 1979. Canadian Arctic Shipping Pollution Prevention Regulations. Canadian Ministry of Transport. Ottawa, Canada.

CHAKRABURTTY, S. et al., 1984. Arctic Marine Design and Construction Handbook, Vol. 1. Prepared for Transportation Development Centre. Transport Canada. Montreal, Canada.

CHRISTENSEN, T. F. and TRYDE, P., 1984. "Extraction of Piles by Repeated Waterlevel Fluctuations." Proceedings IAHR Ice Sympsoium, Vol. 2. Hamburg, New Hampshire.

CHRISTENSEN, T. F. and ZABILANSKY, I., 1985. Review of Experimental Studies of Uplifting Forces Exerted by Adfrozen Ice on Marine Piles. Proceedings POAC 1985.

CLARK, J. I. and GUIGNE, J. V., 1988. "Marine Geotechnical Engineering in Canada." Canadian Geotechnical Journal, Vol. 25.

CLARK, J. I., SHIELDS, D. H., GRAHAM, J., and RYMES, J. E., 1983. "Cold Region Engineering Earthworks." Transportation Development Center, Contract No. 4927. Ottawa, Canada.

CLOUGH, G. W., MOSHER, R., SINGH, Y., and KUPPUSAMY, T., 1987. "Two and Three Dimensional Finite Element Analyses of Cellular Cofferdams." Proceedings of Specialty Conference on Soil–Structure Interaction, Vol. 1.

COMYN, M. I., 1984. "Caisson-Retained Island—First Year of Operation." Proceedings Arctic Offshore Technology Conference. Calgary, Alberta.

COOK, V. B., Co. Limited, 1982. "Air Bubbler Ice Control Experiment in Thunder Bay, Ontario." Draft Report Submitted to Transportation Development Centre, Montreal.

COVENEY, D. B., 1981. "Cutting Ice with High Pressure Water Jets." Conference on Port and Ocean Engineering Under Arctic Conditions. Quebec City, Quebec.

COX, G. F. N., RICHTER-MENGE, J. A., WEEKS, W. F., and MELLOR, M., 1984a. "A Summary of Strength and Modulus of Ice Samples from Multi Year Pressure Ridges." Proceedings International OMAE Symposium. New Orleans.

COX, G. F. N., RICHTER-MENGE, J. A., WEEKS, W. F., MELLOR, M., and BOSWORTH, H., 1984b. "Mechanical Properties of Multi Year Ice, Phase 1: Test Results." CRREL Report CR84-09. Hanover, New Hampshire.

COX, J. R., 1985. Ice Floe Impact Forces on Vertical-Sided Structures. ASCE Proceedings Conference Arctic '85, New York.

CROASDALE, K. R., 1974. "The Crushing Strength of Arctic Ice. The Coast and Shelf of the Beaufort Sea," editors Reed, J. D. and Sater, J. E., Artic Institute of North America, 1974.

CROASDALE, K. R., 1978. "Ice Forces on Fixed, Rigid Structures." Special Report 0-26. International Association for Hydraulic Research. (Published by U.S. Army Corps of Engineers, Cold Regions Research and Engineering Laboratory).

CROASDALE, K. R., 1980a. "Some Implications of Ice Ridges and Rubble Fields on the Design of Arctic Offshore Structures." Proceedings of NRC Workshop on Sea Ice Riding. Technical Memo 134, National Research Council of Canada. Calgary, Alberta.

CROASDALE, K. R., 1980b. "Ice Forces on Fixed Rigid Structures." In Working Group on Ice Forces on Structures, A State-of-the-Art Report. CRREL Special Report 80-26, Hanover, New Hampshire.

CROASDALE, K. R., 1985. "Recent Developments in Ice Mechanics and Ice Loads." Behaviour of Offshore Structures. Elsevier Science Publishers B.V., Amsterdam, Netherlands.

CROASDALE, K. R. and MARCELLUS, N. R., 1978. "Ice and Wave Action on Artificial Islands in the Beaufort Sea." Canadian Journal of Civil Engineering, Vol. 5.

CROASDALE, K. R., MORGENSTERN, N. R., and NUTTALL, J. B., 1977. "Indentation Tests to Investigate Ice Pressure on Vertical Structures. Journal of Glaciology, Vol. 19, No. 81.

CROTEAU, P., ROJANSKY, M., and GERWICK, B. C., 1984. "Summer Ice Floe Impacts Against Caisson-Type Exploratory and Production Platforms." Proceedings International Symposium, Vol. 3, New Orleans, Louisiana.

CURRIER, J. H. and SCHULSON, E. M., 1982. "The Tensile Strength of Ice as a Function of Grain Size." Acta Metall., Vol. 30, No. 8.

CURTIS, D. D., CAMMAERT, A. B., WONG, T. T., and BOBBY, W., 1984. "Numerical Analysis of Impact of Small Icebergs on Semisubmersibles." International Specialty Conference Cold Regions Engineering, Edmonton, Alberta.

DANIELEWICZ, B. W. and BLANCHET, D., 1987. "Measurements of Multi-Year Ice Loads on Hans Island During 1980 and 1981." International Conference POAC Preprint, Fairbanks, Alaska.

DANIELEWICZ, B. W. and CORNETT, S., 1984. "Ice Forces on Hans Island 1983." APOA Project No. 202, Canadian Marine Drilling Ltd. Calgary, Alberta, Canada.

DANIELEWICZ, B. W., METGE, M., and DUNWOODY, A. B., 1983. "On Estimating Large Scale Ice Forces from Deceleration of Ice Floes." Proceedings of the 7th POAC Conference, Vol. 4, Helsinki, Finland.

DANYS, J. V., 1979. "Artificial Islands in Lac St. Pierre to Control Ice Movement." First Canadian Conference on Marine Geotechnical Engineering. April 25–27.

DANYS, J. V. and BERCHA, F. G., 1975. "Determination of Ice Forces on a Conical Offshore Structure." Proceedings International Conference POAC, Vol. 2, Fairbanks, Alaska.

DEJONG, J. J. A. and BRUCE, J. C., 1978. "Design and Construction of a Caisson-Retained Drilling Platform for the Beaufort Sea." Proceedings OTC, Houston, Texas.

DERUITER, H., 1984. "Subsidence of Nanisivik Mines Concentrator Building." Mining Magazine, July.

DERY, J. L., 1981. "Design of Wharves for Winter Navigation in the St. Lawrence River." Proceedings International Conference on Port and Ocean Engineering Under Arctic Conditions POAC, Vol. 1, Quebec City, Quebec.

DET NORSKE VERITAS, 1977. Rules for the Design, Construction and Inspection of Offshore Structures, Oslo, Norway.

DET NORSKE VERITAS CANADA, 1984. State-of-the-Art Description of Methods for Computation of Global and Local Loads in Ice Structure Interaction. Report for Public Works Canada.

DICKINS, D. F. and WETZEL, V. F., 1981. Multi-year Pressure Ridge Study, Queen Elizabeth Islands. Proceedings International Conference POAC, Quebec City, Quebec.

DOMASCHUK, L., MARR, C. S., SHIELDS, D. H., and YOUNG, G., 1983. "Creep Behavior of Frozen Saline Silt Under Isotropic Compression." Proceedings 4th International Conference on Permafrost. Fairbanks, Alaska.

DORONIN, Y. P. and KHESIN, D. E., 1977. "Sea Ice." Translated from Russian. American Publishing, New Delhi, India.

DOUD, J. O., 1978. "Ice Sheet Loads on Marina Piles." Journal of Waterway, Port, Coastal and Ocean Division, ASCE, Vol. 104, No. WW1.

DROUIN, M., 1970. "State of Research on Ice Thermal Thrust." IAHR Proceedings of Symposium. Reykjavik, Iceland.

DROUIN, M. and MICHEL, B., 1971. "Les Poussees d'origine theruique exercees par les converts de glace sur les structures hydrauliques." Rapport S-23, Laval University. Quebec City, Quebec.

DUNNICLIFF, J., 1988. "Geotechnical Instrumentation for Monitoring Field Performance." John Wiley & Sons, New York.

DYKINS, J. E., 1968. Tensile and Flexural Properties of Saline Ice. Proceedings of the International Symposium on Physics of Ice. Munich, Germany.

DYKINS, J. E., 1970. Ice Engineering Tensile Properties of Sea Ice Grown in a Confined System. Naval Civil Engineering Laboratory Technical Report R689. Port Hueneme.

DYKINS, J. E., 1970. Ice Engineering—Material Properties of Saline Ice for a Limited Range of Conditions. U.S. Naval Civil Laboratory Technical Report R720. Port Hueneme.

EAU, 1990. Recommendations of the Committee for Waterfront Structures, Sixth Edition, Erust and John, Berlin, Germany.

EDIL, T. B., ROBLEE, C. J., and WORTLEY, C. A., 1988. "Design Approach for Piles Subjected to Ice Jacking." Proceedings ASCE. Journal of Cold Regions Engineering, Vol. 1, No. 2, June.

EDWARDS, R. Y. and CROASDALE, K. R., 1976. "Model Experiments to Determine Ice Forces on Conical Structures." Symposium on Applied Glaciology, International Glaciological Society. Cambridge, U.K.

ERANTY, E. and LEE, G. C., 1981. "Introduction to Ice Problems in Civil Engineering." Department of Civil Engineering, State University of New York. Buffalo, New York.

ERKIZIAN, H., 1976. "Anchorage Port Survives Nature." Civil Engineering, December.

ETTEMA, R., 1991. "Ice Formations in Navigation Channels." PIANC Bulletin No. 74. Brussels, Belgium.

ETTEMA, R. and HUNG-PIN HUANG, 1990. "Ice Formation in Frequently Transited Navigation Channels." Report 90-40, December. The U.S. Army CRRL. Hamburg, New Hampshire.

EVANS, D. C. B., et al, 1976. "The Kinetic Friction of Ice." Proceedings of the Royal Society of London, Series A, Vol. 347.

EVANS, R. J. and PARMERTER, R. R., 1985. "Ice Forces Due to Impact Loading on a Sloping Structure." Proceedings Conference Arctic '85. ASCE, New York.

EVERS, K. U. and WESSELS, E., 1986. "Model Test Study of Level Ice Forces on Cylindrical Multi-Legged Structure." Polartech 86, Helsinki, Finland.

FENCO ENGINEERS, 1984. Arctic Marine Design and Construction Handbook. Transport Canada Report No. TP 5727E, Vol. 1, Montreal.

FIP, 1978. "Lightweight Aggregate Concrete for Marine Structures." State-of-the-Art Report, Federation Internationale de la Precontraiute. Slough, U.K.

FIP, 1983. Manual of Lightweight Aggregate Concrete, 2nd Edition. John Wiley & Sons, New York.

FITZPATRICK, J. and STENNING, D. G., 1983. "Design and Construction of Tarsuit Island in the Canadian Beaufort Sea." Proceedings OTC, Houston, Texas.

FORIERO, A. and LADANYI, B., 1991. "Design of Piles in Permafrost under Combined Lateral and Axial Load." ASCE Journal of Cold Regions Engineering, Vol. 5, No. 3, September.

FREDERKING, R M. W., 1977. "Plane-Strain Compressive Strength of Columnar Grained and Granular-Snow Ice." Journal of Glaciology, Vol. 18., No. 80.

FREDERKING, R. M. W., 1980a. "Ice Action on Nanisivik Wharf Strathong Sound, N.W.T. Winter 1978–1979." Canadian Journal of Civil Engineering, Vol. 7, No. 3.

FREDERKING, R., 1980b. "Dynamic Ice Forces on an Inclined Structure." In Physics and Mechanics of Ice, edited by P. Tryde. Springer-Verlag, New York.

FREDERKING, R., 1983. "Ice Engineering I." Presented at the Offshore Mechanics and Cold Region Engineering Symposium. University of Calgary. Calgary, Alberta, June 13.

FREDERKING, R. and HAUSLER, F. V., 1978. "The Flexural Behaviour of Ice from In-Situ Cantilever Beam Tests." Proceedings IAHR Ice Sympsoium, Vol. 1, Lulea, Sweden.

FREDERKING, R. and NAKAWO, M., 1984. "Ice Action on Nanisivik Wharf, Winter 1979–1980." Canadian Journal of Civil Engineering, Vol. II.

FREDERKING, R. and TIMCO, G. W., 1983. 'Uniaxial Compressive Strength and Deformation of Beaufort Sea Ice." Proceedings of International Conference POAC, Vol. 1, Helsinki.

FREDERKING, R. M. W. and TIMCO, G. W., 1984a. "Compressive Behaviour of Beaufort Sea Ice Under Vertical and Horizontal Loading." Proceedings of the International OMAE Symposium, Vol. 3.

FREDERKING, R. M. W. and TIMCO, G. W., 1984b. "Measurement of Shear Strength of Granular/Discontinuous Columnar Sea Ice." Journal of Cold Region Science and Technology, Vol. 9.

FREDERKING, R., WESSELS, E., MAXWELL, J. B., PRINSENBERG, S., and SAYED, M., 1986. "Ice Pressure and Behaviour at Adams Island, Winter 1983/84." Canadian Journal of Civil Engineering, Vol. 13.

GAITHER, W. S. and DALTON, R. E., 1969. "All-Weather Tanker Terminal for Cook Inlet, Alaska." Proceedings ASCE, Journal of the Waterways and Harbors Division, May.

GAMMON, P. H., GAGNON, R. E., BOBBY, W., and RUSSELL, W. E., 1983a. "Physical and Mechanical Properties of Icebergs." Proceedings of OTC, Vol. 1, Houston.

GAMMON, P. H., KIEFTE, H., CLOUTER, M. J., and DENNER, W. W., 1983b. "Elastic Constants of Artificial and Natural Ice Samples by Brillouin Spectroscopy." Journal of Glaciology, Vol. 29, No. 103.

GAMMON, P. H., LEWIS, J. C., and MUIR, L. R., 1988. "Iceberg Cutting with a Hot Wire." Cold Regions Science and Technology, No. 15. Elsevier Science Publishers B.V., Amsterdam, Netherlands.

GERSHUNOV, E. M., 1985. "Shear Strength of the Adfreeze Bond and Its Effect on Global Ice Loads Applied to Mobile Offshore Drilling Units Under Arctic Conditions." Offshore Technical Conference Paper Otc 4687, Houston, Texas.

GERSHONUV, E. M., 1986. "Collision of Large Floating Ice Feature with Massive Offshore Structure." ASCE Journal of Waterway, Port, Coastal and Ocean Engineering, Vol. 112, No. 3, May.

GERWICK, B. C., 1985. "Constructibility of Arctic Offshore Structures." ASCE Proceedings of the Conference Arctic '85, San Francisco, California.

GILL, R. J., TSINKER, G. P., and CAMMAERT, A. B., 1983. "Design Criteria for Arctic Ports and Harbours." Proceedings International Conference on Port and Ocean Engineering Under Conditions POAC, Helsinki, Finland.

GIRGRAH, M., 1987. "Wharves in Ice Environment Two Case Histories." Lecture at Seminar on Marine and River Ice Engineering, February. Technical University of Nova Scotia, Halifax, Canada.

GIRGRAH, M. and SHAH, V. K., 1978. "Construction of a Deep-Sea Dock in the Arctic." Proceedings 4th International Conference POAC '77. St. John's, Newfoundland, Vol. 1.

GLADWELL, R. W. M. 1976. Statistical Analysis of Ice Pressure Ridge Distribution in the Southern Beaufort Sea. APOA Project No. 54.

GOLD, L. W., 1984. "Ice Pressures and Bearing Capacity." Geotechnical Eng. for Cold Regions, edited by O. B. Andersland and D. M. Anderson. McGraw-Hill, New York.

GOODMAN, D. J., 1977. "Creep and Fracture of Ice and Surface Strain Measurements on Glaciers and Sea Ice." Ph.D. Thesis, University of Cambridge, U.K.

GOW, A. J., 1977. "Flexural Strength of Ice on Temperate Lakes." Journal of Glaciology, Vol. 19, No. 81.

GRIESBACH, R. J. and KREMER, H., 1973. "Conceptual Study of Marine Terminal Facilities in Arctic." Proceedings 23rd Congress, PIANC, Ottawa.

GROVE, C. S. Jr., GROVE, S. T., and ADIUM, A. R., 1963. "The Theory and Use of Aqueous Foam for Protection of Ice Surfaces in Ice and Snow." Properties, Processes and Applications. MIT Press, Cambridge, Massachusetts.

GUILLERMO, E. C. and LAWRENCE, S. A., 1987. "Design of AIDA Seaport for the De Long Mountain Regional Transportation System at the Chukchi Sea, Alaska." Proceedings International Conference, POAC '87. Fairbanks, Alaska.

GULF CANADA RESOURCES INC., 1983. Rubble Field Study, Issunguak, 1979–1980. Arctic Petroleum Operators Association, Report 171-1, Vol. 1.

GUTTMAN, S. I., PUSKAR, F. J., and BEA, R. G., 1984. "Analysis of Offshore Structures Subject to Arctic Ice Impacts. Proceedings International OMAE Symposium, Vol. 3, New Orleans, Louisiana.

HAGER, M., 1990. Part 1—"Principles for the Dimensioning of Structures Against Ice Load." PIANC Proceedings of the 27th Congress Section 1, Subject 2. Osaka, Japan.

HAGER, M. and KLEIN, E., 1990. Part 1. Principles for the Dimensioning of Structures Against Ice Loads. Part 2. The Struggle Against Ice Using the Elbe River as an Example. Proceedings PIANC 27th Navigation Congress. Osake, Japan.

HALLAM, S. D. and SANDERSON, T. J. O., 1987. "Advances in Ice Mechanics in the United Kingdom." ASME Applied Mechanics, Vol. 40, No. 9.

HAMZA, H. and MUGGERIDGE, D. B., 1979. "Plain Strain Fracture Toughness of Fresh Water Ice." Proceedings 5th International Conferences POAC '79. Trouheim, Norway.

HAMZA, H. and MUGGERIDGE, D. B., 1984. "A Theoretical Fracture Analysis of a Large Ice Floe with a Large Offshore Structure." Proceedings International OMAE Symposium, New Orleans, Louisiana.

HAUSLER, F. V., 1981. "Multiaxial Compressive Strength Tests in Saline Ice with Brush Type Loading Platens." Proceedings IAHR Ice Symposium, Vol. 2, Quebec City.

HAWKES, I. and MELLOR, H., 1972. "Deformation and Fracture of Ice Under Uniaxial Stress." Journal of Glaciology, Vol. 11, No. 61.

HAWKINS, J. R., JAMES, D. A., and DER, D. Y., 1983. "Design, Construction and Installation of a System to Measure Environmental Forces on a Caisson-Retained Island." Proceedings of International Conference POAC, Helsinki, Finland.

HAYLEY, D. W., 1979. "Site Evaluation for Artificial Drilling Island in the Beaufort Sea." First Canadian Conference on Marine Geotechnical Engineering, Calgary, Alberta.

HEYDINGER, A. G., 1987. "Piles in Permafrost." ASCE Journal of Cold Regions Engineering, Vol. 1., No. 2. June.

HNATIUK, J., 1983. "Case Studies on Arctic Offshore Gravity Platforms." Seminar on Offshore Mechanics and Cold Ocean Engineering, Calgary, Alberta.

HOBBS, P. V., 1974. "Ice Physics." Clavendor Press, Oxford.

HOLM, T. A., 1980. "Performance of Concrete in Marine Environment." ACI Publication SP-65, Detroit, Michigan.

HORJEN, I. and VEFSNMO, S., 1984. "Mobile Platform Stability (MOPS) Subproject 02—Icing." MOPS Report No. 15. Norwegian Hydrodynamic Laboratories, STF GO A284002.

HORTON, M. G., 1982. "Designing a Medium Draught Port for Nome." The Dock and Harbour Authority, August.

HUCK, R. W. and HULL, J. R., 1971. "Resonant Driving in Permafrost." Foundation Facts, Vol. 7, No. 1.

HULGAARD, E., 1985. "Examples of Quay Structures in Greenland Placed on Steeply Inclined Rock Surface and Subjected to Ice Forces." Proceedings POAC '85, September Narssarssuag, Greenland.

HUTCHEON, N. B. and HANDEGORG, G. O. P., 1983. "Building Science for a Cold Climate." John Wiley & Sons, Toronto, Canada.

INSTANES, B., 1979. "Coal Loading Pier in Svea, Svalbard." Proceedings International Conference POAC '79. Troudheim, Norway.

IRWIN, G. R., 1957. "Analyses of Stresses and Strains Near the End of a Crack Transversing a Plate." Journal of Applied Mechanics, Vol. 24, No. 3.

ISAACSON, M., 1985. "Iceberg Interaction with Offshore Structures." ASCE Proceedings Specialty Conference Arctic '85.

ISAACSON, M. and CHEUNG, K. F., 1988. "Influence of Added Mass on Ice Impacts." Canadian Journal of Civil Engineering, Vol. 15.

ISAACSON, M. and MCTAGGART, K. A., 1900. "Influence of Hydrodynamic Effects on Iceberg Collisions." Canadian Journal of Civil Engineering, No. 17.

ITOH, Y., YOSHIDA, A., TSUCHIYA, M., and KATOH, K., 1988. "An Experimental Study on Abrasion of Concrete Due to Sea Ice." Proceedings of Offshore Technology Conference, Vol. 2. Houston, Texas, May.

IYER, S. H., 1983. "Size Effects on Ice and Their Influence on the Structural Design of Offshore Structures." Proceedings of International Conference POAC, Vol. 2, Helsinki, Finland.

IYER, S. H., 1989. "A State-of-the-Art Review of Local Ice Loads for the Design of Offshore Structures Special." Fourth State-of-the-Art Report 89-5. U.S. Army Corps of Engineers CRREL. Hanover, New Hampshire.

IZUMIYAMA, K., KITAGAWA, H., KOYAMA, K., and UTO, S., 1991. "On the Interaction Between a Conical Structure and Ice Sheet." Proceedings of the 11th International Conference on Ports and Ocean Engineering Under Arctic Conditions. St. John's, Canada.

JEFFRIES, M. O. and WRIGHT, B. D., 1988. "Dynamic Response of 'Molikpaq' to Ice–Structure Interaction." Proceedings 7th International Conference OMAE, Vol. IV. Houston, Texas.

JIZU, X. and LEIRA, B. J., 1981. "Dynamic Response of a Jacket Platform Subjected to Ice Floe Loads." Proceedings POAC 81, Vol. I, Quebec City, Canada.

JOHNSON, R. C. and BENOIT, J. R., 1987. "Iceberg Impact Strength." Proceedings OTC, Houston, Texas.

JOHNSON, R. C. and NEVEL, D. E., 1985. "Ice Impact Structural Design Loads." Proceedings International Conference POAC, Narssarssuaq.

JOHNSTON, G. H. (ed.), 1981. "Permafrost Engineering Design and Construction." John Wiley & Sons, Toronto, Canada.

JONES, S. J. and CHEW, H. A. M., 1983. "Effect of Sample and Grain Size on the Compressive Strength of Ice." Annals Glaciology, 4.

JORDAAN, I. J., 1987. "Numerical and Finite Element Techniques in Calculation of Ice–Structure Interaction." In 3rd State-of-the-Art Report 87-17 Published by the U.S. CRREL, edited by T. J. O. Sanderson. Hanover, New Hampshire.

JORDAAN, I. J. and MCKENNA, R. F., 1989. "Modeling of Progressive Damage in Ice." In 4th State-of-the-Art Report IAHR published by the U.S. Army CRREL, Hanover, New Hampshire.

JUDGE, A. S., 1973. "Ground Temperature Measurements Using Thermistors." Proceedings of Seminar on the Thermal Regime and Measurements in Permafrost. Technical Memo No. 108. Associate Committee on Geotechnical Research NRC. Ottawa, Ontario (October).

KATO, K. and SODHI, D. S., 1983. "Ice Action on Pairs of Cylindrical and Conical Structures." U.S. Army CRREL, Report 83-25, Hanover, New Hampshire.

KATONA, M. G. and VAUDREY, K. D., 1973. "Ice Engineering—Summary of Elastic Properties and Introduction to Viscoelastic and Non-Linear Analysis of Saline Ice." Naval Civil Engineering Laboratory Technical Report R-797, Port Hueneme, California.

KERR, A. D., 1976. "The Bearing Capacity of Floating Ice Plates Subjected to Static or Quasi-Static Loads." Journal of Glaciology, Vol. 17.

KERR, A. D., 1978a. "Forces an Ice Cover Exerts on Rows or Clusters of Piles Due to a Change of the Water Level." Proceedings of IAHR Symposium on Ice, Part I, Lulea, Sweden.

KERR, A. D., 1978b. "On the Determination of Horizontal Forces a Floating Ice Sheet Exerts on a Structure." Journal of Glaciology, Vol. 20, No. 82.

KERR, A. D., 1980. "On the Buckling Force of Floating Ice Plates." IUTAM Symposium on the Physics and Mechanics of Ice. Springer-Verlag, Berlin.

KERR, A. D., 1986. Response of Floating Ice Beams and Plates with Partial Flooding." Proceedings of the 1st International Conference on Ice Technology. Cambridge, Massachusetts.

KJELDGAARD, J. H., 1977. "Thermal Ice Forces on Hydraulic Structures." A Short Literature Review. SINTEF Report No. STF A77043.

KJELDGAARD, J. H. and CARSTENS, T., 1980. "Thermal in Forces." In the Special State-of-the-Art Report 80-26, edited by T. Carstens. Published by U.S. Army Corps of Engineers CRREL, Hanover, New Hampshire.

KOHNEN, H., 1972. Seismic and Ultrasonic Measurements on the Sea of Eclipse Sound Near Pond Inlet, N.W.T. in Northern Baffin Island. Polarforsehung, Jahrg. 44, Nr. 2.

KORZHAVIN, K. N., 1962. "Action of Ice on Engineering Structures." USSR Academy of Science, Sibirian Branch. Translated from Russian by the U.S. Army CRREL in 1971, Hanover, New Hampshire.

KOVACS, A., 1983. "Characteristics of Multi-Year Pressure Ridges." Proceedings International Conference POAC, Helsinki.

KOVACS, A. and GOW, A. J., 1976. "Some Characteristics of Grounded Floebergs Near Prudhoe Bay." Alasak, CRREL Report 76-34, Hanover.

KOVACS, A. and MELLOR, M., 1971. "Sea Ice Pressure Ridges and Ice Islands." APOA Report TN-122, Calgary.

KOVACS, A., WEEKS, W. F., ACKLEY, S., and HIBLER, W. D., III, 1973. "Stucture of a Multi-Year Pressure Ridge." Arctic, Vol. 26, No. 1.

KOVACS, A., DICKENS, D. F., and WRIGHT, B., 1975. "An Investigation of Multi-Year Pressure Ridges and Shore Ice Pileups." APOA Project 89.

KREIDER, J. R., 1984. "Summer Ice Impact Loads from Multiyear Floes." Proceedings IAHR Ice Symposium, Vol. 2, Hamburg, New Hampshire.

KRY, P. R., 1977. "Ice Rubble Fields in the Vicinity of Artificial Islands." Proceedings of the 4th International Conference on Ports and Ocean

Engineering under Arctic Conditions, POAC '77, Vol. 1. St. John's, Canada.

KRY, P. R., 1978. "A Statistical Prediction of Effective Ice Crushing Stresses on Wide Structures." Proceedings IAHR Symposium on Ice Problems. Part 1. University of Lulea, Sweden.

KRY, P. R., 1980. "Ice Forces on Wide Structures." Canadian Geotechnical Journal, Vol. 17, No. 1.

LADANYI, B., 1983. "Design and Construction of Deep Foundations in Permafrost." North American Practice. Proceedings of 4th International Conference on Permafrost. Fairbanks, Alaska.

LADANYI, B. and JOHNSTON, G. H., 1973. "Evaluation of In Situ Creep Properties of Frozen Soils with the Pressuremeter." Proceedings 2nd International Conference on Permafrost, North American Contribution. Yakutsk, USSR. National Academy of Science, Washington, D.C.

LAFORTE, J.-L. and LAVIGNE, 1986. "Microstructure and Mechanical Properties of Ice Aerations Grown from Supercooled Water Droplets Containing NaCl in Solution." Proceedings 3rd International Workshop on Atmospheric Icing of Structures. Vancouver, British Columbia.

LAINEY, L. and TINAWI, R., 1981. "Parametric Studies of Sea Ice Beams Under Short and Long Term Loadings." Proceedings IAHR Symposium, Quebec City, Canada.

LAINEY, L. and TINAWI, R., 1984. "The Mechanical Properties of Sea Ice—A Compilation of Available Data. Canadian Journal of Civil Engineering, Vol. 2, No. 4.

LANE, S. D., 1988. "Heating Water and Aggregates for Cold Weather Concrete." Concrete International, October, No. 10.

LANGLEBEN, M. P. and POUNDER, E., 1963. "Elastic Parameters of Sea Ice." In Ice and Snow Properties, Processes and Applications, edited by W. D. Kingery. MIT Press, Cambridge, Massachusetts.

LARSEN, P., 1973. "Hydraulic Roughness of Ice Covers." Proceedings ASCE, Journal of Hydraulic Division.

LAU, M. and WILLIAMS, F. M., 1991. "Model Ice Forces on a Donward Breaking Cone." Proceedings the 11th International Conference on Ports and Ocean Engineering Under Arctic Conditions. September, St. John's, Canada.

LAU, M., MUGGERIDGE, D. B., and WILLIAMS, F. M., 1988. "Model Tests of Downward Breaking Conical Structures in Ice." Proceedings International OMAE Symposium, Vol. 4, Houston, Texas.

LAVROV, V. V., 1969. Deformation and Strength of Ice. Arctic and Antarctic Scientific Research Institute, Leningrad; Israel Program for Scientific Translations, Jerusalem, 1971.

LEE, G. C., SHIH, T. S., and CHANG, K. C., 1988. "Mechanical Properties of Concrete at Low Temperature." Proceedings Journal of Cold Regions Engineering, Vol. 2, No. 1, March.

LEGGET, R. F., 1966. "Permafrost in North America." Proceedings International Conference on Permafrost. National Academy of Science, NRC. Publication No. 1287.

LEWIS, J. W. and JOHNSON, R. P., 1980. "St. Lawrence River All-Year Navigation Ice Control System." U.S. Department of Transportation, April.

LINDSTRÖM, C. A., 1990. "Numerical Estimation of Ice Forces Acting on Inclined Structures and Ships in Level Ice" Paper OTC 6445. Proceedings Annual Offshore Technology Conference, Houston, Texas.

LIN'KOV, E. M., 1958. "Izuchenie Uprugikh Svoistv Ledianogo Pokrova v Arktike" (Study of Elastic Properties of an Ice Cover in the Arctic). Vestnik, Leningradskogo Univ. 13.

LISBY, J. and HARTELIUS, H., 1991. "Exposed Pier Structures for Bulk Shipping Systems in Danish Waters." PIANC, Bulletin No. 72.

LIU, W. and MILLER, K. J., 1979. "Fracture Toughness of Fresh Water Ice." Journal of Glaciology, Vol. 22.

LOUSDALE, J. T. and NORRBY, T., 1985. "Electric Heat Tracing Designed to Prevent Icing." Offshore, November.

LOZOWSKI, E. P., STALLABRASS, J. R., and HEARTY, P. F., 1983. "The Icing on an Unheated, Nonrotating Cylinder. Part 1: A Simulation Model." Journal of Climate and Applied Meteorology 22.

MAAGE, M. and HELLAND, S., 1988. "Cold Weather Concrete Curing Planned and Controlled by Microcomputer." Concrete International, October, No. 10.

MÄÄTTÄNEN, M., 1976. "On the Flexural Strength of Brackish Water Ice by In Situ Tests." Proceedings International Conference POAC, Vol. 1, Fairbanks, Alaska.

MÄÄTTÄNEN, M., 1980. "Ice Forces on Fixed, Flexible Structures." CRREL Special Report 80-26, Hanover, New Hampshire.

MÄÄTTÄNEN, M, 1987. "Advance in Ice Mechanics in Finland." ASME. Applied Mechanics, Vol. 40, No. 9, September.

MÄÄTTÄNEN, M., 1989. "Ice-Induced Vibrations of Structures. Self-Excitation." In 4th State-of-the-Art Report IAHR, Published by the U.S. Army CRREL. Hanover, New Hampshire.

MACHEMEHL, J. L., 1983. "Arctic Offshore Engineering—Concepts and Recent Developments." Advanced Project Conference Offshore Northern Sea, Stavanger.

MACKLIN, W. C. and PAYNE, G. S., 1968. "Some Aspects of the Aeration Process." Quarterly Journal of the Royal Meteorological Society, 94.

MAGEAU, D. W. and SHERMAN, M. B., 1983. "Frost Cell Design and Operation." Proceedings 4th International Conference on Permafrost. Fairbanks, Alaska. National Academy Press, Washington, D.C.

MAKKONEN, L., 1989. "Formation of Spray Ice on Offshore Structures." Fourth Special State-of-the-Art Report 89-5. U.S. Army Corps of Engineers CRREL. Hanover, New Hampshire.

MAKKONEN, L. and STALLABRASS, J. R., 1984. "Ice Aeration on Cylinders and Wives." NRC, Technical Report TR-LT-005. National Research Council of Canada.

MANGUS, A. R., 1988. "Air Structure Protection of Cold Weather Concreting." Concrete International, October.

MARCELLUS, R. W., MORRISON, T. B., ALLYN, N. F. B., CROASDALE, K. R., IYER, H. S., and TSENG, J., 1987. "Ice Forces on Marine Structures." Report AES/SAG 1-2:88-5V2 Prepared for Ministry of Supply and Services, Canada, 1988.

MARKIN, K. F., 1977. "Piles Foundations in Permafrost." In Handbook on Construction in Permafrost Zone, edited by Y. Valli, V. V. Dokuchaeu, and N. F. Fedorova. Stroyisdat Publisher, Leningrad, USSR.

MARSHALL, A. L., 1985. 'Behaviour of Concrete at Arctic Temperatures." Proceedings 8th International Conference POAC '85. Marssarssuag, Greenland.

MARSHALL, A. L. and DRAKE, S. R., 1987. "Thermal and Dynamic Response of Concrete at Low Temperature." Proceedings International Conference POAC '87. Fairbanks, Alaska.

MARSHALL, A. L., CROASDALE, F. R., FREDERKING, R. M., SAYED, M., JORDAN, I. J., and NADREAN, J. P., 1989. "Measurement of Load Transmission Through Grounded Ice Rubble." Proceedings of the 10th International Conference on Port and Ocean Engineering Under Arctic Conditions, POAC'89, Vol. 1. Lulea, Sweden.

MAYKUT, G. A. and UNTERSTEINER, N., 1971. "Some Results of Time-Dependent Thermodynamic." (From A. B. Cammaert and D. B. Muggeridge, 1988. "Ice Interaction with Offshore Structures. Van Nostrand Reinhold, New York.)

MCFADDEN, T., 1990. "The Kilpisjärvi Project: Case Study of Cold Regions Research in Finland." Proceedings, ASCE Journal of Cold Regions Engineering, Vol. 4, No. 2, June.

MELLOR, M., 1974. "A Review of Basic Snow Mechanics." Snow Mechanics Symposium. IAHR Publication No. 114. Grindlwald, Switzerland.

MELLOR, M., 1980. "Icebreaking Concepts." CRREL Special Report 80-2, January. Hanover, New Hampshire.

MELLOR, M., 1983. Mechanical Behavior of Sea Ice. CRREL Monograph 83-1, Hanover, New Hampshire.

MELLOR, M., 1986. "Revised Guidelines for Blasting Floating Ice." Special Report 86-10, May. U.S. Army CRREL, Hanover, New Hampshire.

MELLOR, M., COX, G. F., and BOSWORTH, H., 1984. "Mechanical Properties of Multi-Year Sea Ice-Testing Techniques." CRREL Report CR 84-08. Hanover, New Hampshire.

METGE, M., DANIELEWICZ, B. W., and HOARE, R. D., 1981. "On Measuring Large Scale Ice Forces, Hans Island 1980." Proceedings of the 6th POAC Conference Laval University, Quebec City, Quebec.

MICHEL, B., 1970. "Off-Shore Mooring Structure for the Arctic." IAHR Proceeding of Ice Symposium. Reykjavik, Iceland.

MICHEL, B., 1971. "Winter Regime in Rivers and Lakes." U.S. Army Cold Regions Research and Engineering Laboratory Monograph III, Hanover, New Hampshire.

MICHEL, B., 1972. "Effects of Vessel Operation on Sea Ice Growth." In Feasibility Study on Marine Terminal At Hershel Island. Department of Public Works. Ottawa, Canada.

MICHEL, B., 1978. "Ice Mechanics." Les Presses de L'University, Laval, Quebec.

MICHEL, B., 1981. "Advances in Ice Mechanics." Proceedings Sixth POAC Conference. University Laval, Quebec City, Canada.

MICHEL, D. and BERENGER, D., 1975. "Algorithm for Accelerated Growths of Ice in a Ship's Track." Third International Symposium on Ice Problems. Hanover, New Hampshire.

MICHEL, B. and BLANCHET, D., 1983. "Indentation on an 52 Floating in Sheet in the Brittle Range." Annals of Glaciology, Vol. 4.

MICHEL, B. and RAMSEIER, R. O., 1971. "Classification of River and Lake Ice." Canadian Geotechnical Journal, Vol. 8.

MICHEL, B. and TOUSSAINT, N., 1977. "Mechanisms and Theory of Indentation of Ice Plates. Journal of Glaciology, Vol. 19, No. 81.

MILLER, K. J., 1980. "The Application of Fracture Mechanics to Ice Problems." In Physics and Mechanics of Ice. Springer-Verlag, Berlin, Germany.

MINSK, L. D. (ed.), 1983. "Proceedings of First International Workshop on Atmospheric Icing of Structures." Published by the U.S. Army Corps of Engineers CRRL. Hanover, New Hampshire.

MONFORE, G. E., 1954. "Ice Pressure Measurement." USBR Structural Research Laboratory. Report C-662.117.

MOORE, K., WOODHEAD, R., and TUTTLE, H., 1988. "Cold Weather Concreting for Oil Production Plant." Concrete International, October.

MORGENSTERN, N. R. and SMITH, L. B., 1973. "Thaw-Consolidation Tests on Remolded Clays." Canadian Geotechnical Journal, Vol. 10, No. 1.

MORRISON, T. B., MARCELLUS, R. W., ABLYN, N. .F. B., and TSEUG, J., 1988. "Ice Forces on Marine Structures, Vol. 1—Calculations." Ministry of Supply and Services, Canada.

MURAT, J. R., 1978. "La Capacité Portante de la Glace de Mer." Ph.D. Thesis, Research Report EP78-R-49, Ecole Polytechnique, Montreal, Canada.

MURAT, J. R. and LAINEY, L. M., 1982. "Some Experimental Observations on the Poisson's Ratio of Sea Ice." Journal of Cold Region Science and Technology, Vol. 6, No. 2.

NADREAU, J. P. and MICHEL, B., 1984. "Ice Properties in Relation to Ice Forces." Proceeding Second State-of-the-Art IAHR Working Group on Ice Forces, Vol. 4, Chapter 1, Hamburg.

NAKABO, T., ETTEMA, R., and ASHTON, G. D., 1989. "Unconventional Power Source for Ice Control at Locks and Dams." ASCE Journal of Cold Regions Engineering, Vol. 3, No. 3, September.

NAKAWO, M. and SINHA, N. K., 1981. "Growth Rate and Salinity Profile of First–Year Sea Ice in the High Arctic." Journal of Glaciology, Vol. 27, No. 96.

NAWWAR, A. M., NADREAU, J. P., and WANG, V. S., 1983. "Triaxial Compressive Strength of Saline Ice." Proceedings International Conference POAC, Vol. 3.

NEILL, C. R., 1976. "Dynamic Ice Forces on Piers and Piles." An Assessment of Design Guidelines in the Light of Recent Research. Canadian Journal of Civil Engineering, Vol. 3, No. 2.

NESSIM, M. A., CHEUNG, M. S., and JORDAEN, I. J., 1987. "Ice Action on Fixed Offshore Structures." A State-of-the-Art Review. Canadian Journal of Civil Engineering, Vol. 14.

NEVEL, D. E., 1972. "The Ultimate Failure of Floating Ice Sheets." Proceedings IAHR ICE Symposium, Leningrad, USSR.

NIXON, J. F. and MORGENSTERN, N. R., 1974. "Thaw-Consolidation Tests on Undisturbed Fine-Grained Permafrost." Canadian Geotechnical Journal, Vol. 11, No. 1.

NIXON, W. A., ETTEMA, R., MATSUISHI, M., and JOHNSON, R. C., 1993. "Model Study of Cable-Moored Conical Platform." ASCE Journal of Cold Regions Engineering, Vol. 6, No. 1, March.

NOBLE, P. G. and SINGH, D., 1982. "Interaction of Ice Floes with the Columns of a Semi-Submersible." Proceedings 14th Offshore Technology Conference, Vol. 4, Houston, Texas.

NOTTIGHAM, D. and CHRISTOPHERSON, A. B., 1983. "Driven Piles in Permafrost: State-of-the-

Art." Proceedings of the International Conference on Permafrost. Fairbanks, Alaska.

OKSANEN, P., 1980. "Coefficient of Friction Between Ice and Some Construction Materials." Technical Research Centre of Finland, Report 7, Espoo.

OKSANEN, P., 1981. "Coefficient of Friction Between Ice and Some Construction Materials, Plastics and Coatings." Technical Research Center of Finland, Report 7.

OKSANEN, P., 1983a. "Adhesion Strength of Ice." Proceedings International Conference POAC, Vol. 2, Helsinki, Finland.

OKSANEN, P., 1983b. "Friction and Adhesion of Ice." Laboratory of Structural Engineering Technical Research Center of Finland, Publication 10, Espoo.

OSTERKAMP, T. E. and HARRISON, W. D., 1976. "Subsea Permafrost at Prudhoe Bay, Alaska: Drilling Report and Data Analysis." University of Alaska, Geophysical Institute, Report R-245.

OSTERKAMP, T. E. and HARRISON, W. .D., 1981. "Temperature Measurements in Subsea Permafrost off the Coast of Alaska." In the Roger J. E. Brown Memorial Volume. Proceedings 4th Canadian Permafrost Conference. Calgary, Alberta, NRC of Canada, Ottawa.

OSTERKAMP, T. E., HARRISON, W. D., and HOPKINGS, D. M., 1987. "Subsea Permafrost in North Sound, Alaska." Cold Regions Science and Technology No. 14. Elsevier Science Publishers B.V., Amsterdam, Netherlands.

PAIGE, R. A. and LEE, C. W., 1967. Preliminary Studies on Sea Ice in McMurdo Sound, Antarctica, during "Deep Freeze 65", Journal of Glaciology, Vol. 6, No. 46.

PALMER, A. C., GOODMAN, D. J., ASHBY, M. F., EVANS, A. G., HUTCHINSON, J. W., and PONTER, A. .R. S., 1983. Fracture and its Role in Determining Ice Forces on Offshore Structures. Annals of Glaciology, 4.

PARAMESWARAN, V. R., 1987. "Adfreezing Strength of Ice to Model Piles." Canadian Geotechnical Journal, Vol. 24.

PARAMESWARAN, V. R. and JONES, S. J., 1975. "Brittle Fracture of Ice at 77 k." Journal of Glaciology, Vol. 14, No. 71.

PARFENOV, A. F., BALANIN, V. V., and IVANOV, L. V., 1973. "Effect of Ice on Ships and Structures and Maintenance of Their Operation at Subzero Air Temperatures." Proceedings 23rd International Navigation Conference, PIANC, Section II, Ottawa.

PARISET, E., HANSSER, R., and GAGNON, A., 1966. "Formation of Ice Covers and Ice Jams in Rivers." Proceedings ASCE, Journal of the Hydraulic Division, Vol. 92, No. 6.

PARMERTER, R. R. and COON, M. D., 1983. "On the Mechanics of Pressure Ridge Formation in Sea Ice." OTC Paper 1810, Offshore Technology Conference, Houston, Texas.

PATTERSON, D. E. and SMITH, N. W., 1981. "The Measurement of Unfrozen Water Content by Time." Domain Reflectometry: Results from Laboratory Tests. Canadian Geotechnical Journal, Vol. 18, No. 1.

PEKAR, J. W., 1988. "Concreting in Alaska." Concrete International, October, No. 10.

PERDICHIZZI, P. and YASUDA, T., 1978. "Port of Anchorage Marine Terminal Design." ASCE. Proceedings of Specialty Conference on Applied Technology for Cold Environment. Alaska, May.

PERHAM, R. E., 1974. "Forces Generated in Ice Boom Structures." U.S. Army CRREL Special Report 200. January.

PERHAM, R. E., 1977. "Ice and Ship Effects on the St. Mary's River Ice Booms." Proceedings 3rd National Hydrotechnical Conference Canadian Society Civil Engineering. Quebec City, Quebec.

PERHAM, R. E., 1985. Determining the Effectiveness of a Navigable Ice Boom. Special Report 85-17, October. U.S. Army CRREL. Hanover, New Hampshire.

PERMANENT INTERNATIONAL ASSOCIATION OF NAVIGATION CONGRESSES, 1984. "Ice Navigation." Supplement to Bulletin No. 46. Brussels, Belgium.

PEYTON, H. R., 1966. "Sea Ice Strength—Effects of Load Rates and Salt Reinforcement." Arctic Drifting Station Conference. Warren, Virginia.

PEYTON, H. R., 1968. "Sea Ice Forces." In Ice Pressures Against Structures, compiled by L. Gold and G. Williams, NRC Tech. Memo No. 92. Ottawa, Canada.

PHUKAN, A., 1985. "Frozen Ground Engineering." Prentice-Hall, Englewood Cliffs, New Jersey.

PIHLAINEN, J. A. and JOHNSTON, G. H., 1963. "Guide to a Field Description of Permafrost for Engineering Purposes." National Research Council of Canada. Associate Committee on Soil and Snow Mechanics. Technical Memo, 79.

POE, R. W., 1982. "Operation of an Air Curtain Above Lock 8, Welland Canal, For the Opening of the 1982 Navigation Season." St. Lawrence Seaway Authority (internal report).

POLLACK, J., 1985. "Single Point Mooring in Ice Infested Waters." Proceedings of the Conference Arctic '85. San Francisco, California.

PONTER, A. R. S., PALMER, A. C., GOODMAN, D. J., ASHBY, M. F., EVANS, A. G., and HUTCHINSON, J. W., 1983. "The Force Exerted by a Moving Ice sheet on an Offshore Structure." Part 1—The Creep Mode, Journal Cold Regions Science Technology, Vol. 8.

POUNDER, E. R., 1961. "Thermodynamic Considerations on the Use of Air Bubbling Systems in Salt Water." Proceedings of the Syposium on Air Bubbling. Ottawa, Ontario.

POUNDER, E. R., 1965. "The Physics of Ice." Pergamon Press, London.

POUNDER, E. R. and STALINSKY, P., 1961. "Elastic Properties of Arctic Sea Ice." Int. Union of Geodesy and Geophysics, Pub. 54. Helsinki, Finland.

PROCEEDINGS 2ND INTERNATIONAL CONFERENCE ON GEOTEXTILES, 1982. Las Vegas, Nevada. Published by Industrial Fabric Association International, St. Paul, Minnesota.

RALSTON, T. D., 1977a. "Ice Force Design Considerations for Conical Offshore Structures." Proceedings of the Fifth International Conference POAC '79. Trondheim, Norway.

RALSTON, T. D., 1977b. "Yield Criteria for Ice Crushing Failure." Ice Mechanics Worskhop. Calgary, Alberta.

RALSTON, T. D., 1978. "An Analysis of Ice Sheet Indentation." Proceedings IAHR Ice Symposium, Vol. 1, Lulea, Sweden.

RALSTON, T. D., 1979. "Plastic Limit Analysis of Sheet Ice Loads on Conical Structures." Physics and Mechanics of Ice, edited by P. Tryde. International Union of Theoretical and Applied Mechanics. Symposium of the Technical University of Denmark. August. Springer-Verlag, New York, 1980.

RAMSEIR, R. O., 1976. "Growth and Mechanical Properties of River and Lake Ice." Ph.D. Thesis. Fas des Sci, Dep de Genie Civil University, Laval Quebec. Quebec City, Canada.

RICHTER-MENGE, J. A., 1992. US Research in Ice Mechanics—1987–1990. Cold Regions Science and Technology, No. 20. Elsevier Science Publishers B.V., Amsterdam, Netherlands.

ROGGENSACK, J. R., 1975. "Large Scale Laboratory Direct Shear Tests on Ice." Canadian Geotechnical Journal, Vol. 12, No. 2.

ROSE, E., 1947. "Thrust Exerted by Expanding Ice Sheet." ASCE, Vol. 112.

ROWLEY, R. K., WATSON, G. H., and LADANI, B., 1975. "Prediction of Pile Performance in Permafrost Under Lateral Load." Canadian Geotechnical Journal, Vol. 12, No. 4.

SACKINGER, W. M. and SACKINGER, P. A., 1977. "Shear Strength of the Adfreeze Bond of Sea Ice to Structures." Proceedings International Conference POAC, Vol. 2. St. John's Canada.

SAEKI, H., NOMURA, T., and OZAKI, A., 1978a. "Experimental Study on the Testing Methods of Strength and Mechanical Properties for Sea Ice." Proceedings IAHR Ice Symposium.

SAEKI, H., ONO, T., OZAKI, A., and ABE, S., 1978b. "Estimation of Sea Ice Forces on Pile Structures." Proceedings IAHR Symposium on Ice, Vol. 1. Lulea, Sweden.

SAEIKI, H., ONO, T., and OZAKI, A., 1979. "Experimental Study on Ice Forces on a Cone-Shaped and an Inclined Pile Structure." Proceedings International Conference. POAC, Vol. 2. Trondheim, Norway.

SAEKI, H. et al., 1981a. "Experimental Study of the Compressive Strength of Sea Ice and the Ice Forces on a Circular Pile." Coastal Engineering, Japan, Vol. 19, December.

SAEKI, H., OZAKI, A., and KUBO, Y., 1981b. Experimental Study on Flexural Strength and Elastic Modulus of Sea Ice." Proceedings International Conference. POAC. Quebec City, Canada.

SAKHUJA, S., 1985. "Development of a Methodology for the Design of an Offshore Oil Production Platform on the Alaskan Arctic Continental

Shelf." Ph.D. Thesis, University of California, Berkeley.

SANDKVIST, J., 1981. "Conditions in Brash Ice-Covered Channels with Repeated Passages." Proceedings International Conference POAC '81. Quebec City, Quebec.

SAUNDERS, T. F. and TIMASCHEFF, M., 1973. "Ice Effects on Planning, Design and Operation of a Major Oil Terminal." PIANC, Proceedings of 23rd Navigation Congress. Ottawa, Canada.

SCHWARZ, J., 1970. "The Pressure of Floating Ice-Fields on Piles." Proceedings IAHR Symposium on Ice, Paper 6.3, Reykjavik, Iceland.

SCHWARZ, J., 1975. "On the Flexural Strength and Elasticity of Saline Ice." Proceedings IAHR Ice Symposium. Hanover, New Hampshire.

SCHWARZ, J. and WEEKS, W. F., 1977. "Engineering Properties of Sea Ice. Journal of Glaciology, Vol. 19, No. 81.

SCHWARZ, J. et al., 1981. Standardized Testing Methods for Measuring Mechanical Properties of Ice. Journal Cold Region Science Technology, Vol. 4, No. 3.

SERIKOV, M. I., 1961. "Mechanical Properties of Antarctic Sea Ice." Soviet Antarctic Expedition Information Bulletin (English Translation), No. 3.

SHAH, V. K., 1978. "Protection of Permafrost and Ice Rich Shores." Tuktoyaktuk, N. W. T., Canada. Proceedings of 3rd International Conferences on Permafrost. Edmonton, Alberta.

SHEHTMAN, A. N., 1968. "Hydrometeorological Conditions in Icing up of Vessels at Sea." Tranlsated by Naval Scientific and Technical Center (U.K.), Translation No. 2311/71, 1971.

SINHA, N. K., 1981. "Rate Sensitivity of Compressive Strength of Columnar Grained Sea Ice." Experimental Mechanics, Vol. 21, No. 6.

SINHA, N. K., 1982. "Constant Strain- and Stress-Rate Compressive Strength of Columnar Grained Ice." Journal of Materials Science, Vol. 17, No. 3.

SINHA, N. K., 1983a. "Field Test 1 of Compressive Strength of First Year Sea Ice." Annals of Glaciology, Vol. 4.

SINHA, N. K., 1983b. "Field Tests on Rate Sensitivity of Vertical Strength and Deformation of First Year Columnar Grained Sea Ice." Proceedings International Conference POAC, Vol. 1. Helsinki, Finland.

SINHA, N. K., 1984. "Uniaxial Compressive Strength of First-Year and Multi-Year Sea Ice." Canadian Journal of Civil Engineering, Vol. 11, No. 1.

SINHA, N. K. and FREDERKING, R., 1979. "Effect of Test System Stiffness on Strength of Ice." Proceedings International Conference POAC '79. Trondheim, Norway.

SINHA, N. K., TIMCO, G. W., and FREDERKING, R., 1987. "Present Advances in Ice Mechanics in Canada." ASME. Applied Mechanics. Vol. 40, No. 9 (Sept).

SJÖLIND, S.-G., 1984. Viscoelastic Buckling of Beams and Plates on Elastic Foundation." Proceedings, IAHF Ice Symposium 1984, Hamburg, Germany, August 27–31, Vol. I.

SJÖLIND, S.-G., 1985. "Viscoelastic Buckling Analysis of Floating Ice Sheets." Cold Regions Science and Technology, Vol. 11.

SN 76-59, 1959. Standard on Determination of Ice Loads on River Structures. Gosstroy USSR, Moscow.

SN 76-66, 1966. Standard on Determination of Ice Loads on River Structures (Replaced SN 76-59). Gosstroy USSR, Moscow.

SODHI, D. S., 1983. "Dynamic Buckling of Floating Ice Sheets." Proceedings Seventh International Conference on Port and Ocean Engineering Under Arctic Conditions POAC. Helsinki, Finland, April 5–9, Vol. 2.

SODHI, D. S., 1987. "Flexural and Buckling Failure of Floating Ice Sheets Against Structures." In 3rd State-of-the-Art Report 87-17, edited by T. J. O. Sanderson. Published by the U.S. Army CRREL. Hanover, New Hampshire.

SODHI, D. S., 1989. "Ice-Induced Vibrations of Structures." In 4th State-of-the-Art Report IAHR, Published by the U.S. Army CRREL. Hanover, New Hampshire.

SODHI, D. S. and ADLEY, M. D., 1984. "Experimental Determination of Buckling Loads of Cracked Ice Sheets." Proceedings, Third International Offshore Mechanics and Arctic Engineering (OMAE) Symposium, ASME, New Orleans, LA, Vol. III.

SODHI, D. S. and COX, F. N., 1987. "Advances in Sea Ice Mechanics in the U.S.A.." ASME. Applied Mechanics, Vol. 40, No. 9 (September).

SODHI, D. S. and HAMZA, H. E., 1977. "Buckling Analysis of a Semi-Infinite Ice Sheet." Proceedings International Conference POAC, Vol. 1, St. John's, Newfoundland.

SODHI, D. S. and MORRIS, C. E., 1986. "Characteristic Frequency of Force Variations in Cylindrical Structures." Journal of Cold Regions Science and Technology, No. 12.

SODHI, D. S. and NEVEL, D. E., 1980. "A Review of Buckling Analysis of Ice Sheets." In Working Group on Ice Forces on Structures. A State-of-the-Art Report. CRREL Special Report 80-26. Hanover, New Hampshire.

SODHI, D. S., HAYNES, F. D., KATO, K., and HIRAYAMA, K., 1983. "Experimental Determination of the Buckling Loads of Floating Ice Sheets." Annals of Glaciology, Vol. 4.

STAHLE, N. S., 1964. "Protective Coverings for Ice and Snow." Aqueous Foam Studies. U.S. Naval Civil Engineering Laboratory. Technical Report R 340.

STALLABRASS, J. R., 1980. "Trawler Icing." A Compilation of Work Done at NRC. National Research Council Canada. Mechanical Engineering Report ND-56, NRC No. 19372, Ottawa.

STAROSOLSKY, O., 1970. "Ice in Hydraulic Engineering." Norwegian Institute of Technology. Division of Hydraulic Engineering. Report N70-1. Trondheim, Norway.

TABATA, T., 1966. "Studies of the Mechanical Properties of Sea Ice." The Flexural Strength of Small Sea Ice Beams. International Conference on Low Temperature Science, Vol. 1. Sapporo, Japan.

TABATA, T., IWATA, S., and ONO, N., 1968. Studies of Ice Accumulation on Ships. Part 1. Translated by E. R. Hope. NRC Canada. Technical Translation 1318.

TABATA, T., SUZUKI, Y., and AOTA, M., 1975. "Ice Study in the Gulf of Bothnia." Measurements of Flexural Strength. Low Temperature Science, Series A, No. 33.

TIMCO, G. W., 1986. "Indentation and Penetration of Edge-Loaded Freshwater Ice Sheets in the Brittle Range." ASME. Fifth OMAE Symposium, Vol. IV. Tokyo, Japan.

TIMCO, G. W., 1987. "Ice Forces on Multi-Legged Structures." In 3rd State-of-the-Art Report 87-17 by Working Group on Ice Forces, edited by T. J. O Sanderson. Published by the U.S. Army CRREL. Hanover, New Hampshire.

TIMCO, G. W., 1988. "The Influence of Flaws in Reducing Loads in Ice–Structure Interaction Events." Technical Report TR-HY-020, NRC No. 29050, Division of Mechanical Engineering. National Research Council of Canada, March, Ottawa.

TIMCO, G. W. and FREDERKING, R. M. W., 1982. "Flexural Strength and Fracture Toughness of Sea Ice." Journal of Cold Regions Science and Technology, Vol. 8.

TIMCO, G. W. and FREDERKING, R., 1983. "Confined Compressive Strength of Sea Ice." Proceedings International Conference POAC, Vol. 1. Helsinki, Finland.

TIMCO, G. W. and FREDERKING, R. M. W., 1984. "An Investigation of the Failure Envelope of Granular/Discontinuous Columnar Sea Ice." Journal of Cold Regions Science and Technology, Vol. 9.

TIMCO, G. W. and FREDERKING, R. M. W., 1986. "Confined Compression Tests: Outlining the Failure Envelope of Columnar Sea Ice. Journal of Cold Regions Science and Technology, Vol. 12.

TIMCO, G. W. and JORDAN, I. J., 1987. "Time-Series Variations in Ice Crushing." Proceedings POAC '87, Vol. I. Fairbanks, Alaska.

TIMCO, G. W. and PRATTE, B. D., 1985. "The Force of a Moving Ice Cover on a Pair of Vertical Piles." Proceedings Canadian Coastal Conference. St. John's, Newfoundland, Canada.

TRAETTEBERG, A., GOLD, L. W. and FREDERKING, R. M. W., 1975. "The Strain Rate and Temperature of Young's Modulus of Ice." Proceedings IAHR Ice Symposium, Hanover, New Hampshire.

TRYDE, P., 1977. "Ice Forces." Journal of Glaciology, Vol. 19, No. 81.

TRYDE, P., 1989. Appendix B "Ice Engineering" in "Port Engineering" (4th edition) by P. Bruun. Gulf Publishing House, Houston, Texas.

TRYDE, P., 1983. "Report on Vertical Ice Lifting of Piles in a Number of Danish Marines with a

Description of the Lifting Mechanism." Proceedings POAC '83. Helsinki, Finland.

TSALIUK, I. G. and SEMERENKO, I. M., 1981. "Construction of Deep Water Pier in Arctic." Transportnoe Stroitelstvo (Transport Construction), No. 4, April (In Russian). Moscow, USSR.

TSANG, G., 1981. "Fin boom Ice Gate for Ice Control and Winter Navigation." Proceedings International Sympsoium on Ice IAHR. Quebec City, Quebec.

TSINKER, G. P., 1984. "Single Point Mooring for Areas with Landfast Ice and Random Floating Ice Floes." Unpublished Report.

TSINKER, G. P., 1986. "Ice Management for Year-Round Operating Marine Terminals." Proceedings POLARTECH '86. Helsinki, Finland.

TSITOVICH, N. A., 1973. "Mechanics of Frozen Soils" (In Russian). Moscow, USSR. Translated by the U.S. CRREL Translation No. 439. Hanover, New Hampshire.

TSITOVICH, N. A., 1975. "The Mechanics of Frozen Ground," edited by G. P. Swingo. McGraw-Hill, New York.

TSUTAE, S., ITOH, Y., IZUMI, K., ONO, T., and SAEKI, H., 1983. "Estimation of the Compressive Strength of Sea Ice by the Schmidt Test Hammer." Proceedings International Conference POAC, Vol. 2. Helsinki, Finland.

TUNIK, A. L., 1987. Impact Ice Loads on Offshore Structures. Proceedings International Conference on Ports and Ocean Engineering Under Arctic Conditions (POAC '87), Vol. 1. Fairbanks, Alaska.

URABE, N. and YOSHITAKE, A., 1981. "Fracture Toughness of Sea Ice—In Situ Measurement and Its Applications." Proceedings International Conference POAC, Vol. 1, Quebec City, Canada.

URABE, N. and YOSHITAKE, A., 1981. "Strain Rate Dependent Fracture Toughness of Pure Ice and Sea Ice." Proceedings IAHR Ice Symposium, Vol. 2. Quebec City, Canada.

URABE, N., IWASAKI, T., and YOSHITAKE, A., 1980. "Fracture Toughness of Sea Ice." Journal of Cold Regions Science and Technology, Vol. 3, No. 1.

VAN TE CHOW, 1959. "Open-Channel Hydraulics." McGraw-Hill Ryerson Limited, New York.

VAUDREY, K. D., 1977. "Ice Engineering—Study of Related Properties of Floating Sea Ice Sheets and Summary of Elastic and Visco-Elastic analysis." Civil Engineering Laboratory, Naval Construction Battalion Centre, Tech. Report R860, Port Hueneme, California.

VAZIRI, H. and HAN, Y., 1991. "Full-Scale Field Studies of the Dynamic Response of Piles Embedded in Partly Frozen Soils." Canadian Geotechnical Journal, Vol. 28.

VIVATRAT, V. and SLOMSKI, S., 1983. "A Probabilistic Basis for Selecting Design Ice Pressures and Ice Loads for Arctic Offshore Structures." Proceedings OTC, Vol. 1, Houston.

VOELKER, R. P., et al., 1981. "Winter 1981 Traficability Tests of Polar Sea" Vol. II, Environmental Data. U.S. Department of Transportation, Maritime Administration, Office of R. and D., Report MA-RD-940-82018. Washington, D.C.

VOITKOVSKII, K. F., 1960. "The Mechanical Properties of Ice." Idatel'stvo Akademii Nauk USSR, Moscow. Translation, Air Force Cambridge Res. Lab.

WAKE, A., POON, Y-K., and CRISSMAN, R., 1987. "Ice Transport by Wind, Wave, and Currents." ASCE Journal of Cold Regions Engineering, Vol. 1, No. 2.

WALDEN, J. T., HALLAM, S. D., BALDWIN, J. T., and THOMAS, G. A. N., 1987. "Prediction of Multiyear Ice Impact Loads Utilizing Hans Island Data." Proceedings Conference POAC '87, Fairbanks, Alaska.

WANG, QIN-JIAN, 1983. "A Tentative View on Ice Load Applied on Jacket Platforms in Bo-hai Gulf." Proceedings POAC '83, Vol. 2.

WANG, QUIN-JIAN, 1987. "A New Anti-Ice Type of Articulated Single-Point Mooring Structure." Proceedings International Conference POAC '87. Fairbanks, Alaska.

WANG, Y. S., 1979. "Crystallographic Studies and Strength Tests on Field Ice in the Alaskan Beaufort Sea." Proceedings 6th International Conference POAC '81, Vol. 1, Trondheim, Norway.

WANG, Y. S., 1981. "Uniaxial Comparison Testing of Arctic Sea Ice." Proceedings 6th International Conference POAC '81, Vol. 1, Quebec City, Canada.

WANG, Y. S. and POPLIN, J. P., 1986. "Laboratory Compression Tests of Sea Ice at Slow Strain Rates From a Field Test Program." Proceedings International OMAE Symposium, Vol. 4, Tokyo.

WATT, B. J., 1982. "Hydrocarbon Extraction in Arctic Frontiers." Proceedings of the Third International Conference on the Behaviour of Offshore Structures (August). MIT, Cambridge, Massachusetts.

WEEKS, W. F., 1982. "Physical Properties of the Ice Cover of the Greenland Sea." CRREL Special Report 82-28, Hanover.

WEEKS, W. F. and ANDERSON, D. L., 1958. "An Experimental Study of the Strength of Young Sea Ice." Transactions American Geophysical Union, Vol. 39, No. 4.

WEEKS, W. F. and ASSUR, A., 1967. "Mechanical Properties of Sea Ice." CRREL Monograph IIC3, Hanover, New Hampshire.

WEEKS, W. F. and S. F., and ASSUR, A., 1969. "Fracture of Lake and Sea Ice." CRREL, Research Report No. 299, Hanover.

WEEKS, W. F. and MELLOR, M., 1983. "Mechanical Properties of Ice in the Arctic Seas." Proceedings Arctic Technology and Policy." Hemisphere Publishing, New York.

WESSELS, E., 1983. "Ice Loads on Cylindrical and Conical Offshore Structures." Lecture Notes at WEGEMT VII Graduate School "Ships and Structures in Ice", Helsinki, Finland.

WESSELS, E. and KATO, K., 1989. "Ice Forces on Fixed and Floating Conical Structures." Fourth Special State-of-the-Art Report 89-5. U.S. Army Corps of Engineers CRREL. Hanover, New Hampshire.

WILLIAMS, G. P., 1970. "Break-up and Control of River Ice." Proceedings of IAHR Symposium. Ice and Its Action on Hydraulic Structures, Reykjavik, Iceland.

WONG, T. T., MORGENSTERN, N. R., and SEGO, D. C., 1991. "Ice Rubble Attenuation of Ice Loads on Arctic Offshore Structures." Canadian Geotechnical Journal, Vol. 28.

WORTLEY, A. C., 1990. "Ice Engineering for Rivers and Lakes Bibliography." University of Wisconsin, Madison, Wisconsin.

WORTLEY, C. A., 1984. "Great Lakes Small-Craft Harbor and Structural Design for Ice Conditions: An Engineering Manual." University of Wisconsin, Report WIS-SG-84-426, Madison, Wisconsin.

WORTLEY, C. A., 1987. "Design and Use of Floating Docks in Lake Ice Conditions." Proceedings POAC '87, Fairbanks, Alaska.

WRIGHT, B. D., HNATIUK, J., and KOVACS, A., 1978. "Sea Ice Pressure Ridges in the Beaufort Sea." IAHR Symposium on Ice Problems, Lulea, Sweden.

WRIGHT, B., PILKINGTON, G. R., WOLNER, K. S., and WRIGHT, W. H., 1986. "Winter Ice Interactions with Arctic Offshore Structures." Proceedings IAHR. Symposium on Ice, Vol. III, Iowa City, Iowa.

WU, H. C., CHANG, K. J., and SCHWARZ, J., 1976. "Fracture in the Compression of Columnar-Grained Ice," Engineering Fracture Mechanics, Vol. 8, No. 2.

YOUSSEFF, H., 1987. "Cold Regions Engineering: A Century of Canadian Research and Development." Proceedings the CSCE Centennial Conference, Montreal, Canada.

ZAKRZEWSKI, W. P., 1986. "Icing of Ships. Part 1: Splashing a Ship with Spray." NOAA Technical Memorandum, ERL PMEL-66, Seattle, Washington.

ZAKRZEWSKI, W. P., 1987. "Splashing a Ship with Collision-Generated Spray." Cold Regions Science and Technology, 14. Elsevier Science Publishers B.V., Amsterdam, Netherlands.

3

Shiplifts, Marine Railways, Shipways, and Dry (Graving) Docks

by B. K. Mazurkiewicz

3.1 GENERAL INFORMATION ON SHIPBUILDING AND SHIP REPAIR YARDS

3.1.1 Shipyard Layout: Basic Design Considerations

Shipyards are industrial plants located in a suitable water area such as a harbor basin, a bay, or a river, for the building, repair, and maintenance of ships. They are generally classified as shipbuilding yards, which build new ships, and ship repair yards, which are involved mainly in the repair and maintenance of ships in-service. There are also shipyards for both construction and repair of ships. The shipyard equipment depends on the prevailing types of activities; for example, in ship repair yards, shipbuilding will be of secondary importance and vice versa. These secondary importance activities are typically conducted during less intensive primary work periods.

Shipbuilding yards as production plants can be subdivided into different categories, distinguished by such criteria as the material used for building ships, the size of ships, and the range of cooperation with other shipyards and industries.

The ship repair yards are usually distinguished according to the size and type of the ship to be repaired there. These yards are normally designed to handle such operations as preventive maintenance, inspection, and repair of damaged ships.

The basic organizational unit of a shipyard is a specialized workshop. Depending on the specific production process all shipyard workshops can be classified as hull shops, fitting-out shops, machinery shops, and auxiliary shops. The location of these shops in relation to the shipyard's main facilities such as shiplifts, slipways, docks, and fitting-out quays basically depends on a shipyard's production layout.

In shipbuilding yards the typical hull shop includes the plate and shaped steel stockyards, the mold lofts, plate and frame shops, and block fabrication shops with adequate storage space. The final stage of the production process, which takes place in the hull shops, is the block-section assembly. This is typically done on a shipyard main structure that is suitable for launching. Hence, those hull shops where a block-section production process is taking place should be located in close proximity to the shipyard main structures suitable for launching operations.

Naturally, the shiplifts, marine rail shipways, and dry docks are located at the end of the production line.

The fitting-out shops, which should be sited furthest away from the fitting-out quay and/or launching facilities, store the raw materials and prefabricated elements; treatment shops should be placed closer, and the other shops such as pipe-fitting, joiners, locksmith, and carpenter shops, should be sited as close as possible to the fitting-out quay and launching facilities. The outfitting works are typically carried out on a ship partly before launching, and partly after launching at the fitting-out quay.

The arrangement of individual shops relative to the shipyard main structures varies and depends mainly on the local site conditions. The best solution to shipyard layout is to arrange all the workshops in a straight production line resulting in a linear flow of materials and prefabricated components. The latter mitigates transportation problems and increases the yard's productivity. An example of a shipyard layout with a straight-line shipyard arrangement is illustrated in Figure 3–1. Shipyards, as depicted in Figure 3–1, are characterized by a relatively narrow waterfront and deep shipyard territory. Very often the S-shaped, T-shaped, or U-shaped shipyard layout arrangements are used (Mazurkiewicz, 1980).

The adopted design should provide for the efficient integration of the individual production phases (Figure 3–2) and, where possible, the whole yard should be placed under the same roof. Plates and angles should travel on roller tracks from storage via the cleaning–straightening, coating–cutting, and bending stations to the subassembly and assembly stations in the welding zone.

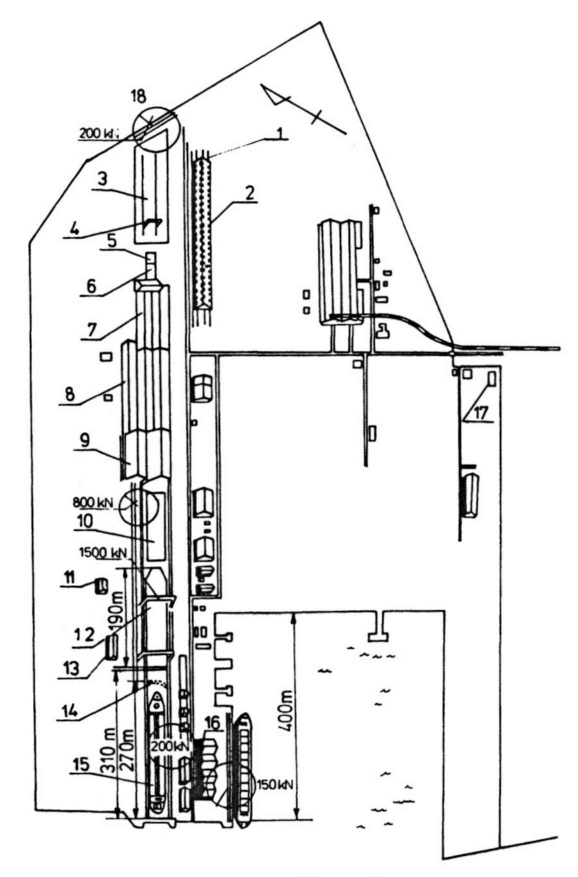

Figure 3–1. Straight line shipyard arrangement. (From Japan Shipping and Shipbuilding, 1965.) 1—angle shot-blast shop, 2—steel structure shop, 3—steel stockyard, 4—200 kN gantry crane, 5—shot-blast shop, 6—straightening-roller shop, 7—marking-cutting shop, 8—parts and processing shop, 9—welding shop, 10—assembly shop, 11—piping shop, 12—bisected-hull building shop, 13—carpenter shop, 14—inner dock gate, 15—repairing and jointing dock, 16—fitting-out shop, 17—power-receiving station, 18—steel-unloading quay.

During this phase the outfitting should be carried out. In shipyards with dry docks (graving docks) prefabricated blocks are assembled while outfitting continues. The next stage, during which the last outstanding piece of equipment is installed and the ship is preparing for trials, is carried out at the fitting-out quay.

In ship repair yards, because of the different technological processes involved, the

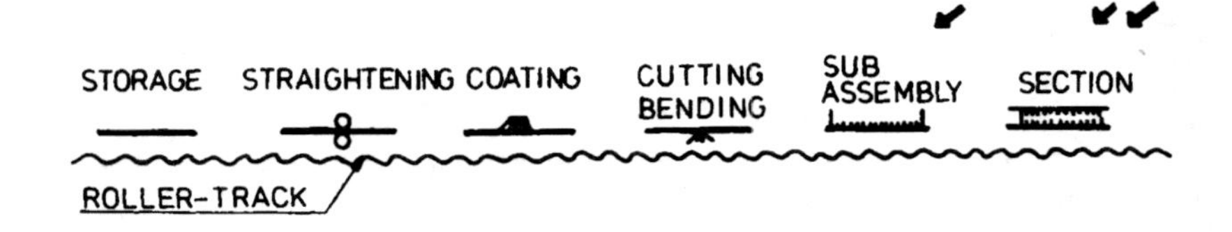

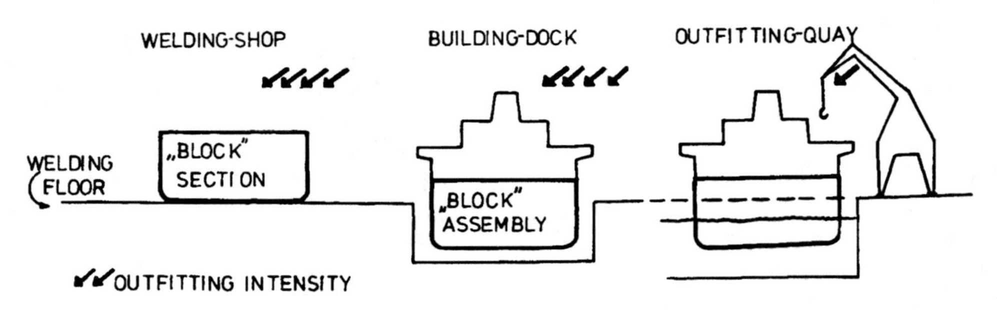

Figure 3–2. Shipbuilding production phases. (From Knegjens, 1968.) 1—Open storage, 2—shop, 3—yard facility for reactor ships, 4—tanker cleaning, 5—repair basin.

layout of the production units is somewhat different from those planned for the shipbuilding yards. Generally the following basic production workshops are required in a ship repair yard: hull repair, maintenance, and paint and ship machinery repair shops.

All these workshops should be located as close as possible to the quays and docks. There are four basic types of ship repair yard layouts: in the form of a pier arrangement, around a basin, on an island, and on a peninsula. The latter layout is the most efficient among the four basic schemes available. An example of a ship repair yard on a peninsula is illustrated in Figure 3–3.

The ship repair yard and its ship handling facilities such as shiplifts, docks, etc., in particular must not be of a single-purpose design, for example, to handle ships of a certain type and size only. Yard design must consider potential changes in ship types and dimensions and therefore its layout and ship handling facilities must be designed with a provision for potential rapid expansion and modification in structure, equipment, and technology.

3.1.2 Shipyard Main Structures: General Specifications

Ship building and repair yards require special facilities or structures generally known as shipyard main structures. These structures, together with cranes and other associated equipment, are used for ship assembly, launching, or docking for repair. These structures include shipways, marine railways, shiplifts, and floating and graving dry docks.

Shipways can be longitudinal, from which a ship is launched parallel to its longitudinal axis, and side or transverse, from which a ship is launched transversely (sideways) to its longitudinal axis. Both side and longitudinal shipways are used only for shipbuilding.

Marine railways can also be longitudinal, allowing docking and then launching of a ship in the direction of the longitudinal axis of the ship, and transverse, allowing docking and then launching of a ship perpendicular to the longitudinal axis of the ship.

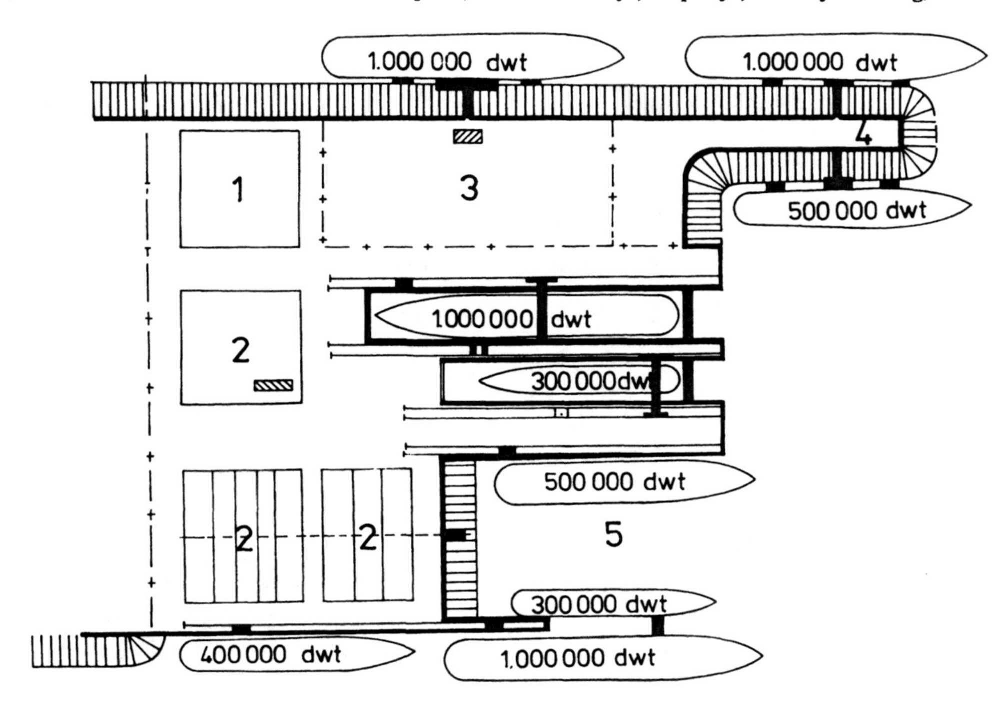

Figure 3–3. Ship repair yard on a peninsula. (From Knegjens, 1968.)

Depending on their track system, marine railways can be equally used for both the launching of newly built ships as well as for docking of ships for repair.

Shiplifts (electric, hydraulic, or pneumatic) are used for both launching of new ships and docking of ships for repair and/or maintenance.

Floating docks (Figure 3–4) are used mainly for ship repair work. However, they can also be used for launching of new ships. The capacity of floating docks typically varies between 1 and 900 MN. For greater flexibility of operation floating docks are usually equipped with dedicated gantry cranes.

Dry (graving) docks, according to their operation, are classified as follows:

- Emergency docks used for docking of ships with or without cargo, however, with greater than normal draught, resulting from serious hull damage
- Repair and maintenance docks for docking of empty ships
- Docks used mainly for ship construction with depth sufficient for launching of an empty or partly equipped hull
- Building and repair docks for ship construction with sufficient depth over the sill for docking ships for repair and/or maintenance work.

The shipyard may have a variety of main structures. The proper choice of number and sizes of these structures for a particular shipyard depends primarily on economic factors. Construction costs will depend on the geological and hydrological conditions of the proposed shipyard site and the contractor's capabilities.

Shipyard main structures also include all kinds of outfitting and unloading quays. The latter do not differ from other harbor quays, a discussion of which is given in other parts of this book. In modern shipyard design practice and particularly in the design of ship

(a)

(b)

Figure 3–4. Floating dock. (a) With transverse transportation system. (Avondale Shipyard, New Orleans, Mississippi.) (b) For ship repair (U.S. Naval Base, San Diego, California).

repair yards, piers often are used in combination with mooring dolphins or sometimes in combination with mooring buoys. This is usually done for reasons of economy. Furthermore, in tidal seas it is possible to dredge an appropriate basin on both sides of a pier, which enables a loaded ship to enter the basin at high tide and subsequently be repaired at low water level.

3.2 SHIPLIFTS

3.2.1 General

The shiplift system of dry docking is the newest innovation in ship transfer. By definition this is a ship elevator. The shiplift concept typically comprises a platform, over which the ship is placed, and a number of lifting mechanisms that raise or lower the platform (Figure 3–5). The number and capacity of the lifting units determine the overall lifting capacity of the shiplift. A schematic layout of typical shiplifts and their transfer system are illustrated in Figure 3–6.

Essentially a shiplift is an elevator type of dry dock. To retrieve a ship, a wheeled docking cradle is rolled onto the platform, which is suspended from a series of wire rope hoists or other lifting mechanisms spaced along each side. The shiplift lifting units are usually mounted on pile-supported foundations slightly above the level of the adjacent shipyard. The platform, with the docking cradle in place, is then lowered into the water. When the platform reaches the required depth, the ship to be dry docked is floated into position above the docking cradle and held with mooring lines attached to the piers. The lifting mechanisms act in a smooth, continuous, and synchronized manner to raise the platform vertically. This lifts the ship to the level of the adjacent land.

In many cases, the ship is then transferred from the shiplift and moved to an onshore berth. When transfer is not required, the ship can be docked directly on the keel and bilge blocks fitted to the shiplift platform. The size, quantity, and spacing of the lifting units along the platform determine its lifting capacity.

Individual elements in the system are simple and straightforward. One of the most important of these is the synchronous lifting mechanisms. Of equal importance is the platform.

The largest shiplift built to date, in terms of overall platform size, is the facility at Todd Pacific Shipyards, Los Angeles Division. It has a gross lifting capacity of 26 400 long tons (263 MN), and its live lifting capacity is 22 000 long tons (253 MN) (Becht and Hetherman, 1990).

According to NEI Syncrolift Inc. (Miami, Florida) designs are also prepared for the shiplift to handle tankers with docking weight of 45 000 tons.

The shiplift can be used both for repair and for construction of vessels. It occupies a smaller waterfront area in comparison with other types of main structures and can be combined with fitting-out berths. It is readily accessible for inspection, repair, and maintenance, and is simple in construction and operation and therefore economical.

The speed of operation and the absence of vertical walls which permits easy access to the ship are additional advantages of a shiplift.

Typically the vertically moving horizontal platform is suspended on cables, chains, or hydraulic components which are drawn up with electric hoisting winches or hydraulic or mechanical lifts. Pneumatic lifts are also used.

The lifting equipment, which is a series of jacks or hoists, is mounted on a pier structure on each side of the pier deck. Miscellaneous mechanical, hydraulic, and electrical systems are usually used for shiplift operation.

Normally the shiplift is connected with a horizontal transfer system to enable the ship to move from the shiplift to the berth.

Figure 3–5. General view of a yard equipped with shiplift and horizontal transfer system. (Courtesy of Pearlson Engineering Co., Inc.)

The location of the shiplift within a shipyard is dependent on available land along the shoreline, availability of the waterfront space, and the needed number of berths. Some typical shiplift layouts are illustrated in Figure 3–6.

The electrical Syncrolift system was most frequently used in the past 3 to 5 years. At present there are 182 Syncrolift units operating in 60 countries. This system consists of three principal components: a structural steel platform with wood decking, a selected

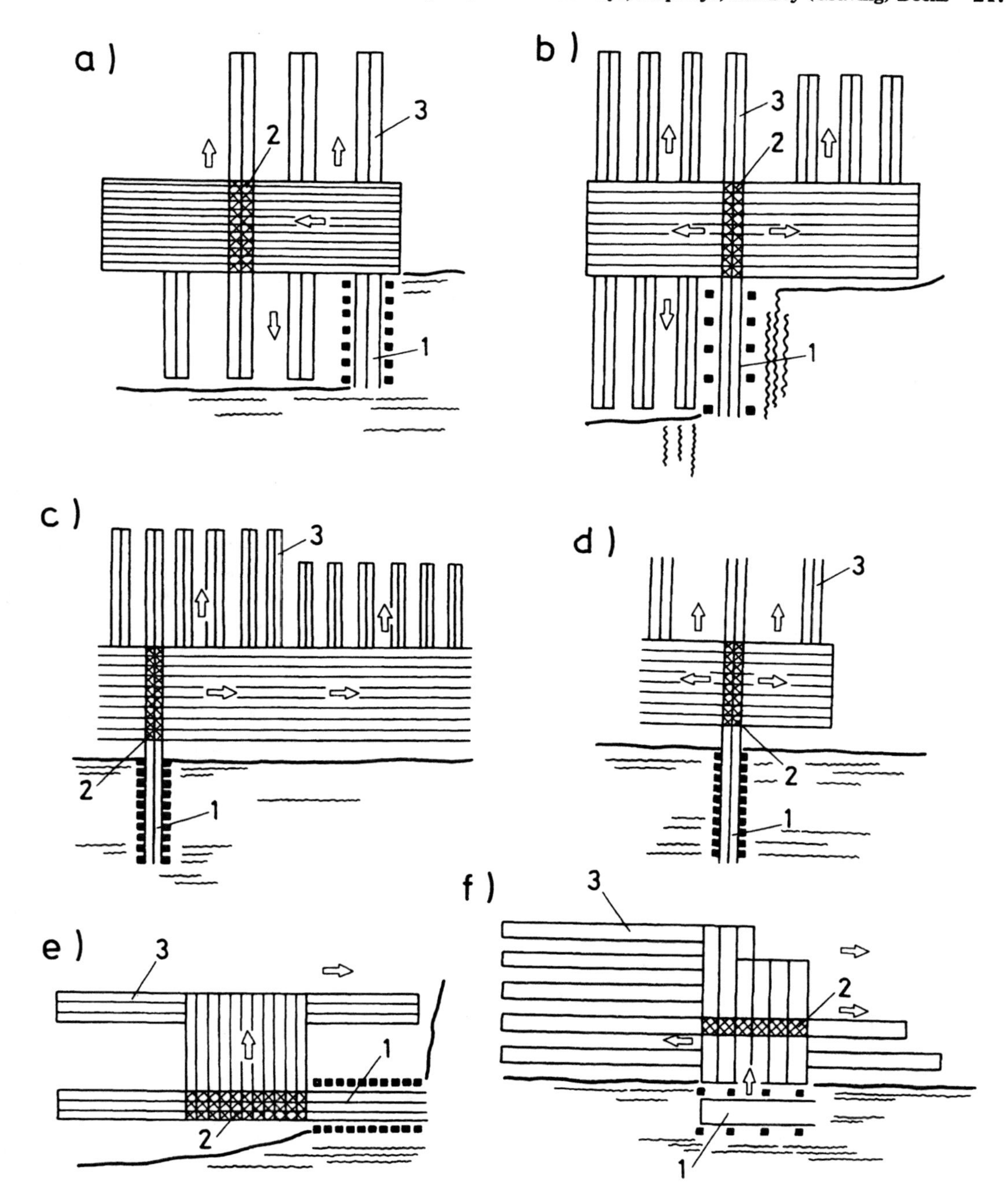

Figure 3–6. Shiplift and transfer system arrangements. (Courtesy of Pearlson Engineering Co., Inc.). (a)–(f) Different locations of shiplifts. 1—Shiplift, 2—transfer carriage, 3—berths.

number of electrically powered wire rope hoists to raise and lower the platform, and an electric motor control center to operate the system. Typically ships are dry docked on the Syncrolift platform using conventional procedures for setting the keel blocks and placing the bilge blocks after grounding. The bilge blocks are moved by a system of chain and pulleys that can position the blocks when the platform is at any elevation along its travel. Usually the blocks are placed on the cradle of the horizontal transfer system (Figure 3–7).

3.2.2 Platforms

Typically the steel platform consists of the main transverse lifting beams, longitudinal and secondary transfer beams, and wooden decking (Figure 3–8). The aforementioned beams are designed depending on ship support arrangements (Figure 3–9); a ship can be supported directly on the platform deck, in which case the whole weight of a ship is transferred to the center of the system, or it can be placed on a system of cradles. In the latter case, the platform is designed for the loads produced by the cradle wheels under the keel, w_1, and under the bilge block, w_2 (Figure 3–10). In each design case the wheel arrangement may be different and therefore the platform should be designed for the worst possible wheel arrangement.

The number of main transverse lifting beams depends on the platform length and the designed load per linear meter. Shiplifts of eight and more hoisting winches have articulated platforms which allows the possibility of independent action of each hoist pair and the main transverse lifting beam. Each main transverse beam is connected to its hoist pair by a multipart wire rope system.

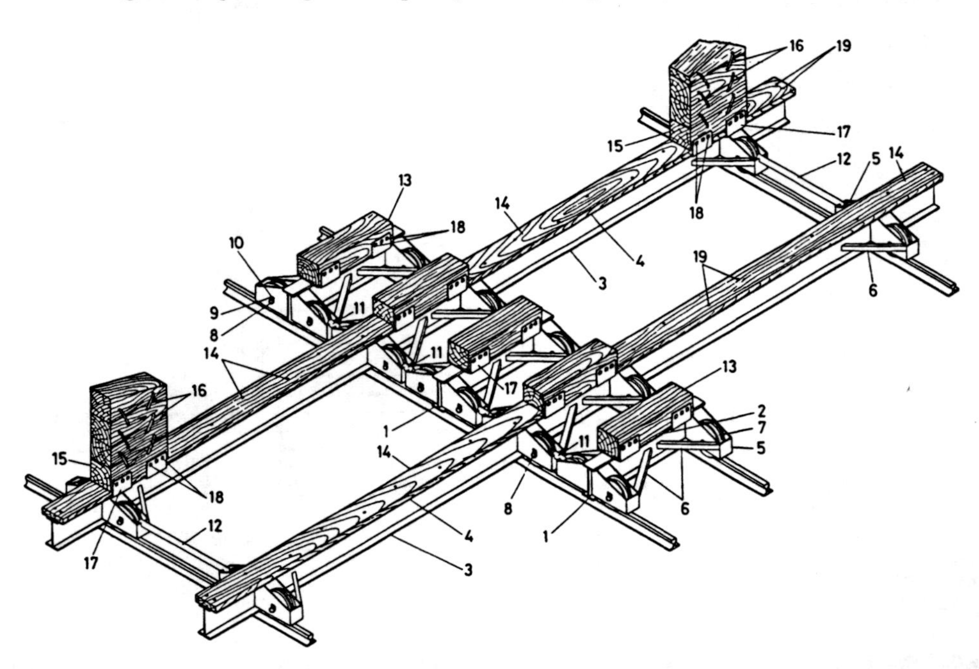

Figure 3–7. System of ship's support on the shiplift. (courtesy of Pearlson Engineering Co., Inc.) 1—Keel block beam, 2—keel block bearing slab, 3—bilge block beam, 4—bilge block bearing slab, 5—wheel housing, 6—wheel housing bracing, 7—wheel, 8—axis, 9—bearing, 10—bearing fastening, 11—connection of keel blocks, 12—connection of bilge blocks, 13—keel block, 14—bilge block sliding surface, 15—bilge block, 16—clamp, 17—fastening steel plates, 18— and 19—bolts.

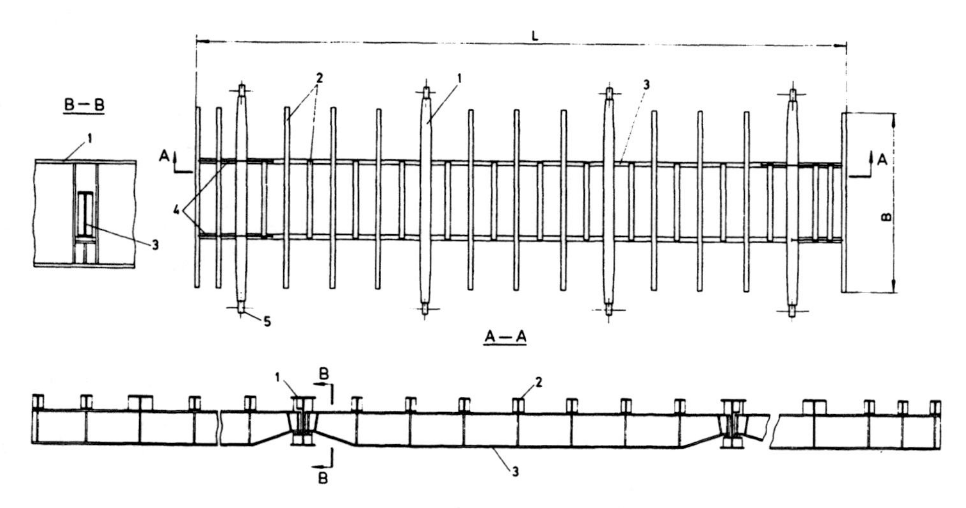

Figure 3–8. Steel shiplift platform (wood decking is not shown). (Courtesy of Pearlson Engineering Co., Inc.) 1—Main transverse lifting beams, 2—intermediate tranverse beam, 3—longitudinal beam, 4—rails of transfer cradle system, 5—beam sheave fastening.

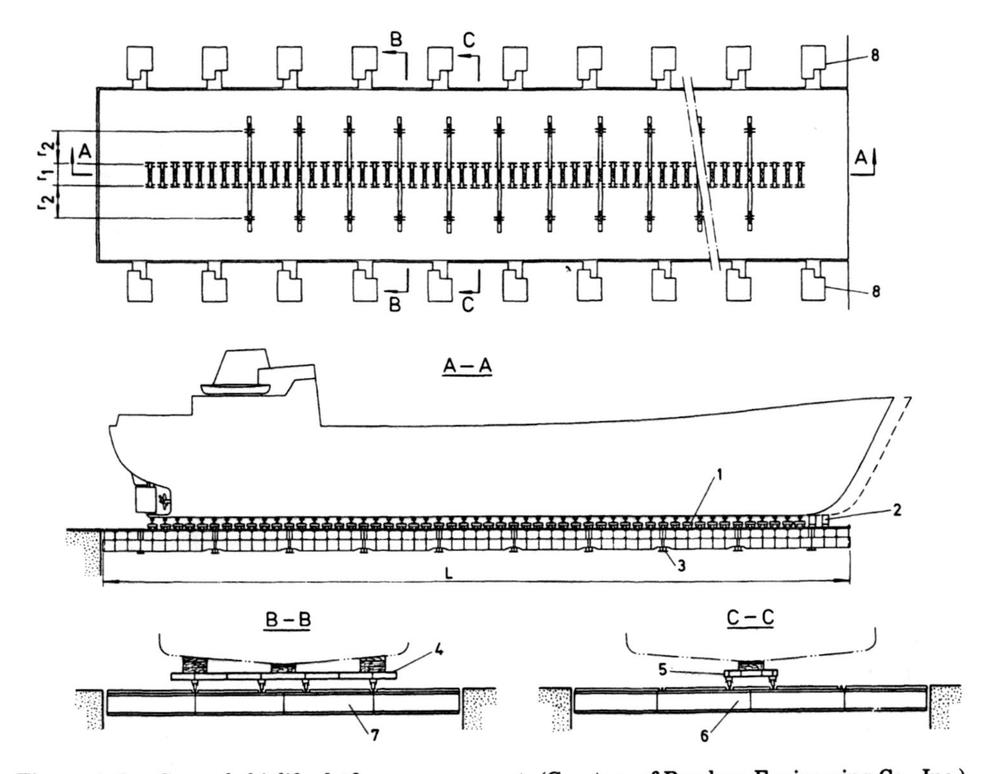

Figure 3–9. General shiplift platform arrangement. (Courtesy of Pearlson Engineering Co., Inc.) 1—Ship supported on end transfer cradles, 2—ship supported on permanent blocks, 3—main transverse lifting beam, 4—bilge block beam, 5—keel block beam, 6—platform beam transferring loads from keel blocks, 7—platform beam transferring loads from bilge blocks, 8—hoists.

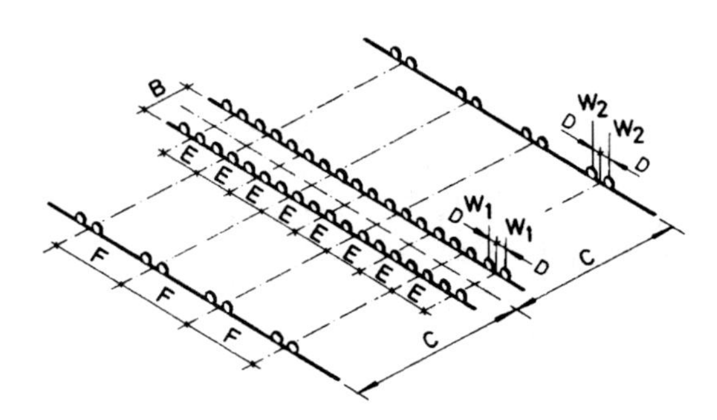

Figure 3–10. Arrangement of wheels of keel and bilge block cradles. (From Mazurkiewicz, 1981.)

Fixed sheaves are mounted on the hoist frame and moveable sheaves mounted on the beam ends. Intermediate beams are supported by longitudinal components held at each end by flexible connections.

The articulated design of the platform enables it to function as a series of supported beams, where each component can be constantly relied upon to carry the local load.

The platform wooden deck typically consists of planks placed on wooden beams bolted to the steel components of the platform.

3.2.3 Hoists

The platform, which supports the vessel during the lifting operation, is normally raised and lowered by electrically controlled hoists and wire rope cables.

Hoists are electric-mechanical synchronized devices designed to operate smoothly while continually monitoring the load distribution along the platform at all the lifting points. The number of hoists and their capacity differ from project to project. Currently Syncrolift Inc. offers hoists in 13 sizes ranging from 610 to 6500 kN lift.

Each hoist is equipped with a simple limit switch activated by an adjustable bracket fixed to its main transverse beam. As the platform rises to the transfer height, each hoist is shut down by the action of the switch. If any wire stretch occurs, it is automatically corrected at each operating cycle of the shiplift. Accordingly, the pier (foundation) settlement can be compensated by adjusting the switch.

Wire ropes are designed to have properties required for a severe marine environment. Every individual wire is galvanized. The wire rope system consists of a six-part reeving. The cable drum is grooved and provides for the maximum amount of travel with a single layer on the drum. Two sheaves mounted on the hoist, together with three sheaves mounted on the main beam, permit reeving the wire rope into a six-part form supported at each hoist. Each hoist assembly consists of an electric motor with a marine-type integral brake, gearing, wire rope drum, ratchet and pawl backstop, sheaves, miscellaneous accessories, and cover (Figure 3–11).

The shiplift operation is made from a control center. There is an ammeter for each hoist that indicates the relative load carried by each hoist during the operation of the shiplift. Safety provisions are incorporated to automatically stop all motors instantaneously if any one motor exceeds the designed capacity. Provision is also made to automatically stop all motors when the platform reaches the upper or lower limits of travel.

3.2.4 Hydraulically Operated Shiplifts

A shiplift may also be equipped with hydraulic synchronous heavy-duty winches (Kessler, 1975).

A representative example of such a type of installation is the shiplift constructed for Schiffswerft Gebr. Schürenstedt KG, Germany. The lifting capacity of this shiplift is 29 400 kN. It is equipped with 12 winches,

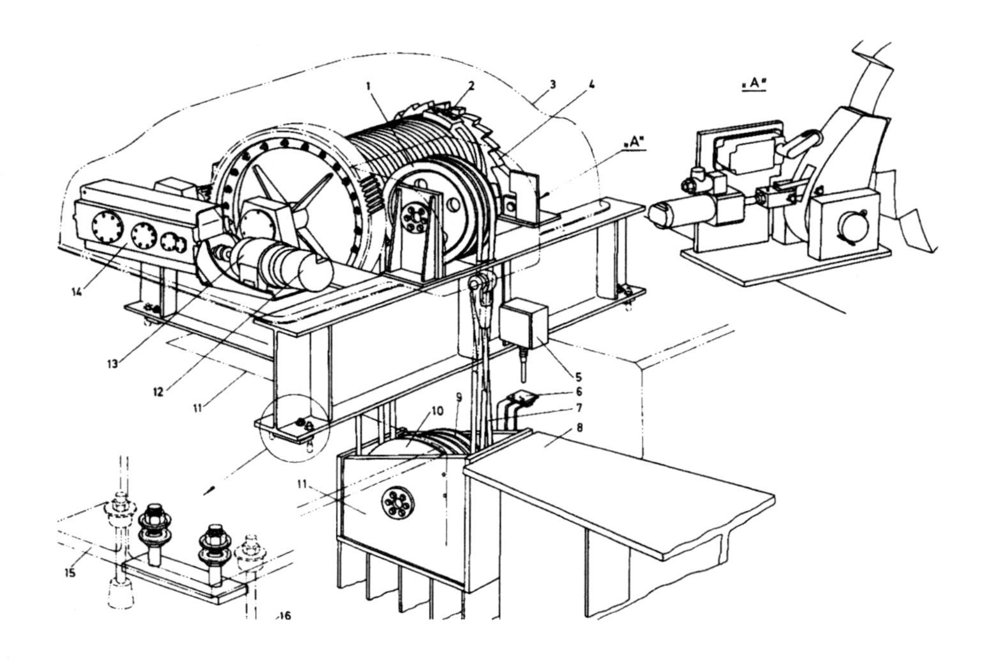

Figure 3-11. Hoist assembly. (Courtesy of Pearlson Engineering Co., Inc.) 1—Sheave mounted on the hoist, 2—rope clamp, 3—hoist cover, 4—pawl system, 5—switch box, 6—limit switch plate, 7—wire rope, 8—main transverse lifting beam, 9—sheave mounted on the beam, 10—housing of the beam sheave, 11—supporting structure, 12—disc brake, 13—motor brake, 14—gearing, 15—base, 16—anchoring.

each having a capacity of 2450 kN; the size of the ship supporting platform is 25 × 58 m.

In hydraulically operated shiplifts the lifting process is carried out step by step by turning the driving wheel by means of raising the work cylinder with the safety lock released, and securing the work cylinder with the safety lock notched-in. The lowering process is carried out in reverse order.

3.2.5 Design

Almost all of the existing shiplifts that can be connected to horizontal transfer systems are constructed in the form of a platform made of steel or wood, suspended from a system of cables or chains, supported on a piled structure constructed around the platform. Once the physical characteristics of the ship have been defined, the correct number, size, and spacing of hoists can be estimated. These characteristics include the weight of the ship (in tonnes per meter of keel length) and the bearing capacity of local soils.

Similarly to the vessel lifting requirements the platform lifting capacity is usually expressed in terms of its tonnes per meter lift capacity. Therefore, to design the shiplift supporting structure, it is essential to know the required overall dimensions of the shiplift, location of the hoisting units, and load distribution along the lifting platform. Knowledge of location of the miscellaneous service lines and switches, control center, and other is normally required. In most cases, it is also necessary to consider the possibility of installing a horizontal transfer system with an articulated link

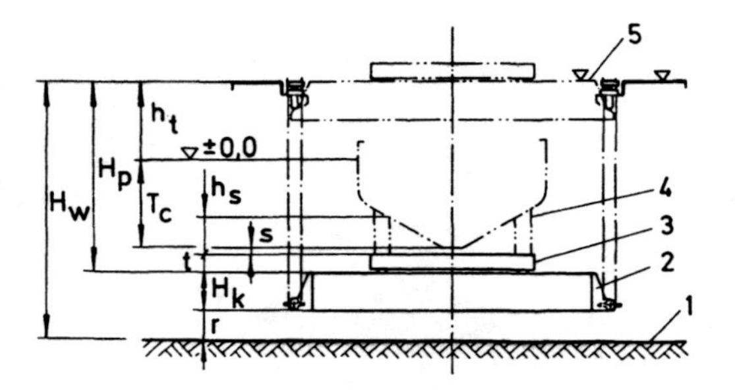

Figure 3–12. Determination of the required depth in the basin at the shiplift location. 1—Depth of basin, 2—shiplift platform, 3—horizontal cradle, 4—ship supports, 5—shiplift platform the highest position. (From Mazurkiewicz, 1981.)

between the shiplift platform and the shipyard berthing area.

The design of the platform supporting structure is preceded by a detailed assessment of the depth of the harbor basin at the shiplift location. The following parameters are normally taken into consideration (Figure 3–12).

- Minimum clearance between the bottom of the platform structure and the bottom of the basin r; it is usually taken as equal to at least 0.5 m.
- The height of the platform structure H_k; the value of H_k varies depending on the type of shiplift, its capacity, and the platform structure; for example, for a platform 30 m wide H_k may reach about 4.0 m.
- Height of a ship transfer cradles t
- Height of keel or bilge blocks h_s from the level of the platform deck or ship transfer cradle (if any) to the ship support level
- Clearance s between the ship bottom and the supporting level of the ship transfer cradle ($s_{\min} = 0.25$ m)
- Maximum draught of the ship to be dry docked, T_c
- The elevation of the shipyard area above the design mean water level in the basin, h_t

Hence, the lifting height (vertical travel) of the platform H_p is computed from the following:

$$H_p = h_t + T_c + s + t \tag{3–1}$$

and the depth of the basin in relation to the elevation of the shipyard area H_w is determined as

$$H_w = H_p + H_k + r. \tag{3–2}$$

Subsequently, in plain view the overall dimensions of the shiplift are defined according to the following parameters (Figure 3–13):

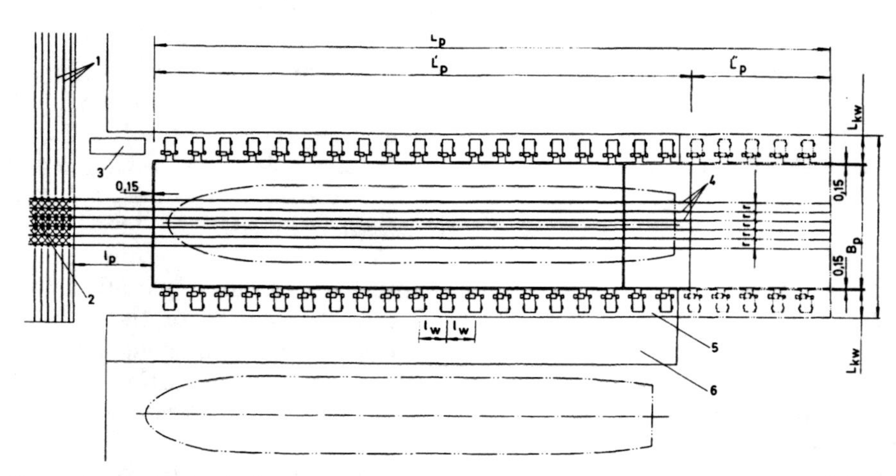

Figure 3–13. Main shiplift parameters. (Courtesy of Pearlson Engineering Co., Inc.) 1—Rails of the side transfer carriage, 2—side transfer carriage, 3—control center, 4—rails of the end transfer cradle, 5—shiplift supporting structure, 6—outfitting quay.

- Required length of the shiplift platform, L_p
- Required width of the shiplift platform, B_p
- Distance between the edge of the platform and the edge of the shiplift supporting structure (pier) (usually 0.15 m)
- Minimum width of the shiplift supporting structure (pier) enabling the setting (foundation) of hoists or jacks, L_{kw}
- Axial spacing of hoists or jacks, ℓ_w
- Spacing of cradle rails in the horizontal transfer system, r_i
- Minimum distance between the front edge of the platform and the transfer carriage in the horizontal ship transfer system, ℓ_p.

As stated earlier, in most shiplift structures the main transverse beams are extended beyond platform width (Figure 3–8). The latter is required for beam sheave fastening, and the extended length of the cantilevers typically corresponds to the sizes of the recesses cut in the pier deck for operation of winches.

To avoid platform jamming and unacceptable horizontal movements the precise location of the winch setting points along the shiplift length is clearly a must.

Some shiplifts, such as hydraulically operated units beside recesses, have holes in the deck structure for chains and chain jacks.

As mentioned earlier, in most cases the shiplift hoists supporting structure is constructed on piled piers. The width of these piers depends on the intended use, for example, for support of winches only, or in combination with fitting-out quays, or cargo handling docks. Piled piers transfer the loads from the hoists or jacks onto the underlying soil via steel and/or reinforced concrete piles of relatively small cross-section.

Use of a piled structure in pier construction is important. This ensures free unobstructed flow of water around the structure especially during lowering of the shiplift platform with a ship on it.

The shiplift supporting structure can be integrated with the shipyard pier or quay wall. In such cases the effect of a new structure on an existing pier or wall must be investigated, for example, whether or not the loads or settlement of the new structure will have detrimental effects on an existing structure.

Several structural schemes that can be considered for the shiplift support structure are illustrated in Figures 3–14 through 3–20.

In most practical cases service lines, for example, electric power, compressed air, gas, oxygen, etc., are located in special ducks placed under the pier deck. However, in some cases service lines are installed in special service galleries (Figure 3–18).

Depending on the deck design the supporting piles can be uniformly distributed along the structure or concentrated at the hoist locations; in the former case the deck slab must be stiff enough to distribute the hoist load uniformly between all piles, and in the latter case it must be sufficiently stiff and large to accommodate all piles needed to support the load at the location of hoist.

It is good practice to protect the bottom under the shiplift platform where a siltation problem can be expected. The latter enables use of the water jet to clean the bottom under the platform in case of substantial accretion of sediments in this area. Bottom protection can also be accomplished by placing a layer of the small size rocks (50 to 100 mm diameter), or by installation of gabions, different kind of fabriforms, and prefabricated concrete slabs.

The shiplift structure must be protected against corrosion. From this point of view all metal parts vulnerable to corrosion such as anchor bolts, hoisting wire ropes, and miscellaneous metal-embedded parts must be galvanized. Furthermore, use of reinforced concrete piles is preferred over that of the regular steel H- or pipe piles.

For a detailed discussion on protection of marine structures from corrosion the reader is referred to Chapter 1.

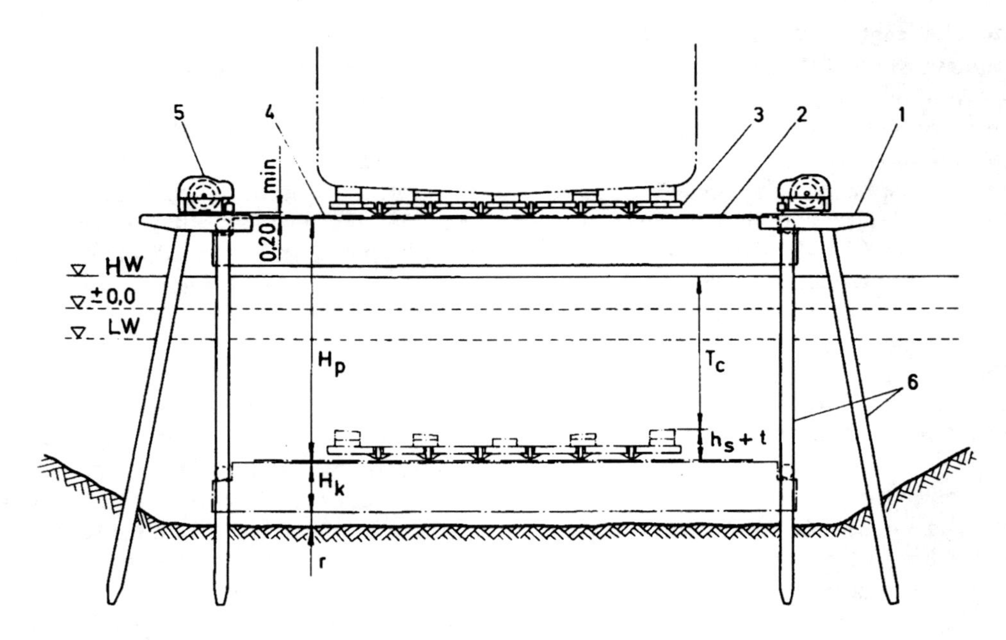

Figure 3–14. Shiplift supporting structures, as independent piers on piles. (Courtesy of Pearlson Engineering Co., Inc.) 1—Concrete cap beam on piles, 2—rail head level of the horizontal transfer system, 3—end transfer cradle, 4—shiplift platform, 5—hist, 6—reinforced concrete piles.

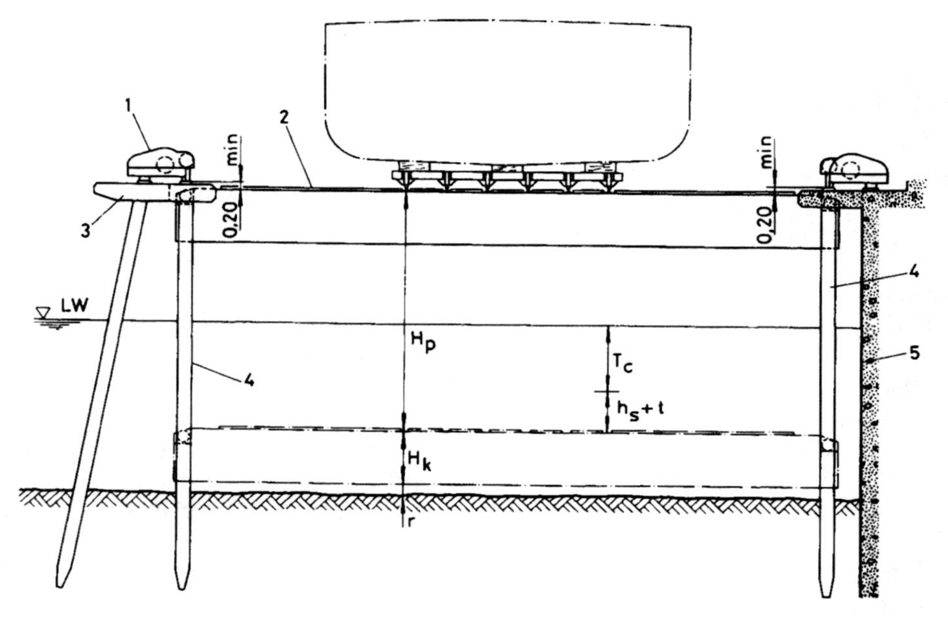

Figure 3–15. Shiplift supporting structure in the form of independent pier on piles and concrete slab on piles and sheet wall. (Courtesy of Pearlson Engineering Co., Inc.) 1—Hoist, 2—shiplift platform, 3—pile cap, 4-concrete pile, 5—sheet wall.

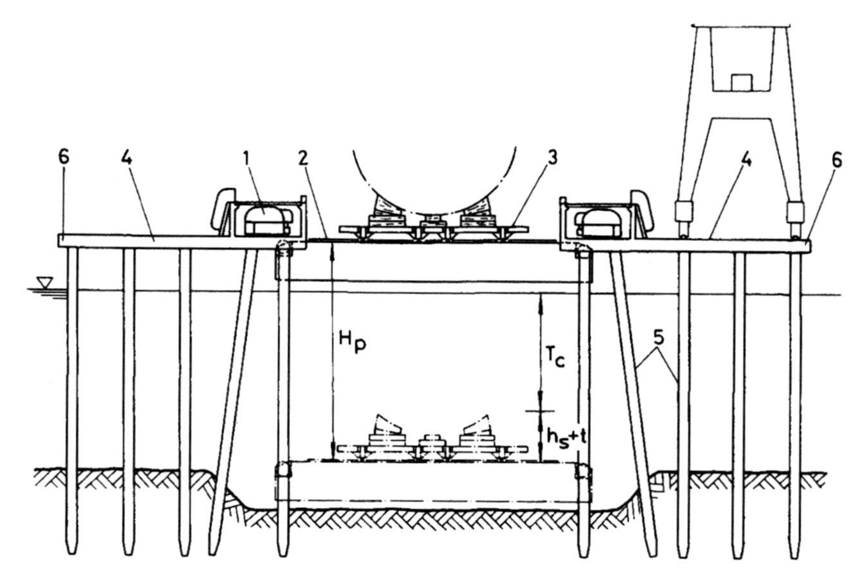

Figure 3–16. Shiplift supporting structure in the form of independent piers on piles with hoists in channels above the pier slab level. (Courtesy of Pearlson Engineering Co., Inc.) 1—Hoist, 2–shiplift platform, 3—end transfer cradle, 4—pier concrete slab, 5—concrete piles, 6—mooring line of the outfitting wharf.

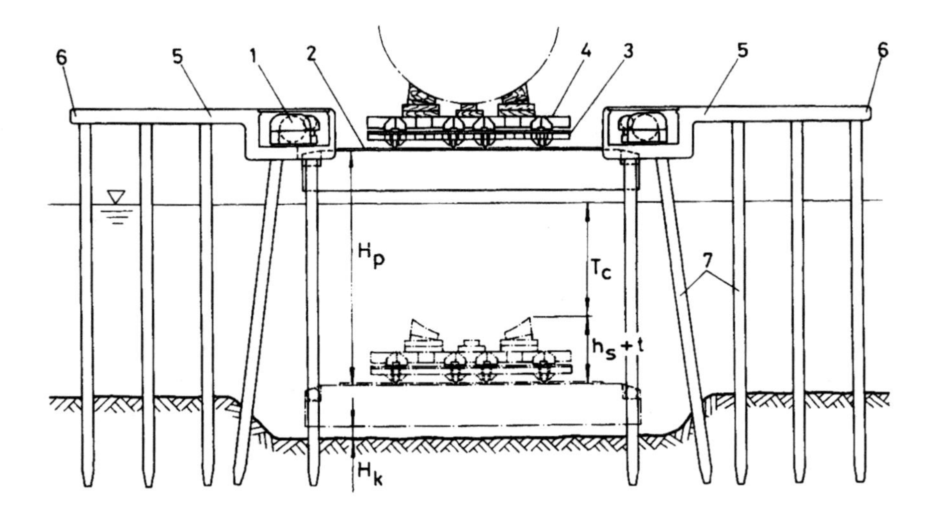

Figure 3–17. Shiplift supporting structure in the form of independent piers on piles with hoists in channels under the pier slab level. (Courtesy of Pearlson Engineering Co., Inc.) 1—Hoist, 2—shiplift platform, 3—lower cradle, 4—upper cradle, 5—pier concrete slab, 6—mooring line of the outfitting wharf.

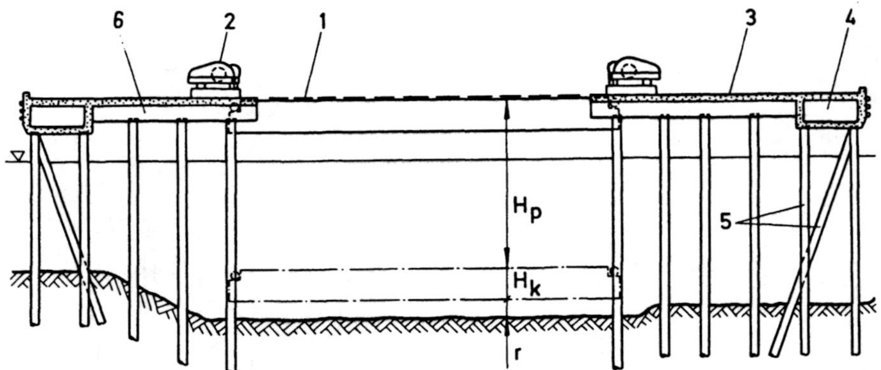

Figure 3–18. Shiplift supporting structure in the form of independent piers used as outfitting wharfs. (Courtesy of Pearlson Engineering Co., Inc.) 1—Shiplift platform, 2—hoist, 3—pier concrete slab, 4—installation channel, 5—concrete piles, 6—Capping beam.

3.2.6 Horizontal Ship Transfer System

The transfer system is used to move a ship from the berth to the shiplift or vice versa. Some typical layouts of horizontal ship transfer systems are depicted in Figure 3–6. The horizontal transfer system includes the following basic elements:

- The end transfer cradle which sits upon rails placed on the shiplift platform (Figure 3–21a)
- The side transfer carriages enabling moving of the end transfer cradles with a ship in the direction perpendicular to the longitudinal axis of the shiplift
- Double-level cradles (Figure 3–21b) enabling moving of the side transfer cradles with the ship from or to the shiplift to or from the dry berths located on both sides of the shiplift
- Rails for the end transfer cradles on both the berths and the transition areas to the berths
- Rails for side transfer carriages
- Retaining walls supporting grade separation between different levels of rails from the side transfer carriage and the end transfer cradles
- Transition structures between the rails of the shiplift and the rails of the shipyard area.

The possible layout arrangements of the shiplift, side transfer carriage, and berths are illustrated in Figure 3–6. From these arrangements it is seen that between the edge of the shiplift platform and the edge of the side transfer carriage there is a transitional section of the end transfer cradle tracks that is supported either on special pile structure, or directly on the solid quay wall (Figure 3–22).

The successful performance of both the end cradles and the side transfer carriages depends on an accurate spacing and parallelism of rails incorporated in both the platform structure and on land. Furthermore, the right angles at rail intersections, as well as uniform level of heads of all rails included in horizontal ship transfer system, must be carefully maintained.

The rail spacing for keel and bilge block cradles is typically equal to 1.5 to 3.0 m, whereas the spacing of rails for side transfer carriages typically varies from 2.0 m to 7.0 m. The latter basically depends on the soil conditions and on the required number of rails and tracks.

The rail supporting structure is also dependent on the soil conditions as well as on the number of rails and tracks in the rail system. It is normally designed in the form of the beams supported on piles, or as a mat foundation supported on a grade. In the latter case, depending on the specific project

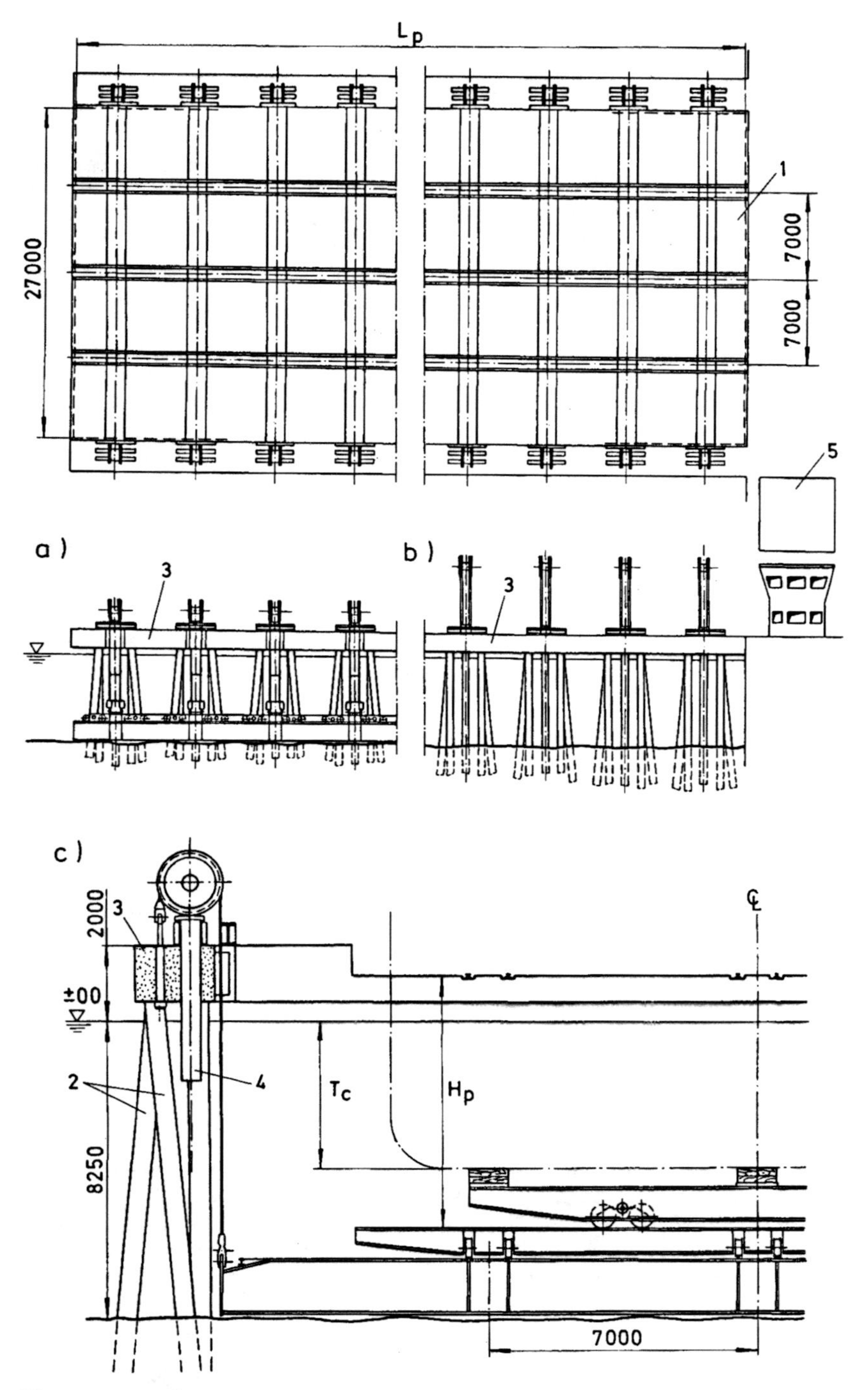

Figure 3–19. Supporting structure of the hydraulic shiplift. (Courtesy of Hyku Shiplift Int.) (a) Shiplift platform in the lowest position. (b) Shiplift platform in the upper position. (c) Cross-section. 1—Shiplift platform, 2—concrete piles, 3—concrete capping beam, 4—housing of the jack cylinder, 5—control center.

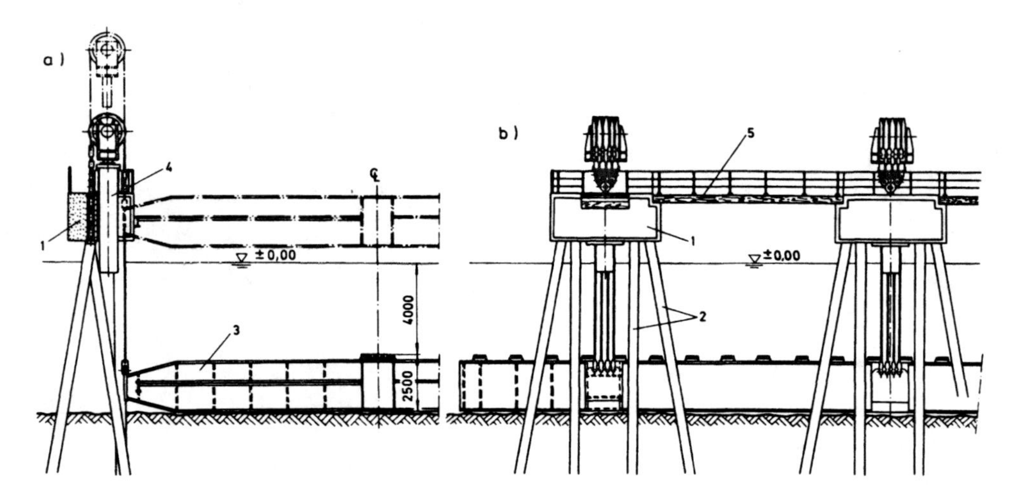

Figure 3–20. Supporting structure of the hydraulic shiplift. (courtesy of Hyku Shiplift Int.) (a) Cross-section. (b) Longitudinal elevation. 1—Concrete block, 2—concrete piles, 3—shiplift platform in the lowest position, 4—cylinder of the hydraulic jack, 5—bridge.

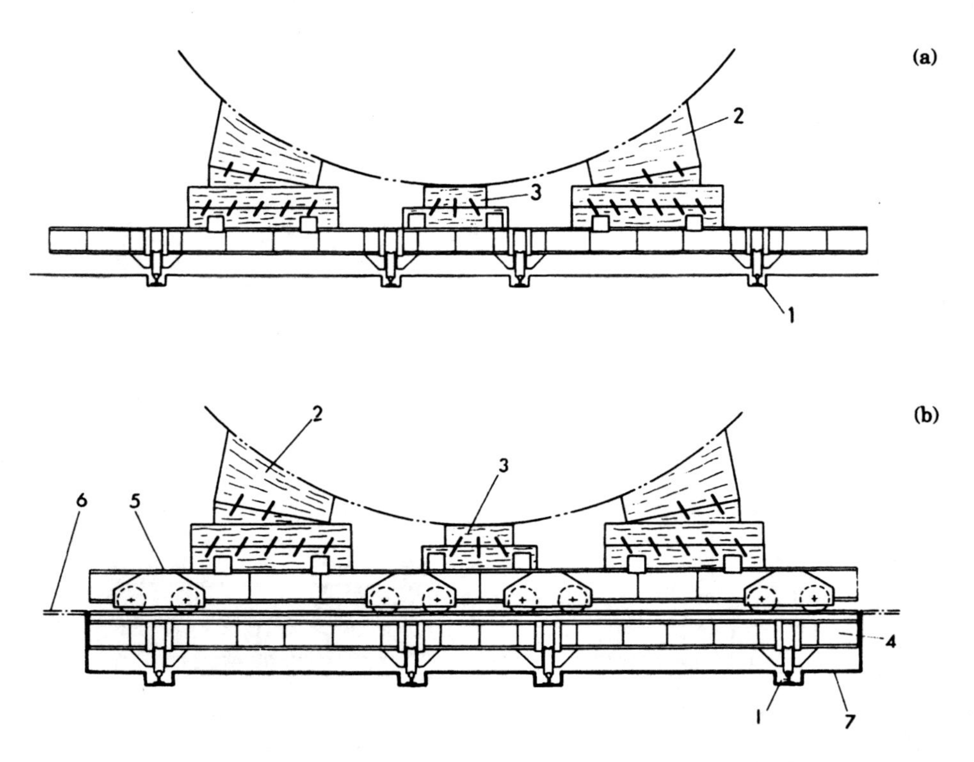

Figure 3–21. Typical transfer cradle. (courtesy of Pearlson Engieering Co., Inc.) (a) The end transfer cradle which sits upon rails placed on shiplift platform. (b) The double-level cradle. 1—Rail in groove, 2—bilge block, 3—keel block, 4—tide transfer carriage, 5—carriage that enables lateral ship movement to the shiplift, 6—rail for lateral movements of the carriage (5) to and from shiplift, 7—recess for side carriage (4).

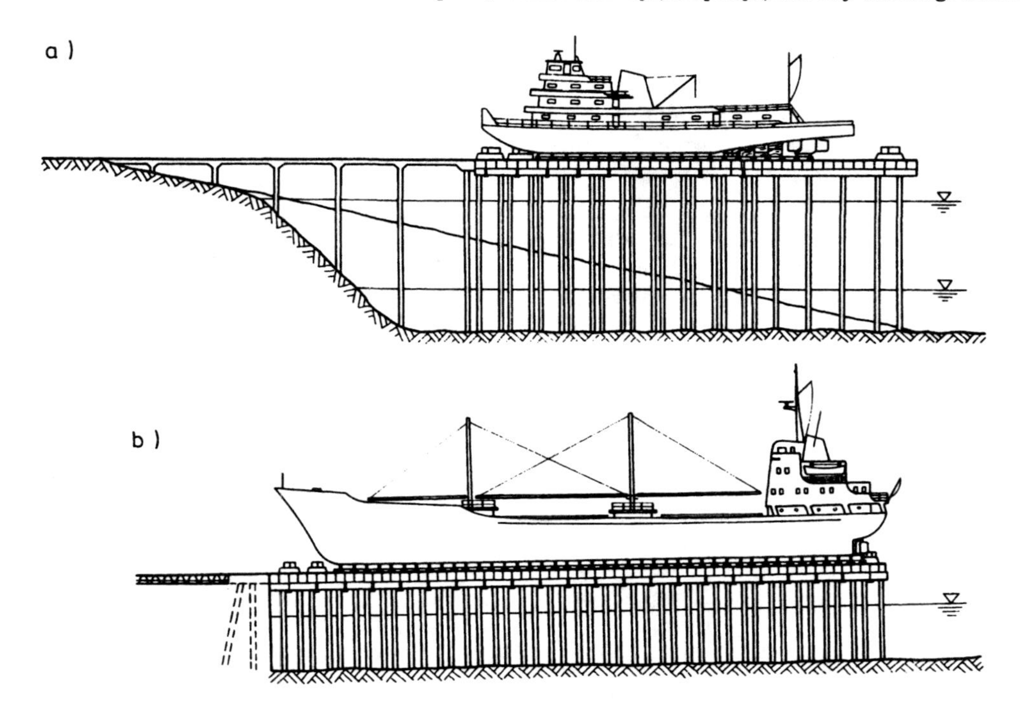

Figure 3–22. Transitional part of the end transfer system. (Courtesy of Pearlson Engineering Co., Inc.) (a) On piled pier. (b) On a quay wall.

conditions, the mat foundation may be designed either as a solid slab which would cover the space within the rail transfer system, or in the case where there is a substantial distance between bilge and keel block cradles, for example, if the distance between the rails of the bilge block and rails of keel block cradles is twice as large as the spacing of rails for the keel block cradles, in the form of continuous slabs in both longitudinal and transverse directions to support the aforementioned cradles on separate foundations.

Where practical, the rails can be placed on continuous or transverse sleepers. When rails are placed on individual tracks, the elements of track structure must be braced together by tie rods or by other means. It should be pointed out that typically loads on the cradle and carriage wheels are relatively small and vary from 50 to 100 kN/wheel. The latter results in a relatively small (about 0.1 MPa) bearing pressure on the subsoil; typically the value of bearing stress on the subsoil governs the differential settlement of rails.

In the berthing areas the pavement is usually made of concrete. The pavement loads, caused by loads from the keel and bilge blocks of a ship under construction or by supporting blocks of a ship under repair, can be significant. As a result the pavement slabs can be quite thick. Subsequently there are no difficulties in placing rails from the horizontal transfer system on them. The rails can be placed in a groove and therefore become flush with the pavement surface, or they can be placed directly on the pavement.

The former arrangement is preferable because it does not prevent traffic from moving freely in all directions. However, grooves can be easily plugged with dirt and debris, and therefore have to be shaped in a way to ease their cleaning; the latter is usually done by water jets. In the case of poor soil conditions the tracks are normally supported on piles.

Of fundamental importance in shiplift operations is the adequate design and construction of the transition structure or device between the shiplift platform and the land structure. This is provided to avoid a substantial difference between the platform and the land-based rail systems during transfer of the ship. In some cases the shiplift platform after reaching the extreme upper level is fixed in this position by special bolts pressed via hydraulic jacks in to corresponding female components installed in the land structure. To avoid heavy shear forces as may be the case when the aforementioned bolts are used, it may be useful to install a short piece of an articulated rail about 0.5 to 1.0 m long pin-joined to both the platform and the land-based rails (Figure 3–23).

3.3 MARINE RAILWAYS

3.3.1 Function and Main Parameters

By definition the marine railway is a system of cradle(s) or carriage(s) that is (are) lowered into the water along an inclined track(s) on a system of rollers or wheels until the vessel to be dry docked can be floated over it.

Marine railways (sometimes referred to as slipways) are mechanical means for lifting a ship out of the water to an elevation above the highest tides for repair and maintenance. They are also used for launching of the repaired or built ship. According to operating principles, marine railways are classified as longitudinal (Figure 3–24) and transverse or side-haul (Figure 3–25); in the former case the ship is lifted out of the water or launched parallel to the railway main axis, and in the latter case it is lifted in a perpendicular direction to the railway main axis. The trans-

Figure 3–23. Articulated connection between platform and land-based rail systems. (From Tsinker, 1983.) 1—articulated rail, 2—pinned connection, 3—bearing connection, 4—platform, 5—bulkhead, 6—rail.

Figure 3–24. Longitudinal marine railway. (From Mazurkiewicz, 1980.)

verse railway is most commonly used at river sites, where there is a wide range of water levels.

In principle at locations where a large and deepwater area is available the transverse marine system is very well suited for launching or lifting of large ocean-going vessels. In general, selection of a railway system is usually done on the basis of careful evaluation of available alternatives and a comparison to the other dry docking systems.

The standard marine railway consists of the following three basic elements:

1. Operational area, that is, the sloped plane with tracks on which carriages or cradles are lowered into the water on a system of wheels or rollers until the vessel to be dry docked can be floated over it.
2. Berths (repair and storage or working areas) to which the ship is brought after the lifting operation is completed, that is, when the carriage or cradle deck is clear of the water leaving the ship high and dry. The berth is designed such that the lifted vessel is positioned at or above the level of the yard. This allows materials and staging to be easily moved into position from the adjacent yard area. In some recent projects berths have been shifted from the marine railway area by introduction of a transfer system which uses the marine railway as the basic lifting and launching facility.
3. Hauling machinery and a system of wire ropes or chains that pull the carriage or cradle and its superimposed load on a system of wheels or rollers.

In general the marine railway is best suited for a capacity range of 1000 to 80 000 kN; theoretically marine railways beyond 80 000 kN could also be technically practical provided that foundation conditions are adequate to take increased loads from the marine railway structure.

Marine railways can be distinguished according to different criteria, such as:

1. The profile of the operational area, which can be straight (one slope), broken (two slopes), or arched.

2. The position of dry berths in relation to the operational area; the berths can be situated in the extension or from both sides of the operational area or relatively far from it. The level of the berths can be the same as the level of the operational area or it can be raised in relation to the operational area.
3. The shape of carriages moving on the rails in the operational area; the carriages can be of triangular or rectangular geometry.

The carriage of a triangular geometry is illustrated in Figure 3–25 (transverse marine railway); in the case of the carriage's rectangular geometry the operational area has two parallel gradients which enables use of the rectangular carriage (Figure 3–26) on straight, broken, or arched slopes.

Examples of use of the triangular and rectangular carriages incorporated into the longitudinal railway system are depicted in Figure 3–27; there in case (a) (triangular carriage) the ship is seated upon a carriage whereas in the horizontal position, and in case (b) (rectangular carriage) the ship is initially placed upon the uppermost carriage and then gradually rests upon all remaining carriages in the process of moving up out of the water. In the process the ship changes its position from horizontal (in floating mode) to

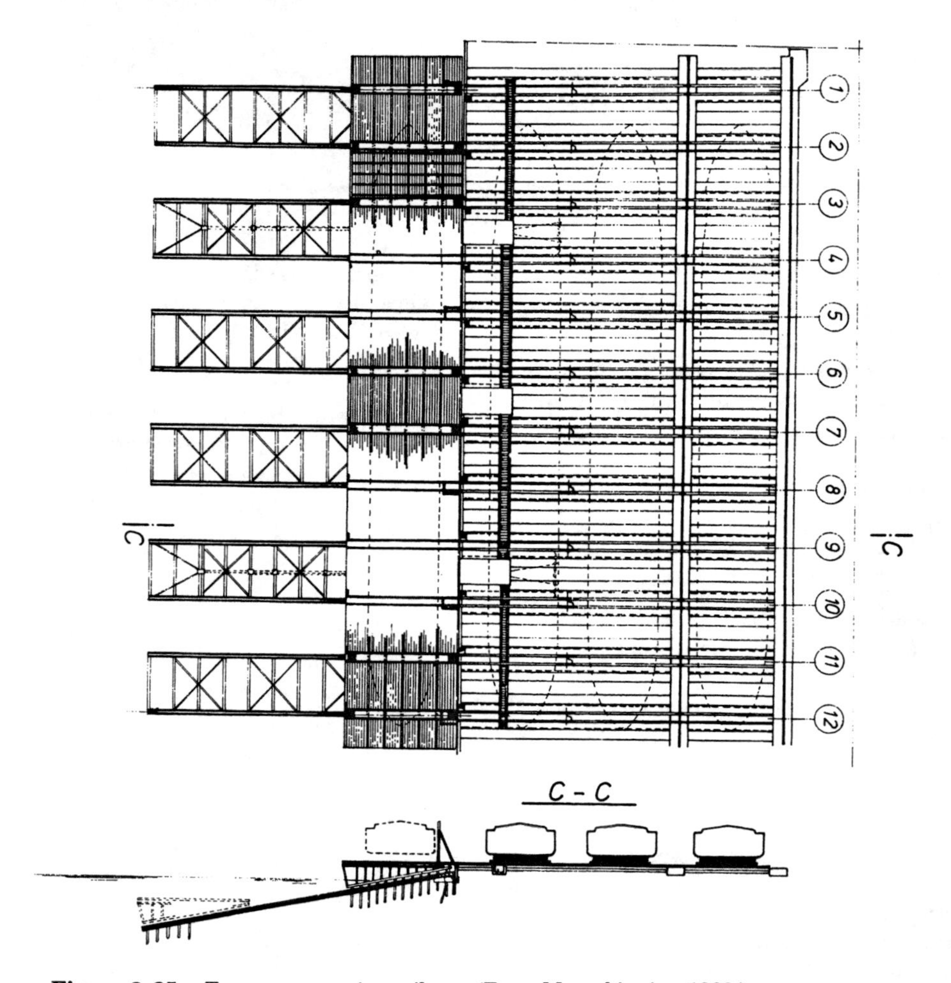

Figure 3–25. Transverse marine railway. (From Mazurkiewicz, 1980.)

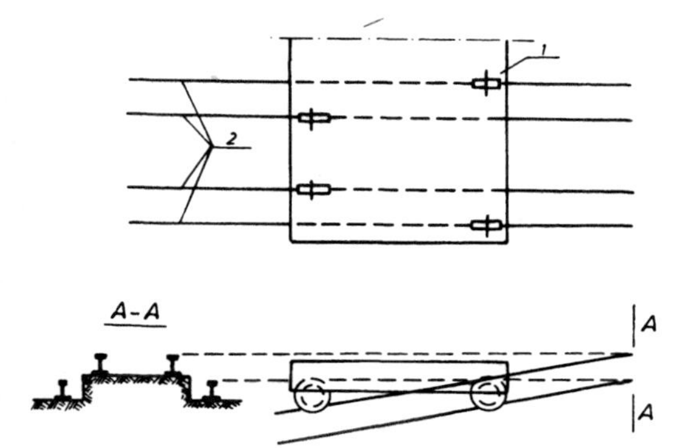

Figure 3–26. System of wheels and rails in a double gradient marine railway. (From Hückel, 1975.) 1—Carriage, 2—rails.

inclined and parallel to the railway track. In a longitudinal marine rail system one large or a number of smaller carriages put in the train pattern can be used.

To move the ship from the rectangular carriage to the berth area a turntable may be employed (Figure 3–28). This allows a change in position of the carriage from inclined to horizontal, after which the carriage, together with a ship seated on it, can be moved to the dry berth area.

The turntable also allows for both a limited amount of vertical movement as well as rotation in the horizontal plane, which brings the carriage onto one of the radially situated tracks leading to a certain designated berthing area.

The minimum depth of water on the sill of a transverse marine railway h_p (Figure 3–29) measured from the mean low water level to the top of the rail normally includes the following allowances:

1. Maximum vessel draught t (where relevant the vessel draught after accident is considered)
2. Underkeel allowance r; for larger vessels $r = 0.5$ m and for smaller $r = 0.2$ to 0.3 m
3. Height of keel blocks h_s
4. Height of the auxiliary (upper) carriage h_{w_2}
5. Height of the main (lower) carriage or cradle h_{w_1}
6. Height of rails fastened on the main carriage or cradle s_2
7. Height of rails or rollers of the marine railway s_1.

$$h_p = t + r + h_s + h_{w_1} + h_{w_2} + s_1 + s_2 \qquad (3\text{–}3)$$

All of the above allowances are also valid for longitudinal marine railways. However, in the latter case it requires proper interpretation for the height of the lower carriage (Figure 3–30).

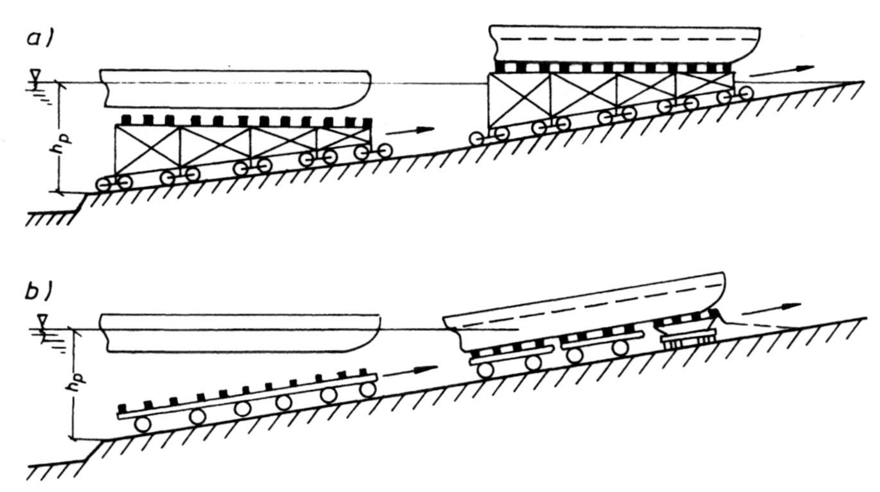

Figure 3–27. Longitudinal marine railways. (From Hückel, 1975.) (a) With a triangular carriage. (b) With a rectangular carriage.

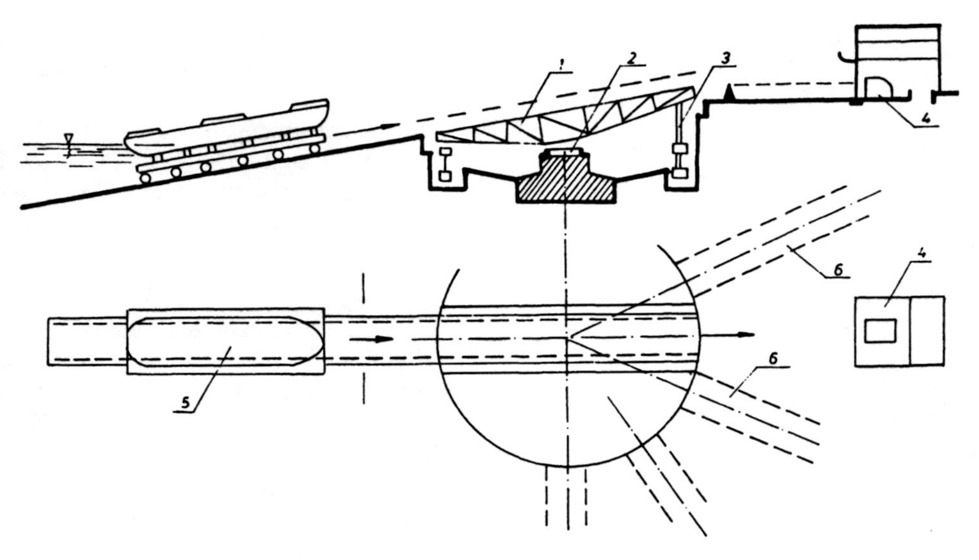

Figure 3–28. Longitudinal marine railway with turntable. (From Hückel, 1975.) 1—Turntable beam, 2—middle ring, 3—carriages with hydraulic jacks, 4—machinery house, 5—carriage with vessel, 6—berths.

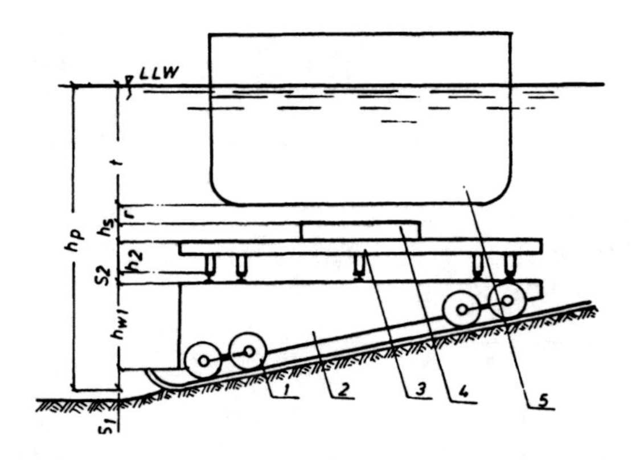

Figure 3–29. Estimation of the depth at sill of the transverse marine railway. (From Hückel, 1975.) 1—Rail of the operational area, 2—main (lower) carriage, 3—auxiliary (upper) carriage, 4—keel block, 5—ship hull.

The length of the marine railway L extended from the sill to the area where a change of slope is taking place, or from the sill to the top edge, can be estimated from the following:

$$L = (h_t + h_p)\tan\alpha \qquad (3\text{–}4)$$

where

h_t = distance from the mean low water level to the higher horizontal part, or higher edge of the operational area

α = angle of inclination of the operational area to horizontal.

Typically transverse marine railways are inclined at 1 : 6 to 1 : 8 (sometimes 1 : 5 to 1 : 10) and longitudinal railways are inclined at 1 : 12 to 1 : 16 (sometimes 1 : 10 to 1 : 25).

The actual length of the marine railway depends on its structure. As an example two typical structural arrangements are presented in Figure 3–31. There the length of the marine railway as is depicted in case (a) is equal to:

$$L = (h_p + h_o)\tan\alpha + \ell_r \qquad (3\text{–}5)$$

where

ℓ_r = allowance, estimated from the following:
$\ell_r = (0.5 \text{ to } 0.8)B + R\tan\frac{\alpha}{2} + (1.0 \text{ to } 2.0\,\text{m})$

where

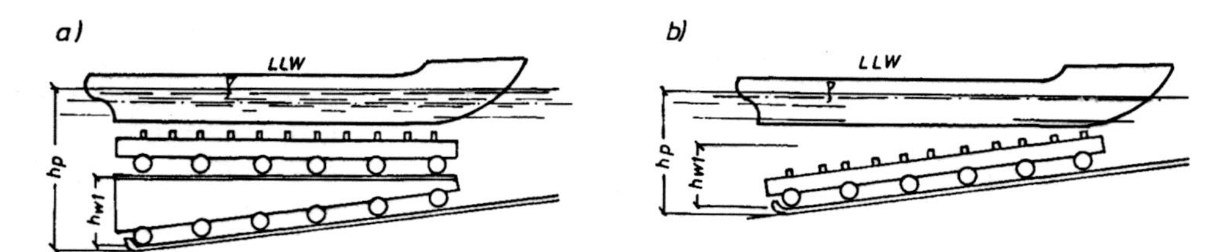

Figure 3–30. Estimation of the depth at sill of the longitudinal marine railway. (From Hückel, 1975.) (a) For triangular carriages. (b) For rectangular carriages.

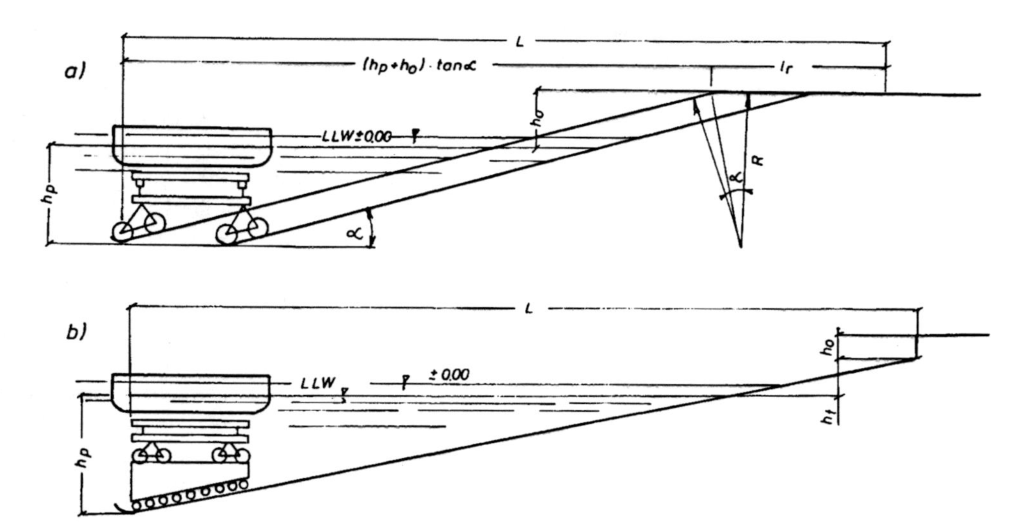

Figure 3–31. Typical marine railway systems. (From Szopowski, 1963.) (a) Transverse double gradient railway with rectangular carriage and two level tracks. (b) Transverse one-gradient railway with main triangular carriage and berths on elongation of the operational area.

B = width of the ship
R = radius of the arch of the rail (usually 10 to 30 m).

In case (b) the length of the marine railway is obtained from the following expression:

$$L = (h_p + h_t)\tan\alpha + h_o. \qquad (3\text{–}6)$$

The width of a transverse marine railway (its operational area) is roughly equal to the length of the largest vessel that shall be dry docked there. The carriage can be a little shorter than the ship length. In the latter case the ship bow and its stern will overhang the carriage.

The width of the longitudinal marine railway is roughly equal to width of the largest ship to be dry docked there.

Modern longitudinal marine railways are usually provided with a service deck over the cradle and with a docking platform for bilge blocks and bilge block winches.

3.3.2 Structural Design and Construction Aspects

The basic structural part of the marine railway is its operational area. It normally includes two or more tracks on which the carriages or cradles move. A small number of tracks are used in longitudinal railways and several tracks are usually required for transverse railways, particularly for large vessels. The tracks run from the yard area at about 1 to 2 m higher than the mean water level to a depth of 5 to 7 m below minimum design water level depending on the design lifting capacity of the marine railway and on the water level changes in the area of the railway. Smooth movement of carriages or cradles on the tracks requires their even assembly with the same inclination. Furthermore, it requires proper foundations for those tracks that do not allow substantial horizontal or vertical displacements.

The simplest structures, particularly those used for dry docking of small vessels, normally comprise an operational area made of a simple track, consisting of two or three rails, fastened to the timber or steel sleepers resting directly on a soil slope or placed upon stone bedding. The rails can be assembled on a steel or reinforced concrete frame and then installed in place on a prepared bedding. After installation the frame is adjusted to the specified levels by placement of additional fill.

If the foundation soil has sufficient strength and is protected against scour then the railway tracks can be placed on a concrete slab or beams constructed directly on the prepared slope. To avoid use of the expensive, cost prohibitive cofferdams the concrete or steel components of the track are usually prefabricated and then installed underwater by divers. The number of prefabricated units should be as low as possible; in case of use of the prefabricated beams they must be joined together by cross beams to form a rigid frame consisting of at least two main beams to which the rails are fastened.

Where the foundation soil in the operational area is poor the railway structure is supported on piles.

An example of a marine railway structure comprised of prestressed prefabricated concrete frames installed on timber piles is illustrated in Figure 3–32. In this example the underwater structure contains six tracks placed 8.5 m apart from each other. Each track consists of two prestressed concrete beams rigidly joined together with transverse bracing. The weight of each prefabricated unit was 450 kN. Each frame was placed at a 1 : 6.2 gradient upon three supports made of timber piles capped by concrete. The prefabricated units were installed underwater by a floating crane.

The marine rail carriages are mechanical units built of steel and designed to fit a particular type of marine railway berthing area, and to accommodate vessels of a certain size. Furthermore, the carriage design is affected by the type of land-based transverse transportation system. A typical marine rail carriage is depicted in Figure 3–33.

The bearing load from the carriages varies according to a certain phase of the ship dry docking operation. An example of the load distribution on a longitudinal marine rail track with rectangular carriages is illustrated in Figure 3–34. As seen from the figure, four load conditions are typically distinguished during the ship dry docking operation. During phase (1) a ship is positioned above the carriage, but is still completely buoyant and only the weight of the carriage is acting on the track structure. During phase (2) a ship is partly buoyant and partly supported on blocks. At this stage the carriage weight plus associated weight of the partly buoyant ship are considered. In phase (3), during which a ship is fully supported on blocks, the weight of the carriage plus full weight of a partly buoyant ship are acting on the track structure. Finally (phase 4), the dry weight of a ship and the carriage are acting on the track system. In load calculations the uplift effect on weight of the submerged carriages is usually ignored.

When triangular carriages or cradles are used two basic methods of load distribution from the wheels or rollers onto the railway tracks are usually considered. The first one is used where load distribution from the vessel

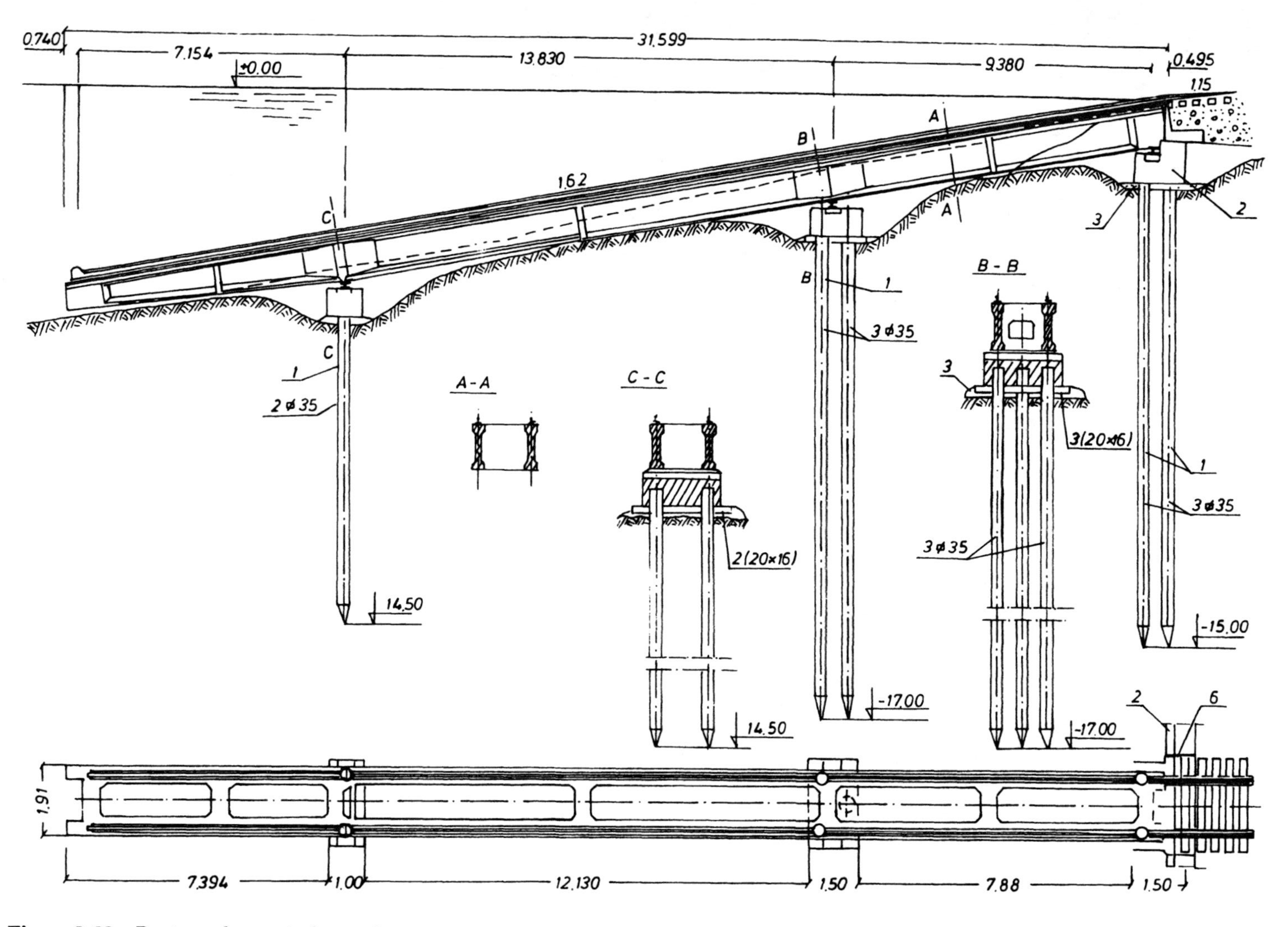

Figure 3-32. Prestressed concrete frame of a transverse marine railway installed on piles. (From Szopowski, 1963.) 1—timber piles, 2—upper support made after final adjustment, 3—concrete, 4—presstressing cables, 5—timber form.

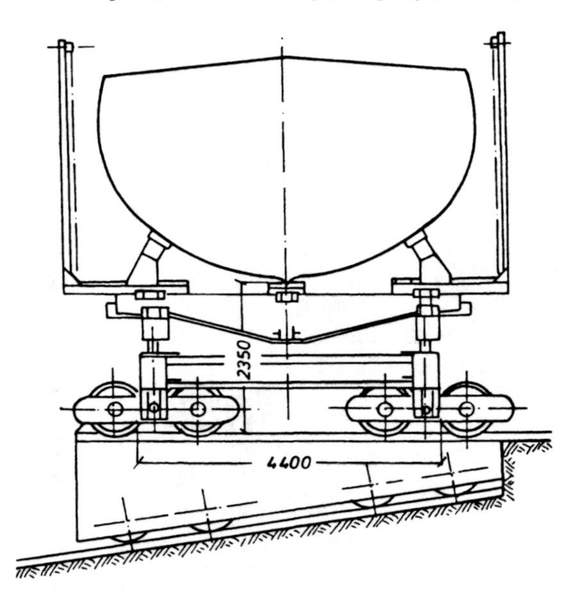

Figure 3–33. Transverse marine railway carriages. (From Mazurkiewicz, 1980.)

is uneven, for example, load from the ocean-going vessels where weight of the machines and the bridge structure are usually concentrated at the stern. In the latter case a trapezoidal load distribution diagram is usually considered (Figure 3–35).

In the case when lifting or launching concerns a vessel with evenly distributed weight a uniform distribution of loads onto the railway track from the ship and the cradle is assumed (Figure 3–36); the uneven distribution of weight of the carriage can be neglected.

Depending on the structural solution the rail supporting system is usually treated as a beam or slab on an elastic foundation, or as a continuous beam on stiff or elastic supports.

In practice the horizontal component of vertical loads can sometimes have a significant value and therefore must be considered in the track system design. Where required the anchor blocks are usually placed at the sill level to resist the aforementioned horizontal loads.

3.4 SHIPWAYS

3.4.1 Functions and Main Parameters

Shipways (sometimes also referred to as slipways) are generally classified as longitudinal, from which a ship is launched parallel to its longitudinal axis, and side or transverse, from which a ship is launched transversely (sideways) to its longitudinal axis. Both side and longitudinal shipways are normally used in shipbuilding only. There are no technical limitations to the size of ships (or other floating units) to be built and launched from a shipway.

Typically the longitudinal shipway is a slab (surface) inclined in the direction of water, on which structures and equipment used for building and launching of ships are placed. The purpose of the shipway structure

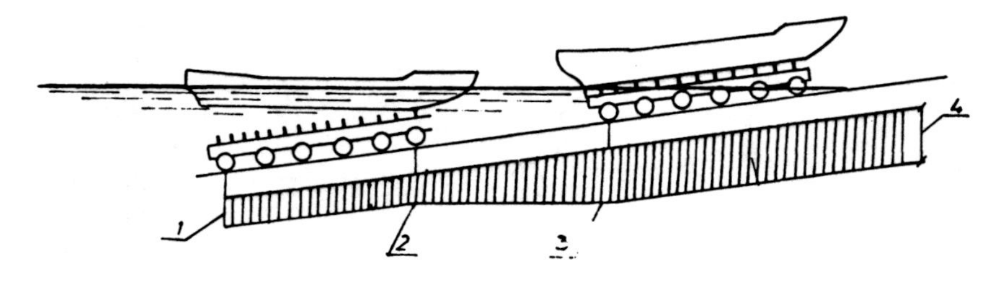

Figure 3–34. Distribution of loads on a longitudinal marine railway with rectangular carriage. (From Hückel, 1975.) 1—Load from the carriage only, 2—loads from part of buoyant vessel and carriage, 3—load from carriage and buoyant vessel, 4—loads from the dry vessel and carriage.

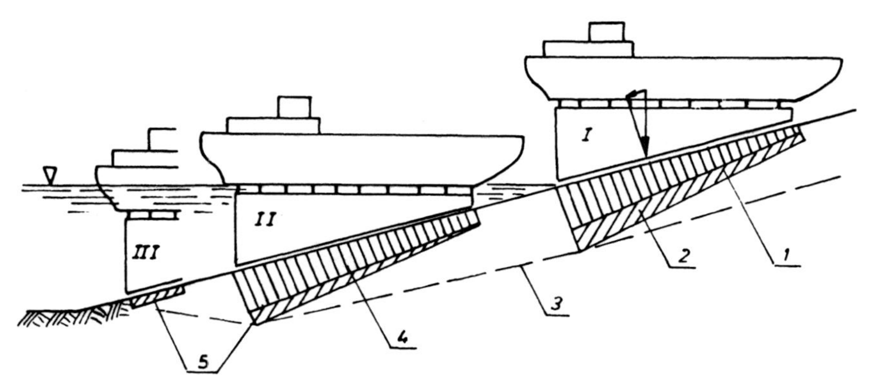

Figure 3–35. Distribution of loads on a longitudinal marine railway with a triangular carriage and an uneven distribution of the weight of the vessel. (From Hückel, 1975.) (a) The vessel and the carriage out of water. (b) The vessel out of water and the carriage under water. (c) The vessel is floating while the carrage is submerged. 1—The pressures from the vessel, 2—the pressures from the carriage without uplift, 3—envelope of maximum pressures, 4—pressures from the vessel, 5—pressures from the carriage with uplift consideration.

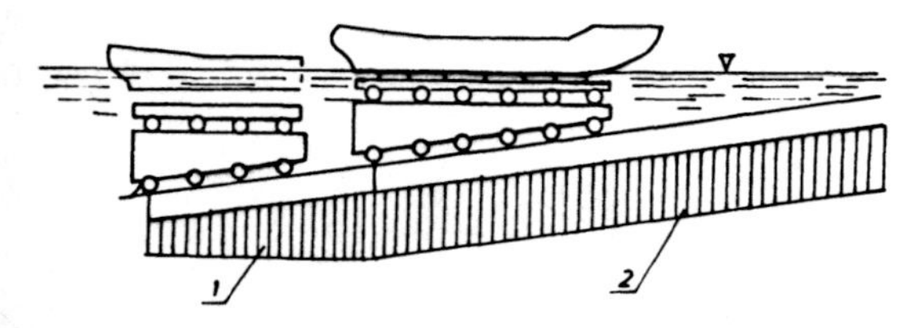

Figure 3–36. Distribution of pressures on a longitudinal marine railway with triangular carriage and uniform distribution of the weight of the vessel. (From Hückel, 1975.) 1—Loads from carriage with consideration of the uplift, 2—loads from dry ship and carriage.

is to transfer loads accumulated during building and launching of ships to foundation soil, and to keep the shipway carrying part at a constant level.

On a longitudinal shipway the ship hull is built on an incline toward the water position and parallel to the longitudinal axis of the shipway. There the ship is usually built with its stern directed toward the shipway sill (Figure 3–37). During construction the ship hull is supported on a set of blocks from which the launching takes place (Figure 3–38).

Longitudinal shipways are usually located perpendicular or askew to the harbor channel or shipyard basin where the length of the water area existing in the front of the shipway is adequate for ship launching. It is usually assumed that the minimum length of this area should be 2.5 to 3.0 times the length of the maximum size ship planned for launching from the shipway in consideration. If this length is smaller than required, special equipment must be installed to prevent the ship from traveling after launching beyond the available space.

The shipway dimensions above all depend on the size of the ship that has to be launched. Therefore during the design phase the future development of the shipyard has to be carefully evaluated and the maximum size of the design ship as well as the ship building method must be established and economically justified.

The shipway is usually divided into two parts, namely, working area of length L_r on which the ships are built, and launching part of length L_w required to bring the ship from the working part into the water.

The maximum length of the working part L_r is computed from the following (Figure 3–39):

$$L_r = L_c + 2a + 2b \qquad (3\text{–}7)$$

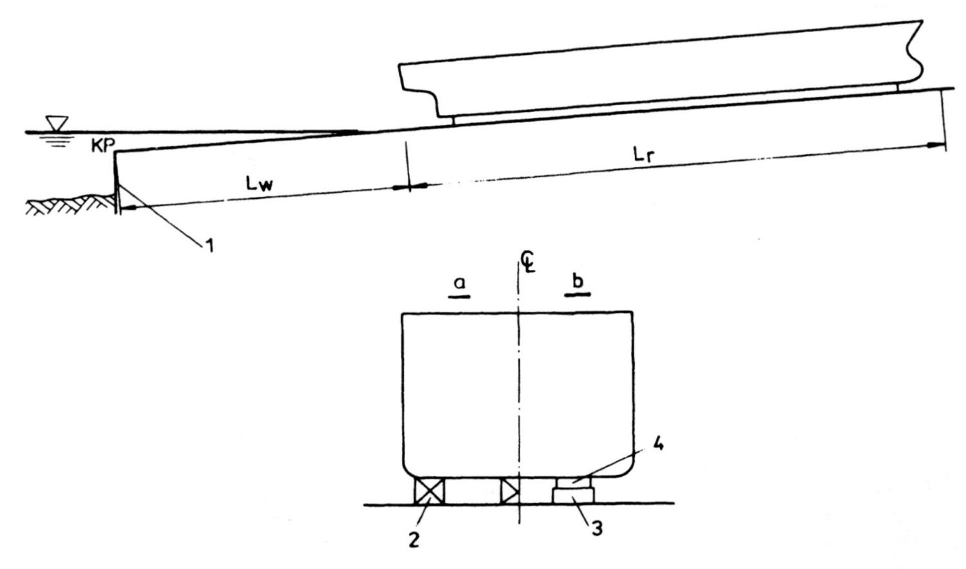

Figure 3–37. Longitudinal shipway. (From Mazurkiewicz, 1981.) (a) ship construction phase. (b) Launching phase. 1—Sills, 2—supporting blocks, 3—launch way, 4—cradle.

Figure 3–38. Launching of a ship from the longitudinal shipway. (Courtesy of Gdynia Shipyard, Poland.)

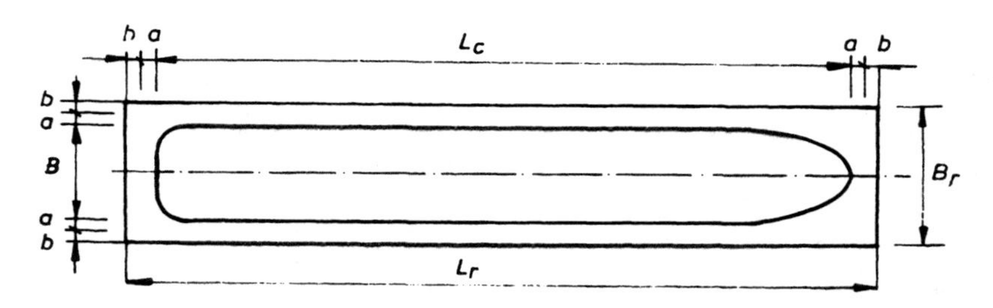

Figure 3–39. Dimensions of a longitudinal shipway. (From Mazurkiewicz, 1981.)

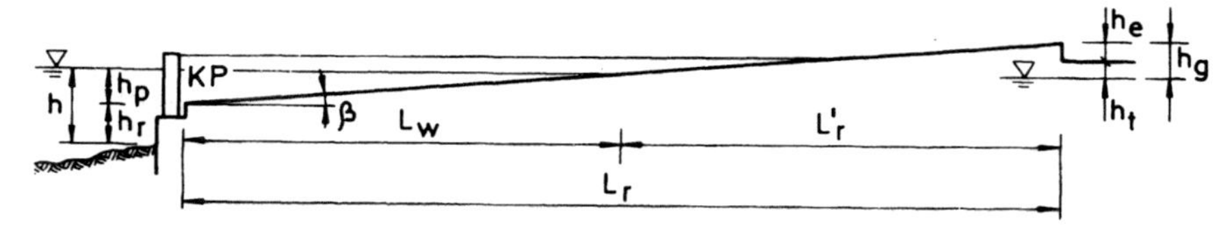

Figure 3–40. Dimesions of a closed (dry dock) longitudinal shipway. (From Mazurkiewicz, 1981.)

where

L_c = maximum length of the ship to be built

a = width of scaffoldings (usually about 2 m)

b = minimum width of the path at the bow or stern of the ship (about 1.5 m).

The width of the working part is equal to

$$B_r = B + 2a + 2b \tag{3–8}$$

where

B = maximum width of the ship to be built.

From the structural point of view it is possible to divide the shipway into a strip on which the ship is supported during the construction phase and sliding ways for use during launching. The former strip is defined as the foundation of the shipway; the width of this structure should be equal to at least the width of the largest ship to be built there in the future.

In practice, however, the same structural slab is used across the whole width of the shipway.

The working part of the shipway very often overlaps its launching part. In the latter case the shipway is ended with a gate placed either on the shipway sill (closed or dry dock shipway) or on its launching part (half-closed or half-dry dock shipway). In fact, because dimensions of the largest ship to be built at the new shipway very often are uncertain, mainly gated shipways are used in shipbuilding industries (Figure 3–40 and 3–44).

The following shipway parameters have the most significant impact on the ship launching procedure: the angle of the sliding part of a shipway β, the geometry (profile) of this part, and the depth of water at the sill.

The decrease in angle β increases the pressure on the shipway sill and accordingly decreases the pressure on the bow sliding bilge during the stern rotation. Naturally,

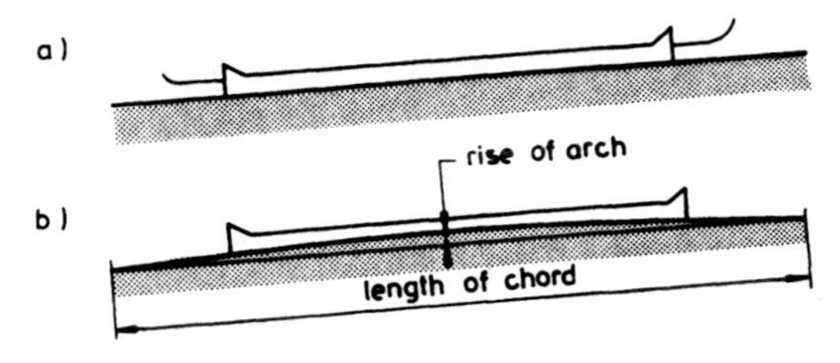

Figure 3–41. Profiles of a longitudinal shipway. (From Mazurkiewicz, 1981.) (a) Straight. (b) Arched.

Figure 3–42. Ship launch from a throw transverse shipway. (courtesy of Gdynia Shipyard, Poland.)

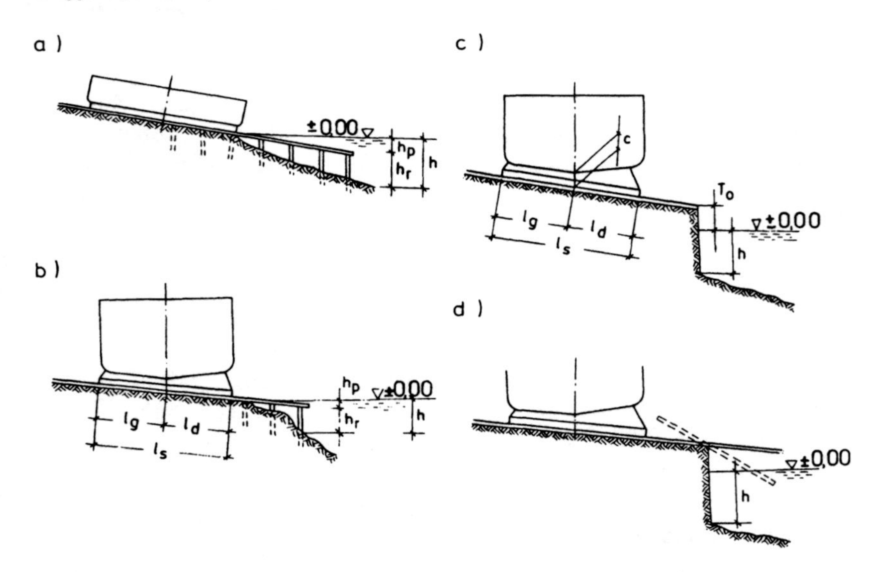

Figure 3–43. Ship launching from a transverse shipways. (From Mazurkeiwicz, 1981.)

increase in β causes a reverse phenomenon. The minimum depth of water above sill h_p affects the value of pressure on the sill, as well as location of the point of stern rotation. Thus the lowest water level above the sill, which makes the launch of a ship of a certain size possible, must be carefully evaluated.

The selection of an adequate inclination of the launchways requires the consideration of many factors, the most important of which is the friction resistance.

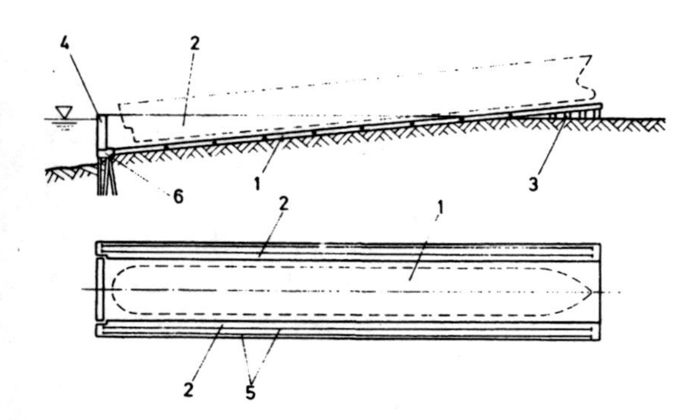

Figure 3–44. Scheme of a closed longitudinal shipway. (From Mazurkeiwicz, 1981.) 1—bottom slab, 2—side wall, 3—bridge, 4—closing gate, 5—crane tracks, 6—sill.

The launchway inclination must be such that the condition $tg\beta > \mu$, where $\mu =$ static friction coefficient, is satisfied. On the other hand too steep an inclination of the launchway may cause a ship to move too fast. The latter may result in large loads on the bow sliding bilge and in substantial increase in bending moment within the ship's hull. Furthermore, if the inclination of the launchway is too large the bow of a large ship will protrude high above the shipyard elevation, causing difficulties in providing the ship with proper lateral support.

For the above reasons, instead of a straight profile arched-shaped launchways sometimes are used (Figure 3–41).

The shipway arch profile (inclination) must be such as to enable a ship to start moving. Depending on the ship size the arch radius can vary from 3000 to 23 000 m with a chord inclination of 44 to 53 mm/m (about 1 : 19 to 1 : 12.5); the chord length may also vary from 90 to 250 m.

On shipways with straight launchways the inclination for small ships is usually assumed as equal to 62.5 to 125 mm/m (about 1 : 16 to 1 : 8) and for large ships this may vary from 37 to 71.5 mm/m (about 1 : 27 to 1 : 14). The working part of a shipway usually has a smaller inclination; for small ships it typically varies from 62.5 to 77 mm/m (about 1 : 16 to 1 : 13), and for large ships it is about 33.3 to 50.0 mm/m.

The length of the launching part of the shipway L_w should be such as to provide adequate depth of water at the ends of launchways. This is required to avoid excessive pressures on the sill, that is, on the frontal edge of the shipway and to prevent the ship from inclination and too fast rotation about its bow. Furthermore, the depth of water should allow the hull to reach the free floatation condition just before leaving the launchways. This length is determined from the following relationship:

$$L_w = h_p / tg\beta \quad (3\text{–}9)$$

where

β = inclination angle of the launching part of the shipway

h_p = the depth of water at the end of launchway.

The end of the launching part of the shipway is usually located at the mean water level. However, if large water level differences occur in the basin at which the shipway is located, the end of this part should be located at mean low water level. Thus the elevation of the upper edge of the shipway working part h_g above the mean water level should be computed from the following:

$$h_g = L_r' \tan\beta \quad (3\text{–}10)$$

where

L_r' = the length of the shipway working part above the mean water level in the water basin.

If the shipyard elevation above the mean water level is equal to h_t the height of the shipway edge h_e above this elevation is equal to

$$h_e = h_g - h_t. \quad (3\text{–}11)$$

The depth of water at the launchway ends, h_p, which would correspond to the water depth at the shipway sill, can be estimated as

$$h_p = 1.25(h_d + h_s) + h_w \quad (3\text{–}12)$$

or

$$h_p = 1.25(D_s/\delta_w L_c B + h_s) + h_w \qquad (3\text{–}13)$$

where

h_d = the forward draught of the largest ship that will be launched from the shipway

h_s = the height of sliding ways (cradle height)

h_w = difference between mean and mean low water levels

D_s = ship lightweight (ship's hull launching weight)

δ_w = block displacement coefficient.

The forward draught of the launched ship hull depends on the ship type, for example, tanker, container ship, navy vessel, or other and its degree of outfitting.

The water depth h at the shipway end should correspond to the draught of the launched ship, together with cradles during the whole launching run, and include an allowance h_r

$$h = h_p + h_r = h_p + 2.0 \text{ (in meters)} \qquad (3\text{–}14)$$

The added depth of water at the sill, h_r, is used to account for sinking of the hull after it leaves the launchways of the shipway. It is estimated from the following empirical relationship:

$$h_r = 0.75(h_d + h_s) + 0.40 \text{ (in meters).} \qquad (3\text{–}15)$$

In practice, design values of $h_r \geq 2.0$ m are usually considered. The best estimate of the required depth of water at the shipway end may be obtained from large-scale physical model tests.

After the ship hull is ready for launch it is shifted onto launching equipment which includes launchways, sliding ways (bilge and cradles), and launch fore and aft poppets.

The longitudinal ship launching is usually made from two launchways; in some instances one, three, or four ways are also used.

On the side shipway the ship is built and launched parallel to the shoreline (Figure 3–42). The water area for side shipways can be smaller than for longitudinal ones. This is largely attributed to the larger water resistance occurring during launching, which practically brings the ship to a standstill just behind the shipway sill. Based on experience it can be assumed that the required minimum width of the water basin for the side shipway should be equal to about 2.5 times the width of the ship to be launched.

The side (transverse) shipways are classified as

1. Flat shipways with launchways of constant slope brought under the water surface to such a depth that the ship becomes buoyant before reaching the launchway end, or with launchways brought just to the surface of the-water
2. Throw shipways with launchways of constant inclination ending above the water level.

Side shipways are generally divided into sliding and roller shipways. On sliding shipways launching takes place after setting up the ship on sliding ways (cradle) with subsequent sliding of the shipways system on a layer of grease placed on the launchway. In the roller shipways the sliding ways are moving on rollers. The method of ship launching affects the speed of its movement which in general must not exceed 7 m/s.

The dimensions of the side shipways depend on the dimensions of the largest ship to be laid down. The length and width of the shipway are calculated according to the formulas given for the longitudinal shipways.

The side (transverse) ship launching has the following advantages.

1. The ship before launching and for launching is set up horizontally. The latter considerably facilitates the laying down of the hull and assembly of equipment.
2. The erection of the launchway foundation, launchways, and sliding ways for side launching is much easier than in the case of longitudinal launchway.

3. Normally the side shipway is more economical than the longitudinal one.
4. From a side shipway the ship can be launched into a much narrower water area than that required for launching from the longitudinal shipway.

The side or transverse launchways can be of the following basic types (Figure 3–43):

1. The shipway with its lower edge located sufficiently deep below the surface of the water to enable the ship to become buoyant before leaving the shipway (Figure 3–43a).
2. The launchways go under the water surface, however, not far enough to enable the ship to run afloat (Figure 3–43b).
3. The launchway end is located from 0 to 0.3 m above the water surface and a "jump down" of the ship is taking place (Figure 3–43c); sometimes the launchways at the shipway edge can rotate (Figure 3–43d).

The first type of launchway is used for launching of the floating units with a large weight per unit length, for example, floating docks. In this case the inclination of the launchways must be larger than that which would be used for longitudinal launching. The latter is attributed to larger water resistance which could stop the vessel on the shipway. Because of substantial lateral strength of the standard floating units the side launching technique does not cause any problems in structural integrity.

The length of the launching part of the transverse shipway and the depth of water at the shipway sill h are estimated in the same manner as for longitudinal shipways. The depth should not be smaller than $h = h_k + 2.0$ (in meters), where h_k = maximum draught of the largest ship to be launched. In this case the additional 2.0 m depth of water is required due to temporary trim of the ship immediately after the launch and to provide for an underkeel allowance.

The second type of shipway is usually used for launching floating units of a relatively small longitudinal strength and small buoyant stability that prevents them from a "jump down" launch. Here the slope of launchways should be larger than that for a longitudinal launch. The side shipways of the second type usually have the underwater part of launchways short enough so that after the center of gravity shifts beyond the shipway edge a rotation of the ship takes place around this edge. The latter causes a concentrated load on the edge that must be considered in a shipways structural design.

For small launching velocities it is required to assume the possible smallest length (lowest position) of sliding ways ℓ_g to prevent blow of the ship into the shipway edge at the return heel after the launch.

$$\ell_g = B/2 + (0.25 \text{ to } 0.3) \text{ (in meters)}. \quad (3\text{–}16)$$

The third type of shipway is used most often where there are no specific ship launching conditions. The inclination of a typical transverse launchway is 200 mm/m (1 : 5) to 83.5 mm/m (1 : 12). The latter normally ensures that the velocities of unequipped ships on the shipway edge equal 3 to 5 m/s, and for equipped ships 5 to 7 m/s. The main problem in sideways launching is the ship heel, which under certain unfavorable conditions, for example, drastic reduction in metacentric height, may cause it to capsize.

The ship launching velocity depends on the launchway length, its slope, and the dynamic friction coefficient. When the velocity increases then the heel angle decreases while the point of the largest draught of the launched ship will be shifted away from the shipway edge. The safety of ship launching is affected by its metacentric height. When the metacentric height decreases the ship heel angle increases and under a certain critical value of metacentric height the ship can capsize. The value of the ship return heel angle also depends on its metacentric height; normally heel angles increases with a decrease of the metacentric height. The latter means that under a certain critical value of metracentric height the ship can hit the shipway edge with the side of its hull. The increased displacement affects the increase of draught and decrease of the underkeel clearance between the lowest point of the launched vessel or the cradle and the bottom of the basin. The increase of the jump down height causes an increase of the heel angle, while the increased water depth and the width of the water area cause a small increase of the return heel angle.

The total width of the sideways shipway depends on the required launching velocity. It consists of the launchway length ℓ_s and the width of a strip between the lower edge of the sliding way and the shipway edge. The width of this strip is usually equal to about 2 to 3 m.

The best design of side or transverse launchways for the optimum launching conditions can be accomplished on the basis of the large-scale model test.

3.4.2 Structural Design and Construction of Longitudinal and Transverse Shipways

Modern longitudinal shipways are constructed as either closed or dry dock shipways. In both cases shipway structural components heavily depend on site geotechnical conditions. In the case of competent soils present at the construction site the shipway structure can be placed directly upon foundation soil. If the founding soils are too weak the structure can be founded upon steel or concrete piles. The structure can also be founded upon improved foundation soils; where economically viable the poor soils can be replaced by good ones.

The closed shipway is equipped with a gate that helps keep the shipway basin dry.

The structural solutions for the underwater part of the longitudinal closed shipways do not differ substantially from those used for dry (graving) docks described in Section 3.5. The basic differences between the two structures are the depth at the sill, inclination of the bottom slab, and the distribution of loads.

By applying the appropriate loads all basic design and construction methods used for the dry (graving) docks can be used for design of the shipway structure.

Dewatering of the shipway basin takes place via pumping station when the shipway gate is in its closed position. Typically the gate structure is made of steel and is designed to allow it to open and close the shipway during all possible water levels in the basin. Floating gates are usually used in the longitudinal closed shipways. During a ship launching the gate is towed away and is moored at the fitting out quay or at any other suitable location.

A practical example of a 46.0 m wide shipway gate is shown in Figure 3–45. The weight of this gate is about 530 tons.

The flooding and dewatering system of a shipway is similar to that used in dry (graving) docks.

The side or transverse shipways are usually constructed as a set of independent launchways placed directly on a prepared natural slope or supported on piles.

A practical example of the piled side throw launching shipway is illustrated in Figure 3–46. It consists of 11 launchways spaced at 15.0 m that are laid partly horizontally and partly with an inclination of 1 : 10. The depth of water at sill $h_p = 0.6$ m. The shipway is equipped with a gantry crane.

Structural calculations for shipways normally include:

1. Determination of the value and distribution of loads acting on the shipway
2. Sizing of structural elements on the basis of bending moments and shear forces distribution.

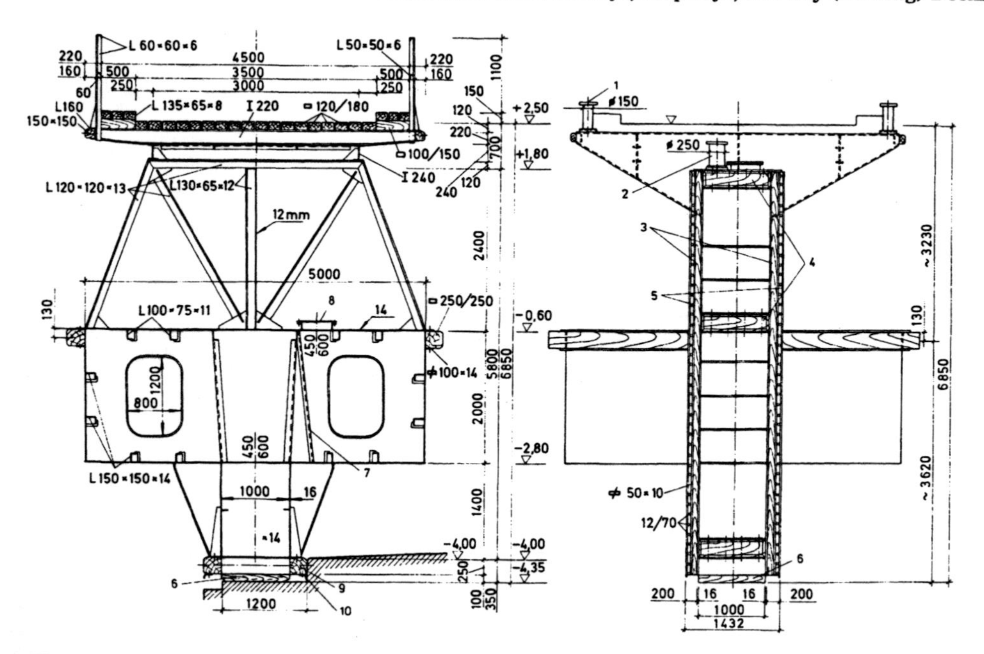

Figure 3–45. Cross-section and end view of the gate of the longitudinal shipway. (Courtesy of PROMOR, Gdansk.) 1—100 kN bollard, 2—150 kN bollard, 3—rebate beams, 4—front fenders, 5—rubber proile, 6—lower fender beam, 7—ladder, 8—cover, 9—side fender beam, 10—rubber profile.

When determining the values and distribution of loads acting on the shipway, two basic kind of loads are considered:

1. Static loads during the period of vessel hull assembly before and after its transfer from permanent blocks on cradles
2. Dynamic loads during launching, which have different phases depending on the transitional position of the hull.

The static loads on the shipway are further divided into two groups:

1. Weights acting on the shipway via cradles and launchways, causing at the same time the deformation of the hull
2. Weights acting on the shipway directly.

The first group includes the weights built in the hull and placed or hung on the hull. They are as follows:

- Dead weight of the hull
- Weight of the equipment installed in the hull while on the shipway
- Weight of the inside and outside hanging scaffolding
- Weight of the launching drags
- Weight of water used for watertightness tests of ship tanks and panels
- Weight of ballast water used for the regulation of the position of forces acting on the hull during launching.

The second group consists of

- Weight of the launchway structure
- Weight of the cradles with cribbing
- Weight of scaffolding erected on the shipway
- Weight of tools and auxiliary equipment
- Weight of hull blocks and sections stored on the shipway structure
- Weight of materials stored on the shipway structure.

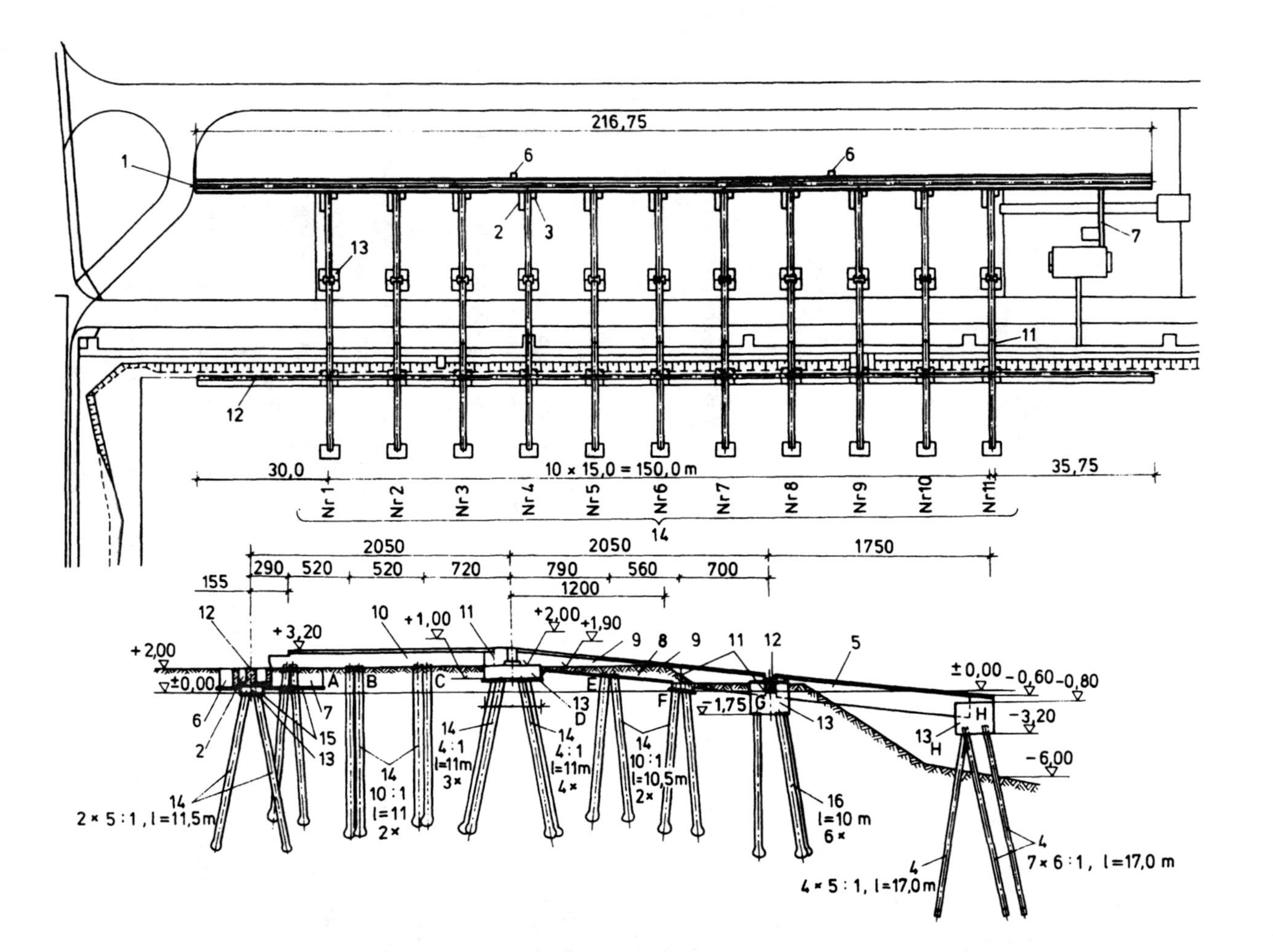

Figure 3-46. Plan and cross section of a throw transverse shipway. (From Mazurkeiwicz, 1981.) 1—Landside crane track, 2—trolley channel, 3—compressed air delivery well, 4—prefabricated reinforced concrete piles 0.35 × 0.35 m, 5—prestressed concrete beam, 6—electrical power supply well, 7—electrical culvert, 8—launch way, 9—inclined beams, 10—transverse beam, 11—expansion joint, 12—crane track beam, 13—support, 14—Franki piles 0.52 m diameter, 15—lean concrete.

The ship launching weight is not uniformly distributed along the hull; for each vessel this distribution is different because, among other reasons, it depends on the equipping rate of the hull.

Shipbuilding practice indicates that the shape, structure, and weight of the hull have a decisive effect on the distribution of blocks supporting the hull during the construction period. In fact the shape of the hull governs the total length of block rows and the magnitude of overhangs of bow and stern.

The distribution of the ship launching weight between all rows of blocks can be estimated either approximately, by ignoring effects of the interaction between the hull, blocks, and shipway structure, or with mathematical precision, by taking into account all the aforementioned factors.

In the former method it is normally assumed that the ship launching weight D_p on every row of supporting blocks (in the direction perpendicular to the ship axis) is equivalent to the part of the distribution curve of the launching weight D_s, contained between the centers of the block rows (Figure 3–47).

The loads on the rows of bilge and keel blocks are determined as indicated in Figure 3–48. There, 2/3 D_p are transferred onto the keel blocks; because of potential overloading caused by wind, it is assumed that each bilge block is loaded by 1/4 D_p. If the ship block coefficient is much smaller than unity it is considered that the whole load is carried on the keel blocks only (Figure 3–49). Accordingly, for dimensioning of the side supports it is assumed that they are loaded by 1/6 D_p.

During the period between the assembly of the hull, and its transfer from blocks onto the cradles the ship launching weight is equally distributed between two launchways; a wind load P_w should be added to the launching weight according to the scheme as indicated in Figure 3–50.

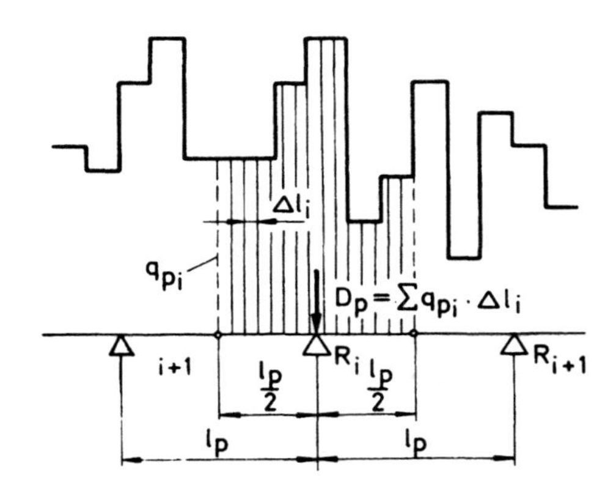

Figure 3–47. Loads on a row of blocks. (Mazurkeiwicz, 1981.)

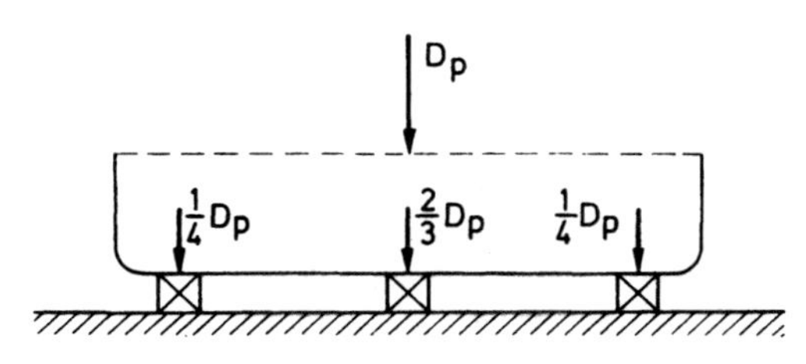

Figure 3–48. Load distribution from ship on keel and bilge blocks during ship assembly. (From Mazurkeiwicz, 1981.)

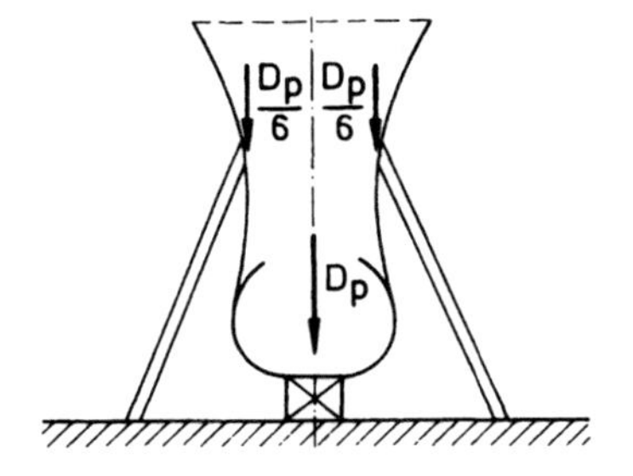

Figure 3–49. Load distribution from ship on keel blocks and on side support in the bow part.

$$P_w = qF_n t/L_s d = Q_w t/L_s d \qquad (3\text{–}17)$$

where

d = distance between axes of the launchways
q = unit wind pressure on the hull
F_n = side surface of the hull
t = the height of the wind resultant force

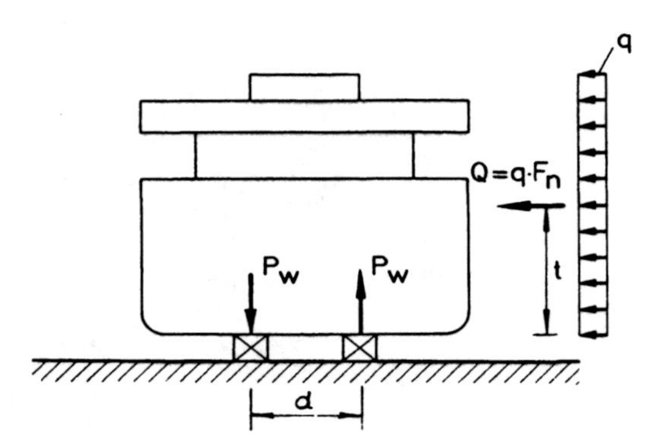

Figure 3–50. Distribution of wind load on a ship hull and launchways.

L_s = effective length of cradles.

The static load of the shipway P_s caused by the vessel of launching weight D_s, placed on cradles, acts at the cradle level and is distributed along the cradle width b_s.

Assume that the hull launching weight D_s is uniformly distributed on the cradles of length L_s consisting of two skids of the width b_p. Then the average load on the launchways P_t is computed from the following:

$$P_t = D_s/2L_sb_p = P_w/b_p. \tag{3–18}$$

Note that for further use in the longitudinal or transverse shipway designs the calculated vertical loads should be resolved into components normal and parallel to the shipway surface.

In practice, however, the shipway inclination angle is rather small, for example, the inclination of a conventional longitudinal shipway varies between 1 : 13 and 1 : 30, and therefore the full vertical load can be considered in proportioning of the shipway structural components.

In the case of transverse shipways the loads are calculated similarly to the above procedure with consideration given to the number of launchways.

The more accurate approach to load calculations takes into account the stiffness (flexibility) of both the hull and the hull supporting structure; the latter produces more realistic load distribution between different structural components (Mazurkiewicz, 1981).

During the launch from the longitudinal shipway the travelling hull causes a progressively changing load that depends on the location of the hull. The load diagrams, as presented in Figures 3–51a through 3–51d, are self-explanatory. For estimation of these loads two groups of calculations are required, namely, the calculation of launching curves and the calculation of loads acting on the shipway during launching. For more details on this subject the reader is referred to Mazurkiewicz (1980).

In transverse shipways the method of load calculation depends on the method of hull launch, for example, in the case of the flat type of shipway, with launchways ending under the water level, the calculations are

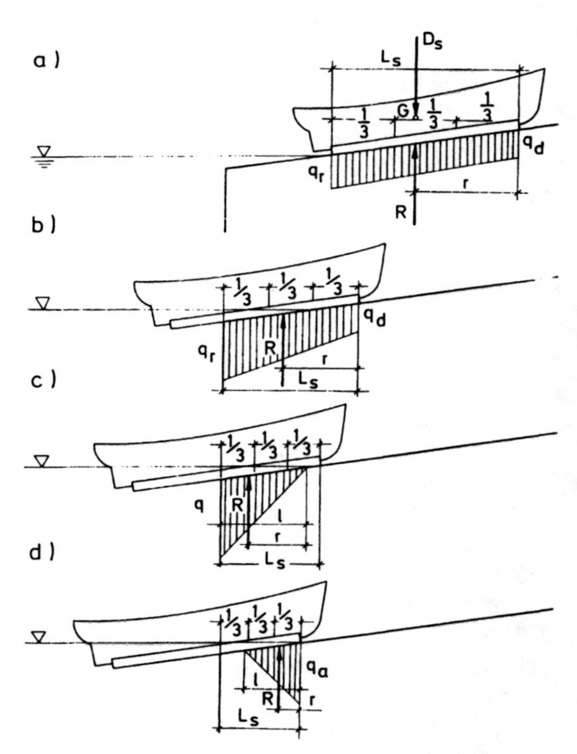

Figure 3–51. Approximate distribution of loads acting on a longitudinal shipway during different launching phases.

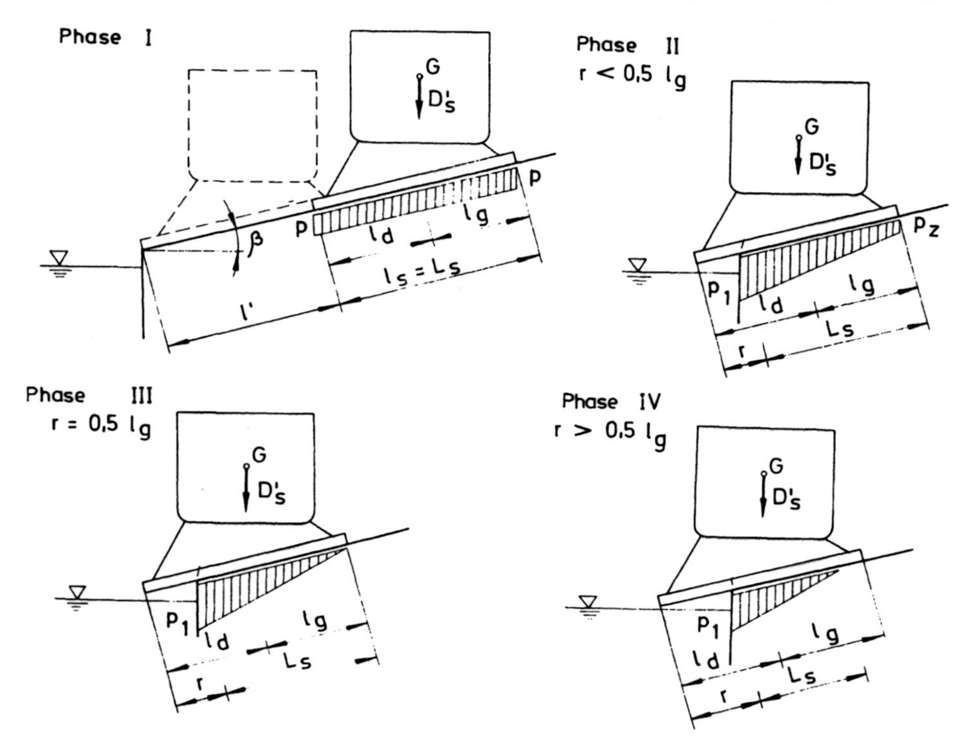

Figure 3–52. Launching phases of a transverse throw shipway.

similar to those used in the longitudinal shipways, and in the case of the throw type shipway several launching phases must be evaluated (Figure 3–52). The load pressure diagrams as indicated in Figure 3–52 are very similar to those used for the design of longitudinal shipways.

3.5 DRY (GRAVING) DOCKS

3.5.1 Functions, Types, and Main Parameters

The dry dock, by definition, is a chamber or basin separated from the adjacent harbor by a gate (Figure 3–53). The dock's basic structure comprises a floor, side walls, head wall, and a gate.

The part of the dock structure located at the entrance, known as the dock head, is used to support the dock gate and installation of any auxiliary devices necessary for docking a ship. The floor of the dock head is constructed as a sill and its side walls are used as vertical sealing faces for the dock gate.

Flooding and dewatering of the dock is carried out by means of pumps and other installations usually located in a pumping station.

When the gate is open, the water level in the dry dock is the same as in the harbor basin. It is then possible to bring the ship in or float her out of the dock. When the gate is closed, the water level in the dock can be lowered by pumping out the water so that the docked ship can settle on supports placed on the dock floor. These supports consists of keel blocks, side blocks, and bilge blocks (Figure 3–54). The number of blocks and spacing between them depend on the type and size of the docked ship.

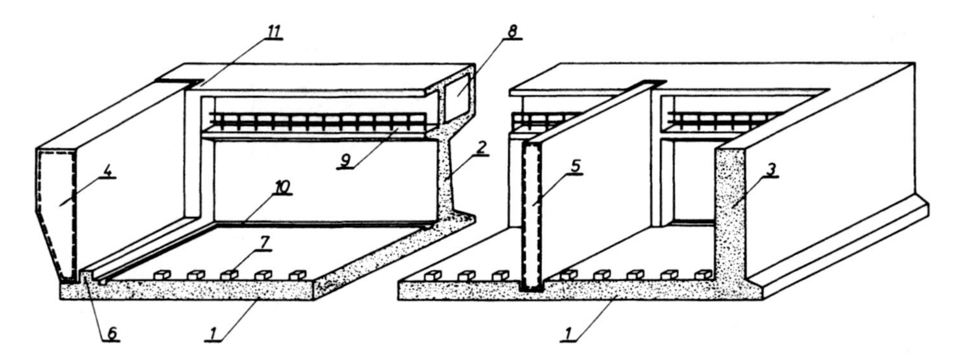

Figure 3–53. General scheme of a dry dock. (From Maxurkiewicz, 1980.) 1—dock floor, 2—side wall, 3—head wall, 4—dock gate, 5—inner gate, 6—sill, 7—keel blocks, 8—culvert, 9—barriers, 10—dewatering channel, 11—gate supports.

Figure 3–54. Keel and movable bilge blocks. (F. C. De Weger Int., Netherlands.)

Long docks can be partitioned into two chambers by inner gates. In some cases a dry dock may have two entrances. The second entrance is especially useful when the dock has an inner gate. Dry docks can be classified as:

1. Emergency dry docks used for docking ships with or without cargo but of greater than normal draught resulting from serious hull damage
2. Repair and maintenance docks used for docking ships without cargo
3. Building docks used mainly for ship construction with sufficient depth for undocking an empty or partly equipped hull
4. Building and repair docks used for ship construction, however, with sufficient depth of sill for docking ships for repair work.

Structurally the dry docks differ from each other basically by the method with which they counteract the hydrostatic uplift force. Accordingly all dry docks may be divided into three basic categories:

1. Heavy or gravity docks, in which the permanent loads are larger than the largest hydrostatic uplift.
2. Anchored docks, in which the largest hydrostatic uplift is totally balanced by the permanent weight of the dock structure and by a proper anchorage in the ground.
3. Drainage docks in which the hydrostatic uplift is reduced by a proper drainage system installed under the dock floor.

Each dry dock is characterized by physical dimensions that define the size of ships that can be built or repaired in it.

These data are related to the following (Figure 3–55):

- Total length of the dock structure L_d
- Effective length of the dock L_u. This length should be up to 3 to 4 m greater than the total length of the ship using the dock.
- Total width of the dock structure S_d
- Effective width of the dock S_u. The effective width is equal to the beam of the largest ship plus an allowance on each side for working space. This allowance normally ranges from 3 to 6 m on each side depending on the dry dock specialization, for example, building or repair.
- The width of the dry dock at the coping level S_t
- The width at dock floor S_b
- The average width S_k which is the average distance between side walls of the dry dock $S_k = 0.5(S_t + S_b)$
- The width at entrance S_w
- The depth at the entrance h_w, that is, the depth at the sill during the design water level. To accommodate a ship of maximum draught T_c the depth of entrance h_w should be equal to the sum of T_c and r, where r is the required underkeel allowance at sill; typically $r = 0.5$ m.
- The depth of the dry dock at the entrance h_{kw}
- The depth of the dry dock at the head wall h_{kt}
- The average depth of the dry dock $h_k = 0.5(h_{kw} + h_{kt})$
- The effective depth of the dry dock h_u, that is the depth of the upper level of the keel blocks measured from the design water level.
- The height of the coping edge h_p, that is the dock freeboard above the maximum design water level; h_p should not be smaller than 0.5 m
- The height of the dock structure H_k
- The height of the dock walls at the entrance H_{cw}
- The height of the dock walls at the head wall H_{cs}
- The height of the dock walls at the sill H_p
- Dock volume V, that is, the volume of the water that must be pumped out of the dry dock for complete dewatering at the design water level.
- The dock capacity factor w_p, that is, the ratio of the displacement of the largest design ship to be served at the dry dock W and the dock volume V. It usually ranges from 0.4 to 0.6; it may be much smaller, however, in wide dry docks.

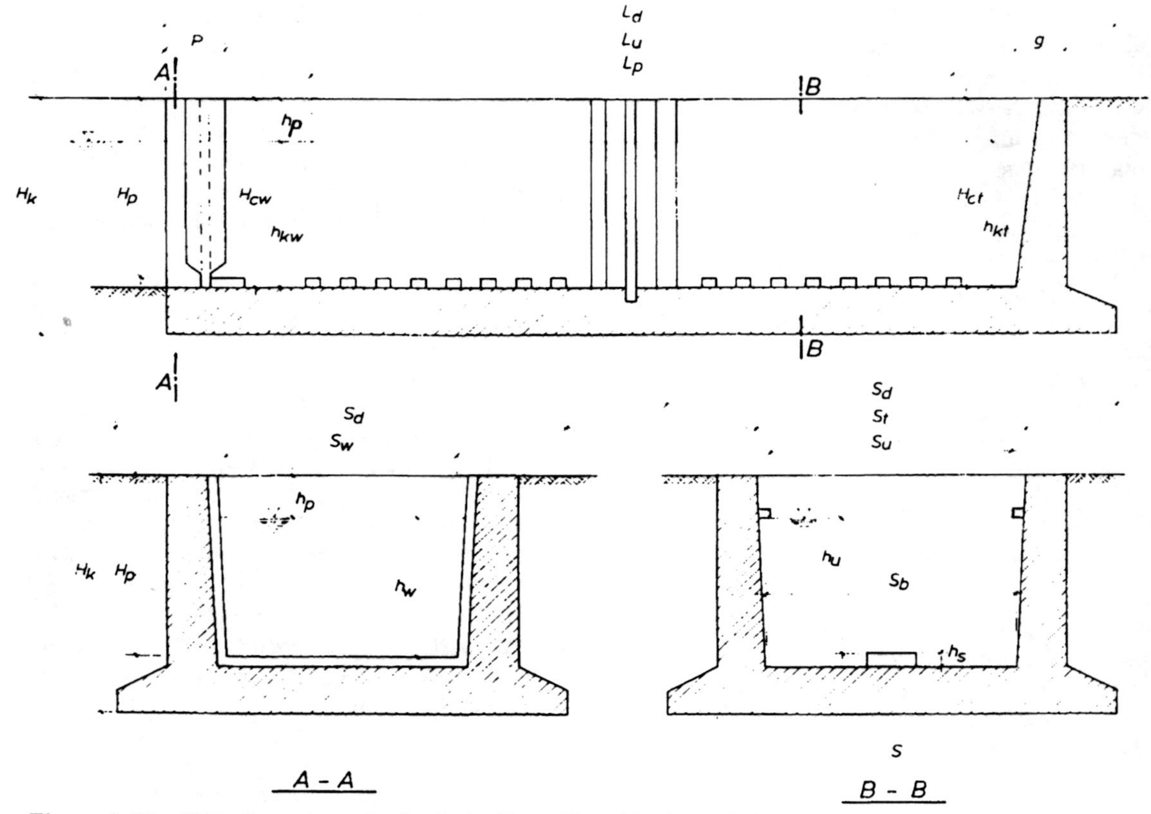

Figure 3–55. Main dimensions of a dry dock. (From Mazurkiewicz, 1980.)

3.5.2 Heavy (Gravity) Dry Docks

Heavy docks are those in which the permanent loads such as dead weight of the structure and weight of soil placed on the structure exceed the maximum hydrostatic uplift.

Structurally these docks can be built either as frames in which the floor slab is rigidly jointed with side walls or as structures in which the floor slab and side walls of miscellaneous constructions are separated. In the latter case the floor slab can interact with the side walls, thus transferring part of the hydrostatic uplift to these walls, or the floor slab can be heavy enough to balance the total uplift load with its own weight.

The most recent state-of-the-art observation on gravity-type docks is given in PIANC (1988).

Each dry dock structure is site specific. The following description of some practical examples of heavy graving dry dock structures built at different shipyards illustrates this statement.

A sheet pile wall assembly and heavy concrete slab comprised the dry dock structure built in Jarrow, England (Figure 3–56). The dock has an effective length $L_u = 183$ m, an effective width $S_u = 26.67$ m, and a depth at entrance $h_w = 7.16$ m. The dock floor—a reinforced concrete slab of a maximum thickness of 3.66 m—was designed to transfer the unbalanced hydrostatic uplift onto the sheet pile walls. A drainage system installed under the slab was connected via a system of pipes

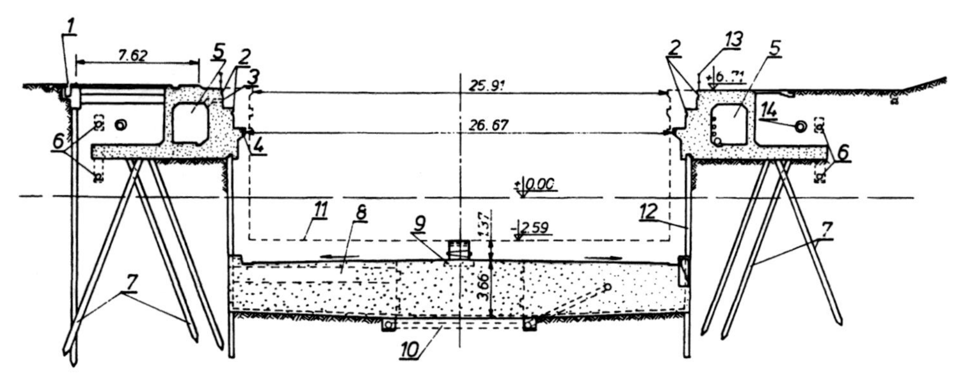

Figure 3–56. Cross-section of the dock in Jarrow, England. (From Hausser et al., 1964.) 1—Trolley channel, 2—inset handrail, 3—cable port, 4—timber fender, 5—service culvert, 6—groundwater drain, 7—reinforced concrete 410 × 410 mm piles, 8—in situ concrete struts, 9—high-grade concrete insert, 10—under floor drains, 11—sill, 12—sheet pile wall (Frodingham No. 5), 13—handrails, 14—surface drain.

directly to the dry dock chamber. When the hydrostatic uplift load exceeds the weight of the floor slab, which may be the case during an exceptionally high water level, the unbalanced load is transferred to the sheet pile walls. Furthermore, during this event water overflows into the dock, thus keeping the pressure on the floor slab within the allowable range.

A heavy dock structure comprised of massive side walls interacting with a massive floor slab was built in Copenhagen, Denmark (Figure 3–57) (Hansen et al., 1956; Knudsen and Larsen, 1956; Birch, 1957, 1958; Larsen et al., 1957). The dock has an effective length $L_u = 218$ m, an effective width $S_u = 30.0$ m, and a depth at entrance $h_w = 8.25$ m.

The 5.3 m thick side walls were made from bottomless prefabricated caissons installed on a gravel bed. The caissons were filled with tremie concrete to a depth of −7 m, consisting of regular concrete laid dry to el −2.15 m and reinforced concrete placed up to the dock coping level (el +2.0 m).

The dock floor slab, 4 m thick at the side walls and 5 m thick along its longitudinal axis, was placed mainly under water. Only the upper portion 0.4 m thick was cast in dry conditions. In total, 35 caissons were used for dock construction. This included 33 typical caissons used for construction of

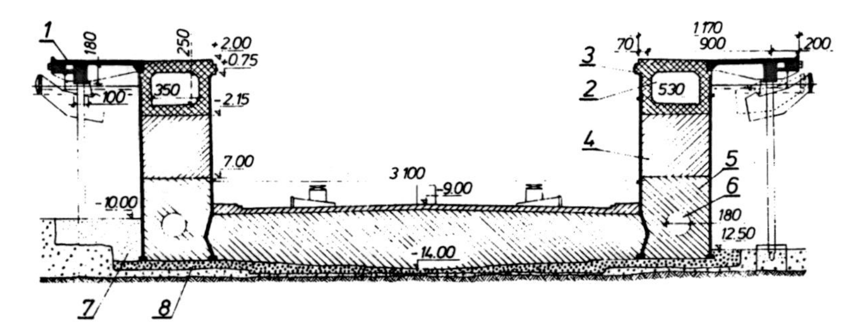

Figure 3–57. Cross-section of the dry dock in Copenhagen, Denmark. (From Knudsen and Larsen, 1956.) 1—Trolley channel, 2—service culvert, 3—reinforced concrete, 4—concrete filling in dry, 5—underwater concrete, 6—oil container, 7—sand fill, 8—gravel bedding.

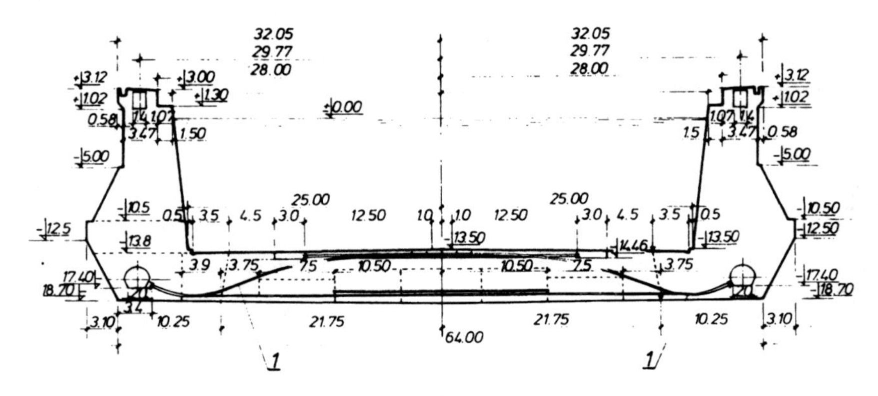

Figure 3–58. Cross-section of Dry Dock No. 8 in Marseilles, France. (From Couprie et al., 1961.) 1—Prestressing cables.

walls and two caissons used for the pumping station. Caissons were made of various heights ranging from 13.65 to 14.55 m. The standard length of all caissons was 14.06 m. The caisson side wall had a recess forming a joint with the dock floor slab. The latter was designed to take any unbalanced hydrostatic uplift loads from the floor slab and transfer them to the side walls.

Side walls, designed as the retaining walls carrying over the reactions from the hydrostatic and soil lateral pressures onto the dock floor slab, were used in Dry Dock No. 8, built in Marseilles, France (Figure 3–58) (Couprie et al., 1961). The dimensions of the dock are: effective length $L_u = 259$ m, width at entrance $S_w = 44.2$ m, and depth at entrance $h_w = 10.7$ m.

A typical cross-section of this dry dock included heavy retaining walls forming the side walls, and a floor slab prestressed with the three groups of cables. All the cables were prestressed by a force of 0.65 MN and placed at a spacing 0.7 m. The floor slab is 5.5 m thick and acts as a beam fixed at side walls.

An example of a ship repair dry dock with a heavy floor slab and very light side walls laterally supported by a special anchor system built at Hebburn, England, is illustrated in Figure 3–59 (Martin and Shirley, 1963). The basic dimensions of this dock are as follows: length $L_u = 259$ m; width at entrance $S_w = 44.2$ m, and depth at entrance $h_w = 10.7$ m. The 7.32 m thick floor slab was reduced in side wall area to about 1.7 m;

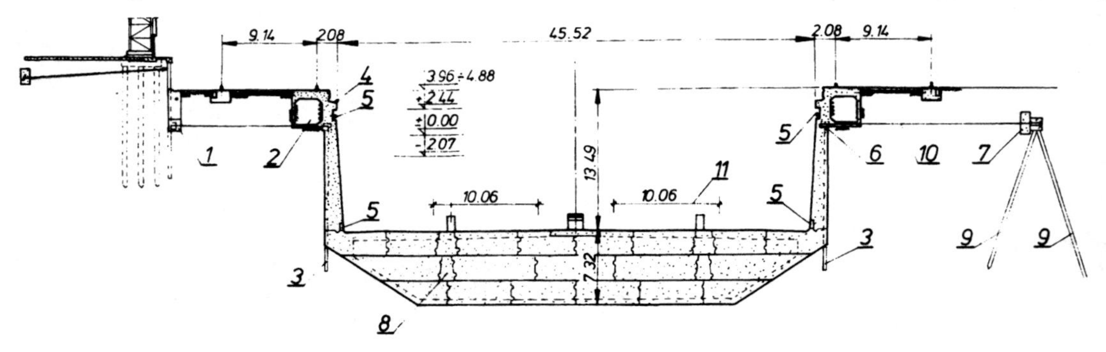

Figure 3–59. Cross-section of the dry dock in Hebburn, England. (From Martin and Shirley, 1964.) 1—Reinforced concrete anchorage attached to sheet piling, 2—service culvert, 3—wide flanged beams 410 × 300 mm and 300 × 300 mm, 4—leading-in trolley rails, 5—dock lights, 6—reinforced concrete waling beam, 7—reinforced concrete anchor beam, 8—contraction gaps, 9—raking reinforced concrete piles, 10—75 mm diameter tie rods, 11—travel of bilge blocks.

however, the weight of the slab of reduced thickness plus the weight of the walls were sufficient to safely resist the design hydrostatic uplift load. The reinforced concrete side walls, with a maximum thickness of 1.68 m, were anchored about 2.05 m below the cope. Anchorage consisted of 76 mm diameter tie rods secured at 1.52-m centers to a reinforced concrete piled structure.

The dry dock at Scaramanga, Greece, is also one with a heavy floor slab (Figure 3–60) (Martin and Irvine, 1972). It was built to the following dimensions: effective length $L_u = 335.3$ m, effective width $S_u = 53.65$ m, and depth at entrance $h_w = 9.25$ m. The adopted design comprised a very thick concrete floor slab with walls constructed of the floated-in concrete caissons installed on a thoroughly levelled foundation.

The floor slab about 8 m thick was made using grout-injected concrete with layers of single-size aggregate consisting of stones 80 to 110 mm in diameter.

The floating-in concrete caisson technique was also used for construction of the cellular type dry dock No. 5 in Genoa, Italy. This facility is used for repairing ships of up to 70 000 DWT (Figure 3–61). The dock was constructed to the following dimensions: effective length $L_u = 250$ m, effective width $S_u = 38$ m, and depth at entrance $h_w = 20.2$ m. The dock structure comprised a prestressed precast concrete caisson sunk on thoroughly levelled stone bedding in the harbor basin. The overall dimensions of this dock are as follows: total length $L_d = 260.49$ m; total width at level of base $S_{d1} = 52$ m; total width at level of cope $S_{d2} = 56.34$ m, and height of structure $H_k = 21.5$ m. The floor slab was made of 15 cells 52 m long and 17.4 m wide. The dock comprised both prefabricated and built in situ components. After installation the caisson was filled with sand.

It should be noted that in general (with few exceptions) the frame type dry docks built from monolithic concrete are not divided by expansion joints and are formed as one unit. Several characteristic examples of frame type dry docks built without expansion joints are given in Szopowski (1963), Hauptmann (1968), Ebert (1950), and Mazurkiewicz (1980).

3.5.3 Anchored Dry Docks

Heavy graving dry docks with a heavy bottom slab to balance the hydrostatic pressure are usually preferred where wide dry docks are required. However, construction of these structures presents many problems, particularly where deep docks are required.

These difficulties arise mainly during the lowering of the groundwater level so the slab can be cast in dry conditions. Furthermore, with an increase in depth, resulting in an increase in hydrostatic uplift, the thickness of the floor slab also increases considerably as does the amount of concrete work and excavation. Owing to the aforementioned

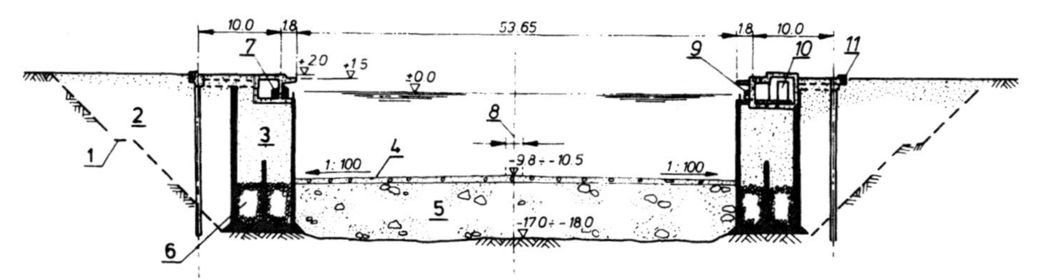

Figure 3–60. Cross-section of the dry dock in Scaramanga, Greece. (From Martin and Irvine, 1972.) 1—Dredged slope, 2—filling, 3—earth fill, 4—reinforced concrete topping placed in the dry, 5—grout intrusion concrete placed under water, 6—stone filling, 7—culvert for electrical services, 8—level keel block strip, 9—piped services, 10—substation, 11—crane beam and trolley channel.

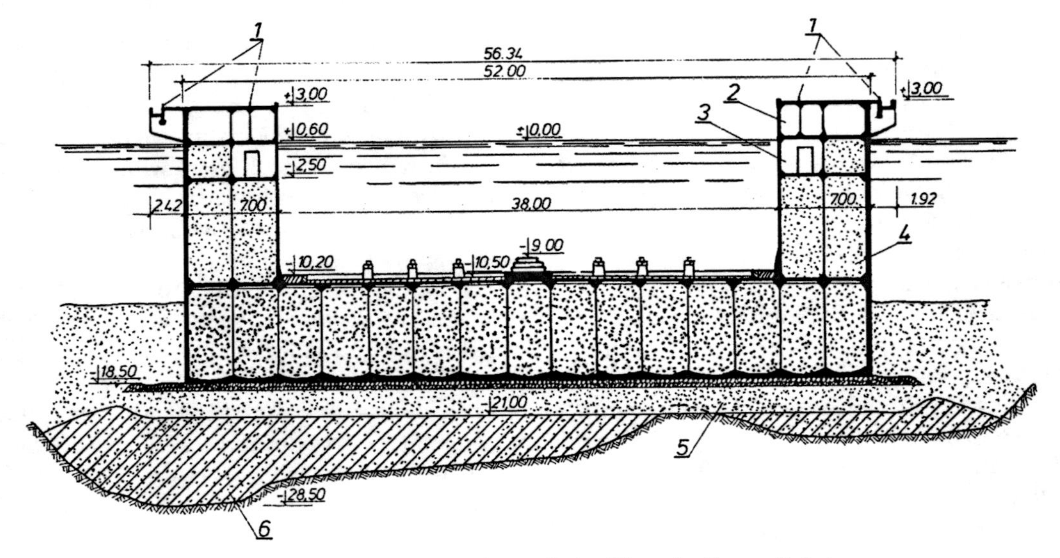

Figure 3–61. Cross-section of the dry dock in Genoa, Italy. (Fincosit, Genoa, Italy.)

problems, construction of the gravity type docks, especially for great depths, becomes too laborious and uneconomical.

For these reasons there is a tendency in modern dock construction practice toward replacing the heavy dock with the anchored and drainage type structure. However, improvements of the foundation soils are sometimes required. Furthermore, in anchored dock strucures a relatively thin floor slab, anchored in to the foundation soil or bedrock, is used. Depending on local foundation conditions the floor can be anchored by ground or rock anchors, or by piles.

Naturally, the anchored docks are much lighter than heavy graving docks; they require less construction materials, less excavation, and thus are much quicker to build, particularly where deep docks are required. Therefore, anchor docks are more economical to construct than gravity docks.

In anchored docks there is always contact between the foundation soil and the floor slab; the contact pressure exists even when the dock is empty, dewatered, and exposed to full hydrostatic uplift. With the floor slab anchored in the ground using post-tensioned steel cables or anchor bolts, a contact pressure between the floor slab and the foundation soil occurs due to both prestressing and weight of the slab. The minimum prestressing force is usually selected to cause the bearing stress on the foundation soil to equal the maximum design hydrostatic uplift load.

A floor slab anchored in the subsoil by means of post-tensioned cables and reinforced concrete blocks of cruciform section was used for construction of the dry dock in Emden Germany (Figure 3–62) (Agatz and Lackner, 1954; Lackner, 1962). Its basic dimensions are as follows: effective length $L_u = 218\,\text{m}$, width $S_k = 32\,\text{m}$, and depth at entrance $h_w = 8.2\,\text{m}$. The concrete floor slab, 2 m thick at the dry dock axis and about 2.5 m thick at the side walls, was anchored by 496 anchoring blocks arranged in cross-section at irregular spacings and in longitudinal section at approximately 3 and 6 m spacings. The side walls are 2 m thick.

The floor slab was anchored by post-tensioned steel cables 46 mm in diameter covered with seven coats of anticorrosive paint; the cables were prestressed with a force of 1.05 MN, while the design maximum load was 0.95 MN. The anchoring blocks were jetted into the soil to various depths from 11 to 14 m.

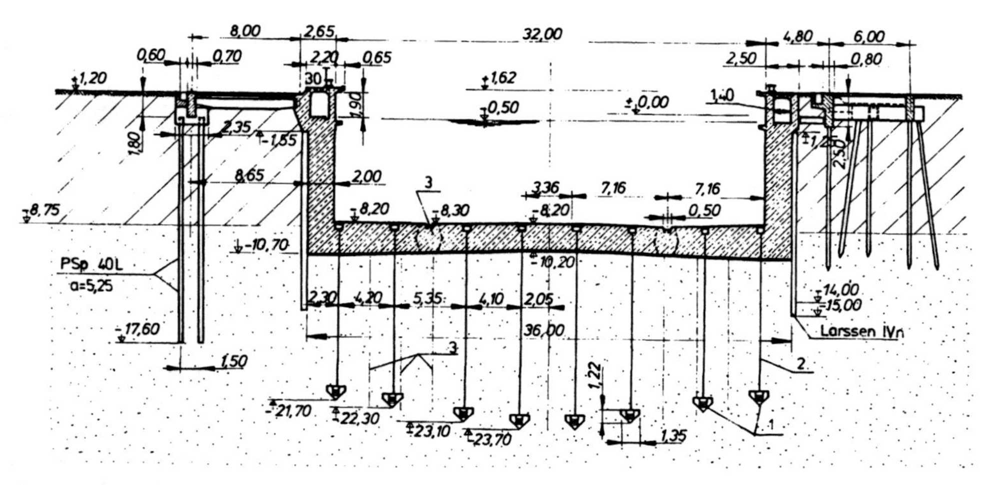

Figure 3–62. Cross-section of the dry dock in Emden, Germany. (From Lackner, 1962.)

The floor slab and anchor system, comprised of prefabricated anchoring blocks and prestressed cables, was also used for the dry dock in Alexandria, Egypt (Lackner, 1962).

Where steel post-tensioned cables or rockbolts are used for anchoring the dock floor they have to be reliably protected from corrosion. Martin and Arnold (1992) described spectacular failure of dry dock floor in Scaramanga, Greece resulting from stress corrosion damage to anchor pile tendons.

A floor slab anchored with Franki piles 0.5 m in diameter was used in dry dock No. 8 constructed in Kiel, Germany (Schenck, 1964; Schenck and Kolloge, 1958). The dock has the following dimensions: effective length $L_u = 285\,\text{m}$, effective width $S_u = 44\,\text{m}$, and depth at entrance $h_{kw} = 7\,\text{m}$. The dock was constructed with side walls designed as retaining walls having wide cantilevers which carried the weight of the overlying soil fill (Figure 3–63). There, the piled foundation was used to ensure uniform settlement of the dry dock structure and for proper transfer of the loads from the keel blocks to ground; piles wer also considered to resist hydrostatic uplift. Thickness of the floor slab varies from 1.1 to 1.5 m; thickness of the side

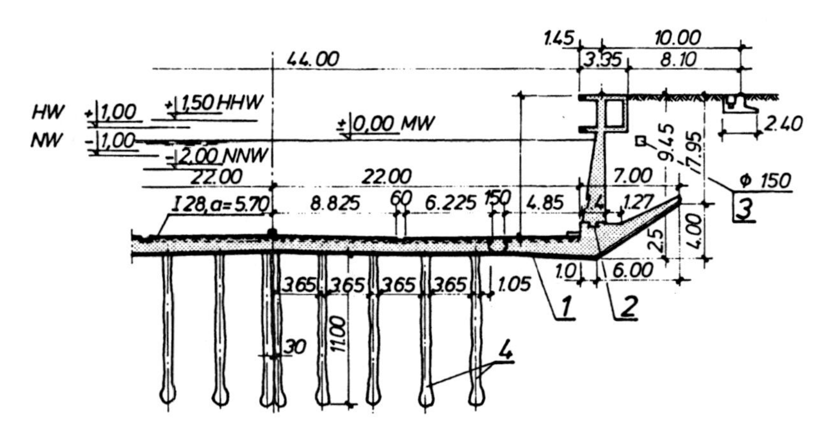

Figure 3–63. Cross-section of dry dock No. 8 in Kiel, Germany. (From Schenck, 1964.)

walls at the base is 1.5 m, and at the bottom of the gallery 0.5 m. The piles were designed to resist tensile load equal to 0.9 MN and compression load of 1.35 MN.

The dry dock with a floor slab placed on steel piles was built at Nobiskrug Shipyard, Germany (Figure 3–64). The dimensions of the dock are: effective length $L_u = 194.3$ m, effective width $S_u = 32.9$ m, and depth of the dry dock $h_k = 7.5$ m.

The dock walls were made from steel sheet piles anchored by steel raking piles. The steel sheet pile wall and the raking piles were connected by a reinforced concrete cap of 1.5 × 1.2 m. The floor slab is 1.2 m thick. To resist load from the keel blocks the foundation steel piles were driven transversely at 2.8 m and longitudinally at 3 m. At dock axis the piles were driven with a spacing of 1.5 m. The piles were connected to the slab by means of steel flat bars, 20 × 80 mm in cross-section, welded to the head of the piles and set into the concrete.

3.5.4 Drainage Dry Docks

There are two basic types of drainage dry docks which differ from each other by the drainage system used. The first type comprises docks in which the drainage system is installed to reduce the water pressure acting on both side walls and the floor slab. To the second type belong docks in which only hydrostatic uplift acting on the floor slab is relieved via drainage system.

In the former type of dry docks the drainage system is installed externally to lower the groundwater level in order to reduce or eliminate completely the hydrostatic pressure on walls and floor slab. The drainage system is linked with a pumping station through a set of collecting pipes or culverts.

In the latter type of dry docks where the drainage system is installed to reduce the hydrostatic uplift load acting on the floor slab only, the flow of groundwater to the drainage system is controlled by a cutoff wall of appropriate depth installed along the dock perimeter. This wall is usually designed to reach an impervious layer of underlying soil, if such exists, in the foundation subsoil, or to be deep enough to substantially reduce hydrostatic pressure gradient.

In both categories of drainage dry docks the thorough preparation of the foundation soil is of paramount importance.

Normally the following foundation conditions should be satisfied.

1. Deformation of the drainage layer placed under the floor slab caused by external loads; for example, dry ships, or weight of water during flooding, should be as small as practical. This layer should also distribute the loads acting on the floor slab onto a larger area of underlying foundation soil.

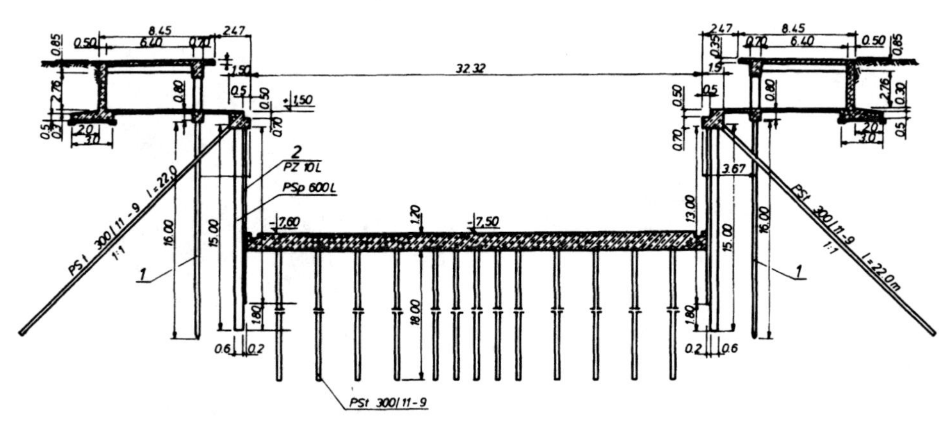

Figure 3–64. Cross-section of the dry dock in Rendsburg, Germany. (From Mazurkiewicz, 1980.)

2. The foundation soil under the layer of filter material, where required, should be well compacted to reduce the stresses acting on deeper and weaker soil stratum.
3. During all phases of the dock floor loading the pore pressures in the foundation soil and in the filter material should not change significantly.

The first and the last conditions are usually fulfilled through proper proportioning of grain sizes of filter material; compaction of the filter layer; and the placing of drainage pipes, collecting culverts, and suction chamber of the pumping station, as required. The second condition usually involves careful preparation and compaction of the subsoil under the filter layer. The latter is particularly important where a layer of a weak soil exists in the foundation subsoil; the weak soil may deform considerably under effects of the heavy loads. Proper compaction of the subsoil has a beneficial effect on the interaction between the floor slab, filter layer, and foundation soil, and normally results in a reduction in slab thickness. The latter, of course, is the result of lesser stresses applied on weak soils. For preliminary design purposes the modulus of deformation of the compacted upper layer of the foundation soil may be assumed as equal to 150 to 200 MPa.

Naturally, the fundamental concern in the drainage dock design is the choice of the proper drainage system. The drainage system installed in the docks of the first type typically consists of a side drainage system designed to relieve the water pressure acting on the dock walls and a lower drainage system designed to relieve the hydrostatic uplift load acting on the floor slab. Both these systems may work independently or can be linked together, for example, the side drainage system may be connected to drainage pipes of the floor system or to the common drainage collecting culverts.

In drainage dry docks of the second type the drainage system typically comprises a filter layer and drainage pipes installed under the floor slab. The latter are connected to the collecting pipes which convey the water to a drainage pumping station.

The amount of water to be drained from docks of the first category may range from 1000 to 5000 m^3/h, and in the case of docks of the second type it may reach up to 500 m^3/h. The dock building practice indicated that the optimum design of drainage dry dock is usually obtained when discharge of the drainage water does not exceed about 150 m^3/h.

The type of drainage system affects the design of the dock structure and its behavior during operation. The following basic factors affect the performance of conventional drainage dry dock:

Thickness and layout of the filter layer
Sizes and layout of drainage pipes
Continuity of the drainage operation.

Water from the drainage system can be discharged during all periods of dock operation or only during dewatering. The former is usually required in dry docks where inner gates are used and cut-off walls separating the individual parts of the dry dock are not provided. A continuous water discharge has the advantage that the filter layer remains in the same working condition all the time and therefore operation of the drainage system has little impact on the bearing capacity of the foundation subsoil.

Design of the dock drainage system normally includes design of the system's layout and its elements such as drainage pipes, inspection wells, and safety valves and sizing of all structural components. To ensure system reliability and longevity it is usually considered that the diameter of the main collecting pipes should be such as to enable discharge of about 10 times more groundwater than the calculated volume.

The floor slab of the drainage dock has a relatively small thickness and therefore low weight. Consequently, it cannot resist the full hydrostatic uplift load; under full uplift load, the dock floor may be lifted up and potentially seriously damaged. To prevent such an occurrence safety valves must be installed in floor slab.

For a drainage dock built on an impermeable foundation soil the floor slab is laid on a filter layer and the penetrating water is drained directly to the main drainage culverts.

As stated earlier, in drainage docks of the second category built on permeable foundation soils a sealing membrane (cut-off wall) around dock perimeter is usually installed. The purpose of this membrane is not only to reduce the amount of inflow of groundwater into the drainage system, but also to protect this system from direct contact with the highly permeable granular fill material placed behind the dock walls. It also can help to control the water level behind the dock walls.

Provided that the subsoil backfill and filter layer permeability parameters are known the amount of water flow penetrating the drainage system can be determined fairly accurately.

Operation of existing drainage dry docks has indicated that the amount of discharged water usually decreases with time. The latter may suggest that the design volume of water to be discharged through the dock drainage system may be considered as conservative.

The following are examples of drainage docks constructed on permeable soils at Puget Sound, Bremerton, USA: (Ross, 1960; Zola and Boothe, 1960; Tate, 1961; Colbert, 1963) and Sparrows Point, Baltimore, USA (Millard and Hassani, 1971; Donnelly, 1971; Hassani, 1973).

The dry dock at Puget Sound Naval Shipyard, Bremerton, (Figure 3–65) has the following dimensions: effective length $L_u = 359.7\,\mathrm{m}$, effective width $S_u = 54.9\,\mathrm{m}$, and depth $h_k = 16.15\,\mathrm{m}$. The reinforced concrete floor slab is 2.13 m thick, increasing near the walls to 3.65 m. Reinforced concrete side walls have a variable thickness from 3.65 m at the floor slab level to 0.84 m under the service culvert.

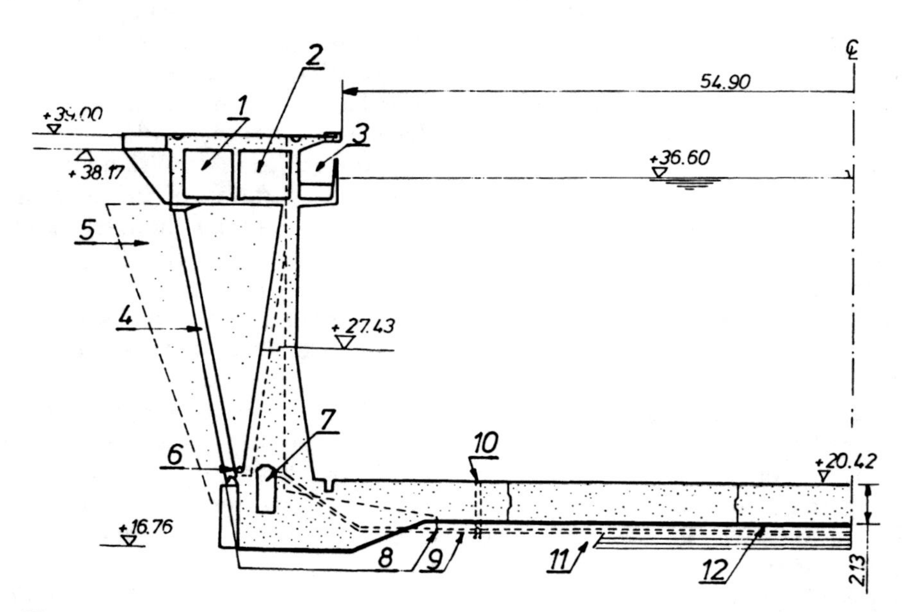

Figure 3–65. Cross-section of the dry dock in Puget Sound, USA. (From Zola and Boothe, 1960.) 1—Electrical culvert, 2—mechanical culvert, 3—services gallery, 4—concrete column, 5—drainage course, 6—300 mm diameter perforated drainage pipe, 7—drainage culvert, 8—piezometer pipes, 9—200 mm drainage pipe, 10—check point, 11—drainage course, 12—100 mm concrete mat.

There the hydrostatic pressure acting on the dock structure is reduced by a continuous discharge of water from the filter layer under the floor slab and behind the side walls. A network of perforated pipes draining the water to the collecting culverts is installed within the filter layer; from culverts the water flows to the suction chamber of the drainage pumping station.

The dry dock at Sparrow Point (Figure 3–66) has the following dimensions: effective length $L_u = 365.8\,\text{m}$, effective width $S_u = 60.4\,\text{m}$, and depth $h_k = 8.7\,\text{m}$. The drainage system of this dock consists of a layer of sand 0.3 m thick and a layer of coarse aggregates 1 m thick. The network of 200 mm diameter perforated pipes is laid in the layers. The drainage pipes are connected to vertical pipes closed with watertight covers. These pipes serve as inspection wells and may be also used for cleaning the drainage pipes in case they become plugged with fine soil particles. These wells also can serve as a pressure relieving system in case of a sudden rise in water pressures in the drainage system.

The next category of drainage docks are those built on permeable soils and sealed by an impermeable layer or by grouting in which case the amount of water discharged from the drainage system is considerably reduced.

A dock of this type was built at Kockums Shipyard in Malmö, Sweden. It has the following dimensions: effective length $L_u = 405\,\text{m}$, width $S_u = 75\,\text{m}$, and depth $h_k = 9\,\text{m}$ (Figure 3–67) (Lidström and Lindskog, 1967; Lidström and Bäckström, 1968). The dock walls comprise a sheet-pile bulkhead and piles capped with a box-like concrete superstructure. The wall structure includes the service culvert and carries one track of the gantry crane. The dock floor is constructed of a reinforced concrete slab 0.8 m thick laid on a 0.5 m layer of macadam. The water is drained through the layer of macadam by three drainage culverts 1 m in diameter each placed parallel to the dock longitudinal axis. The drained water is discharging into the suction chamber of the pumping station.

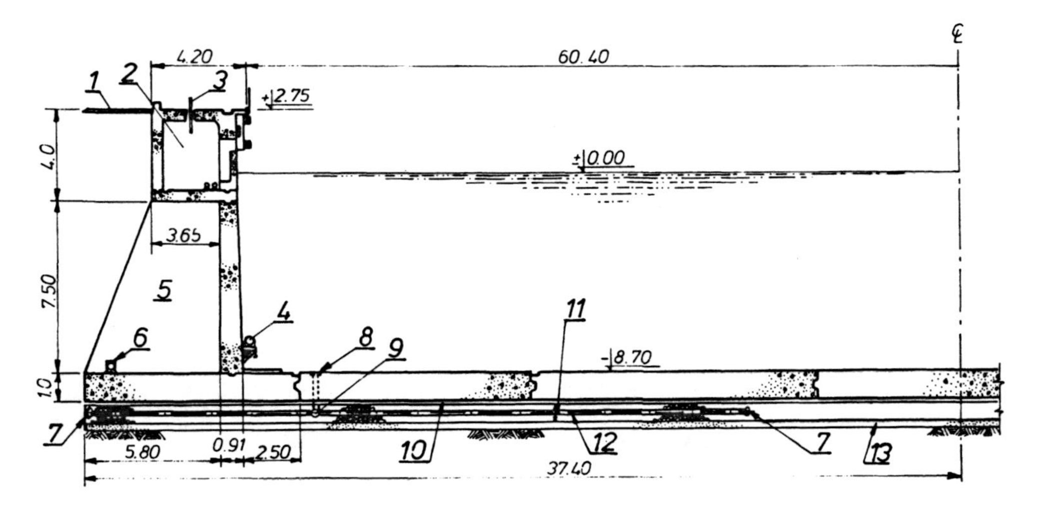

Figure 3–66. Cross-section of the dry dock in Baltimore, USA. (From Millard and Hassani, 1971.) 1—Asphalt paving, 2—service gallery, 3—crane rail conductor collector and support arm, 4—400 mm diameter saltwater main, 5—counterfort, 6—300 mm diameter drain pipe, 7—concrete seal, 8—200 mm diameter one-way valve with hinged cover-plate, 9—300 mm diameter underdrain, 10—vapor barrier (polyethylene sheet), 11—0.9 m drainage filter blanket, 12—200 mm diameter porous under drain pipe at 9 m ctrs, 13—0.3 m sandfill.

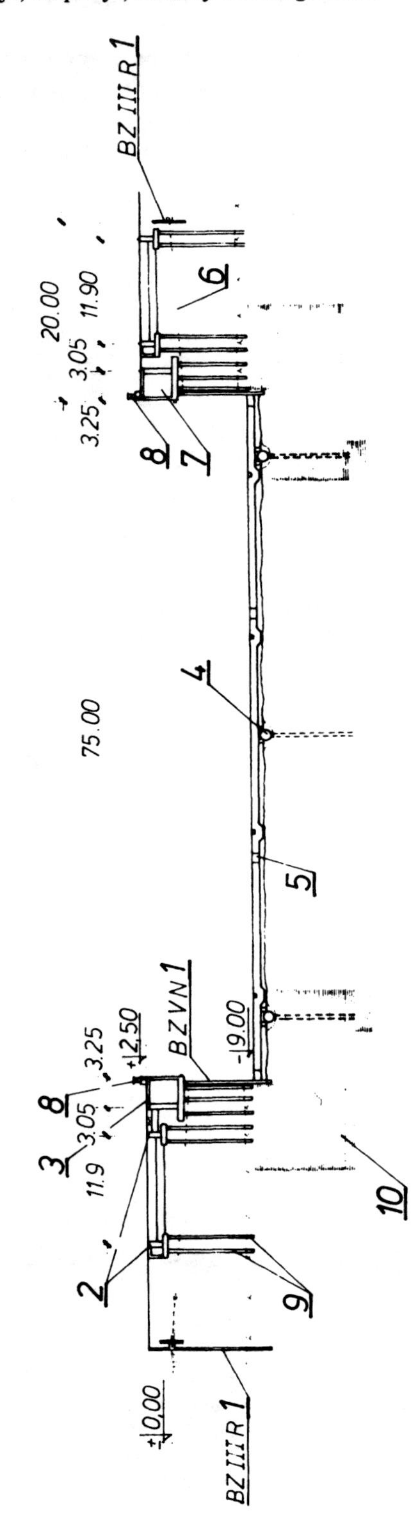

Figure 3-67. Cross-section of the dry dock in Malmö, Sweden. (Skanska, Sweden.) 1—Sheet pile wall, 2—jib crane tracks, 3—gantry crane track, 4—concrete drainage pipe, 5—drainage layer, 6—tie rod, 7—service culvert, 8—750 kN bollard, 9—prefabricated concrete pile 340 mm in diameter, 10—areas of grout injection.

To reduce the sheet-pile bulkhead permeability the sheet-pile walls were driven into a marl layer that exists below the floor elevation. To reduce further the amount of water flowing to the dock, from below the sheet-pile walls, the bottom portion of sheet piles was grouted.

A dock of a very similar construction was built in Gdynia, Poland (Mazurkiewicz, 1976, 1980; Mazurkiewicz and Najder, 1977).

The principle of cutting off the inflow of water into the drainage system with a sheet-pile wall is quite common; it was used in Japan on a dry dock constructed in Tsu, and in the USA in a dry dock at Lorain (Mazurkiewicz, 1980) and at many other locations worldwide.

The next category comprises dry docks built directly on impermeable foundation soil, for example, firm clays or various kind of rocks. Solutions for drainage system design in docks of this category may differ considerably from each other. An example of drainage docks built on impermeable foundation soil is discussed by Mazurkiewicz (1980) and Knudsen (1961).

The last group of drainage docks concerns the docks cut completely or partly in bedrock. Examples of such structures are the dry docks in Gothenburg (Figure 3–68) and Uddevalla, Sweden, and Birkenhead, England (Mazurkiewicz, 1980).

3.5.5 Mechanical Equipment

Basic mechanical equipment required for the dry dock operation includes miscellaneous pumps housed in the pumping station; gates for closing and opening of the dry dock chamber, and cranes used for ship construction, repair, and maintenance.

The pumping equipment used for dock flooding and dewatering is generally installed in a separate building or chamber known as the pumping station.

The amount and type of this equipment depends on the flooding and dewatering system used in the dry dock. The flooding system usually consists of an inlet culvert, flooding valve, and flooding channel (Figure 3–69) whereas the dewatering system comprises an inlet culvert, suction chamber, suction pipes, pumps, discharge pipes, and outlet culvert (Figure 3–70).

The pumping station also has equipment to discharge rain water from the dock floor and surrounding area, to provide ballast water and fire-fighting water, and to discharge drainage water in case of a drainage dock. The dry dock has a system of culverts, and pipelines, which are used to supply and discharge water for different purposes. Naturally, the network of culverts and pipelines in a dry dock requires several pumps, which are usually installed in the pumping station and are operated from a central control room.

The pumping stations are generally located near the dock entrance because the dock floor normally has an inclination in the direction of the dock entrance. This considerably reduces the length of the inflow and outflow culverts required.

The best location for the pumping station is at the dock wall itself; in this case no special structure or building for the pumping station is needed. Furthermore, such location of a pumping station decreases construction costs considerably and reduces the length of the suction and discharge culverts and also of the flooding channels. For this purpose in some cases two pumping stations, located on either side of the entrance, are provided in some cases.

In some projects the flooding equipment was installed in one wall and the dewatering equipment in the other. Subdivision of the two systems seems to be particularly advantageous in the case of twin docks with a common pumping station. More recently it has become common for the flooding valves to be set in the dock gate structure. The latter has many advantages, the main one

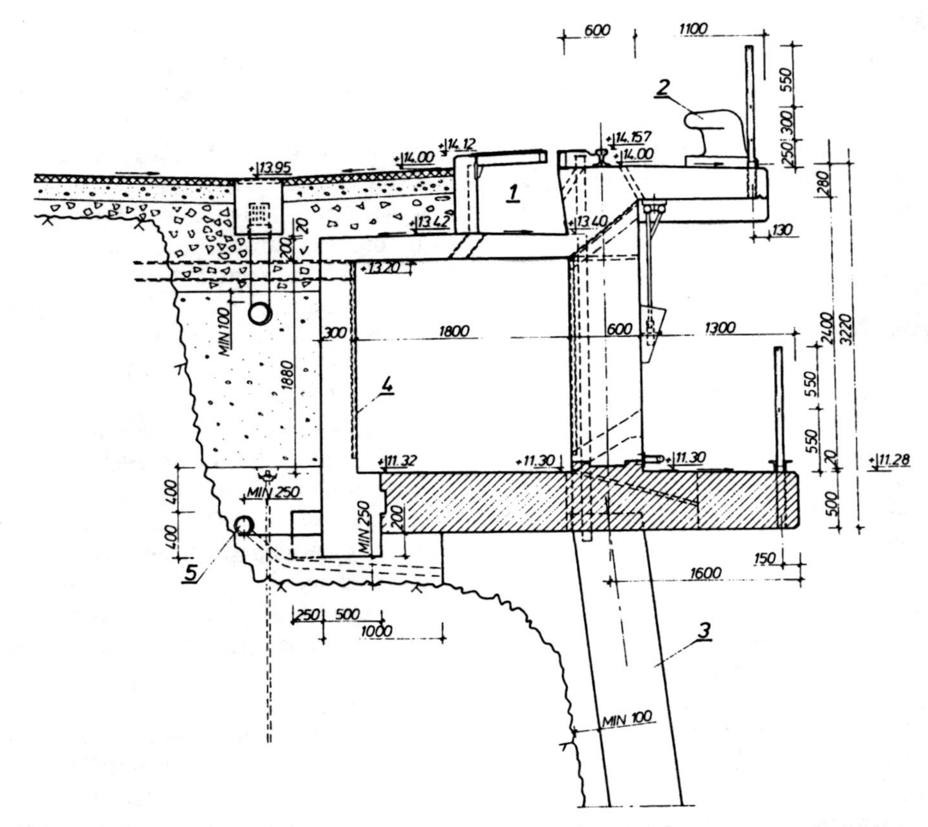

Figure 3–68. Cross-section of the side wall of the dry dock in Gothenburg, Sweden. (Skansa, Sweden.) 1–Trolley culvert, 2—700 kN bollard, 3–reinforced concrete pile 600 × 600 mm at 4.83 center to center, 4—Vemo profile, 5—concrete pipe 150 mm in diameter.

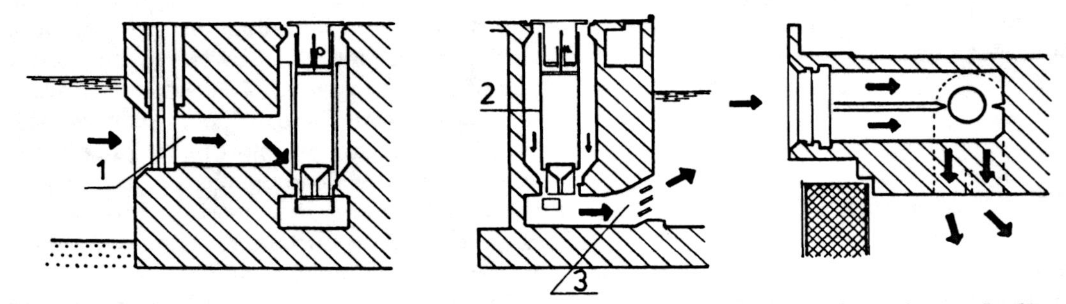

Figure 3–69. Flooding system. (From Mazurkiewicz, 1980.) 1—Inlet culvert, 2—flooding valve, 3—flooding channel.

of which is that the flooding valves can be inspected during gate repair or inspection.

A typical dry dock is equipped with the following pumps:

- Main pumps for dewatering the dry dock
- Auxiliary or seepage pumps for discharge of rain water, ballast water, water from hull watertightness tests, and water seeping into the dock from leaks around the gate
- Sludge pumps for removing any water that seeps into the pumping station, mainly from

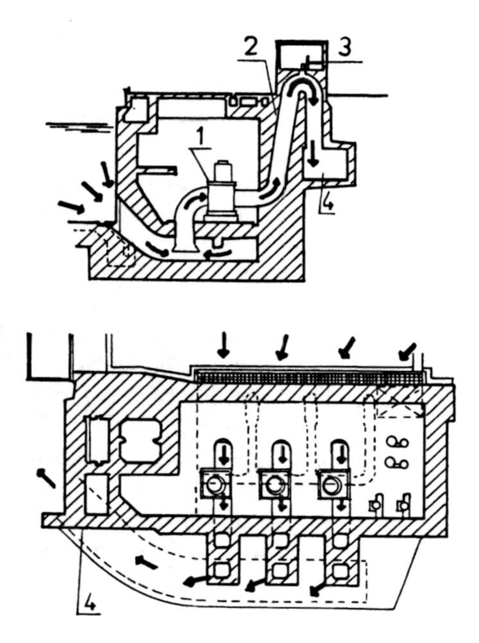

Figure 3–70. Dewatering system. (From Mazurkiewicz, 1980.) 1—pump, 2–discharge pipe, 3—breaking valve, 4—outlet culvert.

pumps or valve glands, and also during the dismantling of pumps for maintenance purposes

- Ballast pumps to supply water to ballast the ship tanks and watertightness tests
- Fire pumps
- Drainage pumps for discharging water from the pressure relief system in the drainage dry dock
- Cooling pumps used to pump cooling water for auxiliary machinery which is kept running while the ship is in the dry dock
- Sewage pumps for discharging sewage from the ship
- Any special or miscellaneous pumps installed in the dry dock pumping station used for special purposes, sometimes unrelated directly to dock operation

In the flooding system three types of flooding valves are used, namely: sluices or sluice valves, installed in the flooding culvert; cylindrical valves placed in separate shafts filled with water and located between the intake and flooding culvert; and separate sluice valves installed in the dock gate.

Pumping water in and out of the dry dock chamber must not be rapid or uncontrolled; the latter assumes that the flooding of the dry dock must not cause either big waves in the dock chamber or produce a sudden impact of water on the ship supports, or on the undocked ship itself. The dewatering of the dock should allow a controlled settling of the ship on the supports.

Construction of the pumping station depends on the dry dock structure, actual location and elevation of certain pumps, as well as type of pumps, for example, horizontal spindle pumps or vertical spindle pumps.

In pumping stations with horizontal spindle pumps the latter are located close to the dock floor level because of their limited suction height. These pumping stations usually require a sizeable floor area.

In the most recently built dry docks pumping stations with vertical spindle pumps were used. Here the driving motor, which sits above the pump, has a common vertical spindle with the pump that results in a significant reduction in the required floor area for the pumping station.

Each pumping station includes a large number of motors of various types. These motors are usually operated from the control center typically located in the pumping station; from there the operator can conveniently control operation of motors and other equipment.

Pumping stations should normally be located within the dock operating area where the shipyard cranes can be used for servicing the pumping station equipment.

The pumping station is one of the most important components in the entire dry dock system. It usually requires the use of special construction methods. This is basically because the pumping station is usually the lowest structure in the entire dry dock system, and thus is exposed to very high concentrated loads resulting from the large weight of the pumps, pipelines, and other equipment installed around pumps.

The following basic principles should be carefully followed for design and construction of the dry dock pumping station.

1. Structural components of a pumping station, such as walls, roofs, and covers, should be designed in a way to prevent cracks, which could cause leakage and corrosion of concrete and/or steel; in general the pumping station should be constructed as a watertight structure.
2. The location of the pumping station and its structure should permit replacement of pumps, motors, and other equipment without removal of structural components such as walls, roofs, or ceilings. For this purpose several openings with sufficiently strong and tight covers should be provided; hoisting equipment, such as overhead cranes or chain lifts, should be installed to handle the aforementioned equipment within a pumping station.
3. All floors of the pumping station should be slightly inclined, to ensure the discharge of water leaking into the station from pumps, valve glands, or from outside. It is essential that several small channels are inset in the floor to discharge the water to the suction chamber of the dewatering pump(s) or sludge pump(s).
4. For safety reasons each pumping station should have at least one entrance directly from the ground level.

3.5.6 Gates

Many types of dock gates are presently in use (PIANC, 1988). They are classified as floating, sliding, mitre, hinged, and flap gates. The purpose of dock gates is to provide for watertightness during all practical working stages with the shortest possible opening and closing time.

Dewatering of the dry dock takes place after the gate is set in place and sealed against the dock head. The dock gate is usually designed to permit closing of the dock chamber at the highest design water level and its opening at least at the mean design water level. The gate structure and its supports must safely withstand the largest water pressure from the water side or, in the case of inner gates, from the flooded part of the dry dock.

Gates must be mechanically reliable and easily accessible for servicing and maintenance with minimum operation and maintenance costs. The gate position during opening and closing must be closely monitored.

All type of gates are required to be sealed against the dock head. Different kinds of wood and hard rubber are usually used for gate seal. Installation of a hardwood seal is time consuming; sealing timbers need to be accurately shaped especially for the quoins to fill in any unevenness of the sealing surfaces. Naturally, the sealing timbers must be strong enough to withstand the full hydrostatic pressure on the gate. The strength of the sealing surfaces of the dock head must also be adequate to resist all design loads.

All types of gates must be protected against corrosion. Steel gates are normally protected by galvanization and/or application of special painting; repeatedly applied thin coats of paint during the dry season will produce an elastic protective film. Concrete gates are usually protected against corrosion by using very dense concrete and epoxy-coated reinforcing bars.

The dock gates must have sufficient buoyancy to allow them to be placed in their seatings, and be removed for maintenance or repair, as well as for moving from the dock entrance to mooring place when the dock is opened. The gate buoyancy is typically achieved through built-in buoyancy tanks.

Gate dimensions are controlled by the gate's effective span and design head of water. In principle the gate is designed as a structure loaded by hydrostatic pressure and supported at three edges and free at one. Where applicable gates are evaluated for effects of wave load and for loads resulting from earthquakes.

Floating gates usually do not have permanent connections with the dock structure; they are normally towed away from the dock entrance after each opening. Sometimes, however, gates may be temporarily joined to the dock structure in order to let heavy rail or road traffic pass over the gate crown, or when bilateral water pressure is exerted upon the gate.

The disadvantages of floating gates are the relatively long time required for dock opening and closing, and the need to use a separate mooring place when the dock is open. In spite of the aforementioned disadvantages, floating gates are still widely used in contemporary dock construction practices (Hassani, 1987; PIANC, 1988).

An example of a steel floating gate made for the dry dock in Belfast, Northern Ireland, is shown in Figure 3–71. It spans a width of 92.96 m and is 10.62 m wide, carrying the loads from both sides. Owing to the possibility of bilateral setting it allows for maintenance without the need to lift the gate out of the seat.

Sliding gates are either of the floating steel or concrete pontoons type, or caissons that sit on a sliding track installed in a special chamber when the dry dock is open. The advantage of the pontoon type sliding gate is that it permits easy docking for carrying out ship maintenance. Sliding gates are used as main or inner gates, and are built in a similar manner as conventional gates. Sliding gates, as compared to floating ones, are more convenient for maneuvering and allow

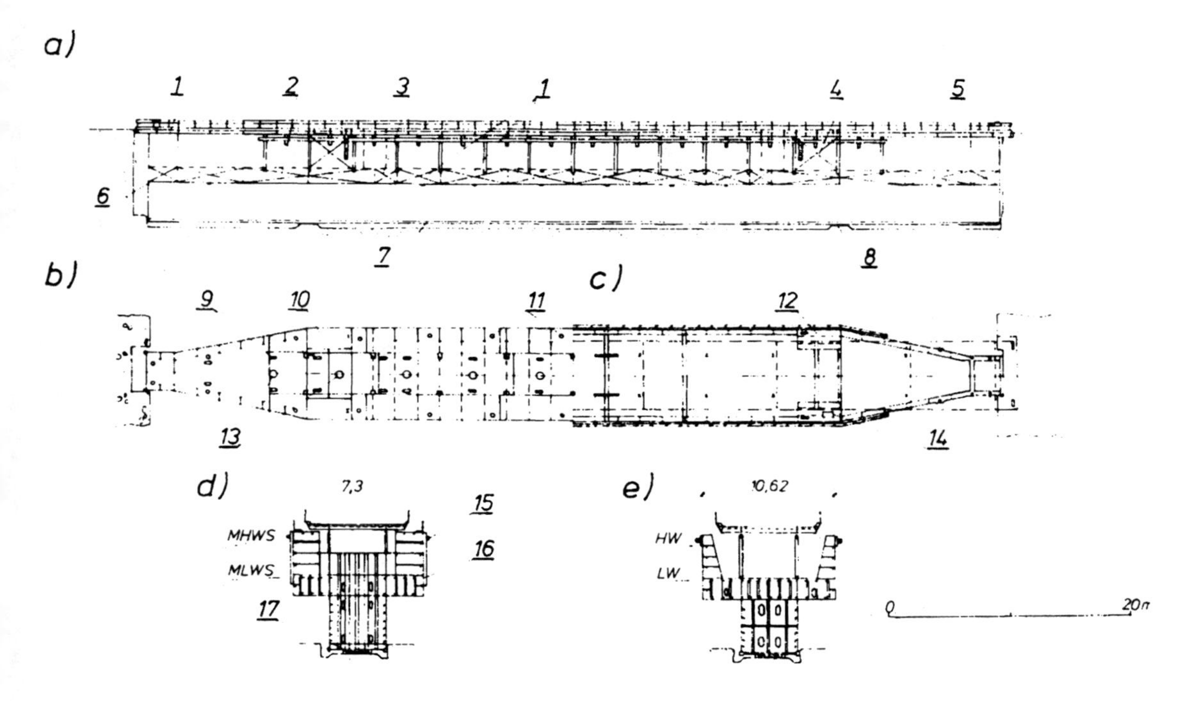

Figure 3–71. General arrangements of floating gate in dry dock at Belfast, Northern Ireland. (From Geddes et al., 1972.) (a) Elevation. (b) Half plan at upper deck of box beam. (c) Half plan at underside of roadway deck. (d) Cross-section at bearing. (e) Typical cross-section. 1— Air cambers, 2—scuttling tank, ballast, 4—pump chamber, 5—bridging, 6—steel meeting face at quoins, 7—greenheart timber meeting face at sill, 8—gate bearing, 9—watertight access manholes, 10—pump chamber, 11—scupper pipes, 12—buoyancy chamber, 13—flooding trunks, 14—pump chamber, 15—305 × 305 mm greenheart timber fender, 16—254 × 204 mm hollow rubber fender, 17—greenheart timber pad.

quicker opening of the gate. Their main disadvantage is the necessity for a special chamber into which the gate is hauled for dock opening. Sliding gates are usually built with a rectangular horizontal cross-section, although gates with an arched cross-section have also been constructed.

During the sliding operation the gates are ballasted in a way that their total reaction on the sliding tracks is minimal and usually does not exceed 100 kN. They usually move on rails placed on the dock sill and on special wheels or rollers fastened to the bottom of the gate. These gates are also designed to slide on smooth surfaces, eliminating the need for wheel or roller maintenance.

An example of a sliding reinforced concrete dock gate installed at the dry dock in Marseilles, France is depicted in Figure 3–72. It has the following dimensions: length 87.35 m, width 15 m, and depth 13.5 m. Structurally this gate is a self-stabilizing prestressed concrete caisson that behaves like a gravity dam. It consists of 28 identical cells, 5.82 × 6.64 m arranged into four ballast chambers.

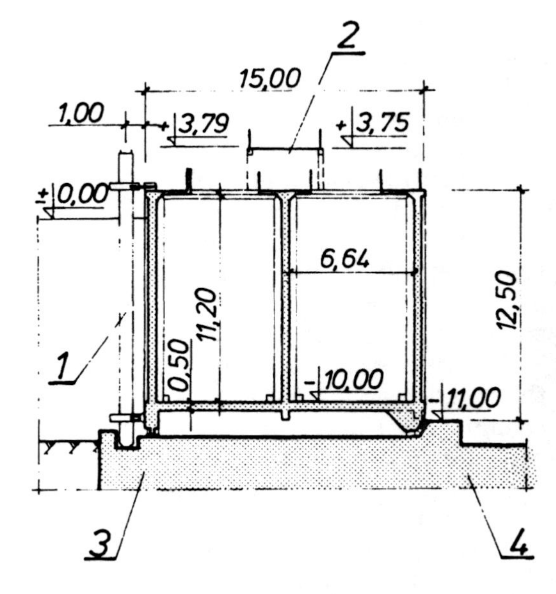

Figure 3–72. Cross-section of the concrete gate of Dry Dock No. 10 in Marseilles, France. (Port of Marseilles Authority, Marseilles, France.) 1—Guiding post, 2—service road, 3—sealing system, 4—bearing panel block.

Mitre gates are composed of two leaves, each rotating about hinges located at the dock side walls. When in the open position both leaves are located in curved recesses formed in the side walls at the dock entrance. The leaves are so shaped that in the closed or mitred position they form an arch across the entrance to resist the water pressure; the gate reaction are transferred to the dock walls. When the gate leaves are in the open position, and enter the recesses made in the side walls, the width of the dock entrance is equal to the width of the dock chamber. For examples of miter gates the reader is referred to PIANC (1986 and 1988).

A large double-leaf reinforced concrete floating gate was installed at the dry dock for building the oil production platforms at Graythorp near Hartlepool, England. It closed a dry dock entrance 122 m wide and 13 m deep. Each half of the gate was 56 m long, 14 m wide, and 14.4 m deep. Internal bulkheads were installed at approximately 6 m center. The gates were floated off their foundations, being guided vertically by the "pin piles" installed at each end of the gate.

One-leaf hinged gates are structurally similar to mitre gates. They are usually built as floating pontoons having a rectangular or arc horizontal section. The pontoon itself rotates about a vertical axis set in to two heavy bearings attached to the dry dock wall.

This gate, owing to its own buoyancy, can be towed to any other place for maintenance and repair.

An example of a hinged gate built at the dry dock in Temse, Belgium is illustrated in Figure 3–73. This prestressed concrete gate closes the entrance to the dry dock 53 m wide with a water depth at sill equal to 9.5 m during a high water level. The gate is 6 m wide and turns on bearings installed in the specially profiled recess made in the dock

Figure 3–73. Hinged concrete caisson gate of the dry dock in Temse, Belgium under construction. (F. C. de Weger Int., Netherlands.)

wall. It may be loaded with hydrostatic pressures on both sides. A steel multiarch type wall was built on the inside part of the gate to resist the differential hydrostatic pressure from inside of the dock chamber when the water level in the harbor is lower than inside of the dry dock.

The flap or falling-leaf type gate (Figure 3–74) consists of a single leaf hinged horizontally at the dock floor (sill) level. The operation of the gate is simple and fast, being raised, or lowered, within a few minutes by means of an electrically driven winch.

The gate is usually of welded steel construction; its typical arrangement consists of a series of horizontal girders spanning between the side jambs of the entrance and vertical stiffeners from gate keel to top deck. This system, however, may vary to suit local design conditions. Within the confines of the skin plating, which covers both sides of the gate, watertight compartments are formed to act as buoyancy chambers. These compartments reduce the loads on the operating gear and on the gate bearings under working conditions.

The hinges are placed in a way that the gate falls from its meeting faces and opens the dry dock entrance when the hydrostatic pressure on both sides of the gate is balanced. A wire rope lead, from the winch, travels via the fairlead pulleys over guide pulleys on the gate to an anchorage point on the far side of the dock, thus providing a double fall. Usually about 5 min is needed to lower or raise the gate.

The ever-increasing width of the dock entrance has led to development of many new gate concepts. One example is the gate system called *PROMOD (Proper Modular Design for Dock Gates)*. This gate consists of a number of typical buoyancy modules supported by props and intermediate plates spanning between the buoyancy modules.

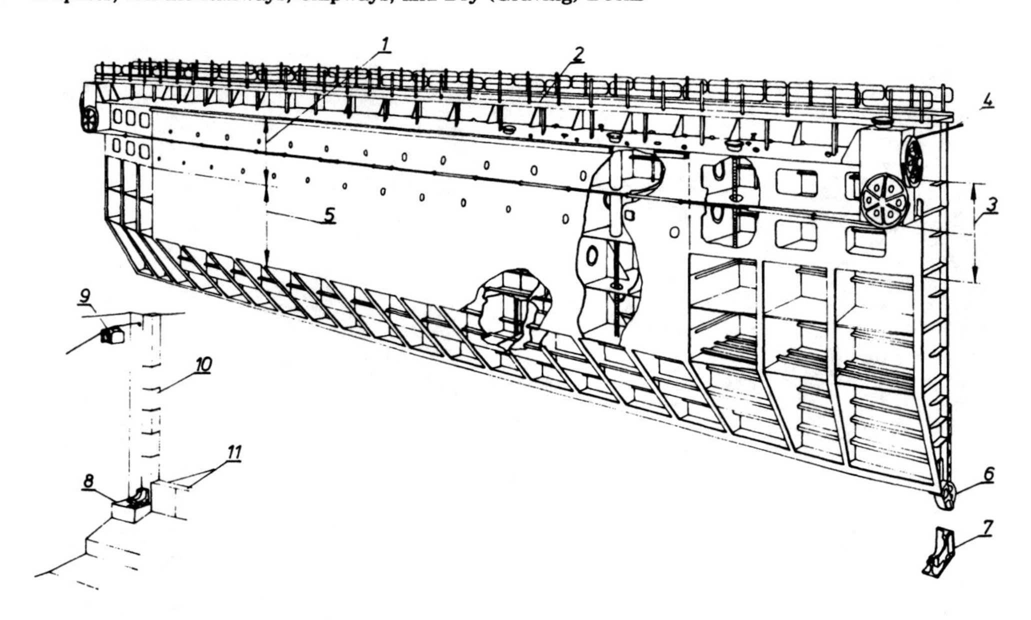

Figure 3–74. General arrangement of a flap gate. (From Mazurkiewicz, 1980.) 1—Free-flood area, 2–removable walkway, 3—tide range, 4—gate haulage rope, 5—buoyancy and water ballast chambers, 6—pivot, 7—bearing, 8—hinge pedestal, 9—wall shave, 10—quoin stones, 11—sill stones.

The hydraulic load from the gate module is transmitted to the prop and from the prop through bearing pads to the sill. The load is not transmitted to the side walls, and consequently the gate structure, which in fact looks like a keyboard, does not depend on the width of the dock.

The PROMOD design for lightweight modular steel gates was adopted for closing the three docks incorporated in the Dubai Dry Docks in the Arabian Gulf. These dry docks—525 m × 100 m, 415 m × 80 m, and 370 m × 66 m—were closed by gates 15 m high. The largest gate with 15 main modules, each incorporating an air chamber and a ballast chamber, was pivoted below the sill level (Figure 3–75).

3.5.7 Cranes

Cranes are among the most important mechanical installations in shipyards and although their function has not changed with time, their characteristics, such as lift capacity and reach, have changed fundamentally. Generally the following types of cranes are used for dry dock service: jib cranes, gantry cranes, mobile cranes, and floating cranes.

Gantry cranes are used mainly in building docks whereas the other cranes can be found in every type of dock. Mobile cranes with a lift capacity of 0.5 MN are widely used to supplement the conventional dry dock equipment and, if necessary, can be positioned inside the dock. Floating cranes with lift capacities of up to 4 MN can be regarded as auxiliary equipment. Apart from the gantry cranes and the different types of jib cranes, various other auxiliary lifting devices are used, for example, fork-lift trucks, hydraulic lifts, pneumatic lifts, and mobile platforms. Different kinds of overhead cranes installed within the assembly halls may travel beyond the workshop area. None of this equipment will affect the structural design of the dry dock but it may determine the design of other features such as the type of ducts (for

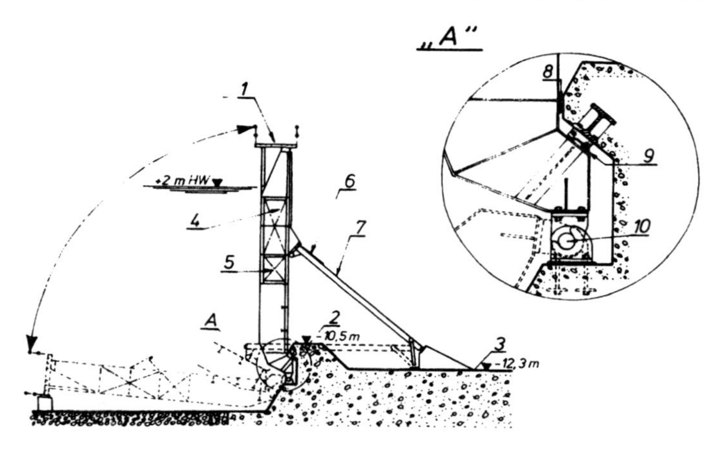

Figure 3–75. General arrangement of gates of the dry dock in Dubai. (T. F. Burns & Partners, England.) 1—Roadway, 2—sill level, 3—dock floor level, 4—air chamber, 5—balancing chamber, 6—sill protection plates, 7—hinged prop, 8—rubber seal, 9—thrust bearing, 10—pivot.

supply or installation), pavements, and power station roofs.

The main task of the gantry crane(s) is to lift heavy ship blocks or sections and move them into the dry dock area. They can either enter the assembly shop or can straddle it or other cranes, for example, jib cranes. However, it is necessary to limit the crane height and its tracks for these that are inside the shop. These gantry cranes that straddle the assembly shop, to hoist finished blocks and sections through the open roof or from the assembly areas equipped with jib cranes (Figure 3–76), are practically not limited in height.

In recent years the capacities of gantry cranes have reached 15 MN and their spans reached 185 m. Modern gantry cranes can be manufactured to span not only the whole width of a dock designed for the construction of 1 000 000 DWT ships, but also to cover a strip along the dock sufficiently wide to be used for the preassembled sections of the ships.

Shipyard cranes require special heavy-duty tracks. The design of crane tracks should take into consideration the following factors:

1. Spacing of wheels and groups of wheels of individual crane bogies
2. Number, spacing, and type of rails
3. Vertical and horizontal loads acting on a crane wheel caused by the different positions of the crane jib or by the position of the lifted weight in relation to the points of support for the gantry crane
4. Overall dimensions of cranes, with special reference to the minimum space between simultaneously operating adjacent cranes
5. Possible arrangements of adjacent cranes during simultaneous operation as well as their size and load distribution along track structure
6. Vertical and horizontal tolerance of the crane rail and the horizontal tolerance of spacing between rails
7. Means of supplying the cranes with electric power
8. Size and distribution of the buffer loads
9. Methods of anchoring the crane during operation and at a standstill, especially during storm conditions.

In order to reduce the required outreach of the jib cranes or spans of the gantry cranes, the track should be located as close to the edge of the dock side wall as practical.

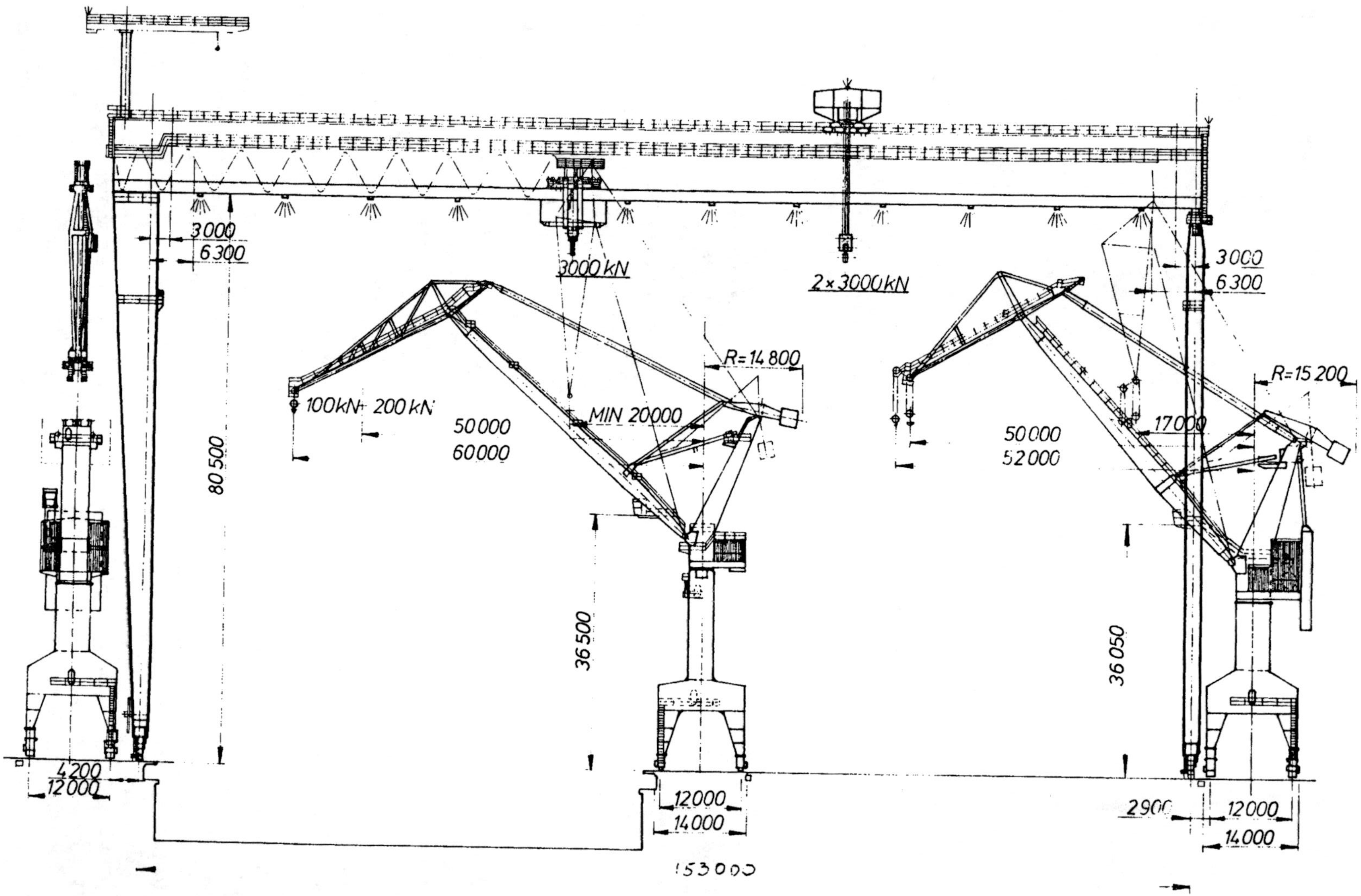

Figure 3–76. Arrangement of shipyard cranes around a dry dock and assembly areas. (From Mazurkiewicz, 1980.)

The dock walls are usually used as foundation for the track of gantry and jib cranes (Figure 3–76).

The typical track of each shipyard crane consists of a rail with fixing devices, foundation structure (beam, slab, or dock wall), and feeding channel.

For more details on crane tracks the reader is referred to Mazurkiewicz (1980).

3.5.8 Structural Design

Throughout their different stages of operation, dry dock structures are subjected to varying loads over very short periods of time. Specifically, rapid changes in loads can occur during dock flooding or dewatering.

There are three basic phases of dock operation during which certain specific external loads and load combinations are acting on the dock structure:

1. *Phase I*: The dry dock is dewatered during periods of repair or cleaning. During these periods the dock walls may be loaded by soil and water pressures and by the reaction of the dock floor. The dock floor is loaded by hydrostatic uplift and the reaction of the subsoil and walls.
2. *Phase II*: The dry dock is dewatered but with a ship resting on the dock floor. The external loads may be the same as in Phase I, but the weight of the ship must be added.
3. *Phase III*: The dry dock is flooded. The hydrostatic pressures on the dock floor and dock walls are added to loads occurring in Phase I.

Furthermore, in all three operational phases loads from the relevant construction and/or operational mechanical equipment must be added to the aforementioned external loads.

Sometimes loads during dock construction phase may control design of certain structural elements.

The weight of the dry dock structure and soil are estimated and used in all design phases according to generally accepted methods.

The hydrostatic pressure acting on dock walls, as well as hydrostatic uplift, depend on the water levels around the dry dock structure and the permeability of the dry dock foundation. For computing the hydrostatic pressures and the uplift load acting on the dry dock structure, the highest possible water level in the surrounding ground or in the harbor observed over a long period of time is considered. When data on the groundwater table are not available, then the highest water level in the harbor in which the dry dock is located is usually used for computing the hydrostatic and uplift loads. Examples of hydrostatic pressures for different foundation conditions are illustrated in Figure 3–77.

If the base of the dry dock can reach strata in which the groundwater is under pressure, the hydrostatic uplift load is calculated for the highest possible static water level.

For safety of dock operation a minimum uplift hydrostatic pressure of 10 kPa should be considered even where the pressure relieving means are provided.

Lateral soil pressures on the dry dock walls are estimated by conventional methods. The type of structure used for the side walls has a major impact on the soil lateral thrust; for example, massive side walls and particularly those rigidly joined with the foundation slab will be exposed to soil pressure "at rest" while flexible walls, or walls that allow for certain movements, will be exposed to active soil pressures.

From inside of the dry dock chamber the floor structure is loaded by the docked ship through the keel, side, and bilge blocks; obviously this load depends on the width, type, and weight of the ship and the layout of aforementioned blocks. It is important to evaluate all possible locations of the ship(s) sitting in a dry dock, specifically in a very wide dock. A study of effects of a ship(s) location within dry dock is normally made for a ship of maximum size, a single ship of

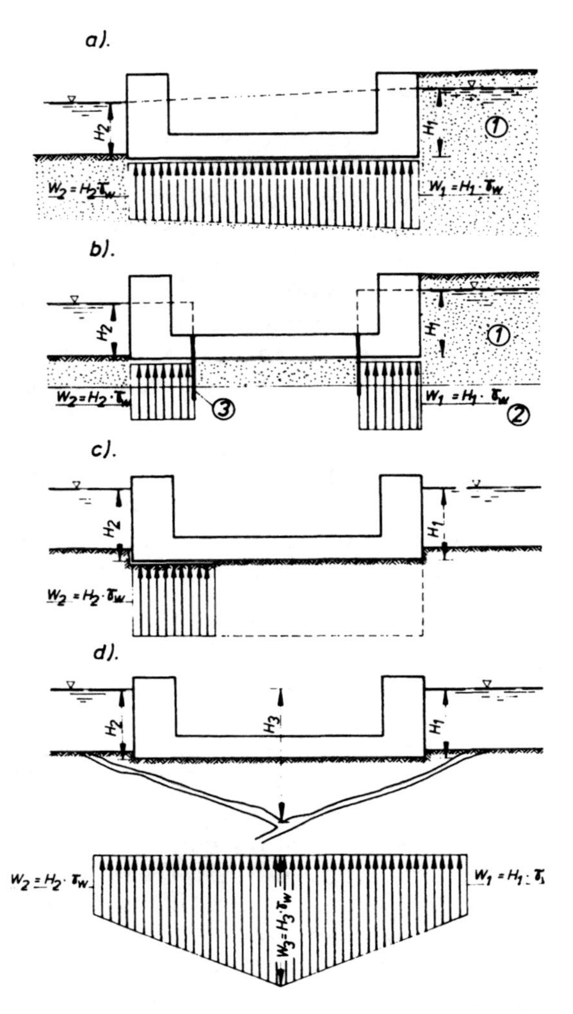

Figure 3–77. Hydrostatic uplift acting on a dry dock structure. (From Mazurkiewicz, 1980.) (a) Free flow of water under the dry dock. (b) Sealing membranes separating the dock floor. (c) Dry dock partly on an impermeable foundation. (d) Dry dock on fissured bed rock. 1—Permeable soil, 2—impermeable soil, 3—sheet pile wall.

maximum weight, several smaller ships sitting parallel to each other and symmetric to the longitudinal axis of a dock, and one or more ships placed asymmetrically to the longitudinal axis.

Determination of the best methods for estimating the loads exerted on the dry dock floor by ships during construction or repair was the subject of many studies, the results of which have led to certain empirical recommendations for the most adverse load combinations that need to be considered in the dock design (Mazurkiewicz, 1980). Naturally, a better design solution is achieved when more detailed and accurate input data for external loads and the foundation soil conditions are available.

It should be noted, however, that despite any design assumptions, no matter how accurate they may be, in real life actual loads may substantially differ from those considered in the dock's original design. If the latter would occur then a detailed analysis should be carried out with the new stress levels in the dock's structural components. The new stress diagrams obtained from these calculations should be compared with those in original designs and obtained from dock watertightness tests.

In addition to the loads resulting from the ships resting on the dock floor, there are also forces caused by the ship during docking and undocking, as well as during flooding and dewatering of the dock. These forces are transmitted from the ship to the dock walls by hauling winches, bollards, and capstans.

Furthermore, additional loads acting on the dry dock structure are those resulting from the storage of various kinds of materials and prefabricated structural components of ships, scaffolding, mobile cranes, and keel blocks that can be stored on the dock floor and/or the surrounding area. These loads should be investigated each time according to the actual conditions existing during building or repair of a particular ship.

Crane loads are obtained from their specifications. These loads are vertical P_d and horizontal H_d; the latter can be perpendicular and/or parallel to the crane track, acting horizontally in the plane of the crane rails.

The most typical load combinations during different phases of dock operation that should be considered in dock structure design are illustrated in Figure 3–78. In Phase I the dock is dewatered and is loaded

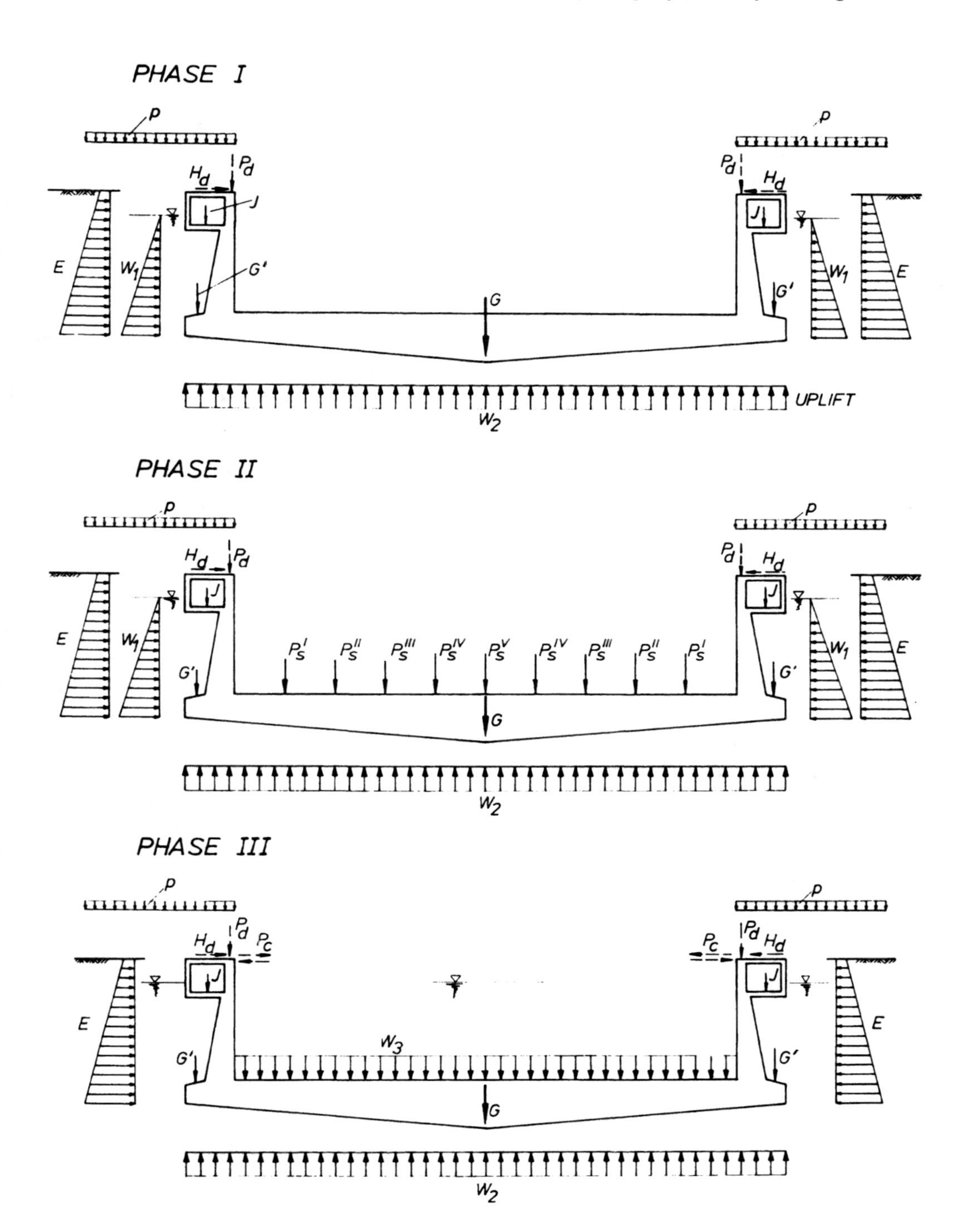

Figure 3–78. Load combinations acting on the dry dock structure. (From Mazurkiewicz, 1980.)

with soil E and hydrostatic pressures W_1 and W_2 acting upon the walls and slab structures, respectively. Weight of the structure G with associated weight of the backfill soil G', and the worst combination of crane loads P_d and H_d, weight of miscellaneous services J, and surcharge load p should also be included.

During the ship construction/repair stage (Phase II) the weight of the ship transmitted to the dock floor through the bilge and keel blocks is added to the loads identified in Phase I. When the dock is flooded (Phase III) the hydrostatic pressure W_1 is balanced by pressure of water from inside the dock; the weight of this water W_3 should be applied to the dock floor and the ship mooring forces P_c should be included in the worst combination of crane, services, and surcharge loads.

The dry dock stability is determined for the largest possible hydrostatic uplift occurring in the dewatered and empty dry dock. Stability calculations are usually carried out per unit length of the dock structure; for this the cross-section with the smallest weight is usually considered. The latter means that the stability of the whole dry dock structure will be somewhat larger than calculated.

The factor of safety is the ratio between resistance and driving forces. In dock stability analysis resistance forces include weight of the dry dock structure, weight of soil resting on the structure, soil friction against dock side walls, and forces produced by the ground anchors in docks with an anchored floor slab; the driving force is hydrostatic uplift.

When the resistance forces consist primarily of the weight of the structure which can be calculated fairly accurately, the dock stability factor of safety can be considered between 1.05 and 1.10. In stability analysis of the anchored dry docks depending on soil conditions at dock site and/or type of anchor, for example, ground or rock anchors or piles a safety factor ranging from 1.2 to 1.5 is usually considered. For drainage dry docks where the floor slab is subjected to reduced hydrostatic uplift, or is resting on an impermeable foundation, depending on the type of dock structure the allowable safety factor is usually used as equal to 1.2 to 1.5.

When the side walls are not separated from the floor slab in one way or the other the dock structure is analyzed as a frame on an elastic foundation, loaded by soil and hydrostatic water pressures, and by loads caused by docking ship, cranes, and other mechanical equipment.

In the latter case, when the side walls are rather heavy and the floor slab is relatively thin the walls can be designed as a separate retaining structure loaded by the aforementioned loads with addition of the reactions from the floor slab, that is, designed as a slab on an elastic foundation fixed at both ends in the side walls.

It should be noted that in modern design practice the finite element method is being used more often for dock structure design. In this case the frame type dock structure is designed as a single three-dimensional unit sitting upon the foundation soil, being surrounded by granular fill and loaded as discussed earlier.

In lightweight docks with floor slabs anchored by ground or rock anchors, the slab is analyzed differently for each basic loading phase: in a dewatered and empty dock with hydrostatic uplift as the main load, the slab is analyzed as a multispan beam or slab supported on elastic supports; and in a dewater dock loaded with docked ships or in a flooded dock, the floor is usually designed as a slab or beam on an elastic foundation; at this stage anchors do not carry any load.

In lightweight docks with floor slabs sitting on piles, the latter are treated as stiff anchors when the dock is dewatered and empty. During aforementioned operational Phases II and III (Figure 3–78) the anchor piles are compressed and therefore the floor slab is designed as supported on piles only; the capacity of each pile should be sufficient to carry out the design load without substantial settlement or deformation. If the piles do not possess sufficient capacity, then the floor slab is designed as if placed on an elastic foundation having an increased coefficient of subgrade reaction resulting from the introduction of piles into the body of soil.

It must be pointed out that the dock structure should be designed in a way to keep its settlement according to acceptable tolerances.

The latter implies that careful investigation of the foundation soil conditions and construction aspects, which may affect the foundation, must be carried out. The optimum solution for the dock structure depends heavily on site stratigraphy, soil geotechnical parameters, and hydrological conditions. The most important soil parameters to be investigated in case of different structural solutions are as follows:

- *Heavy dock*: Soil shear strength and compressibility parameters which control vertical displacements of the dock structure.
- *Anchored dock*: Soil shear strength parameters that supply the bearing capacity of the compression and tension piles and buried anchors.
- *Drainage dock:* Exact depth of the impervious layer, data on the continuity of the impervious layer, horizontal and vertical permeability coefficients, and the actual phreatic radius for a certain amount of pumped water.

REFERENCES

AGATZ, A. and LACKNER, E., 1954. Das neue Trockendock der Nordseewerke Emden GmbH, Emden, Hansa 50/51.

BECHT, P. M. and HETHERMAN, J. R., 1990. "Introduction to Dry Docks." In "Design of Marine Facilities" by J. W. Gaythwayte. Van Nostrand Reinhold, New York.

BIRCH, S., 1957. Caissoner Til en Tordok. Ingenioren September 1957, pp. 1–9.

BIRCH, S., 1958. Beregning af en Dokbund. Nordisk Betong 1/1959, pp. 73–89.

BORZANI, G. and VIAN, P., 1962. Bacino di Carenaggio n.5 del Porto di Genova. L'Industria Italiana del Cemento 5/1962, pp. 1–30.

COLBERT, R. D., 1963. Puget Sound Graving Dock. the Dock and Harbour Authority, February 1963, pp. 309–312.

COUPRIE, P., COURBON, J., MICHE, R., and FABIE, P., 1961. Formes de Radoub Nos 8 et 9 du Port de Marseille. Annales ITBTP December 1961, pp. 1263–1364.

DONNELLY, C. W., 1971. River Yields to Below-Sea-Level Dock Construction. Construction Methods and Equipment, March 1971.

EBERT, P., 1950. Bau Eines Trockendocks im Hamburger hafen. Der Bauingenieur 2/1950, pp. 45–60.

GASCO, L., GRIMALDI, M., and DEL BALZO, E., 1962. New Dry Dock No. 5 at the Port of Genoa. The Dock and Harbour Authority, January 1962, pp. 277–283.

GEDDES, W. G. N., STUROCK, K. R., and KINDER, G., 1972. New Shipbuilding Dock at Belfast for Harland & Wollf Limited. Proceedings Institution of Civil Engineers, January 1972, paper 7448, pp. 17–47.

HANSEN, E., KIRKERUP, C., GUDUM-OLSEN, G., and KNUDSEN, M., 1956. Udrestingen af Burmeister & Wain's nye tordok. Ingenioren 47/1956, pp. 945–951.

HASSANI, J. J., 1973. Very large Graving Docks. ASCE National Structural Engineering Meeting. Meeting Preprint 1938, April 1973, pp. 1–29.

HASSANI, J. J., 1987. "Shipyard Facilities—New and Old Closures for Dry Docks." Journal of Ship Production, Vol. 3, No. 3, August.

HAUPTMANN, J., 1968. Some Problems Concerning the Construction of the Dry Dock for 65 000 DWT Ships at Gdynia Shipyard. Proceedings of the International Havenkongress Antwerpen 1968, Paper 2.

HAUPTMANN, J. and SZOPOWSKI, Z., 1962. Final Design of the Dry Dock Structure at the Shipyard at Gdynia (in Polish). Technika i gospodarka morska 11/1962, pp. 331–334.

HAUSSER, P., GREENWOOD, C., FINLINSON, J. C., and ELLIOT, A. J., 1964. A Comparison of the Design and Construction of Dry Docks at Immingham and Jarrow. Proceedings of the Institution of Civil Engineers, February 1964, pp. 291–324, November 1964, pp. 637–655.

HÜCKEL, S., 1975. Marine Structures, Vol. III (in Polish). Wydawnicto Morski, Gdansk.

KESSLER, H., 1975. Lift-Dock für Rationelle Schifsbewegungen der Neuen Werft. Schiff and Hafen 10/1975.

KNEGJENS, H., 1968. The Layout of the Modern Shipbuilding- and Shiprepair-Yard. Schip en Werf 1/1968, pp. 1–18.

KNUDSEN, M. O., 1961. Modern Shipyard Development. Ingenioren-International Edition, 1/1961, pp. 1–7.

KNUDSEN, M. O. and LARSEN, O. C., 1956. NY Tordock A/S Burmeister & Wain's Maskin-og Skipsbyggeri Refshaleoen. Ingenioren 47/ 1956, pp. 1–24.

LACKNER, E., 1962. Vorgespannt Verankerte Trockendocke. Vorträge der Baugrundtagung 1962 in Essen. Essen: Verlag Wilhelm Ernst & Son, pp. 105–129.

LARSEN, C., MOE, A. J., JORGENSEN, R. B., and PAULSEN, J., 1957. Three New Dry Docks in Denmark. XIX in the International Naval Congress, London, S.II C.2.

LIDSTRÖM, G.and BACKSTRÖM, S., 1968. Cement-och Betongarbeton i Samband med Utforandet av Kockums byggdocka. Cement och Betong 2/ 1968.

LIDSTRÖM, G. and LINDSKOG, L., 1967. Fartygsdocka i Världsformat. Byggnadsindustrin, 19/1967.

MARTIN, G. P. and ARNOLD, A. C., 1992. "Failure of dry dock floor in Scaramanga, Greece, resulting from stress corrosion damage to anchor pile tendons", Proceedings Institution of Civil Engineers, Paper 9990, December, volume 96.

MARTIN, G. P. and IRVINE, K. D., 1972. Graving dock at Scaramanga, Greece. Proceedings of the Institution of Civil Engineers. November 1972, pp. 269–290.

MARTIN, G. P. and SHIRLEY, W. J., 1963. Design and Construction of No. 2 Dry Dock for Vickers-Armstrongs (Shipbuilders) Ltd., Hebburn, Co. Durham, Proceedings of the Instutition of Civil Engineers, Deeber 1094/3, pp. 513–548, October 1964, pp. 431–445.

MAZURKIEWICZ, B., 1976. Structural Solution of the New Dry Dock at Gdynia (in Polish). Technika i Gospodarka Morska 8/1976, pp. 485–487.

MAZURKIEWICZ, B., 1980. Design and Construction of Dry Docks, Clausthal Zellerfeld: Trans Tech Publications.

MAZURKIEWICZ, B., 1981. Shipyard Marine Structures (in Polish). Gdansk: Wydawnictwo Morskie.

MAZURKIEWICZ, B. and NAJDER, T., 1977. Construction of the Dry Dock No. 2 at the Gdynia Shipyard (in Polish). Technika i Gospodarka Morska 4/1977, pp. 229–234.

MILLARD, C. F. and HASSANI, J. J., 1971. Graving Dock for 300 000 Ton Ships. Civil Engineering—ASCE, June 1971.

PERMANENT INTERNATIONAL ASSOCIATION OF NAVIGATION CONGRESSES (PIANC), 1988. "Dry Docks." Supplement to Bulletin No. 63. Brussels, Belgium.

PIANC, 1986. Final Report for the International Commission for the Study of Locks. Supplement to Bulletin No. 55. Brussels, Belgium.

ROSS, F. K., 1960. Dry Dock Floor Being Poured 60 ft Below Puget Sound. Pacific Builder and Engineer, November 1960, pp. 70–72.

SCHENCK, W., 1964. Trockendocks mit Unabhängigen, Pfahlverankerten Sohlen. Messen ihrer Hebungen und Setzungen. Die Bautechnik 9/1964, pp. 289–301, 10/1964, pp. 351–356.

SCHENCK, W. and KOLLOGE, W., 1958. Zwei Neue Trockendocks der Kieler Howaldtswerhe AG, Kiel. Hansa 44/45 November 1958, pp. 2127–2134.

SZOPOWSKI, Z., 1963. Shipways. Marine Railways. In Marine Structures (in Polish). Warszawa: Arkady, pp. 169–210.

SZOPOWSKI, Z., 1980. Structural Solution of a Closed Shipway (in Polish). Inzynieria Morska 1/1980.

TATE, T. N., 1961. World's Largest Dry Dock. Civil Engineering, December 1961.

TSINKER, G. P., 1983. Technical Report on Visit to Shiplift Facilities in Seattle, WA and Vancouver, BC. Unpublished Report for City of Seward, Alaska. Acres Consulting Services, Niagara Falls, Ontario.

ZOLA, S. P. and BOOTHE, P. M., 1960. Design and Construction of Navy's Largest Drydock. Proceedings ASCE, Journal of the Waterways and Harbour Division, March 1960, pp. 53–84.

4

Offshore Moorings*

by J. R. Headland

4.1 INTRODUCTION

Offshore moorings provide temporary or permanent berthing for ships as well as for a wide range of floating structures including dry docks, piers, bridges, and oil drilling/production facilities. Tankers are often moored at offshore moorings during oil transfer operations. Floating dry docks, on the other hand, are normally secured to permanent mooring systems. The following paragraphs summarize the design of offshore mooring systems with an emphasis on the design of ship moorings in nearshore waters (i.e., water depths of 30 m or less).

4.2 OFFSHORE MOORING SYSTEMS

Offshore mooring systems are generally classified as either single-point moorings or multiple-point moorings (i.e., spread moorings). A typical single-point mooring is shown in Figure 4–1. A vessel moored at a single-point mooring is free to swing or "weather-vane" and will tend to align itself bow-on to the prevailing weather conditions. This alignment minimizes the vessel area exposed to wind, waves, and currents which, in turn, minimizes the load on the mooring. As a result, single-point moorings are generally more economical than multiple-point moorings. On the other hand, single-point moorings require ample anchorage area to avoid interference with neighboring navigation channels, structures, and moored vessels.

The most common single-point mooring system is the catenary anchor leg mooring (CALM) shown in Figure 4–1. CALM moorings are comprised of a buoy secured to the seafloor by a number of anchor chain legs. CALMs used in oil industry applications incorporate oil transfer lines that rise from

* In this chapter some materials from the U.S. Navy's Design Manual "Basic Criteria and Planning Guidelines" DM-26.5 are included. The aforementioned manual was prepared for the U.S. Navy by J. Headland.

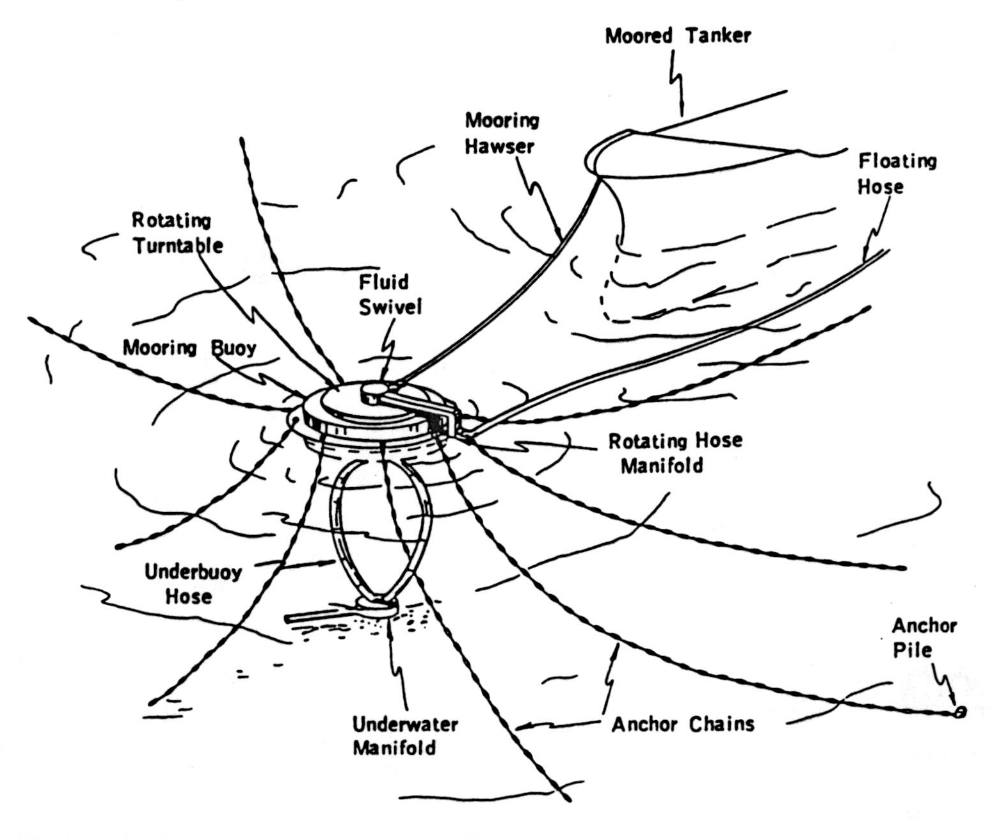

Figure 4–1. Catenary anchor leg mooring. (Flory et al., 1977.)

the seafloor and terminate at the buoy. The vessel may be attached to the CALM buoy by means of a mooring line (i.e., hawser) as shown in Figure 4–1 or by a rigid steel frame (i.e., yoke or turret) as shown in Figure 4–2. A single anchor leg mooring (SALM) consists of a relatively large buoy that is attached by a single anchor chain to a structural base located on the seafloor (Figure 4–3). CALM design will be emphasized in this chapter. A discussion of SALM design can be found in Flory et al. (1977).

Multiple-point moorings are required when there is insufficient area for a single-point mooring or when operations at the mooring cannot tolerate large vessel motions. Selection of a specific multiple-point mooring arrangement depends on site

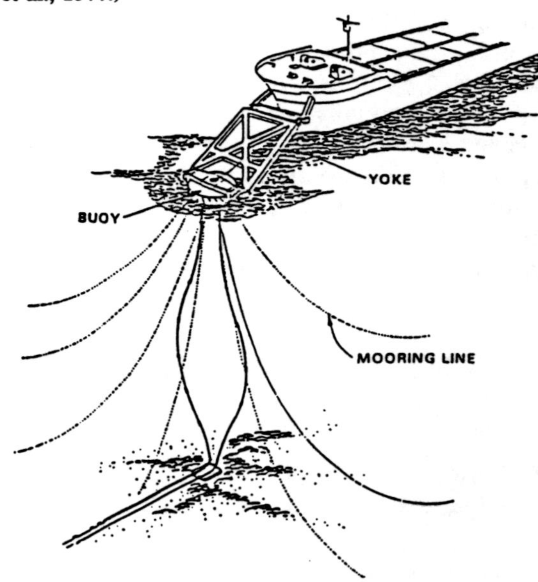

Figure 4–2. Catenary anchor leg mooring with rigid yoke. (American Petroleum Institute, 1990.)

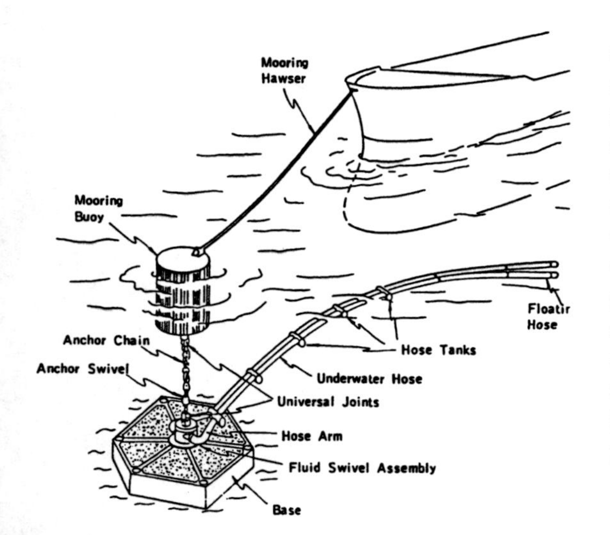

Figure 4–3. Single anchor leg mooring. (Flory et al., 1977.)

conditions, available anchorage area, and mooring use. The bow-and-stern mooring shown in Figure 4–4 can be used when environmental conditions are moderate; however, the moored vessel will still experience relatively large motions. Spread moorings, on the other hand, can be designed for extreme weather conditions and relatively small vessel motions. The number of mooring lines used in a spread mooring is variable and depends on operational and design conditions. Figure 4–5 illustrates a spread mooring arrangement used by the U.S. Navy for fuel loading.

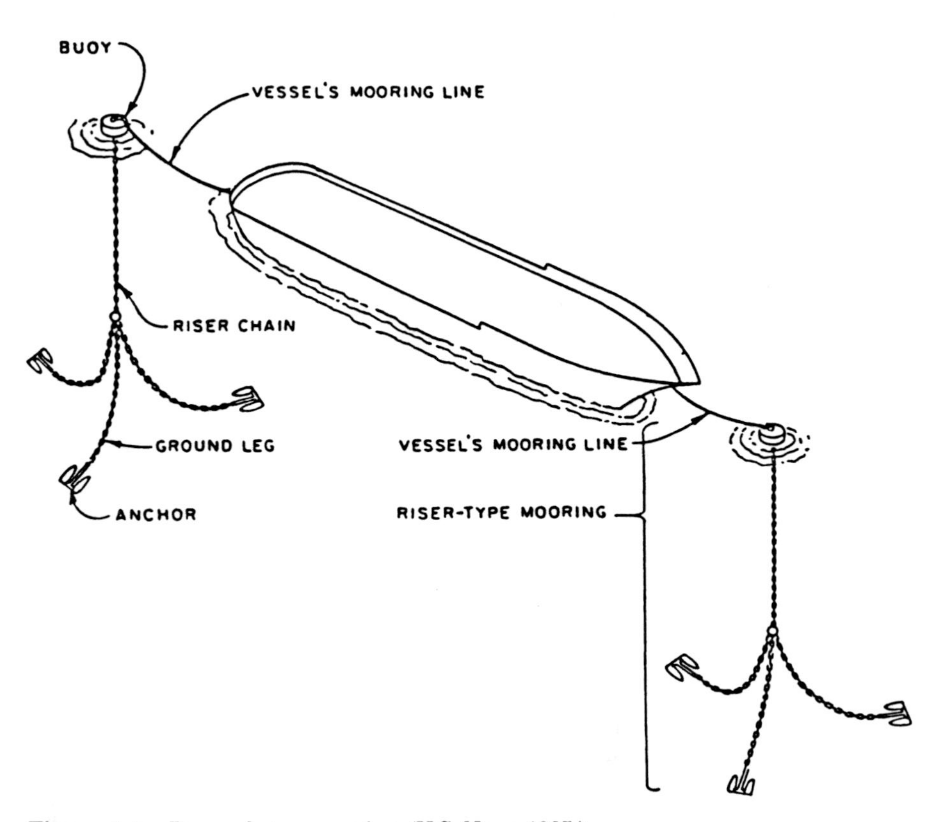

Figure 4–4. Bow and stern mooring. (U.S. Navy, 1985.)

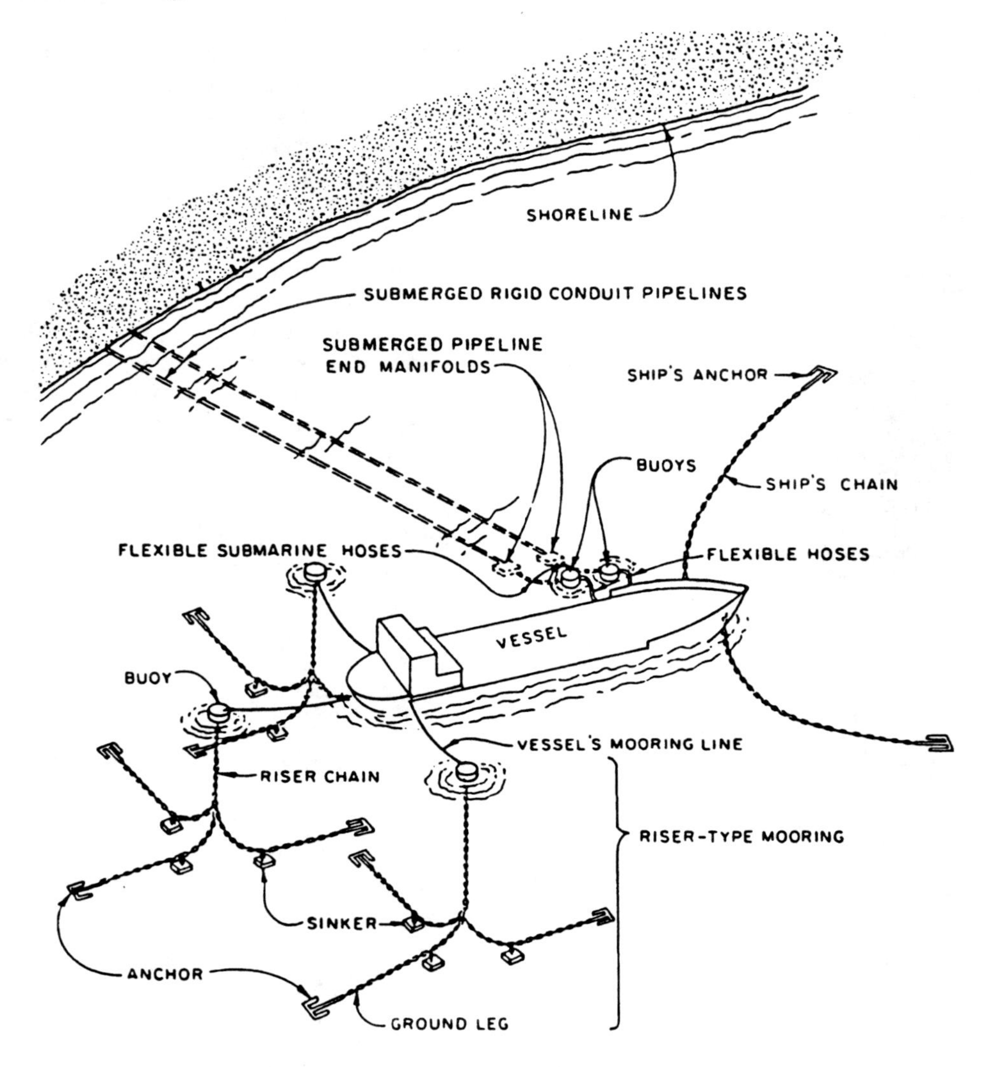

Figure 4–5. Typical spread mooring. (U.S. Navy, 1985.)

4.3 MOORING SYSTEM COMPONENTS

Figures 4–1 through 4–5 identify the principal components of offshore moorings. These components include anchors, sinkers, anchor chain, buoys, and mooring lines (hawsers); each of these is discussed below.

4.3.1 Anchors

Several anchor types are used in offshore moorings including drag-embedment anchors, deadweight anchors, and direct-embedment anchors. Drag-embedment anchors are particularly common in offshore moorings and a variety of commercially available anchor types are shown in Figure 4–6.

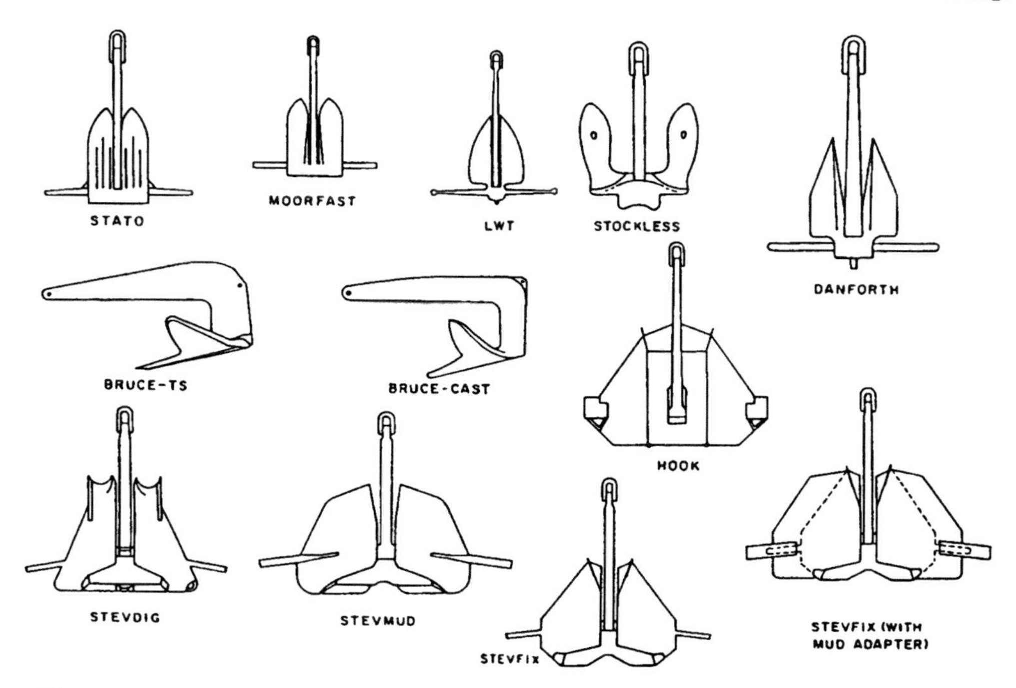

Figure 4–6. Drag-embedment type anchors. (U.S. Navy, 1985.)

The important elements of most drag-embedment anchors are summarized in Figure 4–7 which presents a U.S. Navy Stockless anchor. The anchor shank is used to transfer the mooring-line load to the anchor flukes which bear directly against the bottom substrate. The leading edge of the anchor fluke, called the fluke tip, is designed to penetrate into the seafloor on anchor dragging. Drag-embedment anchors are designed to resist horizontal loading and are incapable of resisting appreciable vertical loading. A near-zero angle between the anchor shank and the seafloor is required to ensure horizontal loading at the anchor and can be achieved by providing sufficient mooring line scope (i.e., ratio of mooring line length to water depth). Methods for designing drag anchors are presented in Tsinker (1986), Berteaux (1992), True (1992), and in Section 4.6.4.

Pile anchors are designed to resist horizontal and vertical loading and usually consist of simple shapes of structural steel fitted with a mooring-line connection. Pile anchors are well suited for short-scope moorings and are often used when seafloor soil characteristics are unsuitable for drag anchors. Figure 4–8 presents some example pile anchor designs. Pipe piles are well suited as anchors because they can sustain loading equally in any direction (although the mooring-line connection may not). Wide-flange sections, on the other hand, possess a weak and a strong axis against bending. Builtup sections may be fabricated with

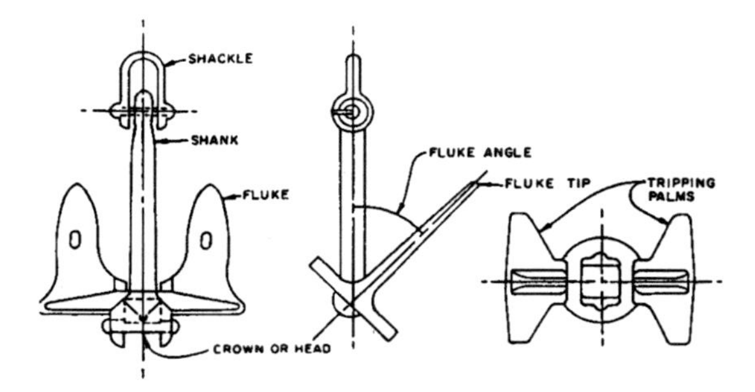

Figure 4–7. Elements of a drag-embedment anchor. (U.S. Navy, 1985.)

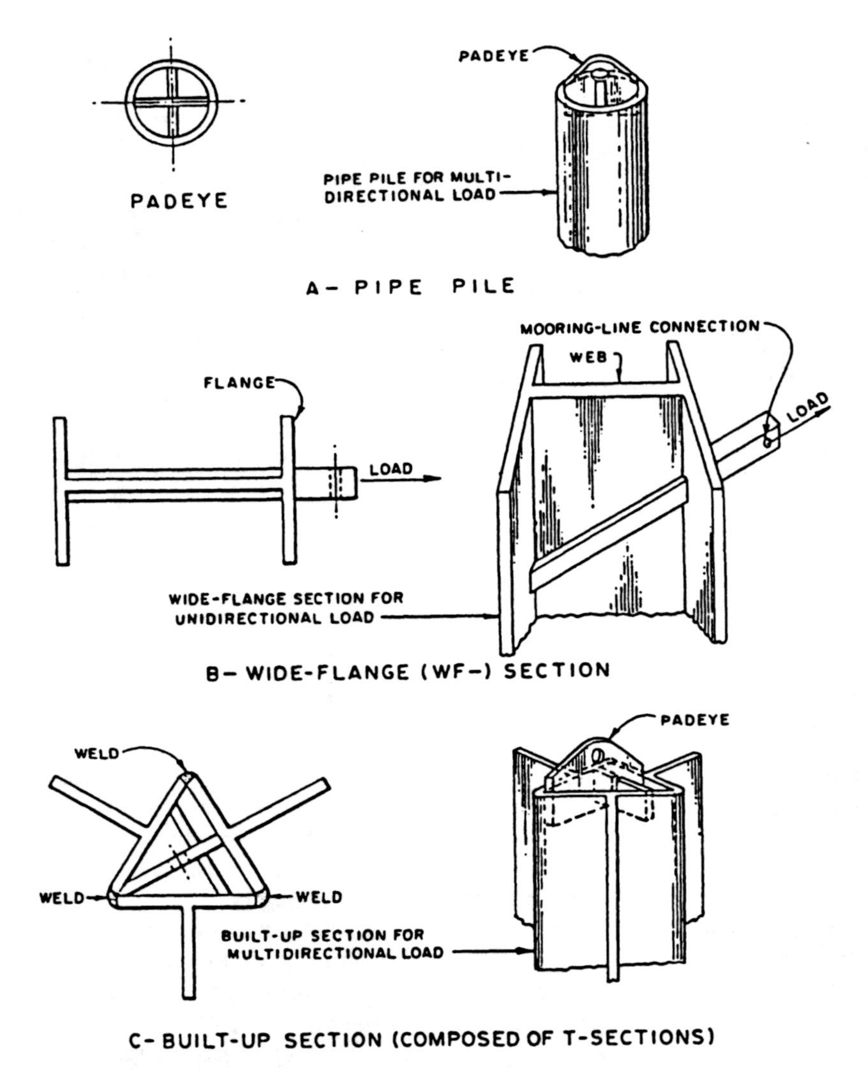

Figure 4–8. Types of pile anchors. (U.S. Navy, 1985.)

other structural shapes to resist either multidirectional or unidirectional loading. Procedures for designing anchor piles are given in Rocker (1985) and API (1987).

A deadweight anchor is a large mass of concrete or steel that resists both horizontal and vertical loading. Deadweight-anchor construction may vary from simple concrete clumps to specially manufactured concrete and steel anchors with shear keys (Figure 4–9). Deadweight anchors are generally larger and heavier than other anchor types and may, for this reason, be uneconomical. Uplift loads on a deadweight anchor are resisted by anchor weight whereas lateral load is resisted primarily by static friction between the anchor block and the seafloor. Lateral capacity can be enhanced with shear keys which bear against the soil substrate. Detailed design procedures for deadweight anchors are given in Rocker (1985) and Tsinker (1986).

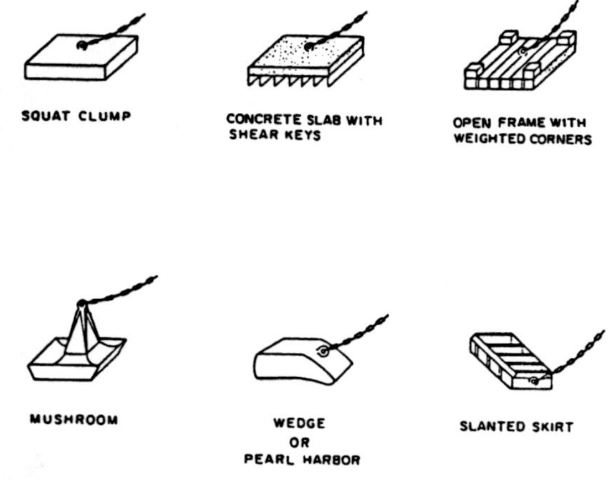

Figure 4–9. Types of deadweight anchors. (U.S. Navy, 1985.)

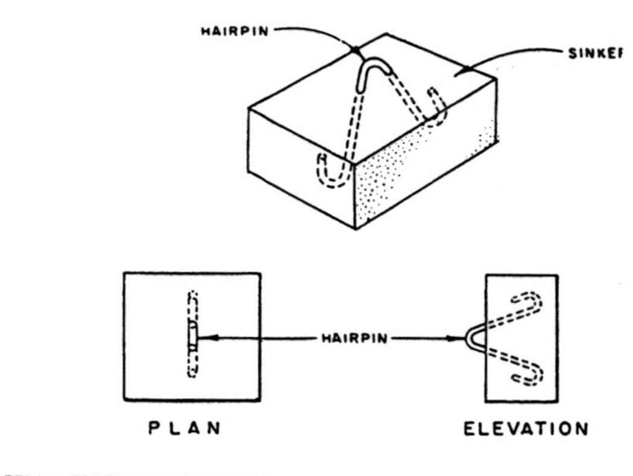

Figure 4–10. Concrete sinker block. (U.S. Navy, 1985.)

A direct-embedment anchor is an anchor that is driven, vibrated, or propelled vertically into the seafloor, after which the anchor fluke is expanded or reoriented to facilitate pullout resistance. Direct-embedment anchors are capable of withstanding both uplift and lateral loading. A discussion of direct-embedment anchors is presented in Rocker (1985).

4.3.2 Sinkers

A sinker is a large weight, made of concrete or steel, and placed on a catenary mooring leg to: (1) ensure horizontal loading at the anchor, (2) enhance mooring line energy absorption, or (3) affect mooring line pretension in a way that can be useful in controlling vessel motions. A typical concrete sinker is shown in Figure 4–10. Dimensions of the sinker depend on desired sinker weight. A discussion of sinkers and energy absorption is presented in Section 4.4.4.3.

4.3.3 Anchor Chains

A stud link chain is normally used in offshore moorings because the chain is strong, highly durable, heavy (which can enhance catenary performance), and has good shock-absorbing characteristics. Wire rope or synthetic lines, which are considerably lighter than chains, are sometimes used in deepwater moorings to reduce the vertical loading on a mooring buoy. On the other hand, wire ropes and synthetic lines require frequent maintenance and/or replacement and are not often used in permanent offshore moorings.

Anchor chain links have a center stud that is designed to hold the link in its original shape under tension and to prevent the chain from kinking when it is piled. There are a variety of grades of stud link chain available for use in moorings including normal, high, and extra-strength chain that are designated by classification societies as grade 1, grade 2, and grade 3 strength, respectively. Before a chain is accepted from a manufacturer, its strength should be established by testing. A break test consists of loading the chain in tension to a designated breaking strength appropriate for that grade and size chain. A proof test, which consists of loading a chain to 70% of its designated breaking strength, is normally applied to each shot of chain prior to acceptance. The characteristic dimension of chain is its diameter, and chain strength is proportional to that diameter. Chain length is generally reported in 15-fathom (about 27.5 m) lengths known as shots.

4.3.4 Buoys

The principal purpose of a buoy is to provide a convenient means for connection of the ship's hawser to the mooring. In the case of a SALM, the buoy also serves to provide a restoring force for the mooring. Buoys must have sufficient buoyancy to support the weight of the subsurface mooring chains (and sinkers). Buoys are designed using basic principles of hydrostatics. Detailed aspects of buoy design can be found in Berteaux (1976) and the American Bureau of Shipping (1975).

4.3.5 Mooring Lines or Hawsers

Hawsers may consists of synthetic ropes, wire ropes, or chains. A detailed discussion of hawser design for single-point moorings can be found in Flory et al. (1977). It should be noted, however, that Flory et al. (1977) recommend nylon hawsers and subsequent testing and design experience point to polyester as the preferred material.

4.4 MOORING DESIGN PROCEDURE

Mooring design consists of three principal steps, namely: (1) mooring layout, (2) evaluation of environmental conditions and associated loads, and (3) mooring component design. Figure 4–11 presents a flow chart outlining a general mooring design procedure. Although the flow chart is by no means applicable to every mooring design, it provides a useful summary of the mooring design process. An important aspect of this process readily apparent in Figure 4–11 is that one often must evaluate several mooring designs prior to the development of a final design concept. This is particularly true in cases where the mooring design is controlled by dynamic loading. The following sections of this chapter summarize the important elements of the mooring design procedure.

4.4.1 Mooring Layout

Moorings are often located at protected sites (e.g., within a harbor) in order to minimize environmental loads. On the other hand, many moorings are located offshore where the ship-mooring system is fully exposed to ocean wave attack. In either case and whenever possible, a mooring should be oriented so that the longitudinal axis of the vessel is parallel to the direction of the prevailing wind, waves, and currents as this practice will minimize the loads on the mooring.

The mooring site, vessel, and mooring configuration are normally given prior to commencement of detailed design. In some cases, however, it may be necessary to review several mooring configurations in order to determine the one most appropriate. Often the designer will have to analyze several vessels for a given mooring configuration. Characteristics such as length, breadth, draft, displacement, broadside wind area, and frontal wind area must be determined for each vessel. The mooring configuration used depends on: (1) mooring usage; (2) space available for mooring; (3) mooring loads; (4) strength, availability, and cost of mooring components; (5) allowable vessel movement; and (6) difficulties associated with maneuvering the vessel into the mooring.

4.4.2 Environmental Site Conditions

Environmental site conditions important to mooring design include bottom soil charac-

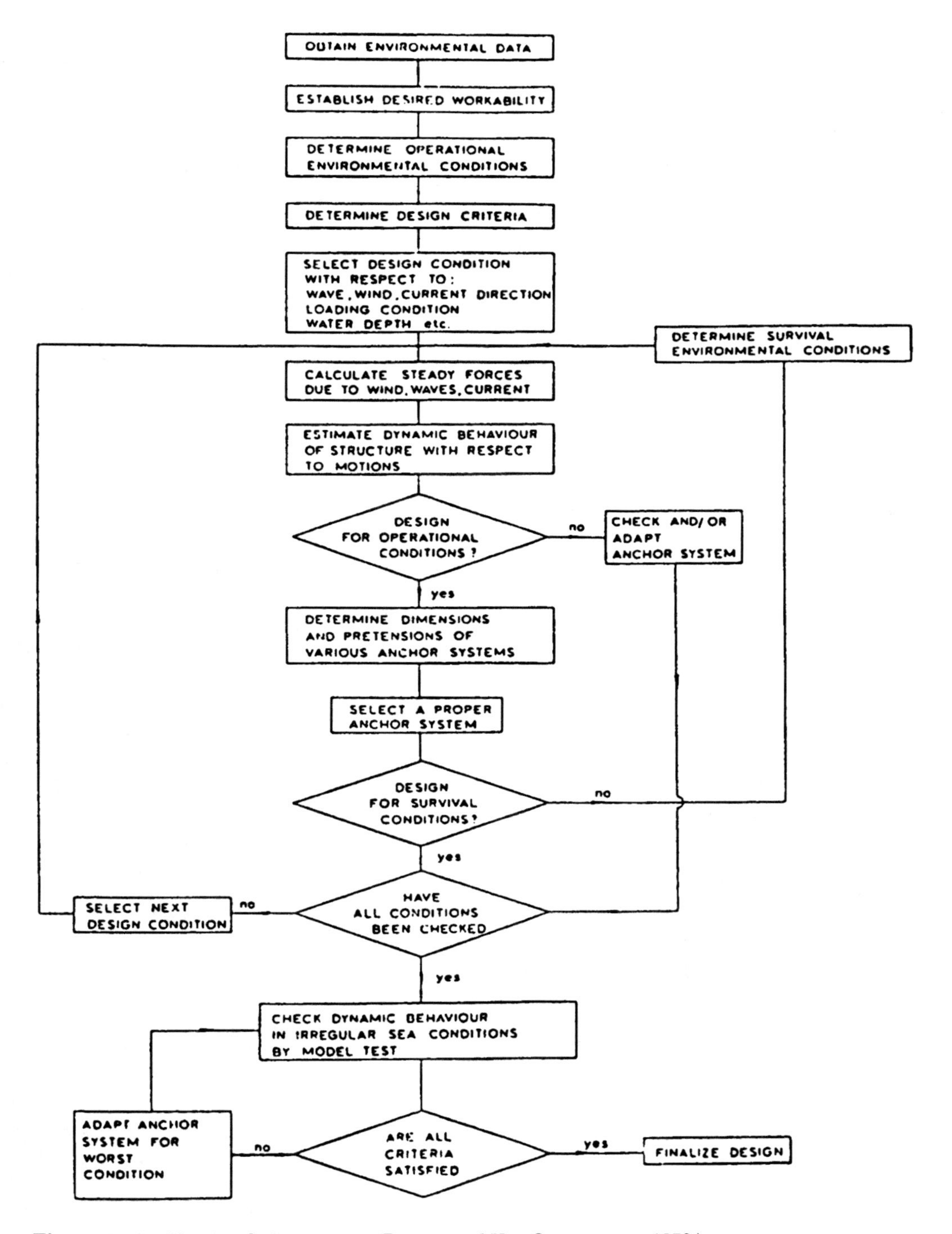

Figure 4–11. Mooring design process. (Remery and Van Oortmerssen, 1973.)

teristics, water depth, winds, currents, and waves. A detailed determination of these conditions is beyond the scope of this chapter; however, a review of important environmental factors is appropriate. Design values of water level, wind, current, and wave characteristics should be selected using probabilistic methods in order to achieve an economical and safe mooring design. Probabilistic analysis of measured environmental conditions taken at or near a site will provide an estimate of how frequently a given condition will occur or be exceeded (probability of exceedence) during the design life of the mooring. The return period of an event, estimated from the probability of exceedence, is defined as the average length of time between occurrences of that event. The concept of statistical return period is useful for determining the design values. For example, a 100-year design windspeed indicates that a windspeed equal to or greater than the 100-year design windspeed will occur, on the average, once every 100 years.

Moorings for offshore oil production facilities are often designed using 100-year return period conditions (American Petroleum Institute, 1990 and American Bureau of Shipping, 1975). The U.S. Navy (1985) recommends that standard ship moorings (which are normally unoccupied during extreme storm events) be designed using 50-year return period conditions. Some offshore moorings are designed on the basis of operational criteria that require that a vessel exit the mooring when weather conditions exceed prescribed threshold values. In such cases, the ship-mooring system is designed for the operational criteria rather than the more severe survival conditions (e.g., 100-year return period event). In any event, the mooring itself (i.e., mooring without a vessel secured to it) is generally designed for a 100-year return period of oceanographic conditions (American Bureau of Shipping, 1975).

4.4.2.1 Soil Conditions

Seafloor soil conditions must be evaluated in order to select and design anchors. Selection of an anchor system is often made on the basis of soil type. Pile or deadweight anchors, for example, would be considered for a rock seafloor whereas drag anchors would be eliminated from consideration. Accordingly, a detailed soil investigation including soil borings and associated laboratory testing is normally required to develop an anchor design. Rocker (1985) gives detailed guidance regarding offshore soil investigations.

4.4.2.2 Water Depth

Mooring-site bathymetry and water-level fluctuations must be investigated to: (1) ensure that there is adequate depth for vessels using the mooring, (2) to determine mooring-line geometry, and (3) to determine wave and current loads on the vessel. Factors contributing to water-level fluctuations include astronomical tides, storm tides, seiche, and tsunamis. Design water levels at a mooring site located in a harbor or estuary are often controlled by the astronomical and/or storm tide. Harbor sedimentation can result in long-term changes in bottom elevation and the potential for sedimentation should be investigated.

4.4.2.3 Wind

Wind is an extremely important factor in mooring design. Design windspeeds are often determined from windspeed records collected either at or in the vicinity of the mooring site. A minimum of 50 years of yearly extreme windspeed data is desired for a good estimate of the 100-year design windspeed. Once obtained, the windspeed data must be adjusted for elevation, duration, and overland–overwater effects in order to represent conditions at the mooring site. These adjustments should be made before the probabilistic analyses are conducted and procedures for making these adjustments are given in U.S. Navy (1985).

Wind measurements are normally corrected to a standard elevation of 10 m (American Bureau of Shipping, 1975; U.S. Navy, 1985; American Petroleum Institute, 1990). Wind-induced loads on a moored vessel depend on the duration. For example, a wind gust with a speed 50% higher than the average windspeed, but lasting only a couple of seconds, may cause little or no response of a moored vessel. On the other hand, a spectrum of windspeeds (i.e., a wind field with a range of speeds and durations) containing energy near the natural period of a vessel-mooring system, can excite the vessel dynamically and result in mooring loads in excess of those computed from the average wind speed. Accordingly, it is necessary to establish a standard wind duration that will provide reliable estimates of steady-state, wind-induced loads on moored vessels. Winds of longer or shorter duration should be corrected to this level. Based on analytical considerations and previous experience, a 1-min duration windspeed is used in the design of single-point moorings for oil tankers (American Bureau of Shipping, 1975; Flory et al., 1977; American Petroleum Institute, 1990). Wind loads on U.S. Navy vessels, which are considerably smaller than oil tankers, are based on a 30-s-duration windspeed (U.S. Navy, 1985).

Probabilistic procedures for determining the return period for various windspeeds based on measured data are presented in U.S. Navy (1985) and Simiu and Scanlon (1978). Changery (1982a, 1982b) reports extreme windspeeds for a number of locations along the East and Great Lakes coasts of the United States based on historical wind measurements. Batts et al. (1980) give hurricane windspeeds for Gulf and East coasts of the United States based on analytical studies of simulated hurricanes. Thom (1973) provides a method for estimating extreme winds at locations throughout the world which can be useful in preliminary planning.

4.4.2.4 Currents

Currents can play a major role in the layout and design of a mooring and current-induced loads on a moored vessel can dominate mooring design. With regard to mooring layout, it is highly desirable to moor vessels headed into the current. Currents may also affect the ability of a vessel to maneuver into the mooring. Tidal currents are the most common type of current in harbors and estuaries. These currents range in speed from less than 1 knot[1] to about 6 knots. Tidal currents are best estimated on the basis of direct measurements. Where measurements are not available, current speeds may be estimated using physical or numerical models. Current speed and direction varies during the tidal cycle and such variations have, in some cases, resulted in significant mooring loads (de Kat and Wichers, 1990). For the most part, however, maximum current values (under extreme tide conditions) are used in design. Currents resulting from river discharge can also be significant. River discharge currents are generally estimated from existing flow records. Wind-driven currents result from the stress exerted by the wind on the sea surface and attain a mean velocity of about 3% to 5% of the mean windspeed at 10 m above the sea surface. The direction of the current is approximately that of the wind and the current speed decreases sharply with depth. Wind-driven currents are seldom a factor in protected harbors, but can be important in the open ocean. Methods for estimating wind-driven currents are presented in Bretschneider (1967).

The probabilistic nature of current speed and direction at a given site should be taken into account particularly if the maximum

[1] 1 knot = 1870 m/h $\simeq$ 0.5 m/s.

currents are associated with a tropical storm event. In areas dominated by currents resulting from astronomical/storm tides, maximum flood and ebb currents under 100-year conditions are often used as a design value.

4.4.2.5 Waves

Wave loading can be an important element of mooring design, particularly when the mooring is located in the open ocean. Moorings located in protected harbors are generally sheltered from waves although a mooring located near the harbor opening may be exposed to sea and swell. Moreover, winds may generate local seas within the harbor of sufficient size to affect the moored vessel. Waves generated by storm activity can contribute significantly to mooring loads and have periods in the range of 6 to 20 s. Long waves, with periods ranging from 20 s to several minutes, are capable of causing oscillations within a harbor (i.e., seiche) which may be a factor in mooring design. Similarly, passing vessels can, in some cases, impart significant dynamic loading to a moored vessel.

Design wave conditions can be estimated from extrapolation of measurements taken at the mooring site. Such measurements, however, are seldom available over sufficient periods of time to make statistically meaningful estimates of 100-year design conditions. Consequently, analytical wave hindcast studies must be made in order to estimate design wave conditions (see, e.g., U.S. Army Corps of Engineers, 1984, 1989). It is best to rely on previous experience to estimate the potential mooring problems associated with seiche and passing vessels.

4.5 STATIC WIND AND CURRENT LOADS

Winds and currents produce static loads on moored vessels that can be separated into longitudinal load, lateral load, and yaw moment (Figure 4–12).

4.5.1 Wind Load

A variety of methods are available for determining wind loads on moored vessels. The following procedure is taken from Owens and Palo (1982) as presented in U.S. Navy (1985). Similar methods can be found in Remery and Van Oortmerssen (1973), OCIMF (1977), Benham et al. (1977), and Gould (1982). In each method, static wind loads are computed using a time-averaged wind speed (e.g., 30-s or 1-min duration windspeed).

Lateral wind load is determined using the following equation:

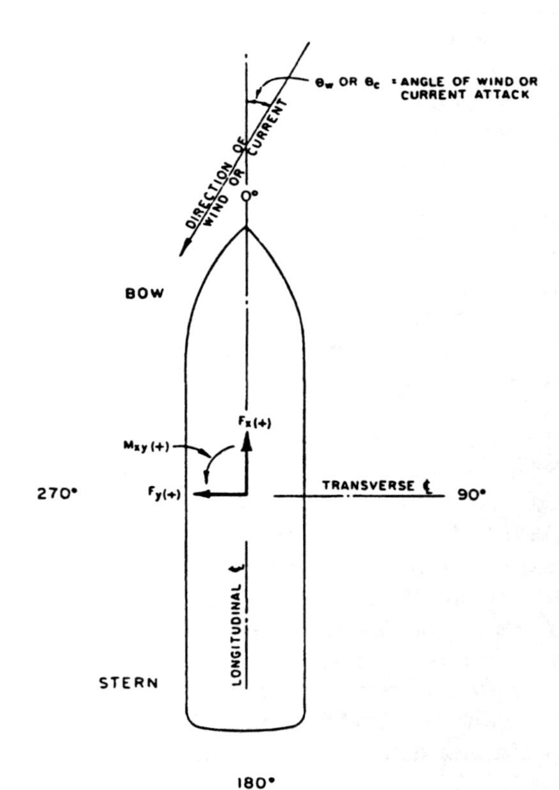

Figure 4–12. Static mooring load definition. (U.S. Navy, 1985.)

$$F_{yw} = \frac{1}{2}\rho_a V_w^2 A_y C_{yw} f_{yw}(\theta_w) \tag{4–1}$$

where

F_{yw} = lateral wind load
ρ_a = mass density of air
V_w = wind velocity
A_y = lateral projected area of ship
C_{yw} = lateral wind-force drag coefficient
$f_{yw}(\theta_w)$ = shape function for lateral load
θ_w = wind angle.

The lateral wind-force drag coefficient depends on the hull and superstructure area of the vessel:

$$C_{yw} = 0.92\,\frac{\left[\left(\frac{V_S}{V_R}\right)^2 A_S + \left(\frac{V_H}{V_R}\right)^2 A_H\right]}{A_y} \tag{4–2}$$

where

C_{yw} = lateral wind-force drag coefficient
V_S = average wind velocity over superstructure
V_R = average wind velocity at 10 m above sea level
A_S = lateral projected area of superstructure
A_H = lateral projected area of hull.

The following formulas are used to estimate the average wind velocities over the vessel superstructure and hull:

$$\frac{V_S}{V_R} = \left(\frac{h_S}{h_R}\right)^{\frac{1}{7}} \tag{4–3}$$

$$\frac{V_H}{V_R} = \left(\frac{h_H}{h_R}\right)^{\frac{1}{7}}. \tag{4–4}$$

The shape function for lateral load, $F_{yw}(\theta_w)$, is given as:

$$f_{yw}(\theta_w) = \frac{\left[\sin(\theta_w) - \frac{\sin(5\theta_w)}{20}\right]}{\left(1 - \frac{1}{20}\right)} \tag{4–5}$$

Longitudinal wind load is determined using the following equation:

$$F_{xw} = \frac{1}{2}\rho_a V_w^2 A_x C_{xw} f_{xw}(\theta_w) \tag{4–6}$$

where

F_{xw} = longitudinal wind load
ρ_a = mass density of air
V_w = wind velocity
A_x = longitudinal projected area of ship
C_{xw} = longitudinal wind-force drag coefficient
$f_{xw}(\theta_w)$ = shape function for longitudinal load.

The longitudinal wind-force drag coefficient varies according to vessel type and characteristics. In addition, a separate wind-force drag coefficient is provided for head wind (over the bow: $\theta_w = 0°$) and tail wind (over the sterm: $\theta_w = 180°$) conditions. The head wind (bow) wind force drag coefficient is designated C_{xwB} and the tail wind (stern) wind-force drag coefficient is designated C_{xwS}. The following longitudinal wind-force drag coefficients are recommended for hull-dominated vessels, such as aircraft carriers, submarines, and passenger liners:

$$C_{xwB} = 0.4$$
$$C_{xwS} = 0.4. \tag{4–7}$$

For all remaining types of vessels, except for specific deviations, the following are recommended:

$$C_{xwB} = 0.7$$
$$C_{xwS} = 0.6. \tag{4–8}$$

An increased head wind-force drag coefficient is recommended for center-island tankers:

$$C_{xwB} = 0.8. \tag{4–9}$$

For ships with an excessive amount of superstructure, such as destroyers and cruisers, the recommended tail wind-force drag coefficient is:

$$C_{xwS} = 0.8. \tag{4-10}$$

An adjustment consisting of adding 0.08 to C_{xwB} and C_{xwS} is recommended for all cargo ships and tankers with cluttered decks.

Longitudinal shape function, $f_{xw}(\theta_w)$, differs over the headwind and tailwind regions. The incident wind angle that produces no net longitudinal force, designated θ_{wz} for zero crossing, separates these two regions. Selection of θ_{wz} is determined by the mean location of the superstructure relative to midships. (See Table 4–1.)

Table 4–1. Selection of θ_{wz}

Location of Superstructure	θ_{wz} (degrees)
Just forward of midships	80
On midships	90
Aft of midships	100
Hull-dominated	120

For many ships, including center-island tankers, $\theta_{wz} \simeq 100°$ is typical; $\theta_{wz} \simeq 110°$ is recommended for warships.

The shape function for longitudinal load for ships with single, distinct superstructures and hull-dominated ships is given below. (Examples of ships in this category are aircraft carriers and cargo vessels.)

$$f_{wx}(\theta_w) = -\cos(\phi) \tag{4-11}$$

where

$$\phi_{(-)} = \left(\frac{90°}{\theta_{wz}}\right)\theta_w \quad \text{for } \theta_w < \theta_{wz}$$

$$\phi_{(+)} = \left(\frac{90°}{180° - \theta_{wz}}\right)(\theta_w - \theta_{wz}) + 90°$$
$$\text{for } \theta_w > \theta_{wz}.$$

The value of $f_{xw}(\theta_w)$ is symmetrical about the longitudinal axis of the vessel. Therefore, when $\theta_w > 180°$, use $360° - \theta_w$ as θ_w in determining the shape function. For example, if $\theta_w = 330°$, use $360° - \theta_w = 360° - 330° = 30°$ for θ_w.

Ships with distributed superstructures are characterized by a "humped" cosine wave. The shape function for longitudinal load is:

$$f_{xw}(\theta_w) = \frac{-\sin\gamma - \dfrac{\sin 5\gamma}{10}}{1 - \dfrac{1}{10}} \tag{4-12}$$

where

$$\gamma_{(-)} = \frac{90°}{\theta_w}\theta_w + 90° \quad \text{for } \theta_w < \theta_{wz}$$

$$\gamma_{(+)} = \frac{90°}{(180° - \theta_w)}\theta_w + \left(180° - \frac{90°\theta_{wz}}{180° - \theta_{wz}}\right)$$
$$\text{for } \theta_w > \theta_{wz}.$$

As explained above, use $360° - \theta_w$ for θ_w when $\theta_w > 180°$. Wind yaw moment is calculated using the following equation:

$$M_{xyw} = \tfrac{1}{2}\rho_a V_w^2 A_y L C_{xyw}(\theta_w) \tag{4-13}$$

where

M_{xyw} = wind yaw moment
ρ_a = mass density of air
V_w = wind velocity
A_y = lateral projected area of ship
L = ship length
$C_{xyw}(\theta_w)$ = yaw moment coefficient.

Typical yaw-moment coefficient curves are presented in Figure 4–13.

4.5.2 Current Load

Methods published by Remery and Van Oortmerssen (1973) and OCIMF (1977) can be used to compute current loads on moored vessels. The methods used to compute current loads described below are taken from U.S. Navy (1985). Lateral current load is determined from the following equation:

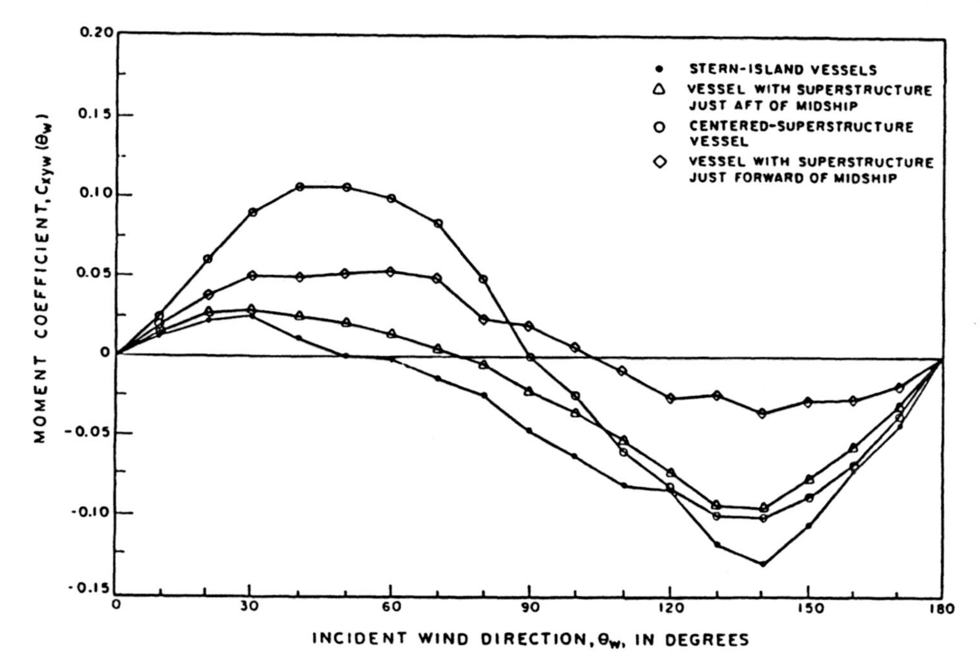

Figure 4–13. Wind Yaw-moment coefficients. (U.S. Navy, 1985.)

$$F_{yc} = \frac{1}{2}\rho_w V_c^2 L_{wL} T C_{yc} \sin(\theta_c) \qquad (4\text{–}14)$$

where

F_{yc} = lateral current load
ρ_w = mass density of water
V_c = current velocity
L_{wL} = vessel waterline length
T = vessel draft
C_{yc} = lateral current-force drag coefficient
θ_c = current angle.

The lateral current-force drag coefficient is given by:

$$C_{yc} = C_{yc}|_\infty + (C_{yc}|_1 - C_{yc}|_\infty)e^{-k\left(\frac{wd}{T}-1\right)} \qquad (4\text{–}15)$$

where

$C_{yc}|_\infty$ = limiting value of C_{yc} for large value of wd/T
$C_{yc}|_1$ = limiting value of C_{yc} for $wd/T = 1$
e = 2.718
k = coefficient
wd = water depth.

Values of $C_{yc}|_\infty$ are given in Figure 4–14 as a function of L_{wL}/B (the ratio of vessel waterline length to vessel beam) (ordinate) and vessel block coefficient, ϕ (abscissa). The block coefficient is defined as:

$$\phi = \frac{35D}{L_{wL}BT} \qquad (4\text{–}16)$$

where

D = displacement in long tons
L_{wL}, B, and T are given in feet.

Values of $C_{yc}|_1$ are given in Figure 4–15 as a function of $C_p L_{wL}/\sqrt{T}$.

The prismatic coefficient, C_p, of the vessel is defined as:

$$C_p = \frac{\phi}{C_m} \qquad (4\text{–}17)$$

where

C_m = midship section coefficient, $C_m = A_{ms}/(BT)$, where A_{ms} = immersed are of midship section

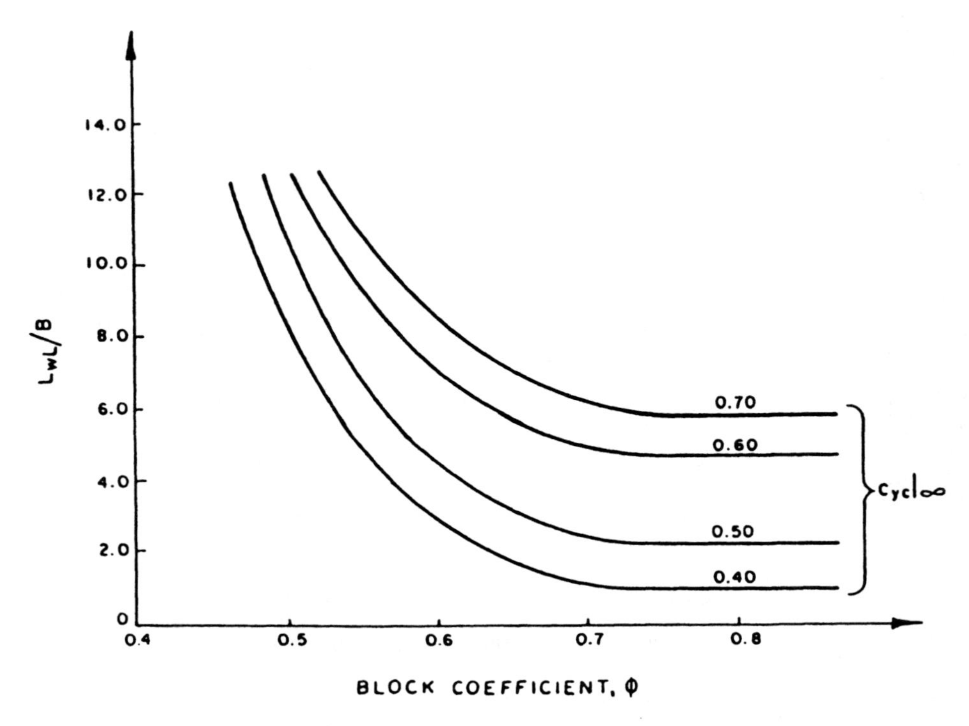

Figure 4–14. C_{yc}—Deepwater limit. (U.S. Navy, 1985.)

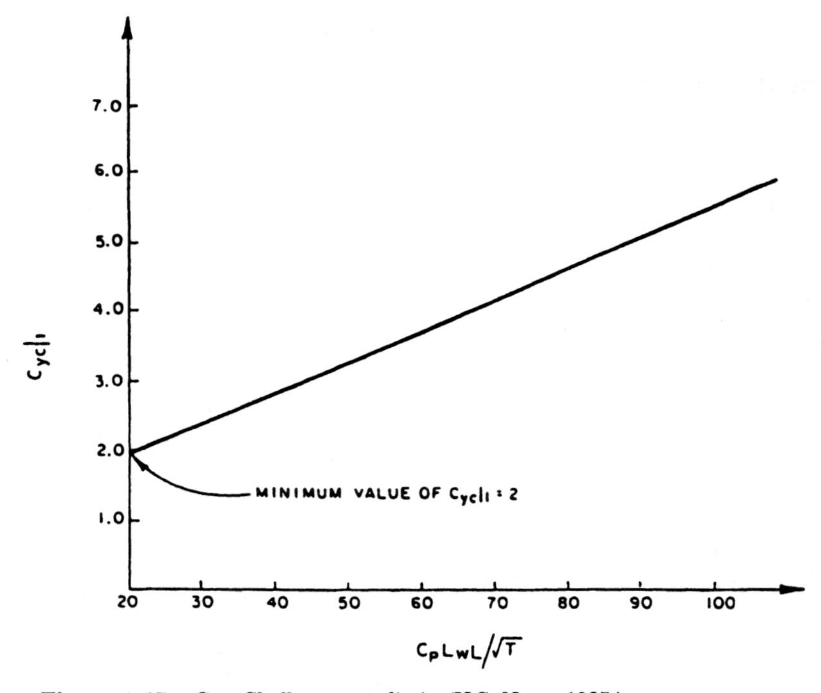

Figure 4–15. C_{yc}—Shallow water limit. (U.S. Navy, 1985.)

The value of the coefficient, k, is given in Figure 4–16 as a function of the vessel block coefficient, ϕ, and vessel hull shape (block-shaped or normal ship-shaped).

Longitudinal current load is determined using the following equation:

$$F_{xc} = F_{x\,form} + F_{x\,friction} + F_{x\,prop} \qquad (4\text{–}18)$$

where

F_{xc} = total longitudinal current load

$F_{x\,form}$ = longitudinal current load due to form drag

$F_{x\,friction}$ = longitudinal current load due to friction drag

$F_{x\,prop}$ = longitudinal current load due to propeller drag.

Form drag is given by the following equation:

$$F_{x\,form} = \frac{1}{2}\rho_w V_c^2 BTC_{xcb}\cos(\theta_c) \qquad (4\text{–}19)$$

where

C_{xcb} = longitudinal current-force form drag.

Friction drag is given by the following equation:

$$F_{x\,friction} = -\tfrac{1}{2}\rho_w V_c^2 SC_{xca}\cos(\theta_c) \qquad (4\text{–}20)$$

where

S = wetted surface area of hull $= 1.7TL_{wL} + 35D/T$ (S in square feet with T, L_{wL} in feet and D in long tons)

C_{xca} = longitudinal current-force friction drag coefficient C_{xca} is computed as follows:

$$C_{xca} = \frac{0.075}{(\log R_n - 2)^2} \qquad (4\text{–}21)$$

where

$R_n = V_c L_{wL}\cos(\theta_c)/\nu$ = Reynolds Number

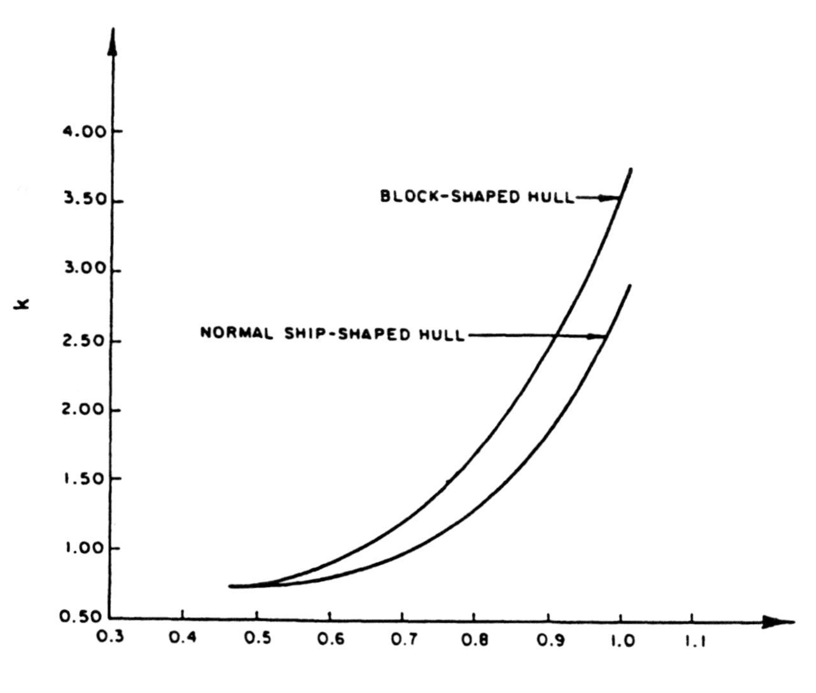

Figure 4–16. Current coefficient k. (U.S. Navy, 1985.)

ν = kinematic viscosity of water (1.4×10^{-5} ft^2/s).

Propeller drag is the form drag of the vessel's propeller with a locked shaft. Propeller drag is given by the following equation:

$$F_{x\,prop} = -\tfrac{1}{2}\,\rho_w V_c^2 A_p \cos(\theta_c) \qquad (4\text{–}22)$$

where

A_p = propeller expanded (or developed) blade area

C_{prop} = propeller drag coefficient = 1.0

A_p is given by:

$$A_p = \frac{A_{Tpp}}{0.838} \qquad (4\text{–}23)$$

where

A_{Tpp} = total projected propeller area

$$A_{Tpp} = \frac{L_{wL}B}{A_R} \qquad (4\text{–}24)$$

Table 4–2 shows the area ratio, A_R, for six major vessel groups. (The area ratio is defined as the ratio of the waterline length times the beam to the total projected propeller area.)

Current yaw moment is determined using the following equation:

$$M_{xyc} = F_{yc}\left(\frac{e_c}{L_{wL}}\right)L_{wL} \qquad (4\text{–}25)$$

where

M_{xyc} = current yaw moment

e_c/L_{wL} = ratio of eccentricity.

Table 4–2. A_R for Propeller Drag

Vessel Type	A_R
Destroyer	100
Cruiser	160
Carrier	125
Cargo	240
Tanker	270
Submarine	125

The value of (e_c/L_{wL}) is given in Figure 4–17 as a function of current angle, θ_c, and vessel type.

4.6 DESIGN OF MOORING COMPONENTS

4.6.1 Selection of Anchor Chain

The maximum mooring-chain tension is always higher than the horizontal load on the chain; however, normally only the horizontal load is known. The maximum tension can be approximated from the horizontal tension as follows:

$$T = 1.12H \qquad (4\text{–}26)$$

where

T = maximum tension in mooring chain

H = horizontal tension in mooring chain.

This equation provides conservative estimates of mooring-chain tension for water depths of 30 m or less.

According to the U.S. Navy (1985), the maximum allowable working load for mooring chain loaded in direct tension is:

$$T_{design} = 0.35T_{break} \qquad (4\text{–}27)$$

where

T_{design} = maximum allowable working load on mooring chain

T_{break} = breaking strength of the chain.

Similarly, the U.S. Navy (1985) stipulates the following for mooring chain that passes through hawse pipes, chocks, chain stoppers, or other fittings that cause the chain to change direction abruptly within its loaded length:

$$T_{design} = 0.25T_{break}. \qquad (4\text{–}28)$$

The American Bureau of Shipping Rules for Classing Single Point Moorings (1975) call for anchor legs to be designed with a

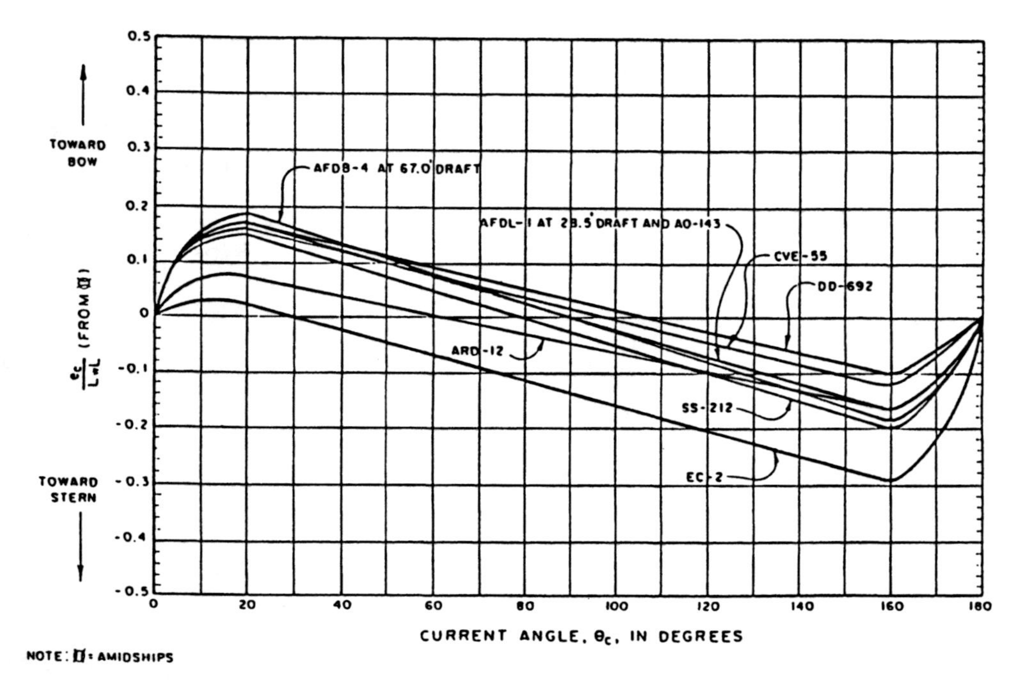

Figure 4–17. E_c/L_{wL} for various vessel types. (U.S. Navy, 1985.)

factor of safety of 3 based on the breaking strength of the chain, i.e.:

$$T_{design} = 0.33T_{break}. \qquad (4\text{–}29)$$

Chains and fittings should be selected with a breaking strength equal to or exceeding T_{break}. The breaking strength of a chain can be found in manufacturers' catalogs. A useful summary of breaking strengths for various types of chains and chain fittings are presented in U.S. Navy (1986). It is common practice to round up to the nearest 6 mm (1/4-inch) size when selecting chain or fittings and it may be desirable to specify the next largest size of chain or fitting if excessive wear is expected.

The weight per shot of chain presented in manufacturers' catalogs is often given in air. The weight of chain in water can be obtained by multiplying the weight in air by 0.87. When tables of actual chain weights are unavailable, the weight of a stud link chain may be approximated as follows:

$$w_{air} = 9.05d^2 \qquad (4\text{–}30)$$

$$w_{submerged} = 8.26d^2 \qquad (4\text{–}31)$$

where

w_{air} = weight of chain (in air) in pounds/foot of length

$w_{submerged}$ = weight of chain (in water) in pounds/foot of length

d = diameter of chain in inches.

4.6.2 Computation of Chain Length and Tension

A chain mooring line supported at the surface by a buoy and extending through the water column to the seafloor behaves as a catenary. Figure 4–18 presents a definition sketch for use in catenary analysis. At any point (x, y) the following hold:

$$V = wS = T\sin(\theta) \qquad (4\text{–}32)$$

$$H = wc = T\cos(\theta) \qquad (4\text{–}33)$$

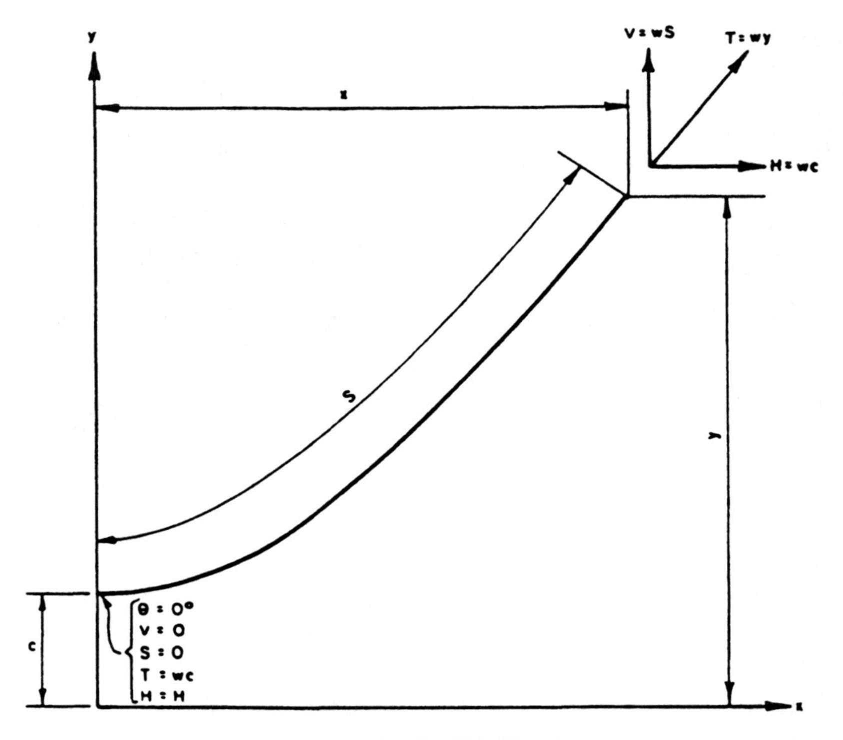

Figure 4–18. Catenary definition sketch. (U.S. Navy.)

$$T = wy \qquad (4\text{–}34)$$

$$c = \frac{H}{w} \qquad (4\text{–}35)$$

where

V = vertical force at point (x, y)
w = submerged unit weight of chain
S = length of curve (chain length) from $(0, c)$ to point (x, y)
T = line tension at point (x, y)
θ = angle of mooring line with horizontal
H = horizontal force at point (x, y)
c = distance from origin to y-intercept.

The shape of the catenary is governed by the following equations:

$$y^2 = S^2 + c^2 \qquad (4\text{–}36)$$

$$y - c\cosh\left(\frac{x}{c}\right) \qquad (4\text{–}37)$$

$$S = c\sinh\left(\frac{x}{c}\right). \qquad (4\text{–}38)$$

Equation (4–38) may be more conveniently expressed as:

$$x = c\,\ln\left(\frac{S}{c} + \sqrt{\left(\frac{S}{c}\right)^2 + 1}\right). \qquad (4\text{–}39)$$

Note that, in the above equations, the horizontal load in the chain is the same at every point and that all measurements of x, y, and S are referenced to the catenary origin.

When catenary properties are desired at point (x_m, y_m), as shown in Figure 4–19, the following equations are used:

$$\sqrt{S_{ab}^2 - wd^2} = 2c\sinh\left(\frac{x_{ab}}{2c}\right) \qquad (4\text{–}40)$$

$$\frac{wd}{S_{ab}} = \tanh\left(\frac{x_m}{c}\right) \qquad (4\text{–}41)$$

$$x_m = x_a + \frac{x_{ab}}{2} \qquad (4\text{–}42)$$

$$x_b = x_m + \frac{x_{ab}}{2} \qquad (4\text{–}43)$$

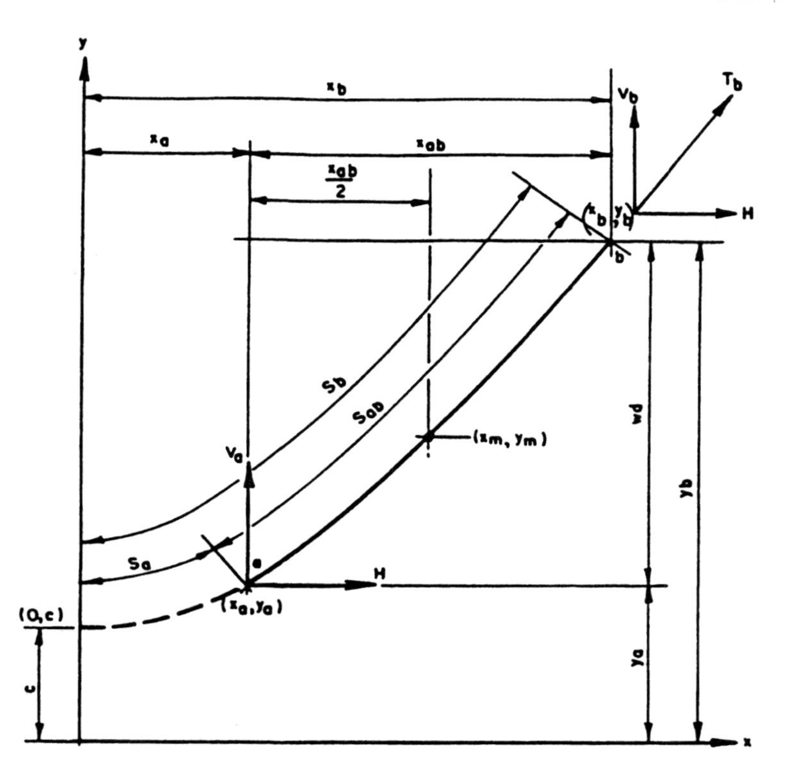

Figure 4–19. Catenary definition sketch. (U.S. Navy, 1985.)

where the terms in the above equations are defined in Figure 4–19. Equation (4–41) is more conveniently written as:

$$x_m = \frac{c}{2}\left[\ln\left(1+\frac{wd}{S_{ab}}\right) - \ln\left(1-\frac{wd}{S_{ab}}\right)\right] \quad (4\text{–}44)$$

4.6.3 Some Applications of the Catenary Equations

4.6.3.1 *Case I*

The known variables are the mooring-line angle at the anchor, θ_a (which is zero: $\theta_a = 0°$); the water depth, wd; the horizontal load, H; and the submerged unit weight of the chain, w. The length of mooring line, S_{ab}; the horizontal distance from the anchor to the buoy, x_{ab}; and the tension in the mooring line at the buoy or surface, T_b, are desired. Procedures for determining these values are outlined in Figure 4–20. Check to determine if the entire chain has been lifted off the bottom by comparing the computed chain length from anchor to buoy, S_{ab}, to the actual chain length, S_{actual}. If the actual chain length is less than the computed, then Case I cannot be used and Case V must be used.

4.6.3.2 *Case II*

The known variables are the mooring-line angle at the anchor, θ_a (or equivalently, a specified vertical load at the anchor, V_a); the water depth, wd; the horizontal load at the surface, H; and the submerged unit weight of the chain, w. This situation arises when a drag anchor is capable of sustaining a small prescribed angle at the anchor, or an uplift-resisting anchor of given vertical capacity, $V_a = H \tan\theta_a$, is specified. The origin of

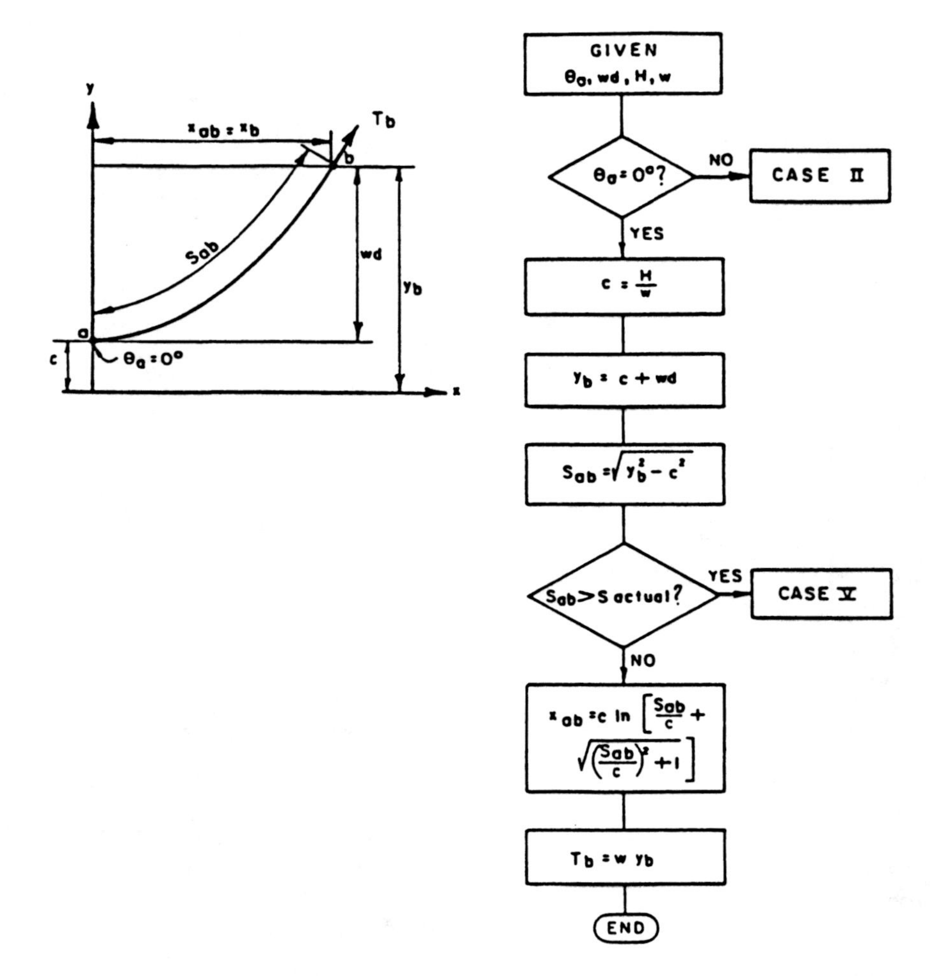

Figure 4–20. Case I. (U.S. Navy, 1985.)

the catenary is not at the anchor, but is some distance below the bottom. The length of the chain from anchor to buoy, S_{ab}, the tension in the mooring line at the buoy or surface, T_b, and the horizontal distance from the anchor to the surface, x_{ab}, are desired. Procedures for determining these values are presented in Figure 4–21.

4.6.3.3 Case III

The known variables are the horizontal distance from the anchor to the buoy, x_{ab}; the water depth, wd; the horizontal load, H; and the submerged unit weight of the chain, w. This situation arises when it is necessary to limit the horizontal distance from buoy to anchor due to space limitations. The length of chain from anchor to buoy, S_{ab}; the tension in the mooring line at the buoy, T_b; and the vertical load at the anchor, V_a, are required. Procedures for determining these values are outlined in Figure 4–22.

4.6.3.4 Case IV

The known variables are the water depth, wd; the horizontal load, H; the submerged unit weight of the chain, w; the angle at

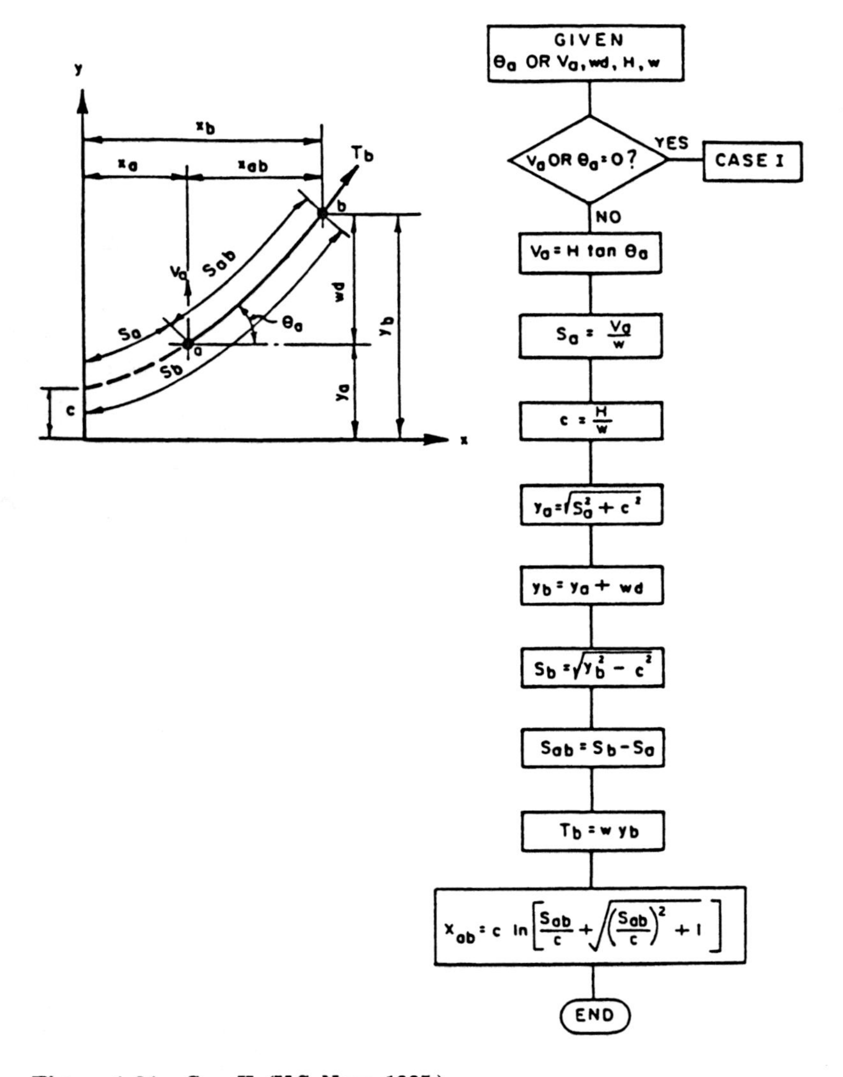

Figure 4–21. Case II. (U.S. Navy, 1985.)

the anchor, θ_a; the sinker weight, W; the unit weight of the sinker, γ_s; the unit weight of water, γ_w; and the length of chain from anchor to sinker, S_{ab}. The mooring consists of a chain of constant unit weight with a sinker attached to it. The total length of chain, S_{ac}; the distance of the top of the sinker off the bottom, y_s; and the tension in the mooring line at the buoy, T_{c2}, are desired. The solution to this problem is outlined in Figure 4–23.

4.6.3.5 Case V

The known variables are the water depth, wd, the horizontal load on the chain, H, the submerged unit weight of the chain, w, and the length of chain from anchor to buoy, S_{ab}.

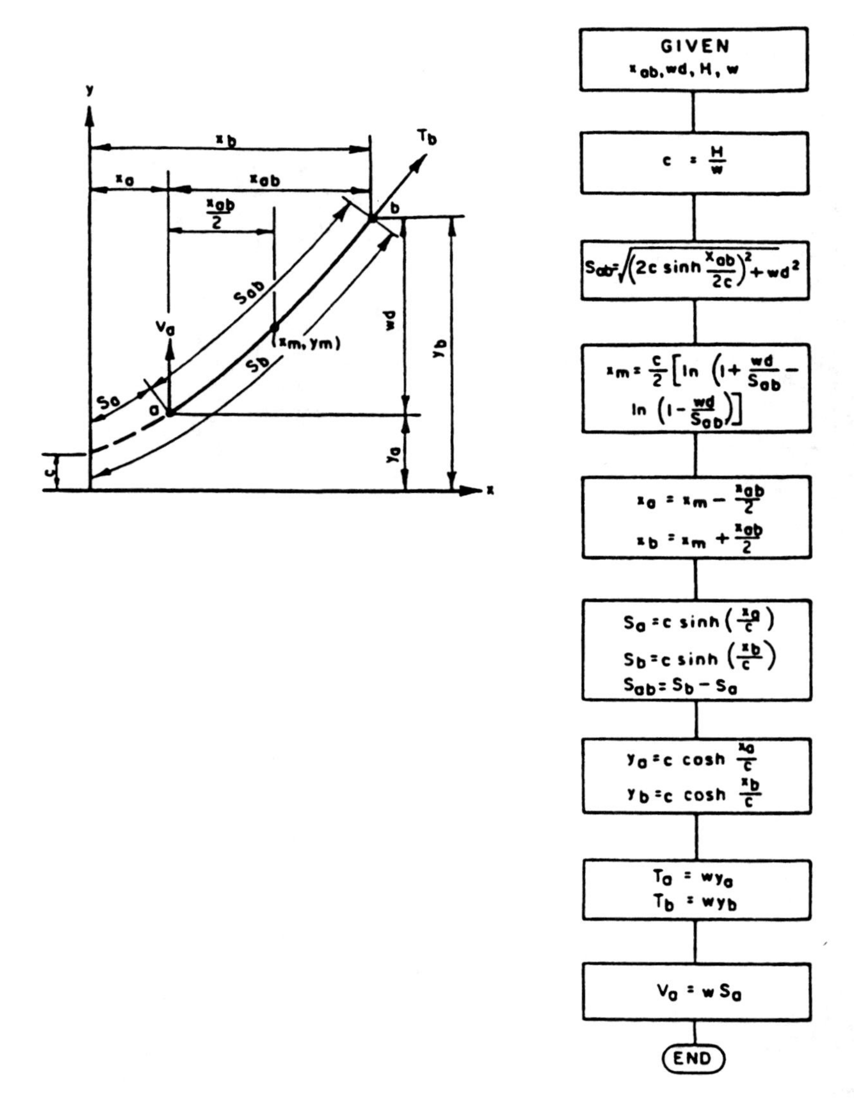

Figure 4–22. Case III. (U.S. Navy, 1985.)

The horizontal load, H, is sufficiently large to lift the entire chain off the bottom, resulting in an unknown vertical load at the anchor, V_a. This situation arises when one is computing points on a load–deflection curve for higher values of load.

The solution involves determining the vertical load at the anchor, V_a, using the trial-and-error procedure presented in Figure 4–24. The problem is solved efficiently using a Newton–Raphson iteration method (Gerald, 1980); this method gives accurate solutions in two or three iterations, provided the initial estimate is close to the final answer.

4.6.3.6 Load–Deflection Curve

On loading, a vessel connected to a mooring will deflect from its initial position in the

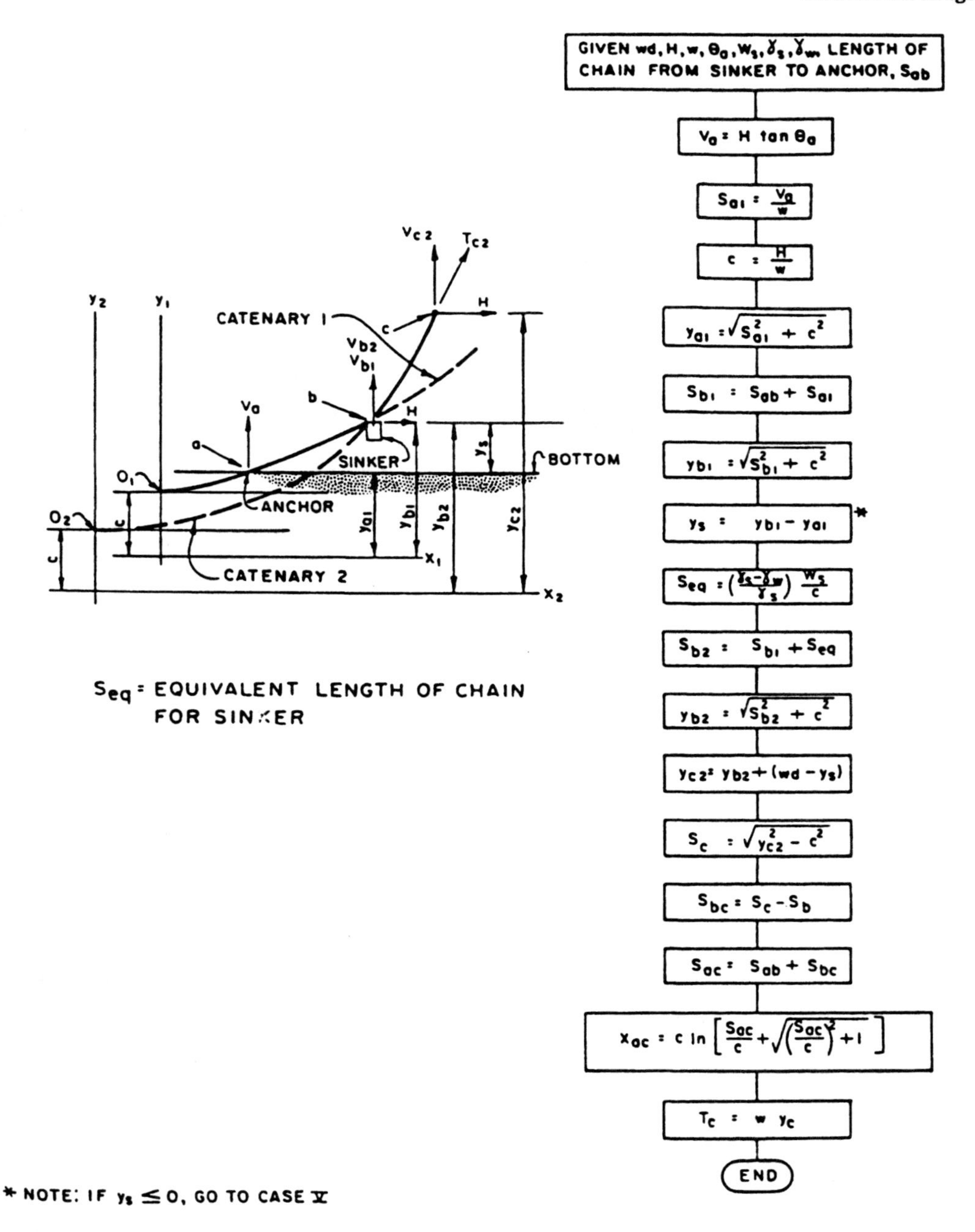

Figure 4–23. Case IV. (U.S. Navy, 1985.)

direction of the applied load. As the vessel moves, the restraining force in the mooring chain will increase from its initial or pretension value. A plot of the restraining force in the catenary mooring chain versus the deflection of the vessel is known as a load–deflection curve and an example load–deflection curve is shown in Figure 4–25. A load–deflection curve can be used to determine vessel movement for a given applied load.

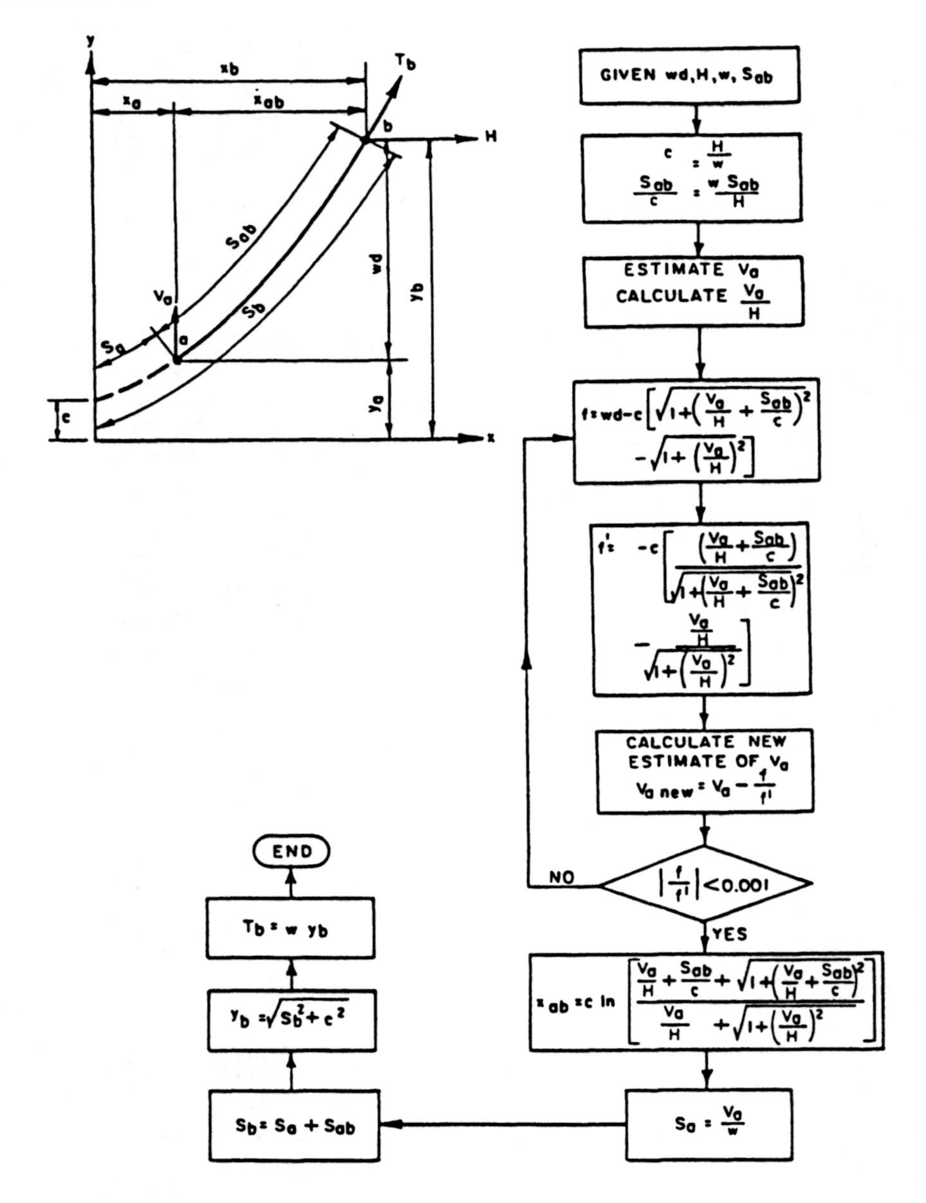

Figure 4-24. Case V. (U.S. Navy, 1985.)

This information is useful for planning a mooring layout and estimating the amount of area required to moor a vessel under normal and design conditions. The load–deflection curve also provides information on the energy-absorbing capability of a mooring. This information is obtained by applying the concepts of work and energy to the load–deflection curve.

The work required to move a vessel from its initial to its equilibrium position under the action of wind and current loading is equal to the increase in potential energy of the mooring-line system. An increase in

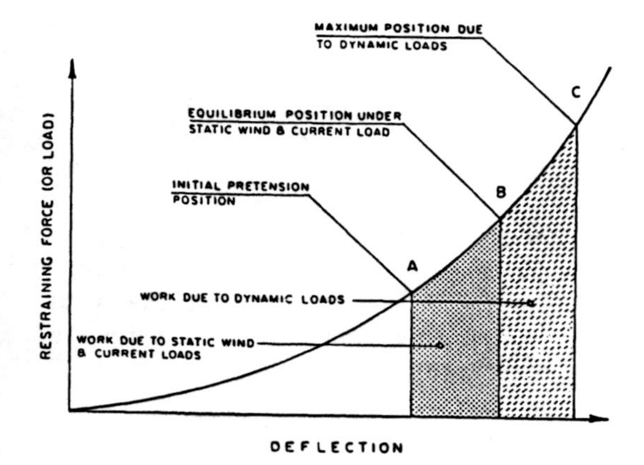

Figure 4–25. Load–deflection curve. (U.S. Navy, 1985.)

potential energy results as the mooring lines and hardware are raised under loading. The principle of work–energy dictates that the work done on the vessel as it is moved from its initial position to its equilibrium position is equal to the area under the load–deflection curve as shown in Figure 4–25. Point *A* denotes the initial position of the vessel resulting from the initial pretension in the mooring line. Point *B* denotes the equilibrium position of the vessel after the static wind and current loads have been applied. The area under the load–deflection curve between Points *A* and *B* represents the work done on the vessel by the static wind and current loads. If dynamic wind, current, or wave loads are also applied, then additional work will be done on the vessel. Point *C* denotes the maximum position due to dynamic wind, current, or wave loads. The area under the load–deflection curve from Point *B* to Point *C* represents the work done on the vessel by dynamic loads. This additional work must be absorbed by the mooring system without allowing the maximum load in the mooring line (Point *C*) to exceed the working load of the mooring line. The maximum dynamic mooring load (Point *C*) is generally difficult to determine. However, where moderate dynamic effects, such as those due to wind gusts, are anticipated, a resilient mooring capable of absorbing work (or energy) is desired.

Sinkers can be used to make a mooring more resilient. Figure 4–26 illustrates how

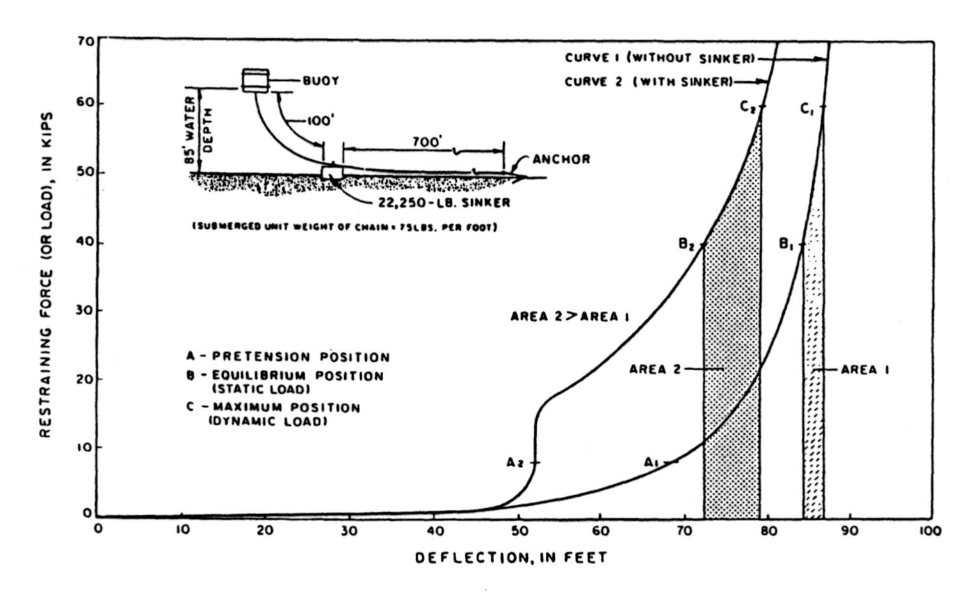

Figure 4–26. Load–deflection curves. (U.S. Navy, 1985.) *Note*: 1 foot = 0.3048 meters; 1 pound = 4.448 newtons.

a sinker can be used to increase the energy absorption of a mooring. Curve 1 is the load–deflection curve for a mooring system without a sinker. Curve 2 is the load–deflection curve for the same mooring system with a sinker added to it. The portion of the load–deflection curve that rises vertically on Curve 2 corresponds to the loads that lift the sinker off the bottom. Points *A*, *B*, and *C* represent the pretension position, equilibrium position under static loading, and the maximum position under dynamic loading, respectively. The sinker is added to the mooring line to increase the energy-absorbing capability of the mooring between Points *B* and *C*. The shaded Areas 1 and 2 under load–deflection Curves 1 and 2 represent the amount of energy absorbed by the mooring without the sinker and with the sinker, respectively. Clearly, the amount of energy absorbed between Points *B* and *C* by the mooring equipped with a sinker is considerably larger than that absorbed by the mooring without a sinker.

4.6.4 Anchor Design

This section provides procedures for designing drag anchors. Procedures for selecting pile, deadweight, and direct-embedment anchors are not included, but can be found in Rocker (1985). The procedure for designing drag anchors consists of determining anchor holding capacity, burial depth, and drag distance. The required holding capacity used in anchor selection should be the maximum computed horizontal mooring-line load.

The soil type, soil depth, and variation of soil type over the mooring area should be known. The soil type encountered in most mooring designs where drag anchors can be used may be classified as either mud or sand. The soil depth is an important consideration because there must be sufficient soil depth for anchor embedment.

A suitable anchor must be chosen from one of the many available drag anchor types. The designer should refer to the manufacturers' literature for detailed information regarding anchor specifications and performance. Once the anchor type has been selected, the anchor size (weight) is chosen to satisfy the required holding capacity. Figures 4–27 and 4–28 provide maximum holding capacity as a function of anchor weight for several anchor types in sand and clay/silt bottoms, respectively. The required maximum holding capacity, H_M, is determined by applying a factor of safety, *FS*, to the horizontal load, *H*:

$$H_M = FSH. \qquad (4\text{–}45)$$

A factor of safety of 1.5 to 2.0 is normally used in design.

When H_M and the anchor type are known, Figures 4–27 and 4–28 provide the required anchor weight (in air) for sand and clay/silt bottoms, respectively.

Anchor holding capacities determined from Figures 4–27 and 4–28 assume there is a sufficient depth of soil to allow for anchor penetration. At some sites, however, there may be a limited layer of soil overlying a hard strata such as coral or rock. The soil-depth requirements (i.e., maximum fluke-tip penetration) for various types of anchors in sand and mud is summarized in Table 4–3.

If the depth of sediment is less than that determined from the above procedures, then the anchor capacity must be reduced. The reduced anchor capacity due to insufficient sediment depth is computed from the following equation:

$$H_{Ar} = fH_a \qquad (4\text{–}46)$$

where

H_{Ar} = reduced anchor capacity
f = anchor capacity reduction factor
H_a = anchor capacity for full sediment depth.

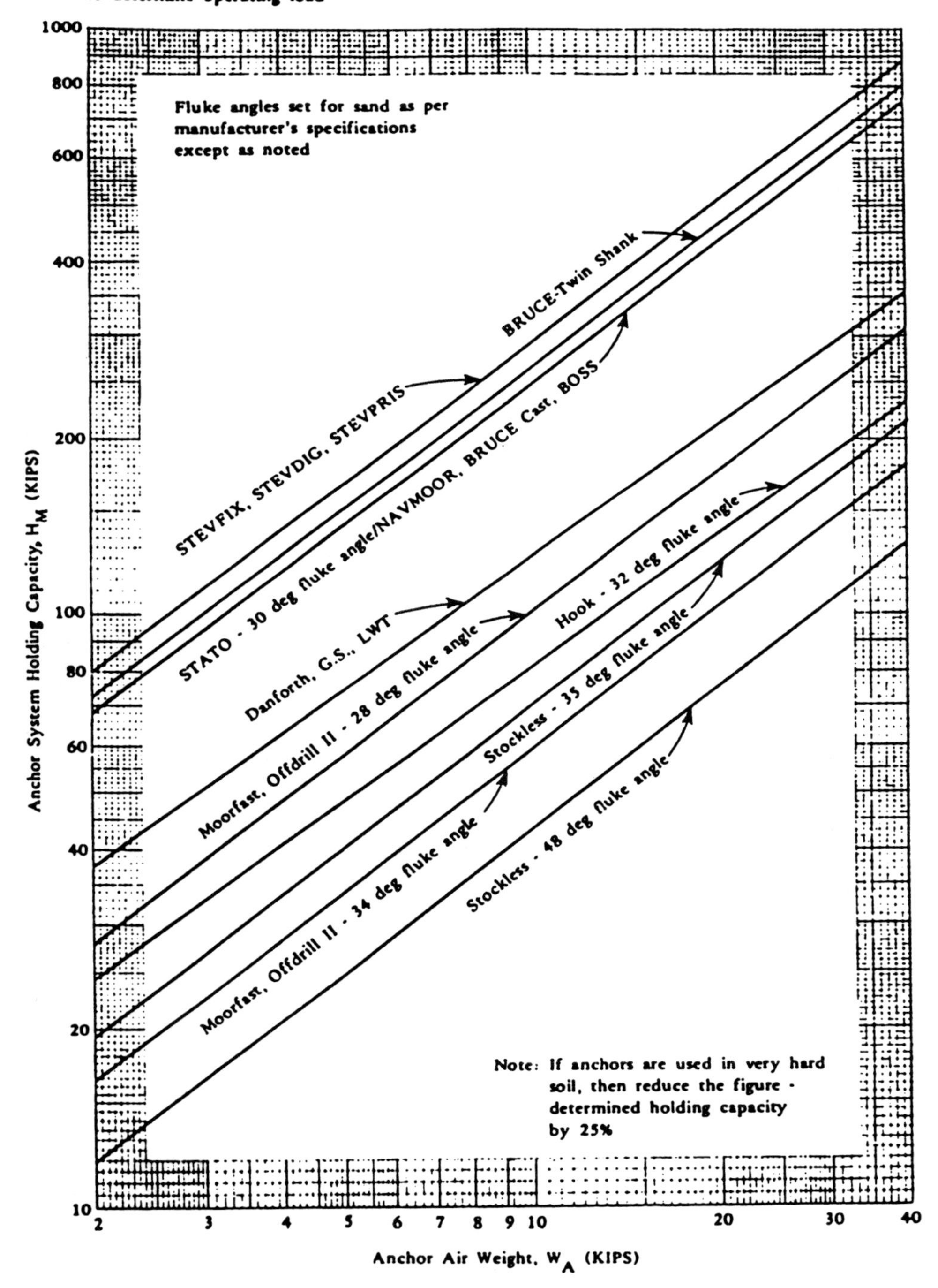

Figure 4–27. Drag anchor holding capacity—sand bottom. (U.S. Navy, 1985.)

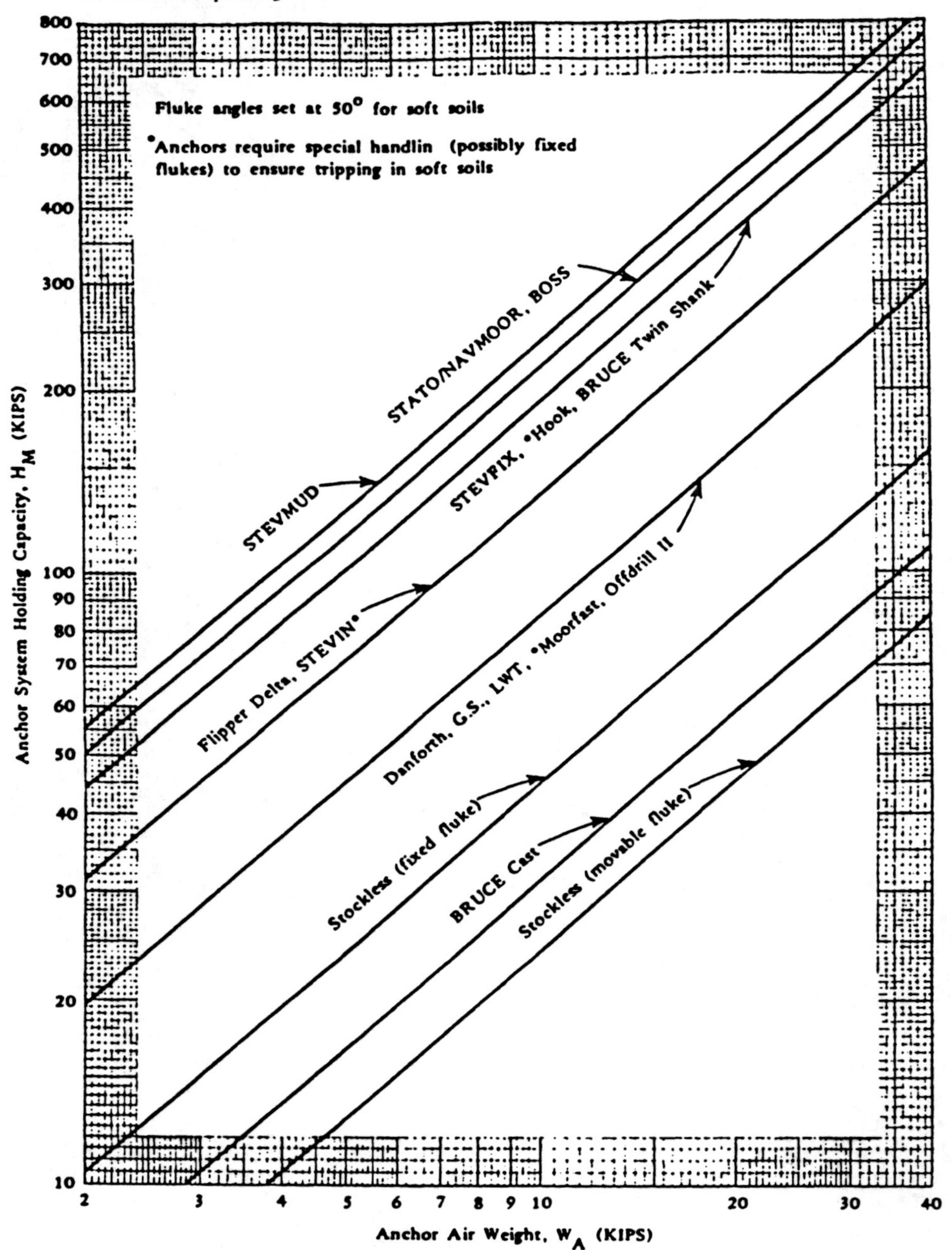

Figure 4–28. Drag anchor holding capacity—Mud bottom. (U.S. Navy, 1985.)

The correction factor, f, is determined using the following equation:

$$f = \frac{actual\ soil\ depth}{required\ soil\ depth} \tag{4–47}$$

where the required soil depths for both mud and sand are determined from Table 4–3.

Anchor holding capacity increases with drag distance; however, for many mooring applications anchor drag distance must be limited. Anchor drag distances in sand for the Stockless and Stato anchors are determined from Figure 4–29, which presents a plot of the percent of maximum capacity (ordinate) versus the normalized drag distance (abscissa). The drag distances indicated in Figure 4–29 are based on a factor of safety of 1.5 for the Stockless anchor and 2 for the Stato anchor. The anchor drag distance in sand for the various commercially available anchors is estimated to be about 3.5 to 4 fluke lengths, corresponding to a factor of safety of 2.

Anchor drag distances in mud can be determined from Figure 4–30. The drag distances for the Stockless (factor of safety of 1.5) and Stato (factor of safety of 2) anchors are indicated in Figure 4–30. Figure 4–31 provides anchor drag distance for various commercially available anchors.

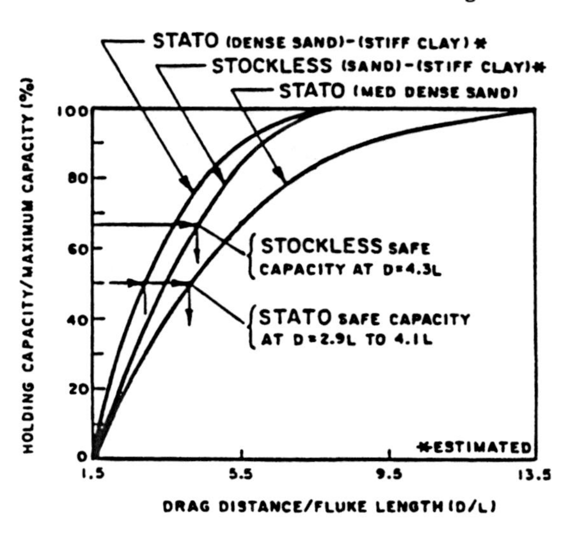

Figure 4–29. Anchor drag distance in sand. (U.S. Navy, 1985.)

Increased holding capacity can be achieved by using combinations of anchors. Methods for using multiple anchors are described in U.S. Navy (1985). Information on high-capacity hollow fluke anchors and special anchors, for example, self-bearing gravity type or suction type anchors is found in Tsinker (1986).

Table 4–3. Estimate Anchor Fluke Tip Penetration

Anchor Type	Fluke-Tip Penetration/Fluke Length	
	Sand/Stiff Clay	Mud (soft silts and clays)
Stockless	1	3 (fixed fluke)
Moorfast		
Offdrill II	1	4
Stato	1	4.5
Stevfix		
Flipper Delta		
Boss		
Danforth		
LWT		
GS (type 2)		
Bruce Twin Shank		
Stevmud	1	5.5
Hook	1	6

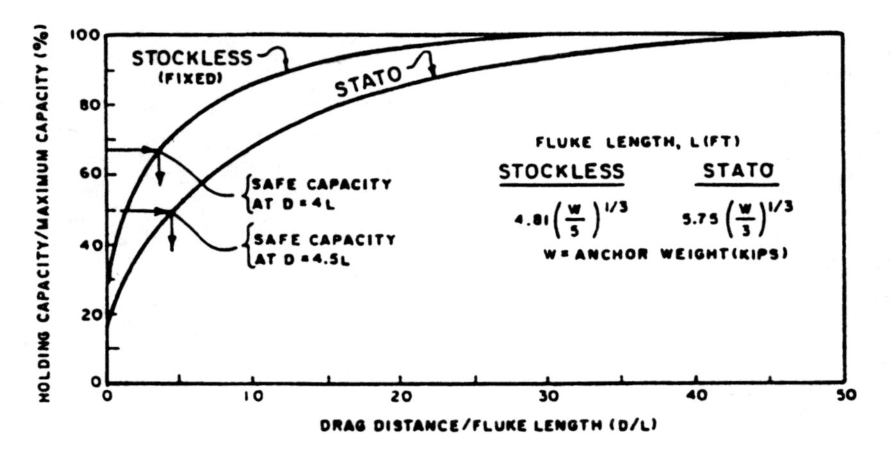

Figure 4–30. Anchor drag distance in mud. (U.S. Navy, 1985.)

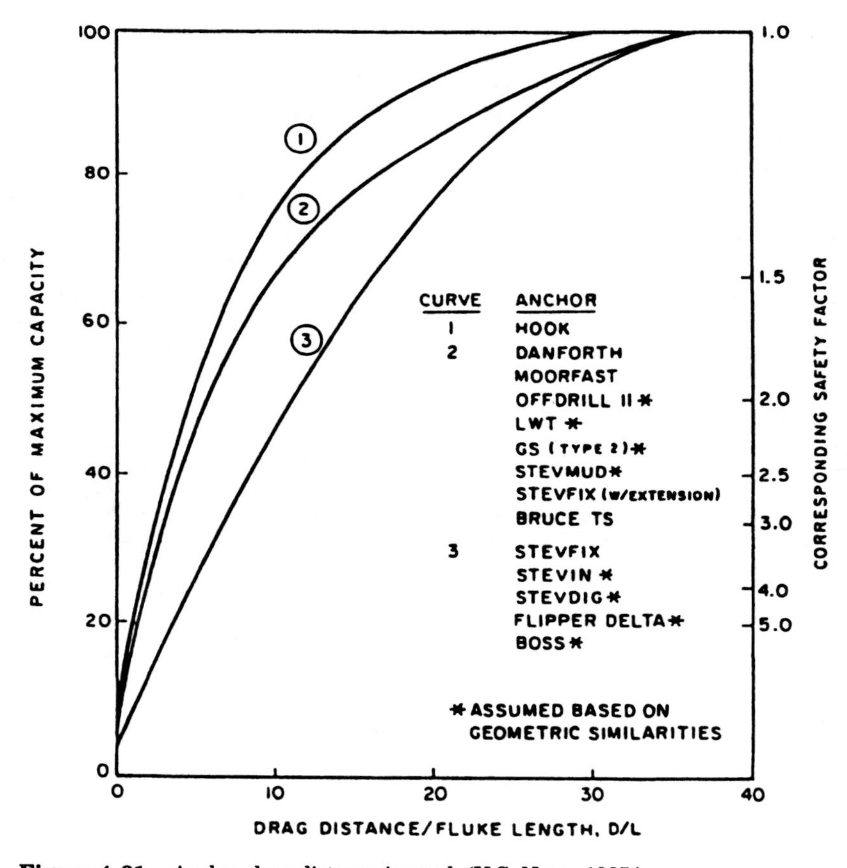

Figure 4–31. Anchor drag distance in mud. (U.S. Navy, 1985.)

4.7 LOADS ON MOORING ELEMENTS

4.7.1 Static Versus Dynamic Analysis

Loads on mooring elements (i.e., mooring lines) can be evaluated by means of static or dynamic analysis methods. Static analysis is appropriate when one is reasonably sure that the ship-mooring system will not experience significant dynamic motions under design environmental conditions. Dynamic analysis, on the other hand, is required when large dynamic motions of the moored-ship system are anticipated. Unfortunately, it is difficult to determine, a priori, whether a ship-mooring system will experience important dynamic loading. This determination must be made for each design problem on the basis of environmental site conditions, vessel size, and mooring arrangement. Generally speaking, one can expect single- or multi-point moorings to experience significant dynamic loads when that mooring is exposed directly to ocean wave attack (i.e., when the mooring is located in the open ocean). Within a harbor or estuary, the ship-mooring system may or may not experience significant dynamic motions depending on the size of the waves in relation to the ship and mooring. Even in the absence of waves one cannot eliminate the possibility of significant dynamic motions of a moored ship system. Single-point moorings, for example, are prone to large dynamic loads when subjected to steady wind and current in the absence of waves. (This matter will be discussed in more detail later.) Furthermore, moored vessels may experience large dynamic loads as a result of the gustiness of the wind (i.e., changes in speed) or abrupt changes in wind direction (wind shift).

Table 4–4 summarizes common static and dynamic mooring problems. The purpose of this table is to assist the designer in identifying general situations where dynamic analysis may be required. The reader is reminded, however, that Table 4–4 is a crude guideline and that the need for dynamic analysis must be made on a case-by-case basis.

4.7.2 Static Analysis

Winds and currents produce a longitudinal load, a lateral load, and a yaw moment on a moored vessel. These loads displace and rotate the vessel relative to its position before the loads were applied. The vessel will move until it reaches an equilibrium position where the applied load is equal to the restraining load provided by the mooring lines. Procedures for determining the

Table 4–5. Common Static and Dynamic Mooring Problems

Static	Dynamic
Multiple-point mooring exposed to moderate wind and current	Single-point mooring exposed to high wind and current
Single-point mooring exposed to moderate wind and current	Single-point or multiple-point mooring exposed to ocean waves (and wind and current)
Floating pier exposed to wind and current	Single-point mooring exposed to sudden shifts in wind or current
	Single- or multiple-point mooring subject to impact of a large floating object
	Floating dock or pier subject to moderate waves or passing vessels

mooring-line loads differ depending on whether the mooring is a single- or a multiple-point mooring. In either case the first step in analyzing static loads on mooring elements is to determine the total longitudinal and lateral loads and total yaw moment on the moored vessel:

$$F_{xT} = F_{xw} + F_{xc} \quad (4\text{–}48)$$

$$F_{yT} = F_{yw} + F_{yc} \quad (4\text{–}49)$$

$$M_{xyT} = M_{xyw} + M_{xyc} \quad (4\text{–}50)$$

where

F_{xT} = total longitudinal force
F_{xw} = longitudinal wind force
F_{xc} = longitudinal current force
F_{yT} = total lateral force
F_{yw} = lateral wind force
F_{yc} = lateral current force
M_{xyT} = total moment
M_{xyw} = wind moment
M_{xyc} = current moment.

4.7.2.1 Single-Point Mooring

Figure 4–32 depicts a vessel secured to a single-point mooring exposed to wind and current. The relative angle between the wind and the current is θ_{wc}. The longitudinal and lateral forces (i.e., F_x and F_y) act through the center of gravity (c.g.) of the vessel and the yaw moment, M_{cg}, acts about the c.g. For static equilibrium, the applied loads must equal the restoring loads of the mooring system:

$$\sum F_x = 0 \quad (4\text{–}51)$$

$$\sum F_y = 0 \quad (4\text{–}52)$$

$$\sum M_{cg} = 0. \quad (4\text{–}53)$$

The vessel will adjust its position around the single-point mooring until the above equations of equilibrium are satisfied. The longitudinal force, lateral force, and yaw moment are a function of the angle between the vessel and the wind, θ_w, and the angle between the vessel and the current, θ_c.

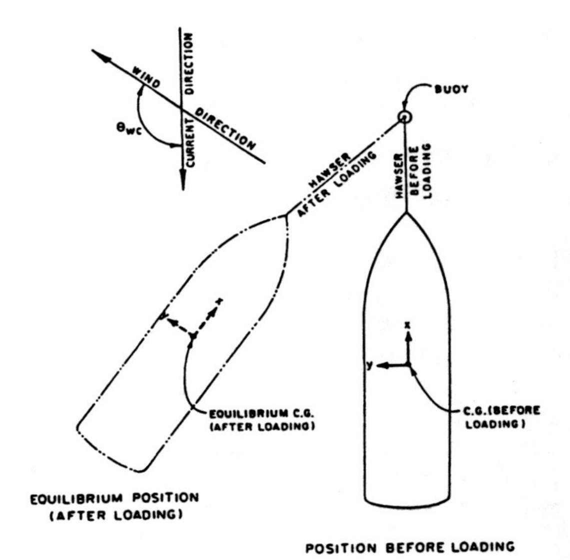

Figure 4–32. Static analysis of a single-point mooring. (U.S. Navy, 1985.)

These angles vary as the vessel achieves its equilibrium position. Computation of the maximum hawser load is a trial-and-error procedure in which the orientation of the vessel is continually adjusted until the point of zero moment is determined. The general procedure for determining the maximum load on a free-swinging (single-point) mooring involves assuming a ship position (θ_c and θ_w) and calculating the sum of moments on the vessel. This process is repeated until the sum of moments is equal to zero. The procedure is tedious and involves a number of iterations for each wind-current angle, θ_{wc} (angle between wind and current). Due to the large values of moment (which can be on the order of 10^5 to 10^7 foot-pounds) it is difficult to determine the precise location at which the sum of moments is zero. As a result, the point of equilibrium (zero moment) is best determined either by computer or graphically. The procedure involves halving the interval (between values of θ_c) for which the moment changes signs. A step-by-step procedure is given in Figure 4–33. An accompanying example plot of sum of moments,

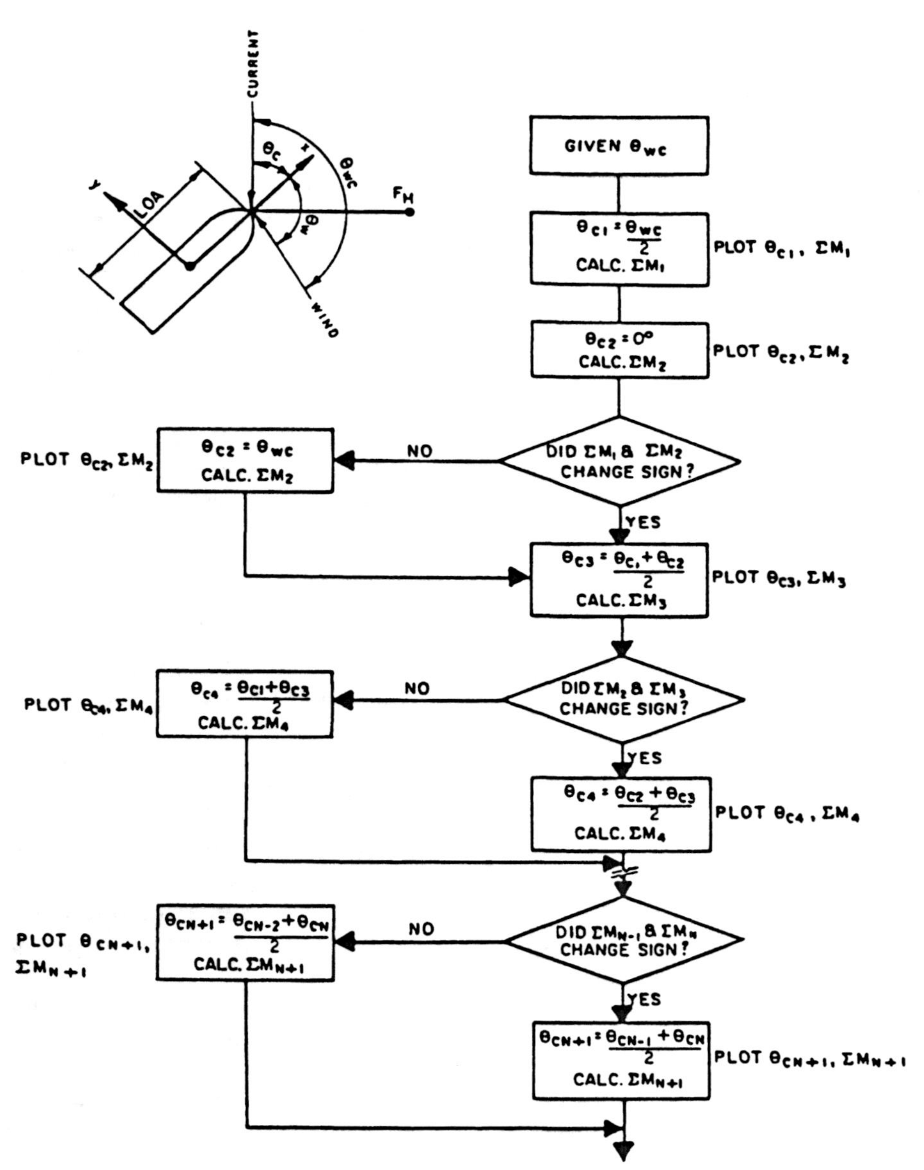

Figure 4–33. Static single-point mooring analysis method. (U.S. Navy, 1985.)

$\sum M$, versus current angle, θ_c, is shown in Figure 4–34.

$\sum M$ is determined as follows:

$$\sum M = M_{xyw} + M_{xyc} - F_{yT}ARM \qquad (4\text{–}54)$$

where

ARM = distance from bow hawser attachment point to center of gravity of vessel ($ARM = 0.48\,LOA$)

LOA = overall length of the vessel.

Once the point of zero moment has been found, the horizontal hawser load, H, is determined using the following equation:

$$H = \sqrt{F_{xT}^2 + F_{yT}^2}. \qquad (4\text{–}55)$$

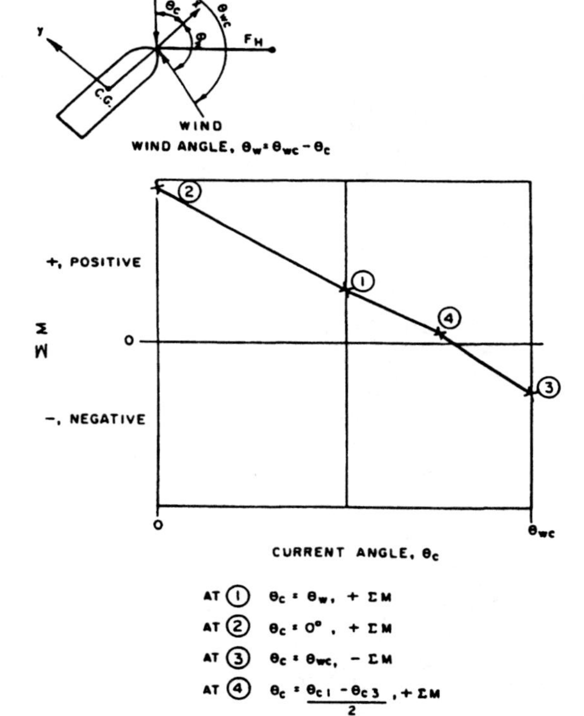

Figure 4–34. Example static single-point mooring analysis. (U.S. Navy, 1985.)

A computer program that performs the above proceudre has been published by Cox (1982).

4.7.2.2 Multiple-Point Moorings

The general procedure for determining the horizontal line loads in a multiple-point mooring is more complicated than the procedure described above for a single-point mooring because one must account for the stiffness of the mooring lines. Figure 4–35 depicts a typical spread mooring both before and after wind and current loads are applied. The vessel is reoriented as the applied load is distributed to the mooring lines. The mooring lines, which behave as catenaries, will lengthen or shorten until they are in equilibrium with the applied loads. Equations (4–51) through (4–53) must be satisfied for static equilibrium to exist. Determining the equilibrium position of the vessel under load is outlined as follows:

Step 1: Determine the total longitudinal load, lateral load, and yaw moment on the vessel due to wind and current.

Step 2: Determine the mooring-line configuration and the properties of each of the mooring lines. Calculate a load–deflection curve for each of the mooring lines using catenary analysis.

Step 3: Assume an initial displacement and rotation of the vessel (new orientation) under the applied load.

Step 4: Determine the deflection in each of the mooring lines corresponding to the vessel orientation.

Step 5: Determine the forces in each of the mooring lines from the above mooring-line deflections.

Step 6: Sum the forces and moments according to the above equations, accounting for all the mooring-line loads and applied wind and current loads.

Step 7: Determine if the restraining forces and moments due to all the mooring-line loads balance out the applied forces and moments due to wind and current.

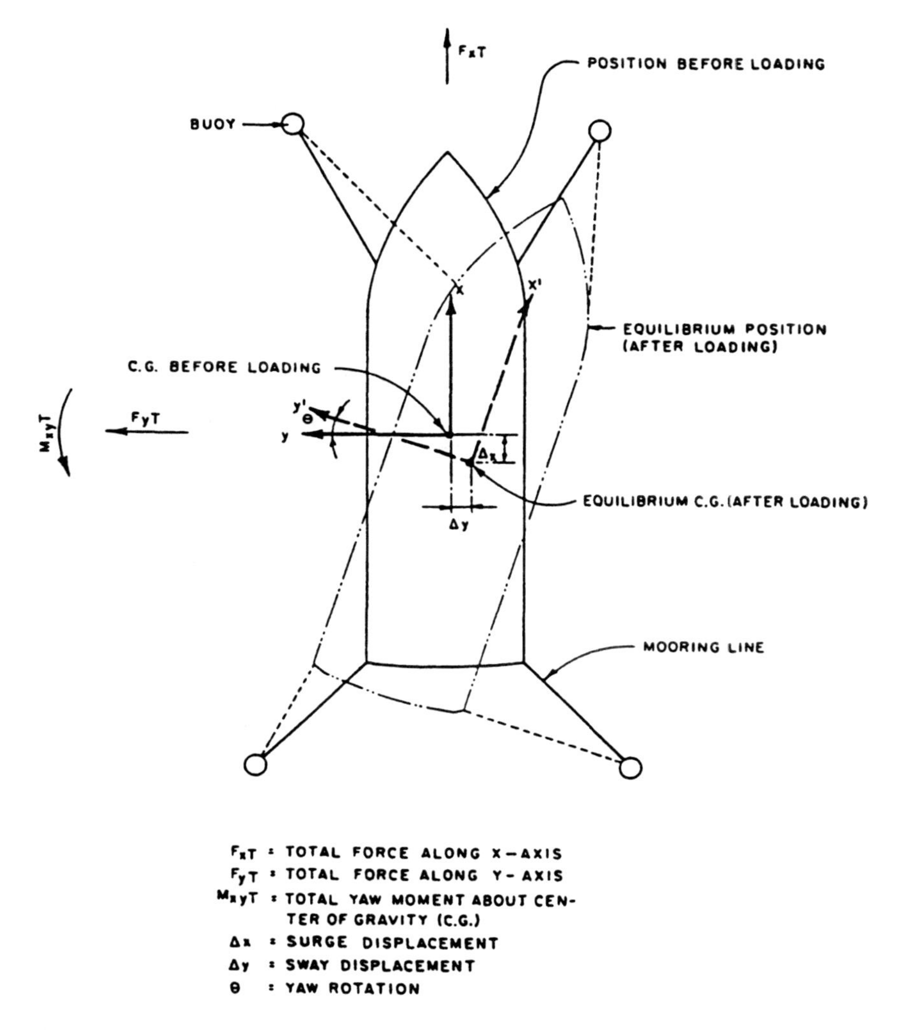

Figure 4–35. Spread mooring static analysis. (U.S. Navy, 1985.)

Step 8: If the above forces and moments do not balance, then the vessel is not in its equilibrium position under the applied load. A new vessel orientation must be assumed.

Step 9: Steps 3 through 8 are repeated until the equilibrium position of the vessel is determined.

Simplified methods for determining mooring-line loads in multiple-point moorings are presented in the following paragraphs. These simplified solutions have been used successfully in the past and are useful for preliminary design. The computer program summarized below and presented in U.S. Navy (1985) is recommended for final design or for preliminary designs where mooring geometries are more complicated than the following simple situations.

Bow-and-Stern Mooring

A force diagram for a typical bow-and-stern mooring is shown in Figure 4–36. In order to facilitate hand computations, a vessel in a bow-and-stern mooring is assumed to

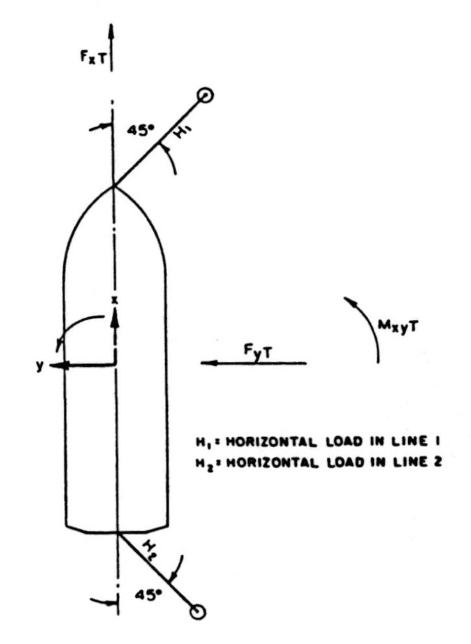

$$\Sigma F_x = 0: \quad F_xT + \cos 45^\circ H_1 - \cos 45^\circ H_2$$
$$\Sigma F_y = 0: \quad F_yT - \sin 45^\circ H_1 - \sin 45^\circ H_2$$

OR

$$H_2 = \frac{F_yT}{2 \sin 45^\circ} + \frac{F_xT}{2 \cos 45^\circ}$$
$$H_1 = H_2 - \frac{F_xT}{\cos 45^\circ}$$

NOTE: DESIGNER MUST BE SURE THAT ALL SIGNS ARE CONSISTENT WITH APPLIED LOADS AND LINE ORIENTATIONS

Figure 4–46. Bow stern mooring analysis. (U.S. Navy, 1985.)

move under applied loading until the mooring lines make an angle of 45° with the longitudinal axis of the vessel, as indicated in Figure 4–36. Horizontal loads in the bow line, H_1, and in the stern line, H_2, are determined by summing the forces in the x- and y-directions. Equations for line loads are given on Figure 4–36. Note that this procedure is an approximation that does not provide a moment balance.

Spread Mooring

A force diagram for a six (6) leg spread mooring is shown in Figure 4–37. The mooring consists of bow and stern mooring lines that resist longitudinal load, and four lateral load resisting mooring lines placed perpendicularly to the longitudinal axis of the vessel. The bow or stern mooring line is assumed to take the longitudinal mooring load entirely. Mooring-line loads may be determined from the equations shown on Figure 4–37.

Four-Point Mooring

A force diagram for a typical four-point mooring is shown in Figure 4–38. Each line in this mooring resists both longitudinal and lateral loads. Mooring-line loads may be determined from the equations shown on Figure 4–38.

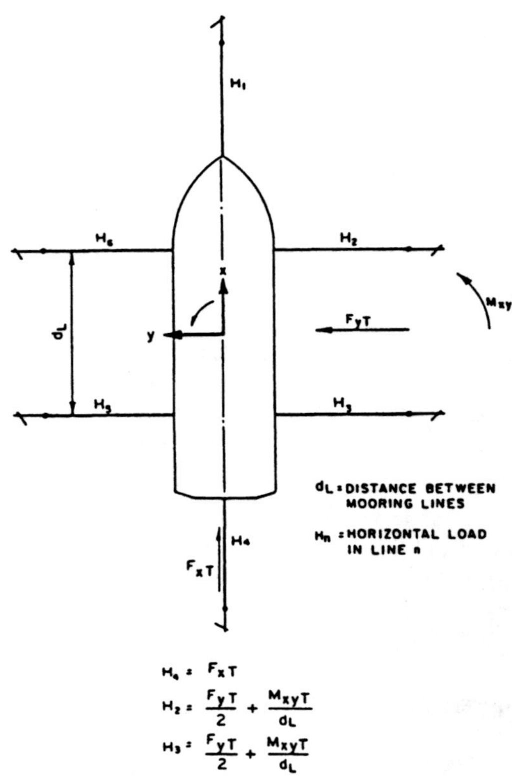

NOTE: DESIGNER MUST BE SURE THAT ALL SIGNS ARE CONSISTENT WITH APPLIED LOADS AND LINE ORIENTATIONS

Figure 4–37. Spread mooring analysis. (U.S. Navy, 1985.)

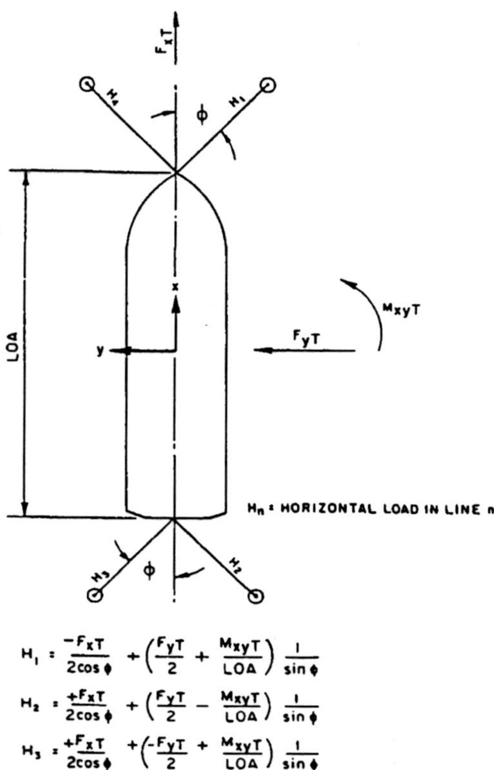

$$H_1 = \frac{-F_{xT}}{2\cos\phi} + \left(\frac{F_{yT}}{2} + \frac{M_{xyT}}{LOA}\right)\frac{1}{\sin\phi}$$

$$H_2 = \frac{+F_{xT}}{2\cos\phi} + \left(\frac{F_{yT}}{2} - \frac{M_{xyT}}{LOA}\right)\frac{1}{\sin\phi}$$

$$H_3 = \frac{+F_{xT}}{2\cos\phi} + \left(\frac{-F_{yT}}{2} + \frac{M_{xyT}}{LOA}\right)\frac{1}{\sin\phi}$$

$$H_4 = \frac{-F_{xT}}{2\cos\phi} + \left(\frac{-F_{yT}}{2} - \frac{M_{xyT}}{LOA}\right)\frac{1}{\sin\phi}$$

NOTE: DESIGNER MUST BE SURE THAT ALL SIGNS ARE CONSISTENT WITH APPLIED LOADS AND LINE ORIENTATIONS

Figure 4–38. Four-point mooring analysis. (U.S. Navy, 1985.)

4.7.2.3 Multiple-Point Mooring Computer Solution

A computer program for solution to multiple-point mooring problems is presented in U.S. Navy (1985). The program can be used to determine forces and displacements in the mooring systems of ships subjected to horizontal static applied loads. The moorings may be composed of mooring lines (hawsers) and anchor chains. The load–deflection characteristics of hawser materials are entered as input and anchor chains are computed as catenaries. The program can also be used to examine moorings with fenders; however, this feature of the program will not be discussed here.

Solutions are obtained iteratively, starting with the ship in an assumed position relative to its mooring points. Reactive loads in the lines and fenders are added successively to the applied forces to obtain the resultant surge and sway forces and yaw moment on the ship. Derivatives of these force components with respect to displacement in surge, sway, and yaw are also computed, and the Newton–Raphson method (Gerald, 1980) is used to get an approximation of the ship displacement that will bring the forces to equilibrium. The process is repeated with the ship in its new position, and continued until the resultant forces are within tolerance.

Figure 4–39 is a definition sketch showing the main dimensional variables used in the program. The coordinate system (referred to as the "global coordinate system") is defined relative to the ship's initial assumed position, with the origin, O, at the ship's center of gravity. The x-axis coincides with the ship's longitudinal axis; location of the y-axis is arbitrary, but it is convenient to locate the y-axis along the transverse axis of the ship. When the ship moves as the result of unbalanced forces, the center of gravity moves to a new location, designated S and referred to as the ship origin (or origin of ship local coordinate system). Three variables are needed to describe the new position: the surge displacement, x, the sway displacement, y, and the yaw angle, θ. Positive yaw is measured from the positive x-axis toward the positive y-axis. The horizontal component of tension in a mooring line, such as that shown as AC in Figure 4–39, can be determined as a function of the horizontal distance between the attachment point (chock) on the ship and the fixed anchor or mooring point. The procedure for computing load–deflection curves is described later. The load–deflection curves are stored as series of load–distance pair. Using simple geometry, the following steps lead to expressions for the mooring-line length and its direction:

$$x_2 = x_c \cos\theta - y_c \sin\theta \quad (4\text{–}56)$$

$$y_2 = x_c \sin\theta + y_c \cos\theta \quad (4\text{–}57)$$

$$x_3 = x_1 - x - x_2 \quad (4\text{–}58)$$

$$y_3 = y_1 - y - y_2 \quad (4\text{–}59)$$

$$r = \sqrt{x_3^2 + y_3^2} \quad (4\text{–}60)$$

$$\cos\theta_3 = \frac{x_3}{r} \quad (4\text{–}61)$$

$$\sin\theta_3 = \frac{y_3}{r}. \quad (4\text{–}62)$$

Variables in the above formulae are defined in Figure 4–39.

The x- and y-components of the force exerted by the mooring line on the ship, and the moments (due to the mooring line) about the ship origin, S, are then given by:

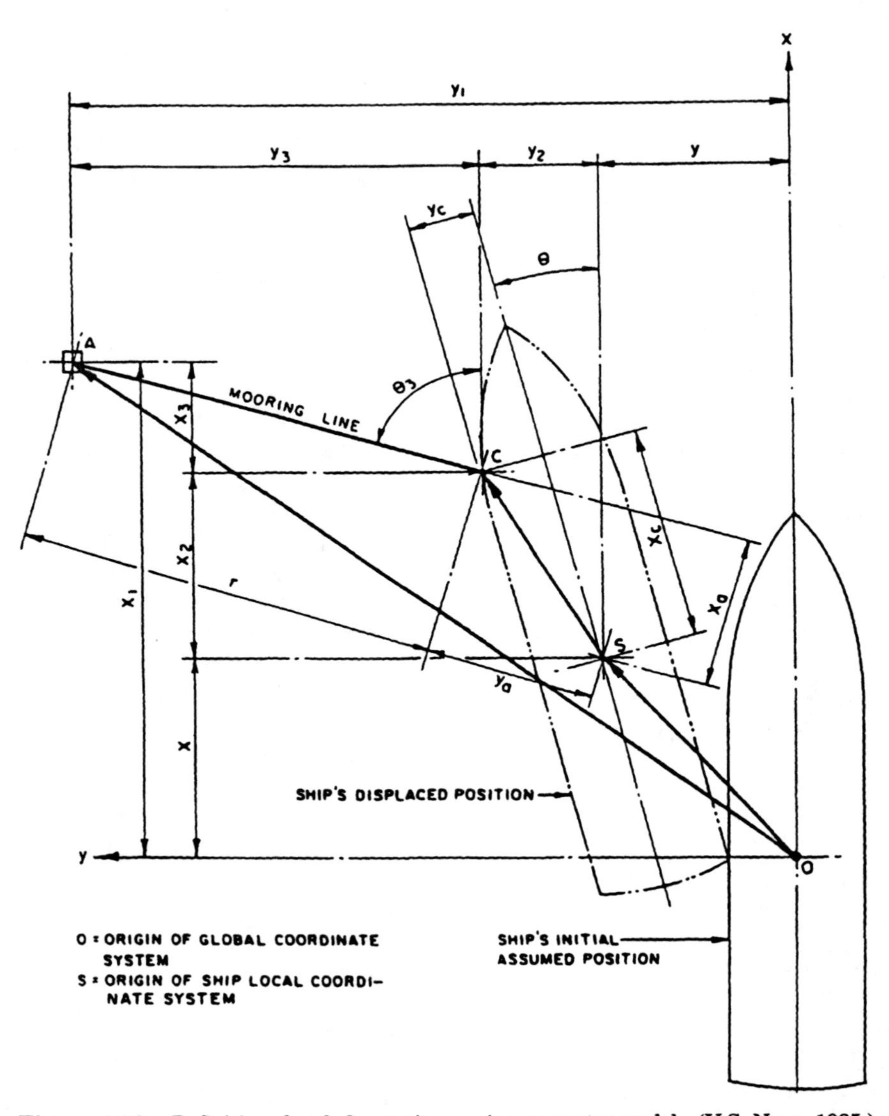

Figure 4–39. Definition sketch for static mooring computer model. (U.S. Navy, 1985.)

$$H = f(r) \tag{4–63}$$

$$F_x = H\cos\theta_3 \tag{4–64}$$

$$F_y = H\sin\theta_3 \tag{4–65}$$

$$M_{xy} = F_y x_2 - F_x y_2 \tag{4–66}$$

where H is the horizontal component of the mooring line force and the other variables are defined in Figure 4–39.

In order to implement the Newton–Raphson method it is also necessary to compute the derivatives of the above force components with respect to x, y, and θ. Expressions for these derivatives are presented in U.S. Navy (1985).

The total surge force on the ship is obtained by summing expressions such as Eqs. (4–64) through (4–66) for each mooring line and adding the applied x-force (due to wind and current). Total sway force and yaw moment are computed in the same way, and the derivative expressions are also summed over all lines. It is assumed in the computation that the applied loads remain constant during changes in ship position and orientation.

The Newton–Raphson method is used to arrive at values of x, y, and θ for which the total force and moment on the ship are zero. In the expressions for the total differential of force components

$$dF_i = \frac{dF_i}{dx}\,dx + \frac{dF_i}{dy}\,dy + \frac{dF_i}{d\theta}\,d\theta \tag{4–67}$$

the differential motions forces are approximated by finite increments as follows:

$$\sum\frac{dF_x}{dx}\,\Delta x + \sum\frac{dF_x}{dy}\,\Delta y + \sum\frac{dF_x}{d\theta}\,\Delta\theta = -F_{xa} - \sum F_x \tag{4–68}$$

$$\sum\frac{dM_{xy}}{dx}\,\Delta x + \sum\frac{dM_{xy}}{dy}\,\Delta y + \sum\frac{dM_{xy}}{d\theta}\,\Delta\theta = -M_{xya} - \sum M_{xy}. \tag{4–69}$$

This set of equations is solved for Δx, Δy, $\Delta\theta$; the ship is moved to:

$$x + \Delta x \tag{4–70}$$

$$y + \Delta y \tag{4–71}$$

$$\theta + \Delta\theta \tag{4–72}$$

and the process is repeated until the computed total force components are all within the desired tolerance.

The expressions given above for mooring-line forces and their derivatives are applicable to hawsers and to anchor chains, both of which run between a mooring point and a fixed chock on the ship.

Load–deflection curves for each mooring line and anchor chain must be available in order to proceed with the iterative computation described above. The characteristics of a hawser as accepted by the programs are illustrated in Figure 4–40.

The line is assumed to be weightless. It runs from the mooring point to a chock on the ship, and there may be additional on-deck length between the chock and the point of attachment. The hawser may be made of elastic steel wire, or of other material for which a dimensionless load–deflection table has been furnished. If the hawser is steel, it may have a synthetic tail. Chocks are frictionless.

Anchor chain load–deflection curves are computed with the aid of catenary equations, as described in Section 4.6.2. The most general system that can be handled by the program is shown in Figure 4–40. It consists of lower and upper sections of chain which can be of different weights, a sinker at the connection point, and a hawser between the ship and the mooring buoy. Hawser characteristics are as described above, except that the total line load and outboard line length are used in place of their horizontal projections (that is, the hawser is assumed to run horizontally between buoy and ship). The buoy, if present, is assumed always to remain at the water surface.

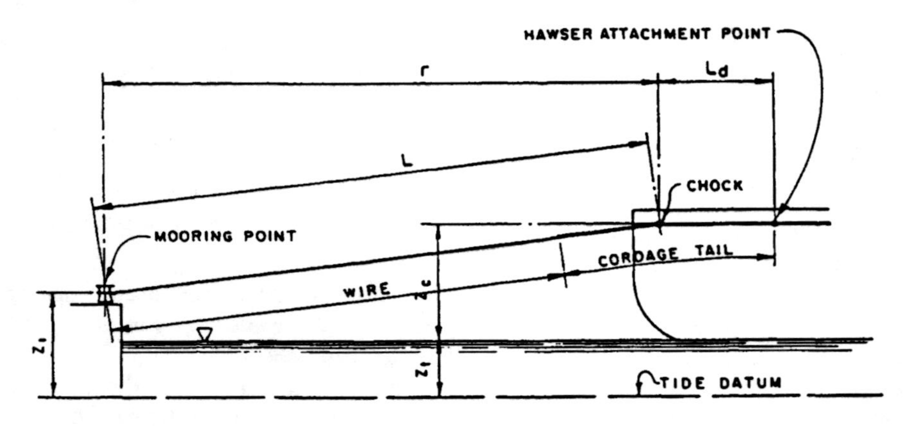

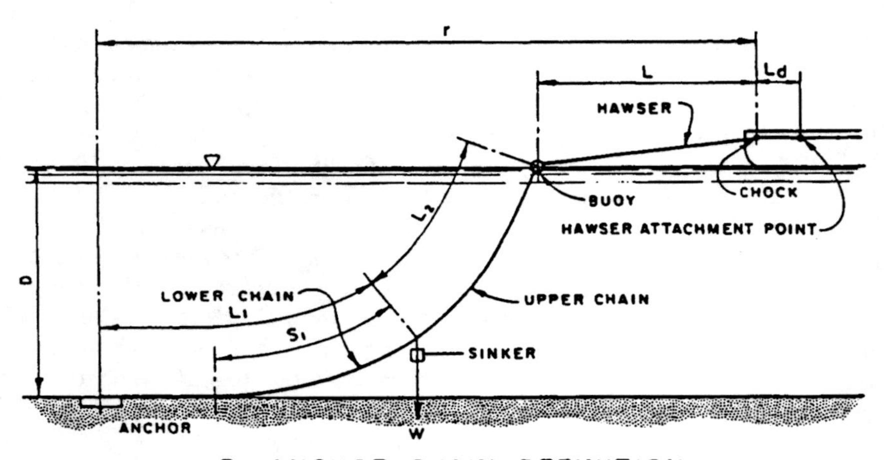

Figure 4–40. Definition sketch for hawser and anchor chain in static mooring computer model. (U.S. Navy, 1985.)

The horizontal length of chain systems subjected to a given horizontal load can be calculated from simple equations once the length of chain raised off the bottom and the vertical force on the anchor are known. Four cases must be distinguished: Case 1—upper chain partly raised, Case 2—upper chain completely raised but sinker on the bottom, Case 3—lower chain partly raised, and Case 4—lower chain completely raised. In computing a load–deflection curve, the four cases are examined in sequence to determine which one prevails. As the load increases, fewer cases need to be considered. A Newton–Raphson method algorithm is used to solve for raised chain length and vertical anchor load in Cases 3 and 4, respectively. The equations used are provided in U.S. Navy (1985).

4.7.3 Dynamic Analysis

An introductory discussion of the dynamics of floating bodies is presented in Chapter 5 of this book in connection with the design of floating breakwaters. This material is not repeated in this chapter; however, the reader is reminded that two basic mathematical approaches are possible; namely: (1) frequency domain analysis and (2) time domain analysis. Frequency domain analysis provides a computationally efficient approach that can be used to estimate first-order wave-induced mooring loads. On the other hand, frequency domain analyses are not as amenable to evaluation of second-order wave drift forces, nonlinear mooring line behavior, and, as discussed in the sections that follow, the slow "fishtailing" motions of single point moorings. Accordingly, time domain analysis is often preferred in design applications. A complete review of computational procedures for performing dynamic analyses of offshore moorings is beyond the scope of this chapter. Instead, the following paragraphs will summarize the dynamic analyses of two typical offshore moorings, namely, a spread mooring and a single-point mooring.

4.7.3.1 Dynamic Analysis of a Spread Mooring

Floating dry docks are normally moored within a harbor and secured by means of anchors and chains in a spread mooring. Because floating dry docks cannot get underway in advance of a storm, their moorings must be designed to withstand extreme events such as hurricanes. Floating dry dock moorings are often designed for wind and current using static analysis methods; however, recent experience indicates that dynamic motions resulting from wind gusts and waves may also be important to design. The following sections, taken from Headland et al. (1989), evaluate dynamic loading on spread moored floating dry docks using a time-domain mooring dynamics model. Although the discussion is limited to a moored dry dock, the analysis techniques are germane to the design of any large floating structure located in a partially sheltered harbor or estuary.

Dynamic analyses were performed on the U.S. Navy AFDM-10 floating dry dock shown in Figure 4–41 moored as shown in Figure 4–42 by 10 mooring legs. The mooring is located in a water depth of about 12.2 m, occupied by a large naval vessel, and ballasted to a draft of 4.9 m. This mooring arrangement and loading situation typifies working conditions for the AFDM-10 dry dock. Geometrical characteristics of the dry dock are summarized in Table 4–5. The mooring lines used to restrain the dry dock consist of a 70-mm chain with a breaking strength of 3.7 MN (825 000 pounds). Sinkers, weighing 44.6 tons (10 000 pounds), are located on each leg. Figure 4–43 presents a load–deflection curve for each mooring leg. Currents at the dry dock location were negligible.

Dynamic mooring analyses were performed using a three-degree-of-freedom time domain model. The numerical model, which is based on work by Van Oortmerssen (1976), integrates equations of motion for surge, sway, and yaw (i.e., x-direction, y-direction, and yaw moment direction in Figure 4–12). The equations of motion are as follows:

Table 4–5. Floating Dry Dock Characteristics

Length	552 ft
Beam	124 ft
Draft	16 ft
Displacement	20 000 long tons
Lateral wind area	62 000 ft^2
Longitudinal wind area	10 800 ft^2

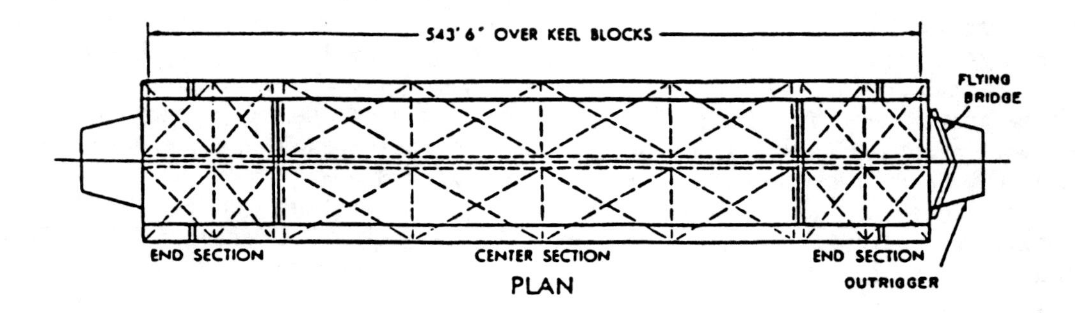

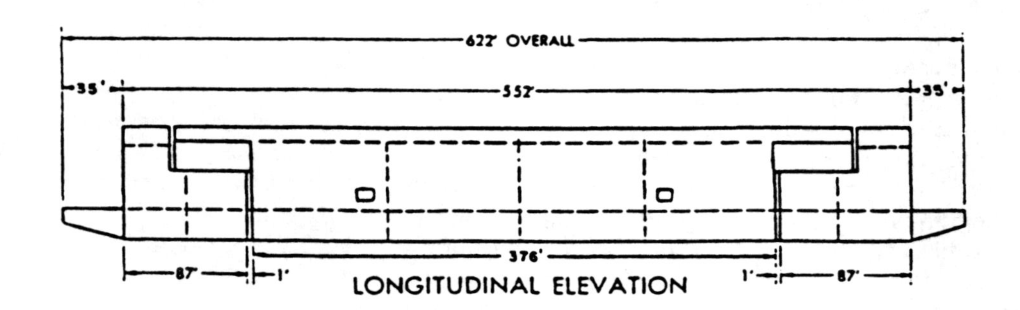

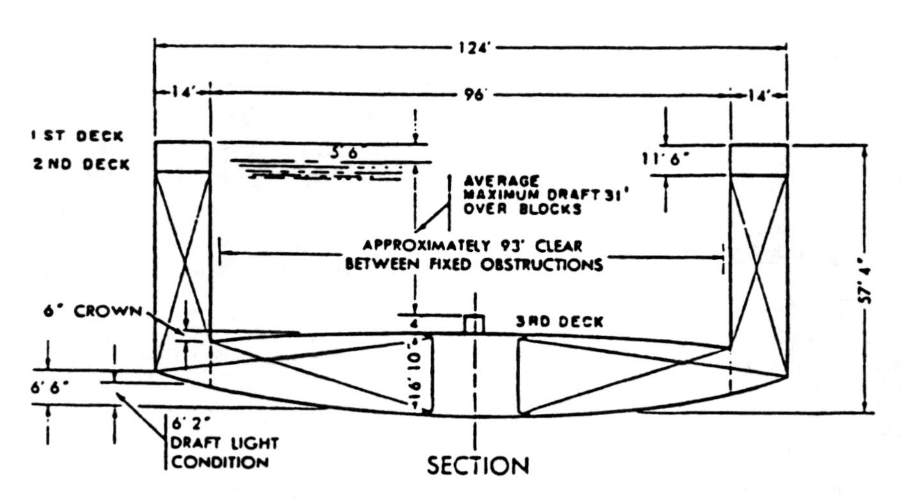

Figure 4–41. AFDM floating dry dock. *Note*: 1 foot = 0.3048 meters.

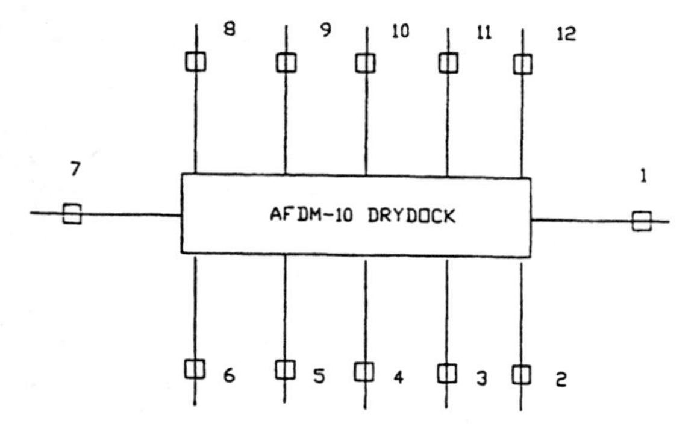

Figure 4–42. Floating dry dock mooring geometry.

$$\sum_{j=1,2,6} [(M_{kj} + m_{kj})\ddot{x}_j + \int_{-\infty}^{t} K_{kj}(t-\tau)\dot{x}_j(\tau)\, d\tau + C_{kj}x_j = F_k(t) \qquad (4\text{–}73)$$

where

k = denotes motion in the i-th mode (1 = surge, 2 = sway, 6 = yaw)

j = denotes motion in the j-th mode

$x_j, \dot{x}_j, \ddot{x}_j$ = acceleration, velocity, and displacement in the j-th mode

$M_{k,j}$ = vessel mass in modes i, j

$m_{k,j}$ = constant inertial coefficient in modes i, j

$K_{k,j}$ = impulse response function for modes i, j

$C_{k,j}$ = hydrostatic restoring force in modes i, j

F_j = arbitrary time forcing function which includes loads from winds, waves, currents, and mooring restraints.

The constant inertial coefficient, $m_{k,j}$, and the impulse response coefficient, $K_{k,j}(t)$, are hydrodynamic reactive coefficients determined from added mass and damping coefficients computed in the frequency domain. Formulas for computing these coefficients are presented in Chapter 5 of this book and are taken from Van Oortmerssen (1976). The added mass and damping coefficients for the subject dry

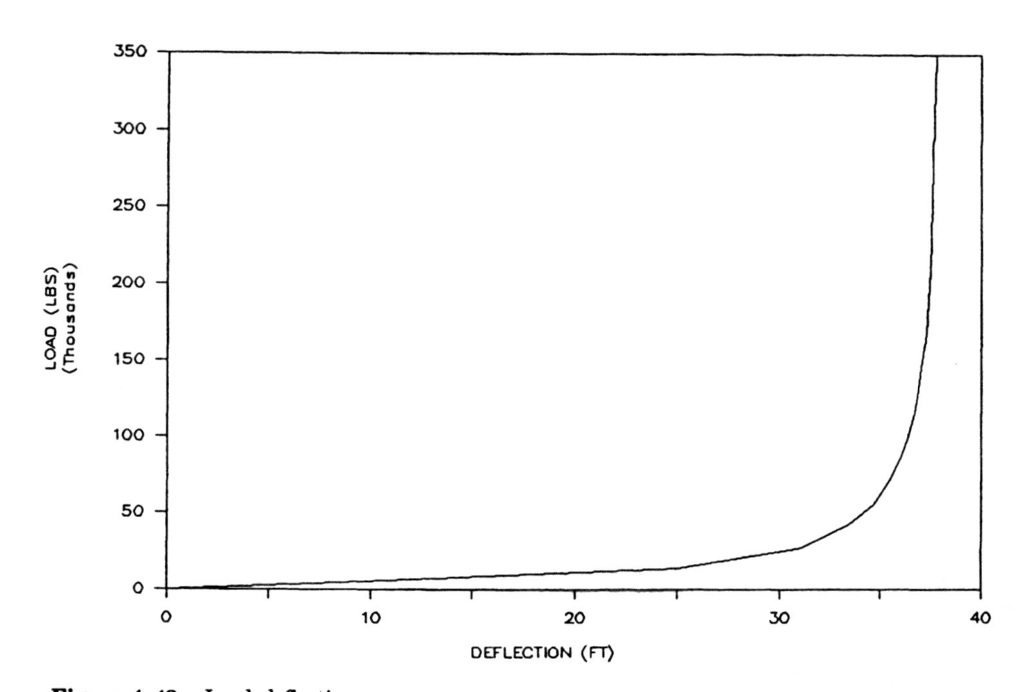

Figure 4–43. Load–deflection curve.

dock were computed using diffraction theory. Similarly, diffraction theory was used to generate first- and second-order unit wave forces. These unit wave forces were used, in combination with a wave spectrum, to generate wave force–time histories. The wave spectrum was computed using the formula published by Goda (1985) for a given significant wave height and peak spectral period [see Eq. (5–9) in Chapter 5]. First- and second-order wave force–time histories were generated using methods developed by Newman (1974) [see Eq. (5–52) in Chapter 5].

Wind force–time histories were generated using spectral simulation techniques and the recently developed wind spectrum reported by Ochi and Shin (1988). The Ochi and Shin spectrum is computed as follows:

$$S(f_*) = \begin{pmatrix} 583 f_* & \text{for } 0 \le f_* \le 0.0 \\ \dfrac{420 f_*^{.070}}{(1 + f_*^{0.35})^{11.5}} & \text{for } 0.003 \le f_* \le 0.1 \\ \dfrac{838 f_*}{(1 + f_*^{.35})^{11.5}} & \text{for } f_* \ge 0.1 \end{pmatrix} \qquad (4\text{–}74)$$

where

$$f_* = 10\,\frac{f}{V_w} \qquad (4\text{–}75)$$

$$S(f_*) = \frac{f S(f)}{C_{10} V_w^2}$$

where

f = frequency of oscillation in hertz
V_w = wind velocity
C_{10} = surface drag coefficient
$S(f)$ = spectral density.

The above spectrum differs from other published wind spectrums inasmuch as it was developed from wind data collected at sea and because it gives a relatively large amount of energy at low frequency. Energy at low frequency has an important bearing on the subject of dry dock mooring because, as will be discussed in the following paragraphs, the natural frequencies of the moored dry dock system in surge, sway, and yaw are relatively low. A typical wind spectrum for a mean hourly windspeed of 50 knots is presented in Figure 4–44. A windspeed time–history, synthesized from the spectrum in Figure 4–44, is given in Figure 4–45. Figures 4–44 and 4–45 both indicate considerable energy at low frequency. Wind force–time histories are developed using wind speed time histories in combination with the static wind drag force formulas presented in Section 4.5.1.

The natural periods of motion of the moored dry dock system in surge, sway, and yaw were determined from numerical extinction tests. The numerical extinction tests were performed in the same manner as would be done with a physical model, that is, the dry dock was displaced from its equilibrium position in a single mode of motion and released. The initial mooring line pretension at the equilibrium position was about 67.0 kN. Motions of the moored dry dock subsequent to release were oscillatory at the natural period of the moored ship system. Results of these extinction tests indicated natural periods for the moored dry

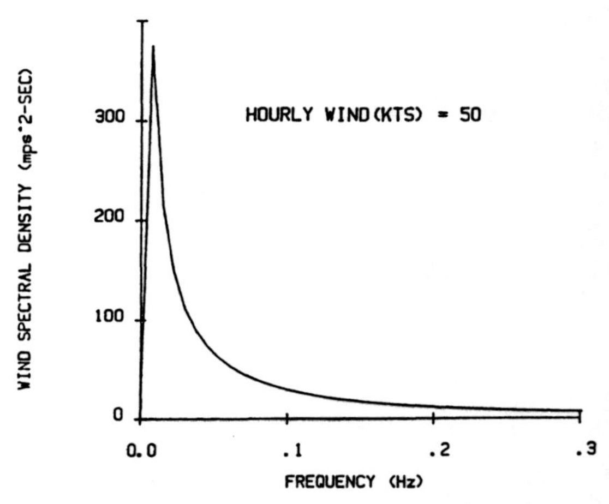

Figure 4–44. Ochi wind spectrum for a 50–knot wind.

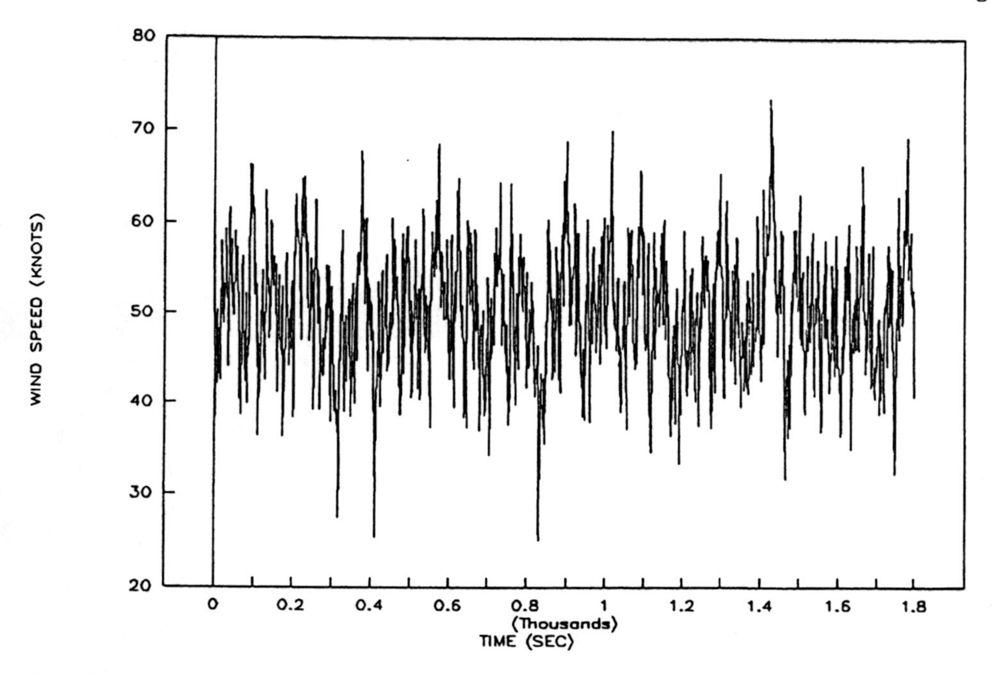

Figure 4–45. Wind time–history. *Note*: 1 knot = 0.515 meters/second.

dock system in surge, sway, and yaw to be 125, 68 and 45 s, respectively.

Five (5) mooring analysis cases were examined each corresponding to a different set of environmental conditions as outlined in Table 4–6. Case 1 consisted of a static mooring analysis for various 30-s wind-speeds, U_{30}. Case 2 consisted of examining various wind spectra for hourly windspeeds, U_{3600}, which corresponded to the 30-s winds examined in Case 1. Cases 3, 4, and 5 examined mooring loads resulting from a mean hourly windspeed of 50 knots and waves having a 6-foot significant height, H_s, and a peak spectral period, T_p, of 5.2 s. The applied wind and wave loading are idealized in Case 3. Specifically, the mean hourly windspeed of 50 knots is idealized as a steady 66-knot, 30-s duration wind while the wave is idealized as a sinusoid with a height of 8 ft and a period of 5.2 s. The 8-ft wave height was chosen because it approximates the 10% exceedence wave height, H_{10}, corresponding to the 6-ft significant height. In Case 4, the wind is still idealized as a steady wind of 66 knots, but the full wave spectrum is simulated. Finally, Case 5 examines the full wind and wave spectrums.

Table 4–6. Dynamic Analysis Cases

Case 1	Steady wind:	$U_{30} = 26–66$ knots
Case 2	Unsteady wind:	$U_{3600} = 20–50$ knots
Case 3	Steady wind: Regular wave:	$U_{30} = 66$ knots $H = 2.45$ m, T = 5.2 s
Case 4	Steady wind: Irregular wave:	$U_{3600} = 50$ knots $H_s = 1.8$ m
Case 5	Unsteady wind: Irregular wave:	$U_{3600} = 50$ knots $H_s = 1.8$ m $T_p = 5.2$ s

Case 1: Static Mooring Analysis

Static mooring analysis was performed using the techniques presented in Sections 4.5 and 4.7.2 of this Chapter. Specifically, a steady 30-s windspeed was used to estimate the total wind load on the floating dry dock. This total load was transferred to the catenary legs of the spread mooring using the static analysis computer model discussed earlier in Section 4.7.2 and presented in U.S. Navy (1985). Static analyses were prepared for several beam-on windspeeds. To simplify presentation of results only beam-on wind conditions are presented. Under such conditions, mooring lines 2 through 6 and 8 through 12 are loaded uniformly and analysis results can be summarized in the form of a single mooring line load. Maximum individual mooring line loads ranged from 29 000 pounds (about 130.0 kN) for a 26-knot, 30-s wind (i.e., 20-knot mean hourly windspeed) to 182 000 pounds (about 810.0 kN) for a 66-knot, 30-s wind (i.e., 50-knot mean hourly windspeed); see Figure 4–46. These loads are well below the 826 000 pound (3.7 MN) breaking strength of the mooring lines.

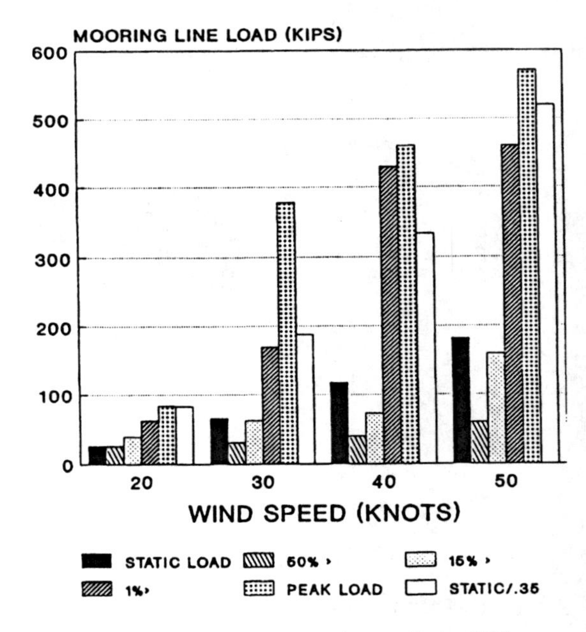

Figure 4–46. Bow stern mooring analysis. (U.S. Navy, 1985.)

Case 2: Unsteady Wind

Dynamic analyses of the moored floating dry dock were performed for wind spectrums having mean hourly windspeeds of 20, 30, 40, and 50 knots. Each dynamic simulation was prepared using beam-on winds, a total simulation time of 1800 s (30 min), and an initial preload of 15 000 pounds (about 67.0 kN). Figure 4–46 summarizes the dynamic analysis results in terms of the 50% exceedence, 15% exceedence, 1% exceedence, and peak mooring line load and compares these to the forces obtained from the Case 1 static analysis results. Figure 4–47 presents a mooring line force–time history for a mean hourly windspeed of 50 knots. As shown in Figure 4–46, the 1% exceedence and peak dynamic mooring line loads greatly exceed the 50% and 15% exceedence mooring line loads. The static mooring line loads from Case 1 are comparable to the 15% exceedence mooring line loads.

Case 3: Steady Wind and Regular Waves

Results from the Case 3 dynamic analysis are presented in Figure 4–48 and a mooring line force–time history is presented in Figure 4–49. Analysis of Figure 4–49 indicates that the period of moored dry dock response is around 16 to 17 s and is about three times longer than the 5.2 s wave period. This subharmonic response is consistent with the findings of Van Oortmerssen (1976) and is attributed to the asymmetric and nonlinear stiffness provided by the mooring lines. In the present case, the mooring system is symmetric in terms of stiffness when there is no applied loading; however, when a large wind load is applied to the vessel the windward mooring lines are considerably stiffer than the leeward mooring lines. Figure 4–48 indicates that the peak mooring line loads are about 490 000 pounds (2.2 MN) or about 2.7

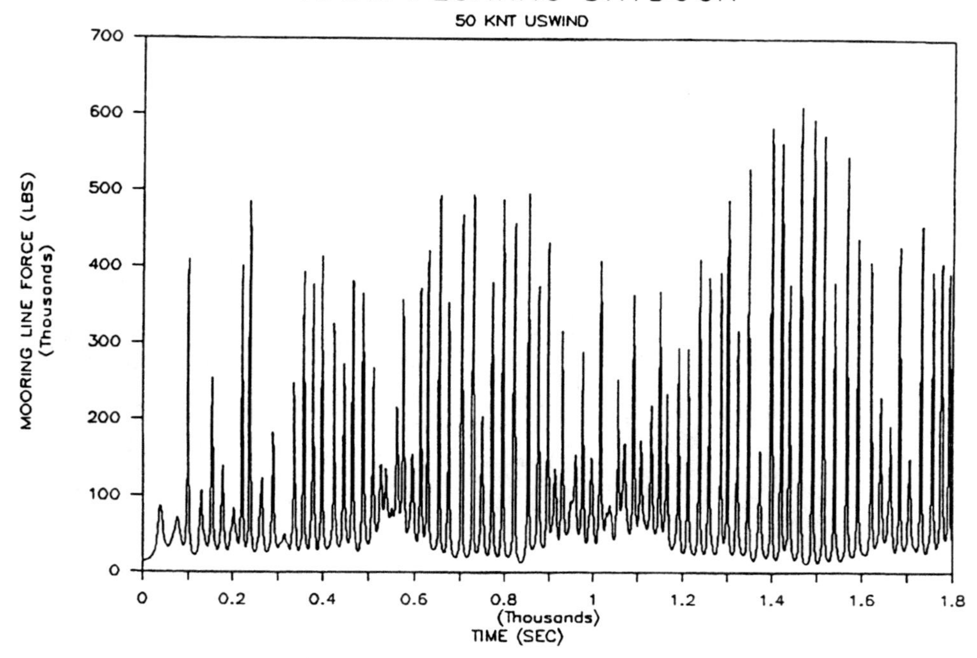

Figure 4–47. Mooring force–time history for 50-knot wind. *Note*: 1 pound = 4.448 newtons.

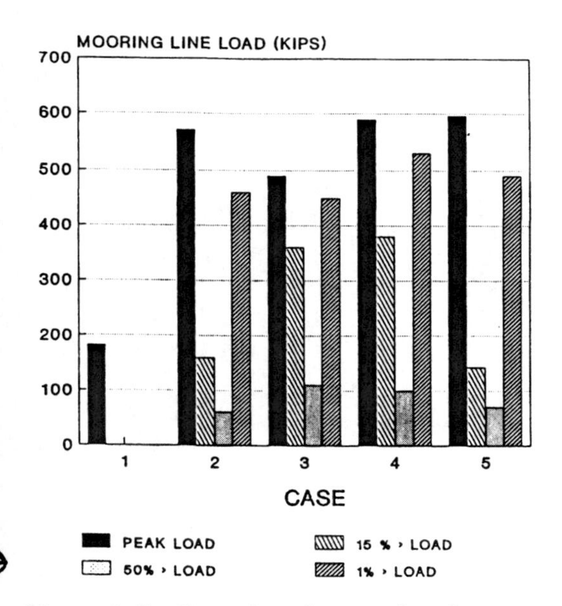

Figure 4–48. Dynamic analyses results—Cases 1 through 5. *Note*: 1 pound = 4.448 newtons.

times the mooring line load associated with static wind loads alone. The 15% exceedence mooring line load is 360 000 pounds (about 1.6 MN) which is roughly twice that associated with unsteady winds alone.

Case 4: Steady Wind and Irregular Waves

Results from this dynamic analysis, which includes the effects of both first- and second-order wave loading, are summarized in Figure 4–48. Analysis of the mooring line force–time history shown in Figure 4–50 indicates that the primary dry dock response is subharmonic at a period equal to about one third the peak spectral wave period. It is clear from Figure 4–50, however, that the dry dock response includes a lower frequency component. Figure 4–48 indicates that the 15% exceedence mooring line loads are similar for irregular and regular wave conditions with a value of around

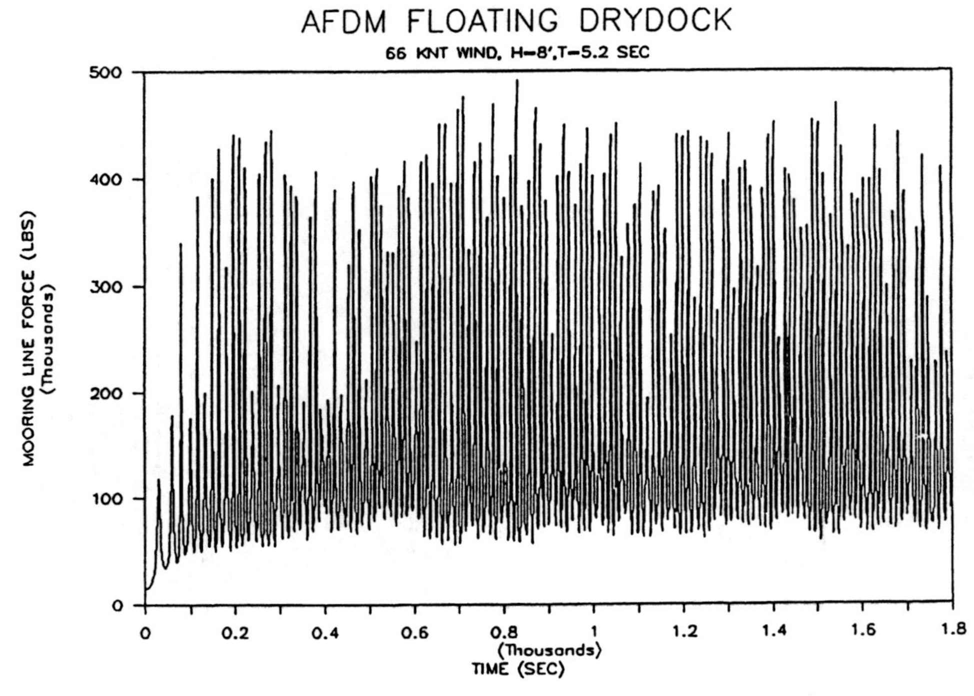

Figure 4–49. Mooring force–time history for Case 3. *Note*: 1 pound = 4.448 newtons.

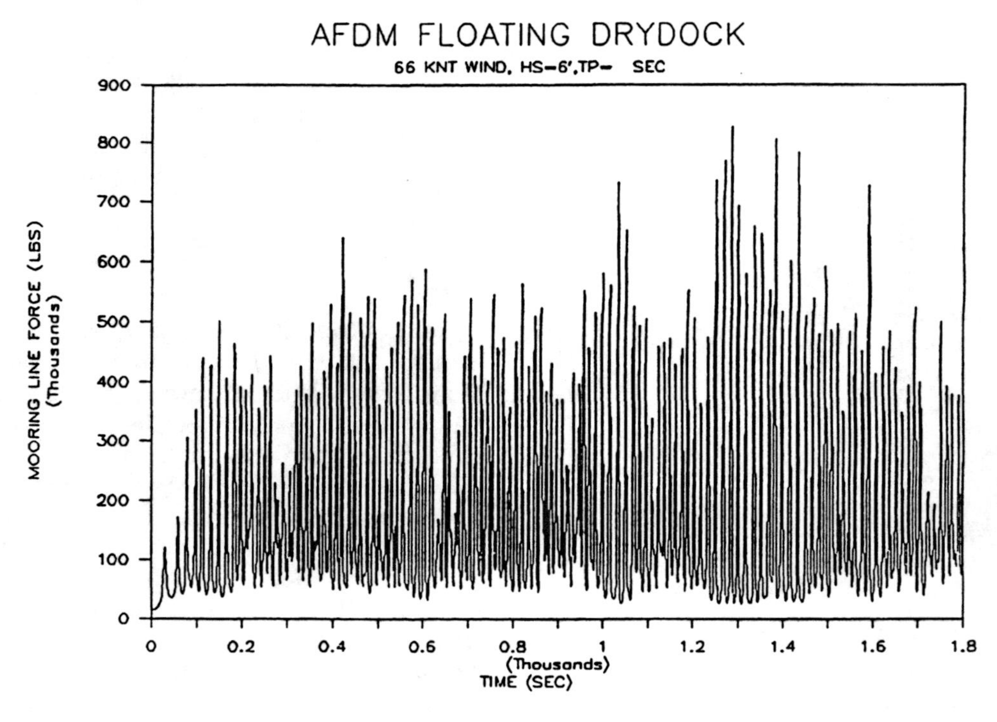

Figure 4–50. Mooring force–time history for case 4. *Note*: 1 pound = 4.448 newtons.

380 000 pounds (1.7 MN). Peak mooring line loads for irregular wave are considerably higher than those simulted for regular waves with a maximum value on the order of 590 000 pounds (2.63 MN).

Case 5: Unsteady Winds and Irregular Waves

Case 5 is more representative of actual conditions than Cases 3 and 4 because the complete wind and wave spectra are simulated. Results of the simulation are summarized in Figure 4–48 and a mooring line force–time history is provided in Figure 4–51. The dry dock response is a mixture of subharmonic response, second-order slowly varying wave drift response, and low-frequency wind response. The peak mooring line loads of 600 000 pounds are comparable to Case 1 (in the absence of waves) and Case 4. It is interesting to note that the 50% and 15% exceedence mooring line forces for Case 5 are considerably less than those for Cases 3 and 4.

In summary, the above paragraphs demonstrate that dynamic wind and wave loading can be an important factor in the design of spread moorings. Physical model tests are generally needed to confirm final designs for important spread mooring installations.

4.7.3.2 Dynamic Analysis of a Single-Point Mooring

As previously mentioned, single-point moorings allow the moored vessel to "weather vane" and align itself with the prevailing environmental conditions (i.e., winds, waves, and/or currents). Preliminary designs of single-point moorings have been made on the basis of the static analysis procedures described in Section 4.7.2 of this chapter;

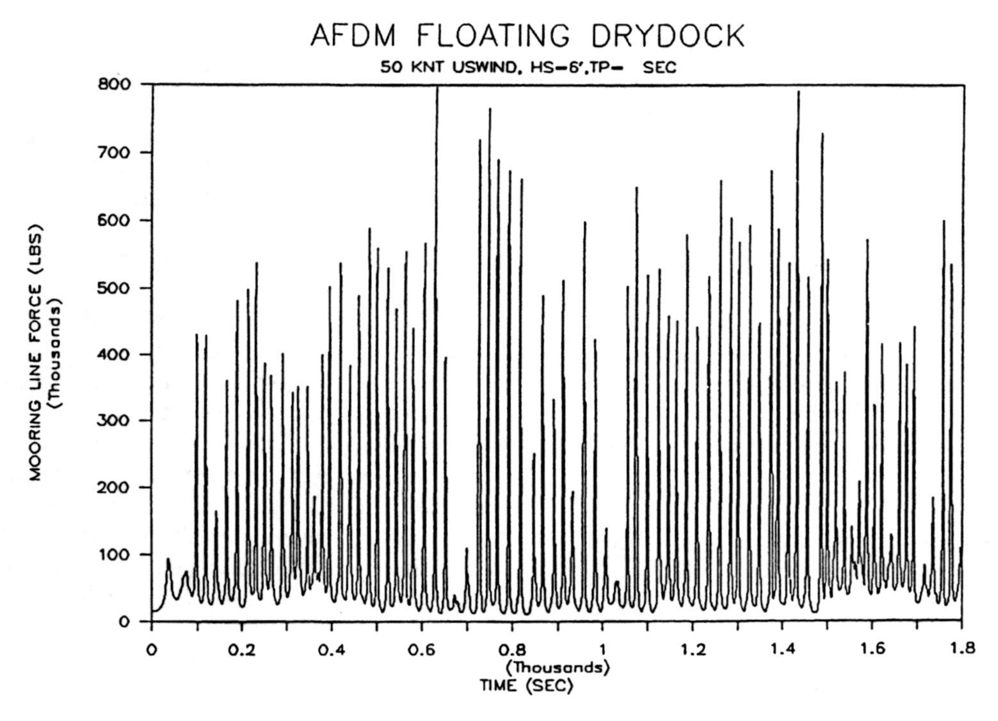

Figure 4–51. Mooring force-time history for Case 5. *Note*: 1 pound = 4.448 newtons.

however, final designs are normally confirmed by physical model testing. On the other hand, recent progress in numerical modeling of single-point moorings has led to an increase in the use of numerical models during design development prior to physical model testing.

The static analysis procedure presented in Section 4.7.2 of this chapter assumes that a vessel secured to a single-point mooring will attain an equilibrium position under wind and current loading. Field experience, physical modeling, and analytical studies have shown, however, that vessels secured to a single-point mooring do not always attain an equilibrium position. Instead, vessels may, depending on a variety of factors, slowly oscillate around an equilibrium position. This phenomenon, called "fishtailing" or "kiting," can result in mooring hawser loads considerably higher than those computed by static analysis methods.

Fishtailing motions of a single-point mooring can be simulated by means of a time domain numerical model. Examples of single-point mooring simulation models can be found in Muga and Freeman (1977) and Wichers (1979, 1988). Wichers (1988) also presents an analytical method that can be used to determine whether a vessel secured to a single-point mooring and subjected to wind and current will experience unstable behavior (i.e., fishtailing motion). Stability analysis generally show:

1. A vessel is more prone to dynamically unstable motions when wind and current are colinear (i.e., act in the same direction).
2. A vessel is more prone to dynamic instability as the bow hawser length increases.

Although the stability analysis gives an indication of the potential for fishtailing motion, one must perform a complete dynamic analysis in the time domain in order to estimate mooring line loads.

Time domain analysis of single-point moorings can be performed using the impulse–response–function formulation presented in Chapter 5 and such an approach was taken by Muga and Freeman (1977). On the other hand, Wichers (1979) has shown that the following formulation, utilizing constant added mass and damping coefficient values, gives results comparable to those obtained from the impulse–response–function approach.

$$(M_{11} + \bar{m}_{11})\ddot{x}_1 = M_{11}\dot{x}_2\dot{x}_3 - b_{11}\dot{x}_1 + F_{1cs} + F_{1w} + F_{1w} + F_{1h} + P \tag{4–76}$$

$$(M_{22} + \bar{m}_{22})\ddot{x}_2 = -M_{22}\dot{x}_2\dot{x}_3 - b_{22}\dot{x}_2 + (F_{2cs} + F_{2cd}) + F_{2w} + F_{2h} \tag{4–77}$$

$$(M_{66} + \bar{m}_{66})\ddot{x}_6 = -b_{66}\dot{x}_6 + (F_{6cs} + F_{6cd}) + F_{6w} + F_{6h} \tag{4–78}$$

where

M_{ii} = mass in mode ii ($i = 1 =$ surge, $i = 2 =$ sway, $i = 6 =$ yaw)
$\overline{m_{ii}}$ = constant added mass at low frequency in mode ii
x_j = displacement in mode j
b_{ii} = constant damping coefficient at low frequency in mode ii
F_{jcs} = steady current drag force in mode j
F_{jcd} = dynamic current drag force in mode j
F_{jw} = wind drag force in mode j
F_{jh} = hawser force in mode j
P = thrust due to astern propulsion.

The steady wind and current drag forces in the above equations are computed using expressions similar to those presented in Section 4.5 of this chapter for static wind and current forces. Unlike the static situation, however, the current forces are computed using a relative current (i.e., superposition of vessel and current velocities). Dynamic viscous current drag forces, which account for the additional force and moment caused by vessel yaw motion in the relative current, are computed as follows (Wichers, 1979):

$$F_{2cd} = \frac{1}{2}\rho_w \beta C_{yc}(90°) \int_{-\frac{L}{2}}^{+\frac{L}{2}} [(V_{cr} - \dot{x}_6\xi)|V_{cr} - \dot{x}_6\xi| - V_{cr}|V_{cr}|)]d\xi \tag{4–79}$$

$$F_{6cd} = \int_{-L/2}^{+L/2} \Delta F_{2cd}\, d\xi \tag{4–80}$$

where

$C_{yc}(90°)$ = steady lateral current drag coefficient
β = empirical coefficient
V_{cr} = relative current in sway
L = ship length
ξ = incremental ship length
ΔF_{2cd} = incremental lateral dynamic drag force.

Details regarding the above expressions can be found in Wichers (1979). Other formulations of the dynamic current forces have been developed by Molin and Bureau (1980), Obokata (1983), and Wichers (1988).

A single-point mooring simulation model, based on the above equations, was used to simulate the motions and attendant bow hawser forces of a 200 000 dead weight ton (200 000 DWT) tanker moored to a single pile by means of an elastic hawser. The vessel and mooring characteristics correspond to the laboratory measurements published by Wichers (1988). The characteristics of the 200 000 DWT tanker are given in Table 4–7 for 25% loading conditions. The bow hawser had an unloaded length of 90 m and a load–deflection curve for the bow hawser is given in Figure 4–52. The tanker was exposed to bow-on wind and current. The wind speed was 60 knots (30.9 m/s) and the current speed was 2 knots (1.03 m/s).

Simulation results are summarized in Figure 4–53 which presents time histories of: (a) longitudinal motion of the bow hawser attachment point—X_a in meters, (b) lateral motion of the bow hawser attachment point—Y_a in meters, (c) yaw motion of the vessel center of gravity—yaw in degrees, and (d) hawser force—*HF* in ton force. The mooring response consists of an oscillatory fishtailing-type motion where maximum mooring line loads exceed 268 metric tons. This value significantly exceeds the maximum hawser force value of 87 metric tons computed from a static analysis. The numerical results presented in Figure 4–53 are in reasonable agreement with the physical model test results of Wichers (1988).

Similar results, also in reasonable agreement with Wichers' (1988) model tests, are presented in Figure 4–54 for the same 200 000 DWT tanker exposed to bow on winds of 60 knots, bow on currents of 2 knots, and a bow on wave spectrum characterized by a significant wave height of 3.9 m and an average wave period of 10.2 s.

In summary, single-point moorings are prone to dynamic motions and attendant mooring loads that can significantly exceed values computed from static analyses.

Table 4–7 Particulars of 200 000 DWT Tanker

Property	Unit	Value
Length	m	310.00
Beam	m	47.17
Draft	m	7.56
Displacement volume	m^3	13 902
Longitudinal wind area	m^2	1897
Lateral wind area	m^2	7785

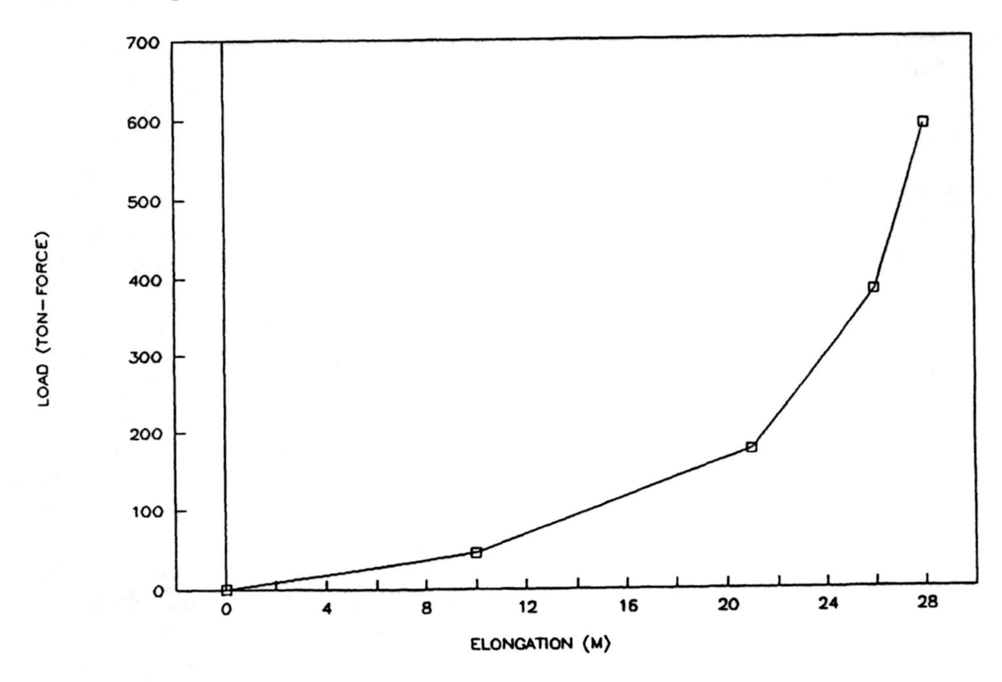

Figure 4-52. Load-deflection curve.

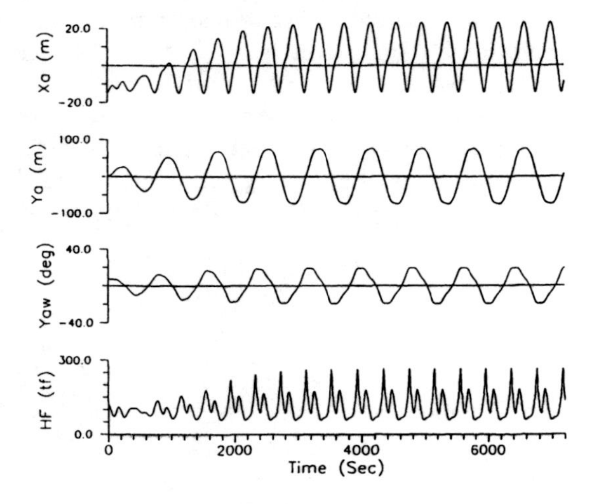

Figure 4-53. Dynamic analysis of a single-point mooring exposed to wind and current.

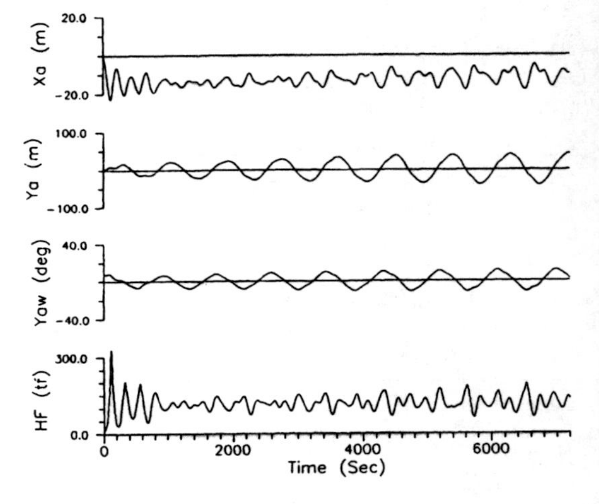

Figure 4-54. Dynamic analysis of a single-point mooring exposed to wind, wave, and current.

Numerical simulation models can be used to estimate dynamic loads on single-point moorings and are particularly valuable during the development of concept designs. Notwithstanding the recent advances in numerical simulation techniques, physical model tests should be conducted for final design of important single-point mooring facilities.

REFERENCES

AMERICAN BUREAU OF SHIPPING, 1975. "Rules for Building and Classing Single Point Moorings," 45 Broad Street, New York.

AMERICAN PETROLEUM INSTITUTE, 1987. "Recommended Practice for Planning, Designing and Constructing Fixed Offshore Platforms," API RP 2A, Seventeenth Edition, April 1.

AMERICAN PETROLEUM INSTITUTE, 1990. "Recommended Practice for Design, Analysis and Maintenance of Mooring for Floating Production Systems," Draft, May.

BATTS, M. E., CORDES, M. R., RUSSELL, L. R., SHAVER, J. R., and SIMIU, E., 1980. "Hurricane Wind Speeds in the United States," NBS Building Science Series 124, National Bureau of Standards, May.

BENHAM, F. A., FANG, S. S., and FETTERS, C. H., 1977. "Wind and Current Shape Coefficients for Very Large Crude Carriers," Offshore Technology Conference, OCT 2729, pp. 97–108, May.

BERTEAUX, H. O., 1976. "Buoy Engineering." John Wiley & Sons, New York.

BERTEAUX, H. O., 1992. "Coastal and Oceanic Buoy Systems." Chapter 9 in Handbook of Coastal and Ocean Engineering, edited by J. B. Herbich Gulf Publishing Company, Houston, Texas.

BRETSCHNEIDER, C. L., 1967. "Estimating Wind-Driven Currents Over the Continental Shelf," Ocean Industry, pp. 45–48, June.

CHANGERY, M. J., 1982a. "Historical Extreme Winds for the United States—Atlantic and Gulf of Mexico Coastlines," NUREG/CR-2639, National Oceanic and Atmospheric Administration, National Climatic Data Center, Asheville, North Carolina, May.

CHANGERY, M. J., 1982b. "Historical Extreme Winds for the United States—Great Lakes and Adjacent Regions,"' NUREG/CR-2890, National Oceanic and Atmospheric Administration, National Climatic Data Center, Asheville, North Carolina, August.

COX, J. V., 1982. "STATMOOR—A Single Point Mooring Static Analysis Program," TN:N-1634, Naval Civil Engineering Laboratory, June.

DE KAT, J. O. and WICHERS, J. E. W., 1990. "Dynamic Effects of Current Fluctuations on Moored Vessel," Offshore Technology Conference, OTC 6219, pp. 183–198, May.

FLORY, J. F., BENHAM, F. A., MARCELLO, J. T., PORANSKI, P. F., and WOEHLEKE, S. P., 1977. "Guidelines for Deepwater Port Single Point Mooring Design," Report No. EE-17E-T77, Exxon Research and Engineering Company, September.

GERALD, C. F., 1980. "Applied Numerical Analysis," Addison-Wesley Publishing Co., Reading, MA, May, 1980.

GODA, Y., 1985. "Random Seas and Design of Maritime Structures," University of Tokyo Press.

GOULD, R. W. F., 1982. "The Estimation of Wind Loads on Ship Superstructures," Royal Institution of Naval Architects.

HEADLAND, J. R., SEELIG, W. N., and CHERN, C., 1989. "Dynamic Analysis of Moored Floating Drydocks," ASCE Ports '89 Specialty Conference, Boston.

MOLIN, B. and BUREAU, G., 1980. "A Simulation Model for the Dynamic Behaviour of Tankers Moored to SPM," International Symposium on Ocean Engineering and Shiphandling, Gothenburg.

MUGA, B. J. and FREEMAN, M. A., 1977. "Computer Simulation of Single Point Moorings," Offshore Technology Conference, OTC 2829.

NEWMAN, J. N., 1974. "The Second Order, Slowly Varying Forces on Vessels in Irregular Waves," Symposium on Dynamics of Marine Vehicles and Structures in Waves, London.

OBOKATA, J., 1983. "Mathematical Approximation of the Slow Oscillation of a Ship Moored to Single Point Moorings," Marintec Offshore China Conference, Shanghai, October.

OCHI, M. K. and SHIN, Y. S., 1988. "Wind Turbulent Spectra for Design Considerations of Offshore

Structures," Offshore Technology Conference, OTC 5736, pp. 461–476.

OCIMF, 1977. "Prediction of Wind and Current Loads on VLCCs," London.

OWENS, R. and PALO, P., 1982. "Wind Induced Steady Loads on Moored Ships," TN:N-1628, Naval Civil Engineering Laboratory.

ROCKER, K., 1985. "Handbook for Marine Geotechnical Engineering," Naval Civil Engineering Laboratory.

REMERY, G. F. M. and VAN OORTMERSSEN, G., 1973. "The Mean Wave, Wind and Current Forces on Offshore Structures and Their Role in the Design of Mooring Systems," Offshore Technology Conference, OTC 1741.

SIMIEU, E. and SCANLON, R. H., 1978. "Wind Effects on Structures: An Introduction to Wind Engineering," A Wiley-Interscience Publication, New York.

THOM, H. C. S., 1973. "Distribution of Extreme Winds Over Oceans," Journal of Waterways, Harbors and Coastal Engineering Division, ASCE, Vol. 99, No. WW1, pp. 1–17.

TRUE, D. G., 1992. "Anchors." Chapter 8 in Handbook of Coastal and Ocean Engineering, edited by J. B. Herbich. Gulf Publishing Company, Houston, Texas.

TSINKER, G. P., 1986. "Floating Ports," Gulf Publishing Company, Houston, Texas.

U.S. ARMY CORPS OF ENGINEERS, 1984. Coastal Engineering Research Center, Shore Protection Manual, Vols. 1 and 2.

U.S. ARMY CORPS OF ENGINEERS, 1989. "Water Levels and Wave Heights for Coastal Engineering Design," EM 1110-2-1414.

U.S. NAVY, NAVAL FACILITIES ENGINEERING COMMAND, 1985. "Fleet Moorings: Basic Criteria and Planning Guidelines," Design Manual DM-26.5.

U.S. NAVY, NAVAL FACILITIES ENGINEERING COMMAND, 1986. "Mooring Design: Physical and Empirical Data," Design Manual DM-26.6.

VAN OORTMERSSEN, G., 1976. "The Motions of A Moored Ship in Waves," Publication No. 510, Netherlands Ship Model Basin, Wageningen, The Netherlands.

WICHERS, J. E. W., 1979. "Slowly Oscillating Mooring Forces in Single Point Mooring Systems," Proceedings of Symposium on Behaviour of Offshore Structures, London.

WICHERS, J. E. W., 1988. "A Simulation Model For A Single Point Moored Tanker," Publication No. 7970., MARIN, Wageningen, The Netherlands.

5

Floating Breakwaters

by J. R. Headland

5.1 INTRODUCTION

The engineering and subsequent construction of the "Bombardon" floating breakwaters was an important episode in the historical development of floating breakwater technology. These floating structures were elements in two artificial "Mulberry" harbors constructed along the coast of France for the D-Day invasion in June 1944. The "Bombardon" floating breakwater, which is discussed in detail in Lochner et al. (1948), consisted of a steel structure in the shape of a Maltese Cross with a 7.6 m beam, a 5.8 m draft, and a total height of 7.6 m. The breakwater units were constructed in 61-m lengths and placed with a 15.2-m longitudinal gap between each unit. The breakwaters were constructed in two lines, roughly 244 m apart, to obtain the desired wave height reduction. Initial breakwater designs were prepared for a wave height of 2.4 m and wave length of 30.5 m, the latter corresponding to a wave period of 4.7 s in the expected water depth of 12.8 m. Final designs were prepared for a wave height of 3.3 m and a wave length of 45.7 m (5.6 s period). In a remarkable engineering achievement, decisions to use the "Bombardon" floating breakwater were made in early 1943 and theoretical analyses, hydraulic model testing, engineering design, construction, and field testing were completed in time for the D-Day invasion in June 1944. The "Bombardon" floating breakwaters provided shelter for invasion troops for the first 12 days of the D-Day operation. Unfortunately, the "Bombardon" breakwaters were destroyed by an unexpected storm (i.e., worst storm in June for 40 years) with wave heights on the order of 4.6 m and wave lengths of 91 m which corresponded to a wave period of 8 s in the actual harbor water depth of about 22 m. Although the breakwaters were destroyed and there was, in hindsight, considerable controversy surrounding decisions made to use floating breakwaters vis-a-vis fixed structures, authorities were pleased with the "Bombardon" performance inasmuch as they provided much needed protection during the initial, critical period of the D-Day invasion. Not surprisingly, conclusions drawn from "Bombardon" floating breakwater development still hold true today and virtually all subsequent floating breakwater development has reinforced these findings:

1. Floating breakwaters offer a feasible and economic alternative to conventional breakwaters in deepwater when exposed to moderate wave conditions (i.e., <2 m).
2. Floating breakwaters can be designed to reduce waves to acceptable levels as long as those waves are not too long. A practical upper limit is on the order to 4 to 6 s and most floating breakwaters are designed for wave periods less than 4 s.
3. Floating breakwater structural units and their mooring systems are vulnerable to catastrophic failure during severe storms when they are needed the most.
4. Floating breakwaters require a relatively high level of maintenance compared to conventional fixed structures.

A variety of floating breakwater types have been developed for commercial use over the past 30 years. Figure 5–1 summarizes some of the floating breakwater types in accordance with the classification of McCartney (1985). There are also a variety of schemes that can be used to moor floating breakwaters. Most floating breakwaters, however, are moored by means of anchors and chains (soft catenary-type mooring) or by guide piles (rigid guide pile-type mooring). With regard to catenary mooring systems, a designer may select from a variety of anchor types including drag anchors, deadweight anchors, and pile anchors. Catenary computations, mooring design, and anchor selection are presented in Chapter 4.

There are far too many floating breakwater systems to cover in this treatise. Because the goal of this chapter is to familiarize practicing engineers with the principles and related mechanics that govern floating breakwater design, emphasis will be placed on those breakwater types that are most easily analyzed by analytical means, namely caisson and pontoon-type floating breakwaters. A more complete survey of floating breakwaters is presented in Hales (1981).

There are four fundamental aspects of floating breakwater design: (1) buoyancy and floating stability, (2) wave transmission, (3) mooring forces, and (4) breakwater unit structural design.

With regard to buoyancy, a floating breakwater must possess sufficient buoyancy to support the weight of the breakwater and its moorings. Furthermore, the breakwater must be rotationally stable in accordance with standard practices of naval architecture. The fundamental goal of a floating breakwater is to attenuate or reduce waves. Accordingly, evaluation of floating breakwater wave transmission constitutes a critical element of design. Moorings, whether constructed of piles or mooring lines and anchors, must hold a breakwater in place and a careful assessment of mooring forces during design storm wave attack must be made in order to ensure the survival of the breakwater. In addition, the breakwater unit itself must sustain the stresses imposed by wave-induced hogging and sagging as well as those related to the moorings. With regard to the structural integrity of the breakwater unit, consideration must be given to both survival and fatigue-related stresses. Waves loading generally dictates the design of a floating breakwater and wave-induced load is emphasized in this chapter. On the other hand, the designer should also evaluate other possible loads such as those associated with currents, water level variations, ice, wind, and vessel impact.

Floating breakwater design is a complicated and iterative process due to the interdependency of each design factor. For example, wave transmission performance depends on breakwater geometry, mass, and mooring properties. Similarly, mooring forces depend on breakwater geometry/mass. Finally, breakwater structural integrity depends on breakwater geometry/mass

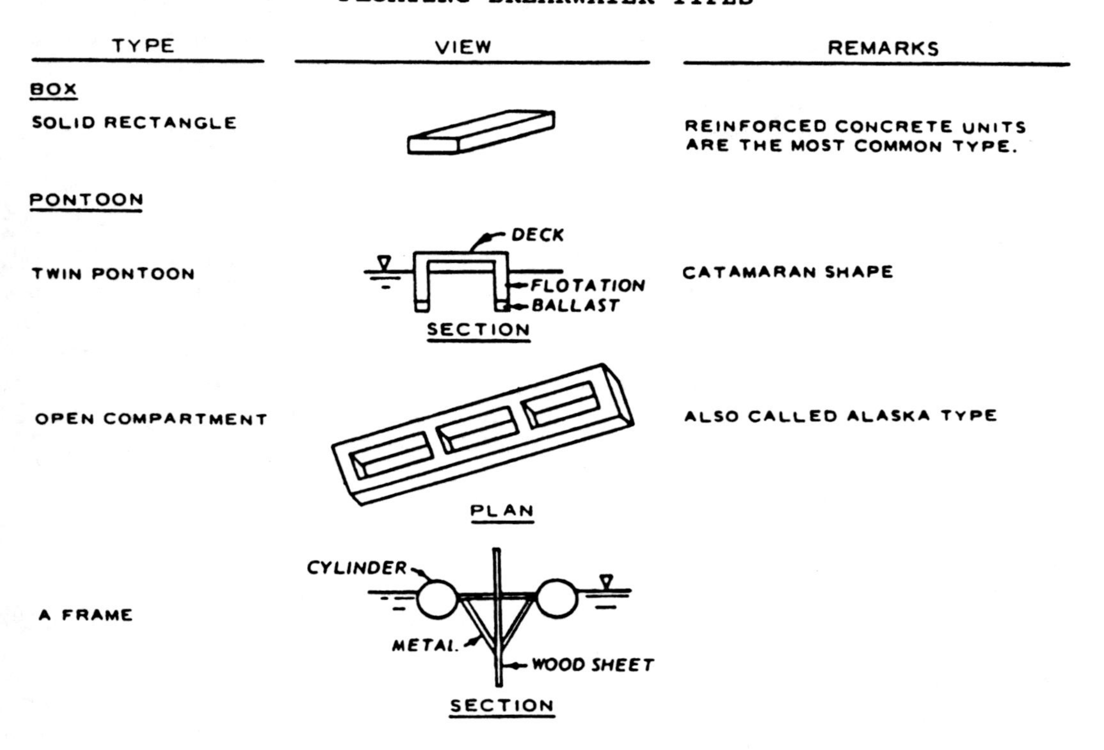

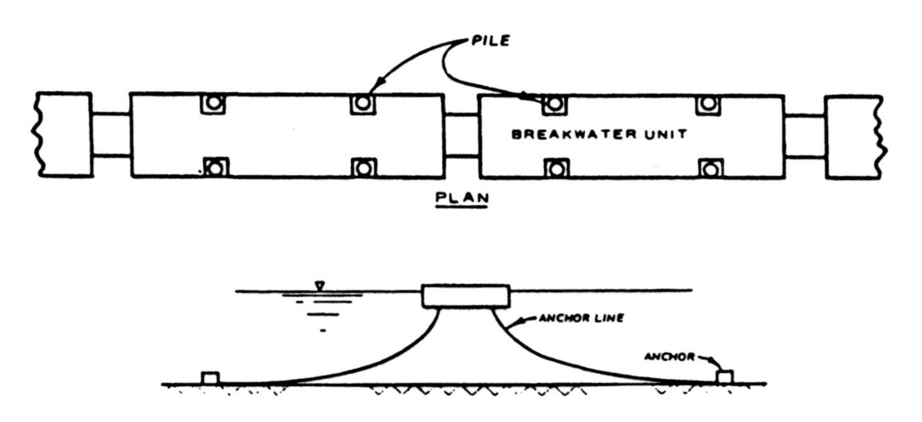

Figure 5–1. Floating breakwater types and typical anchoring systems. (From McCartney, 1985.)

and mooring forces. At the present state of technology, there is neither an integrated nor an accepted methodology for the design of floating breakwaters. Instead, most designs have been prepared in an ad hoc manner using one approach for evaluation of wave transmission, a separate approach for mooring force determination, and yet another approach for breakwater structural integrity. This chapter provides a review of some approaches that have been used to design floating breakwaters.

5.2 WAVE MECHANICS

Linear wave theory is generally used to describe the mechanics of ocean waves. A summary of linear wave theory can be found in numerous texts and an interested reader may consult the U.S. Army Corps of Engineers Shore Protection Manual (1984) or Dean and Dalrymple (1984) for a detailed account of this theory. Only relevant elements of linear wave theory are discussed in this chapter. Linear theory idealizes ocean waves as a sinusoidal free surface fluctuation:

$$\eta(t) = a \sin(t + \epsilon) \tag{5–1}$$

where

η = time varying free surface elevation
a = amplitude which is equal to one half the wave height, H
σ = circular wave frequency
ϵ = wave phase
t = time.

Wave period, T, circular wave frequency, σ, and wave frequency, f, are interrelated as follows:

$$\sigma = \frac{2\pi}{T}; \qquad f = \frac{1}{T}. \tag{5–2}$$

An important characteristic of an ocean wave is its speed, c, and wave length, L, which are related as follows:

$$L = cT. \tag{5–3}$$

Wave length can be determined from the incident wave period, T, the acceleration due to gravity, g, and the water depth at the site, h, using linear wave theory:

$$L = \frac{gT^2}{2\pi} \tanh\left(2\pi \frac{h}{L}\right). \tag{5–4}$$

Equation (5–4) is transcendental in L and generally requires numerical solution. Convenient solutions to Equation (5–2) are presented in tabular form in numerous references (e.g., U.S. Army Corps of Engineers, 1984 and Wiegel, 1964). In the absence of such tables the following explicit approximation to Eq. (5–2) can be used (Nielsen, 1984):

$$2\pi \frac{h}{L} = \sqrt{\left(2\pi \frac{h}{L_0}\right)\left(1 + \frac{1}{6} 2\pi \frac{h}{L_0} + \frac{11}{360}\left(2\pi \frac{h}{L_0}\right)^2\right)}; \qquad L_0 = \frac{gT^2}{2\pi}. \tag{5–5}$$

Real sea states do not consist of a single wave but of a spectrum of waves having a range of heights and periods. The surface elevation of a real sea state can be represented as the superposition of n sinusoidal wave components as follows:

$$\eta(t) = \sum_{n=1}^{\infty} a_n \sin(\sigma_n t + \epsilon_n). \tag{5–6}$$

The resulting wave surface is not a regular sine wave but an irregular or random function of time. Irregular waves and wave spectral theory are the subject of many texts on ocean wave mechanics (see e.g., Dean and Dalrymple, 1984; U.S. Army Corps of Engineers, 1984; Goda, 1985); and will not be discussed in detail here. Nevertheless, a review of a few basic principles is necessary in order to understand the performance of floating breakwaters in actual sea conditions.

A representative wave height known as the significant wave, H_s, is often used in design applications to characterize an irregular sea. The significant wave is the average of the highest one third of all waves present in the spectrum of waves. The statistical distribution of waves in a random sea has been found to follow a Rayleigh Distribution which can be used to estimate various wave parameters as follows:

$$H_{\max} = 0.707 H_s \sqrt{\ln(N)} = (1.6 \sim 2.0) H_s \quad (5\text{–}7)$$

$$H_1 = 1.67 H_s$$

$$H_{10} = 1.27 H_s$$

$$H_{\text{avg}} = 0.63 H_s$$

where

$H_{\max}$ = maximum wave height in the spectrum H_1

H_{10} = waves exceeded by 1% and 10%, respectively, of the wave heights in the spectrum

H_{avg} = average wave height in the spectrum

N = the number of waves in the spectrum and can be estimated by dividing the storm duration by the average wave period

T_s = significant wave period which has been defined as the highest of the upper one third of wave periods. The average wave period is equal to about (0.77 to 0.9) T_s according to Goda (1985).

An irregular sea can be summarized in the form of a plot of wave energy as a function of wave frequency. This plot is referred to as a wave spectrum for the irregular sea condition. An example wave spectrum is provided in Figure 5–2 in which wave spectral energy, $S(f)$, is plotted on the ordinate and wave frequency, f, is plotted on the abscissa. The wave spectrum indicates the relative wave energy at various wave frequencies. The wave period corresponding to the frequency of maximum spectral wave energy is known

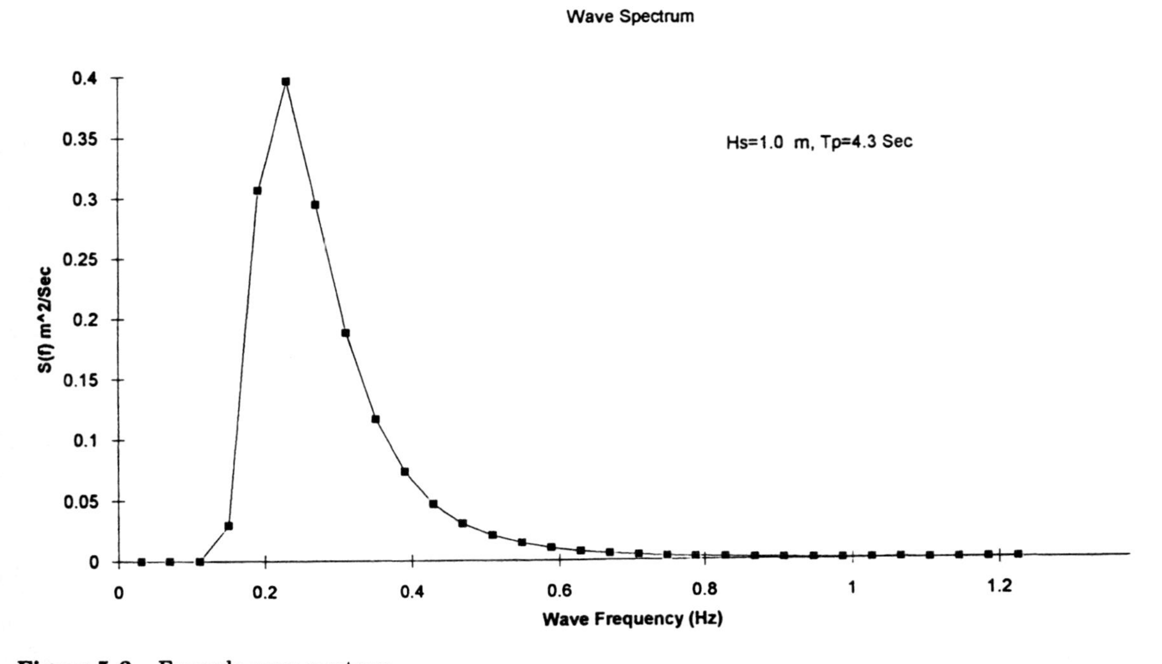

Figure 5–2. Example wave spectrum.

as the peak spectral wave period, T_p, and is often used as a characteristic wave period in design applications. The significant wave period, T_s is, for all practical purposes, equivalent to the peak spectral period, T_p. The U.S. Army Corps of Engineers (1989) recommends: $T_p = 1.05T_s$.

The following theoretical relationship provides a link between the significant wave height and the wave spectrum:

$$H_s = 4\sqrt{\int_0^\infty E(f)\,df}. \tag{5–8}$$

In other words, the significant wave height is related to the area under the spectral energy plot.

It is generally necessary to estimate design wave conditions for a specific breakwater site; however, methods for estimating conditions will not be presented here. The U.S. Army Corps of Engineers (1984) provides detailed methods for hind-casting design wave conditions for a given site and is sufficiently accurate for most floating breakwater design applications. If design values of the significant wave height, $H_{1/3}$, and peak spectral period, T_p, for a design site are known the following expression can be used to develop a plot of the wave spectrum. This expression is given in Goda (1985).

$$S(f) = 0.257H_{1/3}^2 T_s(T_s f)^{-5} \exp(-1.03(T_s f)^{-4}). \tag{5–9}$$

Similar expressions for other types of wave spectrums can be found in Goda (1985), the Shore Protection Manual (1984), and similar texts.

5.3 MECHANICS OF VIBRATION

A floating breakwater can be idealized as a linear vibratory system as will be discussed later. Accordingly, a brief review of vibration mechanics is presented. The dynamic response of a floating breakwater to waves in a single mode of motion (e.g., sway) can be likened to the dynamic motions of a mass, spring, and dashpot subject to a sinusoidal forcing function. In this analogy the mass of the vibratory system is equal to the mass of the floating breakwater, the damping of the vibratory system is equal to the hydrodynamic damping of the floating breakwater, and spring of the vibratory system is equal to the spring of either the moorings or buoyancy or both. The equation of motion for a single degree of freedom system subject to a sinusoidal force is:

$$m\ddot{x} = b\dot{x} + cx = f\sin(\sigma t) \tag{5–10}$$

where

$\ddot{x}, \dot{x}, x$ = acceleration, velocity, and displacement
m = mass
b = damping coefficient
c = spring coefficient
f = amplitude of the applied force
σ = $(2\pi/T)$ = circular frequency of applied force.

Solution to the above equation of motion is:

$$x = |X|\sin(\sigma t + \epsilon) \tag{5–11}$$

where

$|X|$ = amplitude of the motion x
ϵ = phase of the displacement relative to the exciting force.

The amplitude of the motion, $|X|$, is computed from:

(1) The magnification factor

$$\Lambda = \frac{|X|c}{f} \tag{5–12}$$

(2) The natural period of the system

$$\sigma_n = \sqrt{\frac{c}{m}} \tag{5–13}$$

(3) The percent critical damping

$$\zeta = \frac{b}{2m\sigma_n}. \qquad (5\text{–}14)$$

The magnification factor and phase are computed as follows:

$$\Lambda = \frac{1}{\sqrt{\left(1-\left(\frac{\sigma}{\sigma_n}\right)^2\right)^2 + \left(2\zeta\left(\frac{\sigma}{\sigma_n}\right)^2\right)^2}} \qquad (5\text{–}15)$$

$$\phi = \tan^{-1}\left(\frac{2\zeta\left(\frac{\sigma}{\sigma_n}\right)}{1-\left(\frac{\sigma}{\sigma_n}\right)^2}\right). \qquad (5\text{–}16)$$

A plot of the amplification or magnification factor versus frequency ratio (σ/σ_n) is presented in Figure 5–3. This figure shows that there is a relatively large, or magnified, response of the vibratory system for forcing frequencies near the natural frequency. This corresponds to a condition of

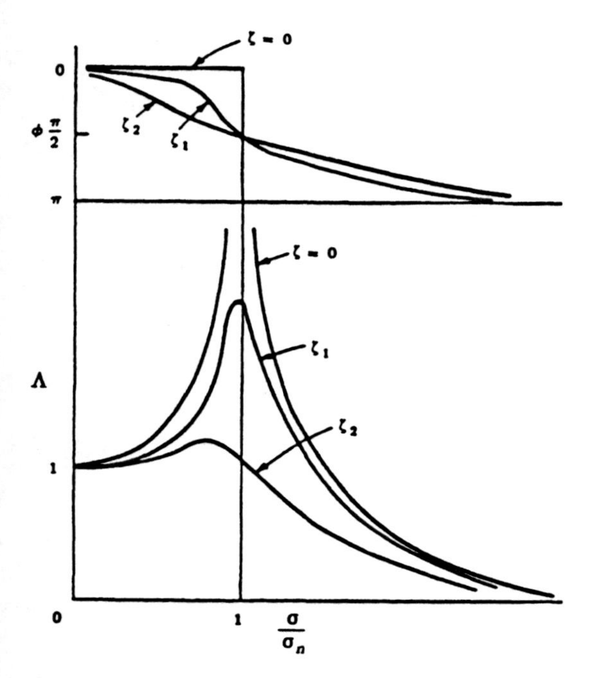

Figure 5–3. Magnification factor and phase for a linear system.

resonance. When the magnification factor is equal to 1 the force seen by the structure is equal to the applied force. This occurs for small frequency ratios (i.e., the natural period of the system is very short compared to the period of the applied force) and corresponds to a condition where the structure is held rigidly. The magnification factor is considerably lower than one for high-frequency ratios where the period of the applied force is much longer than the natural period.

5.4 DYNAMICS OF FLOATING BODIES

Analytical treatment of moored floating breakwaters is complicated and there are two general approaches used in such analyses: (1) frequency domain analysis, and (2) time domain analysis. The former approach is generally used in examination of floating breakwater wave transmission, whereas both methods have been used to examine floating breakwater mooring forces and internal stresses. Frequency domain analysis idealizes the moored breakwater as a linear system characterized by a mass, spring, and dashpot. The analysis is analogous to the vibrations theory discussed above and involves solution of an equation of the following form:

$$(m + a)\ddot{x} + b\dot{x} + cx = F(t) \qquad (5\text{–}17)$$

where

m = breakwater mass
a = breakwater added mass
$\ddot{x}$ = breakwater acceleration
$\dot{x}$ = breakwater velocity
x = breakwater displacement
b = damping coefficient
c = spring constant
$F(t)$ = time varying applied forcing function.

In reality, there is a separate governing equation for each of the six modes of breakwater motion defined in Figure 5–4 (i.e., surge, sway, heave, roll, pitch, and yaw).

Terms on the lefthand side of Eq. (5–17) account for forces imposed on the breakwater from motion in still water (i.e., in the absence of waves). The added mass, a, accounts for hydrodynamic forces proportional to the breakwater acceleration in still water. Similarly, the damping coefficient, b, represents the hydrodynamic forces proportional to the breakwater velocity in still water. The added mass and the damping coefficients are collectively called the hydrodynamic coefficients and are primarily a function of the breakwater shape, motion frequency, and water depth. The spring constant, c, accounts for forces proportional to the breakwater motion displacement and represents either hydrostatic restoring (buoyancy) forces or a linearized mooring restraint, or both. $F(t)$, on the righthand side of Eq. (5–17), is a sinusoidally varying wave force with a frequency equal to that of the incident wave and is a function of breakwater shape, wave amplitude, frequency, and direction. $F(t)$ is computed assuming the breakwater is rigidly held.

Hydrostatic restoring forces are computed from basic principles of naval architecture and are discussed in Tsinker (1986). The hydrodynamic coefficients and the wave forcing function may be computed by either slender body (Frank, 1967) or diffraction (Van Oortmerssen, 1976) hydrodynamic theories. The hydrodynamic coefficients are normally computed for a variety of motion frequencies. The wave forcing function, on the other hand, is computed for a variety of wave frequencies and directions. Equation (5–17) is then solved for each wave frequency and direction of interest. A consequence of this approach is that the vessel responds sinusoidally at a frequency equal to the wave frequency. Because an irregular wave field is characterized by a variety of frequencies and directions, Eq. (5–17) is

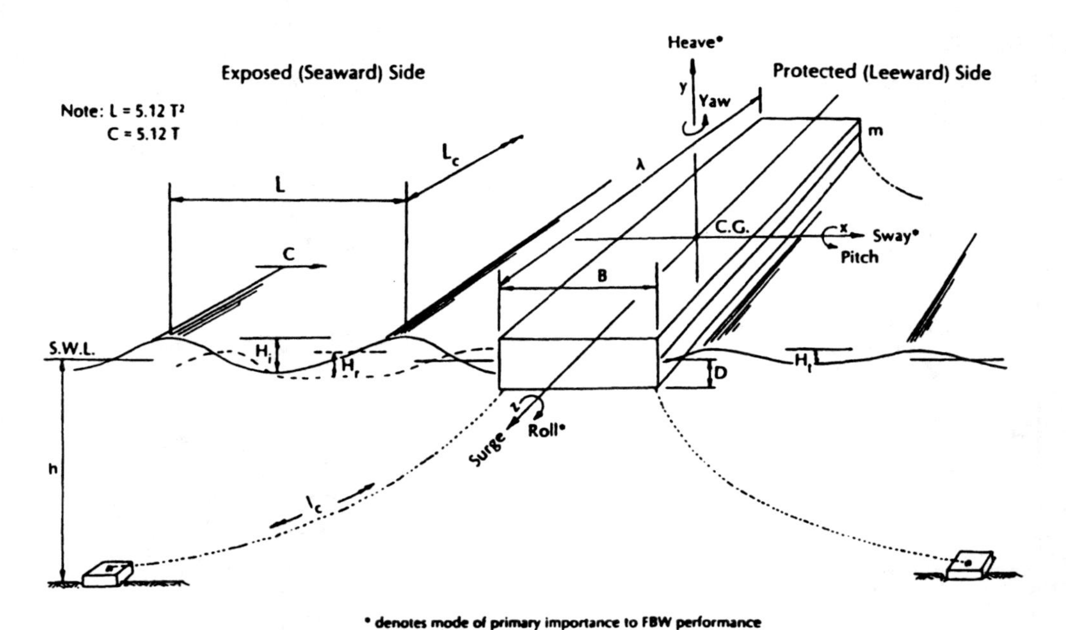

Figure 5–4. Floating breakwater motions.

solved for the range of wave frequencies and directions present in the wave field. Results of such analysis are normally computed for waves of unit amplitude and summarized in terms of a Response Amplitude Operator (RAO). Because the moored-ship system is linear, the motions of a breakwater in a irregular wave field can be determined using concepts of wave spectrum which will be discussed later.

The frequency domain approach has been, and continues to be, widely used as it is simpler and requires less computational effort than time domain analysis. Frequency domain analysis, as has been mentioned, is commonly used in connection with evaluation of wave transmission. Frequency domain techniques can, in the case of mooring analysis, provide misleading results. First of all, frequency domain analysis normally assumes the mooring restraints to be linear, that is, mooring restraint load is a linear function of displacement. Unfortunately, the loads in catenary mooring lines are generally nonlinear functions of breakwater displacement. When the mooring restraints are nonlinear, non-negligible subharmonic motions of the vessel can occur at frequencies that differ from the forcing wave frequency (Van Oortmerssen, 1976). Such motions are not simulated by frequency domain analysis. Second, frequency domain analysis normally accounts only for *first-order wave forces* (i.e., wave forces described above with frequency equal to that of the incident wave) whereas moored structures are subject to slowly varying wave drift forces. The wave drift is a *second-order wave force* of relatively small magnitude and is of no significance in regular waves. In irregular waves, however, differences in wave frequencies result in low-frequency time-varying drift forces. Floating breakwaters moored by means of catenary moorings are often characterized by relatively low natural frequencies in surge, sway, and yaw. Moreover, damping of moored systems is small at low wave frequencies. These facts make catenary moored floating breakwaters prone to low-frequency excitation. The above described frequency domain analysis is generally used to evaluate the first-order wave forces and cannot be used to evaluate breakwater response to both first-order wave and second-order wave drift forces.

The shortcomings of frequency domain analysis are overcome by time domain analysis at the expense of added computational effort. Time domain analysis has been described in detail by Van Oortmerssen (1976) and consists of solving equations of the following form:

$$(m + m')\ddot{x} + \int_{-\infty}^{t} K(t-\tau)\dot{x}\, d\tau + cx = F(t) \quad (5\text{–}18)$$

where

m' = constant inertial coefficient
$K(t)$ = impulse-response function
$F(t)$ = arbitrary forcing function
c = hydrostatic restoring force coefficient

Other terms are as previously defined for Eq. (5–17). Like the frequency domain approach, terms on the lefthand side of Eq. (5–18) account for forces on the ship when moving in still water. In contrast to Eq. (5–17), however, the constant inertial coefficient, m', and impulse-response function, $K(t)$, are independent of motion frequency and represent hydrodynamic forces on the vessel for any arbitrary vessel motion. The righthand side of Eq. (5–18) is an arbitrary force–time function that may include first-order wave frequency forces, second-order wave drift forces, nonlinear mooring restraint forces, forces from ship waves, wind forces, and current forces.

The constant inertial coefficient, m', and impulse–response coefficients cannot be determined directly. Instead, they are determined from the frequency-dependent hydrodynamic coefficients as follows:

$$K(t) = \frac{2}{\pi}\int_0^\infty b(\sigma)\cos(\sigma t)\, d\sigma \qquad (5\text{–}19)$$

$$m' = a(\sigma) + \frac{1}{\sigma}\int_0^\infty K(t)\sin(\sigma t)\, dt. \qquad (5\text{–}20)$$

Van Oortmerssen (1976) describes the numerical solution of Eq. (5–18) which provides time histories of breakwater motions and mooring restraint loads. To the author's knowledge, no attempt has been made to evaluate floating breakwater wave transmission using a time domain approach.

5.5 BUOYANCY AND STABILITY OF FLOATING BREAKWATERS

A floating breakwater must have sufficient floatation to prevent sinking and should be stable against loads that tend to turn the breakwater over. Examination of floating breakwater buoyancy and stability is fundamental and an important aspect of any floating breakwater design. A detailed presentation of the buoyancy and stability of floating structures is given in Tsinker (1986) and Gaythwaite (1990).

5.6 PREDICTION OF WAVE TRANSMISSION

Wave transmission refers to the characteristic of a breakwater, floating or otherwise, to reduce incident waves. The wave attenuating characteristics of a floating breakwater are quantified through the use of a wave transmission coefficient defined as follows:

$$K_t = \frac{a_t}{a_i} = \frac{H_t}{H_i} \qquad (5\text{–}21)$$

where

K_t = wave transmission coefficient
a_i = incident wave amplitude = $H_i/2$
a_t = transmitted wave amplitude = H_t
H_i = incident wave height
H_t = transmitted wave height.

The wave transmission coefficient, K_t, is strongly, but by no means entirely, dependent on the incident wave period, T, and it is common to plot the wave transmission coefficient of floating breakwaters against wave period as shown in Figure 5–5. Many floating breakwater tests are summarized in a plot of K_t versus ratio of the breakwater width or beam, B, to incident wave length, L, as shown in Figure 5–6. Generally speaking, a floating breakwater reduces incident wave energy by either reflecting that energy seward or dissipating that energy through turbulence. The following important relationship, which assumes no energy dissipation, can be derived on grounds of conservation of wave energy:

$$a_i^2 = a_t^2 + a_r^2 \qquad (5\text{–}22)$$

where

a_r = reflected wave amplitude = $H_r/2$.

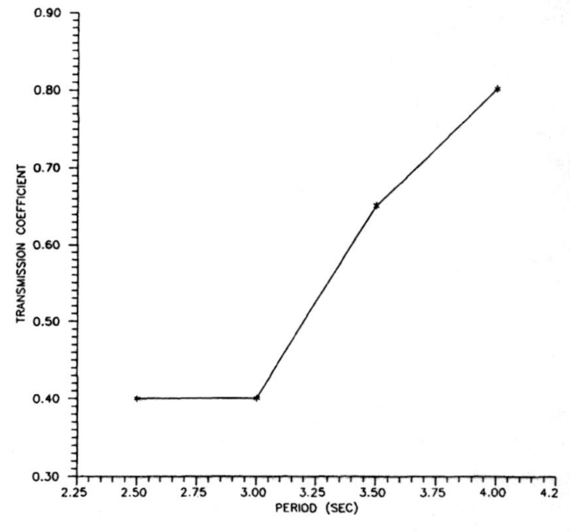

Figure 5–5. Transmission coefficient as a function of wave period.

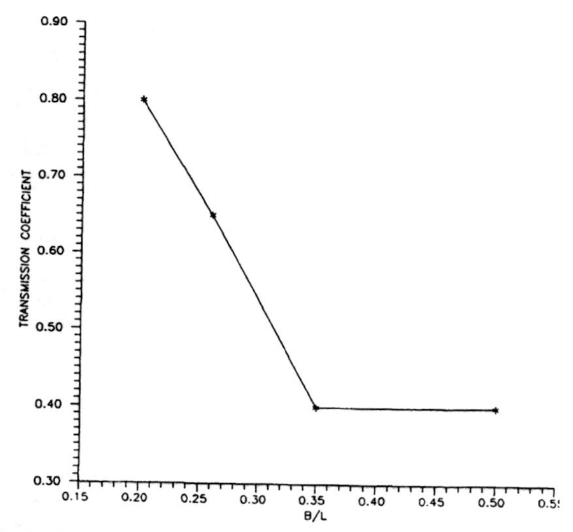

Figure 5–6. Tranmission coefficient as a function of *B/L*.

Equation (5–22) indicates that the transmitted wave amplitude, or equivalently wave height, can be reduced by increasing wave reflection. This is why a larger draft breakwater can, in certain cases, be more effective than smaller draft structure. This is not always the case, however, because wave transmission depends not only on the breakwater geometry but also on the dynamic motions of the breakwater.

Figure 5–5 presents wave transmission results obtained from hydraulic model tests (Carver, 1979) for a rectangular caisson-type floating breakwater with a beam of 4.9 m and a draft of 1.1 m moored by means of chains and anchors in a water depth of 7.6 m. Figure 5–5 clearly shows a dependency of the K_t on the incident wave period, namely K_t is relatively high for long waves and relatively small for short waves. The wave transmission coefficient for this caisson-type floating breakwater is 0.8 for a 4-s wave and only 0.4 for a 2.5-s wave. Figure 5–6 presents wave transmission coefficients for the same structure as a function of B/L and shows that as the incident wave length increases the wave transmission coefficient increases (i.e., B/L decreases). In other words, the ability of the breakwater to reduce waves decreases dramatically as the incident wave length approaches the width of the breakwater. This finding is intuitive; the breakwater does little to reduce waves when the dimensions of the breakwater are small compared to the wave dimensions.

Figures 5–5 and 5–6 are useful in summarizing results of hydraulic model tests for a particular floating breakwater system (i.e., width, draft, mass, water depth, mooring system). Such figures, on the other hands, cannot be used to predict wave transmission for conditions that differ markedly from those tested in the hydraulic model facility. In the past, engineers have *incorrectly* attempted to use test results, such as those presented in Figure 5–6, to predict the performance of a seemingly similar structure. For example, one might conclude from Figure 5–6 that a K_t value of 0.4 would be obtained for a similar rectangular shaped structure as long as the breakwater B/L ratio is equal to 0.4 or less. In the case of the 4.9 m wide structure that is represented in Figure 5–6, the maximum wave length corresponding to a B/L ratio of 0.4 is 12.2 m. One cannot assume, however, that incident waves with a length of 24.4 m can be reduced by 40% if the breakwater width is increased to 9.8 m simply because the B/L ratio of 0.4 is maintained. The reason that this assumption cannot be made is that an increase in breakwater width results in an increase in the breakwater mass and rotational inertia which, in turn, has a dramatic effect on the motion response of the floating breakwater to the incident waves. In short, floating breakwater response and attendant wave transmission depends on more than breakwater width.

Waves transmitted by a floating breakwater are the aggregate result of a complex hydrodynamic interaction between the floating breakwater and the incident wave field. As previously stated, a portion of the incident wave energy is reflected by the structure and some wave energy passes under (and over) the breakwater. Furthermore, a portion of the incident wave field excites motions of the breakwater which, in turn, generate waves in a manner similar to a wavemaker in a hydraulic model. The total transmitted wave is the sum of components that pass under (and over) the breakwater and components that are generated by breakwater motion.

The numerical model of Adee and Martin (1974), which will be discussed in more detail later, actually separates analyses into: (1) determination of waves transmitting past the rigidly held breakwater, that is, "fixed-body" transmission, and (2) determination of wave transmission resulting from breakwater motion. This separation fosters a better understanding of the mechanisms that dominate floating breakwater response at various incident wave periods. Figure 5–7 presents measured and numerically predicted wave transmission characteristics for a catamaran-type floating breakwater. The measured results are from Davidson (1971) and the numerical predictions, which are reasonably close to the measured values, are given by Adee and Martin (1974). The numerical model predictions indicate that wave transmission is dominated by "fixed-body" transmission for beam to wave lengths, B/L, less than 0.15(i.e., wave periods greater than 3.5 s). In other words, when the waves are long compared to the breakwater beam they pass relatively undiminished in height under the breakwater. The "fixed-body" wave transmission drops off quickly for wave periods shorter than 3.5 s and is insignificant for wave periods less than 3 s. Waves generated by body motion, however, affect wave transmission for periods less than 3.5 s. An interesting aspect of the numerically predicted wave transmission curve is evident in Figure 5–7 for a wave period of about 3 s where a relatively low transmission coefficient value is indicated. As previously stated, "fixed body"

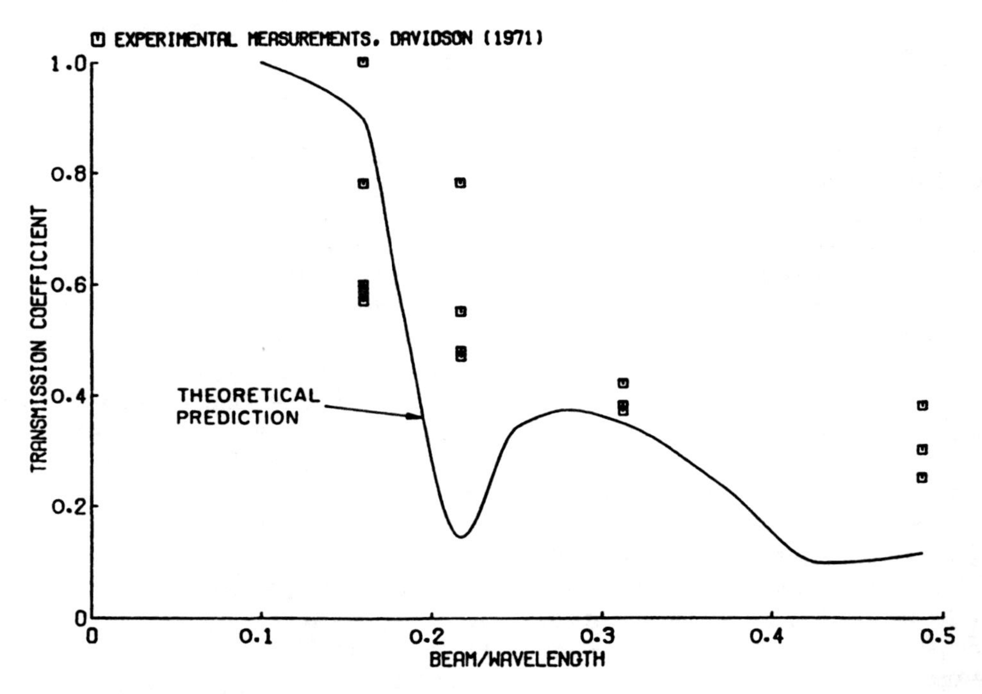

Figure 5–7. Transmission coefficient for catamaran breakwater.

transmission is very low at this wave period; however, the heave and sway motion generated wave transmission are relatively high and heave motion is near resonance. The transmission coefficient is low, however, because the heave and sway generated waves are nearly 180° out of phase and cancel each other out. This phenomenon is not clearly evident in the hydraulic model tests. Wave transmission is governed by sway motion for wave periods less than 2.5 s although the total wave transmission is less than 0.4.

5.6.1 Hydraulic Model Tests

Hydraulic model tests constitute the most reliable means for determining the wave transmission coefficients of floating breakwaters. Normally these tests are based on Froude similarity laws and performed at geometric scales around 1 : 10. Most tests have been performed in regular wave conditions. A few tests, however, have been conducted in irregular waves which are more representative of real wave conditions (Torum et al., 1989). Some tests have been conducted in three-dimensional basins but most have been conducted in two-dimensional wave flumes. Typical hydraulic model results are summarized in Figures 5–5, 5–6, and 5–7. Readshaw (1981) provides a useful summary of hydraulic model test results for various breakwater types.

Hydraulic model tests are highly desirable and should be conducted prior to final design on any large-scale project. On the other hand, hydraulic model tests are expensive and should be viewed as a means for confirming a final breakwater scheme. It is therefore desirable to approximate breakwater performance before conducting hydraulic model tests. Such approximations can be made with previous model test results; however, as stated above it is difficult to apply results of previous hydraulic models tests to new floating breakwater designs unless the conditions for the new development correspond to those tested in the hydraulic model. In the absence of appropriate model test results, analytical methods can be used to estimate floating breakwater performance.

5.6.2 Simplified Analytical Methods

Several investigators have developed simplified methods for determining the wave attenuating characteristics of various shaped structures. Some of these methods can be used to make a rough estimate of floating breakwater transmission coefficients and are presented below. It should be noted, however, that the following methods assume a rectangular floating breakwater that is rigidly held, that is, the breakwater does not move and generate waves. As a consequence these methods should not be used for general design purposes. Nevertheless, they are useful in preliminary investigations of floating breakwater arrangements. Moreover, these methods are reasonably accurate for conditions where the floating breakwater mooring is relatively rigid as is often the case for a breakwater moored by means of guide piles.

Macagno (1953) presented an analytical formula for computing the wave transmission coefficient of a rectangular breakwater of width, B, draft, D, rigidly moored in a water depth, h, and subjected to incident waves having a period, T, and wave length, L:

$$K_t = \frac{1}{\sqrt{1 + \left[\dfrac{\pi B \sinh\left(\dfrac{h}{L}\right)}{L \cosh\left(2\pi \dfrac{(h-d)}{L}\right)}\right]^2}}. \qquad (5\text{–}23)$$

A simpler formula has been presented by Carr (1952) for shallow water wave conditions ($H/L < 0.04$):

$$K_t = \frac{1}{\sqrt{1+\left[\left(\frac{\pi B}{L}\right)\left(1+\frac{D}{h-D}\right)\right]^2}}. \quad (5\text{–}24)$$

Example 1

A floating breakwater with a beam, B, of 4.9 m, and a draft, D, of 1.1 m is rigidly moored in a water depth of 7.6 m and is subjected to an incident wave with a height, H, of 0.91 m and a period, T, of 3.5 s. This breakwater is identical to that tested by Carver (1979) and summarized in Figures 5–5 and 5–6. Compute the height of the transmitted wave using both the Macagno and Carr formulas.

Step 1: Compute wave length, L using Eq. (5–5).

$$L_0 = \frac{gT^2}{2\pi} = \frac{9.8(3.5)^2}{2\pi} = 19.1\,\text{m}$$

$$2\pi\frac{h}{L} = \sqrt{\left(2\pi\frac{h}{L_0}\right)}\left(1+\frac{1}{6}2\pi\frac{h}{L_0}+\frac{11}{360}\left(2\pi\frac{h}{L_0}\right)^2\right)$$

$$2\pi\frac{7.6}{L} = \sqrt{\left(2\pi\frac{7.6}{19.1}\right)}\left(1+\frac{1}{6}2\pi\frac{7.6}{19.1}+\frac{11}{360}\right.$$

$$\left.\left(2\pi\frac{7.6}{19.1}\right)^2\right)$$

$$2\pi\frac{7.6}{L} = 2.5432.$$

Solving for L gives $l = 18.8$ m.

Step 2: Compute K_t using Macagno's theory [Eq. (5–23)].

$$K_t = \frac{1}{\sqrt{1+\left[\frac{\pi B \sinh\left(2\pi\frac{h}{L}\right)}{L\cosh\left(2\pi\frac{(h-D)}{L}\right)}\right]^2}}$$

$$= \frac{1}{\sqrt{1+\left[\frac{\pi 4.9 \sinh\left(2\pi\frac{7.6}{18.8}\right)}{18.8\cosh\left(2\pi\frac{(7.6-1.1)}{18.8}\right)}\right]^2}} = 0.66$$

Step 3: Compute K_t using Carr's Theory [Eq. (5–6)].

$$K_t = \frac{1}{\sqrt{1+\left[\left(\frac{\pi B}{L}\right)\left(1+\frac{D}{h-D}\right)\right]^2}}$$

$$= \frac{1}{\sqrt{1+\left[\left(\frac{\pi 4.9}{18.8}\right)\left(1+\frac{1.1}{7.6-1.1}\right)\right]^2}} = 0.727$$

Step 4: Compute Transmitted Wave Heights using Eq. (5–1).

The transmitted wave height according to Macagno's theory is:

$$H_t = K_t H_i = 0.66(0.91) = 0.60\,\text{m}.$$

The transmitted wave height according to Carr's Theory is:

$$H_t = K_t H_i = 0.727(0.91) = 0.66\,\text{m}.$$

It should be noted that Carr's Theory is not valid for this example as the value of h/L is 0.4 and exceeds the shallow water limit of $h/L = 0.04$. Strictly speaking, Carr's formula should not be used for the conditions presented in this example.

5.6.3 Numerical Models

There are several numerical models available for predicting the performance of floating breakwaters. A literature survey of available floating breakwater numerical models has been conducted by Bando and Sonu (1985). The Ito and Chiba (1972) and the Adee and Martin (1974) numerical models are discussed below.

5.6.3.1 Ito and Chiba (1972) Numerical Model

Ito and Chiba (1972) present a numerical model for determining the wave transmission characteirstics of a rectangular floating breakwater. The details of this model have been presented by Tekmarine Inc. (1986), who suggest that the model be used as a quick diagnostic tool. The model is two-dimensional, assumes incident waves are beam-on, and accounts for breakwater response in heave, sway, and roll (Figure 5–8). The model utilizes linear wave theory and, as shown in Figure 5–9, separates the hydrodynamic domain into two computational regions. The interior region is beneath the breakwater and is bounded by the sides and bottom of the breakwater and the seafloor. The exterior region extends from the interior/exterior boundary to infinity on the seaward and leeward sides of the breakwater. Expressions for velocity potentials are derived for the interior and exterior regions. For those not generally familiar with hydrodynamic theory or the specific notion of velocity potential it may be useful to think of a velocity potential as a function whose spatial derivative in any direction is a velocity in that direction. In other words, it is a convenient mathematical expression that contains information concerning the velocity of a specified flow problem. The velocity potential for a particular flow problem is normally developed from hydrodynamic equations of continuity of mass, momentum, and energy and boundary conditions appropriate to a particular problem. Ito and Chiba (1972) develop separate velocity potentials for the interior and exterior hydrodynamic regions shown in Figure 5–9.

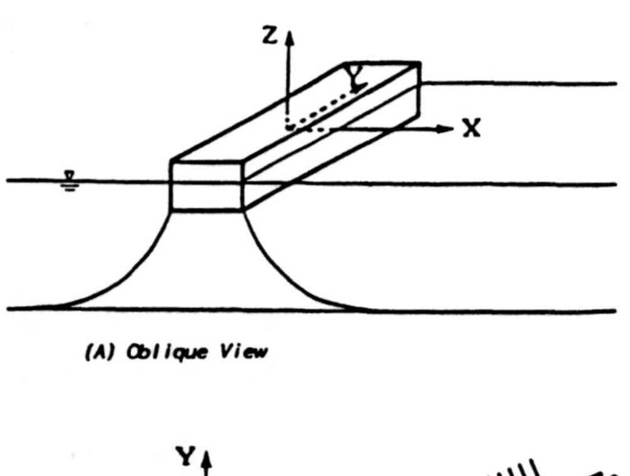

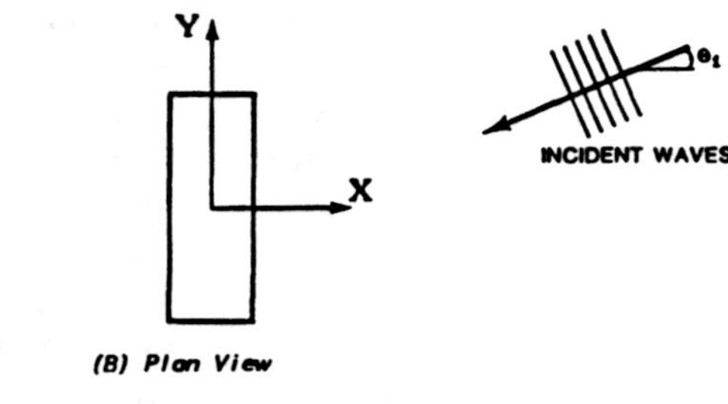

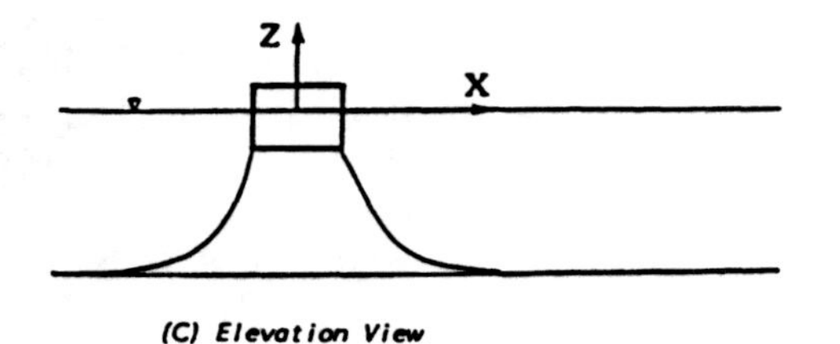

Figure 5–8. Definition sketch—Ito and Chiba model.

Numerical solutions for wave transmission, floating breakwater motions, and mooring line forces are obtained by setting up a system of five simultaneous equations. Three of these are equations of motion for sway, heave, and roll in the following form:

$$\sum_{i=1}^{3}[(m_{ij}+a_{ij})\ddot{x}_i+(K_{ijh}+K_{ijm})x_i]=F_j(t)$$
$$\text{for } j=1,2,3. \tag{5–25}$$

Subscripts 1, 2, and 3 denote motion (or acceleration) in sway, heave and roll as shown in Figure 5–8. M_{11} and M_{22} are the breakwater mass per length and M_{33} is the roll mass moment of inertia per unit length. Added mass terms a_{11}, a_{22}, a_{33} are for sway, heave, and roll motions, respectively. The effects of moorings are simulated as linear springs with the terms K_{11m}, K_{22m}, K_{33m} representing restraints in sway, heave, and roll, respectively. The hydrostatic restoring

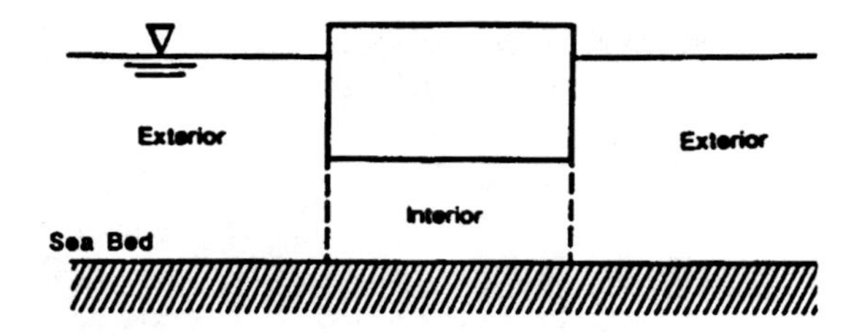

Plan View

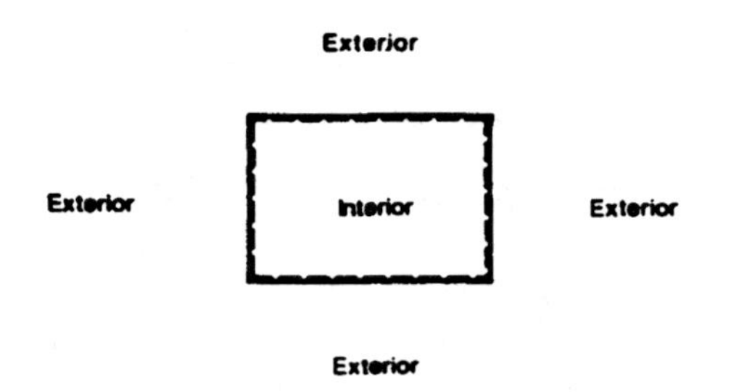

Figure 5-9. Computation Domain—Ito and Chiba model.

force in heave and moment in roll are represented by the terms K_{22h} and K_{33h}, respectively. It will be noted that no damping terms are included in the equations of motion. The hydrostatic restoring force and moment are defined as follows:

$$K_{22h} = -2\rho gB \tag{5–26}$$

$$K_{33h} = -2\rho gDBGM \tag{5–27}$$

where GM is the breakwater metacentric height. Highly idealized expressions for added mass coefficients, which are independent of motion frequency, are used in the model:

$$a_{11} = 0 \tag{5–28}$$

$$a_{22} = 2p \frac{a}{3(h-D)} (B^2 + (h-D)^2) \tag{5–29}$$

$$a_{33} = 2\rho \frac{B^3}{(h-D)} \left(\frac{B^2}{45} + \frac{(h-D^2)}{9}\right). \tag{5–30}$$

The righthand terms of the above equations, that is, F_1, F_2, and F_3, represent the total wave forces in sway, heave, and roll, respectively. Wave forces are derived by integrating wave pressures on the front, bottom, and lee sides of the breakwater and are given, because the breakwater is assumed rectangular, in terms of the incident, reflected, and transmitted wave amplitudes. For example the equation for the sway force, F_1, is:

$$F_1 = \rho gD \frac{\sinh(kh) - \sinh(k(h-D)}{kD\cosh(\text{kh})} (a_i + a_r - a_t)\sin(\sigma t) \tag{5–31}$$

where

$$k = 2\frac{\pi}{L}, \qquad \sigma = \frac{\pi}{T}$$

where ρ is the water density.

The equations of motion are solved in the frequency domain assuming the motion to be sinusoidal:

$$x_i(t) = |x_i|\sin(\sigma t + \epsilon_i) \quad \text{for } i = 1, 2, 3 \tag{5–32}$$

where
$|x_i|$ = i-th motion amplitude
ϵ_i = i-th motion phase.

The three equations of motion contain five unknowns, namely, x_1, x_2, x_2, a_r, and a_t. Accordingly, two additional equations are necessary in order to solve for all of the unknowns. These additional equations are derived by matching hydrodynamic boundary conditions along the interface between the interior and exterior computational regimens. Specifically, two additional expressions are obtained from the interior and exterior velocity potentials assuming continuity of pressure and velocity across the interior/exterior interface. Details of this derivation can be found in Ito and Chiba (1972) or Tekmarine (1986) and are not included herein. It is remarked, however, that the matching of velocity potentials for continuity of velocity involves the use of a depth-averaged velocity under the breakwater. This is a very crude approximation especially considering that most floating breakwaters are placed in deepwater and exposed to relatively small waves. Under such conditions, wave-induced water particle velocities drop off dramatically with depth and one would not expect a depth-averaged velocity to adequately represent hydrodynamic conditions. Nevertheless, this assumption permits a substantially simplified solution methodology. The Ito and Chiba (1972) method will be compared to other prediction techniques later. It suffices to say that crude assumptions notwithstanding, the model has been shown to compare reasonably with physical model and field measurement results. Detailed methods for solution of the governing equations can be found in Ito and Chiba (1972) or Tekmarine (1986).

5.6.3.2 Adee and Martin (1974) Model

The floating breakwater model of Adee and Martin (1974) is similar in some respects to the Ito and Chiba (1972) model. Specifically, the model is two-dimensional, uses linear wave theory, assumes beam-on wave conditions, and solves linearized equations of motion in the frequency domain. A definition sketch for the Adee and Martin (1974) model is given in Figure 5–10. The model uses the following governing equations of motion which incorporate linear damping terms:

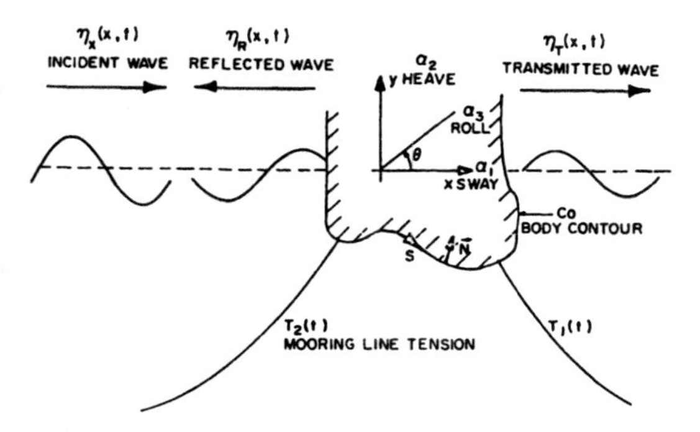

Figure 5–10. Definition sketch—Adee and Martin model.

$$\sum_{i=1}^{3} [(m_{ij} + a_{ij})\ddot{x}_i + b_{ij}\dot{x}_i + (K_{ijh} + K_{ijm})x_i] = F_j(t) \quad (5\text{–}33)$$

where b_{ij} are the damping coefficients in the j-th direction due to velocity in the i-th direction and the other variables are as previously defined. It will be noted, however, that the added mass and damping coefficients are computed as a function of wave/motion frequency in the Adee and Martin (1974) method. As with the Ito and Chiba (1972) model, frequency domain solutions to the equations of motions are assumed to be of the following sinusoidal form:

$$x_i(t) = |x_i| \sin(\sigma t + \epsilon_i) \quad \text{for } i = 1, 2, 3 \quad (5\text{–}34)$$

where
$|x_i|$ = i-th motion amplitude
ϵ_i = i-th motion phase.

Substitution of the assumed solution into the equations of motions leads to a system of simultaneous linear algebraic equations that are solved for motion amplitudes and phases. Adee and Martin's hydrodynamic techniques for computing the added mass and damping coefficients, the applied wave forces, and reflected and transmitted wave amplitudes are considerably more sophisticated (and computer intensive) than the Ito and Chiba (1972) method. The approach is known among naval architects as the "Frank-Close Fit" method and was originally developed by Frank (1967). One of the primary advantages of this method is that it can be used to evaluate floating breakwaters of arbitrary shape. Furthermore, there is no need to make any assumptions regarding the depth-averaged velocity under the breakwater as is done with the Ito and Chiba model.

The "Frank-Close Fit" method computes the total velocity potential along the breakwater surface and at selected points seaward and leeward of the structure. Inputs required by the "Frank-Close fit" method are incident wave period (or length) and the breakwater geometry. The total wave potential consists of the incident wave potential; the diffracted wave potential (that is, "fixed body" potential which includes effects of wave reflection and transmission); and the potential resulting from sway, heave, and roll motion.

$$\overset{\Phi}{\phi_{\text{total}}} = \overset{\Phi}{\phi_{\text{indicent}}} + \overset{\Phi}{\phi_{\text{diffracted}}} + \overset{\Phi}{\phi_{\text{motion}}} \quad \text{for } i = 1, 2, 3. \qquad (5\text{–}35)$$

The added mass and damping coefficients for the breakwater are computed from the motion potentials whereas the wave forces on the breakwater are computed from the incident and diffracted potentials. Surface amplitudes at desired points (e.g., seaward and leeward of the breakwater) can be obtained from the total potential:

$$a = -\frac{1}{g}\frac{\overset{\Phi}{\phi_{\text{total}}}}{\partial t}. \qquad (5\text{–}36)$$

Computational details of the model, which are too involved for this presentation, are discussed in Adee and Martin (1974).

5.6.4 Comparison of Predictive Techniques

The following paragraphs compare the above analytical methods for estimating wave transmission with the hydraulic model tests of Carver (1979). Carver (1979) examined several breakwater types; however, the present discussion will focus on the single breakwater geometry summarized in Figures 5–11 and 5–12. Specifically, the prototype breakwater module was characterized by a 4.9 m beam, a 1.1 m draft, and a length of 29.3 m. The breakwater was moored in 7.6 m of water by means of four anchor chains as shown in Figure 5–12. The unit weight of each anchor chain was 179.5 N/m

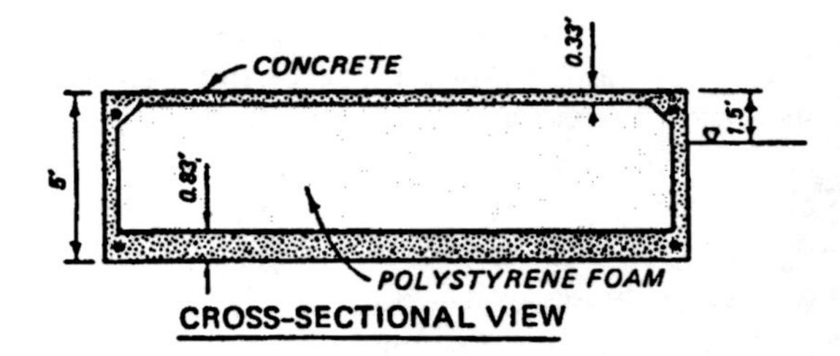

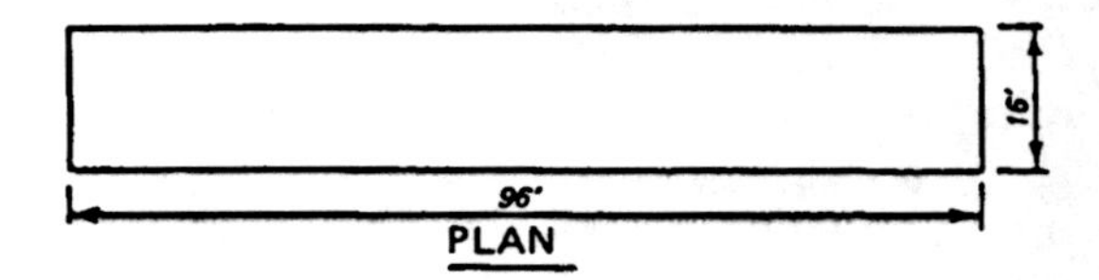

Figure 5–11. Floating breakwater geometry.

in water. The breakwater units were exposed to regular waves with heights ranging from 0.5 to 1.1 m and wave periods ranging from 2.5 to 4.0 s. The model experiment was conducted at a geometric scale of 1 : 10. Wave transmission coefficients for the breakwater geometry presented in Figures 5–11 and 5–12 have already been summarized in Figures 5–5 and 5–6. Computations were prepared using both

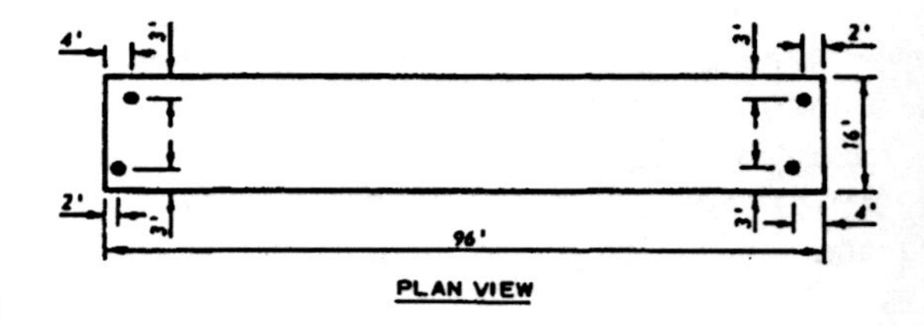

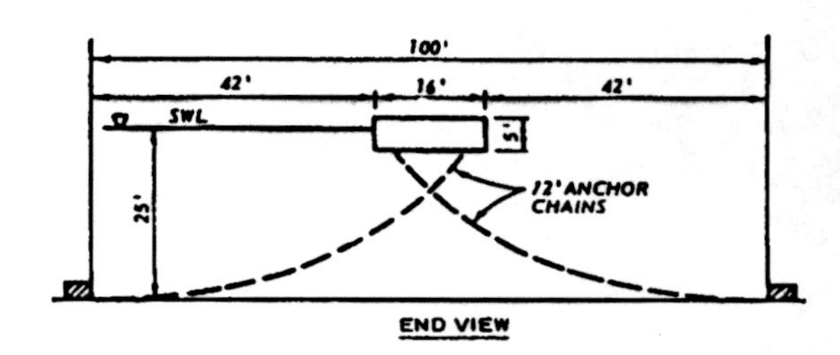

Figure 5–12. Floating breakwater geometry.

the simplified analytical methods of Macagno (1953) and Carr (1952) as well as the more complex numerical models of Ito and Chiba (1972) and Martin and Adee (1974). Wave transmission coefficients obtained from each of these analytical computations are presented in Figure 5–13 along with the measurements recorded in the physical model tests. The purpose of this comparison is simply to illustrate the use of various predictive methods for a specific breakwater arrangement. There is no intent on the part of the author to draw conclusions as regards the general predictive capability of any of the analytical methods on the basis of this single comparison.

Before considering these results, the reader is reminded that the simplified methodologies of Macagno (1953) and Carr (1952) should not be expected to provide particularly accurate predictions because these methodologies assume the breakwater to be rigidly held. In fact, the breakwater was secured by a relatively soft mooring system that impeded, but by no means prevented, breakwater motion. Furthermore, the Carr (1952) method assumes shallow water wave conditions (i.e., $h/l < 0.04$) and the wave conditions in the hydraulic model tests represented deep or intermediate water conditions (i.e., $h/L = 0.3$ to 0.78) for all wave periods. Despite the shortcomings of the Macagno (1953) theory it predicts wave transmission coefficients which are remarkably close to the measured results for wave periods of 3 s or higher. The Macagno theory underpredicts wave transmission for a 2.5-s period and it appears that the theory would underpredict wave transmission for shorter wave periods. This would be expected inasmuch as wave transmission for short period waves would be governed by breakwater motions neglected in the Macagno (1953) theory. The Carr (1952) theory overpredicts wave transmission for the range of tested wave periods.

Numerical model predictions of wave transmission are compared with the model test results in a separate plot shown in Figure 5–14. Wave transmission curves

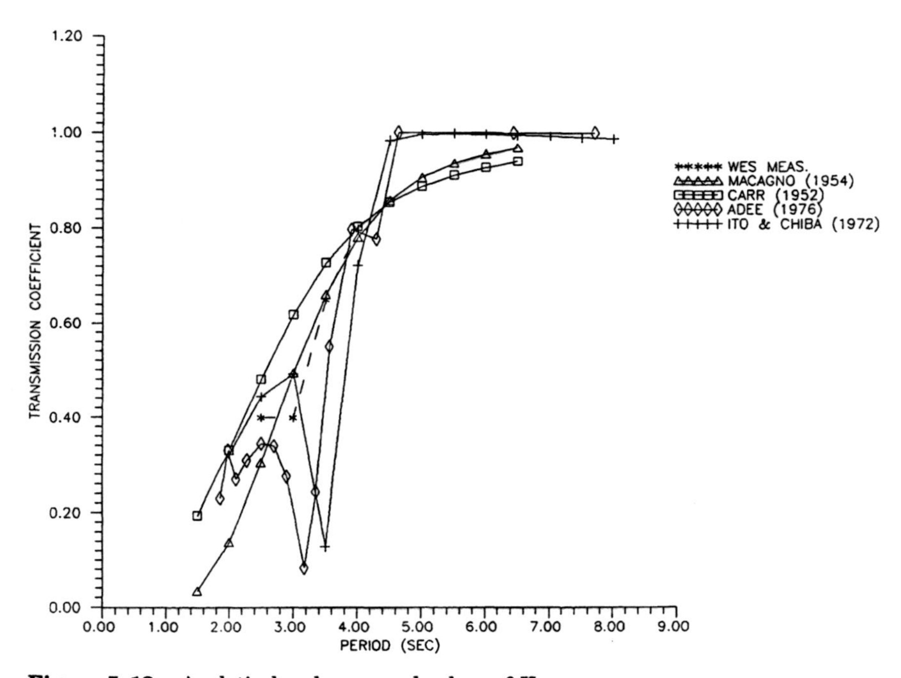

Figure 5–13. Analytical and measured values of K_t.

obtained from the Ito and Chiba (1972) and Martin and Adee (1974) numerical models both show characteristic dips in transmission coefficients in the region of heave resonance (i.e., 3.3 to 3.5 s). This dip in the transmission curve is not evident in the hydralic model test results and in this sense the numerical predictions do not correspond closely with the measurements. The lack of a dip in the hydraulic model transmission curve may be the result of wave overtopping which is neglected in the numerical models. Both numerical models predict a sharp rise in wave transmission coefficient values for wave periods higher than the period of heave resonance. For example, both numerical models give transmission coefficients on the order of 0.8 for a 4-s wave period. Similarly, the wave transmission coefficients predicted by the numerical models increase from relatively low values at heave resonance to values on the order of 0.3 (Martin and Adee model) to 0.5 (Ito and Chiba model) for the wave period range from 2 to 3 s. In summary, numerically predicted wave transmission coefficients compare reasonably well with the hydraulic model test results for wave periods of 2 to 3 s and 4 s. The numerical models do not compare well with the measured results for periods of 3 to 3.8 s where the numerical models predict breakwater heave resonance. This notwithstanding, it is clear that the numerical models are sufficiently accurate to serve as a useful planning tool. With regard to the relative performance of the numerical models, neither of the models provide substantially better predictions of wave transmission for the case examined.

5.6.5 Computation of Wave Transmission for Irregular Waves

The above paragraphs show that floating breakwater wave transmission is highly dependent on wave period. Although it is useful to examine the general nature of a floating breakwater wave transmission in

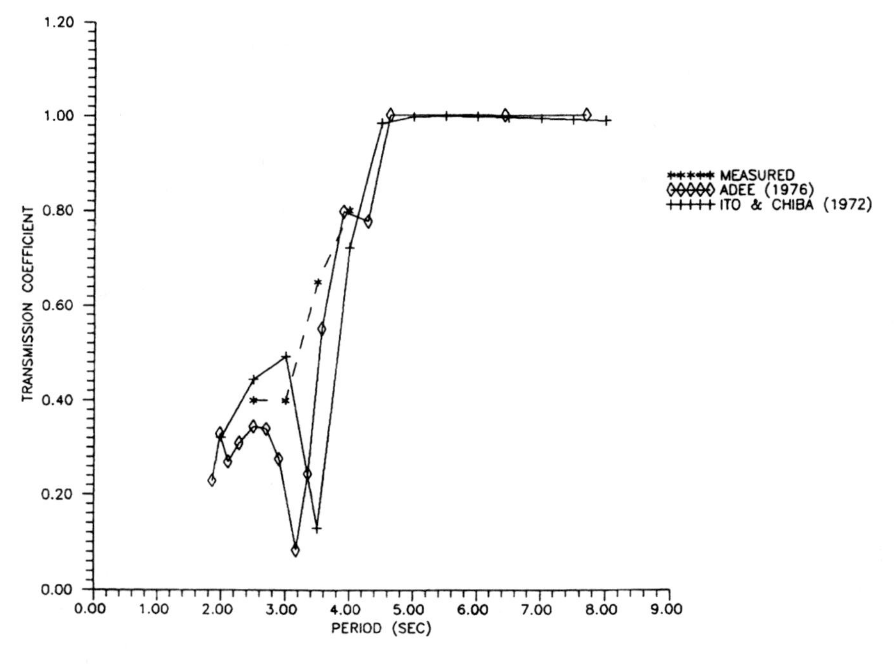

Figure 5–14. Analytical and measured values of K_t.

regular waves, a designer should also examine wave transmission for irregular waves. Fortunately, wave transmission coefficients determined on the basis of regular wave conditions can be used, in combination with a wave spectrum, to provide an assessment or irregular wave transmission. Specifically, a transmitted wave height spectrum can be obtained from the incident wave spectrum and the wave transmission coefficient for a given frequency as follows:

$$S_t(f) = [K_t(f)]^2 S_i(f). \tag{5–37}$$

Once the transmitted height spectrum has been computed one can compute the transmitted significant wave height as follows:

$$H_{t/13} = 4.0\sqrt{\int_0^\infty S_t(f)\, df}. \tag{5–38}$$

Furthermore, one can define a significant transmission coefficient as follows:

$$k_{t1/3} = \sqrt{\frac{\int_0^\infty S_t(f)\, df}{\int_0^\infty S_i(f)\, df}} \tag{5–39}$$

Example 2

Compute the spectrally based wave transmission coefficient for the rigidly moored breakwater examined in Example 1 using a significant wave height, H_s, of 0.91 m and a peak spectral period, T_p, of 3.5 s using Macagno's formula. Computations for this example were made with a microcomputer.

Step 1: Compute the wave spectrum corresponding to the given H_s and T_p using Eq. (5–9). The resulting wave spectrum is given in Figure 5–15.

Step 2: Compute the wave lengths and attendant wave transmission coefficients corresponding to a range of wave frequencies using Eq. (5–23). The computed wave

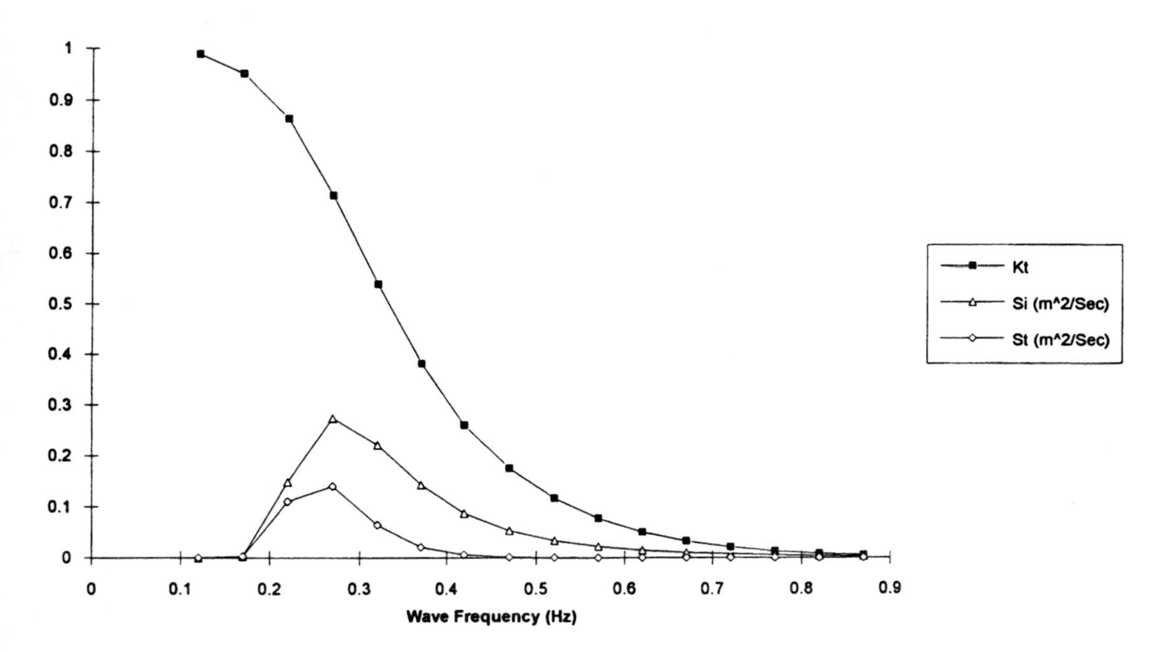

Figure 5–15. Incident and transmitted wave spectra.

transmission coefficients are shown in Figure 5–15.

Step 3: Compute the transmitted wave spectrum using Eq. (5–37) and the attendant significant height of the transmitted wave using Eq. (5–38). The transmitted wave spectrum is shown in Figure 5–15 and the computed value of the transmitted significant wave height is 0.53 m.

Step 4: Compute the significant wave transmission coefficient using Eq. (5–39). The computed value is 0.58.

The reader will note that in this example the spectrally based transmission coefficient is 0.58 which is somewhat lower than the 0.67 value computed using a wave period equal to the peak spectral period. Furthermore, the significant height of the transmitted wave is lower (0.53 m) than that computed in Example 1 (0.61 m). Hence, for the case examined a lower transmitted wave height is obtained using a spectral approach than would be obtained by using an incident regular wave with a height equal to the significant height having a period equal to the peak spectral period.

5.7 PREDICTION OF MOORING FORCES

Accurate determination of mooring forces is an important aspect of floating breakwater design. The fact that a floating breakwater is a highly efficient wave attenuator is no consolation if the mooring system fails to keep the breakwater in position during survival storm conditions. Furthermore, mooring characteristics have a direct bearing on wave transmission performance and breakwater structural design.

It was mentioned in the introduction to this chapter that floating breakwaters are often moored with either a catenary-type or guide pile-type system (Figure 5–1). Catenary mooring systems have significant cost advantages over guide pile systems in deepwater (i.e., greater than 10 m). Furthermore, catenary mooring systems are softer than guide pile systems which generally result in lower mooring forces. This can be seen in Figures 5–16 and 5–17, taken from Davidson (1971), which present mooring forces obtained from hydraulic model tests for a guide pile and a catenary mooring system, respectively. A comparison of Figures 5–16 and 5–17 indicates that the mooring forces for the guide pile system are about 10 times greater than those for the catenary mooring system. Clearly, the differences in mooring forces have a profound effect on the mooring system design as well as the internal stresses experienced by the breakwater.

5.7.1 Hydraulic Model Tests

As with wave transmission, hydraulic model tests serve as the most reliable means for predicting mooring forces. Early hydraulic model studies of floating breakwater mooring forces were performed in regular waves. Example results of such studies have already been presented in Figures 5–16 and 5–17. Advances in the offshore industry have indicated that regular wave tests seldom provide a conservative estimate of wave forces for catenary-type mooring systems. This stems from the fact that regular wave tests do not simulate second-order wave drift forces. This matter has been discussed previously and will be discussed in more detail in the following section. It suffices to say here, however, that model tests for wave mooring forces should be conducted using irregular waves. As previously stated in the context of wave transmission, model tests are expensive and should be conducted for final design. Accordingly, simple methods for estimating mooring forces are needed to prepare preliminary designs prior to model testing.

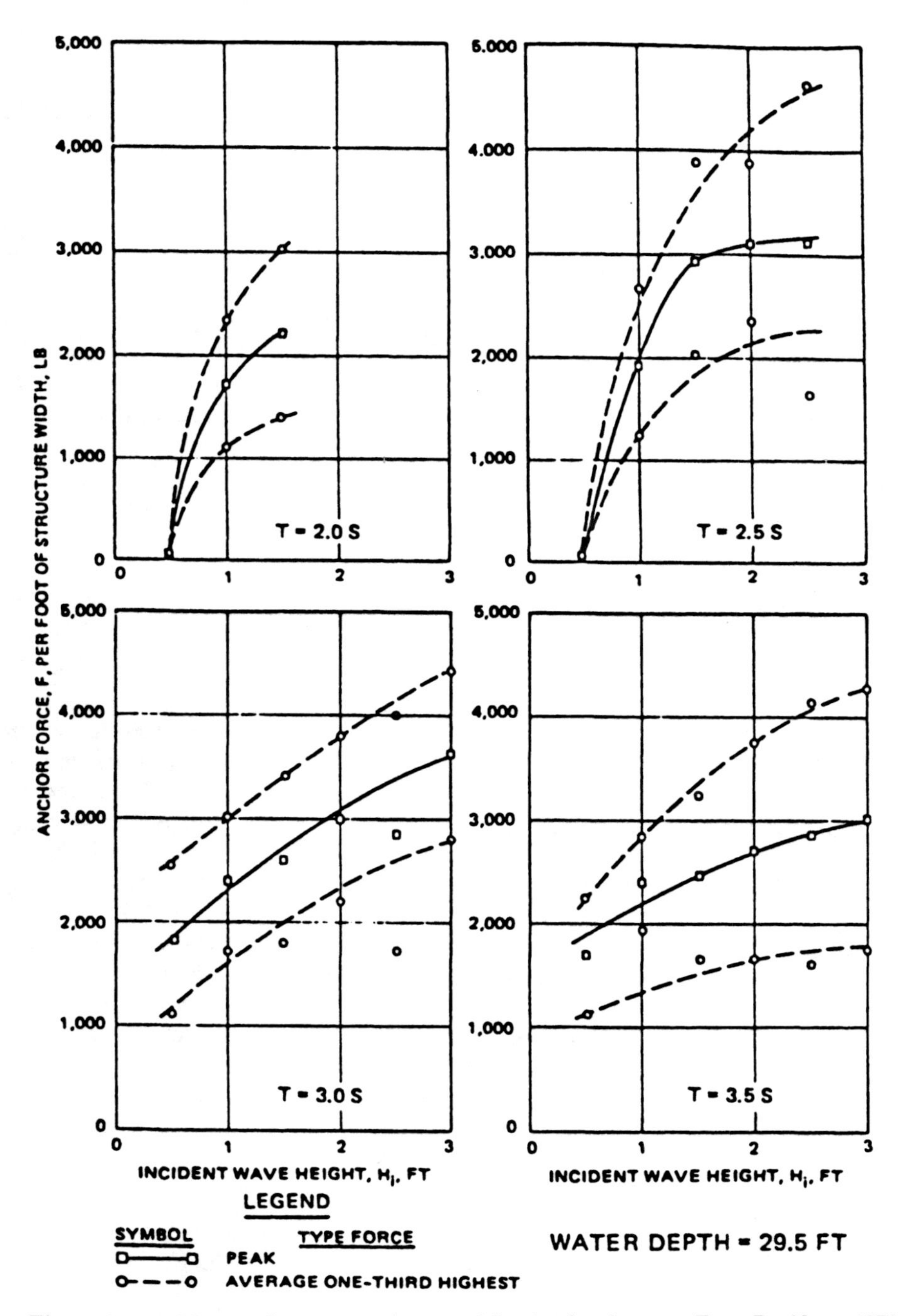

Figure 5–16. Mooring forces on a pile moored floating breakwater. (From Davidson, 1971.)

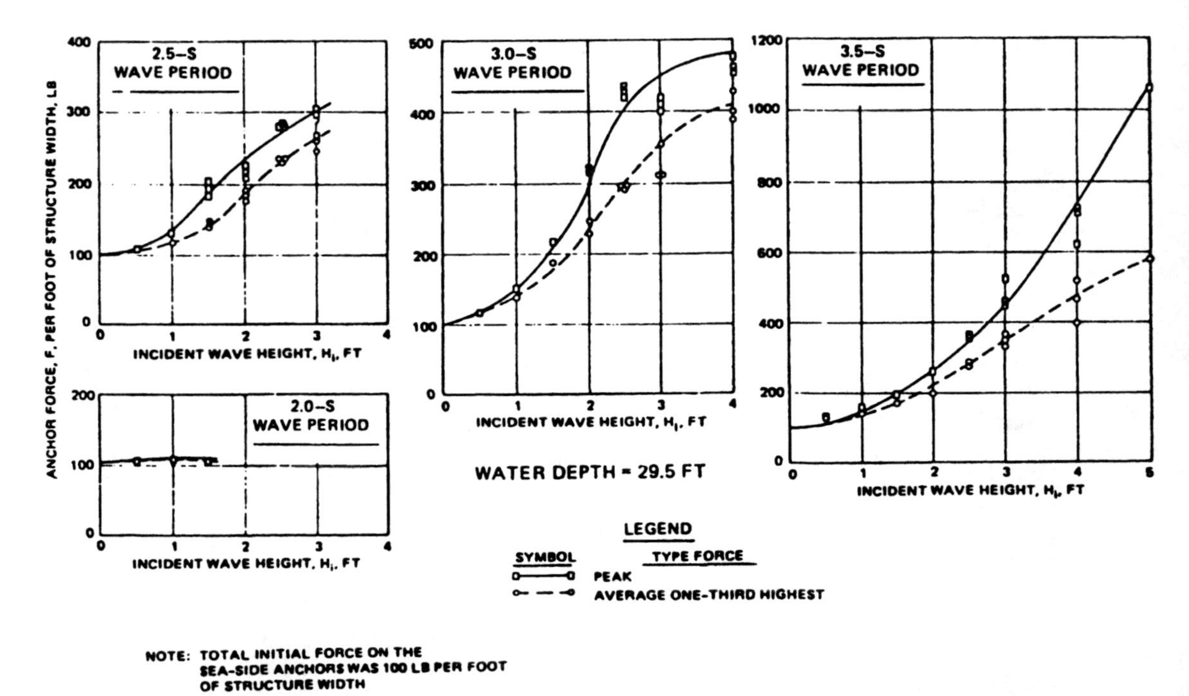

Figure 5–17. Mooring forces on a catenary moored floating breakwater. (From Davidson, 1971.)

5.7.2 Simple Analytical Methods

Simple analytical methods for computing first-order wave forces on a moored floating breakwater can be divided into those that account for breakwater motion and those that do not. Methods that assume the breakwater to be rigidly moored ignore breakwater motions, are relatively easy to apply, and assume that the entire wave-induced load is seen by the moored breakwater system. Referring to the lefthand limit of Figure 5–3, these methods assume a magnification factor of 1 and are most applicable to stiff guide pile-type mooring systems where the natural period is very short compared to the period of the incident waves. This rigid mooring analyses provides extremely conservative estimates of mooring loads when the actual natural period of the moored breakwater system is long compared to the period of incident waves. This is often the case for catenary-type mooring systems where the natural period on sway is much longer than the incident wave period. Under these conditions the magnification factor is much less than 1; see righthand limit of Figure 5–3. Accordingly, the rigid mooring system assumption is seldom used for the design of catenary-type mooring systems. Care must always be taken to ensure that the period of incident waves does not coincide with the natural period of the moored breakwater system. Under such conditions, the rigidly moored assumption cannot be made as the breakwater will experience large resonant motions and attendant mooring forces.

5.7.2.1 Wave Forces on Rigidly Moored Breakwater

Hooft (1982) presents a relatively simple method for computing wave forces on a rigidly held floating box. The maximum lateral (i.e., sway) force on a rigid box of length,

L_s; beam, B; and draft, D, resulting from a wave of amplitude, a, acting at an angle μ, is computed as follows:

$$Y_a = Y_{1a} + Y_{21a} \quad (5\text{–}40)$$

$$Y_{1a} = \rho g a \left[\frac{4\sin(\frac{1}{2}Bk\sin\mu)\sin(\frac{1}{2}L_s k\cos\mu)}{k^2\cos\mu}\right] \quad (5\text{–}41)$$

$$x\left[\tanh(kh) - \frac{\sinh k(h-D)}{\cosh(kh)}\right]$$

$$Y_{21a} = a a_{22}\left[\frac{4\cos(\frac{1}{2}Bk\sin\mu)\sin(\frac{1}{2}L_s k\cos\mu)}{k\cos\mu}\right]$$

$$\times\left[\frac{\sigma^2\cosh k\left(h-\frac{D}{2}\right)}{\sinh(kh)}\right]\sin\mu. \quad (5\text{–}42)$$

The term a_{22} is an added mass term equal to:

$$a_{22} = \rho\frac{\pi}{4}D^2. \quad (5\text{–}43)$$

The computed wave force actually varies sinusoidally with a period equal to the incident wave. The above formulas give the maximum value or amplitude of the sinusoidally varying wave force. The above formulas are singular for beam-on (90°) and bow-on (0°) wave attack; hence, it is recommended that an incident wave angle of 89.9° be used in lieu of 90° for beam-on waves and an incident wave angle of 0.1° be used in lieu of 0 degrees for bow-on waves.

Example 3

Compute the wave force on the floating breakwater presented in Example 1 subject to beam-on waves. Assume the breakwater to be rigidly held and use breakwater dimensions and incident wave characteristics given in Example 1. Use a breakwater length $L_s = 22.9$ m and a water density of 1025 kg/m^3.

From Example 1: $H = 1.1$ m, $T = 3.5$ s, and $L = 18.8$ m.

Step 1: Compute added mass term from Eq. (5–43):

$$a_{22} = \rho\frac{\pi}{4}D^2 = 1025\frac{\pi}{4}(1.1)^2 = 974.1\ \text{kg/m}.$$

Step 2: Compute Y_1 from Eq. (5–40):

$$k = 2\frac{\pi}{L} = 2\frac{\pi}{18.8} = 0.3342\ \text{1/m}$$

$$\sigma = 2\frac{\pi}{3.5} = 1.795\ \text{rad/s}$$

$$Y_{1a} = 1025(9.8)\frac{1.1}{2}$$

$$\left[\frac{4\sin\left(\frac{4.9}{2}0.3342\sin 89.9\right)\sin\left(\frac{22.9}{2}0.3342\cos 89.9\right)}{0.3342^2\cos(89.9)}\right]$$

$$\times\left[\tanh[(7.6)0.3342] - \frac{\sinh[0.3342(7.6-1.1)]}{\cosh[0.3342(7.6)]}\right]$$

$$= 170683\ \text{N} \simeq 170.7\ \text{kN}$$

$$Y_{21a} = \frac{1.1}{2}974.1$$

$$\left[\frac{4\cos\left(\frac{4.9}{2}0.3342\sin 89.9\right)\sin\left(\frac{22.9}{2}0.3342\cos 89.9\right)}{0.3342\cos(89.9)}\right]$$

$$\times\left[1.795^2\frac{\cosh\left[0.3342\left(7.6-\frac{1.1}{2}\right)\right]}{\sinh[0.3342(7.6)]}\right]$$

$$= 45634\ \text{N} \simeq 75.6\ \text{kN}$$

$$Y_a = Y_{1a} + Y_{21a} = 170.7 + 45.6 = 216.3\ \text{kN}$$

$$\frac{Y_a}{L_s} = 9.4\ \text{kN per meter of breakwater}$$

5.7.2.2 Wave Forces on Nonrigidly Moored Floating Breakwater (One Degree of Freedom Frequency Domain Model)

Simple analytical methods for computing forces on a floating breakwater moored by means of guide-pile or catenary-type moorings are highly approximate. This is especially true of catenary-type moorings where loads are affected by second-order wave drift forces and the nonlinear behavior of mooring lines. Accordingly, these simple methods should not be used for design of catenary-type moorings. Moreover, the methods should be applied with care to semirigid guide pile systems, especially when those systems are forced near resonance. Despite some shortcomings, the approximate analyses are useful in providing insight into the behavior of floating breakwater moorings. These analysis methods are based on mechanics of vibrations theory and will be illustrated for regular waves by example.

Example 4

Assume the floating breakwater analyzed in the previous example is moored by means of several catenary-type mooring lines that provide a total spring constant, k, in the sway direction of 14.6 kN/m. Using the regular wave conditions presented in Example 1 (i.e., $H = 0.91$ m and $T = 3.5$ s) compute the total wave-induced force transferred to mooring system. Ignore the effects of damping.

Step 1: Compute the natural frequency (and period) of the moored breakwater in sway using Eq. (5–13):

$$\text{Assume } a_{22} = 10\% \text{ of mass } (m)$$

$$m + a_{22} = \rho L_s BD + a_{22}$$

$$= 1025(22.9)4.9(1.1(1 + 0.1) = 139\,168\,\text{kg}$$

$$\sigma_n = \sqrt{\frac{c}{(m+a)22}} = \sqrt{\frac{14.6(1000)}{139\,168}} = 0.324 \text{ rad/s}$$

$$T_n = \frac{2\pi}{\sigma_n} = 19.4\,\text{s}$$

Step 2: Compute the magnification factor using Eq. (5–15) with critical damping coefficient, ζ, equal to zero:

$$\sigma = \frac{2\pi}{T} = \frac{2\pi}{3.5} = 1.795 \text{ rad/s}$$

$$\Lambda = \frac{1}{\sqrt{\left(1 - \left(\frac{\sigma}{\sigma_n}\right)^2\right)^2 + \left(2\zeta\left(\frac{\sigma}{\sigma_n}\right)^2\right)^2}}$$

$$= \frac{1}{\sqrt{\left(1 - \left(\frac{1.795}{0.324}\right)^2\right)^2 + 0}} = 0.034.$$

Step 3: Compute the maximum total wave-induced force on the mooring system. According to Example 3, the maximum lateral wave force on the rigidly held breakwater, $Y_a = 174.5$ kN. The wave load transferred to the mooring system is:

$$F_m = \Lambda Y_a = 0.034(216.3) = 7.3 \text{ kN}$$

Because the period of the incident wave is much lower than the natural period of the moored breakwater system, only a portion of the rigidly moored wave force is transferred to the mooring system.

Example 5

Repeat Example 4 for a guide-pile mooring system with a total spring constant, k, in the sway direction of 573 kN/m. Again, ignore the effects of damping.

Step 1: Compute the natural frequency (and period) of the moored breakwater in sway using Eq. (5–13):

$$\text{Assume } a_{22} = 10\% \text{ of mass } (m)$$

$$m + a_{22} = \rho L_s BD + a_{22}$$

$$= 1025(22.9)4.9(1.1)(1 + 0.1) = 139\,168\,\text{kg}$$

$$\sigma_n = \sqrt{\frac{c}{m + a_{22}}} = \sqrt{\frac{1616.6(1000)}{139\,168}} = 3.41$$

$$T_n = 2\frac{\pi}{\sigma_n} = 1.84\,\text{s}.$$

Step 2: Compute the magnification factor using Eq. (5–15) with critical damping coefficient, ζ, equal to zero:

$$\sigma = \frac{2\pi}{T} = \frac{2\pi}{3.5} = 1.795 \text{ rad/s}$$

$$\Lambda = \frac{1}{\sqrt{\left(1 - \left(\frac{\sigma}{\sigma_n}\right)^2\right)^2 + \left(2\zeta\left(\frac{\sigma}{\sigma_n}\right)^2\right)^2}}$$

$$= \frac{1}{\sqrt{\left(1 - \left(\frac{1.795}{3.4}\right)^2\right)^2 + 0}} = 1.384$$

Step 3: Compute the maximum total wave-induced force on the mooring system. According to Example 3, the maximum lateral wave force on the rigidly held breakwater, $Y_a = 174.8\,\text{kN}$. The wave load transferred to the mooring system is:

$$F_m = \Lambda Y_a = 1.384(216.3) = 299.4 \text{ kN}$$

The period of the incident wave is close to the natural period of the moored breakwater system; accordingly, the wave forces transferred to the mooring system exceed the rigidly moored wave force.

5.7.2.3 Extensions of One Degree of Freedom Model to a Wave Spectrum

The above examples illustrate simple (and highly approximate) methods for estimating wave forces on a floating breakwater. In each of these examples, wave forces are computed for a single, regular wave. This one-dimensional methodology can, however, be extended to account for a spectrum of waves. This extension consists of computing: (1) wave forces, (2) added mass and damping coefficients, and (3) magnification factors for a range of wave periods (i.e., frequences) and a unit wave amplitude of 1 m. A plot of the mooring force per unit wave amplitude, F_m, as a function of wave frequency is known as a Response Amplitude Operator (RAO). The RAO can be combined with the incident wave spectrum to provide a mooring force spectrum:

$$S_f(f) = (F_m(f))^2 S(f). \qquad (5\text{–}44)$$

This expression is similar to that given in Eq. (5–37) for the transmitted wave height spectrum. Various force statistics can be computed from the force spectrum as follows:

$$F_s = 2\sqrt{\int_0^\infty S_f(f)\,df} \qquad (5\text{–}45)$$

$$F_{\max} = 0.707 F_s \sqrt{\ln(N)} \qquad (5\text{–}46)$$

where F_s is the significant mooring force and $F_{\max}$ is the maximum mooring force, N is the number of waves in the spectrum.

The reader should note that it is particularly important to include a damping term when the mooring system is forced by waves having a period near the natural period of that system. Unfortunately, there is no really easy method for objectively estimating the damping coefficient. Normally damping terms are computed using numerical theories [e.g., Frank (1967) or Van Oortmerssen (1976)]. However, the following equation (Hooft, 1982) can be used to estimate damping from the applied wave forces [i.e., Eq. (5–40)]:

$$b_{yy}(\omega) = \frac{k^2}{8\pi\rho g C_g} \int_0^{2\pi} \left[\frac{Y_a(\mu,\omega)}{\zeta a}\right]^2 d\mu. \qquad (5\text{–}47)$$

The term C_g is the wave group velocity:

$$C_g = \frac{c}{2}\left(1 + \frac{2kh}{\sinh(2kh)}\right) \qquad (5\text{–}48)$$

where c is the wave speed as defined in Eq. (5–3).

Equation (5–47) for damping can be solved by computing wave forces from Eq. (5–40) for a range of wave periods and directions. Such computations are rather laborious but can be performed easily on a microcomputer. The percentage of critical damping at a given frequency can be computed from Eq. (5–14).

For the purposes of illustration, the above methodology is applied to a small floating breakwater moored by means of guide piles in a water depth, h, of 4.6 m. The dimensions of the breakwater are: $B = 3.7$ m, $L_s = 30.5$ m, $D = 0.61$ m. The mooring system consists of five steel pipe piles: 40.6 cm outer diameter and 1.91 cm wall thickness, steel yield strength $F_y = 24\,821$ N/cm^2, driven in medium sand with a length to point of fixity of $4.6 + 3.3 = 7.9$ m and a maximum load capacity per pile of 67.4 kN. The spring constant per pile is 531.5 kN/m which gives a total spring constant in sway, $k = 5 \times 531.5 = 2657.5$ kN/m. The floating structure is exposed to beam-on wave conditions with a significant wave height of 0.46 m and a peak spectral period, T_s, of 1.5 s (spectrum given in Figure 5–18).

Figure 5–19 presents a plot of the applied (or rigidly moored) wave force and the damping coefficient in sway for a unit amplitude wave and various wave frequencies. The magnification factor and the percent of critical damping are presented in Figure 5–20. The mooring force RAO for a single pile (vis-a-vis that for the entire mooring system) is presented in Figure 5–21. The natural period of the moored breakwater system in sway is 1.1 s which is within the range of incident wave periods. The mooring force spectrum for a single pile is given in Figure 5–22 which also presents the significant and maximum mooring forces. The maximum wave force per pile is 76 kN which exceeds the maximum allowable load per pile.

5.7.3 Numerical Models

The above analytical models neglect a number of important factors that can be very important in practice. First, the simplified methods are based on a one-dimensional analysis of sway motion and neglect the

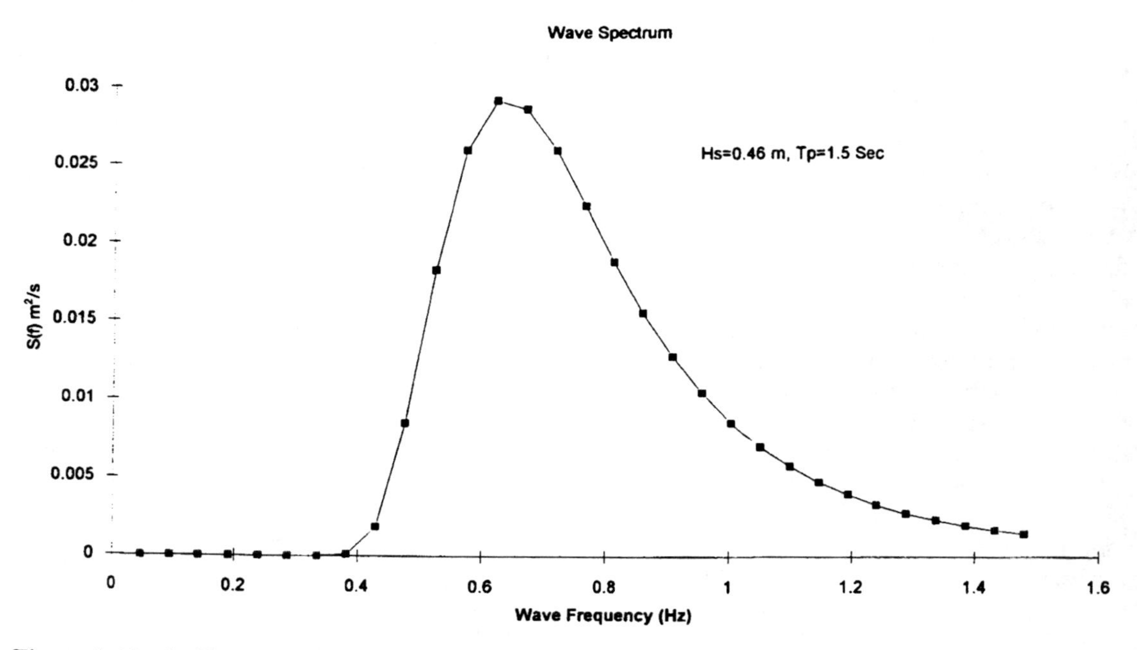

Figure 5–18. Incident wave spectrum.

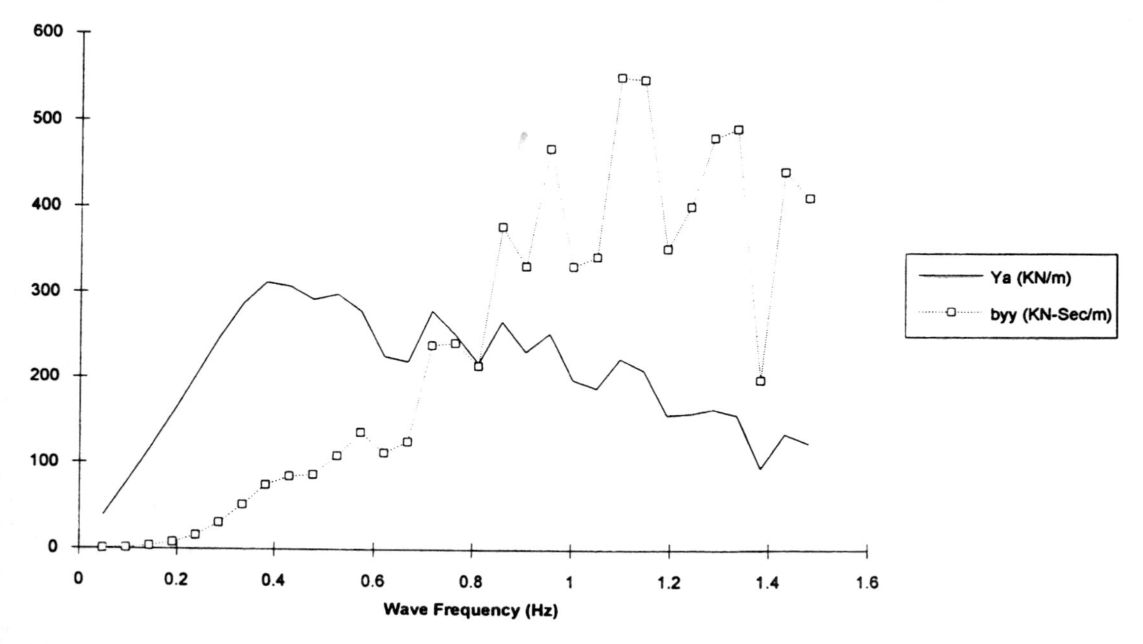

Figure 5–19. Applied wave force and damping force.

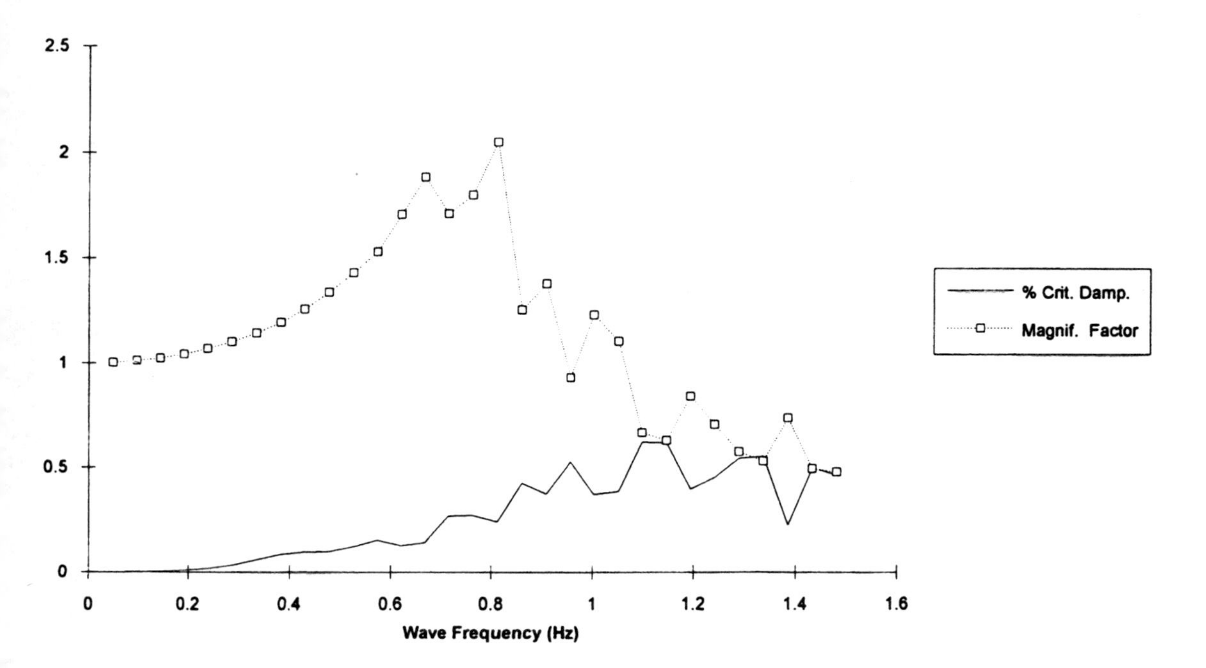

Figure 5–20. Magnification factor and percent critical damping.

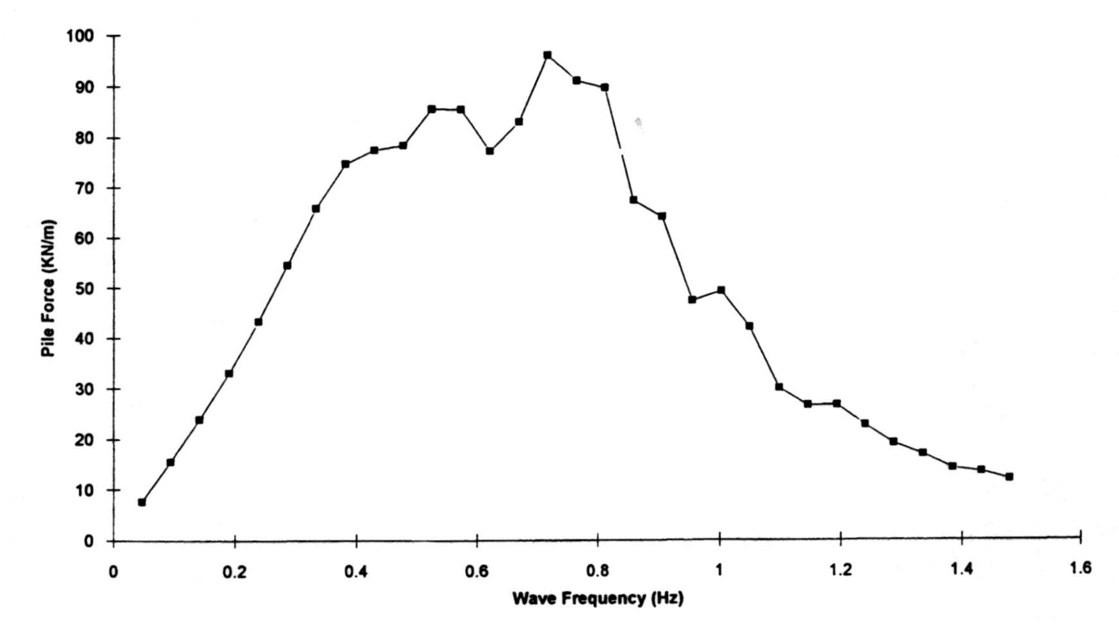

Figure 5–21. Pile force per unit wave height.

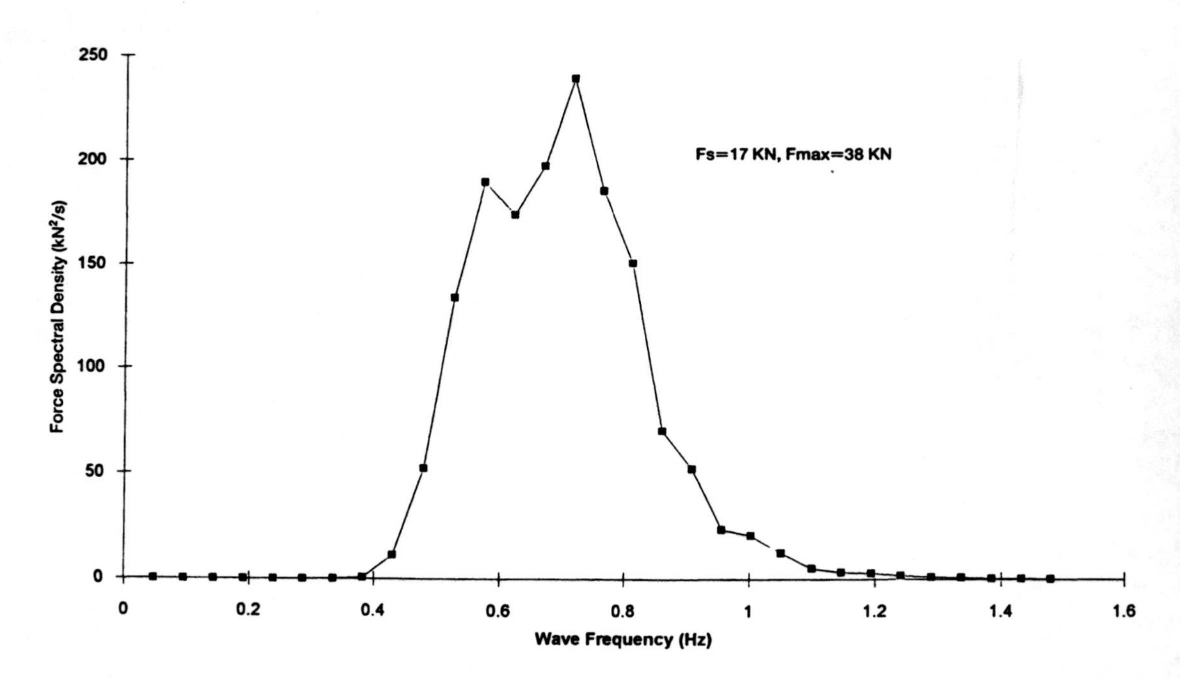

Figure 5–22. Mooring force spectrum.

other five modes of motion, namely: surge, heave, roll, pitch, and yaw (see Figure 5–4). Second, the simplified analyses neglect the nonlinear behavior of catenary mooring systems. Finally, the simplified analyses neglect second-order wave drift forces that can dominate mooring loads for catenary-type mooring loads. Numerical models, which account for some of the factors neglected in the simplified analyses, can be used to provide a better estimate of mooring forces for design purposes.

For the purposes of illustration, floating breakwater mooring loads measured in field and laboratory experiments will be compared to results obtained from three numerical models; namely: (1) the one degree of freedom (1DOF) model described above, (2) a spectral implementation of the Ito and Chiba ((1972) three degrees of freedom (3DOF) model, and (3) an impulse–response function (IRF) time domain model.

The U.S. Army Corps of Engineers initiated the Floating Breakwater Prototype Test Program in 1981. A concrete caisson-type floating breakwater was constructed, installed, and monitored on an exposed site in the Puget Sound as part of this program. Results of this study are presented in Nelson and Broderick (1986). The general arrangement of the floating breakwater installation is shown in Figure 5–23. The floating breakwater was moored in an average water depth of about 13.7 m and was restrained by a mooring system comprised of five mooring lines that provided a (measured) linearized spring constant of about 4.4 kN/m. The breakwater has a beam of 4.9 m, a draft of 1.1 m, and a length of 45.7 m. The breakwater is essentially the same structure tested by Carver (1979) and examined in Section 5.6.3.

In an effort to corroborate the findings of the field experiment, the U.S. Army Corps of Engineers, Coastal Engineering Research Center sponsored a series of 1 : 10 scale physical model tests. These tests are summarized by Torum et al. (1989).

The Ito and Chiba (1972) numerical model (i.e., 3DOF model) has already been described in the context of wave transmission in Section 5.6.3. This model is based on three degrees of freedom linearized frequency domain analyses, and can be used to estimate mooring line forces. Results from the 3DOF model are compared to 1DOF model results in Figure 5–24 for the Corps of Engineers field/model experiments in the form of a mooring line RAO. It can be seen from Figure 5–24 that the 3DOF model gives a mooring line force per unit wave height which is consistently lower than the 1DOF results for wave frequencies of less than 0.08 Hz and at a wave frequency of 0.2 Hz. The 0.2 Hz wave frequency approximates the roll natural frequency of a moored breakwater system.

5.7.3.1 Comparison of 1DOF and 3DOF Models to a Physical Model

Response amplitude operators obtained from the Ito and Chiba (1972) and 1 DOF method were used in combination with a wave spectrum characterized by a significant height of 1 m and a peak spectral period of 4.3 s which corresponds to one of the conditions tested in the physical model. The wave spectrum corresponding to this wave condition was computed from Eq. (5–9) and is presented in Figure 5–25. The mooring line force spectra obtained from both the 1DOF and 3DOF models are shown in Figure 5–26. The computed values of the significant and maximum mooring line force statistics are provided in Figure 5–26. It is remarked that the significant and maximum values correspond to mooring line forces above the 18 kN load to which the mooring system was pretensioned. The maximum mooring line force measured in the laboratory was 15 kN (above the initial pretension of 18 kN). Figure 5–26 indicates that the 3DOF model gives significant and maximum wave force values that are higher than

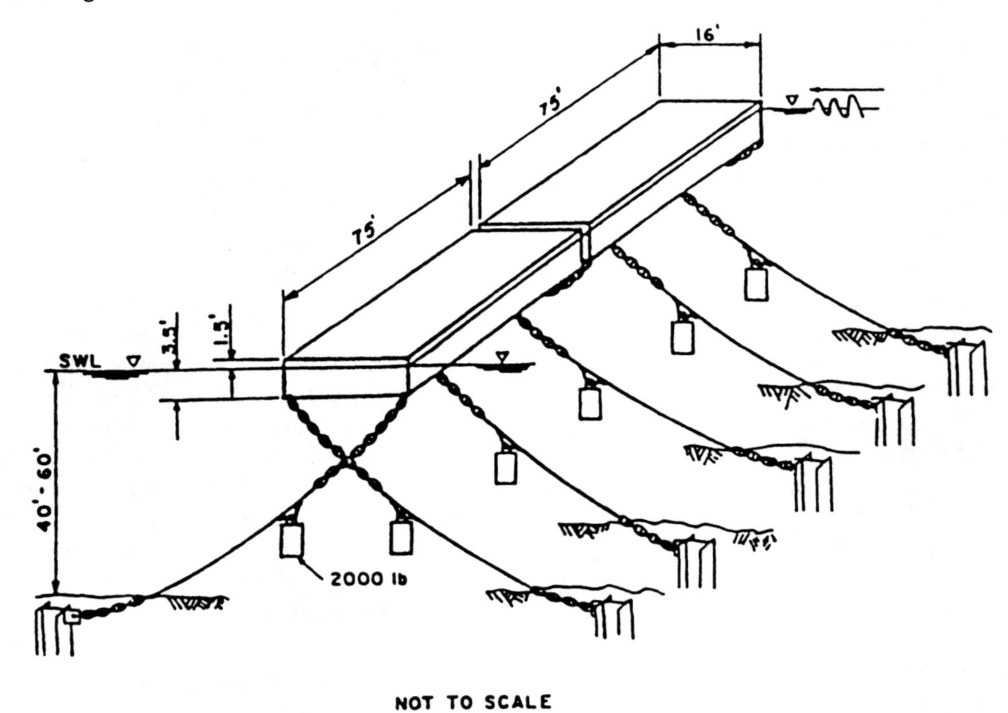

Figure 5–23. Puget Sound floating breakwater.

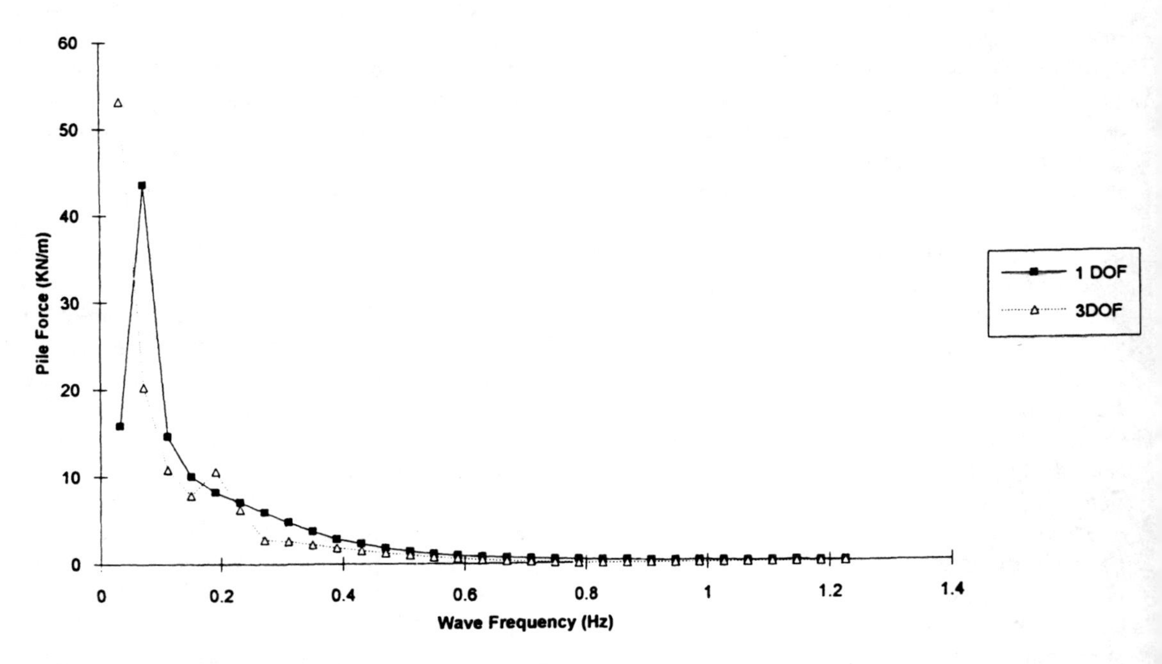

Figure 5–24. Mooring force RAO (IDOF and 3DOF).

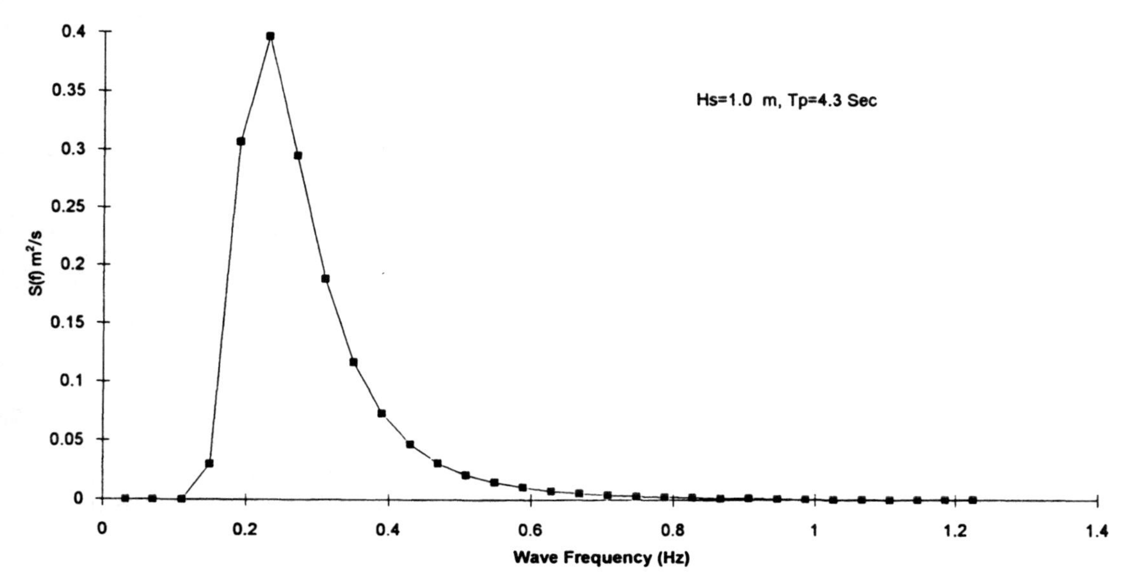

Figure 5-25. Incident wave spectrum.

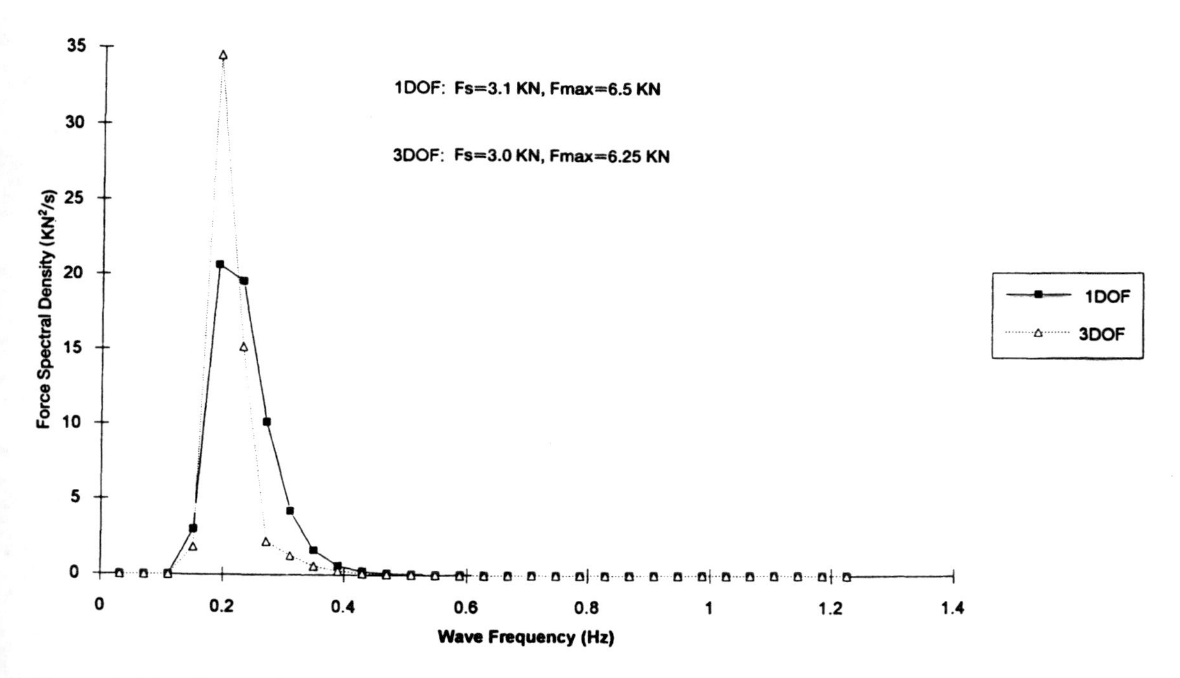

Figure 5-26. Mooring force spectrum.

the 1DOF model. Both the 3DOF and 1DOF models give maximum mooring line loads less than that recorded in the laboratory.

5.7.3.2 Comparison of 1DOF and 3DOF Models to Field Measurements

Analyses were prepared for a significant wave height of 0.39 m (1.29 ft) and a peak spectral wave period of 2.75 s which corresponded to one of the conditions measured in the full-scale COE field tests. The computed wave spectrum for this wave condition is presented in Figure 5–27. The 1DOF, 3DOF, and field measurement results are presented in the form of a mooring line force spectrum in Figure 5–28. Figure 5–28 also presents computed values of the significant and maximum wave forces and the maximum measured wave force. In this case, the 1DOF model gives slightly higher forces than the 3DOF model. The maximum mooring line load recorded in the field was 6 kN (subtracting out the mooring line pretension). Peak mooring line forces from the 1DOF and 3DOF models are 23 and 14% of the measured maximum value, respectively.

5.7.3.3 Time Domain Numerical Model

The shortcomings of linearized frequency domain analyses are evident in Figures 5–26 and 5–28 which show that such methods generally can underpredict floating breakwater mooring line forces. This is consistent with the findings of previous investigators. Miller et al. (1984), for example, found that frequency domain approach adequately predicted floating breakwater motions in heave and roll, but could not predict the low-frequency sway motions which generally dominated mooring forces. Similarly, Adee (1974) concluded that mooring forces developed from frequency domain analysis must be increased substantially in order to provide an estimate of actual mooring forces. Figure 5–29 presents an example of force–

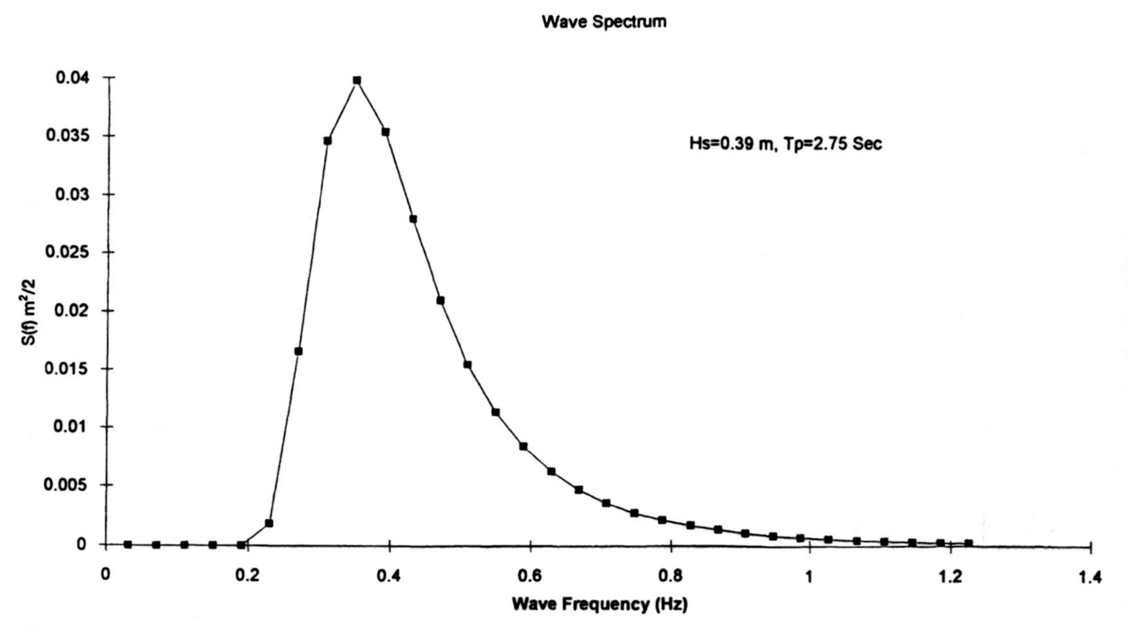

Figure 5–27. Incident wave spectrum.

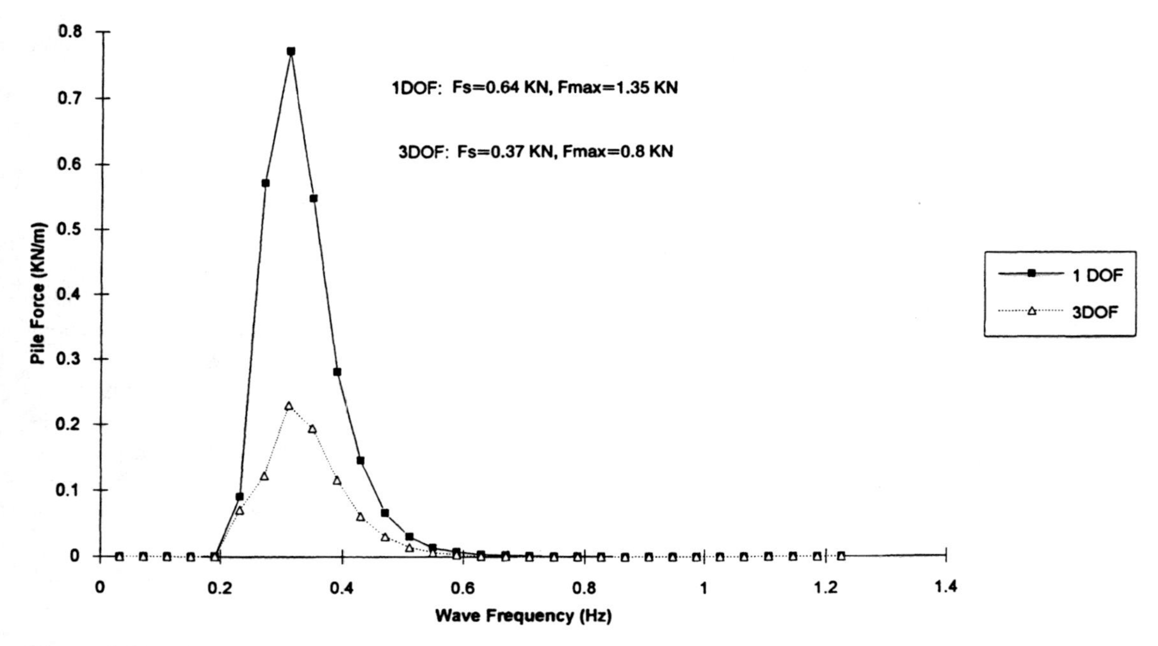

Figure 5–28. Mooring force spectrum.

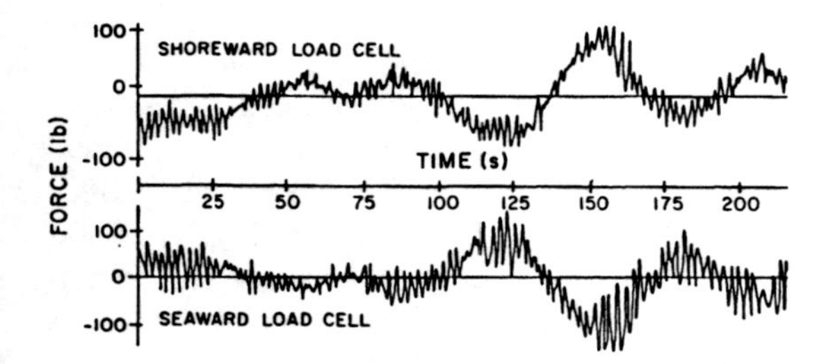

Figure 5–29. Example force record.

time history for a floating breakwater installation at Tenakee Alaska. Incident waves had periods ranging from 1 to 2 s, however, mooring line forces were dominated by long period oscillations with periods of approximately 55 to 60 s. A review of the dynamic characteristics of the floating breakwater mooring system indicates that the recorded sway motion was close to the sway natural period. Similar prototype measurements were reported near Kumamoto, Japan [see Tekmarine (1986)]. Clearly, any numerical analysis of floating breakwater mooring line forces must account for both high- and low-frequency motions. The following paragraphs outline a numerical model (Headland and Vallianos, 1990) that accounts for first- and second-order floating breakwater motions and attendant mooring line forces. Following a description of this numerical model, the model will be compared with the model and field measurements described above.

The governing equations of motion, which account for motion in six degrees of freedom, are as follows:

$$\sum_{j=1}^{6} [(M_{kj} + m_{kj})\ddot{x}_j] + \int_{-\infty}^{t} [K_{kj}(t-\tau)\dot{x}_j(\tau)\,d\tau + C_{kj}x_j] = F_k(t) \qquad (5\text{–}49)$$

where

x_j = motion in the j-th mode
M_{kj} = inertia matrix
C_{kj} = hydrostatic restoring force matrix
K_{kj} = impulse response function matrix

m_{kj} = constant added mass matrix
$F_k(t)$ = arbitrary external force in k-th mode due to waves, mooring line reactions, viscous damping, winds, currents, etc.
k = denotes mode of motion, $k = 1, 2$, ... 6 [i.e., 1 (surge), 2 (sway), 3 (heave), 4 (roll), 5 (pitch), 6 (yaw)].

The inertia matrix, M_{kj}, and hydrostatic restoring force matrix, C_{kj}, are computed using standard methods of naval architecture. The impulse–response function matrix, K_{kj}, and the constant added mass coefficient, m_{kj}, are computed as follows:

$$K_{kj} = \frac{2}{\pi} \int_0^\infty b_{kj}(\omega) \cos \omega t \, d\omega \qquad (5\text{–}50)$$

$$m_{kj} = a_{kj}(\omega^*) + \frac{1}{\omega^*} \int_0^\infty K_{kj}(t) \sin \omega^* \tau \, d\tau \qquad (5\text{–}51)$$

where
a_{kj} = frequency-dependent added mass
b_{kj} = frequency-dependent damping coefficient.

The above frequency-dependent hydrodynamic coefficients and wave transfer functions may be computed using either strip-theory or diffraction theory hydrodynamic analysis. The applied force, $F_{kw}(t)$, resulting from waves can be written as:

$$F_{kw}(t) = \sum_0^n f_k^{(1)}(\omega_n) a_n \cos(\omega_n t + \epsilon_n + \epsilon_k)$$
$$+ \sum_0^n \sum_0^m f_k^{(2)}(\omega_n) a_n a_m \cos((\omega_n t + \epsilon_n) - (\omega_m t + \epsilon_m)) \qquad (5\text{–}52)$$

where
$f_k^{(1)}(\omega_n)$ = first-order wave transfer function in the k-th mode for n-th wave component
$f_k^{(2)}(\omega_n)$ = second-order wave transfer function in the k-th mode for n-th wave component
ϵ_k = phase of first-order wave transfer function in the k-th mode for n-th wave component
a_n = wave amplitude for n-th wave component
ϵ_n = phase of n-th wave component
ω_n = frequency of n-th wave component.

The above formula can be used to simulate first- and second-order wave force time histories resulting from an incident wave spectrum composed of n-waves having amplitudes a_n and frequencies, ω_n. The incident wave spectrum is represented in the above formulation by the wave amplitudes and phases (i.e., a_n and ϵ_n) which correspond to wave frequencies, ω_n. Wave amplitudes and frequencies are determined from an incident wave spectrum using the "equal area" method described by Borgman (1969).

It should be noted that the wave force formulation presented in Eq. (5–22) is valid only for long-crested, unidirectional wave conditions. The present version of the model cannot be used to assess mooring loads from directional spectra. The physical model test results of Torum et al. (1989) indicate that directional spreading has little effect on mooring line forces when the breakwater is exposed to beam seas. Beam seas normally control mooring design.

The first- and second-order wave transfer coefficients were computed using the Frank-Close Fit method. Second-order wave transfer functions were computed using methods based on the work of Gerritsma and Beukelman (1972) and Newman (1967). As will be discussed later, the drift forces computed using this method were found to overestimate the sway motions and attendant mooring line forces of physical model and prototype floating breakwaters. This overprediction was believed to be due to the fact that in both the physical model and in the field, breakwater units were heavily overtopped by waves. In an effort to account for overtopping, wave drift forces were computed using the following expression derived by Longuet-Higgins (1977):

$$f_2^{(2)} = \frac{1}{4}\rho g(a^2 + a_R^2 - a_T^2)\left(1 + \frac{2kh}{\sinh(2kh)}\right) \quad (5\text{–}53)$$

where
a = incident wave amplitude
a_R = reflected wave amplitude
a_T = transmitted wave amplitude
k = wave number
h = water depth.

The reflected wave amplitude can be estimated from the incident and transmitted wave amplitudes as follows:

$$a_i^2 \approx a_R^2 + a_T^2. \quad (5\text{–}54)$$

When measured or computed values of the incident, reflected, and transmitted wave amplitudes are available, Eq. (5–53) can be used to estimate the second-order wave drift forces. Wave overtopping conditions are implicitly accounted for by virtue of the measured reflected and transmitted wave amplitude.

Numerous studies have shown that potential theory damping terms for horizontal motions (i.e., surge, sway, and yaw) are negligible in comparison to viscous damping terms at low frequency. Accordingly, an additional term is added to the time-varying applied force, $F_k(t)$, to account for viscous damping effects. Both nonlinear and linear viscous damping formulations were investigated for sway motions. In accordance with the findings of Wichers (1988), initial efforts were directed toward evaluation of a quadratic damping force formulated as follows:

$$F_{2D}(t) = -\tfrac{1}{2}\rho C_{2D}|\dot{x}_2|\dot{x}_2 \quad (5\text{–}55)$$

where
$F_{2D}(t)$ = nonlinear damping force in sway
ρ = water density
A_2 = lateral area of breakwater
C_{2D} = viscous drag coefficient in sway.

After considerable analyses, it was concluded that the above nonlinear formulation provided very little damping for typical floating breakwater and was abandoned in favor of a linear formulation as follows:

$$F_{2D}(t) = -B_{2D}\dot{x}_2 \quad (5\text{–}56)$$

where
B_{2D} = linear viscous damping coefficient.

A similar formulation has been shown to successfully predict *surge* damping characteristics of vessels secured to single-point moorings in still water (Wichers, 1988).

Mooring line loads are computed from the instantaneous position of the mooring attachment point on the breakwater and from static load deflection curves developed for each line on the basis of catenary analyses. The total mooring restoring forces and moments acting on the center of gravity of the breakwater in each mode of motion are computed by summing the force (and moment) contributions from each mooring line. These forces are then added to the $F_k(t)$ term on the righthand side of Eq. (5–49). Although Torum et al. (1989) suggest that such effects may be important at high frequency, no attempt has been made to incorporate dynamic oscillations of the mooring lines in the numerical model.

5.7.3.4 Comparison of Time Domain Model with a Physical Model

A series of numerical computations were performed in order to determine the natural periods of the moored breakwater system in each of the six modes of motion. This numerical computation is known as an extinction test and consists of displacing the center of gravity of the breakwater in some mode (e.g., sway) and releasing the structure. The breakwater oscillates in the sway mode at the sway natural period until resisting forces eventually dampen the motion out. The extinction tests revealed that the natural periods of motion in sway and yaw were 20, and 14 s, respectively. These periods are relatively long compared to the natural periods in heave, roll, and pitch which were found to be 3.2, 2.4, and 3.4 s, respectively.

The linear hydrodynamic coefficients and wave transfer functions for the caisson-type floating breakwater were computed in the frequency domain and converted to time domain functions using the techniques described above. Unfortunately, given the nature of the physical modeling program described by Torum et al. (1989), there is no way to validate these quantities directly. A systematic series of physical model tests would be required to evaluate the added mass, damping, and wave transfer functions separately. Hence, one can only compare the final output of the physical model (i.e., breakwater motions and mooring line forces under given wave conditions) to that simulated by the numerical model.

Mooring analyses were prepared for a variety of the wave conditions tested in the physical model. As previously mentioned, second-order wave transfer functions were initially computed without accounting for wave overtopping. The resulting mooring dynamic simulations significantly overpredicted breakwater motions and attendant mooring line forces. A systematic variation of the second-order wave transfer functions and viscous damping coefficients demonstrated that the floating breakwater response was relatively sensitive to the second-order wave transfer functions and relatively insensitive to the viscous damping coefficient as long as the linear damping formulation was used. On the basis of these results, the original method for computing the second-order drift forces was abandoned in favor of the method of Longuet-Higgins (1977) presented in Eq. (5–53). In order to apply Eq. (5–53) it is necessary to know the incident, reflected, and transmitted wave amplitudes for each incident wave frequency. Fortunately, breakwater transmission characteristics were presented in Torum et al. (1989). Hence, the second-order wave transfer functions could be estimated directly from Eq. (5–53) and (5–54).

Example numerical model results are presented in Figure 5–30 which presents time histories of water surface elevation, sway, heave, roll, and mooring line forces for beam-on waves with a significant wave height, H_s, of 1 m and a peak spectral wave period, T_p, of 4.3 s. Figure 5–30 is actually a portion of the predicted time history as the numerical model was run for a total time of 960 s. With the exception of roll, the numerical model provided a good prediction of breakwater motions. Specifically, the peak values of sway, heave, and roll measured in the physical model were 3 m, 0.7 m, and 10.04°, respectively. The maximum values of sway, heave, and roll predicted by the numerical model were 2.8 m, 0.64 m, and 25°, respectively. The maximum mooring line load predicted by the numerical model was 31.8 kN and was comparable to 32.9 kN measured in the physical model tests. Figure 5–30 indicates several facets of the breakwater response, namely: (1) heave and roll responses are at the incident wave frequency, (2) sway response and mooring line forces are dominated by low-frequency motion near the natural period of sway motions, and (3) roll motion is overpredicted by the numerical model. Similar results were obtained for other conditions tested in the physical model tests.

The numerical model results described above are generally consistent with the conclusions presented by Torum et al. (1989), who concluded that: (1) floating breakwater sway response was dominated by low-frequency motions and (2) mooring line forces were governed by low-frequency sway motion, wave frequency roll, and heave motions. However, spectral plots of mooring line forces presented in Torum et al. (1989) indicate peak spectral energy at near the

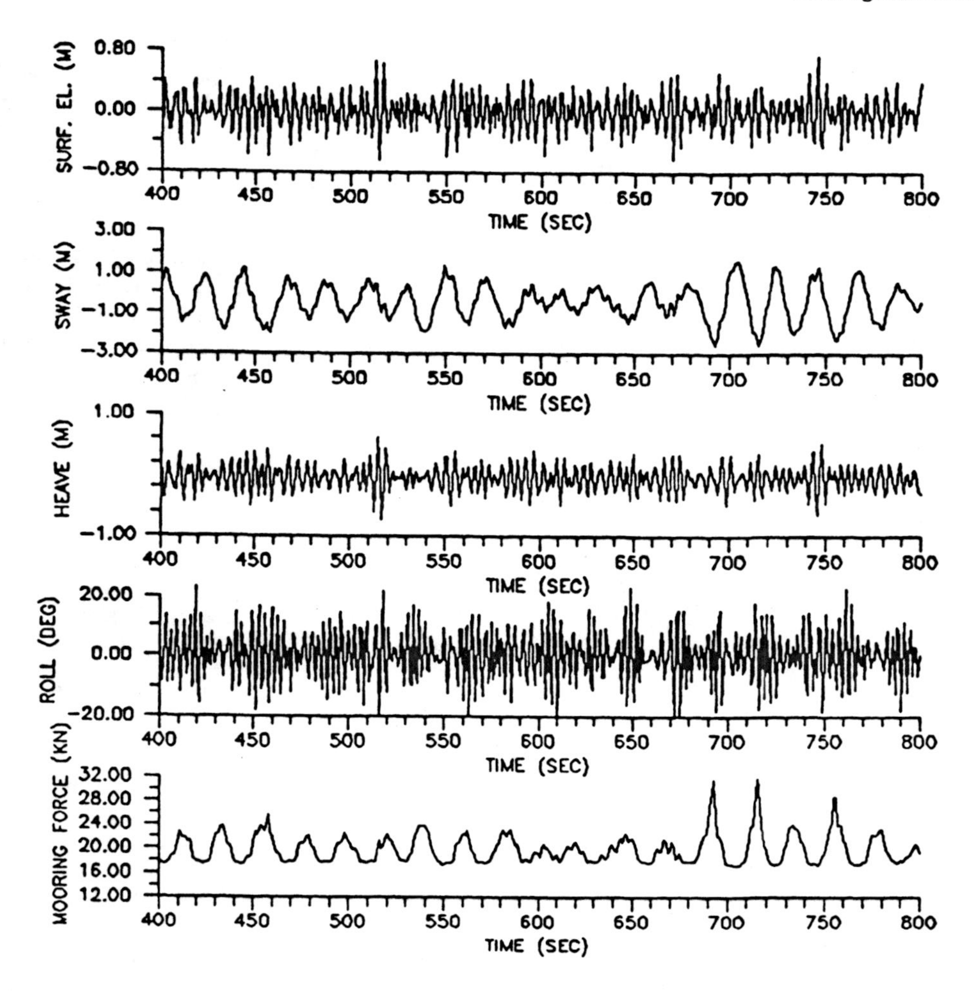

Figure 5–30. Example time domain numerical model results.

wave frequency with a secondary peak at low frequency. This led Stansberg et al. (1990), in a summary of the Torum et al. (1989) report, to conclude that mooring line forces were on average governed by heave and roll motions with extreme events strongly affected by low-frequency sway peaks. The numerical model presented herein indicates that mooring line loads are dominated by horizontal sway motions. Additional analyses will be required to evaluate the apparent discrepancies between the motion and mooring line force spectra presented in Torum et al. (1989).

5.7.3.5 Comparison of Time Domain Model with Field Measurements

Hydrodynamic coefficients and wave transfer functions were computed in the same manner as described above for the phsyical model. It should be noted that the wave transmission coefficients measured in the physical model tests were used to estimate the second-order wave transfer function for the prototype structure. Mooring line load-deflection characteristics were estimated on the basis of the field measurements described in Nelson and Broderick (1986).

Example numerical model results are presented in Figure 5–31 for an incident significant wave height, H_s, of 0.46 m (1.29 ft) and a peak spectral period of 2.75 s. The simulation results presented in Figure 5–31 for the prototype breakwater are similar to those presented in Figure 5–30 for the scale model breakwater. Specifically, breakwater sway response and attendant mooring line force are primarily at low frequency whereas the breakwater response in heave and roll is at a frequency corresponding to the incident wave. The maximum mooring line load measured for these wave conditions was 28.2 kN. The maximum mooring line load simulated by the numerical model was 27.0 kN. Similar results were found for other comparisons of the numerical model to prototype measurements.

The above time domain numerical model provides reasonable estimates of floating breakwater mooring line forces for use in planning and design. The model does not obviate the need for physical model tests but serves as a planning tool prior to physical modeling of final floating breakwater mooring configurations.

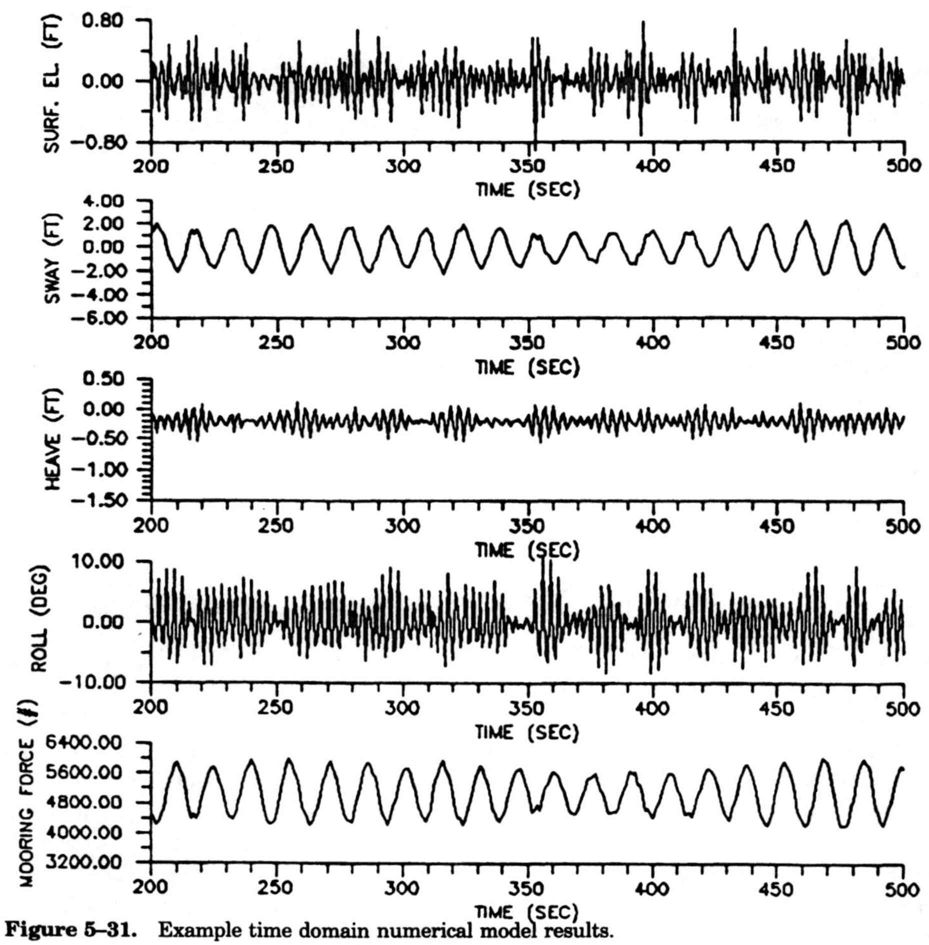

Figure 5–31. Example time domain numerical model results.

5.8 STRUCTURAL DESIGN OF FLOATING BREAKWATERS

Structural design of floating breakwater modules should be in accordance with the basic principles of naval architecture. These principles dictate that the floating breakwater be capable of sustaining the loads resulting from:

- Wave-induced shear, bending, and torsion
- Mooring line/pile loads
- Distributed and concentrated deck loads
- Thermal loads
- Loads imparted during launching and towing.

The cross-sectional geometry (i.e., section modulus) of a floating breakwater module is often governed by wave-induced bending moments along the long axis of the breakwater. Transverse bending between mooring points and torsional loads are also important in design.

The following paragraphs focus on methods for evaluating wave-induced structural loads. Details regarding other types of structural loading can be found in Tsinker (1986) or Gaythwaite (1990).

5.8.1 Hydraulic Model Tests

Hydraulic model tests are seldom used to evaluate the internal stresses of floating breakwaters. This stems from the costs of such physical model testing as well as the difficulties associated with modeling hydraulic and structural phenomena simultaneously. As a result, structural design of floating breakwater modules is generally based on an analytical assessment of wave-induced stresses.

5.8.2 Simplified Methods

Despite the complex nature of a ship hull in waves, it has been standard practice in naval architecture to reduce analysis of this dynamic phenomenon to a quasi-static problem. This practice assumes that the floating body is supported statically by a simple sine wave. The conditions that give the maximum wave-induced bending moments, and in turn generally govern the section modulus of a floating breakwater, are known as wave hogging and sagging (see Figure 5–32). Hogging refers to a condition where the crest of a wave, with a wave length equal to the longitudinal length of the structure, is located at the center of the structure. Conversely, sagging occurs when wave crests are located at each end of the structure. Computation of wave-induced shears and moments under conditions of hogging and sagging is based on a simple hydrostatic force balance. Specifically, the hydrostatic force at any point on the floating breakwater is equal to the local buoyant force per unit length, b_x, minus the total weight (dead plus live load), W_x, per unit length of breakwater. The floating breakwater is then reduced to a nonuniformly loaded beam where the shear force, Q_x, is determined from the following:

$$Q_x = \int (b_x - W_x)\, dx. \tag{5–57}$$

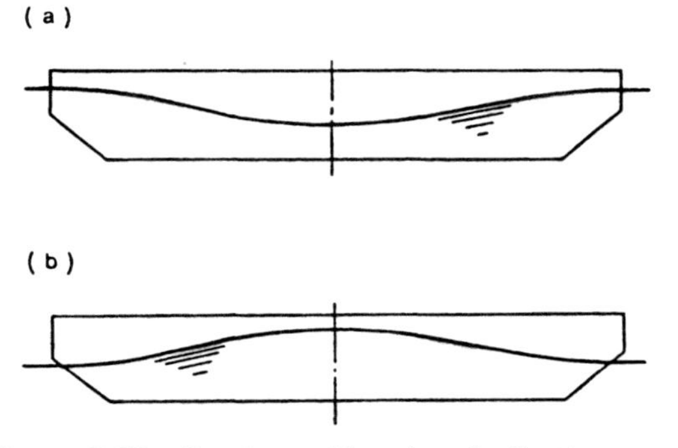

Figure 5–32. Sagging and hogging of a floating structure. (From Tsinker, 1986.)

The shear force having been computed, the bending moment, M_x, is determined from the following:

$$M_x = \int Q_x\,dx. \tag{5–58}$$

Computation of the wave-induced shears and bending moments on a floating body under conditions of hogging and sagging is highly dependent on the height and length of the wave chosen for design.

Generally, the wave length is taken as being equal to the length of the structure because this will result in the maximum hogging/sagging moment. Unless there are compelling reasons to do otherwise, this practice is recommended for design floating breakwaters particularly if the breakwater will be towed through open ocean to the installation site. In fact, Muller in Gerwick et al. (1975) has shown that hogging/sagging moments for oblique wave attack approaching 30° from beam-on conditions are nearly equal to that resulting from wave attack along the longitudinal axis of the structure. Accepting the practice of taking the wave length equal to the structure length leaves selection of the appropriate wave height to the discretion of the designer. At one time, the wave height for ocean applications was taken as the wave length, L, divided by 20 and more recently, the following formula has been suggested (Muckle, 1975):

$$H = 0.607\sqrt{L}. \tag{5–59}$$

The above formula assumes wave height and length are in meters. Similarly, ship design classification societies, such as the American Bureau of Shipping (ABS), have developed guidelines for determining design wave height values in the open ocean. If the breakwater module is not towed through the open ocean, then the designer must estimate the highest waves to which the breakwater will be exposed. The H_1 statistical wave height should be used in design.

Example 5

Compute the wave hogging/sagging shears and bending moments on a rectangular floating breakwater (4.9 m width, 1.1 m draft, and 45.7 m length) exposed to a wave with an H_1 height of 2.5 m. Assume the wave surface to be sinusoidal.

Step 1: The wave surface profile is given as follows:

$$h_w = \frac{H}{2}\cos\frac{2\pi x}{L} = \frac{2.5}{2}\cos\frac{\pi x}{22.9}.$$

Step 2: Compute shear. For sagging conditions, the buoyancy per unit length is:

$$b_x = \rho W\frac{H}{2}g\cos 2\pi\frac{x}{L} = 1025(4.9)\left(\frac{2.5}{2}\right)$$
$$9.8\cos 2\pi\frac{x}{22.9} = 61\,526\cos\pi\frac{x}{22.9}.$$

The wave shearing force is given by:

$$Q_w = \int 61\,526\cos\frac{\pi x}{22.9}\,dx$$
$$= \frac{61\,526(22.9)}{\pi}\sin\frac{\pi x}{22.9}.$$

Because the shearing force is zero at $x = L/2$, $A = 0$. Hence:

$$Q_w = 448\,478.5\sin\pi\frac{x}{22.9}.$$

Step 3: Compute bending moment. The wave-induced sagging moment is:

$$M_w = \int 448\,478.5\sin\frac{\pi x}{22.9}\,dx$$
$$M_w = -\frac{448\,478.5(22.9)}{\pi}\cos\frac{\pi x}{22.9} + B.$$

Because the bending moment is zero at $x = 0$,

$$B = \frac{448\,478.5(22.9)}{\pi} = 3\,269\,093$$
$$M_w = 3\,269\,093\left(1 - \cos\frac{\pi x}{22.9}\right).$$

The maximum bending moment occurs at the center of the breakwater (i.e., $x = 22.9\,\text{m}$); therefore the maximum sagging bending moment is:

$$M_w = 6\,538\,186\ \text{N-m} = 6538\ \text{kN-m}.$$

The hogging bending moment would have the same value, but opposite sign, of the sagging bending moment. A more detailed discussion of wave hogging/sagging is given in Tsinker (1986).

5.8.3 Numerical Models

Georgiadis and Hartz (1982) have developed a finite element numerical model for dynamic response of continuous floating structures in short-crested waves. The model, which was originally developed for use in floating bridge applications, has also been used to evaluate wave-induced stresses in floating breakwaters (England, 1985). In this model, the floating breakwater is schematized as a group of finite beam bending or torsional elements. Sway, heave, and roll motions are modeled separately, that is, motions are not coupled. Breakwater elements are connected by bending or torsional elements that simulate either flexible or rigid connectors. The breakwater (i.e., series of beams) is supported by buoyancy and is restrained laterally by discrete elastic springs at the ends and middle of each element.

One of the compelling features of the Georgiadis and Hartz (1982) model is that it accounts for wave directionality even though just three degrees of freedom are simulated. This is an important consideration when determining the structural loading along the longitudinal axis of the breakwater. Loads at various locations along the breakwater (i.e., nodes) are simulated using a "Monte Carlo" approach. In this Monte Carlo approach the stochastic or statistical nature of the directional wave loading is evaluated by computing the breakwater response to a large number of applied nodal load wave loads. The applied nodal loads are simulated by means of a rather complicated theory involving wave coherence (which is a measure of the correlation between wave surface elevations at various locations). Details of this theory are described in Georgiadis and Hartz (1982). The statistical breakwater response (e.g., mean and maximum values) are determined from the results of the Monte Carlo simulations. The Georgiadis and Hartz (1982) model was compared to the U.S. Army Corps of Engineers field measurements by England (1985), who found that the model compared favorably to measured results and, on this basis, recommended that the model be used in design applications.

5.9 ADDITIONAL ASPECTS OF DESIGN

Detailed structural design procedures are beyond the scope of this text. However, much attention must be given to the optimum length of breakwater units, adequate connections between breakwater units, spacing of anchor lines or piles, construction materials (i.e., concrete, steel, or timber), provision of cathodic protection, fatigue, and long-term maintenance costs. The reader should refer to the cited literature for more insight into these important design issues.

REFERENCES

ADEE, B. H., 1974. "Analysis of Floating Breakwater Mooring Forces," Ocean Engineering Mechanics, Vol. 1, pp. 77–92.

ADEE, B. H. and MARTIN, W., 1974. "Theoretical Analysis of Floating Breakwater Performance, Proceedings of the Floating Breakwater Conference, University of Rhode Island, Kingston, Rhode Island, pp. 21–40.

BANDO, K. and SONU, C. J., 1985. "Evaluation of Numerical Models for a Floating Breakwater," Report to U.S. Army Corps of Engineers Coastal Engineering Research Center.

BORGMAN, L. E., 1969. "Ocean Wave Simulation for Engineering Design," ASCE WWH, November.

CARR, J. H., 1952. "Mobile Breakwaters," Proceedings of the Second Conference on Coastal Engineering, Berkeley, California, pp. 281–295.

CARVER, R. D., 1979. "Floating Breakwater Wave-Attenuation Tests for East Bay Marina, Olympia Harbor, Washington; Hydraulic Model Investigation," Technical Report HL-79-13, U.S. Army Engineer Waterways Experiment Station, CE, Vicksburg, Mississippi, July.

DAVIDSON, D. D., 1991. "Wave Transmission and Mooring Force Tests on Floating Breakwater, Oak Harbor, Washington; Hydraulic Model Investigation," Technical Report HL-71-3, U.S. Army Engineer Waterways Experiment Station, CE, Vicksburg, Mississippi, April.

DEAN, R. G. and DALRYMPLE, R. A., 1984. "Water Wave Mechanics for Engineers and Scientists," Prentice-Hall, Englewood Cliffs, New Jersey.

ENGLAND, G. G., 1985. "Testing and Analysis of a Floating Breakwater," Masters Thesis, Structural and Geotechnical Engineering and Mechanics Program, Department of Civil Engineering, University of Washington, Seattle, Washington, 1967.

FRANK, W., 1967. "Oscillation of Cylinders in or Below the Free Surface of Deep Fluids," Naval Ship Research and Development Center Report, 2375, Washington, DC.

GAYTHWAITE, J. W., 1987. "Floating Breakwaters for Small Craft Facilities," Civil Engineering Practice, Journal of the Boston Society of Civil Engineers Section/ASCE, Vol. 2, No. 1.

GAYTHWAITE, J. W., 1990. "Design of Marine Facilities for Berthing, Mooring and Repair of Vessels," Van Nostrand-Reinhold, New York.

GEORGIADIS, C. and HARTZ, B. J., 1982. "A Finite Element Program for Dynamic Response of Continuous Floating Structures in Short-Crested Waves," International Conference on Finite Element Modeling, Shanghai, China.

GERRITSMA, J. and BEUKELMAN, E., 1972. "Analysis of the Resistance Increase in Waves of a Fast Cargo Ship," International Ship Building Progress, Vol. 19, No. 219, September.

GERWICK, B. C., JR., ed., 1975. "Proceedings of the Conference on Concrete Ships and Floating Structures," Berkeley, California.

GODA, Y., 1985. "Random Seas and Design of Maritime Structures," University of Tokyo Press.

HALES, L. Z., 1981. "Floating Breakwater: State-of-the-Art, Literature Review," TR 81-1, U.S. Army Coastal Engineering Research Center, CE, Fort Belvior, Virginia, October.

HEADLAND, J. R. and VALLIANOS, L., 1990. "Dynamic Analysis of Floating Breakwater Mooring Systems," 22nd Coastal Engineering Conference, Delft, The Netherlands.

HOOFT, J. P., 1982. "Advanced Dynamics of Marine Structures," John Wiley & Sons, New York.

ITO, Y. and CHIBA, S., 1972. "An Approximate Theory of Floating Breakwaters," Report of the Port and Harbor Research Institute, Ministry of Transport, Japan, Vol. 11, No. 2, pp. 138–213. (in Japanese).

LOCHNER, R., FABER, O., and PENNY, W. G., 1948. "The Bombardon Floating Breakwater," The Civil Engineer in War, Vol. 2, Docks and Harbours, The Institution of Civil Engineers, London, England.

LONGUET-HIGGINS, M. S., 1977. "The Mean Forces Exerted by Waves on Floating or Submerged Bodies with Applications to Sand Bars and Wave Power Machines," Proceedings Royal Society of London, A. 352, pp. 463–480.

MACAGNO, E. O., 1953. "Fluid Mechanics Experimental Study of the Effects of the Passage of a Wave Beneath an Obstacle," Proceedings of the Academic des Sciences, Paris, France, February.

MCCARTNEY, B. L., 1985. "Floating Breakwater Design," ASCE Journal of Waterways, Port, Coastal and Ocean Engineering, Vol. 111, No. 2, March.

MILLER, R. W., CHRISTENSEN, D. R., NECE, R. E., and HARTZ, B. J., 1984. "Rigid Body Motion of a Floating Breakwater: Seakeeping Predictions and Field Measurements," Water Resources Series, Technical Report No. 84, University of

Washington, Department of Civil Engineering, February. 1984.

MUCKLE, W., 1975. "Naval Architecture for Marine Engineers," London: Newnes Butterworths.

NELSON, E. E. and BRODERICK, L. L., 1986. "Floating Breakwater Prototype Test Program: Seattle, Washington," MP 86-3, U.S. Army Coastal Engineering Research Center, CE, Vicksburg, Mississippi, March.

NEWMAN, J. N., 1967. "The Drift Force and Moment on Ships in Waves," Journal of Ship Research, March.

NIELSEN, P., 1984. "Explicit Solutions to Practical Wave Problems," Proceedings of the 19th Conference on Coastal Engineer, Houston, Texas.

PINKSTER, J. A., 1980. "Low Frequency Second Order Wave Exciting Forces on Floating Structures," Publication No. 650, NSMB, Wageningen, The Netherlands.

STANSBERG, C. T., TORUM, A., and NAESS, S., 1990. "On a Model Study of a Box Type Floating Breakwater," PIANC, Osaka, Japan, May.

READSHAW, J. S., 1981. "The Design of Floating Breakwaters: Dynamic Similarity and Scale Effects in Existing Results," Proceedings of the Second Floating Breakwater Conference, University of Washington, Seattle, Washington, pp. 99–120.

TEKMARINE, INC., 1986. "Development and Verification of Numerical Models for Floating Breakwaters," Report to U.S. Army Corps of Engineers Coastal Engineering Research Center.

TORUM, A., STANSBERG, C. T., OTTERA, G. O., and SLATTELID, O. H., 1989. "Model Tests on the CERC Full Scale Test Floating Breakwater, Final Report, Norwegian Marine Technology Research Institute, MT51 89-0153, June.

TSINKER, G. P., 1986. "Floating Ports, Design and Construction Practices," Gulf Publishing Co., Houston, Texas.

U.S. ARMY CORPS OF ENGINEERS, COASTAL ENGINEERING RESEARCH CENTER, 1984. Shore Protection Manual, Vols. 1 and 2.

U.S. ARMY CORPS OF ENGINEERS, 1989. "Water Levels and Wave Heights for Coastal Engineering Design," EM 1110-2-1414, July.

VAN OORTMERSSEN, G., 1976. "The Motions of a Moored Ship in Waves," Publication No. 510, Netherlands Ship Model Basin, Wageningen, The Netherlands.

WICHERS, J. E. W., 1988. "A Simulation Model for a Single Point Moored Tanker," Publication No. 797, MARIN, Wageningen, The Netherlands.

WIEGEL, R.L., 1964. "Oceanographic Engineering," Prentice–Hall, Englewood Cliffs, New Jersey.

6

Marinas*

6.1 GENERAL

The word "marina" is generally used to describe a recreational boat facility, commonly referred to as small-craft marina; it is a water-dependent facility generally located on the body of water to enable boaters to obtain the shortest possible way to cruising waters. Sometimes the term "small-craft harbor" is used as a synonym for the word "marina" to describe facilities that harbor and service recreational boats. The former term, however, is more applicable in a broader spectrum of usage, for example, for description of facilities such as small fishing ports or basins to harbor small commercial boats. A marina is a place where boating people can keep their yachts and motorboats and obtain essential supplies such as fuel, food, and drinking water. It provides direct access to each boat, and adequate depth of water at all times, car parking, toilet facilities, technical services, shops, accommodation, and other amenities. Traditionally, small-craft moorings are found on coast lines, estuaries, lakes, and river banks. The increasing prosperity of the world population has resulted in an increased popularity and need for recreational facilities such as small-craft marinas. Many small-craft marinas have developed in the last two decades, and it is obvious that many more will be built in the future.

Estimates indicate that already in 1983 13.2 million pleasure craft were operating in United States waters alone, as compared to some 8 million boats in 1960. The pleasure boating community, which comprised 61.8 individuals in 1983, was served by 5919 marinas, boat yards, and yacht clubs with waterfront facilities. According to the National Marina Manufacturers Association (1880), in 1980, approximately one of every seven

* This is a reviewed and expanded version of Chapter 14 "Small Craft Marinas," by G. Tsinker previously published, in Handbook of Coastal and Ocean Engineering, Vol. 3 (1992), edited by J. B. Herbich. Gulf Publishing Company, Houston Texas.

people in the United States was engaged in some type of recreational boating activity and approximately one in every 20 owned some type of boat. In 1987, these numbers had risen to approximately one in three and one in 15, with 14 835 000 recreational boats owned. In 1988 as many as 72 000 000 Americans were using the nation's waterways more than once per year. The escalation in boat ownership by larger numbers of the population has created a substantial demand for marina facilities. Expenditures for new and used equipment, fuel, launching fees, repairs, boat club memberships, insurance, and other boating-related services amounted to almost $9.4 billion (Brinson, 1985). In Sweden, a country with a population of about 8.3 million, approximately 300 000 pleasure craft were sold between 1975 and 1985, and in 1985 the total amount of existing pleasure craft in this country was estimated to exceed 800 000 (Lellky et al., 1985). Similar developments are seen all over the world. This has created a tremendous demand for berth areas and harbors along seashores, inland waterways, and lakes, and this demand is still growing.

In recent years, however, some developments of small-craft marinas have been curtailed by environmental groups and local residents concerned with the effects of large-scale marinas on the quality of rivers, lakes, estuaries, and ocean shorelines. The latter is basically associated with all kinds of impacts on the environment; for example, water pollution during marina construction and operation, visual pollution due to unaesthetical planning and design, noise, destruction of wildlife habitat, and others.

There are many other practical problems associated with marina development.

One of the problems facing a marina developer is to obtain a site large enough and that can be sheltered with a minimum need for breakwaters, requiring little or no dredging, especially maintenance dredging, and close enough to a large population center and that is easily accessible. For example, in the United States a normal figure for the water area required is about 200 m^2 per boat (Brunn, 1989). The minimum size of land area may be approximately 80% of the water area, or about 160 m^2 per boat. Chaney (1961) recommended that for services such as driveways, parking for cars and boat trailers, boat storage, service areas, accommodations, and restaurant and recreational areas, a land area equal to about 125% of the basic area should be considered. Tobiasson and Kollmeyer (1991) presented the rule of thumb pie chart for land areas versus water area utilization. According to this chart about 32% of the total area should be designated to boat slips and docks, about 24% to channels and fairway access to slips, about 30% for land-based boat storages, haul-out equipment, etc., about 6% for building and support faculties, and about 8% for car parking area. The current trend toward marina developments as mixed use complexes with additional public recreation and commercial oriented services typically calls for a 50/50 land/water area requirements ratio, or even one slightly favoring the land area, where extensive dry storage, public recreation and large destination resort commercial uses and their related parking are present (Corrough et al., 1990). It should be noted that in some cases, nearby communities are able to offer parking space, shopping, and hotel accommodations for boaters, and therefore the required land area can in those cases be reduced accordingly. Utility services, water supply, solid waste collection and removal, collection and removal of waste water and oily waste for boats, removal of floating debris, safety of operations, security systems, servicing of boats, winter storage of boats and in some cases of floating berths, and collection of rental fees are all basic concerns and problems a marina owner/developer must deal with. Naturally, it must be recognized that each project will have its own peculiarities and characteristics and developers will all have their own preconceived ideas. Each marina is site specific and therefore must be planned, developed, constructed, and operated to serve a specific market; no two situations will ever be exactly the same.

The decline or change of waterborne commerce in some urban areas left once busy harbor and waterfront facilities with unattended piers, docks, and infrastructure.

Examples of decaying relics of such once busy facilities are found elsewhere in North America and worldwide. In the past 20 to 25 years, however, these "out-of-business" urban waterfronts became the most desired sites for construction of modern marinas.

The advantages of urban waterfront sites is that recreational boating facilities constructed there are very close to their users; they also provide a general public access to the waterfront and to means of local water transportation.

Revitalized waterfront or abandoned port areas affect the resurgence of the adjacent offices and residences (or vice versa) and make the marina facility there a valuable asset. On the other hand there is a price to be paid for marina development in an urban environment. The latter may include such important issues as development of the modern marina facility within an old and therefore obsolete infrastructure; rehabilitation and modernization of the existing docking facilities, such as piers and quay walls orginally designed to serve large merchant ships; operation of small crafts in deepwater berths, exposed to waters from passing commercial and recreational maritime traffic; and last but not least security in the urban environment.

Another modern concept of marina called "marina village" has been developed in the mid-1980s in the United Kingdom (Hirsh and Lacey, 1990). It proved to be very successful, and is described as a scheme where predominantly residential development is allied closely with a marina.

The marina villages satisfy both demands in that they provide residential accommodation with marine berthing as an integral part of the scheme. The latter proved to be an excellent enhancement of property values.

In addition the "marina village" complex can provide an ideal situation for leisure-oriented developments such as public houses and restaurants.

In all cases, the marina designer must consider legislation concerning the protection of historical areas of particular interest. For example, ancient remains are considered to be a historic heritage of national interest. They belong to the state, even when found on private property and, in most countries, are protected by law. This law typically states that the necessary investigations, documentation, and protective measures will be made at the expense of developers. Ancient cultures usually flourished along coasts and shores of inland waters and traces of old harbors and ancient settlements as well as shipwrecks have been found on marina sites. Whenever a site is expected to be of archaeological interest or when a shipwreck of historical/archaeological interest is discovered, an investigation and inventory must be made in order to evaluate the importance of the find. Those finds of historical or archaeological value must be salvaged.

At present there are numerous sources of information on marina design. Most of them will be pointed out later in this chapter. The purpose of this work is to put together in concise form the common sense guide to prime requisites of small-craft marina design and construction aspects, based on the most recent information on the subject available.

The emphasis here is basically on the structural design and construction aspects of the marina-associated facilities.

General information on marina development, which includes financial, operational, maintenance, and management considerations, is given in Tobiasson and Kollmeyer (1991), Abraham (1991), Morishige (1991), and others. Wortley (1989) provided a comprehensive list of publications on docks and marinas.

Most recently ASCE Ports and Harbors Task Committee prepared a Draft Progress Report on planning and design guidelines for small craft harbors. This report includes four chapters on marine planning and environmental considerations, entrance design and breakwaters, inner harbor structures, and economics and finance (Proceedings ASCE Specialty Conference PORTS '92).

6.2 THE ENVIRONMENTAL DESIGN PROCESS

Environmental conflicts in marina development arise mainly from the need for dredge and fill operations that may alter or destroy the wildlife habitat, change circulation patterns, introduce suspended sediments that eventually smother aquatic plants and shellfish beds, or release toxic substances when polluted bottom sediments are set free. Water can stagnate in deep troughs on the bottom. The degradation or permanent loss of habitat could therefore threaten valuable natural resources such as fish. Also, it may affect migratory bird breeding and feeding areas.

Boating can also have various negative effects on the natural environment. For example, motorboats can cause water pollution, mainly in harbors. At winter storage areas there is a risk of pollution from chemicals and paints. In shallow waters boats are capable of stirring bottom sediments. Noise from motorboats can be a nuisance, particularly in an otherwise undisturbed natural environment. Sensitive shoreline vegetation can easily be worn out at popular mooring areas. Fauna, mostly birds, can be directly disturbed, especially during their breeding periods.

New marinas should therefore be located with consideration to prevailing environmental conditions and never at important fish reproduction areas or bird mass breeding grounds. Furthermore, dredging operations for the construction of marinas should be avoided during the breeding period of existing wildlife. Dredged materials, consisting of clay or granular materials such as sand, gravel, or stone and rocks, should not be dumped on a bottom that has loose sediments, mud, and silt. Dredging or mud dumping activities should not be carried out in areas of aquaculture. Neither should shellfish be consumed that has been exposed within the last year to suspended particles from these activities. The increasing public desire to attain a high-quality environment, coupled with increasing institutional permit complexities, rising cost of construction, and project time constraints demonstrates the need for guidelines to achieve environmentally sound projects.

The successful completion of the environmental study is a milestone for any project, and this is especially true for a marina, because of the inherent difficulties in coping with stringent environmental requirements. Like the engineering design process, environmental design typically involves a preliminary and a final environmental review. During the preliminary review stage, an early assessment will be made to determine whether the marina project can be designed to be environmentally acceptable and still be economically attractive to the developer. Legislation affecting the development of shorelines often results in restrictive environmental regulations with regard to water quality levels within a marina basin, the impact on flora and fauna, the impact on adjacent land uses and utilities, the impact of construction activities such as dredging and dumping, traffic generation, and shoreline stability.

The water quality level within a marina basin is usually the main area of environmental concern. It depends heavily on water circulation within the basin. In the past, little or no attention has been given to water circulation within marina basins. The optimization of marina basin geometry was generally based on the development of maximum berthage. This usually resulted in the construction of square, rectangular, or trapezoidal-shaped marina basins, often with sharp corners. Studies by Nece and Richey (1975) revealed that these shapes do not represent the optimum configuration for water circulation within marina basins. They found that dead spots and sharp corners within the basin interior are responsible for poor water quality level at these locations. Restricted flushing and circulation characteristics at sharp corners of man-made marina basins could result in significantly depressed dissolved oxygen levels and excessively elevated water temperatures, common to stagnant water. Subsequently, substantial changes in ambient water quality levels would adversely affect local and transitory marine life.

From a tidal mixing and flushing capacities point of view, oval and round-shaped marine basins are much superior to those with sharp corners. Natural bays and inlets with good circulation and high water quality can be used as a pattern for sound design. Design and construction of the curvilinear-shaped Point Roberts marina (Figure 6–1) set a good precedent for the design and

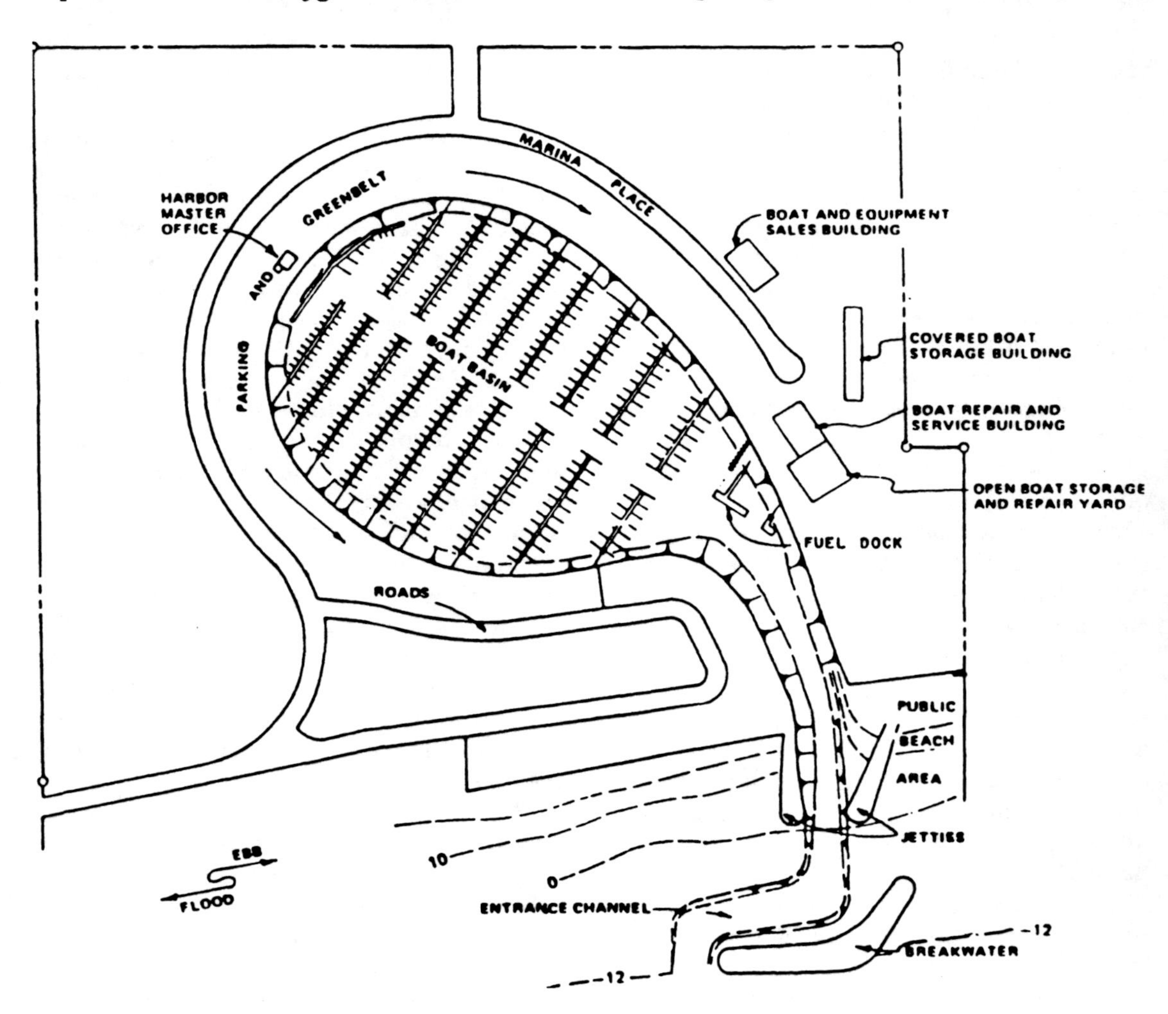

Figure 6–1. Layout of a small-craft curvilinear marina. (From NAVFAC DM-25.5, 1981.)

construction of this kind of marina (Nece, 1976; Layton, 1979, 1991). In addition to the oval shape of the basin, the entrance to the Point Roberts marina has been hydraulically designed in such a manner as to produce a flooding tide jet with sufficient momentum to cause large-scale rotating vortex systems within the basin. Coupled with the oval shape of the basin interior, this produced a strong internal circulation system.

A preliminary environmental review process, during which all basic environmental data are collected and evaluated, should tell the designer whether to proceed to the second stage of design, abandon the project altogether, or whether more information is required in order to arrive at one of the two preceding decisions.

Environmental issues and legislation have become increasingly complex and to avoid costly professional liabilities the designer must carefully address all relevant problems. Because most shoreline-related projects, which also include small-craft marinas, usually are highly controversial, legal counsel experienced in environmental law must be involved at the preliminary environmental review. This will provide the assurance that the environmental impacts of a project are adequately addressed and presented for public review. Impacts of some recent environmental legislations on marina design are discussed by Gustin and Neal (1990) and Obern and Hoy (1992).

The project must be approved by the appropriate regulatory agencies before attempting any construction activities. Failure to obtain the required approvals may result in costly construction delays, litigations, or fines.

Regulatory agencies in most countries are basically of three levels: state (federal), provincial (state in the United States), and municipal. The former is mostly concerned with major issues, such as water quality, protection of fish and wildlife habitats, effects of dredging and sludge disposal, etc.; the second group of agencies are basically concerned with the same group of problems as aforementioned, however, within bounaries of a particular province, plus with traffic and transportation issues; the latter group of agencies is concerned with issues of local coastal and wetland protection, but also with project conformity with local master planning, impact on local community, traffic, and parking issues, and economic and social benefits to the community.

The project approval system usually works in a way that approval at one level should be obtained before approval at another level; the time frame to acquire all approvals may last from several months to several years.

Public review is the next very important stage. During this stage there are usually numerous differences of opinion expressed and conflicts between designers and reviewing agencies usually arise from disagreement with regard to the interpretation of public policies, project environmental impacts, and proposed mitigation measures. Designers and project developers must therefore be prepared to modify the project to eliminate the most controversial issues.

The successful completion of the environmental review gives the green light for the final project design and construction, which consists of preparing architectural and engineering plans and specifications. The job should be awarded only to qualified contractors and a competent construction management company should be retained to coordinate the project. This would relieve the developer of the burden of construction management, provide project quality assurance, and help to keep the project on schedule as well as minimize construction costs.

6.3 SITE SELECTION

A potential marina site should be located according to the local zoning regulations, or where the local master plan can be rezoned without substantial interference with local community development plans.

It is obviously pointless to build small-craft facilities where they will not be used adequately. However, even if there is only a limited demand when the marina is built, often it can in itself create a demand. In most developed countries, any facility for water recreation within a reasonable distance of population centers is likely to find a ready clientele, provided it is not too expensive and is developed to supply a reasonably wide range of activities, such as yacht cruising and racing, canoeing, rowing, waterskiing, diving, swimming, and fishing.

The choice of a site for a marina typically involves the following obvious factors:

1. A new site should be located where there is a demonstrated demand for marine facilities.
2. The site must provide safe navigation access to cruising waters.
3. The site has to be adequately accessible on the land side, that is, have access roads for use by boat owners to conveniently reach their crafts.
4. The site must have enough protected water area or low land that can be dredged to required navigable depth.
5. The size of available land must be sufficient for a parking area, service structures, roads, and auxiliary facilities, including land for potential expansion of the marina.
6. There must be utility services to the site, such as electric power, portable water, telephone, gas, and sewerage.
7. The site must be outside of a wetland or resource protection area, area of restricted historic preservation, and areas designated for construction of new ports.
8. It must not be adjacent to public beaches.

In addition, environmental and sometimes sociological factors may affect site selection.

Because very few potential marine sites can meet all the above criteria, the impact of deviation from one or more of these criteria on the project viability must be evaluated.

Site assessment usually includes field sampling, laboratory analyses and other testing to investigate specific environmental concerns. During the site assessment the following are typically addressed: potential environmental significance of previous activities, environmental compliance history, and indications of potential environmental concerns based upon current or recent observations at the site and surrounding area. The prime objective of site assessment is to confirm or reject suspected contamination based on real data obtained from laboratory analyses of samples taken from the site.

Furthermore, if the existing marina is contemplated for further expansion all the above criteria should be evaluated to assess their impact on new expanded marina viability and potential for further modernization.

The most important factor for site selection is navigability of the approaches to the marinas for cruising waters. The approaches to the marina should be located so as to permit speedy and nonobstructed passage of boats to the marina in case of emergency, for example, storm, fire, or other. According to their location, marina sites could be classified as those located at riversides, river mouths, open sea coastline, and dredged low land. Two examples of a typical small-craft marina location are shown in Figure 6–2.

Rivers often offer excellent water courses for small-craft cruising. Good basins for marinas are frequently found along or just behind a river bank. In the case of a marina basin location along a river bank, it must be protected by a floating boom from floating debris, ice, and occasionally small boats that have gone out of control. Because this type of arrangement is usually adjusted to fluctuations in the water level of the river, it must have an articulated bridge system to obtain convenient access to the land side. Sometimes channels, located behind the river bank, are excavated to gain access to

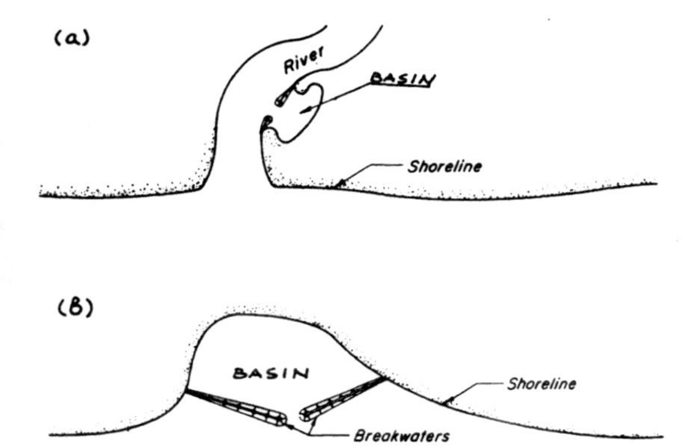

Figure 6–2. Examples of typical small-craft marinas. (a) Typical marina build inside a river mouth. (b) Typical open coast marina built seaward/lakeward from shoreline and protected by breakwaters.

the marina. In some instances the basin needs to be expanded and deepened. In both cases, the site must be located in an area where it can cope with extreme high and low river levels, both with respect to slip anchorage and access from the land.

Minimal water level fluctuations and current velocities, as well as lack of substantial waves, are typical of river-mouth locations. However, the site (basin and approach channel) may be shoaled by sediments moved by river current or tidal action. Many off-river basins have entrance shoaling, as suspended sediments carried by the river are deposited on the bottom in the quiet entrance waters. In that case, maintenance dredging will be required.

The effect of altering the natural regimen of the marina basin must be carefully studied. For example, if the marina basin is formed by excavating lowland, or as an entirely excavated harbor with entrance to the sea or a tidal river, problems of erosion and siltation are likely to arise. Possible salination of surrounding areas must also be considered and questions may arise regarding drainage and flood plains. These factors require in-depth study by a qualified coastal engineer.

Currents are the major exposure problem to be reckoned with in the process of the marina site selection. In general cross-currents of more than 0.5 m/s (1 knot) are considered too swift to sustain a safe and viable marina; under these conditions small craft tend to become less maneuverable especially when they need to turn broadside to current flow.

The latter may present an operational problem particularly where the boat must maintain the most control, namely at the entrance to the marina, in narrow confines between rows of berthed boats, in fairways, and during a docking maneuver such as entering the confine slip.

Sites at the open sea shoreline require protection from wave action which is usually achieved by breakwater construction. To reduce the cost of the marina breakwaters, the entrance to marina basins is usually situated at the minimal depth, which is defined as the depth at low water equal to the maximum draft of the design boat plus sufficient under-keel clearance. This leads to the following important considerations:

1. The danger that the marina entrance may often be on or closer to the shore than the wavebreaking contour, which means that sometimes entry to the marina basin will be impossible.
2. If the marina site is located on a sandy or muddy coast, or where there is much seaweed, sand bars may form at the entrance or within the basin.
3. Because of the limited depth, the maximum possible wave will have broken before it reaches the basin. Therefore, the breakwater profile can be considerably less massive than in the case of a deepwater marina.

Good sites for marinas are often found in natural harbors along the sea shoreline shielded from the prevailing waves by natural terrain features such as islands and shoals. Site selection there will include consideration of occasional wave episodes from directions other than those of the normal wave regimen to which the site may be exposed. The study of wave refraction must be considered to ensure that open sea swell from certain directions does not penetrate the marina basin. In all of these cases, adequate and suitable water and land areas are essential for successful marina operation.

The waterside (basin) requirements depend on the design, number, and size of boats to be accommodated and on the size of the required maneuvering area. Landside requirements largely depend on the planned function of the marina. Generally, the size of this landside area will be approximately as indicated in Section 6.1. However, if a recreational center is required, the landside area may reach several times the size of the water area. In recent years boating as a recreational activity has increased tremendously and successful marinas demonstrate a continuous tendency for expansion of their recreational facilities.

During the site selection process it is prudent to keep in mind the existing regulations concerning the public's rights along waterfront properties. The latter may typically include provision of public walkways or promenades, open to the general public, boat launching ramps and hoisting facilities, provision of scenic overlooks, dockage for public water transportation and for otherwise displaced commercial fishermen, construction of public fishing piers, and others. The public access, however, should not be viewed as a major stumbling block; while in many cases the public access rights will be easy to implement, where security is a legitimate concern the regulators will not force unreasonable public access demands.

Some potential marina sites may be found in close proximity to traffic lanes or within harbors frequented by commercial maritime traffic. The effects of this traffic that passes in close proximity to the marina such as vessel wake, or vessel suction have to be investigated at the project feasibility design stage.

Although maritime law generally restricts ship maneuvers in narrow navigable channels and confined water areas and in some instances stipulates use of tug boats to assist vessel maneuvers still vessel wakes or high-velocity currents or short-term large wakes produced by propeller thrusts from powerful tugs may have a detrimental impact on marina operation. Examples of the aforementioned effects on marina location are illustrated in Figure 6–3.

Normally, marina permit applicants would be advised that construction of small-boat berthing facilities located in close proximity to commercial navigation channels might be exposed to unacceptable wake waves and that the attenuation of this action is the responsibility of the marina developer. Safety of operation of the marina located adjacent to the commercial waterway areas is another concern of the marina operation.

Seagoing larger vessels with the bridge aft positioned and specifically those loaded with containers and long trains of river barges pushed by tugs may have restricted observation in the area forward of the vessel (blind spots) that comprise their ability to see small crafts ahead of them.

Hence, it is prudent to avoid siting marines where passing traffic may have blind spots in areas where boats are leaving and/or entering the marina.

Last, but not least, during the process of site selection it must be recognized that in urban areas, and specifically in industrial areas, soil and water may be significantly contaminated. Therefore the marina developer needs to be aware of the possibility of encountering hazardous wastes or contaminated soils as he purchases or develops industrial lands. To protect himself from the potential for environmental clean-up and liability, project delay, cost overruns, adverse publicity, etc., the developer should undertake water and soil screening investigation prior to property acquisition or major project developments. This investigation must be assisted by a qualified consul-

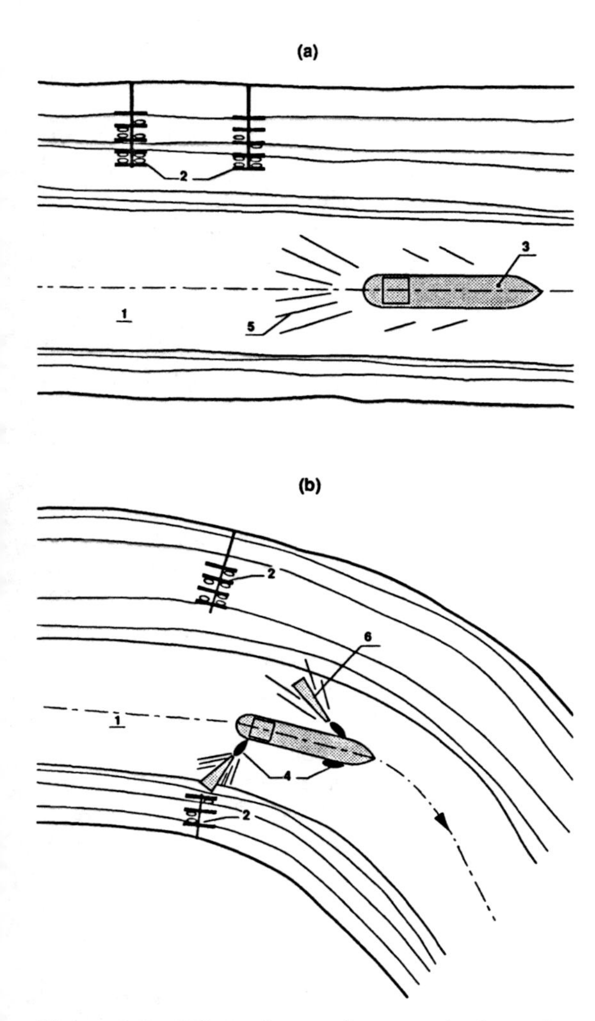

Figure 6–3. Effects of a vessel on a marina located in a narrow waterway. (a) Strait channel. (b) River bend. 1—Navigation channel, 2—marina docks exposed to ship wakes, 3—ship, 4—tugs helping ship maneuvering in a narrow river band, 5—wake, 6—high-velocity propeller thrust from maneuvering ship.

tant experienced in soils and groundwater investigation, with knowledge of hazardous waste regulations and experience in field safety procedures and sampling procedures related to environmental chemical contamination. The site investigation is usually conducted in the following order. First the site history is reinvestigated. The latter includes review of past ownership, and site potentials for contamination by chemicals or spillage.

Next, soil investigation and chemical analysis are conducted. During this stage soil samples are taken from the surface and from some depth at different locations. Then the laboratory chemical analyses are performed to screen the soil for the presence of the heavy metals, PCBs, and other unacceptable chemicals and compounds.

If investigation will reveal the presence of contaminants of concern then environmental agencies must be notified and a decision on further steps of land development should be made.

Several case histories on soil quality testing in ports are discussed by Farr (1986).

Finally, it must be noted that marinas must blend in with the landscape. In most cases the development of marinas does not cause problems from a landscape point of view as, by its nature, it is not obtrusive and a collection of yachts and motorboats is usually visually very attractive.

6.4 SITE CONDITIONS

The natural elements of the site, such as local weather conditions (i.e., precipitation, wind, and fog), ice conditions, waves, tides, currents, shoaling factors, geotechnical conditions, as well as project impact on the environment and sociological factors all must be investigated.

6.4.1 Weather Factors

6.4.1.1 Precipitation

Factors such as maximum probable rainfall or snowfall present no serious problems for small-craft marina operations, although all necessary surface drainage measures have to be considered in marina planning. Drainage facilities have to be designed to be capable of draining or diversion of the design amount of surface water produced by a maximum probable rainfall. In regions where snowfall is heavy, landbased structures must be designed to carry the design snow load.

6.4.1.2 Wind

Nearly every location in the world is subject to what is termed the prevailing wind, that is, a wind blowing from one general direction of the compass for a major portion of the year. Prevailing winds are not necessarily the strongest winds. Very often winds of greater intensity, but that occur less frequently, come from other directions. The direction, frequency, and intensity of winds at a particular location over a period of time are represented graphically by a wind rose. The force of winds is classified in accordance with the Beaufort Scale.

In most regions, wind data are readily available and windloading of structures is specified by local building codes. Where wind could reach hurricane strengths, the water- and landbased structures must be designed to be able to withstand the unusually heavy forces. Heavy wind may affect water levels in the marina basin, raising or lowering the water level substantially in a matter of a few hours. The rise and fall of the water level due to heavy wind conditions can be computed by using relevant guidelines or by examining historical records for the area. In hurricane areas, where wind destructive forces could be beyond practical design factors of safety, some structures of a secondary importance could be designed as "sacrifical," or with sacrifical elements, for example, doors and/or windows, to avoid complete collapse.

Other wind effects, such as the generation of waves or the movement of sand in dune areas that may shoal the basin or the entrance to the marina, must be considered. The former usually requires the construction of breakwaters, whereas the latter requires the implementation of some sand stabilization measures such as land stabilization by planting grass or construction of sand fences.

To date, however, because of lack of extensive wind tunnel tests the realistic wind load on small-craft marinas is unclear, and in today's marina engineering practice it still remains more art than science. Wind load is usually related to the marina characteristics, for example, layout, berth system, boats' shapes and sizes, and others. The recent Australian standard for marina design recommends a range of drag coefficients C_D to be applied to calculate the wind load acting in a different direction. Some representative values are given in Table 6–1.

Some practical recommendations and guidelines on determination of wind load in small-craft harbors are given in ASCE (1969) and in the U.S. Army Corps of Engineers (1984a). A survey of recent design wind factors affecting marina planning is given in Nichol (1990). The shielding effect of one boat or one row of boats by another is the topic of considerable debate among marina designers. Caution, however, must be exercised in assumptions on the shielding effect produced by first, second, etc., rows of boats. Tobiasson and Kollmeyer (1991) suggested that wind pressure on the second row of boats can be considered with a factor of 0.5, and on the third row and beyond a factor of 0.3 could be appropriate. Although use of these values may be considered as somewhat conservative this may lead to a safer design.

Table 6–1. Drag Coefficients Recommended by Australian Standard on Marina Design

Vessel or Structure		Drag Coefficient C_D
Vessels:	bow to wind	1.1–1.2
	stern to wind	1.6–2.0
	beam to wind	1.3–1.6
Tubular piles		1.2
Rectangular members		2.0

From Abraham, 1991.

On the other hand Hunt (1991) suggested that wind load should be used as 100% for the exposed first row of boats and an additional 10% for each shielded boat should be added.

6.4.1.3 Fog

Fog reduces visibility and is a serious navigational problem. Many marina locations have occasional foggy conditions and for this reason channels in a small-craft marina should be as straight as possible. At these locations, navigation in dense fog must be aided by installation of a proper number of marker buoys and other channel marking devices.

6.4.2 Ice

In northern climates ice is a serious problem in the operation of small-craft marinas. In areas with moving ice sheets, marinas must be located in protected areas, because these ice sheets, if unchecked, may not only crush boats but also badly damage marine structures (Figure 6–4). Protection is best provided by locating the entrance to the marina oriented away from the direction of the prevailing wind or current. This will encourage ice floes to move out of the marina during breakup. If possible, the marina should be located as close as possible to an industrial complex so that any available waste heat may be used. Although thin ice formation cannot damage boats, even in protected marinas boats are usually removed from the water during the winter. In protected marinas thick, unbroken ice sheets forming around piles that support marina piers may lift these piles when the water rises, and thus bring the whole structure out of alignment. Repeated freezing and thawing may eventually jack up piles completely out of the ground. In large natural basins, wind-driven ice floes may crash onto marine structures as the ice melts in spring. This

Figure 6–4. Damage to the marine floating docks produced by the wind-driven random ice floes.

may cause considerable damage to these structures. In areas where ice presents problems, special precautions must be taken. These measures, as well as ice load upon marine structures, are discussed in detail in Chapter 2.

In Finland small-craft marinas have been built with considerable success with piers and quays with a width of 1.5 to 3 m supported by wooden batter piles (Kivekas and Sarela, 1985). Batter piles provide better stability in the foundation soil. When water fluctuates steadily, the ice attached to the shore, to a wall of a solid type construction, or to a dense row of piles will break easily at that location when the water changes level. However, in tidal zones ice could easily build up on vertical surfaces of structures that are fixed on the bottom, thus creating a destabilizing buoyancy force or an additional load on the foundation.

Wortley (1991) suggested the most comprehensive design criteria and specific design guidance for ice engineering for marinas located in the Great Lakes.

He proposed a winter condition characteristics classification system for the harbors located on the Great Lakes. The latter include ice conditions, water fluctuation, temperatures, winter durations, and others. Wortley also pointed out the ice sheet confinement effects on the ice loads within a marina basin. As discussed in Chapter 2 the ice loads can be mitigated by miscellaneous means. The most frequently used methods for ice control within the marina basin are bubbler and miscellaneous flow developing systems. Both systems if correctly employed may be effective in both fresh and salt waters. Again, where warm industrial water is available it may be effectively used for ice suppression. Naturally, the ice conditions at the specific marina site affect selection of the dock system. For example, where the heavy ice conditions are expected the use of the bottom fixed structures of solid type construction, or the appropriately designed piled structures would be an obvious choice. Alternatively, at sites with relatively short and mild winter seasons and stable ice conditions the floating berth system may work quite well. Also, as usual, a combination of both systems, for example, bottom fixed access piers and floating finger piers, may be used where ice conditions demand so. In the latter case finger piers may be removed for the winter season if required.

6.4.3 Waves

Natural phenomena such as waves may be cuased by winds, tides, earthquakes, or by disturbances caused by moving vessels. The marina designer is mostly interested in waves produced by wind and moving vessels as they have the most effect on site selection and marina design. Waves that reach coastal regions expend a large part of their energy in the nearshore region. As the wave nears the shore, the wave energy is dissipated through bottom friction and percolation. When deep-water waves reach shallow water, where the depth is equal to about one and one quarter of its height, it will usually break, although it may break in somewhat deeper water, depending on the strength of the wind and the condition of the bottom. Breaking waves affect beaches and man-made coastal structures. Thus, shore protection measures and the design of coastal structures depend on the ability to predict wave form and fluid motion beneath waves, and on the reliability of such predictions. For details on wave fundamentals and classification, as well as a variety of wave theories and recommendations for the prediction of wave height and period, see the related literature, for example, U.S. Army Corps of Engineers "Shore Protection Manual" (1984).

Passing ships may generate substantial waves that are sometimes of greater length than wind waves. This is particularly true for small-craft marinas located on rivers, where passing deep-draft ships or barges may generate damaging waves. The height of waves generated by moving vessels is dependent on the vessel speed, draft, shape of the hull, depth of water, and blockage ratio of ship to channel cross-section. The effect of waves will depend on the height of the wave generated and the distance between the ship and the project site. As a rule of thumb it can be assumed that the wave height is equal to twice the amount of vessel squat. The wave height at the river bank is then computed using refraction and diffraction techniques. The wave length is equal to approximately one third of the vessel length (U.S. Army Corps of Engineers, 1984a). If ship-generated waves are considered to be the design wave, model tests or prototype measurements are needed to verify or adjust the predictions. For detailed discussion on ship generated waves the reader is referred to Sorensen and Weggel (1984) and Weggel and Sorensen (1986). Additional information on the possible impact of vessels wakes may be obtained from Canfield et al. (1980) and Kurata and Oda (1984); a useful discussion on the subject is provided by Tobiasson and Kollmeyer (1991).

Naturally, marina sites need to be protected from adverse wave effects. Some sites have naturally protected entrances from the main water body. This protection may be provided by one or more islands that shield the entrance from waves by reducing wave height, or by shoals where the waves break. If the site does not have natural protection against wave action, then the construction of breakwaters or other wave energy dissipating devices at the entrance or inside the marina must be considered to reduce waves to an acceptable height. The criteria for acceptable wave actions are that the significant height of any wave episode should not exceed about 0.5 to 1.5 m in the entrance channel, and 0.25 to 0.5 in the berthing areas, depending on the craft using the marina. In calculating the load transmitted to the marina-based structures well-recognized theories are usually used. For the preliminary wave load estimation a minimum load of 2.0 kN/m is recommended.

6.4.4 Tides

The rise and fall of water levels at different locations around the world varies from very minimal to up to 16.2 m in the Bay of Fundy area (the world's highest tide). Tide tables are readily available for most parts of the world. The Admiralty Tide Tables cover major ports in the United Kingdom and elsewhere, while the United States Coast and Geodetic Survey lists the tides for the major harbors in the United States and other parts of the world. When using the tide data of the United States Coast and Geodetic Survey, it must be kept in mind that they give the times and heights of high and low waters and not the times of the flood or slack water. For locations on the ocean coast there is usually little difference between the time of high or low water and the beginning of ebb or flood current, but for locations in narrow channels, landlocked harbors, or along tidal rivers, the time of slack water may differ by several hours from the time of high or low water. The predicted times of slack water and tidal current velocities are given in tidal current tables published by the United States Coast and Geodetic Survey; one for the Atlantic Coast of North America, and another for the Pacific Coast of North America and Asia. If a chosen marina location is remote from any of the stations listed in the tables, it is possible to interpolate predictions for the site under consideration from values given in the tables for the two nearest stations. Interior surge, and wind- and boat-induced waves should be considered in conjunction with the tidal fluctuations. This information is necessary for determining the required depth of the access channel and basin and the elevation of the land portion of the marina. It should be noted that consideration must be given to the draft of the design boat to be used at the facility, and to the effects of squat, roll, pitch, and heave of this boat in the approach channel and in maneuvering areas.

Tides and tide-like effects, for example, water level change in inland lakes and rivers due to spring and fall flood, often play an important role in water quality control. The current producing exchange of water between the marina basin by water fluctuation action may be essential to the marine ecology and the prevention of stagnation conditions. As noted earlier in this chapter, water circulation is an important component in marina design and could be accomplished by the effective use of the tidal prism of the water. In inland lakes and rivers water fluctuates in a slower cycle, and although it occurs too slowly to produce substantial water exchange effects, these effects have to be taken into account for the design.

6.4.5 Currents

Currents are essentially horizontal movements of the water. At coastal locations, currents or flow tides or freshets, moving at only a few tenths of a knot, generally cause no serious problems to marine operations. However, in swiftly moving rivers (with a speed of several knots), where seasonal floods are expected, or in large open bodies of water, where wind-generated current may be damaging to the marina, marinas should be located in protected locations, for example, secluded inlets, bays, or lagoons, or breakwaters have to be installed. Apart from the possibility of direct interference with marina operation, currents may also present other adverse functional effects such as scouring, deposition of sediments, and increased corrosion rates.

Currents may cause changes in wave effects and impact of ice and flotsam (floating debris), as well as hamper construction operations. In tidal estuaries the current can be expected to reverse. The value of tidal current velocity for many locations around the world may be obtained from tables published annually by the National Oceanic and Atmospheric Administration. Depending on location as well as importance and cost, current velocity measurements may be considered for the project.

6.4.6 Shoaling

A principal cause of shoaling at entrances to marina basins is littoral drift, which is mainly the result of wave and/or current action. Any structure that interferes with wave or current action would cause abnormalities in the wave or current pattern and could substantially affect the shoaling process. Dunham and Finn (1974) suggested the following example. If the unprotected approach channel is dredged through a beach into an inner basin, the wave impinging on either side at the mouth will be refracted in such a way as to cause changes in the wave pattern approaching the lips of the channel. If the approach of the prevailing waves is normal to the shore, the initial effect will be a movement of the littoral material from the lips inward along each flank of the channel, thus eroding the lips and shoaling the inner channel fed by material from the beach on either side of the entrance. Unless tidal currents are strong enough to maintain an opening against the forces tending to shoal the entrance, the channel will soon be blocked. Where the prevailing wave approach is oblique to the shoreline, sediments being transported along the shore by littoral currents will be interrupted at the channel opening near the updrift lip and that lip will soon begin to accrete. As the wave-induced longshore current again begins to "feel" the shore downdrift of the channel mouth, it attempts to reacquire its sediment load. As a result, the downdrift lip of the channel will erode at about the same rate as the updrift lip accretes, and the channel mouth will migrate in the downdrift direction. In each of these cases, the forces of nature are attempting to reestablish the littoral balance that was present before the channel was excavated. The previous example is an oversimplified version of an extremely complex process, and excludes consideration of the effects of sandbar formation, eddy currents, and tidal channel meandering.

The problem may also stem from the construction of the marina basin or its entrance channel through dredging; there the natural siltation mechanism may simply return the sediments to the excavated area, to restore this area to its natural state.

The customary solution to entrance shoaling is the construction of jetties along each flank of the channel from the lips of the mouth seaward beyond the breaking zone. The structural features of the jetties must be such that the materials will not be washed through or over the structure into the channel. A typical section of a sand-tight, rubble-mount jetty is shown in Figure 6–5. If the littoral transport from one direction predominates and the entrance is stabilized by jetties, accretion will occur along the updrift shore and erosion along the downdrift shore.

The entrance to off-river marinas is often subject to shoaling because of sediment deposition in the quiet water area and to eddy currents that might be created by the entrance configuration and the flowing water in the river. Although shoaling cannot be prevented, it is often reduced by proper entrance design. For example, a flat area on the downstream lip of the entrance could be provided from which a dragline can excavate deposits from the bottom of the entrance channel and cast them into the river downstream of the entrance (Figure 6–6). The entrance must be kept as narrow as practical to permit such an operation, and a training dike at the upstream lip is helpful in reducing the deposits.

A river never takes a straight course from its source to the sea. Geological conditions cause the river to take an alignment with one or several forms, among which meandering and divided flow are common. River flow continuously erodes materials from concave bends where the current concentrates; the material is then deposited on convex bars further downstream, where the current is slower. The marina designer must study the proposed site carefully to determine if this natural channel migration and its related movements of bottom sediments post a threat, or if the effect can be overcome by maintenance efforts. Some useful discussion on this subject is given in Fitzpatrick et al. (1985).

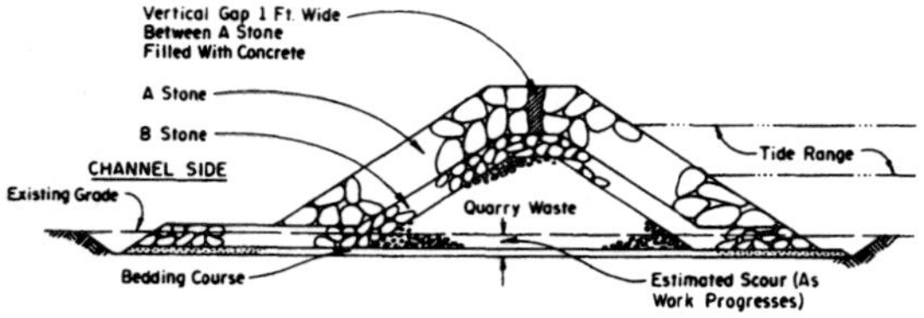

Figure 6–5. Typical cross-section of a rubble-mound jetty. (From Dunham and Finn, 1974.)

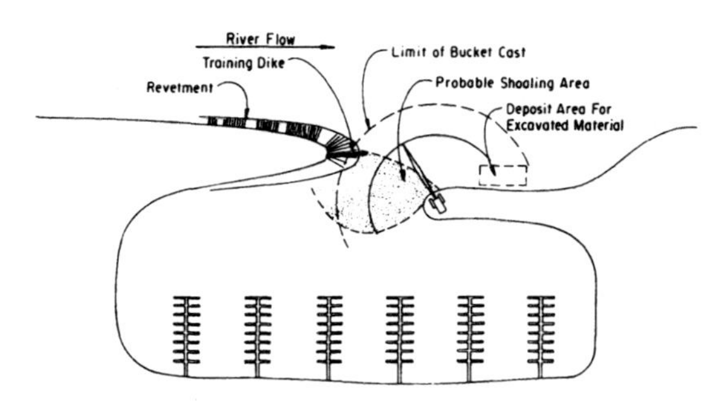

Figure 6–6. Maintenance of entrance to off-river marina basin with land-based equipment.

6.4.7 Geotechnical Conditions

Because in most cases of marina construction some excavation/dredging is required to achieve the correct size and depth and the desired shape of the basin, knowledge of the basin's geotechnical condition is essential to determine the best method of soil/rock excavation. Moreover, data on soil engineering properties at various locations of the selected site are required to design different kinds of marina structures, such as anchorages, piled berths, bulkheads, gravity-type retaining walls, foundations, and other.

Subsurface conditions at a site may be relatively uniform or extremely variable and will largely determine the complexity of the problems to be faced, both in basin excavation and the design and construction of the marina structures. The subsurface investigation must therefore be of sufficient extent to provide enough information for a thorough understanding of the interaction of the proposed structures and supporting soil or rock on which to base a safe and economical design.

The site investigation should be carried out to such an extent that the entire zone of soil or rock affected by changes caused by the excavation or the construction will be adequately explored. A variety of methods can be used for the subsurface investigation to determine the soil properties critical in the design. In particular, it is good practice, whenever possible, to combine field and laboratory tests for strength and compressibility. The properties of soils can be determined from laboratory tests on samples recovered from boreholes. These samples are analyzed to determine the best method of moving material, that is, by dredging, excavation, or blasting. For soil moving purposes, disturbed samples usually are sufficient for the visual identification and classification of the materials encountered, as well as identification by means of grain size.

To obtain a basic knowledge of the engineering properties of the materials that will have an effect on the structure design, a limited number of undisturbed samples, such as those obtained with thin-wall samplers or by other means, will be required. The number of samples taken should be sufficient to obtain information on the shear strength and consolidation characteristics of each major stratum. The results of the soil explorations and laboratory testing are usually presented in the form of a geology and soils report. Apart from boring methods, there are other methods of soil exploration such as the standard penetration test, the seismic method, the electrical resistivity method, and the sounding and probing method. Details on these methods, as well as on boring methods, are given in the Foundation Engineering Handbook, edited by Winterkorn and Fang (1975) and in other related literature.

Recently the flat-plate dilatometer test has been introduced in North America (Marchetti, 1980; and Schertmann, 1988) and this test is gaining more acceptance for foundation design. It has proved to be useful for site investigations and can be classified as a logging tool that is easy to use and provides a range of empirically predicted soil parameters from two measurements. In some instances a satisfactory delineation of the mud–sand or the sand–rock interface may be obtained simply by probing with a steel rod or jet pipe. During the site exploration, seismic conditions must be evaluated and all marine facilities must be designed according to the seismic design data. These data can usually be obtained from the local building codes.

It must be specifically noted that soils containing significant amounts of organic materials, either as colloids or in fibrous form, are generally weak and will deform excessively under load. Such soils include peat and organic silts and clays typical of many estuarine, lacustrine, or fluvial environments. Such soils are usually not satisfactory as foundations even for very light structures because of excessive settlements that can result from loading the soil, and must be replaced by good quality materials. Some geotechnical work is required for the search of material sources required for the construction, for example, rocks for breakwaters, jetties, and revetments; fill material for construction of retaining structures; and aggregates for concrete works. If a developed quarry close to the site is not available, then the cost of transporting the required materials from distant quarries, the use of locally made concrete blocks of different shapes, a geotechnical exploration for a potential new quarry, or the development of an alternative design method must be compared.

All materials used for the construction must be sound: rock should be hard enough and free from lamination and of such characteristics that it will not disintegrate from the action of air, weather, or handling and placing conditions; fill materials must be of adequate structural quality. This may be obtained locally or from nearby areas. Sometimes dredged materials from the basin excavation are suitable for fill. Loose sandfill can be compacted to required density by flooding or vibration. Cohesive soils require mechanical compaction. In most cases quality aggregates for concrete such as gravel and sand can be found locally; if not, they will have to be brought in from available sources.

6.4.8 Sociological Factors

Sociological factors, such as the relation of the prospective marina to adjacent development, related recreation, and transportation facilities, must be examined as part of the investigation program for site selection. Any site selected for a small-craft marina within city boundaries or close to them should be functionally as well as aesthetically compatible with other developments in the area. In some cases zoning restrictions may rule out marina construction at some highly desirable location, and conversely may prescribe the location in accordance with a municipal master development plan. Sometimes these sites can be located in areas not acceptable from a sociological point of view, for example, when they are in close proximity to a waste disposal area.

A marina is a recreational center and it is therefore highly desirable to have other types of recreational facilities, compatible with boating activities, located in close proximity to the marina. Examples of these are swimming pools, scuba diving clubs, rowing courses, gymnasiums, golf courses, tennis courts, and hiking and bicycle trails.

Although most boaters and marina employees will commute to and from the site in their own cars, some of them, as well as transient boaters from other marinas, still require some kind of public transportation. An effort must therefore be made to include a stop at the marina complex on the scheduled route or routes of a local bus or rail system. Where the marina is isolated, some other options, for example, a minibus shuttle service, may be considered.

6.5 LAYOUT PLANNING

6.5.1 Objectives and General Principles

Once the site has been selected and environmental factors, geotechnical conditions, and other related aspects of marina functioning have been investigated, the marina layout can be planned. The following primary objectives of this stage of marina planning should be considered:

1. The entrance should be safe to enter by all craft using the marina, and under all conditions of currents, wind, and waves that may

occur in the water body served under which the craft can navigate. The access channels and fairways must be wide and deep enough to accommodate the anticipated peak-hour traffic of design craft, without undue hazard to navigation.

2. The marina basin must be large enough to accommodate the planned number of berths for all classes of designed boats without encroaching on established clearance standards, fairways, and turning basins.
3. The landside area should be sufficiently large to accommodate all administrative and service facilities and the whole complex of marina structures should be aesthetically pleasing.
4. Where future expansion is anticipated, adequate undeveloped land and/or water areas must be available for all future developments.

All of these objectives and requirements must be considered during the planning process and during preparation of the marina master plan, which should include the layout of both water and land areas for the entire site. The master plan should be arranged in such a way that the marina can be built in stages or increments, with each increment or combination of increments constituting, as nearly as practicable, a complete and well-balanced facility.

The marina layout is usually developed on the basis of a preconceived number of ships which are assumed to generate the rate of return sufficient to make the project profitable. Although there is a general perception that a minimum of 300 boat slips is needed for a financially viable marina it will be a poor way to begin a project without an appropriate feasibility study. Once the optimum number of slips is rationalized then land and water areas requirements should be established. The latter is discussed in Section 6.1.

From the safety of operation viewpoint the entrance to the marina and/or access to fairway if practical should be sited where a rising or setting sun could not blind the incoming or outgoing boaters.

Also where possible sharp entrances to a navigation channel should be avoided and the visual distance from the marina activites to adjacent navigation channels should be as great as possible. Again, from safety of operation viewpoint the boat traffic should enter or leave the marina at the smallest angle possible to the navigation channel.

The proper siting of the various components of a small-craft marina is a prerequisite for the functional soundness of the overall plan. A schematic typical layout of a complete marina complex is shown in Figure 6–7.

Larger craft should generally be berthed near the entrance to the marina. This is because they are less affected by residual wave action entering the basin; this action is usually greater near the entrance. They also need more maneuvering space, which is provided near the entrance, where there is a larger volume of traffic, and, because of their larger drafts, they require a deeper channel and basin. The inner parts of the basin, where small boats are accommodated, can be shallower if they are not used by larger craft.

Commercial small craft usually fall in the same category as large private recreational craft as far as their water area requirements are concerned. The berthing areas for commercial and recreational craft should generally be as far away from each other as possible because of different adjacent land use requirements. Charter boats for sport fishing must have adjacent facilities for selling their services, for controlling the boarding and debarking of clients, and for parking cars. Rental boats should be located in the commercial fleet area and not mixed with private recreational craft. The car parking area for rental boat clients should be separated from the slip rental area, but it may be shared with visitors using facilities in the marina complex.

Parking lots for the berthing basins should be located so that no parking space in any lot is more than about 150 m from

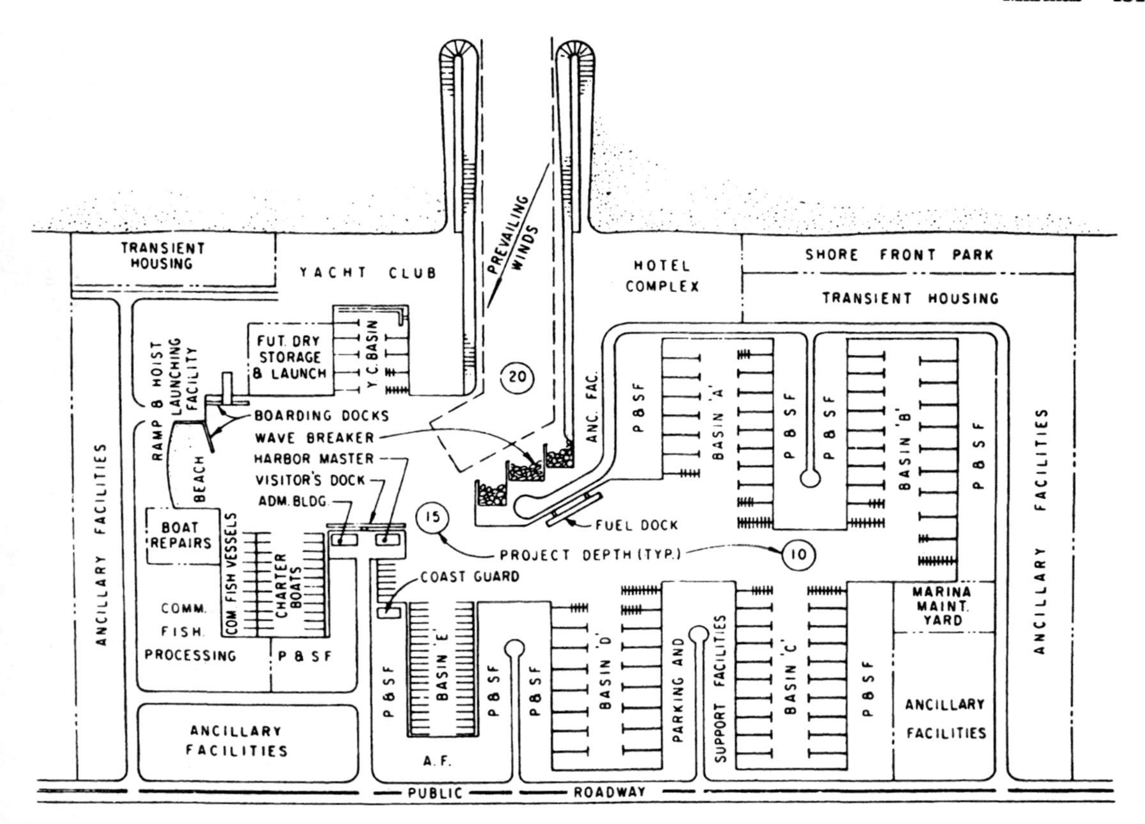

Figure 6–7. Schematic layout of a small-craft marina. (From U.S. Navy Corps of Engineers Engineering Manual EM 1110-2-1615, 1984.)

the head of the pier for the particular lot it is intended to serve. Parking lots for the ancillary facilities should be adjacent to parking lots for the berthing basins so that, under peak conditions, overflow from one lot can be absorbed by the other. As recommended by the Permanent International Association of Navigation Congresses (PIANC) (1979) coastal small-craft marinas should normally be provided with a minimum of 1.2 car spaces per berth and inland marinas with 0.70 spaces per berth, a car space being 5 m × 2.5 m. Access to car spaces should not be less than 5 m wide. In some cases, for example, when a marina is near available parking facilities, or when it caters mainly to boats in transit, the size of the parking lot may be reduced accordingly. The average size of craft and the habits of users should also be considered. Parking lots must be adequately drained and surfaced. An expensive surfacing, such as asphalt or concrete, does not need to be used; a porous surfacing, for example, gravel or crushed stone, would be quite adequate.

Trailered craft should have a separate entrance, or be launched directly into the marina water without using the marina. The latter may be necessary to prevent a conflict situation between owners of trailered and berthed craft if they are using the same fairway. If trailered craft must share the same protected waters with berthed craft, then the launching area should be as near the entrace as possible and physically separated from the berthing areas so that vehicle traffic to berthing areas and trailer traffic to the launching area do not merge.

The best location for a boat fuelling dock is in a well-protected area near the entrance. The adjacent land must be suitable for buried fuel storage tanks and easily accessible for fuel supply vehicles and, of course, the fuel station should not be placed in a location where it interferes with the traffic flow or constitutes a fire hazard because of its proximity to other marina facilities or to berthed craft. The marina administration area should be located near the entrance, where owners of visiting craft can easily come ashore to obtain information.

The repair facility may include various shiplift systems, large hoists, a dry dock, and other devices for launching and retrieving the largest craft. Off-season dry storage facilities should be located in a remote area of the marina; in fact, it may be located elsewhere beyond the marina complex. In many places parking lots are used in the off-season for dry storage in lieu of providing a separate facility for this purpose. Locations for other potential facilities, such as Coast Guard dock, restaurants, hotel complex, shopping areas, water-oriented facilities, and others can be located as shown in Figure 6–7.

Typically, the waterside part of the marina consists of the following elements: entrance channel, turning basin, berth area, and perimeter stabilization structure.

6.5.2 Entrance Channel

A marina's entrance channel is an artificial or natural waterway of perceptible extent that forms a connecting link between the marina basin and the adjacent navigable body of water. The entrance channel normally follows the course of the deepest bottom contours, which requires the least initial construction dredging. Currents also often follow this path, which is desirable for navigation. An alternative location would be the shortest route to deepwater. The entrance channel must be suitable for navigation, and for channel alignment the direction of predominant wind and waves and their impact on navigation must therefore be considered. Channels dredged through shoals or sandbars tend to shoal rapidly, and these locations therefore must be avoided. The inside of river bends must also be avoided because of the high shoaling rate at these locations.

In some cases breakwaters or jetties[1] paralleling the channel may be required to maintain a desired alignment and their design may require a physical model investigation. Breakwaters and jetties, if required, will provide protection to the entrance channel and to other basin elements. Several breakwater layouts and their associated costs are usually considered at the marina planning stage to determine the optimum arrangement. In most cases during the marina planning process model studies (physical or mathematical) are employed to determine the optimum entrance configuration and to predict wave heights at different locations within basin. Allowable wave heights are site specific; the acceptable wave heights depend on boat size and type of moorage, that is, pier or anchorage.

The entrance channel must be so oriented and protected as to avoid the direct intrusion of elements tending to disturb the water in the basin. As noted earlier, this is usually achieved by the effective location of protective structures. Boat traffic to and from the marina basin requires a certain minimum width of entrance to permit safe operation under different kinds of unfavorable conditions, for example, fog, darkness, and heavy

[1] In United States usage a jetty is structure perpendicular to the open sea coast and extended into a body of water which is used to prevent sholing of a channel by littoral materials, and to direct and confine the stream of tidal flow. Jetties are built at the mouth of a river or tidal inlet to help deepen and stabilize a channel. In British usage a jetty is synonymous with a wharf or pier.

wind. Under ordinary conditions, the minimum width of the entrance should not be less than 20 to 25 m, or about 4.5 to 5.0 times the beam of the widest boat to be used there. This allowance will permit boats to pass safely when sailing at low speed. When it is likely that boats will pass each other frequently in the entrance, additional width might be required. Although no criteria have been established for determining the width of an entrance channel to a large marina, where boat traffic is a controlling factor, a good practice is to provide a navigable width of about 100 m for the first 1000 boats, plus an additional 25 to 30 m for every additional 1000 boats berthed in the marina, including the daily launching capacity of operational ramps and hoists (Dunham and Finn, 1974).

PIANC (1991) suggested that minimum widths of entrance channel ranging between 20 and 30 m, except in the case of the large harbors (those with more than 2000 berths), should be considered.

Every entrance has its own special characteristics that may modify the tentative entrance width determined by the aforementioned general recommendations. For example, a short reach of constricted channel with more area for maneuvering at either end can be considerably narrower than would be desirable for a long channel of uniform width. Where boating characteristics of the marina users spread the daily entrance use pattern uniformly over several hours, rather than concentrating on a few peak hours, a narrower entrance may be adopted. In some instances the need of the narrower entrance for better marina protection from wave action may override the congestion consideration; in that case an exceptionally narrow entrance can be provided and its use restricted in some manner during peak hours. To design a safe entrance, the marina planner must exercise a considerable amount of judgement in determining the entrance to the marina, as well as entrance channel orientation and configuration. A typical layout and changing cross-sections of channel flank protection is shown in Figures 6–8 and 6–9. An entrance channel flanked by retaining walls is shown in Figure 6–10.

The channel leading to the marina should be at least twice the entrance width, with a minimum as determined by Eq. (6–1), Figure 6–11, but no less than 30 to 35 m.

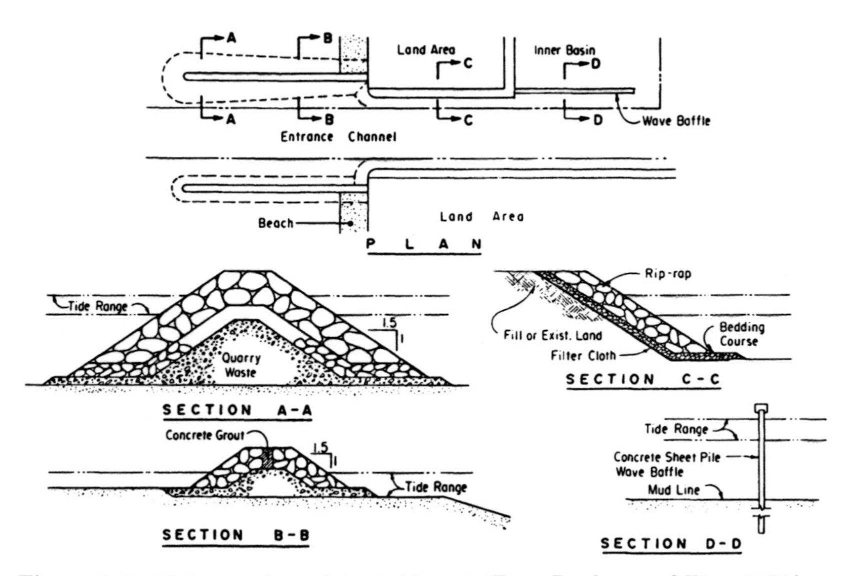

Figure 6–8. Entrance channel, typical layout. (From Dunham and Finn, 1974.)

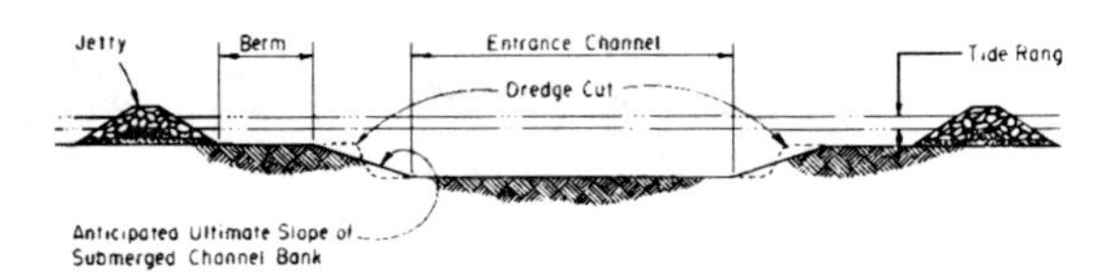

Figure 6–9. Typical dredge cut through entrance channel. (From Dunham and Finn, 1974.)

$$B_c = nS + 2.4B + 1.6(n - 1)B \qquad (6\text{–}1)$$

where

B_c = the channel width at the design low water level

n = the number of sailing boats in a row

B = the beam of the largest design boat

S = projected length of a sailboat mast height.

The height of a sailboat mast S normally depends on the boat's length and for

Figure 6–10. Entrance channel flanked by retaining walls. St. Catharines, Ontario.

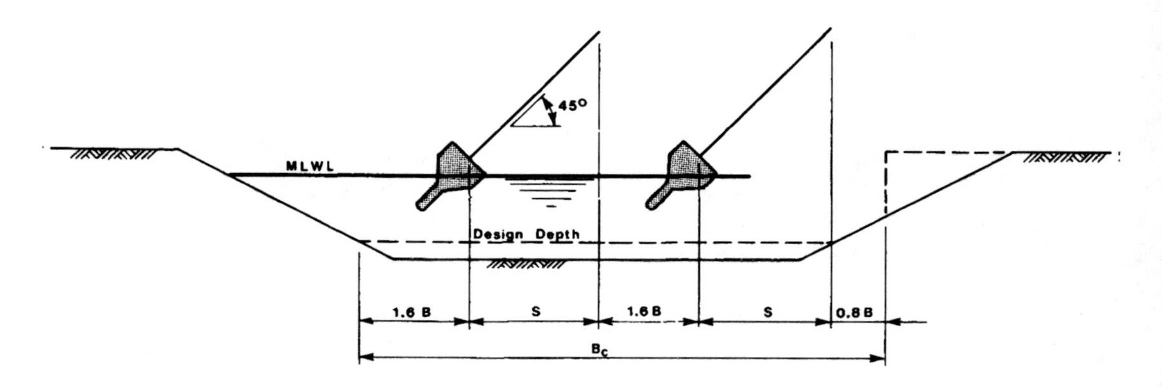

Figure 6–11. Design width of entrance channel. (From Prokofiev and Krivov, 1985.)

preliminary design an upper limit of mast height to boat length ratio varying between 1.3 and 2.3 may be used; the smallest value may be related to boat length equal to about 6.0 m and the larger to boat length equal to about 20.0 m. The aforementioned ratios, although helpful, must be treated with caution because boat parameters are changing constantly. A more detailed investigation is usually required for the specific marina final design where sailboats must transit through bridges or where other height impediments to boat passage may exist; then the vertical clearance criteria (usually 1.0 to 1.5 m) may limit specific marinas to use by boats of certain limited lengths.

The channel depth must be adequate for boat draft and trim (Z_1), squat (Z_3), wave conditions (Z_2), and safety clearance (Z_4). Additional depth is allowed in construction due to dredging inaccuracies (Z_6) (Figure 6–12). Sometimes overdredging (Z_5) may also be included as an advance maintenance procedure. The depth of the channel is usually measured from a design low-water figure. Because the wave action is less pronounced at the basin location, the entrance channel at that side may not have to be as deep as at the entrance location.

While in motion, a boat has a tendency to draw more water astern than when stationary. This phenomenon is known as squat. For channel design purposes and assuming small craft moving at reasonable speed (2 to 4 knots), the squat value (Z_3) is usually taken as equal to 0.3 to 0.5 m. The same value of small-craft squat is used for the design of the moorage (anchorage) area and for the turning basin. The amount of additional channel depth to be provided for squat can be approximated using the procedure recommended in the U.S. Army Corps of Engineers Engineering Manual No. 1110-2-1615 (1984a). This procedure suggests the

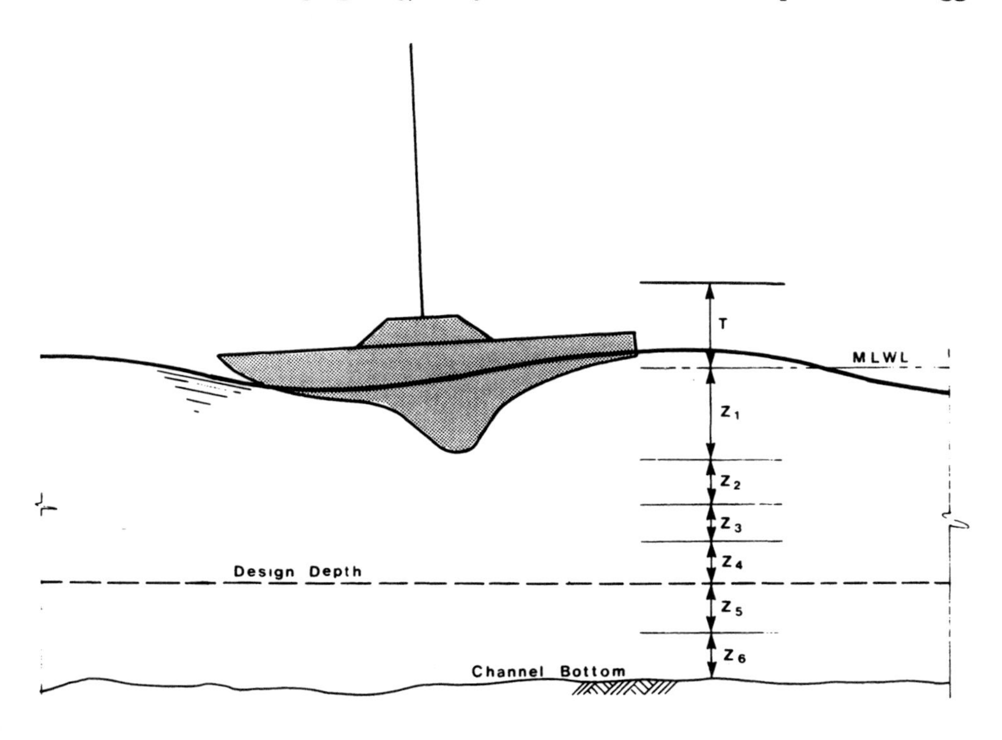

Figure 6–12. Depth of entrance channel. MLWL—minimum low water level, T—tide, Z_1—draft and trim, Z_2—wave allowance, Z_3—squat, Z_4—safety clearance, Z_5—advance maintenance dredging, Z_6—dredging tolerance.

following steps to be taken in computing the value of squat.

1. Determine blockage ratio of the submerged cross-section of the boat to the channel cross-section from

$$s = A_s/WH_c \qquad (6\text{–}2)$$

where

A_s = the craft submerged cross-section in square feet

W = the average width of the channel in feet

H_c = the depth of the channel in feet.

A semiconfined channel, that is, one in which the top of the dredged channel side slope is underwater, is assumed to have the same cross-section as a confined channel. This assumption will produce conservative results.

2. Determine the Froude number (F) from Eq. (6–3)

$$F = V_s/(gH_c)^{0.5} \qquad (6\text{–}3)$$

where

V_s = craft speed in feet per second

g = gravity acceleration = 32.2 ft²/s.

3. Apply the calculated values of s and F to Figure 6–13 to obtain d, a dimensionless squat.

4. Using the d value, compute squat Z_3 from

$$Z_3 = dH_c. \qquad (6\text{–}4)$$

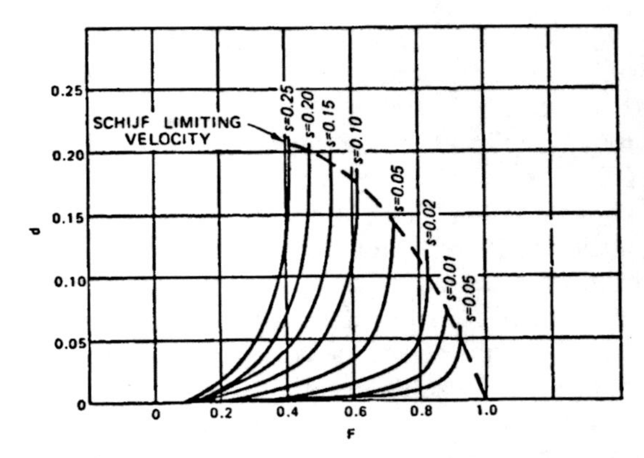

Figure 6–13. Dimensionless squat as a function of Froude number. (From Committee on Tidal Hydrualics, Report 3, 1965.)

It must be considered that in a two-way channel the value of squat will be greater, because the total blockage ratio for more than one passing boat is larger. In unconfined waterways and in open bodies of water, squat is much less than in confined channels, because the submerged cross-section of the vessel becomes a very small percentage of the waterway cross-section.

The channel depth increase for wave action (Z_2) is generally taken as equal to one-half the design wave height for small recreational craft. Boat motion can be determined by prototype observation and, if economically justifiable, by physical models or vessel-simulator mathematical models. The largest value of wave allowance obtained is to be used in channel design. Safety clearance (Z_4) depends on bottom characteristics. Usually a value of 0.5 to 0.6 is used for channels with soft bottoms, and 0.8 to 0.9 for channels with hard bottoms, that is, rock or coral. In the latter case, the additional allowance is to compensate for the greater damage expected if a boat strikes a hard bottom.

Allowance for dredging tolerance (Z_6) is always recongized as a practical need to cover possible inaccurances in channel dredging. Depending on site conditions, the value of dredging tolerance usually ranges from 0.25 to 1.0 m. Dredging tolerance is not considered in determining the theoretical depth of the entrance channel; however, it is indicated in contract specifications.

Finally, advanced maintenance (Z_5) is another factor to be considered. In channels where shoaling is continuous, overdredging is used as a means of prevention dredging to provide reliable channel depth over a longer period of time. Advanced maintenance consists of dredging the channel deeper to prevent the accumulation and storage of sediments. The value of advanced dredging depends on the estimated rate of channel shoaling. Several depths are usually considered to optimize advanced maintenance dredging.

6.5.3 Fairways

The fairways as a part of marina layout must be designed to provide as much room as necessary to allow safe boat maneuvering under existing environmental parameters. In practice the actual sizing of a marina fairway is based on a certain rule of thumb according to which the fairways width is equal to the design boat length multiplied by a certain coefficient. In the past the coefficient was used as equal to 1.25. At present the general rule of thumb most frequently used assumes that the clear distance between boat extremities located on both sides of the fairway should be taken as equal to 1.5 times the longest boat length; California State Department of Boating and Waterways Guidelines for Small Craft Harbor Design (1984) recommended that clear width of fairway, which is defined as the space between oppositely located berth slips, should be equal to 1.75 times the length of the longest boat; PIANC (1991) suggested that the minimum clearance between adjacent berths should be 1.5 to (1.75 to 2.0) times the overall length of the design boat. In general the greater the distance, the better condition for boat maneuvering and better accommodation for potential errors in maneuvering. In locations where fairway width is limited the finger piers may be installed at an angle to the access pier, thus providing for wider fairway and making it possible to accommodate longer boats. The latter berth layout assumes that the boat is expected to enter and leave the basin in the same direction and therefore the boat will not need to turn around before heading into or out of the berth.

6.5.4 Turning Basin

A turning basin is generally provided to allow boats to change direction without having to back up for long distances. The size and depth of the basin depend on the size, draft, and maneuverability of boats using the basin. It should be large enough to allow the turning of small recreational craft without backing up. This figure can be obtained based on practical experience. Boats that might have an unusually large size for the marina in question may need to maneuver forward and reverse several times in order to turn. For average conditions, a width of water area for turning and entering and leaving slips equal to 2 to 2.5 times the length of the longest boat is usually considered reasonable. If a large number of single-screw boats is contemplated, this width is used as equal to 2.5 boat lengths. There are occasions when the waterfront is very limited or a predominance of twin-screw boats is ensured, in which case a water width of only two boat lengths would be required and acceptable. Turning basins located in close proximity to boat ramps may require additional space to allow for waiting areas for several boats while the ramp is occupied.

PIANC (1991) suggested that in the case of particular environments and where navigation continues all year round a minimum diameter of 50 m can be considered as reasonable.

The depth of water in a turning basin must be consistent with the entrance channel depth. Allowances for squat, wave action, and safety clearance must be taken into account. A squat value of about 0.2 m is usually adequate.

6.5.5 Berth Areas

When designing marinas in general and berths in particular, it is important to know as closely as possible the number and size of craft that will use them. It will then be possible to arrange berths economically, both for depth and length. This is important, as the optimum development of the water area is essential if the marina is to be financially successful. At the present time there is a great variety of small craft in use, and reliable statistics on which to base an estimate are generally not available. A small craft classification proposed by PIANC (1965) is shown in Table 6–2. At the present time, the growing popularity of multihull boats such as trimarans and catamarans is quite obvious. These boats are typically 4.5 to 12.0 m long and 0.6 to 1.5 m deep. They are designed either for small sporting cruises (those up to 8 m long) or for ocean navigation (those reaching 10 to 12 m in length). The publicity brought about by the great sponsored races is no doubt partly responsible for this development.

The total area available to the marina development often places a restriction on the number of boats that the marina can accommodate, as well as on the size and scope of the ancillary activity it can support. Small-craft basins vary in size from a handful of boats to those catering to several thousand. For example, in the United States there are marinas designed to accommodate more than 5000 craft. The most recent example of facilities of this magnitude is one built on Lake Michigan offering 1500 slips to boaters (Civil Engineering, November, 1990). However, unless there are special circumstances, it is recommended that marinas should have between 300 and 1200 berths (PIANC, 1979). Below 300, capital funding and running costs are likely to be prohibitive. It is unlikely that a harbor needing extensive breakwater protection will be viable unless it is constructed for at least 500 craft. More than 1200 craft massed together tends to lead to a marina losing its individuality.

The criteria applying to the design of the approach channel and turning basin generally apply also to the berth area. In addition, the effect of bottom depth on the structures of the berthing system, that is, bottom fixed, floating, or anchorage by anchor lines to the bottom, should be considered. Because the cost of the structure increases with the depth of the water, it is customary to berth

Table 6–2. Classification of Pleasure Boats as Proposed by the Permanent International Association of Navigation Congresses (PIANC) 1965

Class	Overall Length L (m)	Subclasses
I	$L < 5$	Motor boats Sailing yachts Motor/Sailing
II	$5 < L < 8$	Motor boats with living quarters Motor boats without living quarters Sailing yachts with living quarters Sailing yachts without living quarters Motor/sailing with living quarters Motor/sailing without living quarters
III	$8 < L < 15$	Motor boats Sailing yachts Motor/sailing
IV	$L > 15$	Motor boats Sailing yachts Motor/sailing

the larger and deeper-hulled craft near the entrance and to decrease the berth area depth in incremental steps away from the entrance. Thus, a narrower fairway and smaller turning area will be required for smaller craft. Hence the amount of water area required for the berth area will be decreased. Also, the shallower back berth area reduces pier and slip construction costs. The depth of the berth area can be determined by using the same approach as for determining the depth of the turning basin.

6.5.6 Berth System

The berth system is designed to accommodate the craft and keep it safe while not in use, as well as to provide access to and from shore. There are two basic methods to hold a boat. The first, known as anchor moorage, is to moor a boat (or cluster of boats) by anchorage to the bottom in such a way that it cannot strike another moored boat or other nearby fixed object while swinging within the confines of its mooring line or lines under all designed weather conditions (Figures 6–14 and 6–15). The second method is to moor the boat against a fixed or floating dock which can be a single pier, marginal wharf, or outboard end of a multiple-slip dock (Figures 6–16 to 6–18).

Typically the method of boat handling is to be decided by the designer in close cooperation with the marina developer. The type of moorage, as well as the berth system to be used at a specific site, should be selected on the basis of mooring requirements, layout compatibility, installation requirements, structural adequacy, maintainability, and serviceability during normal and extreme climatic events. Furthermore, marina aesthetics, life expectancy, and last but not least economics must be considered.

6.5.6.1 Layout

Figure 6–19 shows some of the most frequently used types of layouts for permanent berth installations. The merits and disadvantages of these are presented in Table 6–3. The Type C arrangement is the most popular in small-craft marina construction. It is generally preferred for its economy of space, convenience of use, and ease of boarding. Some details on Type C general arrangement, finger piers, layout of mooring lines with one and two boats docked, and recommended clearances are shown in Figure 6–20 and presented in Table 6–4. Furthermore, when determining the clear width between finger piers $b = (w - 1.2)$ an additional 0.6 to 1.2 m should be added to each boat beam for proper clearance and fendering. Typical relationships between boat length and its beam are depicted in Figure 6–21. The amount of clearance provided depends on site exposure to wind, waves, currents, and on related local experience. It also depends on water area available for the marina development. However, it must be realized that having boats moored too close may cause operational problems resulting in both structure and boat damages. If practical a minimum space between boats of 1.0 m is recommended. Additional information on the relationship between the boat length and the clear space between adjacent finger piers may be obtained from charts in Figure 6–22. The latter are based on information provided by Tobiasson and Kollmeyer (1991).

PIANC (1991) suggested the minimum lateral clearance of a boat during mooring operations as follows: 0.5 m for boats smaller than 7.5 m, 0.75 m for boats between 7.5 and 12.0 m, and 1.0 m for boats exceeding 12.0 m.

It must be mentioned that it is prudent to configure a berthing layout with a variety of clear widths between adjacent finger piers. The latter provides the marina operator with some degree of flexibility in assigning

(a)

(b)

Figure 6–14. Anchor moorage. (a) Escoridido Marina (San Diego, California). (b) Small Craft Moorage, Singapore Harbor.

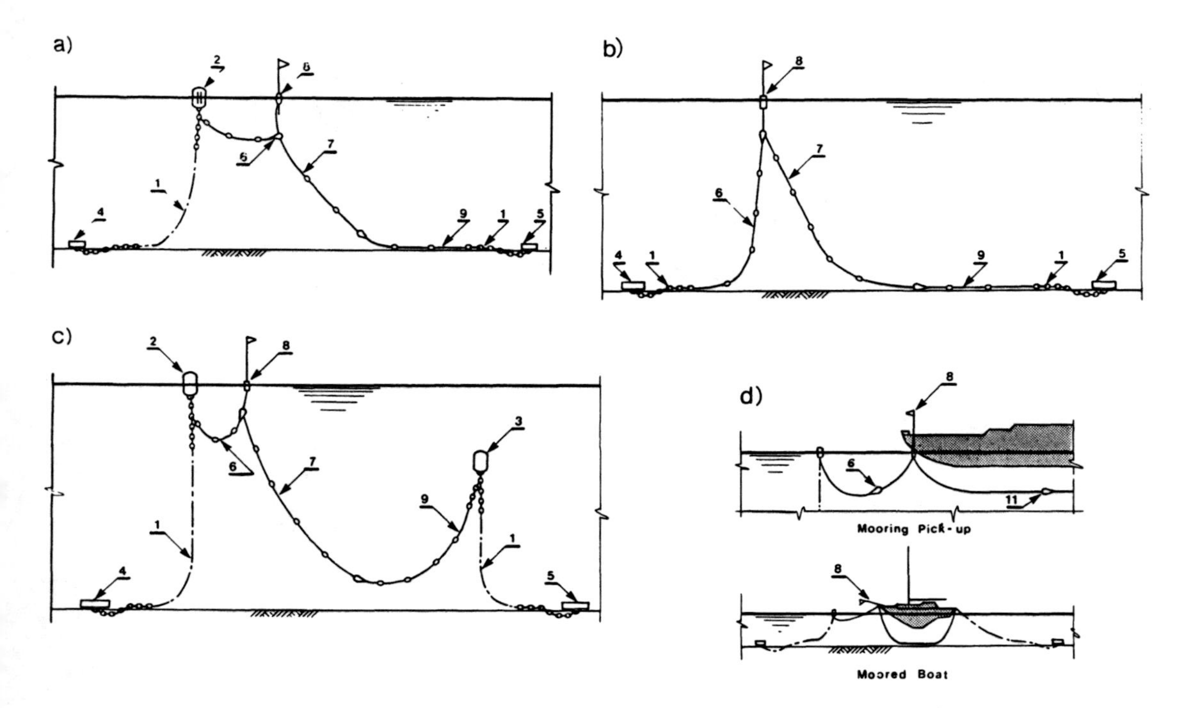

Figure 6–15. Two-point moorages at Catalina Point, Santa Catalina Island. (From Nichol et al., 1986.) (a) Typical mooring. (b) Typical shallow water mooring. (c) Typical deep water mooring. (d) Securing of boat to two-point mooring. 1—Chain, 2—mooring buoy, 3—underwater stern buoy, 4—bow mooring weight, 5—stern mooring weight, 6—bow hawser, 7—spreader line, 8—pickup pole assembly, 9—stern hawser.

Figure 6–16. Bottom fixed marginal wharf.

Figure 6–17. Floating berths.

slips that are most appropriate for the boats under consideration.

Where multihull boats are to be served in a conventional marina, these boats can be docked at a specially provided boating area, designed to serve this type of vessel, or alternatively they can be moored alongside marginal docks or piers without projected finger piers; these docks and piers may also be used by varying beam boats.

If multihull boats represent a significant percentage of the boats using the marina then extra wide slips may be included into facility layout. It should be pointed out that unless specifically designed hauling equipment is available the multihulls are difficult to haul out. Because of their light weight the multihull boats are wind sensitive. Therefore, a reliable mooring and fender system must be provided at multihull berths.

Naturally, the number of berths per hectare that may be accommodated in a basin varies according to the layout arrangement, as well as the size of craft. An average small-craft basin normally accommodates anything from 75 to 125 boats per hectare. Structurally, berth structures can be either bottom fixed, floating, or a combination of both. The latter usually consists of fixed main piers and floating finger piers.

6.5.6.2 Anchor Moorage

Anchor moorage is seldom used in permanent marina facilities. This is because this method is wasteful of sheltered water space even with two-point (bow and stern) moorings, and moreover it requires the use of service boats for access. It is recommended for temporary installations pending construction of permanent berthing facilities, for refuge harbors, and for destination harbors for pleasure boats where wave height could reach 1.5 m or more. The advantages of anchor moorage are as follows:

1. They do not use slip structures.
2. In most cases, they do not require expensive all-weather coastal protection structures.

(a)

(b)

Figure 6–18. Combination of fixed and floating structures. (a) Niagara-on-the-Lake, Ontario. (b) San Diego, California.

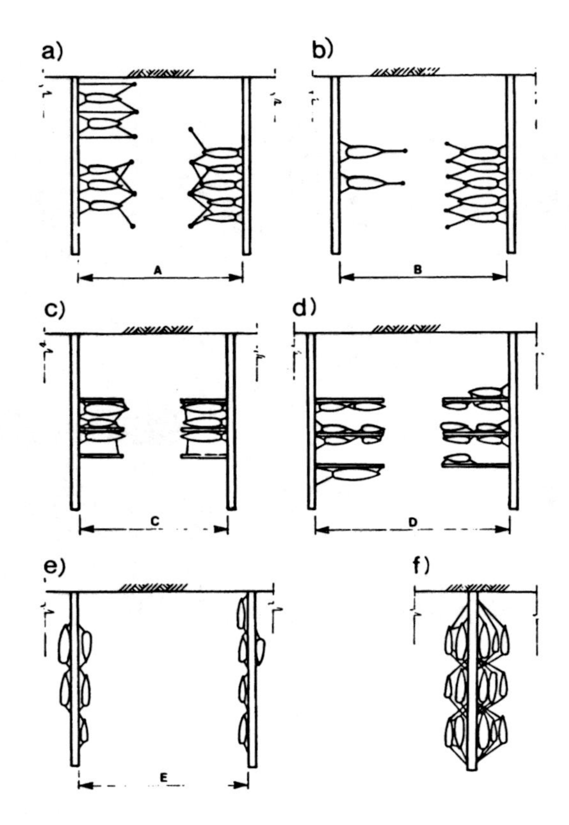

(g)

Figure 6–19. Typical layout of permanent berth installations. *Note*: Layout (e) (Small craft harbor at Hong Kong) is similar to schemes (a) and (b). In former a boat is moored at the pier located mooring accessories and at offshore piles. Note round fenders at anchor piles.

Table 6.3. Berth Layouts (Figure 6–19)

Ref.	Type of Mooring	Advantages	Disadvantages	Remarks
a	Stern to pier, bow to anchor piles	Economy	Not as convenient for operation as alongside berth	
b	Stern to pier, bow to anchors or buoys	Economy	Not suitable with large tide range. Danger of propellers being entangled in head warps	Suitable for large yachts in basins with little tide range where the gangways can be attached to stern.
c	Alongside finger piers or catwalks, one craft on each side of finger	Convenient for operation		
d	Alongside finger pier with more than one craft on each side of each finger	Convenient for operation, allows for flexibility in accommodating crafts of different lengths	Fingers must be spaced wider apart than in Type C layout. This, however, may be compensated for by larger number of craft between piers.	Fingers may be long enough for two or three craft. If more than three, then provision should be made for turning at the root of the berth.
e	Alongside quays or piers	Convenient for operation, allows for flexibility in accommodating craft of different lengths		
f	Alongside quays or piers. Up to 3–4 craft at rest	Economy	Crew from outer craft must climb over inner berthed craft.	

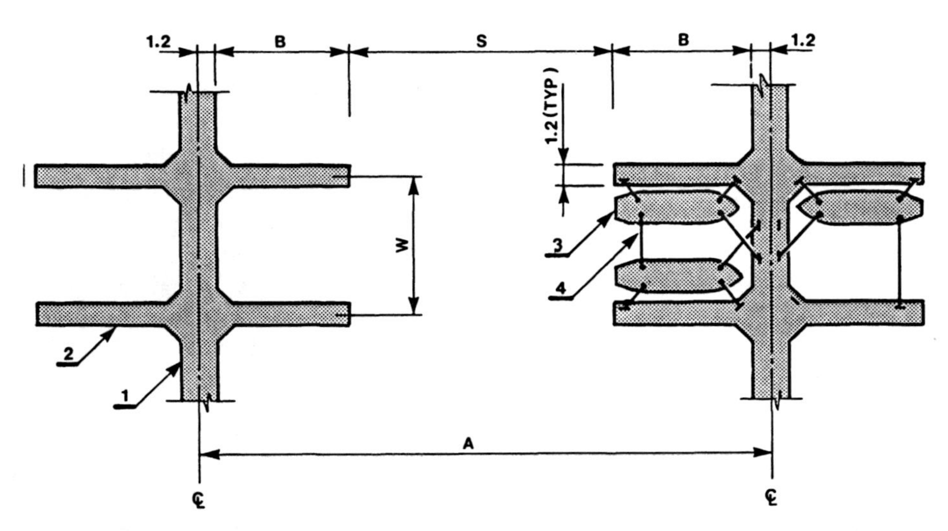

Figure 6–20. Typical pier (Type C) arrangement. 1—Main pier, 2–finger pier, 3–craft, 4—mooring line.

Table 6–4. Recommended Sizes of a Typical Finger Pier Arrangement (Layout Type 'C', Figure 6–20)

Length of Berth (B) (m)	Width (W) (m)	Slip (S) (m)	Center to Center of Piers (A) (m)
6	6	10.5	24.9
9	8.5	17	37.4
12	9.5	21	47.4
15	11	30	62.4
18	12.5	36	74.4
21	14.5	42	84.4
24	17	48	98.4

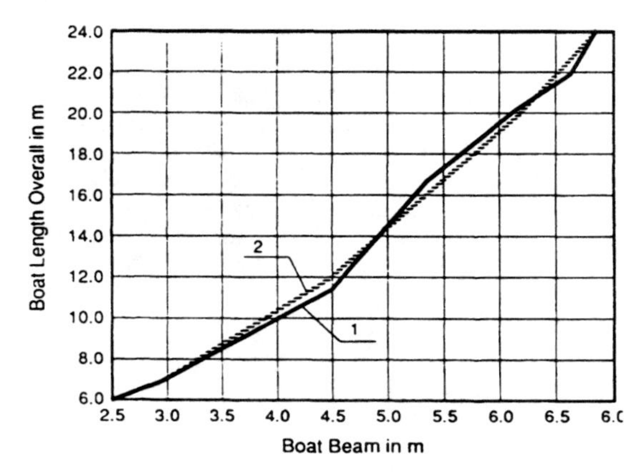

Figure 6–21. Motorboat beam versus length. 1—Based on manufacturer's product (1989), 2—average.

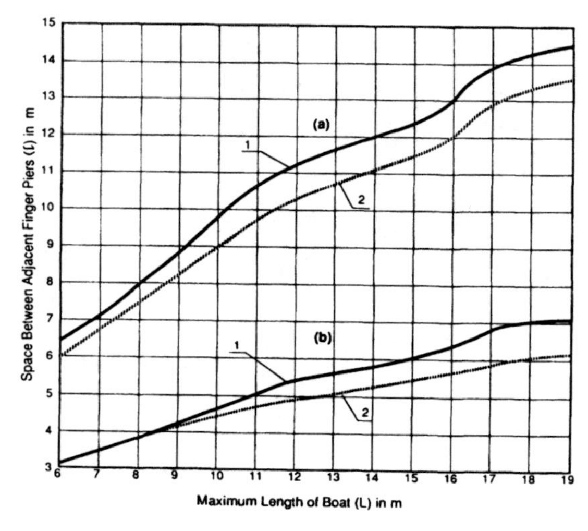

Figure 6–22. Recommended (1) and minimum (2) space between adjacent finger piers as a function of maximum boat length based on 1989 manufacturer's data. (a) Two boats per slip. (b) One boat per slip.

3. The cost of the mooring system components and installation are low because these components are readily available or relatively inexpensive to fabricate.
4. The system is easily inspected and maintained.

The mooring configuration (layout) is essentially controlled by the geometry of the basin, water depth, predominant direction of critical waves and winds, and spacing requirements for the one- or two-point moorings. The two-point mooring, which is basically a bow and stern mooring, is used more frequently for the following reasons:

1. It provides a much higher mooring density than is possible with a single-point mooring system, thus allowing for the most efficient use of available mooring space.
2. It provides for better potential survivability under storm conditions.

Successful installations of anchor moorings at the Catalina Point (Santa Catalina Island) and Isthmus Cove destination harbors are discussed by Nichol et al. (1986). Santa Catalina Island is located approximately 40 km (25 miles) south of the Los Angeles–Long Beach Shoreline area. The predominant wind direction there is from the northwest, although storms can induce strong winds from all directions. The wind can produce waves of over 1.8 m in height and directly impinge on certain mooring areas.

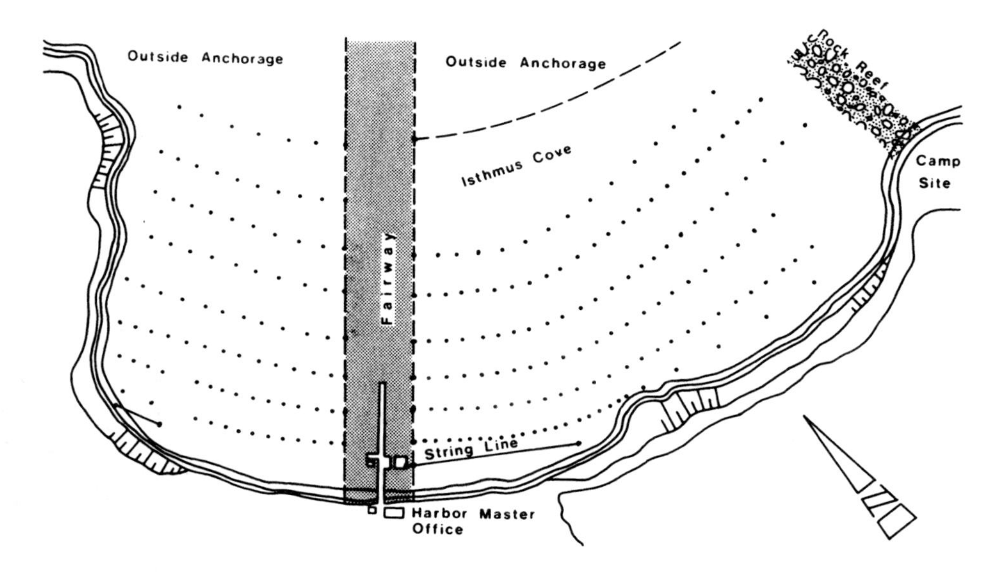

Figure 6–23. Santa Catalina Island Isthmus Cove mooring layout. (From Nichol et al., 1986.)

The mooring system that has been installed at the above locations consists of bow and stern weights, bottom chains and hawsers, a spread line, a pickup pole, and up to two buoys (Figure 6–15). This arrangement is altered somewhat depending on the water depth and length of the boat and is described as the regular type (Figure 6–15a), the skiff/dinghy type (Figure 6–15b), and the deepwater type (Figure 6–15c).

The regular type of mooring is installed to serve boats shorter than 12 m and for depths of water between 3 and 12 m. Skiff/dinghy mooring is used for lightweight skiffs and small ski boats generally under 6 m long, and is installed in the shallower waters. Deepwater mooring is used in water depths of 12 to 25 m.

The size of all components varies according to different loading conditions existing at the specific locations. Mooring buoys are used with regular and deepwater mooring types to offset the weight of the mooring chains and to allow simple hand mooring. Mooring buoys are not required for skiff/dingy mooring because of the limited length of chains used in the mooring lines. Deepwater mooring has a submerged stern buoy to allow the boater to pull the stern hawser to the stern cleat without having to lift the entire weight of the chain. Figure 6–15d shows how boats are secured to a two-point mooring. The average mooring density at small-craft marinas at Catalina Island for an average 12-m water depth is about 20 boats, with a length of 9 to 12 m per hectare of water area. The layout at Isthmus Cove marina is shown in Figure 6–19. The anchor moorages at Escoridido, California and Singapore are shown in Figure 6–14. A typical layout of an anchor moorage system consists of rows of individual moorings running roughly parallel to the shore at the marina landside. Smaller craft are usually placed closer to the shore in the shallower depth whereas larger boats are moored in deeper water. The result of this procedure is increased spacing between moorings and generally wider fairways in deeper water.

6.5.6.3 Fixed Piers

Usually fixed piers are less expensive than floating piers, but are typically limited to marinas in which water is not deeper than approximately 5 to 6 m and where the water surface does not fluctuate more than about 1.0 m. If the surface of the water fluctuates more than 1.0 m, then special adjustments for the structure should be provided to ease boarding during low water levels. In general PIANC (1991) recommended the maximum pier deck elevation above design water level as follows: 0.8 m for boats smaller than 7.5 m, 1.2 m for boats between 7.5 and 12.0 m long, and 1.5 m for boats longer than 12.0 m.

At deepwater locations the cost of bottom-fixed piers is usually prohibitive. A fixed pier by definition is a structure resting on or embedded into the foundation material. It is generally a structure that offers a significant resistance to vertical and horizontal forces imposed upon it. The pier structure depends on bottom subsoil conditions, for example, high bedrock formation will preclude use of piles. In the latter case the pier deck may be resting on gravity-type structures such as timber or concrete cribs, floating-in caissons, and other similar structures or supported on socketed-in columns. In most cases construction of the aforementioned structures will be cost prohibitive. On the other hand soft and loose subsoil stratas may inhibit piled foundations, unless the piles are long enough to reach a competent soil layer; under these bottom soil conditions fixed-type structures may also be cost prohibitive. The most economical bottom-fixed piled structure may be expected where competent foundation soil is present, water depth is not excessive, and ice conditions are such that they do not preclude structure serviceability.

Backfilled structures of miscellaneous constructions may also be used as bottom-fixed docking structures. For economy in the marina basin layout and to save construction costs, piers must not be wider than required for safe pedestrian traffic. This usually calls for main piers 1.2 to 2.4 m wide. The latter allows for two hand carts to pass in opposite directions. Finger piers need to be no wider than 0.8 to 1.0 m. All kinds of structural materials such as pressure-treated timber, concrete, and steel are used in fixed berth construction. These materials must comply with updated local standards. Timber is the most popular material in small-craft marinas. Timber piles used for marina construction should preferably be pressure-treated with creosote or creosote coal-tar. Quality pressure treatment extends the life of the structure, which more than justifies the cost of timber impregnation. All light ferrous hardware, such as nails, bolts, etc., should be hot-dipped galvanized. To increase life span of the structure the engineer should consider a design that precludes, or at least minimizes, deterioration of main structural components, for example, corrosion of fasteners and rot of piles and deck structure. For this an adequate surface water runoff should be provided and the tops of piles should be capped. The designer should be aware that rot in timber and corrosion of fasteners is usually prevalent around mooring hardware. If properly designed and maintained the design life of a timber pier structure may be 25 years and beyond. A typical timber pier construction is shown in Figures 6–24 and 6–25.

In areas where timber is scarce or costly, a reinforced concrete construction is frequently used. A typical solution for concrete piers made from prefabricated components is shown in Figure 6–26. Both regular reinforced concrete and prestressed concrete elements are used for pier construction. Piers constructed from metal component and steel piles are generally too costly and therefore seldom used for marina construction. Normally steel components should be protected from corrosion either by hot-dipped galvanization, or by applying different kinds of protective coatings. In addition, cathodic protection can be used for metal protection against corrosion. Details on corrosion protection for steel are given in

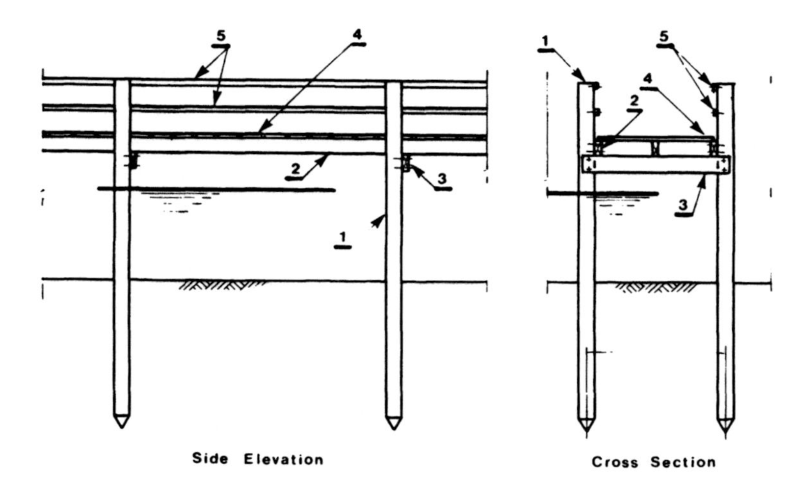

Figure 6–24. Typical bottom-fixed timber pier. 1—pile, 2—stringer, 3—cross beam, 4—decking, 5—handrails.

Figure 6–25. Timber built pier for small crafts. (Florida.)

Chapter 1. Well-maintained steel- or concrete-built piers may exhibit a life span of over 50 years.

In all structural designs, the dock or loading platform elevations should be approximately 0.3 to 0.5 m above the design high water level. If the craft can directly approach and hit the pier, then this pier must be equipped with a fender system extending to at least 0.3 m above the design low-water level.

Utility lines are usually kept below the deck, where they are protected and readily available for inspection and maintenance.

Fixed pier structures are usually designed to sustain all applicable environmental forces, that is, wind, wave, current, ice, as well as design dead and live loads that can

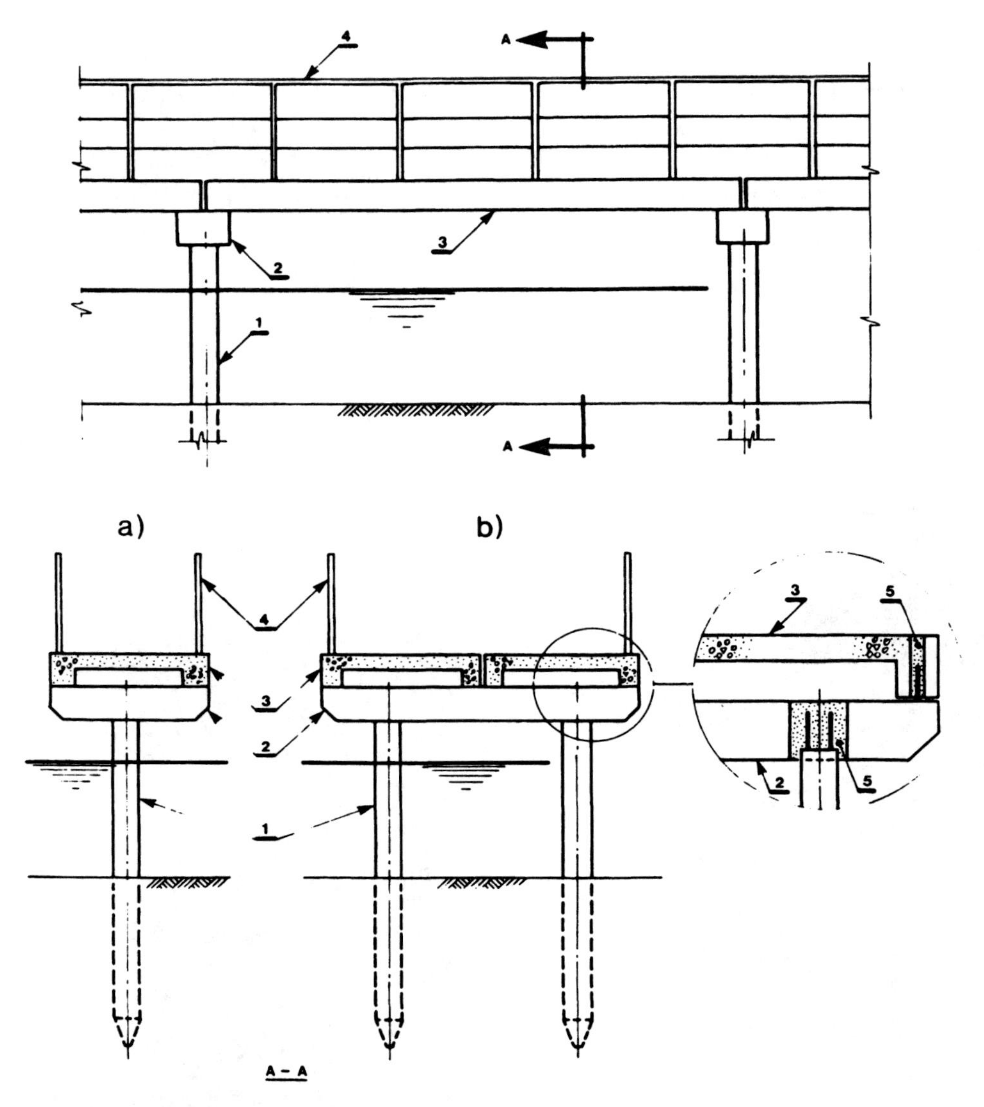

Figure 6-26. Typical concrete pier. (a) Narrow main pier or finger pier. (b) Wide main pier. 1—pile, 2—precast seat, 3—precast span, 4—guardrail, 5—grout.

be assumed as uniformly distributed loads used as prescribed by local codes and regulations, but in no case should be smaller than 2.5 kPa for access piers and 1.5 kPa for finger piers. Furthermore, the structure should be reviewed for a minimum concentrated load of 2 kN applied at any location of the dock structure. Docks that are accessible for vehicular transport should be designed for the appropriate load produced by the heaviest vehicles that can operate there. Although difficult to estimate, craft impact and mooring forces should also be considered. In these calculations the boat approach speed must not be considered less than 30 to 50 cm/s and an approach angle 10° to 20° may be assumed as practical. Where the boat is moored alongside the marginal berth rather than finger pier an approach angle of 45° might be considered as more representative to compute impact force. For preliminary calculation of impact load the boat weight can be used as represented in Figure 6–27.

In cold climates, ice is the largest problem to deal with in fixed berth marinas. Small-craft marinas are customarily built in protected areas, away from moving ice masses. Therefore, from a structural design standpoint, the ice dynamic forces produced by relatively small wind-driven ice floes, as well as by static forces, both vertical and horizontal, should be considered. Ice that forms around piles can apply substantial buoyant and down loads when water levels fluctuate. These piles therefore can fail either in tension or compression. Stationary ice in protected marinas exerts horizontal forces on piling and on other bottom fixed structures. In the United States and Canada, depending on location, these forces are assumed to be between about 700 kN/m^2 and 2800 kN/m^2 (Wortley, 1979). Where a stable ice cover exists, it will respond thermally to changes in ambient temperature. When an ice sheet warms up, expansion will shove pilings and other bottom fixed structures about. It is believed, however, that ice thermal expansion forces against relatively flexible piers in small-craft marinas are less than mooring forces for which the piers are usually designed. However, thermal ice forces against relatively stiff structures such as bulkheads or any type of gravity-type retaining walls could be substantial and these structures must therefore be designed accordingly. Depending on location, values for ice thermal expansion forces in North America are usually taken as equal to 150 kN/m to 300 kN/m. The importance of the structure to the overall project is also a factor in selecting the design value of ice pressure.

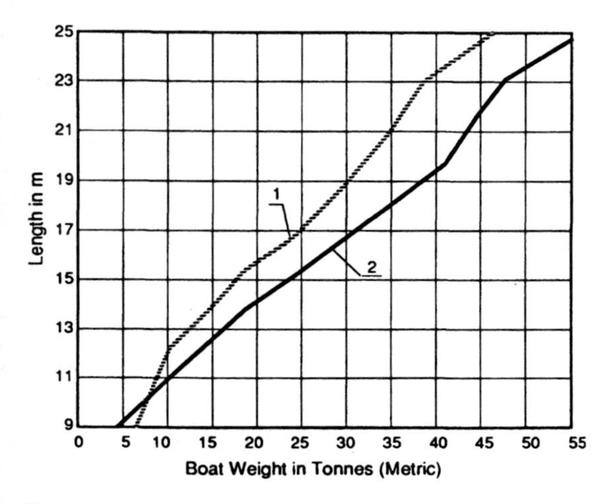

Figure 6–27. Average weight of recreational boats with half fuel and water on board but without owner's accessories, based on manufacturer's literature. 1—Sailboats, 2—motor boats.

Various methods are used for ice control (Tsinker, 1986a). For small-craft marinas, bubblers, ice booms, or warm water from industrial waste can be used for this purpose. Bubblers are systems that discharge air at some depth, usually the basin bottom. As the air bubbles rise to the surface, they entrain the warmer water trapped at the bottom during surface freezing. This may prevent surface water from freezing or inhibit additional ice formation. Booms are usually installed to retain moving ice. They can be composed of large timbers, small pontoons, or oil drums connected to wire ropes and anchored across an area to be protected from moving ice. Warm water, if available, is an effective method of ice control. It is particularly efficient when used in combination with the bubbler system. A point to consider by those designing for cold regions is that cold air on top of the warmer water creates fog. Hence, a thin layer of ice through which boats can move is often preferable to completely open water. For a detailed discussion on ice static and dynamic forces acting on marine structures and ice control methods, see Chapter 2.

6.5.6.4 Floating Piers

Where the water level fluctuates between 0.5 and 1.5 m, a floating pier system is an optional solution; however, at locations where the tidal range exceeds 1.5 m, the use of floating piers is almost unavoidable; despite the cost, the floating system, particularly at locations with a tidal range of more than 1.5 m, provides better moorage for the craft and is more convenient and safer for boating and leaving boats during extremely low water levels. Floating dock systems are often used in modern marinas.

Nowadays the industry that specializes in the manufacture of floating dock systems worldwide has developed a high-strength, low-maintenance, and cost-effective product. Today, a great variety of floating dock systems are available. All of them are designed to suit a certain condition and only the right one, suitable for the specific site, must be used to meet the specific site design criteria.

Materials

Modern floating dock systems offer a wide range of material types that include solid and glued-laminated wood, steel, aluminium, plain and reinforced concrete, fiberglass, and other plastic materials. Each material and each type of pier design has its own merits and disadvantages; the choice of one or the other must be made considering specific site factors, intended project use, projected life span, and cost considerations.

Wood in its various forms is a traditional material used in marina dock construction. If properly designed and treated with chemicals dock components made of wood are durable, easy to repair, and offer a reasonably long life expectancy. Some exotic types of wood, for example, ekki (or "zobe" in French) imported from Africa, although expensive, do however offer extraordinary potentials for floating docks.

Ekki is a very dense and heavy (about 1300 kg/m^3), even-grained, free of knots, and extremely strong kind of wood; its bending, tension, and shear stresses correspondingly are 25.5, 23.4, and 2.04 MPa. For more details on ekki see Tsinker (1986b).

Disadvantages of natural species of wood include potential for rot and mechanical abrasion, as well as a tendency to splinter, warp, and otherwise distort. These disadvantages are less pronounced in dock components fabricated from glued-laminated wood.

Steel is another material most frequently used for fabrication of floating docks. It is readily available in a variety of plates and sections, and is easy for cutting, shaping, welding, and bolting. As a structural material steel offers great flexibility in dock structural design. As mentioned earlier, corrosion is perhaps the only drawback to the use of steel in marinas and especially in saltwater environments. Therefore different kinds of protection from corrosion must be used as discussed earlier. Steel is also an excellent heat and electrical conductor and therefore where appropriate a dock structure made of steel must be properly grounded, and where required insulated.

Concrete is a relatively inexpensive, strong, and durable material that is highly resistant to effects of the marine environment. It is labor intensive; however, relatively unskilled labor is required to form dock components. Stringent control of proportioning of concrete aggregates, water/cement ratio, and placement and curing procedures is a key to successful functioning of docks comprised of concrete components. Crack control is a very important issue in concrete structural design; in most cases, width of cracks must be limited to certain values that prevent corrosion of reinforcement, and otherwise to not permit water to get frozen within cracks. It is also useful to consider epoxy-coated rebars for fabrication of floating dock components such as floats. Fiber-reinforced and lightweight concretes are gaining acceptance as floating dock construction material.

Aluminum is a high strength, lightweight versatile structural material. In recent years aluminum has received wide acceptance for fabrication of some major components (basically docking elements) of the floating dock systems.

It is readily available in a variety of rolled and extruded shapes. Certain alloys of aluminum are relatively resistant to corrosion. In general aluminum is left unpainted. If not properly designed aluminum may be subject to fatigue and cracking. Similarly to steel, aluminum is an excellent heat and electrical conductor and therefore docks made of aluminum must be properly isolated from electrical devices installed on them, and grounded; where required an insulation should be provided.

Fiberglass is extensively (and successfully) used for fabrication of floats for floating docks. It is a light, strong, durable, and easy to repair material. Floats made of fiberglass can be easily incorporated into the pier decking system. Fiberglass may be collared and formed in a variety of shapes and sizes to meet specific needs. It can also be used for fabrication of deck systems.

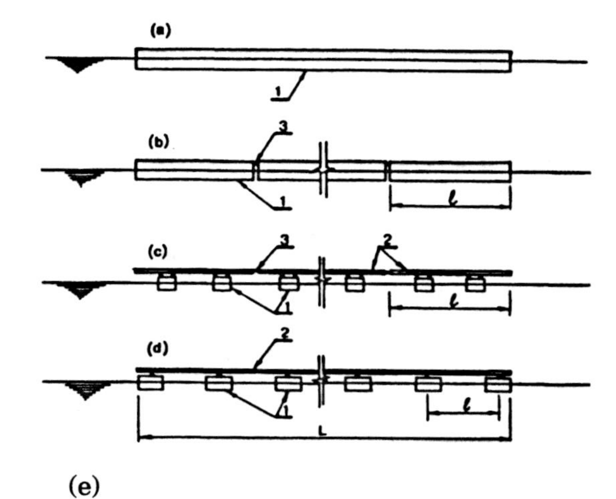

Figure 6–28. Typical structural arrangements of a floating pier. (a) Single pontoon. (b). Several pontoons joined by pivots. (c) Small pontoons (floats) bridged by single-span decks. (d) Small pontoons (floats) spanned by a continuous deck. 1—Float (pontoon), 2—deck, 3—pivot. (e) Combination of schemes (c) and (d).

Floating Dock Structural Arrangements

There are four basic structural layouts that are commonly considered for construction of floating piers (Figure 6–28). They are as follows:

1. One long pontoon
2. Several pontoons joined by pivots at deck level
3. A series of small pontoons (floats) spanned by a number of single-span decks
4. A series of small pontoons (floats) spanned by a continuous deck.

Each of the above structural arrangements has its own merits and drawbacks, and again, is site specific. For example, scheme (1), which represents a single-unit pontoon usually of a robust steel or concrtete construction, would better suit for docking of relatively large crafts, such as large yachts, small commercial fishing boats, and similar, rather than small motor boats. Scheme (2) is a modified version of a scheme (1); it can be site-assembled from prefabricated pontoons trucked to the site. Schemes (3) and (4) are usually used in small-craft marinas. They differ from each other basically by deck structure and by sizes and number of floats required to support decking and associated live loads. A combination of both systems, for example, three or four floats in one section is also used. In most cases scheme (4) represents a more economical system. For details on floating pier systems the reader is referred to Tsinker (1986b).

Bertlin (1972) suggested that floating piers preferably should be no less than 2.5 m in width, although 1.8- to 2.0-m wide piers are frequently used. Although finger piers are narrower than main piers, they are usually of the same construction as main piers. Finger piers are connected with the main pier by means of simple hinges readily available at very reasonable cost. One of these hinges is seen in Figure 6–29. Floating pier systems should ride with the deck at about 0.5 m above the water surface without live loading to provide ease of boarding. The lower permissible limit to prevent wave overtopping is taken as equal to 0.25 to 0.3 m. Therefore the float's capacity to carry dead and full live loads must be designed accordingly. It is important that all exposed edges of the floating pier be equipped with fenders to protect the craft tied alongside. A typical rubber bumper is shown in Figure 6–30. Used rubber tires are often used as fenders. They have proved to be inexpensive and reliable for the protection of small boats and piers. The piers should be designed so that they will tilt less than 15° when the total design load is applied to half the width. It is possible to reduce the width of the pontoon in the case of Type C and D layouts (Figures 6-19 and 6–20) by making the finger piers (or part of them) act as an integral part of the main pier.

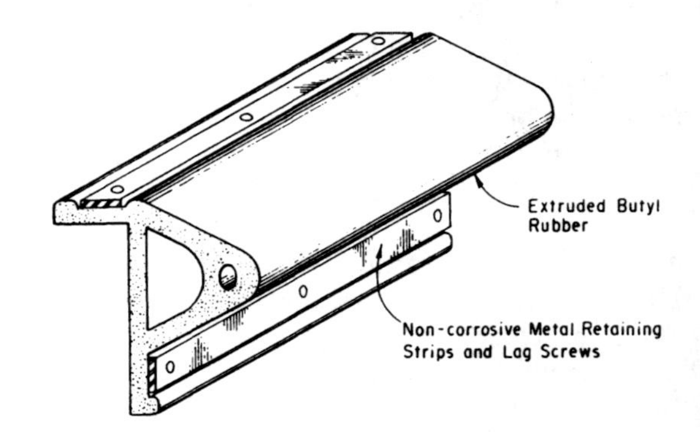

Figure 6–30. Typical extruded rubber fender. (From Dunham and Finn, 1974.)

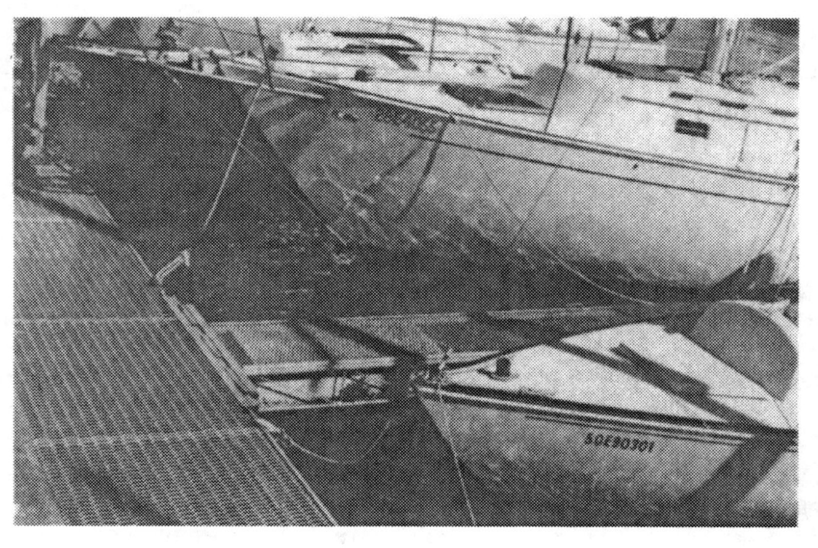

Figure 6–29. Aluminum grating used in floating pier decking system. (St. Catharines, Ontario.)

Floats

Numerous types of floats have been used in floating berth construction. Most of them are described by Chaney (1961), Dunham and Finn (1974), and by the American Society of Civil Engineers (1969). Among them are different kinds of metal drums, steel pipe floats, rectangular steel, concrete and plastic pontoons, styrofoam floats, and polyurethane floats (Figures 6–31, 6–32, and 6–33). One of the most common lightweight solids used for flotation is expanded close-cell polystyrene, with an average density of approximately 32 kg/m^3. This material is impervious to water. It should be pointed out that, according to the recent environmental legislation devices such as floats, buoys and similar that include polystyrene require the encapsulation of any submersible polystyrene foam (Obern and Hoy, 1992). Polyurethane is preferred over polystyrene by some marina designers because of its resistance to hydrocarbons and the ease with which it can be formed into a protective shell or molded into any shape. Its higher cost and greater susceptibility to oxidation, however, have limited its general use as a flotation material. It should always be covered with an oxidation-resistant material. Several other foams exist, but they seem to attract a rapid growth of marine plant and animal life in seawater. These aquatic growths are not considered too objectionable, except for their appearance and the danger of fouling propellers: most growths periodically die and drop off without damage to the float. Exposed foams, however, are subject to vandalism (because they can be cut easily with a knife), and to degradation by various marine organisms, fish, and birds. On the whole, foams are durable, for example, one styrofoam installation has been in existence in Los Angeles harbor for 16 years, with little sign of deterioration (Brunn, 1989).

The relative softness of the synthetic materials and their attraction for marine life can be alleviated by using exterior protection. An epoxy coating that bonds firmly and forms a tough, flexible surface has been used in some areas, and has been found to attract less marine life and be impervious to hydrocarbons. A better but most costly protection is sheeting with a fiberglass-reinforced polyester resin. Floats made from synthetic materials are very light.

One serious objection to lightweight floats is their responsiveness to wave action, which

Figure 6–32. Floating pier consisting of lightweight metal decking supported on corrugated tubes.

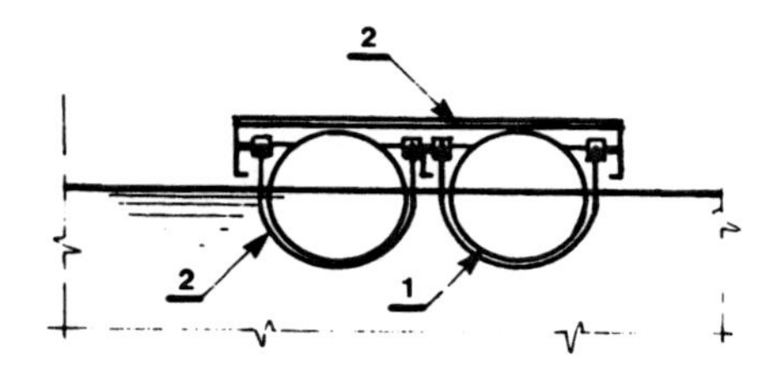

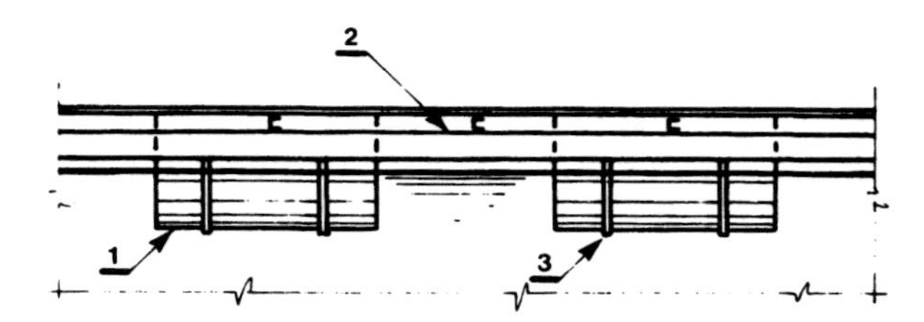

Figure 6–31. Floating pier. 1—Metal drum (alternatively, metal pipe), 2—framing, 3—U-bolt.

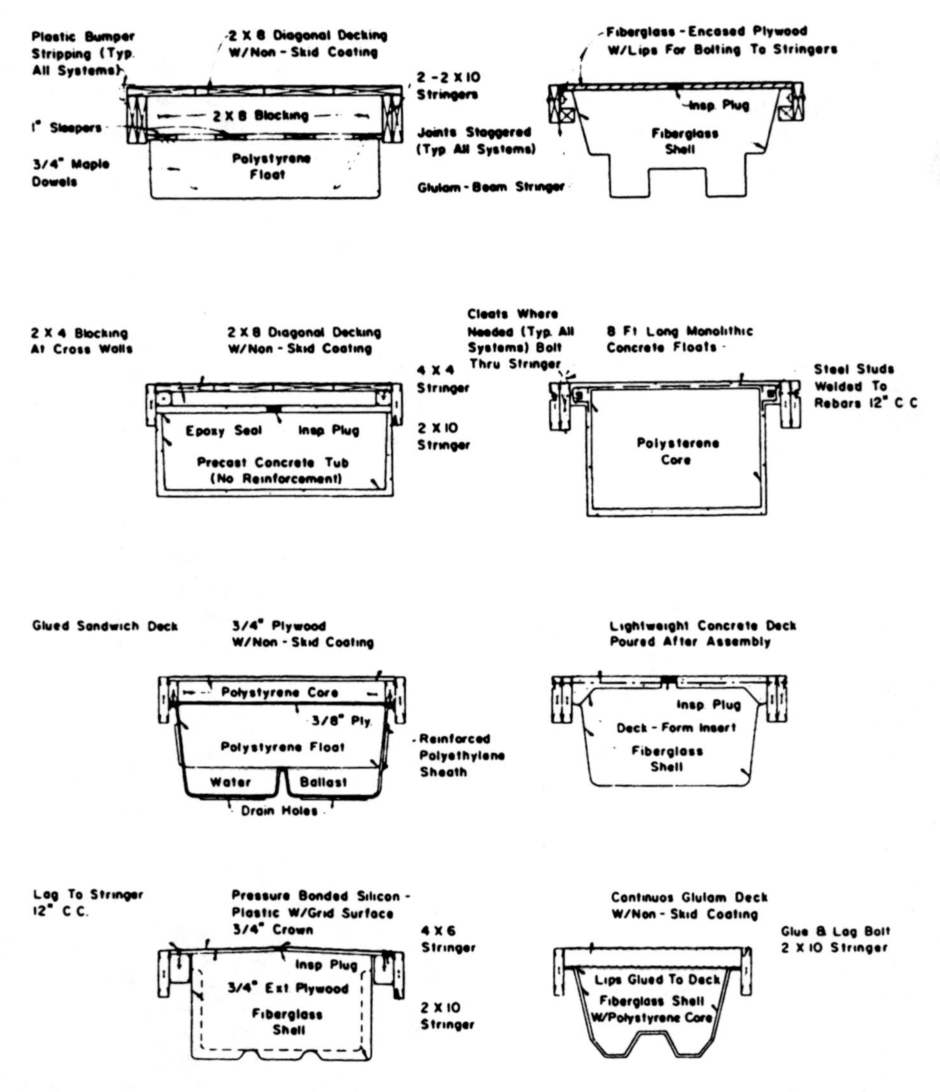

Figure 6–33. Various types of floats for floating piers. (From American Society of Civil Engineers, Report on Small-Craft Marinas, 1969.)

causes them to roll and pitch with the waves. Some manufacturers now offer to coat or ballast lightweight floats with concrete. This adds considerably to the dead load and calls for a proportionate increase in flotation material; however, concrete coating does provide an armored surface that may prolong the life of the pier.

A great variety of waterproof shells on the market are used for floats. All of them have one common characteristic: they can be ballasted with sand or water to achieve corrective levelling of the dock after installation. However, these shells are vulnerable to leakage and loss of buoyancy if the shell becomes permeable for any reason and therefore they must be provided with inspection plugs for periodic checking.

In general all shells can be classified as synthetic molded shells, fabricated metal floats (pontoons), and hollow concrete floats (pontoons).

The most commonly used synthetic-shell floats are those made of fiberglass-reinforced polyester resin. They are light, tough, strong, and unaffected by hydrocarbons, saltwater, or any of the common contaminants found in a small-craft harbor. However, the coatings, in which glass fibers and resins are sprayed into an external form and then rolled smooth, tend to be resin-rich, to contain air bubbles, and to fail through cracking and pinhole leaks. Each float should therefore be pressure-tested. Exterior steel dyes and interior solid-rubber dyes can force all the resin out of the wall, except the one needed to bond the glass fibers. Other synthetic materials may also be used for float construction.

A few manufacturers offer prefabricated steel and aluminum floats. Metal shells are usually folded and welded into rectangular units. Thin-gauge steel sheets with stiffening baffles for greater strength are normally used to manufacture floats. Pipes and corrugated tubes with sealed ends are sometimes used as an alternative to rectangular flotation units (Figure 6–32). Preservative coatings are usually applied to both sides of all corrodible metals. Floats are made compatible with modular deck units. Metal floats are generally used in freshwater, but seldom in seawater because of their high corrosion rates in a saltwater environment. Certain alloys are being offered that may overcome this problem.

Manufacturers of concrete floats point to the maintenance-free performance of concrete construction and to the sense of stability that the heavy mass of concrete gives to the floating pier. Successful concrete floats have been made both with and without reinforcement. When no reinforcing is used, the float must be designed so that at no point will the allowable tensile strength of the concrete be exceeded. Great care must be taken in transporting and launching these floats to prevent their cracking, either through impact stress or from temperature changes. Once in the water and decked over, there is little danger of damage except by accident.

Low-alkali cement is normally used for the concrete float's construction, and concrete must develop a compressive strength not less than 30 MPa at 28 days. When concrete floats are reinforced, a galavanized wire mesh is normally used. Because of the thin walls, the reinforcement is kept to a small diameter and placed near the center of the section. Its chief purpose is to moderate temperature differentials and hold the unit together, even if it cracks. Because such cracks frequently heal themselves by hydration of unset cement that is always present in concrete, the reinforcement is structurally useful.

In recent years floats built from recycled, foam-filled rubber tires have been used in the construction of small-craft marinas (Tsinker, 1986b). In these floats, expanded polystyrene beads pumped into automobile tires are protected with metal disks placed at the top and bottom of the float.

Large independent floats normally only need to be cradled, that is, prevented from sliding laterally by outer stringers and cross struts. This system allows for quick removal for maintenance or replacement, when necessary. If significant waves are expected, floats must be strapped to their saddles. Figures 6–34 and 6–35 show various kinds of deck framing and float connections. The framework and the decks are generally made of pressure-treated wood. However, in recent years galvanized steel and aluminum have been used frequently for framing. Aluminum grating systems are often used for decking (Figures 6–29 and 6–35). Although these are more costly, they are durable and the cost of maintenance is substantially reduced. All hardware used in both wood and metal construction should be

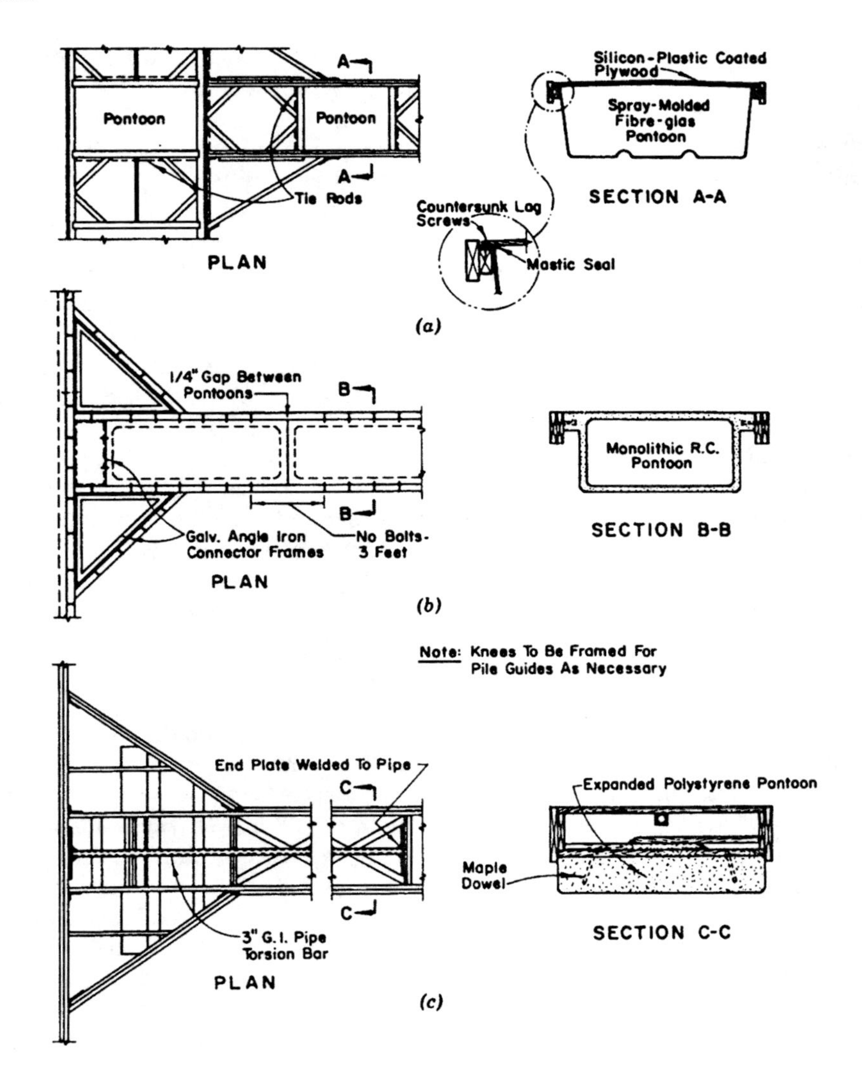

Figure 6–34. Floating pier—typical deck framing. (From American Society of Civil Engineers, Report on Small-Craft Marinas, 1969.)

of rust-resistive alloy and preferably hot-dipped galvanized.

Design Loads

Similarly to the fixed berth system the floating piers should be designed for all environmental loads, such as wind, wave, current, and ice, as well as dead and live loads. Because of concerns with floating dock safety of operation that is usually associated with its buoyancy and stability and also depends on marina activities the following minimum design live loads are recommended:

- Not busy, small residential marinas: 1.5 kPa
- Middle size and large commercial marinas: 2.0 kPa

Figure 6–35. Floating pier consisting of lightweight deck system supported on close-cell polystyrene floats. (St. Catherines, Ontario.)

- Marinas that may be used for assembly, boat shows, events, and other: 2.0 to 3.0 kPa

Furthermore, pier structural strength and stability should be reviewed for concentrated load of 2 kN applied at any location of the floating berth system. With a full live load or with a concentrated load applied at the end of a finger pier a minimum freeboard of 15 cm must be ensured.

Similarly to the bottom fixed pier design, the floating pier must also be designed for impact by incoming boats. Additional useful information on floating pier design loads is provided in Hunt (1991), Morishige (1991) Abraham (1991), and Wortley (1991).

Mooring Systems

The mooring system is an integral part of any floating pier installation. A variety of structural types of mooring systems are used in modern floating pier engineering (Tsinker, 1986b). However, in modern marina developments in most cases floating piers are usually moored in place by anchor piles or an underwater cable system. The effective mooring system includes minimum anchor points to provide for attractive, cost-effective, and reliable marina docking arrangements. In relatively shallow water anchor piles, which require firm but penetrable substrata, are the simplest and most commonly used type of pier anchorage. The depth of penetration into the soil, pile diameter, and the number of piles required depend on the forces acting on the pier depth of water and geotechnical parameters of soil substrata. The reliable anchor pile design is an engineering task based on rigorous investigation of pier–pile and pile–soil interaction. If required design parameters are not available the preliminary anchor pile design may be based on existing precedents and an educated guess. The details on pile–soil interaction are given in appropriate chapters of this work and in most standard texts on foundation engineering.

The use of piles for mooring systems requires the provision of an adequate connection between the pile and the pier. The most commonly employed pile guide system includes loose metal hoops.

Metal hoops at anchor piles are commonly used to guide the pier in a vertical direction with variation in water level. Sometimes sliding type connections or roller type connections (Figures 6–36 and 6–37) are used. The latter type of connection provides smooth vertical movements and reduces vertical forces on piles. Rollers are usually made of plastic material which reduces noise which can be a problem with metal on metal connections. To eliminate axis corrosion they are usually made of stainless steel. In non-rolling sliding type connections metal on metal contact may produce noise and therefore must be avoided; in this type of connection a rubbing element made of timber or plastic material is normally provided. Naturally the pile guides and part(s) of a floating pier to which the pile guides are attached, as well as the entire pier system, must be of robust construction to resist the mooring forces transferred to anchor piles. Pile guides are usually bolted or in some instance welded to the dock system; naturally the strength of fasteners must be compatible with the strength of the attachment system.

When a marina is constructed in deep water, or where large water-level fluctuations occur, floating piers are usually anchored in place with steel wire ropes or chains. Sometimes synthetic ropes are used for the same purpose. In general, pier anchorage by means of underwater anchor cables provides for better aesthetics of the marina.

An anchor cable essentially is a catenary suspended between floating pier and anchor placed on the bottom. Typically, the projected line of this catenary is equal to (7.0 to 10.0) H, where H = depth of water.

As the floating berth system can tolerate only limited movements between each dock, or between floating and fixed parts of the combined berth system, the mooring cables must be tensioned in order to prevent excessive displacement of the docks. For details on cable mooring system design the reader is referred to Tsinker (1986b).

At the bottom mooring cables are secured to conventional anchors, or anchors of special designs. The holding power of the numerous traditional and special anchors is discussed by Tsinker (1986b); some details

Figure 6–36. Roller-type sliding connection between floating pier and anchor piles. (Bangkok, Tailand.)

Figure 6–37. Sliding type connection between floating pier and anchor pile. (London, U.K.)

are given in Chapter 4 and in other related sources.

It must be pointed out that the vertical component of mooring force must not be overlooked in floating pier design, and particularly in buoyancy and stability analysis.

In a soft bottom conventional anchors may be used. Sometimes sinkers are attached at the midpoint of the anchor cables to maintain a nearly horizontal pull on the anchor, and in the case of a steep anchor cable slope sinkers should be added to the system near the cable attachment to the floating dock, so as not to foul boats that move in close by. If the whole floating system may move a considerable distance in and out with changes in water level, hand- or motor-operated winches are employed to keep the dock in place.

Access Ramp

Access to the floating pier usually is obtained through an articulated ramp hinged to the fixed structure at one end and resting on the pier deck at the other. The ramp structure (frame and railings) may be of wood or metal, or a combination of both, designed to support a live load used for the design of the main pier structure (Figure 6–38). The wearing surface of the ramp should be designed to provide safe walking in a sharply inclined position.

Because a floating pier has a tendency toward rolling or some sideways movement, which can severely overstress the hinges at the upper end of the ramp, the latter must be provided with slotted bearings to accommodate the pier motion. Sliding ends are used for ramps weighing less than 1.5 kN. Heavier ramps should always be provided with rollers and an apron. The ramp is usually narrower than the pier. Extra flotation must be provided in the floating pier under the outboard end of the ramp so that, at the point of contact, the deck rides at the same height as the adjacent segments of the deck.

Hardware

These include miscellaneous deck and float connections, hinges between adjacent dock sections and main pier and finger piers, mooring cleats, and other. All hardware used must be fabricated to provide for a long-lasting performance. For example, nails and other fasteners must be hot-dipped galvanized, and hinges and cleats should be protected from corrosion in one way or the other. Cleats must be selected on the basis of expected mooring forces, which must be well defined during the dock design process. They must be securely fastened to the pier frame (not just to the decking). Typically the diameter of a cleat horn should not be less than twice the diameter of the mooring line that passes around the cleat; smaller diameter cleat horns will

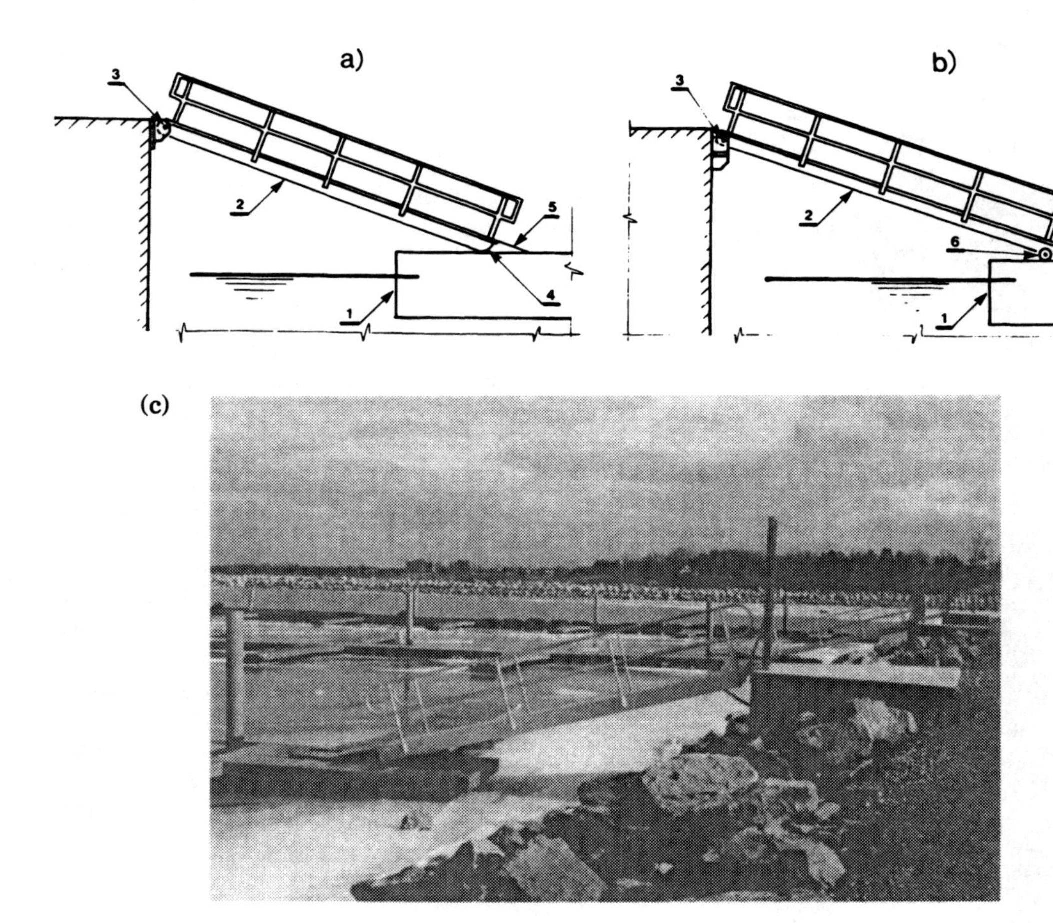

Figure 6-38. Articulated access ramp. (a) with sliding end. (b) End with roller. 1—Pier, 2—ramp, 3—slotted bearing hinge, 4—sliding end, 5—apron, 6—roller.

create overstress in the mooring line and increase potential for premature line failure.

6.5.7 Floating Pier Design

As stated earlier the floating deck is designed to resist all design loads by providing an adequate buoyancy and stability while maintaining an adequate free board criteria. Basic aspects of floating body buoyancy and stability are discussed in Tsinker (1986b).

The pier's structural integrity depends on its ability to resist the following loads:

- Deck load and wave-induced bending moments and shear forces
- Hydrostatic and hydrodynamic loads
- Mooring forces due to effects of environmental loads
- Boat impact load
- Launching and towing loads where applicable.

Consideration is usually given to corrosion, fatigue, and miscellaneous mechanical deteriorations of dock components.

Several characteristic examples of floating pier systems design are provided in the following sections. Work examples of floating piers of miscellaneous design are given in Tsinker (1986b).

6.5.7.1 Chain of Pontoons with Individual Deck Sections

A system of a chain of floats (pontoons) with individual deck sections is statically determinate. Design of a pier consisting of floats joined by individual deck sections normally begins with establishing approximate sizes of floats and a typical space between them. Float sizes determine the draft under dead and live loads and also the design gradient of the span. In the case of central deployment of the deck span on floats (as shown in Figure 6–39), preliminary sizes of the float could be verified by using two influence lines: the reaction line of a single float (a) and the reaction line of two adjoining floats (b); these floats are assumed to be straight walled, and therefore their draft is directly proportional to the load exerted on them. Loading line (a) in Figure 6–39 with dead and live loads defines the maximum load on a float (A_{max}). Hence, the float's maximum draft T_{max} is determined as follows.

$$T_{max} = \frac{A_{max}}{\gamma F} \tag{6–5}$$

where

γ = specific weight of water
F = area of float waterplane.

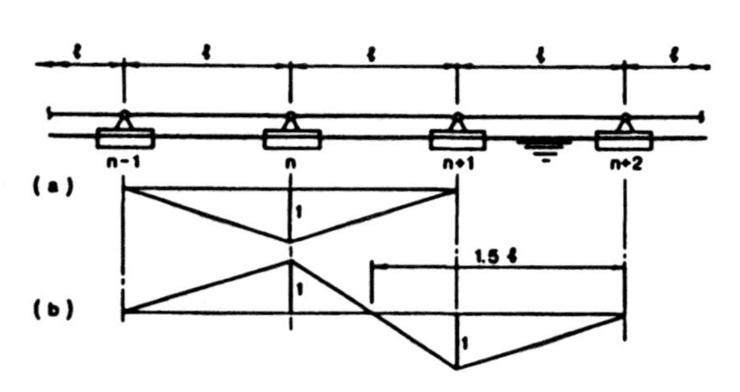

Figure 6–39. Statically determined chain of pontoons with intermediate spans and influence lines.

Using line $A_{n+1} - A_n$ [line (b) in Figure 6–39], determine the deck gradient i. If all spans and floats are identical then

$$i = \frac{A_{n+1} - A_n}{\gamma F \ell} \tag{6–6}$$

where

ℓ = span between two adjacent floats
A_n, A_{n+1} = subsequent reactions of floats n and $n+1$.

The maximum deck gradient can be obtained by loading 1.5ℓ of the deck with live load. Since the area of one part of the influence line $\omega = 0.75/\gamma F$ does not depend on ℓ, the span gradient produced by the uniformly distributed load also does not depend on span but rather on the intensity of this load and on the float's waterplane area. Hence:

$$i_q = q\omega = \frac{0.75q}{\gamma F} \tag{6–7}$$

where

q = uniformly distributed load.

Assuming the permissible value of pier gradient i_{max} and knowing the design value of the uniformly distributed load q, the preliminary value of the float's waterplane area F_{min} could be established.

$$F_{min} = \frac{0.75q}{\gamma i_{max}}. \tag{6–8}$$

In turn, the deck- span gradient produced by concentrated load i_p depends not only on the magnitude of the load and water plane area but also on the span length:

$$i_p = \frac{P}{\gamma F \ell} \tag{6–9}$$

and therefore

$$F_{min} = \frac{P}{\gamma \ell i_{max}}. \tag{6–10}$$

It should be mentioned that the value of F_{min}, determined for the intermediate floats by Eq. (6–10), sometimes may not be sufficient for the first float.

Assume that the scheme shown in Figure 6–39 represents an access pier linked to the shore with an articulated ramp. This would mean that support $n-1$ (land abutment) is fixed. In this case, the influence line i_n would appear as line $A_n(a)$, but with ordinates decreased $\gamma F \ell$ times. Then the area of this influence line can be expressed as follows.

$$\omega' = 1/\gamma F. \tag{6–11}$$

Note that ω' is 33% larger than the area-of-influence line for intermediate floats.

6.5.7.2 Redundant Pier Systems (by A. Mee)

For some linked pier systems, the resultant forces acting on an individual pontoon element cannot be determined by applying equilibrium conditions only. Such a pier system is described as being statically indeterminate or redundant. An example of such a system is shown in Figure 6–40a.

In this figure, the pontoons are pinned together without intervening components, and the loading is applied to the deck surface. If, for example, the system is subjected to a point load acting at one of the joints, the pier will deform as shown in Figure 6–40b. Unknown shearing forces act at the connections, as shown for a typical pontoon in Figure 6–41. These forces cannot be determined using static equilibrium conditions alone, as the buoyant forces acting on the pontoon are indeterminate. To solve for the forces, it is necessary to make use of the relationship between pontoon loads and pontoon displacements. These will be introduced later.

Two additional examples of redundant pier systems are shown in Figures 6–42 and 6–49. Both of these systems consist of a

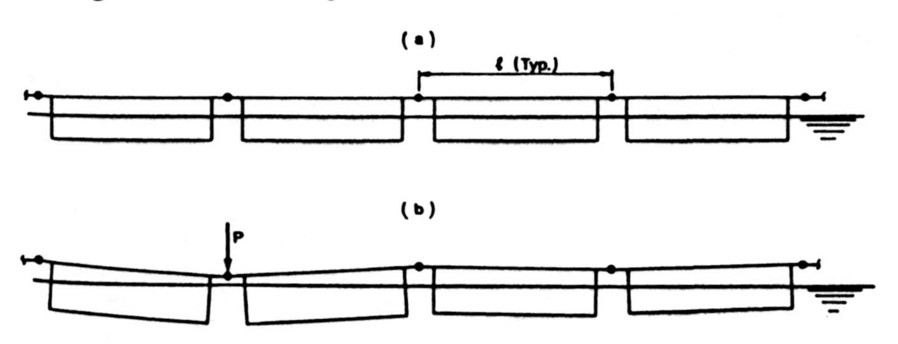

Figure 6–40. Linked pontoon system.

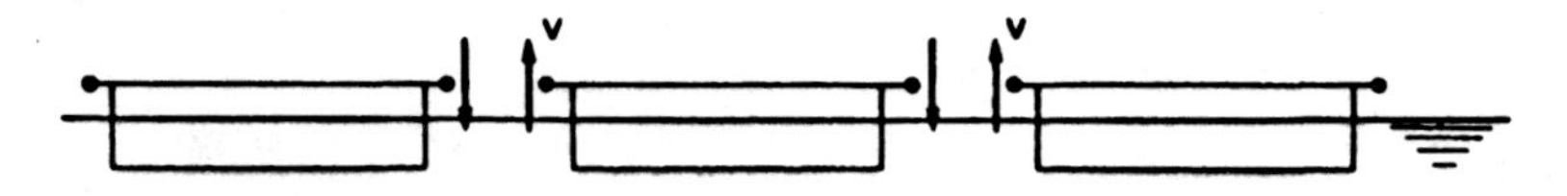

Figure 6–41. Individual pontoon forces.

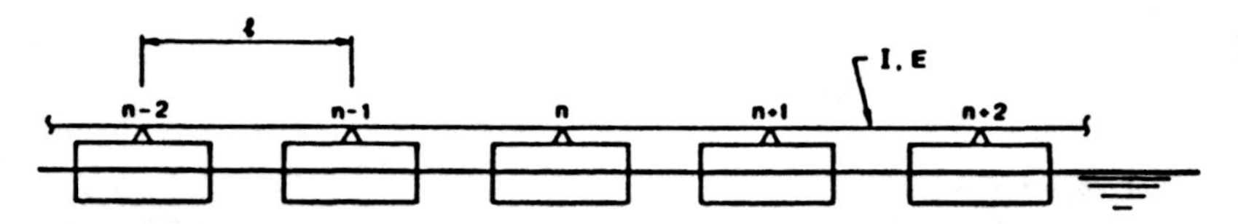

Figure 6–42. Redundant pier systems.

continuous deck beam supported on individual floats (pontoons). The beam is only pinned to the supports. In the second system, the deck is supported from two positions on each pontoon.

Methods for Static Analysis

In the past, only hand-calculation methods were available for the analysis and design of floating piers. The design engineer often had to adopt a number of simplifying assumptions to make the manual analysis more tractable. For example, geometric and component similarities would be assumed, boundary support conditions would be simplified, and the number of alternative loading conditions would be condensed.

The digital computer has eliminated most of the drudgery of manual calculations and has reduced the effort expended on analysis in the design process.

There are many structural frame analysis programs available commercially. In addition, many design firms have developed their own frame-analysis programs. Obviously, a program can be written on an ad hoc basis for a particular pier system. However, it will be shown here how a particular pier system can be idealized to make use of existing frame-analysis programs.

Some effort will be devoted here to manual methods, which are still useful in the preliminary design of pier systems, and which will serve as an introduction to the computer procedures.

Assumptions for Analysis

That which is readily available in many structural-analysis texts (i.e., Przemieniecki, 1968; Livesley, 1975) will not be repeated here. However, a brief summary of the basic assumptions will be given.

Many different schemes are available for structural analysis, but only the so-called displacement or stiffness method will be discussed here, because most computer-oriented methods are based on this approach.

In the stiffness method, the complete structure is regarded as an assembly of elements. For example, in pier-system analysis, the structure can be represented with simple beam and spring components. The overall properties of the system can then be determined by the relatively simpler properties of its idealized parts. The simple components are interconnected at joints or nodes. This joining corresponds to the forcing of displacement compatibility between the elements.

The stiffness properties of the individual elements relate the element nodal displacements and nodal forces and are derived using element geometry and the stress–strain equation of the component material. External loads acting on the structure are replaced by statically equivalent nodal forces.

Note that the terms force and displacement are used in a general sense, implying moment and rotation components in addition to simple loads and translations.

In an analysis based on the stiffness method, the nodal displacements are the basic unknown quantities. Equilibrium relations are developed for the nodes, and these equate the external loads to the sum of the nodal forces acting on the elements that meet there. Substitution is made using the nodal compatibility conditions and the element nodal force–deformation relationships. The result is a set of equations relating the nodal displacements to the applied nodal loads. The equations are solved for the displacements, and the element forces can then be recovered using the individual element stiffness.

Implicitly in this procedure is the assumption of linearity. The behavior of an element or structure is said to be linear if all displacements and internal forces vary in direct proportion to the applied loads. This assumption, which is reasonably accurate for pier structures subjected to normal working loads, is important for two reasons. First, the analysis is much simpler, as the final set of simultaneous equations are linear, and as standard procedures are available for their solution. Second, solutions can be superimposed so that different load cases can be combined.

There are two basic causes of nonlinearity. The first is the behavior of the constituent material. The response of mild steel, for example, is linear until yielding occurs. The second cause of nonlinearity is associated with changes in geometry. If distortion of the structure changes the equilibrium equations, nonlinearity will result.

Individual pontoons floating in water respond linearly to load, provided that the pontoon waterplane area does not change with displacements.

Response of a Single Pontoon

The overall behavior of a pier system depends on the load–deflection response of a single pontoon. Here, the necessary relationships will be derived and a structural analogy developed.

Consider a pontoon floating in water, as shown in Figure 6–43. The plan at water level is shown in (a); a typical cross-section is shown in (b). It is assumed that the individual pontoon does not deform in a rolling mode (rotation about pier longitudinal axis). The deflections can then be described in terms of a vertical displacement u and a rotation v about the axis x–x only. Note that any applied loads can be similarly resolved into the resultant system made up of vertical force U and moment V. In addition, the deflections are assumed small enough that the waterplane area does not change its shape under pier deformation. The response can, therefore, be considered linear.

For vertical load only:

$$\gamma A u = U \tag{6–12}$$

where

γ = specific weight of water
A = waterplane area.

For moment:

$$\gamma I v = V \tag{6–13}$$

where

I = moment of inertia of the waterplane area about the axis x–x.

Equation 6–13 neglects pontoon weight and buoyancy and is a simplified version of the behavior under moment (Tsinker, 1986b). The correct relationship is obtained

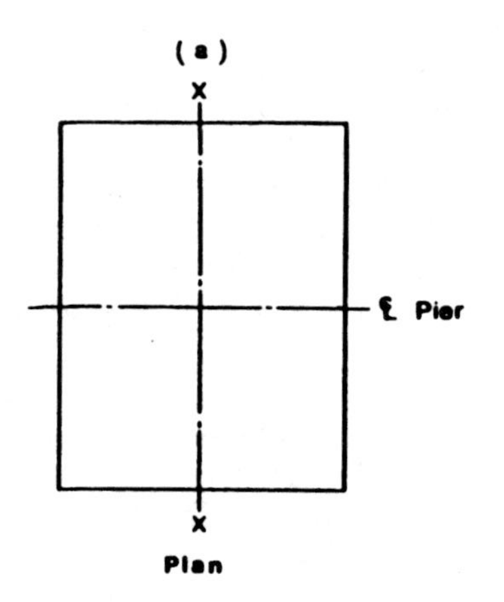

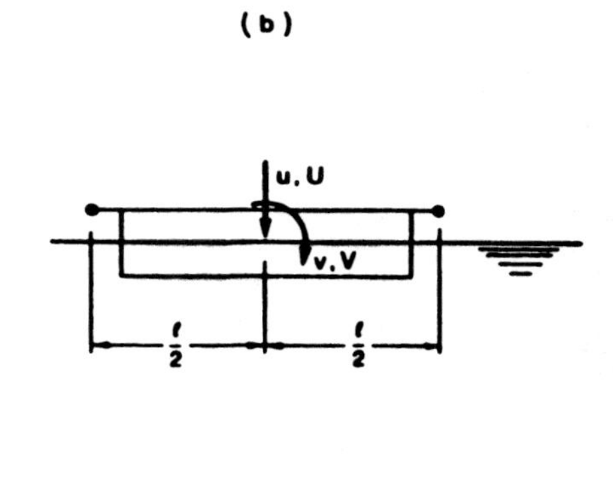

Figure 6–43. Individual pontoon.

if I is replaced by $I - wa$. Here, w is the displacement and a is the distance between the center of gravity and center of buoyancy.

The term wa can generally be ignored as a first approximation and this is done here.

Equations 6–12 and 6–13 can be summarized in matrix form:

$$\begin{bmatrix} \gamma & A & 0 \\ 0 & \gamma & I \end{bmatrix} \begin{Bmatrix} u \\ v \end{Bmatrix} = \begin{Bmatrix} U \\ V \end{Bmatrix}. \tag{6–14}$$

The previous equations represent the hydrostatic response of the pontoon to arbitrary deformation. An analogous response can be achieved by the use of structural components. Consider the arrangement shown in Figure 6–44. The pontoon, idealized as a rigid component, is supported by linear and rotational springs with stiffness values s and r, respectively. The response would then be given by the matrix equation:

$$\begin{bmatrix} s0 \\ 0r \end{bmatrix} \begin{Bmatrix} u \\ v \end{Bmatrix} = \begin{Bmatrix} U \\ V \end{Bmatrix}. \tag{6–15}$$

Thus, if $s = \gamma A$ and $r = \gamma I$, complete equivalence between Eqs. (6–14) and (6–15) will be established.

Although Eq. (6–15) is simple, it is not in the most convenient form for manual structural analysis. Instead of central vertical and rotational deflections, it is better to use the two vertical displacements at the connecting nodes for describing the deformation of the pontoon.

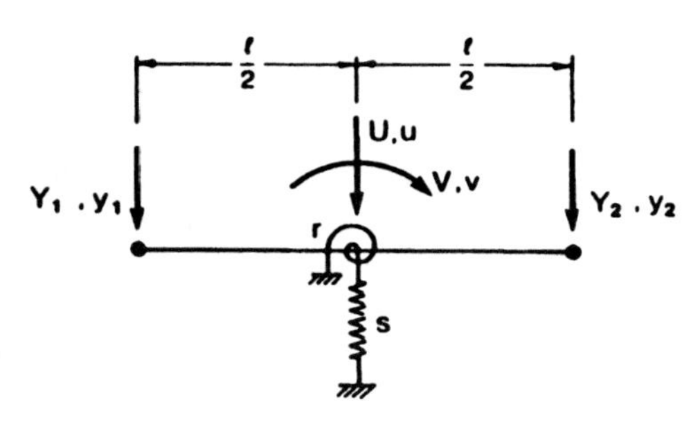

Figure 6–44. Structural components equivalent to pontoon.

The two sets of displacements are related by the expression

$$\begin{Bmatrix} u \\ v \end{Bmatrix} = \begin{bmatrix} 1/2 & 1/2 \\ -1/\ell & 1/\ell \end{bmatrix} \begin{Bmatrix} y_1 \\ y_2 \end{Bmatrix}. \tag{6–16}$$

Similarly, the nodal loads are related by the expression

$$\begin{Bmatrix} Y_1 \\ Y_2 \end{Bmatrix} = \begin{bmatrix} 1/2 & -1/\ell \\ 1/2 & 1/\ell \end{bmatrix} \begin{Bmatrix} U \\ V \end{Bmatrix}. \tag{6–17}$$

Equation (6–15) can be combined with Eqs. (6–16) and (61–17) to give

$$\begin{bmatrix} \frac{s}{4}+\frac{r}{\ell^2} & \frac{s}{4}-\frac{r}{\ell^2} \\ \frac{s}{4}-\frac{r}{\ell^2} & \frac{s}{4}+\frac{r}{\ell^2} \end{bmatrix} \begin{Bmatrix} y^1 \\ y^2 \end{Bmatrix} = \begin{Bmatrix} Y_1 \\ Y_2 \end{Bmatrix} \tag{6–18}$$

The stiffness properties of a simple beam element, as shown in Figure 6–45, also will be used later. The values are derived in many structural analysis texts and are listed here:

$$\frac{2EI}{\ell^3} \begin{bmatrix} 6 & 3\ell & -6 & 3\ell \\ 3\ell & 2\ell^2 & -3\ell & \ell^2 \\ -6 & -3\ell & 6 & -3\ell \\ 3\ell & \ell^2 & -3\ell & 2\ell^2 \end{bmatrix} \begin{Bmatrix} y_1 \\ \theta_1 \\ y_2 \\ \theta_2 \end{Bmatrix} = \begin{Bmatrix} V_1 \\ M_1 \\ V_2 \\ M_2 \end{Bmatrix}. \tag{6–19}$$

Hinged Pontoon System

Consider a section of the hinged pontoon system shown in Figure 6–46. For the analysis, assume that all pontoons have the same dimensions and buoyancy properties. Pontoons are linked through pin joints, such as $n-1$, n, and $n+1$, which transfer vertical shear forces but no bending

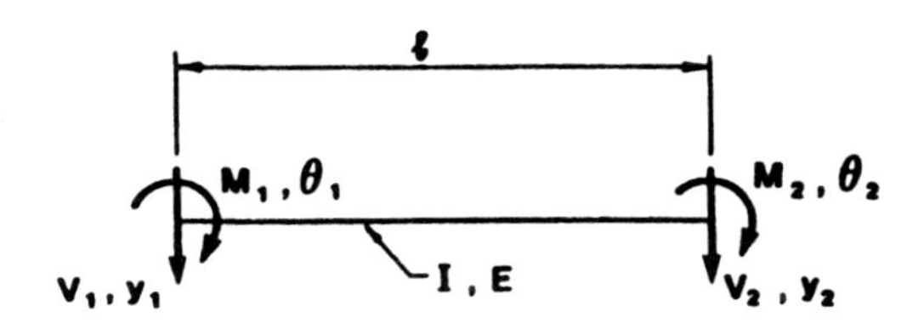

Figure 6–45. Simple beam element.z

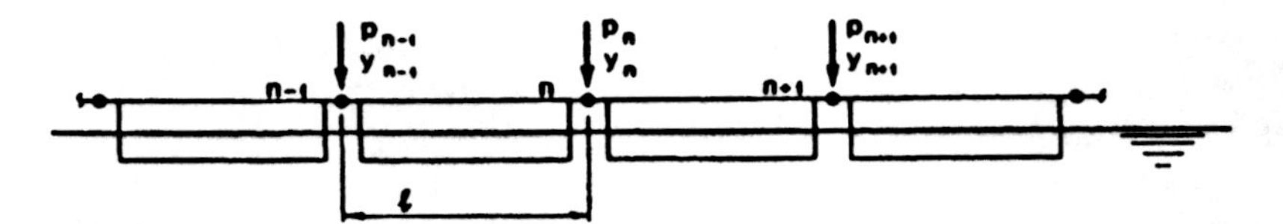

Figure 6–46. Hinged pontoon system.

moments. The pontoons are considered rigid, so the deformation of the system can be described completely in terms of nodal vertical displacements, such as y_{n-1}, y_n, and y_{n+1}. Any loading applied to a particular pontoon can be resolved into two equivalent vertical point loads acting at the pontoon's nodal positions. For example, if the pontoon has a uniform superimposed load of intensity w, the point loads would be given by $p = w\ell/2$.

Thus, the applied loading system is represented in terms of nodal loads, such as P_{n-1}, P_n, and P_{n+1}, respectively. A free body diagram showing the forces acting on joint n is shown in Figure 6–47. For vertical equilibrium of the joint, it is necessary that $Y_{2n} + Y_{1n} = p_n$.

Substituting for Y_{2n} and Y_{1n} from Eq. (6–18):

$$k_2 y_{n-1} + k_1 y_n + k_1 y_n + k_2 y_{n+1} = p_n$$

or

$$k_2 y_{n-1} + 2k_1 y_n + k_2 y_{n+1} = p_n \qquad (6\text{–}20)$$

where

$$k_1 = \frac{s}{4} + \frac{r}{\ell^2}$$

and

$$k_2 = \frac{s}{4} - \frac{r}{\ell^2}.$$

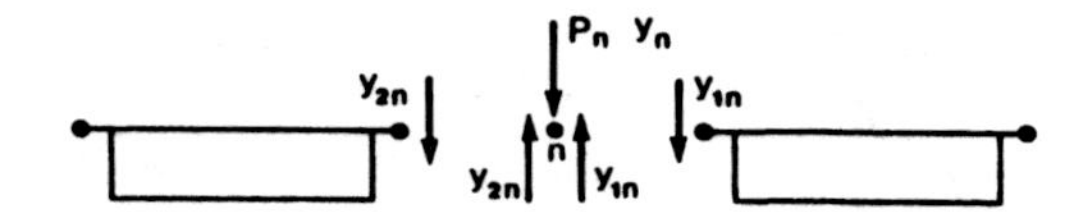

Figure 6–47. Free-body diagram of joint.

If Eq. (6–20) was written out for all joints n and assembled in matrix form, the result would be

$$[K]\{y\} = \{p\}$$

where

$[K]$ = the overall stiffness matrix of the structure
$\{y\}$ = the vector of nodal displacements
$\{p\}$ = the vector of nodal loads.

If the boundary conditions were incorporated into the previous equation, the displacements $\{y\}$ could be solved for directly. Such an approach may be used if the stiffness coefficients do not form a regular pattern, which would occur if the individual pontoons had different geometry. For a regular system of pontoons, however, a different approach is possible. Equation 6–20 becomes

$$y_{n-1} + 2\frac{k_1}{k_2} y_n + y_{n-1} = \frac{p_n}{k_2}. \qquad (6\text{–}21)$$

Note that

$$\frac{k_1}{k_2} = -\frac{1 + \dfrac{4I}{A\ell^2}}{1 - \dfrac{4I}{A\ell^2}} > 1.$$

Equation (6–21) is a difference equation of second order, with constant coefficients. A discussion of difference equations and their solutions may be found in standard texts (Brand, 1966; Hildebrand, 1968) and will not be elaborated on here.

If $P_n = 0$, the general solution of Eq. (6–21) would be

$$y_n = A(-1)^n ch\phi n + B(-1)^n sh\phi n \qquad (6\text{–}22)$$

where

$$ch\phi = \frac{k_1}{k_2} \quad (6\text{–}23)$$

sh = the hyperbolic sine, sinh
ch = the hyperbolic cosine, cosh
A, B = arbitrary constants.

For complete results, a particular solution dependent on the applied loading must be added. The arbitrary constants must be determined from end conditions. For example, if the pontoon has a free end at its maximum node k, the required condition is

$$y_{2k} = 0$$

or, in terms of displacement,

$$k_2 y_{k-1} + k_1 y_k = 0.$$

An alternative condition occurs if the end is pinned. In this case, for node k, say, the condition is simply

$$y_k = 0.$$

Consider as a practical example the pontoon system between nodes 0 and k, with a point load p applied at node m. The loading system can be written as

$$p_n = p\delta_n^m \quad (6\text{–}24)$$

where p is the load, and

δ_n^m = delta function, which has the property

$$\begin{aligned} \delta_n^m &= 1 \text{ if } n = m \\ \delta_n^m &= 0 \text{ if } n \neq m. \end{aligned} \quad (6\text{–}25)$$

The boundary conditions are simple supports at nodes 0 and k, requiring $y_0 = 0$ and $y_k = 0$.

The complete solution for deflections is given by

$$y_n = \frac{p(-1)^{n-m}}{k_2\, sh\phi\, sh\phi k}[sh\phi n\, sh\phi(k-m) - sh\phi k\, sh\phi h\, H(n-m)] \quad (6\text{–}26)$$

where the step function having the following properties has been used:

$$\begin{aligned} H(n-m) &= 1 \quad \text{if } n \geq m \\ H(n-m) &= 0 \quad \text{if } n < m. \end{aligned} \quad (6\text{–}27)$$

The nodal deflections can be used to calculate the shears, Y_n, at the joints. This results in:

$$Y_{1n} = \frac{p(-1)^{n-m}}{sh\phi k}[sh\phi k\, ch\phi(n-m)H(n-m) - ch\phi n\, sh\phi(k-m)]. \quad (6\text{–}28)$$

In Eqs. (6–27) and (6–28), $k > m > 0$.

A similar problem occurs if both ends of the pontoon system are free. In this case, the deflected shape is given by

$$y_n = \frac{p(-1)^{n-m}}{k_2\, sh\phi\, sh\phi k}[-sh\phi k\, sh\phi(n-m)H(n-m) + ch\phi k(k-m)ch\phi n]. \quad (6\text{–}29)$$

The shears are given by

$$Y_{1n} = \frac{p(-1)^{n-m}}{sh\phi k}[sh\phi k\, ch\phi(n-m)H(n-m) - ch\phi(k-m)sh\phi n]. \quad (6\text{–}30)$$

Computer Analysis of Linked Pontoons

A number of alternative models are available to represent the rigid pontoon when a two-dimensional-frame computer analysis is being carried out. The simplest model is perhaps that shown in Figure 6–48.

Two equal beams are rigidly linked and supported on two springs, one linear and

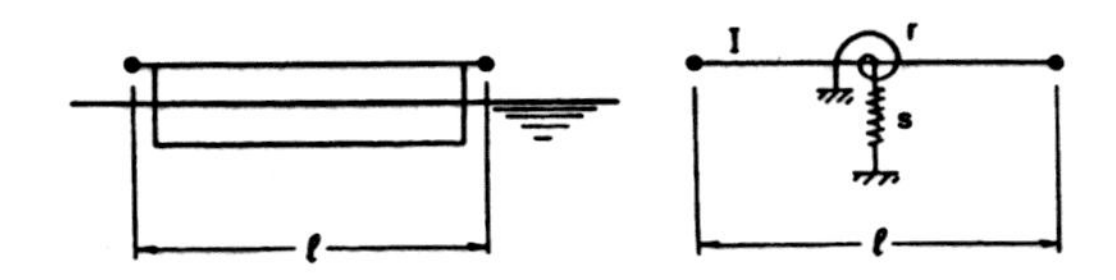

Figure 6–48. Model for computer analysis.

one rotational. The linear-spring constant should have the value $s - A\gamma$; similarly, for the rotational-spring constant, the value should be $r = I\gamma$, where A, and I, and γ are as previously defined. The beams should have a sufficiently large moment of inertia so that they can be considered rigid. This can be achieved by making the beam inertia

$$I_b \approx 10^4 \frac{\ell I \gamma}{3E}.$$

The formula implies that the deflection of the beam as a cantilever is about 10^{-4} times that of a rigid component supported by a rotational spring with stiffness value γI.

Continuous Beam

Consider the pier system shown in Figure 6–49. The deck or beam bridging is continuous over the pontoons, which are pinned to the deck. However, the pontoons supply only vertical reactions and no rotational stiffness. The analysis of the system response is considerably more complex than that of the previous example, as the beam is continuous. As with a typical beam, four boundary conditions need to be satisifed, two at each end. If the structure is regular, a closed-form solution is possible, but arbitrary loading will not be treated, as it is generally too complicated. Instead, the system response, infinite in extent, to unit vertical load and moment will be determined. Subsequently, it will be shown how these elementary solutions can be combined for more complicated loading and boundary conditions.

The solutions given are similar to those found for a beam on continuous elastic support, which have been extensively treated (Hetinyi, 1946).

Consider the equilibrium of an unloaded joint n, shown as a free body in Figure 6–50. For moment equilibrium:

$$M_{2n} + M_{1n} = 0. \tag{6–31}$$

Substituting the stiffness terms given in Eq. (6–19):

$$\frac{3}{\ell}(y_{n-1} - y_{n+1}) + (\theta_{n-1} + 4\theta_n + \theta_{n+1}) = 0. \tag{6–32}$$

For vertical equilibrium of the joint n:

$$V_{2n} + V_{1n} + ky_n = 0. \tag{6–33}$$

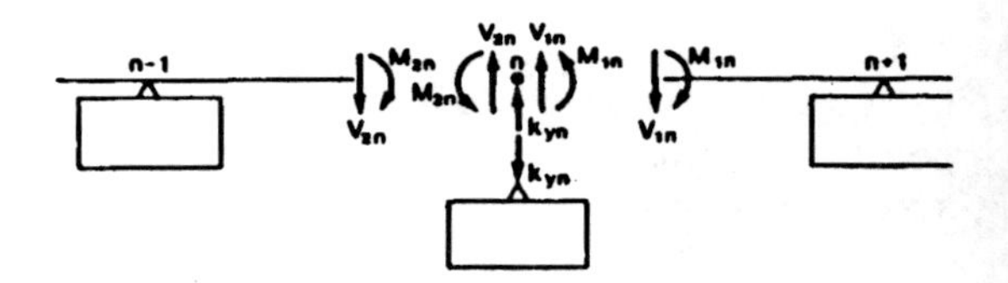

Figure 6–50. Free-body diagram of joint.

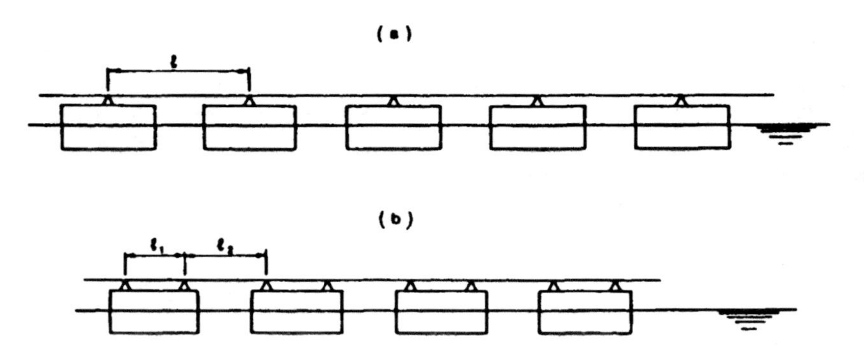

Figure 6–49. Continuous beam over pontoons.

Substituting from Eq. (6–19):

$$\frac{2}{\ell}(-y_{n-1} - 2y_n - y_{n+1}) + \frac{\ell^2 k}{6EI} y_n + (-\theta_{n-1} + \theta_{n+1}) = 0. \quad (6\text{–}34)$$

Using similar equilibrium equations for points $n-1$ and $n+1$, the nodal-slope variables can be eliminated, thereby deriving a fifth-order equation in the deflection:

$$y_{n-2} - (4-\alpha)y_{n-1} + (6+4\alpha)y_n - (4-\alpha)y_{n+1} + y_{n+2} = 0 \quad (6\text{–}35)$$

where

$$\alpha = \frac{\ell^3 k}{6EI}. \quad (6\text{–}36)$$

The general solution of Eq. (6–36) is

$$y_n = Ap^n \cos n\phi + Bp^n \sin n\phi + Cp^{-n} \cos n\phi + Dp^{-n} \sin n\phi \quad (6\text{–}37)$$

where

$$\cos\phi = [1/2(2+\alpha) - 1/4(24\alpha + 3\alpha^2)^{0.5}]^{0.5} \quad (6\text{–}38)$$

and

$$p = \frac{1}{\cos\phi}\left[1 - \frac{\alpha}{4} \pm 1/4(\alpha^2 - 16\alpha + 4(24\alpha + 3\alpha^2)^{0.5})^{0.5}\right] \quad (6\text{–}39)$$

where for a real ϕ value, $\alpha < 24$. The sign in Eq. (6–39) is chosen so that $|p| < 1$.

These relationships are derived from

$$2\cos\phi\left(p + \frac{1}{p}\right) = 4 - \alpha \quad (6\text{–}40)$$

and

$$p^2 + \frac{1}{p^2} + 4\cos^2\phi = 6 + 4\alpha. \quad (6\text{–}41)$$

A, B, C, and D are arbitrary constants to be determined from boundary conditions.

By using symmetry arguments, the solution can be confined to positive values of n. It follows that, because $|p| < 1$, the terms involving p^{-n} must be eliminated, as they do not tend to zero as n tends to infinity, which is a physical necessity. Hence

$$y_n = Ap^n \cos n\phi + Bp^n \sin n\phi$$

and the constants A and B must be determined from the conditions at station $n = 0$. In particular, the two separate systems shown in Figure 6–51 will be examined. In both cases, only half the system need be considered, as the deflections are symmetrical about the origin for case (a) and asymmetrical for case (b). In both cases, only the solution valid for positive values of n will be derived.

For case (a), the boundary conditions at station $n = 0$ are

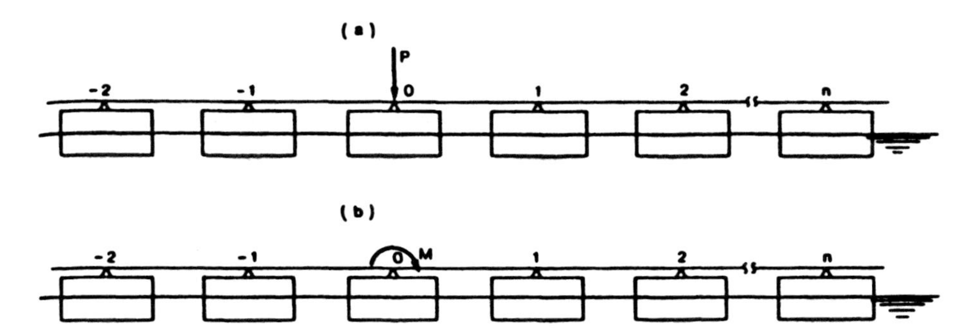

Figure 6–51. Symmetric (a) and asymmetric (b) beam systems.

$$\theta_0 = 0$$

and

$$v_{10} = 1/2(P - ky_0).$$

The shear v_{10} can be determined from the summation

$$v_{10} = \sum_{n=1}^{\infty} ky_n.$$

Similarly, the moment at the origin can be determined from

$$m_{10} = \sum_{n=1}^{\infty} k\ell ny_n.$$

Using the prior equations and the identities

$$\begin{aligned}
\sum_{n=1}^{\infty} p^n \sin n\phi &= \frac{1}{F_1} p \sin\phi \\
\sum_{n=0}^{\infty} p^n \cos n\phi &= \frac{1}{F_1} (1 - p\cos\phi) \\
\sum_{n=1}^{\infty} p^n n \sin n\phi &= \frac{1}{F_1^2} (1 - p^2) p \sin\phi \\
\sum_{n=1}^{\infty} p^n n \cos n\phi &= \frac{p}{F_1^2} ((1 + p^2)\cos\phi - 2p)
\end{aligned} \tag{6–42}$$

where

$$F_1 = 1 - 2p\cos\phi + p^2$$

and

$$|p| < 1.$$

The solution for the deflection can be determined:

$$y_n = \frac{2}{3}\frac{p}{k}\frac{F_1}{F_3} p^n \left[\frac{F_4}{F_2}\cos n\phi + \frac{F_5}{p\sin\phi}\sin n\phi.\right] \tag{6–43}$$

The slope, shears, and moments can be derived from Eq. (6–43), the component equilibrium equations, and the condition at the origin. This leads to

$$\theta_n = \frac{PF_1 p^n}{2k\ell F_2}\left[\frac{F_1}{p\sin\phi}\sin n\phi\right] \tag{6–44}$$

$$v_{1n} = \frac{Pp^{n+2}}{F_3F_2}\left[F_6\cos n\phi + \frac{F_7}{p\sin\phi}\sin n\phi\right] \tag{6–45}$$

$$m_{1n} = \frac{P\ell p^{n+2}}{F_3F_2}\left[2F_8\cos n\phi - \frac{F_9}{p\sin\phi}\sin n\phi\right] \tag{6–46}$$

where

$F_2 = (1 - p^2)$
$F_3 = 1 + 2p\cos\phi + p^2$
$F_4 = 1 + p\cos\phi + p^2$
$F_5 = 1/4(1 + 4p\cos\phi + p^2)$
$F_6 = F_3 - 4$
$F_7 = F_2(1 - p\cos\phi) - 2p^2\sin^2\phi$
$F_8 = F_3 + 1$
$F_9 = 3F_3 + 2.$

For case (b), the boundary conditions at station $n = 0$ are

$$y_0 = 0$$

and

$$m_{10} = \frac{M}{2}\sum_{n=1}^{\infty} kn\ell y_n.$$

If the appropriate identity is used in Eq. (6–42), the solution is

$$y_n = \frac{M}{2k\ell}\frac{F_1}{F_2} p_n \left[\frac{F_1}{p\sin\phi}\sin n\phi\right] \tag{6–47}$$

$$\theta_n = \frac{MF_1^2 p^{n-2}}{6k\ell^2}\left[\frac{F_4}{F_2}\cos n\phi - \frac{F_5}{p\sin\phi}\sin n\phi\right] \tag{6–48}$$

$$v_{1n} = \frac{Mp^n}{2\ell}\frac{F_1}{F_2}\left[\cos n\phi + \frac{F_{10}}{p\sin\phi}\sin n\phi\right] \tag{6–49}$$

$$m_{1n} = \frac{Mp^n}{2}\left[\cos n\phi - \frac{F_{11}}{p\sin\phi}\sin n\phi\right] \tag{6–50}$$

where

$F_{10} = p(\cos\phi - p)$
$F_{11} = \frac{F_1^2}{12F^2}.$

Beams of Finite Length

The combination of solutions derived for the beam of infinite length can be used to determine the forces and displacement in beams of finite length. Consider the example shown in Figure 6–52.

The beam has a pinned joint at the right-hand end, node n, say. For the applied loading, the solutions given for beams of infinite length can be used to calculate the shear force and bending moment just to the right of node n. Assume these are given by v_{1n}^{w} and m_{1n}^{w}, shown in Figure 6–52b. By applying at node n additional external forces V and M of appropriate magnitude, the beam shear and moment can be reduced to zero, which corresponds to the required pinned joint condition. The values of V and M are determined as follows.

Just to the right of node n, let the shear and moment, due to a unit vertical load at node n, be given by v_{1n}^{v} and m_{1n}^{v}, respectively. Thus, for arbitrary vertical load V, the beam actions would be Vv_{1n}^{v} and Vm_{1n}^{v}. The end forces acting on the beam section to the left of an including node n are then $V(-1+v_{1n})$ and Vm_{1n}^{v}, respectively. Similarly, for load M, the end forces are given by Mv_{1n}^{m} and $M(-1+m_{1n}^{m})$, where v_{1n}^{m} and m_{1n}^{m} are the shear and bending moment due to unit applied moment. Thus, for zero end forces in the beam, just to the right of node n in the original system (Figure 6–52) the following equations must be satisfied.

$$V(-1+v_{1n}^{v})+Mv_{1n}^{m}+v_{1n}^{w}=0 \qquad (6\text{–}51a)$$

$$Vm_{1n}^{v}+M(-1+m_{1n}^{m})+m_{1n}^{w}=0. \qquad (6\text{–}51b)$$

These two equations can be solved simultaneously for the unknown values V and M. The coefficients in Eq. (6–51) are derived from Eqs. (6–45), (6–46), (6–49), and (6–50) by putting $n=0$ and $P=1$ or $M=1$.

Obviously, similar principles can be applied to a beam with a pinned joint at its lefthand end. Unfortunately, for a beam of finite length, the condition at one end will have an effect on the forces at the other end. Thus, for strict equilibrium, it would be necessary to set up four simultaneous equations involving the shears and moments at each end. However, as can be seen from a typical internal-force expression such as Eq. (6–45), the term p^n reduces the effect of an applied force quite rapidly away from its point of action. Thus, in most practical cases of finite-length beams, the actions at one end have a negligible effect on the other end, and the boundary conditions can be handled independently. For both manual and computerized numerical examples developed by the writer of this section the reader is referred to Tsinker (1986b).

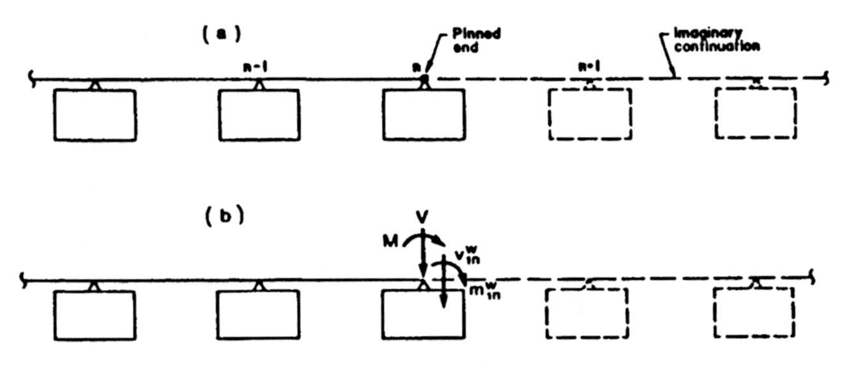

Figure 6–52. End condition for finite-length beam.

6.5.7.3 Floating Pier Dynamic Stability

Normally the floating object if unrestrained is subject to six degrees of freedom of movement. Of these six, three modes, for example, heave, roll, and pitch, where heaving is vertical motion, rolling is rotation about its longitudinal axis, and pitching is rotation about its transverse axis, are of particular significance in floating pier design.

In a real site condition, all three motions can be generated simultaneously, and all three are true oscillatory motions. If the pier is displaced from its equilibrium position by some force, when that force is removed the pier will oscillate until the motion is damped out.

It should be noted that in most cases pitching motion usually is not important, because of the relatively long length of a floating pier structure compared to the length of the incident waves. Pitching motion, however, may have a significant effect on a pier deck structure specifically when the latter is designed in the form of a continuous beam spanned between separate floats (pontoons).

In turn the pier rolling may be affected by mooring restraint which increases damping thus reducing maximum amplitudes.

Rosen and Kit (1985) recommended the following limits of acceptable motions within small-craft harbors as dictated by the human response to the motion:

- Maximum linear accelerations $\cong 0.4\,\text{m/s}^2$
- Maximum angular accelerations $\cong 2^\circ/\text{s}^2$
- Maximum peak roll to roll motion $\simeq 6^\circ$.

Rolling and pitching are similar in that they are both rotations. However, the added, or hydrodynamic, mass of the pier can be ignored with rolling, as it is likely to be relatively small; but this mass cannot be neglected for pitching. Rolling and pitching periods (T_R and T_p) can be determined by Eqs. (6–52) and (6–53), respectively.

$$T_R = 2\pi \frac{r_t}{(g(\rho_t + d/2))^{0.5}} \tag{6–52}$$

and

$$T_p = 2\pi \frac{r_\ell}{(g(\rho_\ell + d/2))^{0.5}} \left[\frac{w_1 + w_2}{w_1}\right]^{0.5} \tag{6–53}$$

where

r_t and r_ℓ = pier (pontoon or float) transverse and longitudinal radii of gyration

ρ_t and ρ_ℓ = pier (pontoon or float) transverse and longitudinal metacentric radius above the pier (pontoon or float) center of buoyancy

d = pier (pontoon or float) draft

g = gravitational acceleration

w_1 = weight of pier (pontoon or float)

w_2 = added weight (weight of the entrained water).

For a discussion on w_2 consult Tsinker (1986b).

As seen from Eqs. (6–52) and (6–53), rolling and pitching periods are inversely proportional to the square root of $\rho + d/2$. Therefore keep periods long, and with small accelerations this value should be small.

The designer has more control over the rolling period of a pier than over the periods of the other motions. In piers having a large value of $\rho_t + d/2$, reducing the value could give longer rolling periods.

The designer has very little control over the natural pitching period, as the value of $(\rho_\ell + d/2)$ is very large, and as any relatively small changes made in $p_\ell + d/2$ will have a negligible effect.

However, as floating piers and specifically those used in small craft hrbors are usually deployed in sheltered locations, pier pitching is unlikely to occur. Therefore, the problem of pier pitching is more theoretical than practical.

In the case of the pier of constant rectangular cross-section heaving freely in still water, the period of heave motion, for example, the time for one complete oscillation (T_H in seconds), can be determined by Eq. (6–54).

$$T_H = 2.83(d)^{0.5} \tag{6–54}$$

where

d = pier draft, in meters.

The waves create a vertical force on the pier, and this force is a simple harmonic function of time. Heaving motion is thus generated, and its magnitude basically depends on the ratio of the heaving period of the pier (T_H) to the encounter period of the waves with the pier (T_E) sometimes called the "turning factor" n. The latter is found from the following ratio:

$$n = T_H/T_E. \tag{6–55}$$

The resonant condition occurs when n approaches unity. Under the resonant condition, the maximum amplitudes of heave will occur.

The pier should be designed to avoid resonant condition (large amplitudes), and under given site conditions this can be done only by altering the pier draft so as to change the encounter period.

For more information on floating pier dynamic stability the reader is referred to standard texts on naval architecture.

6.5.8 Perimeter Structures

6.5.8.1 Perimeter Stabilization Structure

Marina basin perimeter structures can be divided into two groups. The first group includes structures whose objectives is to protect marinas' water area from penetration by heavy waves, ice, or by floating objects such as floating trees or lost control small crafts. These structures may take the form of the bottom founded or floating breakwaters, miscellaneous wave barriers, wave attenuators, or ice booms.

The second group includes structures built inside marinas with the objective of perimeter stabilization.

The basic objectives of perimeter stabilization is to retain the shape of the marina basin, and to prevent damage to the perimeter by external forces. External forces are those that tend to erode or damage the perimeter structure by means of wave action, currents, ice action, etc. The configuration of the land-based perimeter structures may be designed as a rivetted or natural slope, vertical wall, or partly sloped/vertical structure (Figure 6–53). Because inside marina structures usually are not exposed to large waves, strong currents, or severe ice conditions, the degree of required bank stabilization is usually minimal. In many marinas the bank protection is not required at all. In fact, quite often a natural bank presents a very pleasing appearance and does not need any protection. Armoring is required in most cases to protect the natural bank, or the constructed slope from scour to boat-wake waves (Sorensen, 1973). The most common slope armoring is rip-rap, precast concrete

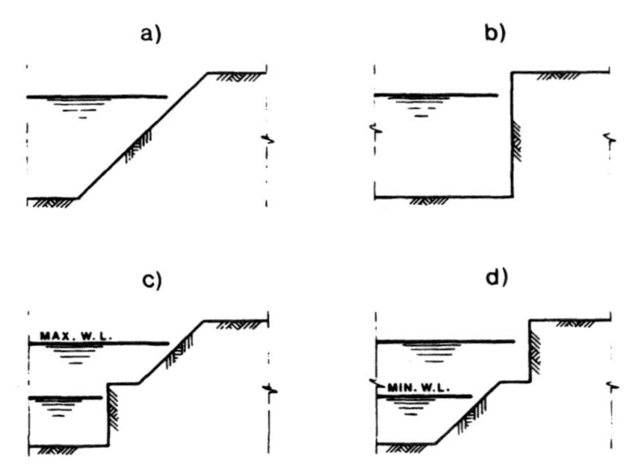

Figure 6–53. Typical perimeter structure configurations. (a) Slope, (b) vertical, (c) partly vertical, (d) partly sloped.

block units, and concrete paving (Figure 6–54). Gabions, placed upon a properly levelled slope, also represent a viable and economical solution to bank protection.

The bank protection is usually designed to sustain effects of wake waves and the design waves are usually those generated by fire boats on call, which can produce wake waves up to about 1.0 m in height. A filter cloth or selected filter course of about 0.15 to 0.2 m, made of gravel or crushed stone, is usually designed to underline armor material.

For guidelines for design and construction of flexible revetments incorporating goetextiles the reader is referred to PIANC (1992).

Vertical walls, although less economical than rivetted slopes, are used where land and/or water areas are limited or costly, because a vertical configuration of a basin perimeter eliminates the waste of space occupied by the slope. However, high wave reflections from vertical walls will be experienced and may cause unfavorable wave conditions inside the basin.

Of the many types of vertical perimeter walls currently in use, the least costly to build are anchored sheet-pile bulkheads. They are built where the foundation soil permits pile driving. Timber, steel, and concrete piles of great varieties of cross-section are used for the manufacture of sheet piles. Most of the problems reported with anchored bulkheads are caused by a wrong design or poorly constructed anchorages. Less frequently, reported bulkhead failures are attributed to heavy overstressing of sheet piles, or sheet-pile wall undermining

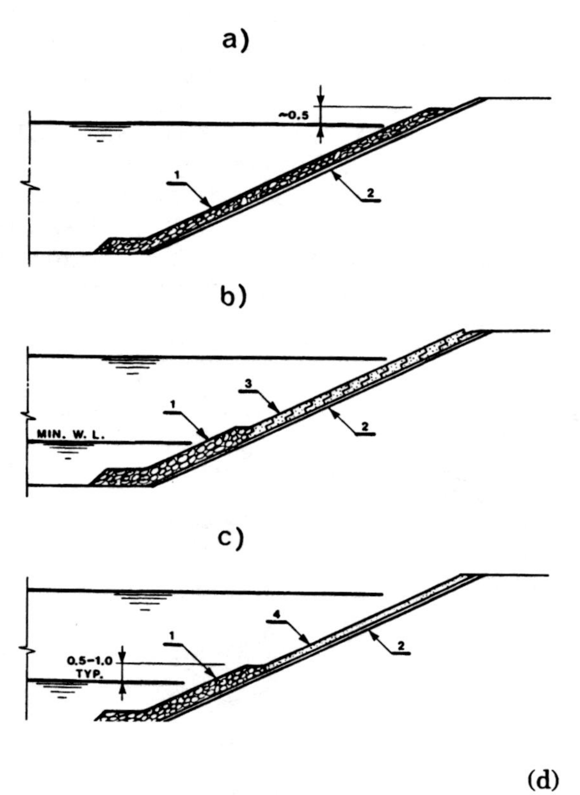

Figure 6–54. Typical slope armoring. 1—Rip-rap, 2—filter cloth, 3—interlocking blades, 4—concrete paving.

by overdredging or by propeller-induced scour. Sheet-pile bulkheads are usually considered as "flexible" structures because of their ability to deform slightly. The problems involved in the design of "flexible" retaining structures are somewhat more complicated than those applying to regular retaining walls. Soil pressure is the main force acting against the sheet-pile wall. The magnitude of this pressure depends on the physical properties of the soil and the kind of interaction of the soil/structure system.

Movement (deflection) of the structure is a primary factor in the development of earth pressures. However, this problem is highly indeterminate. Earth pressures are affected by the time-dependent nature of soil strength, such as consolidation due to vibration, groundwater movement, soil creeping, and chemical changes in the soil.

Construction sequences also have considerable impact on sheet-pile wall performance.

Two basic anchored bulkhead design procedures are in common use today. They are: (1) the free earth support method, and (2) the fixed earth support method. The first one is based on the assumption that the soil into which the lower end of the sheet pile is driven is incapable of producing an effective restraint from passive pressure to the extent necessary to induce a negative bending moment. The sheet pile is driven just deep enough to ensure stability, assuming that the maximum possible passive pressure is fully mobilized. The second method is based on the assumption that the sheet pile is driven deep enough so that the soil beneath the dredge line provides the deflected shape of the pile that reverses its curvature at the point of contraflexure and eventually becomes vertical. Consequently, the bulkhead acts as a built-in beam subjected to positive and negative bending moments. Soil pressure is the principal load acting against any type of retaining structure, including anchored sheet-pile bulkheads.

The nonhomogeneity of soils resulting in uncertainty in the properties of the backfill, and uncertainty to some degree in the mechanism of soil–structure interaction, presents problems in determining of soil pressure against the anchored sheet-pile bulkhead. That is probably why the aforementioned problems cannot be solved with mathematical precision and why one cannot expect to obtain from computation results more than an approximate accuracy in relation to reality. The effect of an error in anchor system design, leading to excessive deflection of sheet piles, is seen in Figure 6–55.

The gravity-type wall built for perimeter stabilization is usually limited to foundations that do not permit pile driving. This type of

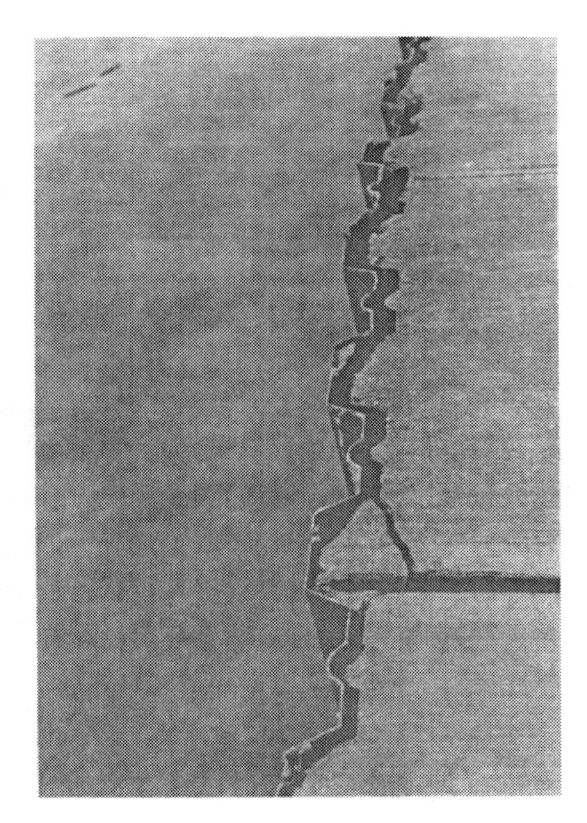

Figure 6–55. Excessive deflection of sheet piles due to a poorly designed anchor system.

construction in wet is generally more costly than sheet-pile bulkhead structures. Where construction in dry conditions is possible, the simplest solution is a masonry cast-in-place concrete gravity wall. For construction underwater where depth of water does not exceed 3 to 4 m the gabion type wall may represent a viable and economical solution (Figure 6–56). Owing to its flexibility this type of wall may be built upon foundation soil directly without stone bedding. If granular backfill is used then filter material must be installed to prevent the backfill from being washed off.

Gabion retaining walls have several advantages. They are free draining and, if proper filtering exists they are usually not subjected to unbalanced hydrostatic pressures due to

(a)

(b)

Figure 6–56. Vertical perimeter wall constructed from gabions (Courtesy of Maccaferri Gabions Ltd.)

water fluctuation. These walls can accommodate substantial settlement without failure, although excessive settlement can produce an unattractive appearance. Gabions represent a useful technique for remote areas, as the wire baskets can be easily brought in by hand and filled with local stone.

The disadvantages of gabion walls are price and durability; they are labor-intensive to construct and their long-term durability is always in question.The latter is basically attributed to gabion basket wire deterioration, particularly in tidal and splash zones.

Even polyvinyl chloride (PVC)-coated wire is subjected to deterioration at gaps in the coating. The latter can be created by the placement of rough angular material. A damaged basket will eventually develop a hole and start losing stone which in turn can compromise wall integrity.

Gabion wall is a gravity type retaining structure and its analysis include review of stability against sliding, overturning and bearing failure. Because these walls are flexible then active soil pressures are appropriate for determination of lateral thrust against the wall exterior. For calculation of soil lateral thrust some gabion manufacturers recommend using of angle of wall friction equal to the angle of internal friction of the backfill. More conservative is value of wall friction, δ, will be $2/3\phi$, where $\phi =$ angle of internal friction. If a relatively stiff, smooth geotextile is used as a filter on the back of the wall then reduced value of δ should be considered.

Partly vertical or partly sloped configurations (Figure 6–57) are generally used as a compromise between sloped and vertical configurations.

6.5.8.2 Protection from Waves

Protection from waves is usually accomplished by construction of the bottom-founded vertical face breakwaters of miscellaneous constructions, rubble mound breakwaters, floating breakwaters, and fixed and floating wave attenuators of miscellaneous designs.

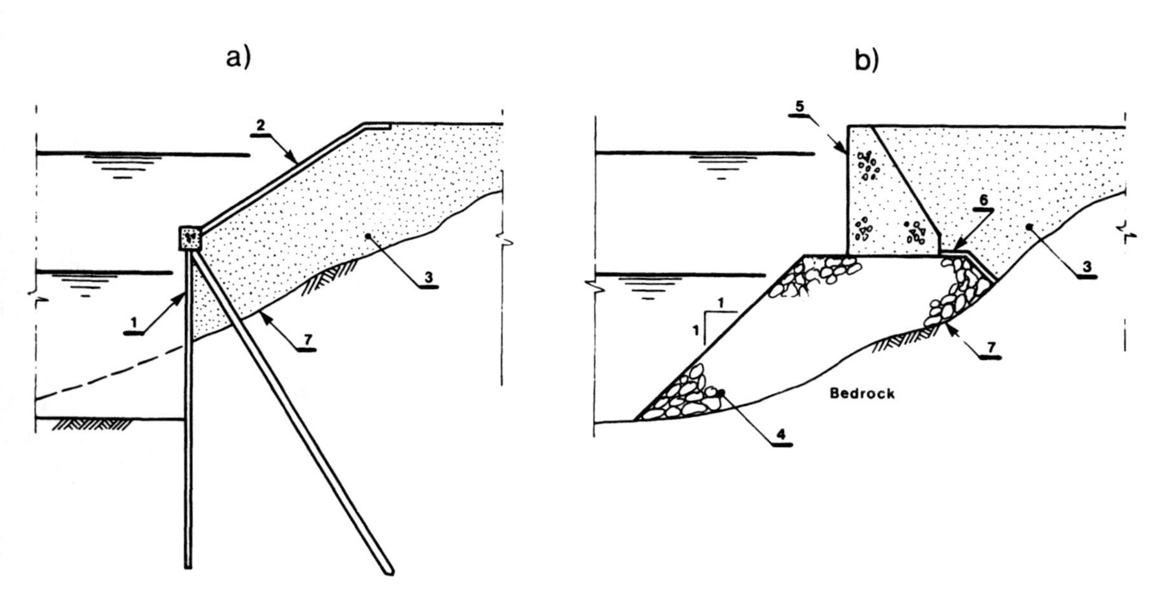

Figure 6–57. Partly vertical (a) and party sloped (b) configurations of perimeter construction. 1—Sheet-pile bulkheads, 2—reveted slope, 3—backfill, 4—rip-rap, 5—gravity-type retaining wall, 6—filter material, 7—natural slope.

The subject of floating breakwater design is covered in detail in Chapter 5.

The type of wave protection structure used in marina construction is largely dictated by wave climate outside the marina water area, the allowable height of waves inside the marina perimeter, bottom conditions, and availability of suitable materials at or near the site of construction. There are two basic types of bottom-founded breakwaters: the impermeable vertical solid-type construction breakwaters and the sloping mound-type structures. The combination of both types is also in common use.

Solid Vertical Barriers

These usually take the form of blockwork walls, floated-in concrete caissons, steel sheet pile cells, concrete cylinders, or timber or concrete cribs filled with rocks.

Use of solid, vertical face type breakwaters is practically limited to a water depth of no more than 15 to 20 m, and where used in deep water they are usually founded upon a bedrock foundation.

Wave energy absorption by a solid wall type breakwater can be increased by using different techniques. One of these is use of perforated caissons. This type of breakwater was pioneered by the Department of Public Works, Canada and National Research Council of Canada in the early 1960s, when the department ordered the feasibility study of perforated breakwater to the site of Baie Comeau, Quebec (McNamara Engineering Ltd., 1961; Jarlan, 1961). In some cases riprap placed along the wave active surface of the vertical breakwater helps to reduce wave carryover ability and reduce wave energy to be absorbed by the breakwater itself.

Rubble-Mound Breakwaters

These are usually made by dumping natural rocks or man-made concrete blocks of different geometry. The size of rocks or weight of man-made concrete blocks depends on available equipment to handle these elements, flatness of exposed wave slope, and, of course, on wave characteristics.

In some cases the presence of ice may have a significant impact on the shape of the exposed slope of the breakwater and size of the armour blocks placed to protect breakwater slopes. Typically rubble-mound breakwater consists of the core made up from a regular run-of-quarry rock, protected by heavy rocks or concrete man-made blocks. Because of the permeable nature of this type of structure the armor layers are usually underlaid by single or several layers of filter material. An example of placement of large rocks for breakwater protection is shown in Figure 6–58.

Typically construction of rubble-mound breakwaters in relatively deep water in

Figure 6–58. Placing heavy armor rocks for protection of small-craft harbor breakwater.

combination with exposure to high waves necessitates the use of heavy protective armor units. Under the aforementioned site conditions three slope geometry breakwaters with a berm appeared to be effective in reducing wave run-up and in increasing stability of the breakwater armor layer (Herbich et al., 1963; Kuzucu, 1973).

Bruun and Johannesson (1976), Bruun and Günbak (1976), and Ergin et al. (1989) described the hydraulics of S-shaped rubble-mound breakwaters. The aforementioned authors recommended use of S-shaped breakwater geometry for its similarity to an equilibrium beach slope instead of continuous single slope, for an increased stability and economy.

It should be noted that being a permanent and quite visible part of the marina landscape the poorly designed and constructed rubble breakwaters may become a "visual pollution" to the area. To reduce the possibility of this type of damage to the environment the breakwater's crest elevation should be as low and as narrow as practical. The latter can be achieved by supplementing a conventional breakwater of any design with a submerged (reef) breakwater placed offshore and operating in tandem with the principal structure.

This type of protection from waves was designed for construction of the marina located in the southern end of Lake Michigan. The U.S. Coast Guard did not consider the reef as a navigation hazard, provided it has adequate signage, and the concept itself was readily accepted by environmental interest groups as a new fish habitat area (Cox, 1991).

Similarly, wave energy dissipation effect can be achieved by construction of trench-type breakwaters (Berenguer et al., 1990) and others like it (Madrigal and Valdes de Alarcon, 1992). This type of construction includes a trench, located just beyond the top of the edge exposed to the wave slope. The waves that are overtopping the breakwater's crest are trapped within the trench. The dissipation of wave energy is caused by friction with the inner and outer armor layers, crest and the crest rear slope, turbulence inside the trench, and splash against the vertical wall at the rear end of the trench. Several breakwaters of this type have been successfully used in Spain. According to Berenguer et al. (1990) reduction in crest elevation up to 30% has been achieved.

A detailed discussed on breakwater design procedure is given in Bruun (1985 and 1989), Quinn (1972), Van der Meer et al. (1991), U.S. Army Corps of Engineers "Shore Protection Manual" (1984b), PIANC (1992), Oumeraci et al. (1994), and in a great many other related publications.

In-depth discussion on stability of the seaward slope of berm breakwaters and effects of such parameters as seaward slope geometry, wave characteristics, storm duration, spectral shape, rock size, shape and grading, depth of water, wave attack, and others is offered by Van der Meer (1992).

Breakwater Layout

Because breakwaters usually are very expensive to build several layouts as well as several structural types should be evaluated in order to arrive at the most economical solution. The optimum solution will be based on establishment of acceptable wave heights in various basin areas; for example, larger wave heights may be acceptable in moorage areas for larger fishing vessels, whereas a 30 cm wave may be the maximum acceptable at a boat ramp. Therefore wave penetration studies are required to investigate the expected wave conditions in all basin areas, as well as in an entrance/exit channel(s).

Wave climate inside the marina basin depends on wave refraction and diffraction, as well as breakwater overtopping. The best breakwater layout is obtained from physical model studies; a good mathematical model can also produce credible results.

Properly placed and angled breakwaters will result in smaller and less expensive structures; the failure to place breakwaters (or wave attenuator devices) at the proper boundaries may result in failure of wave attenuation goals.

While planning the breakwater(s) layout the marina designer must also be concerned with quality of water and flushing of the basin, and therefore layouts that may prevent proper water circulation in and out of the marina basin must be avoided. In some cases to improve water circulation within the marina basin the flow pipes are installed within breakwaters. These pipes can be positioned in a way to conduct well aerated surface water to the bottom of a marina basin and thus to create turbulent mixing in addition to direct bottom water replacement. Alternatively, if basin flushing either by the direct flow of water caused by a river or by tidal current is not feasible then mechanical means such as pumps, propellers, or air bubbler systems sometimes may be effective to prevent poor water quality caused by sluggish water or by the total absence of water movements.

Sheet-Pile Breakwaters, Wave Boards, and Wave Fences

These structures are built with objectives similar to solid vertical barriers. They are usually used where design wave parameters allow for construction of light type structures to dissipate wave energy, or substantially quail the waves

Sheet-pile type breakwaters may be constructed in the form of free-standing walls, or walls supported by battered piles (Figure 6–59). In the former case the reinforced concrete sheets, or reinforced concrete tubular piles of miscellaneous constructions (steel sheet piling included), are propped either on one side or on both by piles (usually by concrete piles). Unsupported walls are used where foundation soils are competent enough to provide reliable lateral support for sheet piling. Depending on bottom soil geotechnical conditions, wave parameters, and sheet-pile construction, for example, flat planks or tubular piles, the height of the unsupported breakwater may be up to 6 to 8 m. The height of the propped sheet-pile wall may be much higher.

An example of a propped sheet-pile breakwater constructed at the Fisherman's Wharf on San Francisco Bay is discussed by Thuet (1987). There the interlocking precast prestressed concrete sheet piles were used for construction of the 460 m long breakwater. Sheeting was supported by precast prestressed batter piles and capped with cast-in–place reinforced concrete cope. Batter piles were used on both sides of this breakwater, except for a 130 m long section which

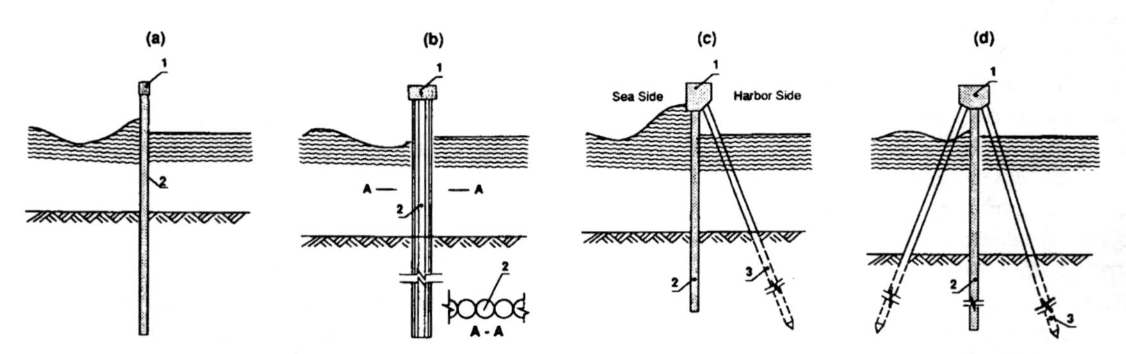

Figure 6–59. Sheet-pile breakwater. (a) Free-standing sheeting made from steel or concrete sheet piles. (b) Free-standing wall composed of steel or concrete large-diameter cylindrical piles. (c) Sheet-pile wall supported by piles on one side only. (d) Sheet-pile wall supported on both side.

had batter piles on the harbor side only. Two types of sheet piles 76 × 76 cm and 62 cm thick and 102 cm wide were used. The batter piles 61 × 61 cm were driven into the bottom at a slope 5 (horizontal) on 12 (vertical). The size, spacing, and length of both sheet piles and batter piles were designed to resist action of the design wave 1.8 m high, and seismic and other horizontal loads. A differential head of about 3.0 m was assumed across the breakwater to account for possible scour and/or liquefaction on the "sea" side and a trough on the "lee" side.

Sheet piles and batter piles were brought together by a cast-in-place 3.05 m wide and 1.22 m high concrete cap.

Wave Board and Wave Fences

These are the lightened version of sheet-pile type breakwaters. They represent a compromise between environmental requirements for water quality inside the marine basin, and at the same time providing some wave protection for the site. The objective of these structures is basically to quail the waves by intercepting the wave energy in part and reflect them seaward rather than completely dissipating wave energy. Wave boards and fences are built in the form of either vertical or horizontal planks attached with predetermined spacing to a support system, which usually consists of piles and walers.

The spacing between boards basically depends on the amount of wave energy allowed to penetrate a marina basin. Usually a certain space is left between the facing and the bottom to allow for water circulation; the latter provides for better flushing action and water renewal within the marina area, and also allows for unrestricted migration of fish and other migrating marine animals. The width of boards, spacing between them, as well as the depth to which they are extended from the minimum water level are governed by the design wave characteristics; the longer the wave period, the greater wave energy can pass between wave boards and beneath the facing.

Floating Wave Attenuators

Under certain site conditions, for example, deepwater, these structures may be a good choice for wave attenuation; they can offer protection of the marina basin, while providing unimpeded flushing of the basin and unrestricted migration of fish and migrating marine animals. Being a kind of a floating breakwater these structures are available in a variety of designs; some of them are made from barges and pontoons ballasted deeper into the water and anchored in place in one way or the other at the desired location, and the others operate using a variety of wave suppressing techniques. Floating wave attenuators are normally designed to provide protection from waves and wake experienced under normal conditions. The structure and its mooring system, however, must be able to survive the extreme condition.

In general, if a site is vulnerable to heavy wave conditions, then the floating attenuators may not be a suitable solution. Examples of the floating wave attenuators of miscellanous designs are found in almost all proceedings of specialty conferences dealing with navigation, ports, and small-craft marina design and construction.

6.6 DREDGING OF THE MARINA BASIN: SOME ENVIRONMENTAL ASPECTS

6.6.1 General

The need for new and/or maintenance dredging is common for construction and operation of not only the marinas, but also of many other harbor facilities. It involves the removal of subaqueous soil materials or sediments to provide adequate depth of water for safe movements of commercial and recreational vessels. Dredging is performed by either onshore based or floating equipment designed to reach the bottom soil. The latter include mechanical dredges such as clamshell, dipper, bucket-ladder, and other types of dredges and hydraulic dredges, such as plane suction, dustpan, and cutter head dredgers. The principles of operation and advantages and disadvantages of these machines are discussed by Turner (1992). The underwater excavation may also be carried out from the shore by dragline and/or backhoe cranes. It should be noted that usually use of shore-based equipment (dragline in particular) results in high turbidity.

The most common equipment used in marina construction is barge-mounted cranes with a mechanical clam shell bucket, or small hydraulic dredge. The excavated sludge is transported from the place of dredging activities to the site of disposal in traditional dump barges, self-propelled barges, or by floating and/or land pipe lines.

The beneficial effects of dredging operations result from the primary dredging objectives such as creation of a basin of desirable parameters, improvements of navigation and waterflow conditions, removal of contaminated soils and sediments, and others.

It may however produce an adverse environmental effect that includes turbidity at the dredging and disposal points, overflow from barges or hoppers creating turbidity in the receiving waters, adverse effects on fish habitat, interference with navigation, noise, and others.

At present the dredging dilemma is disposal of the sludge from basin dredging activities, which in some cases became a precarious problem from the environmental protection viewpoint. Fear of the contaminated "toxic sludge" created, not always warranted, "not in my backyard" mentality. Dredging and disposal of contaminated materials according to a 1989 report by the U.S. National Research Council's Marine Board became an "environmental issue of national importance due to decades of using coastal water for waste disposal." According to the same report one third of the nation's polluted sites are marine or in estuarine waters. In some instances similar conditions exist in Canada.

At present any dredging activities are unthinkable without a thorough evaluation of its impact on the environment.

The latter usually include sampling, testing, and evaluation of contamination of dredged sludge and the effect of dredging and sludge disposal on the environment.

The sludge disposal alternatives include open water and land sites; the latter may be confined or unconfined.

The subject of sludge disposal is beyond scope of this work and for the relevant information an interested reader is referred to Barnard (1978), Cullinane et al. (1990), Van Wijck et al. (1991), Palermo (1992), and PIANC (1990).

This work is primarily focused on the problem of containment of the turbidity at points of dredging and underwater disposal.

6.6.2 Turbidity Created by Dredge and Underwater Disposal

Public awareness of the detrimental effects of adding pollutants to the natural environment has increased dramatically during the last two decades and many governments all over the world have passed laws and set up special agencies to prevent degradation of the environment. One form of environmental contamination is the transport of the dredge-related sediments into seas, lakes, and rivers.

One of the major concerns about dredging, filling, trenching, and open-water disposal of dredged materials that churn up the sediments and cause the finer particles to go into colloidal suspension involves the possible environmental impact associated with the resuspension and subsequent dispersion of fine-grained material. The resulting turbidity may be quite persistent and locally affect the microorganisms' growth and reproductivity. This is particularly true when dealing with dredging of the contaminated materials. The latter is significant considering the fact that the vast majority of the potentially toxic chemical contaminants present in bottom sediments are associated with a fine-grained fraction that is most susceptible to dispersion (Bannon et al., 1976).

The larger particles, fine sands, and silts may have a disastrous effect on oyster beds, both natural and artificial.

The material suspended during dredging or open-water disposal of dredged sludge is often referred to as turbidity and the dense near-bottom suspensions are commonly called fluid mud or bluff.

It should be noted that suction dredging generally causes less turbidity than bucket dredging, and large clamshell buckets cause less disturbance than a bucket ladder dredge. Dredged material suspensions are quantitatively classified by their concentration of suspended solids expressed in milligrams or grams per liter, percent solids (by weight), or bulk density. For more information on this subject the reader is referred to Barnard (1978). It should be noted that dredgers usually describe a slurry in terms of percent solids by volume, where the volume of in situ sediment (including both the solids plus water) excavated during an operation is divided by the volume of slurry pumped; this fraction is then converted to a percentage.

Turbidity is commonly used to describe the cloudy or muddy appearance of water. There are a variety of instruments and techniques that have been used to evaluate or quantify turbidity (McCluney, 1975). The turbidity plumes generated by dredging and/or dredging operations are usually caused by low concentrations of silt and clay-size particles (with diameters of less than 0.03 mm) or small flocs (i.e., masses of agglomerated particles) that settle independently at very slow rates through the water column.

The concentration of suspended solids in a turbid water column is usually classified as low-density and high-density fluid mud.

The former is characterized by the mud concentration in a water column equal to about 5 to 20 g per liter and the latter is considered when the solids concentration exceeds 200 g per liter.

Low-density fluid mud may be stationary or may freely flow outward and is characterized by randomly oriented particles that settle at "hindered" rated (Einstein and Krone, 1962). As solids concentrate up to 200 g per liter the hindered settling process apparently ends and self-weight consolidation begins (Einstein and Krone, 1962; Migniot, 1968).

The high-density fluid mud due to a high degree of solid concentration does not flow as freely as low-density fluid mid.

The nature and extent of dredged material dispersion around a dredging or disposal site are controlled by the characteristics of dredge material such as size distribution, solids concentration and composition, type of dredge equipment, cutter configuration, operational procedure, and, of course, hydrological and hydrodynamic conditions (i.e., waves, currents).

The turbidity of the water column to a large degree depends on a number of factors among which are: the discharge rate, character of the dredge material slurry, depth of water, hydrodynamic regimen, and discharge configuration. The nature and persistence of turbidity plumes are controlled largely by the settling rates of the material suspended in the water column.

Low concentrations of silt and clay (with diameters of less than 0.03 mm) settle very slowly, causing large, persistent turbidity plumes. If the suspended particles are coarse-grained or composed or large flocs, they will settle relatively rapidly; the suspended solids concentrations in the resulting plume will be relatively low and decrease very rapidly with distance from the discharge point.

Turbidity plumes will be relatively persistent in freshwater, because fine-grained particles at low concentrations do not readily form flocs (Wechsler and Cogley, 1977). However, the degree of flocculation increases very rapidly as salt concentrations increase from 0 to 10 g/ℓ and remains essentially constant at concentrations between 10 g/ℓ and seawater concentrations of 35 g/ℓ (Migniot, 1968). The clay mineralogy of the dredged sediment may exert a subtle effect on settling behavior but its importance is relatively small due to the presence of naturally occurring organic material which apparently coats the clay particles. In fact, in saltwater the settling rate of the suspended material generally increases as the organic content increases (Wechsler and Cogley, 1977). Regardless of the sediment/water composition, settling rates generally increase with increasing solids concentrations up to approximately 10 to 20 g/ℓ (Migniot, 1968; Wechsler and Cogley, 1977); at higher solids concentrations the particle settling rates are "hindered" or reduced due to contact with adjacent sediment particles or flocs.

The particle settling rates, slurry discharge rate, water depth, current velocities, and the diffusion velocity (describing horizontal dispersion) all interact to control the characteristics of the turbidity plume during the disposal operation (Schubel et al., 1978). As the current velocity increases, the plume (as defined by a specified level of suspended solids in excess of background) will grow longer. As stated by Barnard (1978), with increasing depth of water in the disposal area, the average concentration of suspended solids in the plume will tend to decrease. As the dredge size increases or particle settling rates decrease, the plume size and suspended solids concentrations will tend to increase. In addition, as the diffusion velocity increases for a given current velocity, the plume becomes longer and wider, while the solids concentrations in the plume decrease. (However, if there is no resuspension of bottom sediment, the total amount of solids in the plume will remain the same.) Finally, with a decrease in diffusion velocity or particle settling velocity, or an increase in water depth, the length of time required for the plume to dissipate after the disposal operation has ceased will increase.

A relatively simple method for predicting plume characteristics has been proposed by Barnard (1978). He developed a turbidity plume model that includes various parameters that have been referred to earlier.

6.6.3 Silt Curtain

Dispersion of turbid water resulting from dredge operation or open water pipeline disposal can be physically controlled by specially designed silt curtains that provide a barrier either down-current or around the working/disposal areas (Tsinker, 1994). Generally the silt curtain is made from a flexible, nylon-reinforced polyvinyl chloride fabric that is maintained in place by flotation segments at the top, ballast chain along the bottom, and an anchor system placed along the curtain. A tension cable is typically built into the curtain immediately above the flotation segments and in some instances also installed above the bottom chain (Figure 6–60).

The problems associated with curtain design are usually of hydraulic or material use nature.

Hydraulic problems are typically related to the water-body characteristics such as currents, waves, and fluctuation of water levels. High wind also proved to be a problem (Johanson, 1976a,b).

With regard to the curtains themselves, difficulties included rapid deterioration of

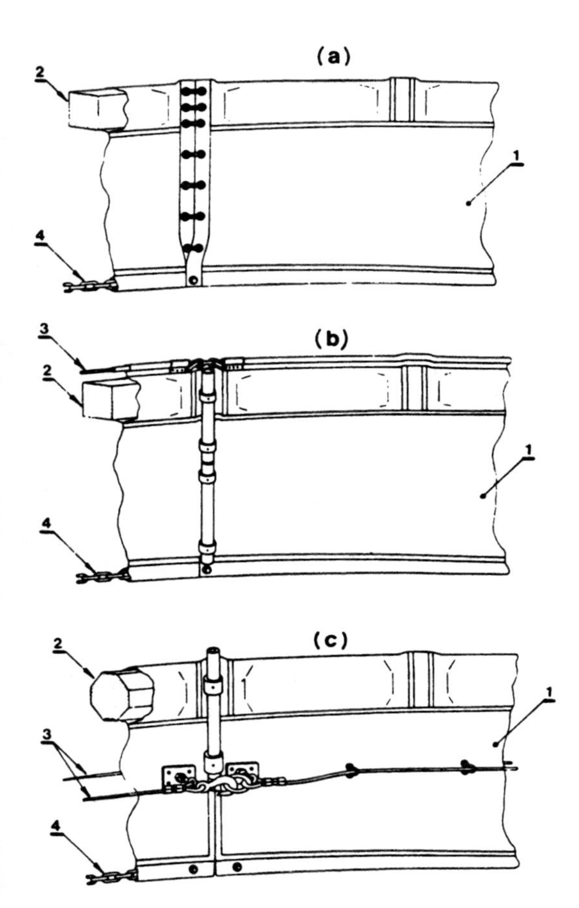

Figure 6–60. Turbidity containment curtain (containment system, Florida, Catalog). (a) Lightweight curtain for installation under no substantial current conditions. (b) Middleweight curtain for installation in rivers, open lakes, and exposed shorelines with moderate current moving in one direction. (c) Heavyweight curtain for installation in rivers, open lakes, and exposed shorelines with heavy current moving in any direction. 1—Polyester reinforced vinyl fabric, 2—flotation made from expanded polystyrene, 3—top cable, 4—bottom chain.

the fabric and marine growth on the fabric (Ekey, 1970), damage by boat traffic and marine animals, silt buildup on the skirt pulling the barrier under the water, flotation failure or inadequate flotation, parting of the fabric seams (Johanson, 1976a,b), inadequate weighing, improper weighing due to rough sea bed (McLuckie, 1981), and build up of sediment in shallow water resulting in prolonged intermittent turbidity as it was subjected to constant agitation (Gerner, 1971).

Silt curtains have been in use for the past two decades (Gerner, 1971; Ekey, 1970; Johanson, 1976a,b; Brown, 1978; Barnard, 1978; Suits and Minnitti, 1989). Kouwen (1990) reported results of model tests conducted on 12 different curtain configurations placed in a narrow channel.

The existing experience indicates that a properly designed silt curtain may reduce turbidity levels in the water column outside the curtain by as much as 80 to 90%. However, the effectiveness of the curtain may decrease with increasing current velocity, for example, if velocity exceeds 5 cm/s (Barnard, 1978).

The following basic options with respect to curtain layout geometry are usually considered (Figure 6–61):

1. On a river, where the current does not reverse and dredging or dredge discharge operations are taking place in the middle of a river an open U-shaped curtain layout is used (Figure 6–61a). In this case, however, the anchored ends of the curtain must be long enough to prevent leakage of turbid water around the curtain.
2. At locations with reverse currents a closed configuration of a curtain layout will be required. It can be of circular, elliptical, rectangular, or other configuration as dictated by local conditions (Figure 6–61b).
3. On a river, or on a shoreline, where dredging or dredge discharge operation are taking place in the relatively close proximity of the bank/shoreline the curtain can be anchored in a semicircular or trapezoidal configuration (Figure 6–61c).

As stated earlier the silt curtain is used to contain turbidity in a water column during dredging or dredge disposal operations. With respect to the aforementioned operations the silt curtain is usually designed as extended

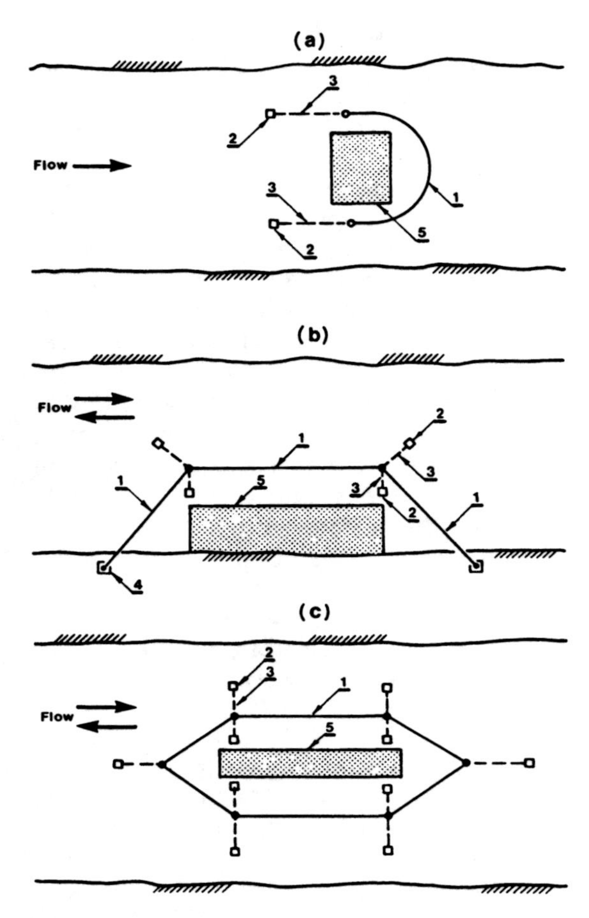

Figure 6–61. Silt curtain. Typical layouts (Tsinker, 1994). (a) In the middle of the river with current moving in one direction. (b) In the river (channel) with the current moving in one or two directions. Curtain is placed at the river bank. (c) The same as (b); however, curtain is placed in the middle of the river (channel). 1—Curtain, 2—bottom anchor, 3-anchor cable, 4—bank (shore) located deadman, 5—dredged area.

all the way from the surface of water to the bed of the waterbody, or as a partial depth curtain. The former, if not provided with a pressure-relieving mechanism, may be exposed to substantial loads which may induce excess stresses on a curtain structure. These stresses may become critical and induce a failure. On the other hand the partial depth curtain, although automatically allowing excess pressure to escape beneath the curtain, would also, however, allow sediments to escape.

The solution is found in providing the curtain which slack and a flap at its bottom part (Figure 6–62). The curtain of such design would normally experience smaller stresses than one "fixed" to the bottom. The stress level in silt curtain fabric as well as vertical components of the anchor forces will greatly depend on the depth (sag) of curtain "sail" which can be mathematically described as a catenary.

This type of curtain is recommended for use in dredging of the contaminated soils type operation. For this type of application, it is important that the curtain extends to the very bottom of the water body as the majority of the turbid cloud generated by the dredging operating moves along the bed (Herbich and Brahme, 1984).

The use of a partial depth curtain would be more appropriate in dredge disposal operations. Any sediment that escapes beneath such a curtain would be coarser material which would settle relatively quckly, thus having little detrimental effect (Johanson, 1976). Barnard (1978) provides a detailed set of recommendations for the design and installation of the partial depth silt curtains based on an evaluation of silt curtain performance under various field conditions. These recommendations are incorporated in the MTO Drainage Manual, Chapter F. (Ministry of Transportation of Ontario, 1985.)

6.6.4 Curtain Design

A silt curtain, when installed in a river, can be considered as a sudden contraction. The curtain itself will represent a floating flexible barrier exposed to hydrostatic and hydrodynamic loads due to potential differences in water levels within the containment system and outside of it ($f_{\Delta h}$), development of the impulse-momentum forces on the upstream

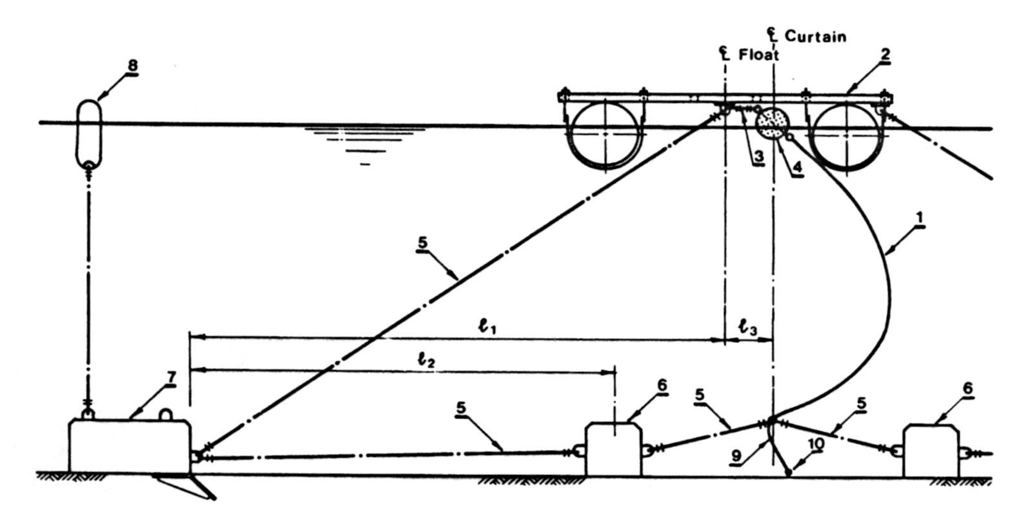

Figure 6–62. Silt curtain with pressure-relieving flap (From Tsinker, 1994). 1—Curtain, 2—anchor flotation, 3—cable connecting curtain to flotation, 4—polystyrene flotation, 5—achor cable, 6—sinker, 7—self-bearing concrete anchor, 8—marker buoy, 9—flap, 10—chain.

portion of the system (f_{iM}), lift-off (suction) forces in area of contraction (f_s), and drag forces along the curtain (f_D) (Figure 6–63).

The solutions for the above forces are given in any standard text on fluid mechanics and hydraulic engineering. Essentially the site characteristics where the curtain is to be employed plays a major role in the design requirements of the floating barrier. Current velocity, tidal and wave action, and sometimes wind characteristics will dictate the amount and type of flotation and anchoring devices required. A typical design diagram of the curtain extended to the river bottom with a flap at the lower end is depicted in Figure 6–64.

As shown there, the lift-off (suction) forces only are considered in curtain design.

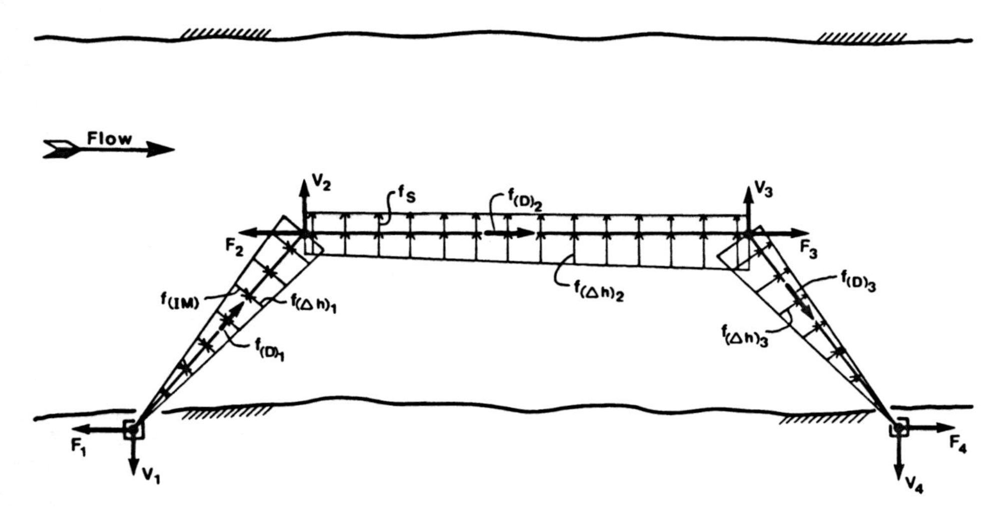

Figure 6–63. Silt curtain. Load diagram (From Tsinker, 1994).

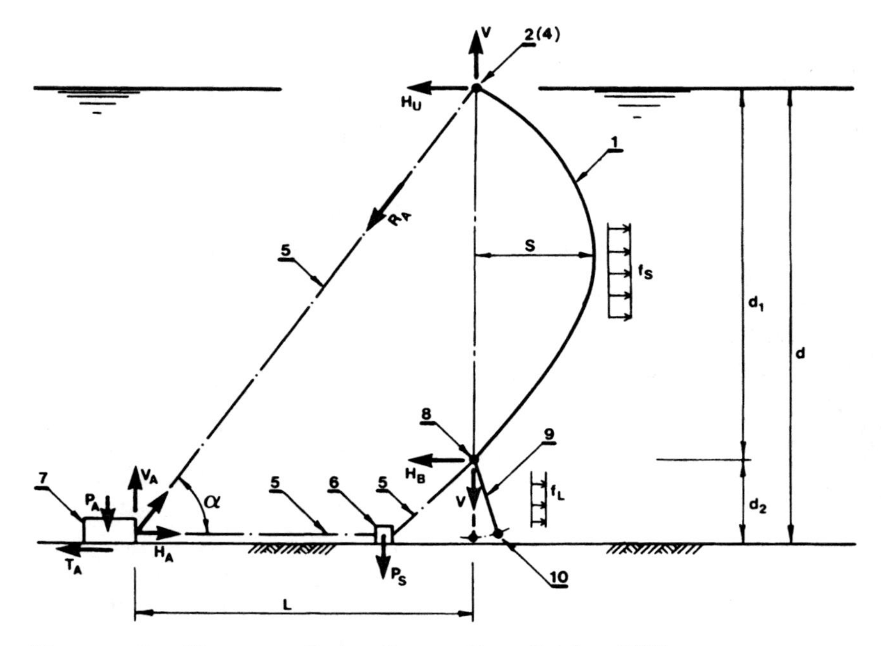

Figure 6–64. Silt curtain. Design diagram (From Tsinker, 1994).

This is because differential hydrostatic pressure is effectively eliminated by installation of a flap at the bottom of the curtain.

Subsequently forces H_u and H_b, which respectively are horizontal components of the reaction forces at the top of the curtain and at the lower wire rope (8), can be approximately determined from the following:

$$H_u = 0.5f_s d_1 \quad (6\text{–}56)$$

$$H_b = 0.5f_s d_1 + f_L d_2 \quad (6\text{–}57)$$

where f_s, d_1, d_2, and f_L are defined in Figure 6–64 and $(f_L d_2)$ is the force required to lift off the flap just a couple of millimeters off the ground to relieve the excess of hydrostatic pressure behind the curtain due to unbalanced head.

Vertical component of the reaction force V can be determined from the catenary equation

$$V = f_s d_1^2/8s \quad (6\text{–}58)$$

where

s = deflection of the curtain (Figure 6–64).

Typically the curtain stays afloat with the help of a continuous flotation incorporated into the curtain fabric. Therefore buoyancy of this flotation must be sufficient to support the weight of the curtain along with vertical component of the reaction force V, as well as provide a sufficient freeboard to ensure the turbid water would not overflow the curtain.

At the bottom part of the curtain vertical component of the reaction force V should be balanced by weight of the lower wire rope (8) and bottom chain (10). If this would happen to be insufficient then a sinker (4) should be added to the system.

The stress level in the curtain fabric is determined by the combined effect of horizontal (H) and vertical (V) components of the reaction force.

Typically the curtain is made from polyester or nylon reinforced vinyl or similar geotextile fabric. Although its strength requirements are determined by the hydraulic forces as discussed earlier, in some cases geotextile parameters may be governed by the specific site environment condition, for example, presence of aquatic plants, algae, animals, chemicals, and temperature (freeze/thaw action) (Johanson, 1976b).

Other design requirements include triple-sewn or heat-sealed seams (Gerner, 1971); a suitable maintenance and monitoring program should be in place to ensure that no formidable gaps exist between the curtain and the bottom. The metal parts of a curtain are usually galvanized.

The important component of a silt curtain is its anchoring system. An improper design of this system typically will lead to failure of the silt curtain. The anchor system design must be given full attention.

In a conventional anchor system the projected length of the anchor cable (L) is usually equal to 8 to 10 d, where d = depth of water. This is dictated by the need to eliminate the vertical component at the anchor to enable the conventional steel anchor to perform satisfactorily. For example, if the depth of water is equal to say 4.0 m then the minimum projected length of anchor cable $L = 40.0$ m. Therefore with added length of the conventional anchor plus some additional space between the anchor and the dredge which would normally be required for safety or dredging operation the total projected length of the "dead" zone will be equal to about 43.0 to 44.0 m. In some practical cases this can be prohibitive. In the case of limited space for example, excavation in a relatively narrow river, the designer will face the dilemma of how to reduce the anchor cable projected length. The latter is usually accomplished by use of sinkers, or introduction of gravity-type anchors.

A special anchor, which combines the properties of conventional anchors with those of gravity-type anchors, is depicted in Figure 6–65. This self-burying system proved to be practical and very efficient (Tsinker and Vernigora, 1980; Tsinker, 1986b). It develops resistance against the pull force by digging into the bottom soil. The anchor continues to dig in until sufficient soil resistance is developed and equilibrium between the pulling out forces and soil passive resistance is reached. As a result of digging into the bottom soil the anchor may tilt; however, for better stability this tilt normally should not exceed 30°. For more information of this type of anchor design the reader is referred to Tsinker (1986b).

In general the conformity of the bed will indicate the type of anchoring to use.

A gravity-type anchor must be heavy enough to resist full hydraulic load H exerted upon curtain system that is equal to

$$H_A = d_1 f_s + d_2 f_L. \qquad (6\text{–}59)$$

To satisfy the anchor sliding stability the required submerged weight of anchor (P_A) is determined from the following equation.

$$P_A = H_A F_{sl}/f + V_A \qquad (6\text{–}60)$$

where

f = friction coefficient between anchor material and the bottom. For a concrete block the following values of f are usually used: 0.25 for clay, 0.4 to 0.5 for sand, and 0.45 to 0.55 for rocky type of foundation material.

$F_{s\ell}$ = safety factor against sliding; $F_{s\ell} = 1.25$ to 1.3 may be considered sufficient.

Note that the upper and lower cables to which anchor cables are attached should be analyzed as a catenary correspondingly loaded with H_u and H_b. The span for these catenaries will be equal to the distance between adjacent anchors, and the design length of curtain cables should allow for the design deflection. Connection between curtain cables and the anchor cables is best made via sling links.

To ensure silt curtain stability in a body of water the anchors should be placed from both sides of it. This will protect the curtain

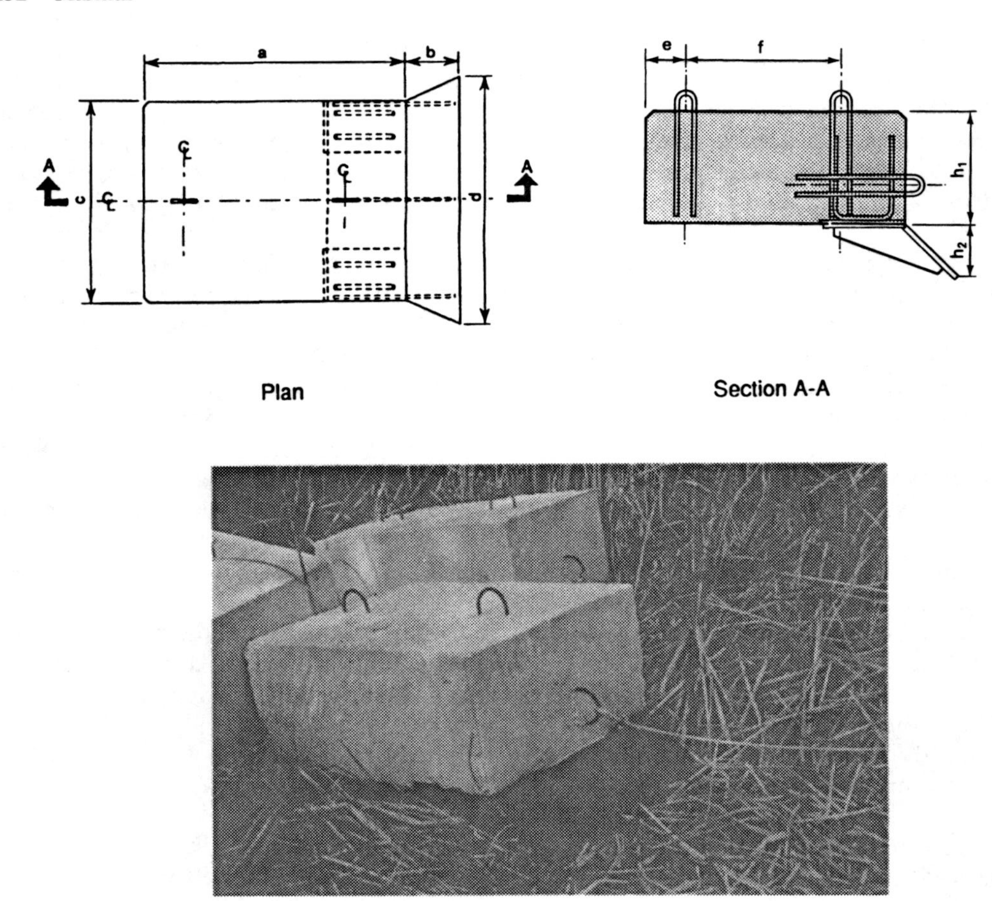

Figure 6–65. Self-bearing anchor.

from being displaced by occasional forces such as wind or waves. A typical silt curtain layout is depicted in Figures 6–66 and 6–67.

Because the vertical component of anchor force (R_A) normally cannot be absorbed by the curtain flotation unit, special floats should be added to the system.

Standard steel drums which are usually used for transporting the oil products, paints, etc., can be used for float construction, provided they are properly sealed and cleaned to prevent pollution of the body of water. These drums are usually readily available at low cost (or at no cost at all) sometimes from local retailers. Any kind of other standard commercially available floats may also be used.

6.7 DRY BERTHS

The extraordinary development of pleasure boating and the increasing difficulty of enlarging marinas for wet berthing by means of expansion of existing marina basins or by developing new marinas, particularly in highly populated areas, have created a situation where a great many boat owners store their craft at home or in some other location

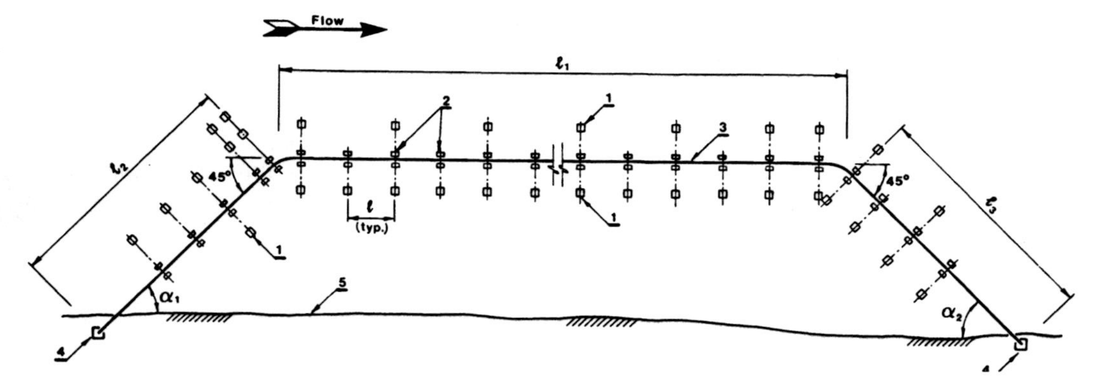

Figure 6–66. Silt curtain. Typical arrangement. 1—Anchor, 2—anchor float, 3—curtain, 4—deadman, 5—shoreline.

Figure 6–67. Upstream part of silt curtain exposed to heavy current. (Welland River, Welland, Ontario.)

and bring them to the waterfront on trailers pulled by their own cars. They launch their boats and haul them upon return to the marina at dry berths especially allocated for this purpose or at locations assigned for periodical removal of pleasure craft for maintenance, that is, repair, hull cleaning, and application of anti-fouling paint. Due to the congestion of small-craft marinas, particularly during the summer season, dry berths quite often are placed outside the marina basin. Dry berths usually include a land area for parking cars and trailers, a launching apparatus (most of them often are simple launching ramps or some kind of mobile crane), and certain facilities such as restrooms, a gas station, small repair shop, etc. Sometimes boaters prefer to keep their craft in dry storage which could be part of a dry berth. Dry berthing is particularly suitable for small-sized motor boats. As land on the coast becomes increasingly scarce and expensive, boats can be stored in storage racks with several levels (Figure 6–68). Advantages and disadvantages of this type of storage facilities are discussed in the PIANC (1980) report on dry berthing of pleasure boats (1980). Sailboats are usually stored using special cradles (Figure 6–69). The most common type of equipment used in marinas and dry berths for transferring boats between land and water are mobile cranes, forklift trucks, mobile marine hoists, lift slips, derricks, monorails, and various kinds of vertical-lift platforms. Mobile cranes are usually too expensive for dry

Figure 6–68. Storage rack for small motorboats. (Florida)

Figure 6–69. Sailboats in dry storage (Niagara-on-the-Lake, Ontario).

berth operations alone, and can be justified only when used in conjunction with some other marina-related operation. A stiff-leg derrick is usually less costly, but it has rotational problems. Derricks are better used to handle larger boats. Smaller boats are normally handled by monorail installations which can rotate about its supporting column similar to the one depicted in Figure 6–70. Forklift trucks used for boat launching have a fork ladder with the capability of hyperextending down over a perimeter wall to the depth required to place the forks under the hull of a floating boat (Figure 6–71). Because of the limited downward reach, forklifts are best suited for use in marinas with little water level fluctuation. In marinas with large tidal ranges, they can be used to pick up boats at high tide and the rest of the time be used for repair and maintenance work.

Figure 6–71. Forklift truck.

In general lifting equipment (facilities) up to a 50-ton capacity is considered as useful.

Most marinas provide launching ramps with or without mobile marine hoists. Figures 6–72 and 6–73 show the first type of launching ramp as well as a mobile marine hoist. Launching ramps can be very large in size and can thus accommodate many crafts. Their longitudinal section usually has a slope of between 1/10 and 1/15. In tideless areas it is preferable to design a steeper slope (up to 1/5) on the part under water to permit a more rapid floating of the craft and to reduce the bowing distance. The surface of the ramp should be paved down an elevation of about 1.5 m below extreme low water level; the top should be rounded over on a 6-m vertical curve until it becomes nearly level at about 0.5 m above extreme high water level. Any part of the ramp placed under water is usually precast in sections of a size that can be lowered in place onto a carefully prepared gravel bed about 0.2 to 0.3 m thick. Depending on the number of boat launchings and retrievals the launching ramp may include a single or several lanes. Usually it is assumed that one lane can be adequate for 100 operations (launchings and retrievals combined) per day. Typical productivity of the properly designed launching ramp is about 10 min per boat for launching or retrieval.

Figure 6–70. Monorail hoist.

Inside covered storage areas, monorails are more often used for hauling small and medium-sized craft and for their ground

Figure 6–72. Mobile marine hoist.

Figure 6–73. Launching ramp.

transfer. Different kinds of lifting platforms have been used, either of a very elaborate type with a transfer system for large craft, for example, Syncrolift, or of a very simple type for small craft that are transported to neighboring land areas by means of trailers. Some additional information on vertical-lift platforms is found in Chaney (1961) and discussed in Chapter 3. Numerous other types of equipment have been proposed for the hauling of small craft. Today, however, forklift trucks and mobile marina hoists are the basic equipment that are generally used in small-craft marinas. This is attributed to the large handling capacity of these vehicles, their speed of handling, and their flexibility in adapting to the marina site and to its subsequent development.

6.8 SERVICES AT BERTH

6.8.1 Water Supply

In modern marias, water and power must be distributed not only to the land-based facilities, but also to the berths. An adequate supply of drinking water should be provided, preferably at water supply points on piers. The water pressure must be sufficient to maintain a full flow when a quarter of the water points are in use. Precautions should be taken to protect the piping system during freezing conditions. Hose reels should be provide for stowing hoses when not in use to prevent obstructing piers. The water outlet at each berth is usually considered as useful for any class of boats.

6.8.2 Electric Power

A well-equipped small-craft marina should have electric power supply on piers. Current-operated earth leakage circuit breakers that will cut the supply when a leakage current of 30 milliamps has occurred for 30 milliseconds should be installed in each circuit, in addition to the usual fuses and circuit breakers. Each circuit should be fused separately and grounded through a ground core in the supply cable. Any metal (non-current-carrying) in the distribution box must be grounded. Cables should be insulated and sheathed with PVC. If underground, they must be armored. Distribution boxes should be located on the main piers, never on finger piers. Outlet sockets should be installed at each berth and must be weatherproof with spring-loaded covers and the voltage and current should be shown on each box. The design of all these services must comply with local codes and regulations. A general lighting system installed over the entire marina area is generally considered necessary.

6.8.3 Lifesaving Apparatus

Lifebuoys and safety ladders should be provided on piers and enable anyone who has fallen into the water to climb out.

6.8.4 Communication System

Communication services are usually limited to the landside facilities. Public telephones are provided on the marina premises. This usually does not present any problems. As the demand for telephone and/or television connections at individual berths is likely to be limited, it is better to provide independent connections rather than a universal network. The presence of the post office and telefax facilities is considered as useful but not necessary.

6.8.5 Fire Fighting

In small-craft marinas the risk of fire is great due to the presence of gas on board motorboats and yachts. The capacity of some fuel tanks on board small craft is likely to be more than 250 to 300 liters. The smallest accident can therefore cause a serious fire. All necessary precautions must therefore be taken. For example, gas cylinders should be stored upright in a cage or in a properly ventilated building; gas installations and storage areas should be clearly marked and must not be located where gas can accumulate; propane and butane cylinders should not be interchanged; naked lightbulbs must be prohibited; and the storage of gas cylinders or cans and tanks containing fuel in storage lockers ashore must not be allowed.

When planning fire fighting equipment and arrangements, the availability and proximity of local authority fire fighting equipment and the accessibility to the berth should be taken into account. Fire points on piers usually contain a small-diameter (20 to 25 mm, with a handle control nozzle of at least 50-mm diameter) fire hose, or hose reel tubing of sufficient length to reach within 6 m of each developed part of the marina basin as well as a fire extinguisher containing 5 kg of dry powder (or other suitable extinguisher). Pier structures should have a minimum fire resistance of half an hour. Where piers are extended more than 15 m into the basin, fire hydrants spaced at about 30-m intervals must be included in the water distribution system on the docks. Alternatively, fireboats must be provided.

Fuel supply is separated from the berth and is limited to a specially allocated space at the landside portion of the marina because of the fire hazard for piers as well as for other land-based structures. A minimum of two fire hydrants must be installed at a fuel dock location.

In North America several important reference standards that may be used in designing fire protection systems are given in the National Fire Protection Association's publications NFPA 303, "Marinas and Boatyards," and "Fire Protection of Vessels During Construction, Repair and Lay-Up."

6.8.6 Pollution Prevention

Toilet facilities should be provided at convenient points (preferably so that no berth is more than 150 m from a toilet). Some countries have introduced legislation to the effect that any vessels, including small craft, should be equipped with marine sanitary devices—not always holding tanks. Local regulations and practices must be studied and, if necessary, facilities should be provided for pumping sewage ashore from small-craft sewage tanks. An ample number of containers and garbage cans should be located at convenient locations for disposal of hydrocarbons, oil, and miscellaneous garbage.

6.8.7 Navigation Aids, Tide Levels, Draft Marks

Navigation aids should be provided with yachts in mind and tide levels and draft should be large (300 mm) in SI units and be clearly visible in the approach channel.

REFERENCES

ABRAHAM, C. H., 1991. A New Australian Standard for Marinas. ASCE Proceedings of the First International Conference World Marina '91. Long Beach, California.

ACRES INTERNATIONAL LIMITED, 1991. Welland River Dredging. Demonstration Project. Unpublished Report for Atlas Specialty Steel. Niagara Falls, Ontario.

AMERICAN IRON AND STEEL INSTITUTE, 1981. "Handbook of Corrosion Protection of Steel Pile Structures in Marine Developments." Washington, D.C.

AMERICAN SOCIETY OF CIVIL ENGINEERS, "Small Craft Harbors." New York, New York.

BANNON, J. M. et al., 1976. "Selective Analytical Partitioning of Sediments to Evaluate Potential Mobility of Chemical Constitutents During Dredging and Disposal Operations." Technical Report D-76-7. U.S. Army Engineers Waterways Experiment Station, CE. Vicksburg, Mississippi.

BARNARD, W. D., 1978. Prediction and Control of Dredged Material Dispersion Around Dredging and Open-Water Pipeline Disposal Operations. U.S. Army Engineers Waterways Experimental Station Environmental Laboratory. Vicksburg, Mississippi.

BERENGUER, J. M., NAVERES, V. S., and SAMPOS, S., 1990. "Trench-Type Breakwaters for Marinas." PIANC Proceedings 27th Congress, S.1-5. Osaka, Japan.

BERTLIN, D. P., 1972. "Marine Design and Construction." Proceedings of a Symposium on Marinas and Small Craft Harbours. Southampton, U.K.

BRAND, L., 1966. "Differential and Difference Equations." John Wiley & Sons, New York.

BRINSON, R. J., 1985. "Recreational Harbor Development in the United States: The Port Authority Role." Proceedings of 26th Congress of Permanent International Association of Navigation Congresses, Section II, Subject 5. Brussels, Belgium.

BROWN, J., 1978. "Fabric Silt Barriers: Faster to Erect, Cheaper than Piling." Engineering and Contract Record, Vol. 8.

BRUUN, P. (ed.), 1985. "Design and Construction of Mounds for Breakwaters and Coastal Protection." Elsevier, Amsterdam, The Netherlands.

BRUUN, P., 1989. Port Engineering, Vols. 1 and 2. Gulf Publishing Company, Houston, Texas.

BRUUN, P. and GÜNBAK, A. R., 1976. "New Design Principles for Rubble-Mounds." Proceedings of the 15th International Conference on Coastal Engineering, Honolulu, Hawaii.

BRUUN, P., and JOHANNESSON, P., 1976. "Parameters Affecting Stability of Rubble Mounds." ASCE Journal Waterway, Harbor and Coastal Engineering, Vol. 102, No. 2.

BUDIN, A. Y. and DEMINA, G. A., 1979. Quays Handbook (in Russian). Stroyisdat Publishing Co., Moscow, USSR.

CANFIELD, F. E., RAY, R. E., and ECKERT, J. W., 1980. "The Possible Impact of Vessel Wakes on Bank Erosion." Report No. CG-W-1-80. U.S. Coast Guard, U.S. Department of Transportation, Washington, D.C.

CHANEY, C. A., 1961. Marinas—Recommendations for Design, Construction and Maintenance, Second Edition, National Association of Engine and Boat Manufacturers Inc., New York.

COMMITTEE OF TIDAL HYDRAULICS, 1965. Evaluation of the Present State of Knowledge of Factors Affecting Tidal Hydraulics. Report No. 3.

CORROUGH, J. C., KLANCHIK, F. A., and TAYLOR, D., 1990. "Trends in the Planning, Design, and Use of Support Facilities for Sport and Pleasure Navigation in the U.S." PIANC Proceedings of 27th Congress, Subject 1 S.1–5. Osaka, Japan.

COX, J. C., 1991. "Reef Breakwater Design for Lake Michigan Marina." ASCE Proceedings of the First International Conference World Marina '91." Long Beach, California.

CULLINANE, M. J., AVERETT, D. E., SHAFER, R. A., MALE, J. W., TRUITT, C. L., BRADBURY, M. R., and PEQUEGUAT, W. E., 1990. "Contaminated Dredged Material. Control, Treatment and Disposal Practice." Noyes Data Corporation Park Ridge, New Jersey.

DUNHAM, J. W. and FINN, A. A., 1974. Small-Craft Harbors: Design, Construction and Operation. U.S. Army Corps of Engineers, Coastal Engineering Research Center. Special Report No. 2.

EINSTEIN, H. A. and KRONE, R. B., 1962. "Experiments to Determine Modes of Cohesive Sediment Transport in Salt Water." Journal of Geophysical Research, Vol. 67, No. 4.

EKEY, P. W., 1970. "The Big Diaper." American Highways, Vol. 49, No. 4.

ERGIN, A., GÜNBAK, A. R., and YANMAZ, A. M., 1989. "Rubble-Mound Breakwaters with S-Shape Design." ASCE Journal of Waterway, Port, Coastal and Ocean Engineering, Vol. 115, No. 5.

FARR, A. K., 1986. "Soil Quality Testing in Ports: Why?" ASCE Proceedings Specialty Conference Ports '86, Oakland, California.

FITZPATRICK, J. B., TANNER, R. G., and TSINKER, G. P., 1985. "Alluvial Rivers and Their Impact on Layout and Design of Dock Facilities." Proceedings of the 26th Congress of Permanent International Association of Navigation Congresses. Brussels, Belgium.

GERNER, J. A., 1971. "Siltation Control Using the 'Florida Diaper'." Highway Focus, Vol. 3, No. 4.

GUSTIN, J. D. and NEAL, L. A., 1990. "Site Assessments," Civil Engineering, August.

HERBICH, J. B. and BRAHME, S. B., 1984. "Turbidity Generated by a Model Cutterhead Dredge." Dredging and Dredged Material Disposal, Vol. 1.

HERBICH, J. B., SORENSON, R. M., and WILLENBROCK, A. M., 1963. "Effect of Berm on Wave Run-up on Composite Beaches." ASCE Journal Waterway and Harbor Division, Vol. 85, No. 3.

HETINYI, M., 1946. "Beams on Elastic Foundation." University of Michigan Press.

HILDEBRAND, F. B., 1968. "Finite Difference Equations and Simulations." Prentice Hall, Englewood Cliffs, New Jersey.

HIRST, M. and LASEY, P., 1990. "Village Marina Concept, United Kingdom." PIANC Proceedings 27th Congress Section 1, S.1-5. Osaka, Japan.

HUNT, F. G., 1991. Performance Specification for Floating Docks. Proceedings of the First International Conference World Marina '91. Long Beach, California.

JARLAN, G. E., 1961. "A Perforated Vertical Wall Breakwater." The Dock and Harbour Authority No. 42 (April).

JOHANSON, E. E., 1976a. "Silt Curtains for Dredging Turbidity Control." Proceedings of the Specialty Conference on Dredging and its Environmental Effects," edited by P. A. Krenkel, J. Harrison, and J. C. Burdick III. ASCE, January 26–28.

JOHANSON, E. E., 1976b. "The Effectiveness of Silt Curtains (Barriers) in Controlling Turbidity." Dredging: Environmental Effects and Technology, Proceedings of WODCON VII, July 10–12.

KIVEKAS, L. K. and SARELA, K., 1985. "Boating and Marina Building in Finland." Proceedings of the 26th Congress of Permanent International Association of Navigation Congresses, Section II, Subject 5. Brussels, Belgium.

KOLLMEYER, R. C., 1991. "Concept and Principles of Operation of Floating and Fixed Wave Attenuators." Floating Structures Design Conference at University of Wisconsin, June. Medison, Wisconsin.

KOUWEN, N., 1990. "Silt Curtains to Control Sediment Movement on Construction Sites." Report MAT-90-02. Ontario Ministry of Transportation. The Research and Development Branch. Downsview, Ontario.

KURATA, K. and ODA, K., 1984. "Ship Waves in Shallow Water and Their Effects on Moored Small Vessel." ASCE Proceedings Coastal Engineering Conference, Chapter 28.

KUZUCU, S., 1973. "Wave Run-Up on Rubble-Mound Breakwater with Composite Slopes." Thesis presented to the M. E. Technical University in partial fulfilment for the degree of M.S., Anchara, Turkey.

LAYTON, J. A., 1979. "Design and Construction of a Curvilinear Marina." Proceedings of the American Society of Civil Engineers Specialty Conference on Design, Construction, Maintenance, and Performance of Port and Coastal Structures. Alexandria, Virginia, March, Vol. 1.

LAYTON, J. A., 1991. Case History of the Point Roberts Marina. ASCE Proceedings of the First International Conference World Marina '91. Long Beach, California.

LAYTON, J. A. and HANSEN, J. D., 1977. "Environmental Design Process: Small-Craft Harbors." Proceedings of the American Society of Civil Engineers Specialty Conference Ports '77. Long Beach, California, March 9–11, 1977.

LELLKY, H., HARTELIUS, K., and BRANDSTROM, C. E., 1985. Paper, Proceedings of the 26th Congress of Permanent International Association of Navigation Congresses, Section II, Subject 5. Brussels, Belgium.

LIVESLEY, R. K., 1975. "Matrix Methods of Structural Analysis." Second Edition. Pergamon Press, New York.

MADRIGAL, B. G. and VALDES DE ALARCON, J. M., 1992. "Influence of Superstructure Geometry on the Behavior of Vertical Breakwaters: Two Case Studies." PIANC Bulletin No. 76, Jan–Feb–March.

MARCHETTI, S., 1980. "In Situ Tests by Flat Dilatometer." American Society of Civil Engineers Journal of the Geotechnical Engineering Division, Vol. 106, GT3.

MCCLUNEY, W. R., 1975. "Radiometry of Water Turbidity Measurements." Water Pollution Control Federation Journal, Vol. 47, No. 2.

MCLUCKIE, R., 1981. "Floating Silt Barriers for Ontario." IR71, Research and Development Branch, Ontario Ministry of Transportation and Communications, Toronto, Canada, September.

MCNAMARA ENGINEERING LTD., 1961. Report on Study of Perforated Concrete Breakwater for the Department of Public Works of Canada, January. Toronto, Canada.

MIGNIOT, C., 1968. "A Study of the Physical Properties of Various Very Fine Sediments and Their Behavior Under Hydrodynamic Action." (Translation from French), La Houille Blanche, Vol. 23, No. 7.

MORISHIGE, I., 1991. "Japan New Standards for Excellence in Marina Design." ASCE Proceedings of the First International Conference World Marina '91. Long Beach, California.

NATIONAL MARINA MANUFACTURERS ASSOCIATION (NMMA), 1988. America's Boating Business—NMMA Boating Industry Report. Washington, D.C.

NECE, R. E., 1976. Hydraulic Model Study: Point Roberts Marina. Prepared for the CH2M Hill, Bellevue, Washington, by Harris Hydraulic Laboratory, University of Washington, Seattle.

NECE, R. E. and RICHEY, E. P., 1975. Hydraulic Model Study of a Proposed Marina Basin in the Northwest Corner of Penn Cove, Whidbey Island, Washington. Prepared for the Innova Corporation, Seattle, Washington, by Harris Hydraulic Laboratory, University of Washington, Seattle.

NECE, R. E., FALCONER, R. A., and TOSHIRO, T., 1976. "Platform Influence on Flushing and Circulation in Small Harbors." Proceedings of the Fifteenth Coastal Engineering Conference of the American Society of Civil Engineers. Honolulu, Hawaii, July 11–17, 1976, Vol. IV.

NICHOL, J. M., 1990. "Wind Design Factors for Small Boat Mooring Facilities, A Survey of Practice." PIANC, Bulletin No. 68, Brussels, Belgium.

NICHOL, J. M., BATTALIO, R., NOLHAN, R., BOUDREAU, R., and BOMBARDO, D., 1986. "An Example of a Destination Harbor for Pleasure Craft." Permanent International Association of Navigation Congresses Bulletin No. 55.

OBERN, D. and HOY, V., 1992. "Impacts of Environmental Legislation on Boating Facilities," ASCE Proceedings Specialty Conference PORTS '92, Seattle, Washington.

OUMERACI, H., VAN DER MEER, J., and FRANCO, L., 1994, editors, "Vertical Breakwaters", Special Issue of Coastal Engineering Vol. 22, Elsevier, Amsterdam, Netherlands.

PALERMO, M. R., 1992. "Dredge Material Disposal." Chapter 6, in Handbook of Coastal and Ocean Engineering, Vol. 3, edited by J. B. Herbich. Gulf Publishing Company, Houston, Texas.

PERMANENT INTERNATIONAL ASSOCIATION OF NAVIGATION CONGRESSES, 1965. "Problems Arising from Increasing Use of Yachts and Other Small Boats for Sport and Recreation." Proceedings of 21st Congress, Section 1, Subject 6, Stockholm, Sweden.

PERMANENT INTERNATIONAL ASSOCIATION OF NAVIGATION CONGRESSES, 1979. Standards for the Construction, Equipment and Operation of Yacht Harbors and Marinas with Special Reference to the Environment. Supplement to Bulletin No. 33, Brussels, Belgium.

PERMANENT INTERNATIONAL ASSOCIATION OF NAVIGATION CONGRESSES, Third International Commission for Sport and Pleasure Navigation, 1980. Report on Dry Berthing of Pleasure Boats, either for Maintenance or Complementary to Wet Berthing—Both the Technical and Financial Aspects. Supplement to Bulletin No. 37, Vol. III. Brussels, Belgium.

PERMANENT INTERNATIONAL ASSOCIATION OF NAVIGATION CONGRESSES, 1984. Report of the International Commission for Improving the Design of Fender Systems. Supplement to Bulletin No. 45. Brussels, Belgium.

PERMANENT INTERNATIONAL ASSOCIATION OF NAVIGATION CONGRESSES, 1990. Management of Dredged Material from Inland Waterways. Report Working Group No. 7 of PTC 1. Supplement to Bulletin No. 70. Brussels, Belgium.

PERMANENT INTERNATIONAL ASSOCIATION OF NAVIGATION CONGRESSES, 1991. Guidance on Facility and Management Specification for Marine Yacht Harbours and Inland Waterway Marinas with Respect to User Requirements. Report of Working Group No. 5. Supplement to Bulletin No. 75. Brussels, Belgium.

PERMANENT INTERNATIONAL ASSOCIATION OF NAVIGATION CONGRESSES, 1992a. Analysis of Rubble Mound Breakwaters. Supplement to Bulletins 78/79. Brussels, Belgium.

PERMANENT INTERNATIONAL ASSOCIATION OF NAVIGATION CONGRESSES, 1992b. "Guidelines for Design and Construction of Flexible Revetments Incorporating Geotextiles in Marine Environments." Supplement to Bulletins 78/79. Brussels, Belgium.

PROKOFIEV, Y. A. and KRIVOV, A. K., 1985. "Experience of Design and Construction of the Sailing Sport

Center in Tallinn." Proceedings of the 26th Congress of Permanent International Association of Navigation Congresses. Brussels, Belgium.

PRZEMIENIECKI, J. S., 1968. "Theory of Matrix Structural Analysis." McGraw-Hill, New York.

QUINN, A. DEF., 1972. "Design and Construction of Port and Marine Structures. McGraw-Hill, New York.

ROSEN, D. and KIT, E., 1985. A Simulation Method for Model Study of Small Craft Harbors. ASCE Proceedings 19th Coastal Engineering Conference.

SCHERTMANN, J. H., 1988. "Dilatometers Settle In." American Society of Civil Engineers Journal of Civil Engineering. March 1988.

SCHUBEL, J. R., et al., 1978. "Field Investigations of the Nature, Degree and Extent of Turbidity Generated by Open-Water Pipeline Disposal Operations." Technical Report D-78-30, July. U.S. Army Engineer Waterways Experiment Station, CE, Vicksburg, Mississippi.

SORENSEN, R. M., 1973. "Water Waves Produced by Ships." American Society of Civil Engineers Journal of the Waterways, Harbors and Coastal Engineering Division. Vol. 99, No. WW2, Paper No. 9954.

SORENSEN, R. M. and WEGGEL, J. R., 1984. "Development of Ship Wave Design Information." ASCE Proceedings Coastal Engineering Conference, Chapter 216.

SUITS, L. D. and MINNITI, A., 1989. "Prototype Turbidity Curtain for the Westway Highway." Geosynthetics, Geomembranes, and Silt Curtains in Transportation Facilities. Transportation Research Board, NRC. Washington, D.C.

THUET, D., 1987. "Fisherman's Wharf Breakwater." Concrete International, May.

TOBIASSON, B. O. and KOLLMEYER, R. C., 1991. "Marinas and Small Craft Harbors." Van Nostrand Reinhold, New York.

TSINKER, G. P., 1986a. "Ice Management for Year-Round Operating of Marine Terminals." Proceedings of the International Offshore and Navigation Conference POLAR TECH '86. Helsinki, Finland.

TSINKER, G. P., 1986b. "Floating Ports." Gulf Publishing Company, Houston, Texas.

TSINKER, G. P. and VERNIGORA, E. B., 1980. "Floating Piers." Proceedings ASCE Specialty Conference PORTS '80, Norfolk, Virginia.

TURNER, T. M., 1992. "Shallow-Water Dredging." Chapter 5, in Handbook of Coastal and Ocean Engineerings, Vol. 3, edited by J. B. Herbich. Gulf Publishing Company, Houston, Texas.

U.S. ARMY CORPS OF ENGINEERS, 1984a. "Hydraulic Design of Small Boat Harbors." Engineering Manual No. 1110-2-1615. Washington, D.C.

U.S. ARMY CORPS OF ENGINEERS, 1984b. Shore Protection Manual. Three volumes. Washington, D.C.

VAN DER MEER, J. W., 1992. "Stability of the Seaward Slope of Berm Breakwaters." Coastal Engineering, Vol. 16, No. 2, January. Elsevier Science Publishers B.V., Amsterdam, The Netherlands.

VAN DER MEER, J. W., BURCHARTH, H. F., and PRICE, W. A. (eds), 1991. Coastal Engineering, Vol. 15, March. Special Issue—Breakwaters. Elsevier, Amsterdam, The Netherlands.

VAN WIJCK, J., VAN HOOF, J., and SMITS, J., 1991. "Underwater Disposal of Dredged Material: A Viable Solution for the Maintenance Dredging Works in the River Scheldt." PIANC, Bulletin No. 73, April to June.

WECHSLER, B. A. and COGLEY, D. R., 1977. "Laboratory Study Related to Predicting the Turbidity-Generation Potential of Sediments to be Dredged." Technical Report D-77-14, November. U.S. Army Engineer Waterways Experiment Station, CE, Vicksburg, Mississippi.

WEGGEL, J. R. and SORENSEN, R. M., 1986. "Ship Wave Prediction for Port and Channel Design." ASCE Proceedings Specialty Conference Ports '86.

WINTERKORN, H. F. and FANG, H. Y., 1975. Foundation Engineering Handbook. Van Nostrand Reinhold, New York.

WORTLEY, C. A., 1978. "Ice Engineering Guide for Design and Construction of Small-Craft Harbors." Advisory Report No. WIS-S6-78-417. University of Wisconsin Sea Grant College Program. Madison, Wisconsin.

WORTLEY, C. A., 1979. "Small-Craft Harbor Structures Design for Lake Ice." Proceedings of Conference on ASCE Design, Construction, Maintenance and Performance of Ports and

Coastal Structures. Alexandria, Virginia: March 14–16, 1979, Vol. 1.

WORTLEY, C. A., 1989. "Docks and Marinas Bibliography." University of Wisconsin Sea Grant Advisory Services. Medison, Wisconsin.

WORTLEY, C. A., 1991. "Ice Engineering Design for Marinas." ASCE Proceedings of the First International Conference World Marina '91. Long Beach, California.

7

Bridge Pier Protection from Ship Impact

7.1 INTRODUCTION

On May 9, 1980 during an intense early morning thunderstorm, low visibility, and high wind conditions, the empty 40 000 DWT bulk carrier MV "Summit Venture" collided with one of the anchor piers of the twin Sunshine Skyway Bridge across Tampa Bay, Florida. Ship impact caused a 396-m section of the southbound main span to collapse with subsequent loss of 35 lives. The most recent catastrophic event of this kind in the United States has occurred in September, 1993 in Mobile, Alabama. There the barge hit the bridge pier that allegedly resulted in displacement of the rail included in the train track.* This was followed by derailment of the passenger train, which resulted in the loss of 45 lives. Unfortunately these tragic events are only two of many. A long list of similar accidents is given by Blaauw and Buzek (1988) and many other investigators.

Frandsen (1983) notes that annual rates of catastrophic collisions increased from 0.5 bridges/year in the decade 1960 to 1970, to 1.5 bridges/year in the period 1971 to 1982. Accidents at bridges are common occurrences in certain areas. For instance, New Jersey Transit Authority and Conrail's bridge across the Raritan River have had well over 100 collisions in the period 1970 to 1981 alone (Heming, 1981).

According to AASHTO (1991) in the period 1965 to 1989, an average of one catastrophic accident per year involving bridge collision by merchant vessels have been recorded worldwide. More than 100 lives have been lost in these accidents and very large economic losses were incurred in repair/replacement costs, lost vessels, disruption of transportation service, and other damages. More than half of these accidents occurred in the United States.

Many factors account for the rise in number of ship–bridge collisions. Increased global trade expansion, resulting from economic development and increased demands for oil, coal, minerals, grain, etc., is an important contributor. As a result, an increased number of larger merchant ships are now in service. The world fleet has increased three times and worldwide sea-borne tonnage has increased by more than 255% (U.S.

* By the time this book was written the final report on the Mobile, Alabama accident has not yet been made public.

Department of Commerce, 1978; McDonald, 1983). During the same period, the number of bridges in the United States has increased by one third (Knott, 1987). The increased number of bridges and, in particular, poorly sited bridges is another factor that contributes to ship–bridge collision.

Many bridges in place today have been in existence for more than 50 years and now seriously restrict navigation because their spans are too short to be safe for transit of modern large merchant vessels.

All these factors contribute to the growing number of catastrophic ship–bridge collisions in recent years. The loss of lives, property damage, and economic losses due to ship collision with bridges in recent years far exceed those of bridge accidents caused by earthquakes, winds, and waves combined (U.S. National Research Council, 1983).

In the last decade, intensified studies of ship collision with bridges have been initiated in an attempt to examine the risk and consequences of collision accidents and to improve the basis for the treatment of the problems involved. Recent studies performed all over the world were presented and discussed at the International Association for Bridge and Structural Engineering (IABSE) colloquium on "Ship Collision with Bridges and Offshore Structures," held in Copenhagen, Denmark in 1983, sponsored by the American Society of Civil Engineers Structural Congress 1987, held in Orlando, Florida, in 1987, and most recently at the 27th PIANC Congress held in Osaka, Japan (1990).

Most studies conducted to date have emphasized the need for developing uniform engineering criteria regarding ship–bridge collision. The criteria proposed include measures to avoid or at least to reduce the risk of collision, for example, bridge siting and layout, with due consideration given to navigation conditions, geometry of the waterway, and aids to navigation.

Other design criteria deal with bridge design to withstand ship impact, the use of a protective system to absorb impact energy or to deflect the ship from a collision course, as well as with some general design recommendations for reducing the risk and consequences of a collision.

Some industrialized countries have established national regulations concerned with the safety of navigation and the determination of ship impact loadings from collision events involving ships and barges. Also, efforts regarding the establishment of the international design guidelines for vessel collision have been initiated by the International Standardization Organization (ISO) and by IABSE.

In Sweden design guidelines have been developed on the basis of the research work carried out for the Oresund crossing (Olnhausen, 1966). Subsequently, the Swedish regulations were developed into the Common Nordic Regulations (Nordic Road Engineering Federation, 1980). These regulations are recognized in Sweden as the national standard for the design of new bridges and review of all other bridges placed across navigable waterways and shipping channels (Olnhausen, 1983). In recent years Nordic Regulations underwent some changes in Norway (Norwegian Public Road Administration, 1986) and in Denmark (Danish Road Directorate, 1984).

Some European countries that depend heavily on their river and canal transportation systems have established their own regulations to ensure the safety of navigation (French Ministry of Public Works, 1971; German Traffic Ministry, 1988).

In Japan, a very comprehensive investigation with respect to ship collision with bridge piers has been conducted in connection with several recent major bridge projects (Iwai et al., 1983; Kuroda and Kita, 1983; Matsuzaki and Jin, 1983; Ohinishi et al., 1983).

The Canadian Coast Guard (1982) published a study on the vulnerability of bridges placed across Canadian navigable waters to collision by ships. This study concluded that there are a large number of unprotected bridges in Canada with a high risk of being damaged.

The study also produced guidelines with respect to achieving better understanding of the collision mechanism between ship and bridge pier, and to design of bridge pier protection systems.

In the United States the Louisiana Manual (Criteria, 1985) and Interim Proposal (1988) put together by the Special Task Force of the AASHTO Bridge Committee give collision forces for each class of waterways as a function of the water depth at the location of the pier and, in the case of deep-draft waterways, as a function of the design ship selected and a typical average speed. Once the force is determined, its distribution on piers with respect to their distance from the navigation channel has to be found.

The United States National Research Council (1983) published a report on ship–bridge collisions. This report recommended the development of a national standard for the design and construction of bridges to resist ship collision (with the exception of criteria for fenders to protect railroad bridges). Such a standard has been developed recently and is published as a Guide Specification by the AASHTO (1991).

The American Railway Engineering Association (AREA, 1986) also developed specifications for the design of bridge pier protection systems at bridges over navigable waterways. The emphasis of this guideline seems to be oriented toward the protection of bridges from small collision forces, which can be absorbed by the commercially available conventional fender systems. The specification, however, is general in nature and lacks specific guidelines on how to estimate the risk of ship impact and impact forces.

The International Organization for Standardization (ISO, 1987) published a working draft of regulations for ship impact on bridges from river and canal vessel traffic. The document, which is the result of the collective work of engineers and scientists from more than eight participating European countries, recommends some typical impact loads as a function of the type of waterway and mass of a ship.

The aforementioned colloquium held by IABSE in 1983 in Copenhagen, Denmark has established a working commission for the preparation of an international guideline or a state-of-the-art report.

It should be noted that in some cases of catastrophic ship–bridge collisions, the greatest loss of life has resulted from the continuation of the highway traffic after the bridge has been severed. Following the investigation of the Sunshine Skyway Bridge collapse, the U.S. National Transportation Safety Board (1980) recommended that motorist warning systems be installed to detect highway bridge-span failures and to warn motorists. These system components are generally grouped into the following three categories:

1. Devices to detect hazards, either environmental or man-made
2. Devices to verify hazards or problems
3. Traffic interrupting devices.

Furthermore, the safety of navigation in the vicinity of a bridge can be ensured by having in place well-maintained standards and, if required, electronic navigation systems.

The former typically include ranges on inbound/outbound channels, additional buoys, radar reflectors and lights on all buoys near the bridge, high-intensity light beacons on the bridge structure, sound devices (fog horns) on the bridge, and a RACON device on the bridge structure for improved radar image of a ship.

The use of advanced electronic navigation systems both on-board and on-shore (bridge) has shown significant reductions in the probability of aberrancy by pilots under similar conditions.

The safety evaluation of a bridge typically involves risk analysis of vessel collision and selection and design of pier protection alternatives. The economic analysis of the design of a bridge over a navigable waterway would normally include consideration of the construction of unprotected piers, which are able to withstand direct impact of the design ship, versus use of a protective system designed to protect the piers from ship impact.

7.2 RISK ANALYSIS OF VESSEL COLLISION

Any bridge pier that can be reached by a passing ship is at risk of being hit. A collision event can be characterized by its probability of occurrence and by its consequences. The probability of a collision taking place and risk analysis are affected by a variety of factors, as noted below.

- Physical features of the site, such as geometry of the navigation channel (waterway), proximity and nature of banks, the geometry of the bridge, pier shape and accessibility to ships, and distance between piers
- Hydraulic and weather conditions. From the studies done by the U.S. and Canadian Coast Guards, it was concluded that most vessel accidents with bridges were due to currents (Heming, 1981; Canadian Coast Guard, 1982). Depth of water in the navigation channel and around a pier, and tide, ice, and wind conditions are additional contributing factors affecting the vulnerability of a bridge.
- Vessel traffic, traffic management, and aids to navigation. These are essential in the bridge environment, particularly for bridges over approaches to busy ports and harbors, and at changes in configurations and alignment of navigation channels.
- Vessel characteristics. For example, large fully loaded ships with very little under-keel clearance can produce the highest impact forces. On the other hand, ships sailing in ballast (or empty), or ships that have a large wind-exposed area such as container or ferry ships, are particularly sensitive to wind and wind gusts and may wander far outside the navigation channel.
- Human error (misjudgment, negligence, etc.). Evidence indicates that 60% to 85% of all vessel collisions are a result of pilot error (Knott, 1987).
- Mechanical failure, for example, engine or steering failure.

Because most of the factors involved are uncontrollable and random in nature, a probabilistic approach to risk analysis is typically used (Rowe, 1982; Prucz and Conway, 1987). As denoted by Rowe (1982) "risk evaluation" is a rather subjective process for arriving at an "acceptable" level of risk, for example, determining how to manage the estimated risk. The latter involves such a highly emotional and moral issue as estimation of the cost of loss of human lives. General approaches to formulation of the risk assessment model can be found in Fujii et al. (1974), Fujii (1983), Macduff (1974), Larsen (1982), Knott and Bonyun (1982), Kuroda and Kita (1983), and in a report by COWIconsult Inc. (1987).

One means for ensuring that the risk associated with the construction and/or operation of a bridge over a navigable waterway is reasonably low and is acceptable to the community is the development of the risk acceptance criteria (Rowe, 1979). Typically the risk acceptance criteria are set up on the basis of the existing standards and regulations. These can include safety codes set on the basis of probability of a specified consequence. In the absence of specific regulations, the risk acceptance criteria can be established on the basis of criteria previously developed for similar projects. However, as addressed by Fairley (1981), the assessment of the catastrophic risks is a very difficult task.

There are a number of typical categories of risk that are associated with the construction and/or operation of bridges over navigable waterways. These include: owner's risk, bridge user risk, third party risk, risk of service disruption, and risk of negative impact on environment.

Owner's risk includes potential loss of revenue and cost of reconstruction. A rational way to decide the owner's risk will be to base the decision on analysis of cost-effectiveness of risk reduction. An alternative would be to establish an acceptable risk level based on similar precedents.

User risk covers impact to the user which includes loss of lives and other damages to user. The relevant criteria can be found in U.S. Department of Transportation (DOT) (1986), NCDCB (1978), COWIconsult (1987a,b), and AASHTO (1991).

Typically, the user risk criteria are based on a small percentage of the variance of the historic accident rate on equivalent facilities. The year-to-year variation in fatality rates provides a historic average and a standard deviation. It is assumed that for a specified year with fatality rates several standard deviations from the average corrective action will be initiated. Variations close to the average, less than one standard deviation, are considered acceptable on a revealed preference basis.

Third party risk involves risks to those who receive no direct benefit from the use of the bridge. It includes damages to the passing ships and persons on ships due to ship collisions or grounding due to increased complexity of navigation caused by a bridge. Some relevant criteria on third party risks are found in Rowe Research and Engineering Associates, Inc. (1985), Chicken and Harbison (1986), Chicken and Hayus (1986), and Vinck et al. (1987).

Risk of service disruption involves socioeconomic costs depending on the duration of disruption.

Examples and relevant data can be found in the Modjeski and Masters (1985) report to the Louisiana DOT and the FHA.

Risk of negative impact on environment includes oil spill or toxic gas release as a result of a ship–bridge collision. Various types of risk assessment models have been developed for ship collisions with bridges. In practice all of them are based on the assumption that the annual frequency of vessel collision with the bridge (A_f) is dependent on a value (N) related to the number, size, and type of vessels expected to transit the bridge and to the loading and sailing conditions of the vessel, the probability of vessel aberrancy (P_a) which is the statistical probability that a vessel will stray off-course and threaten the bridge, the geometric probability of a collision between an aberrant vessel and a bridge (P_g), and the probability of bridge collapse due to a collision with an aberrant vessel.

A description of the existing and future *traffic of vessels* that will transit the bridge along with physical dimensions and sailing conditions of vessels (loaded, empty, in ballast), speed, and type of cargo is required for a risk analysis. The speed is a very important element of the data because it affects a vessel's momentum and associated impact force during a collision with a bridge structure. The speed may vary depending on ship type, channel geometry, and current condition. This information is usually obtained from local pilots and ship masters of the vessels using the waterway.

Two very important parameters in ship–bridge collision risk analysis that greatly affect accuracy of computation are probability of ship aberrancy (P_a) and geometric probability (P_g).

Probability of aberrancy (P_a) is a measure of the risk that a vessel is not under proper control which may be a result of human error, mechanical failure, or adverse environmental conditions. Most of the statistics indicate that human errors and adverse environmental conditions are the major reasons for accidents.

The aberrancy rate is typically computed using accident data (collisions, groundings, rammings) in the waterway, and statistics of the frequency of vessel traffic during the same period of time. The aberrancy rate estimated by various authors around the world for ships varies, depending on previously stated conditions, between 0.5×10^{-4} and 2×10^{-4} and in some cases up to 5×10^{-4} to 7×10^{-4}. The aberrancy rate of barges may be two to three times that estimated for ships in the same waterway.

Geometric Probability (P_g) is typically defined as the conditional probability that a vessel will collide with a bridge structure given that it has lost control (i.e., it is aberrant) in the vicinity of the bridge. The probability of such an occurrence depends on the waterway's geometry, depth of water, location of bridge piers, span clearance, sailing path and size, draft and maneuvering characteristics of a vessel, and environmental conditions.

Similar to the probability of aberrancy, the methods used to determine the geometric probability vary significantly among various researchers. A relatively simple method for computing of a geometric probability (P_g) has been proposed by Fujii (1983) and by Fujii et al. (1974). The proposed method is based on statistical data of ship groundings and collisions with oil drilling platforms in several Japanese waterways. Accordingly, the geometric probability is calculated by dividing the width of the pier (D) plus the beam of the vessel (B) by the width of the waterway (W) as shown below.

$$P_g = (D + B)/W. \qquad (7\text{–}1)$$

Macduff (1974) proposed a model for computing the geometric probability of a ship hitting the wall of the channel or groundings based on statistics of the straits of Dover and the English Channel. According to Macduff, the geometric probability can be determined by Eq. (7–2).

$$P_g = 4T/\pi W \qquad (7\text{–}2)$$

in which T = stopping distance of the ship, and W is the waterway width as in Eq. (7–1).

More elaborate and relatively complex models have been proposed by COWIconsult (1987a and b), Modjeski and Masters Consulting Engineers (1985), and Greiner Engineering Sciences (1985).

Edectech Associates (1985) have developed a design procedure for evaluating the risk of grounding in restricted waterways for various layouts of short-range aids to navigation systems. Edectech provides useful data for the evaluation of the probability of aberrancy and geometric probabilities. These data may be particularly useful for those waterways where historical data are not available.

For risk analysis, Prucz and Conway (1987) described the occurrence of a collision in probabilistic terms and proposed a rather simple model for ship–bridge pier collision risk analysis. It should be noted that the proposed model considers only the bridge piers and not the superstructure. As the superstructure and approach spans in particular can also be hit by an aberrant vessel, in general they should be included in the risk analysis.

In the case of very important bridges or in special cases an individual approach to risk analysis is required.

Typically the results of risk analysis should yield an assessment of the annual frequency of a ship collision with various parts of the bridge structure and their annual frequency of collapse. Using these data, an appropriate design vessel can be selected. Subsequently the pier protection scheme alternatives, for example, a pier that is able to withstand a ship impact force versus protective schemes, can be developed and evaluated for cost effectiveness.

According to Prucz and Conway (1987) "... A more complete probability based design approach is justifiable only in specific cases of major bridges where sufficient data on the local conditions, including vessel traffic and history of collision accidents are available". The general approach used by Prucz and Conway in the formulation of the proposed risk assessment model is similar to the approach employed by Fujii et al.

(1974), Knott and Bonyun (1982), Larsen (1982), and Macduff (1974). It assumes that a certain number of passing ships will be out of control in the vicinity of a bridge and that some of those uncontrollable vessels will collide with a bridge pier.

The probability of a ship hitting a given pier i (P_i) is expressed as follows:

$$P_i = P_{gi} P_a \qquad (7\text{–}3)$$

where

P_{gi} = geometric probability

P_a = probability of vessel aberrance.

In order to adjust the value of P_a for a given location, a correction factor $F_c = 2.5$ to 4.0 to the basic probability of vessel aberrancy is introduced. The value of F_c to be used in analysis is dependent on vessel traffic density, and the history of vessel accidents in the area.

P_{gi} can also be determined by the following formulation (7–4) which is based on a Cauchy probability density function, modified to take into account the waterway width and the fact that not all collision accidents will result in severe damage:

$$P_{gi} = \frac{X_{max}(D_i + B)}{6(X_{max}^2 + X_i^2)} \qquad (7\text{–}4)$$

where

D_i = width of pier i

B = width of vessel

X_i = distance from the navigational channel centerline to pier i

X_{max} = maximum distance from the navigational channel centerline to the waterway's edge. If $X_{max} > 365$ m, then $X_{max} = 365$ m should be used.

For the case where there is a curvature in the waterway in the vicinity of the bridge, then in formula (7–4) X_i = distance from a tangent to the navigational channel centerline, before the curvature to pier i, and X_{max} = half the distance from the location of curvature to the bridge centerline.

The larger of the two values of P_{gi} obtained should be used for the pier considered. In fact Eq. (7–4) accounts for an expected higher concentration of aberrant vessels closer to the navigational channel centerline than further away from it.

Based on the assumption that, out of a total number of vessels passing the bridge per year (N_v), a certain percentage (N_{ci}) are capable of exerting a collision force larger than or equal to the lateral resistance of the pier, the annual probability of failure of the pier is expressed as follows:

$$P_i = F_c P_a P_{gi} N_{ci} N_v. \qquad (7\text{–}5)$$

If the failure of one pier is considered to result in failure of the whole bridge, then the annual probability of bridge interruption (P) will be computed as follows:

$$P = F_c P_a N_v \sum_{i=1}^{N} P_{gi} N_{ci} \qquad (7\text{–}6)$$

where

N = total number of piers exposed to vessel collision.

Equation (7–6) assumes statistical independence and that the pier probabilities of failure (P_i) are very small compared to unity.

Due to the assumptions made in the above formulations and usually limited data available to support these assumptions, the probability values obtained by the described risk assessment model should be treated with caution when related to the absolute vulnerability of a bridge. For more information on a subject matter the interested reader is referred to AASHTO (1991).

7.3 DESIGN VESSEL SELECTION

The collision force is determined by a design vessel, which in turn is determined by the intensity of the ship traffic, the waterway geometry in relationship to bridge line, and navigational aspects.

The design vessel size in most cases is determined on the basis that the annual probability (P) of the pier in question being hit by a ship larger than the design vessel is very small. For example, Nordic Road Engineering Federation (1980) recommends that the annual probability of such an event should be 0.02 (i.e., a return period of 50 years). Alternatively, this document recommends that for easily navigable water, the design vessel size shall be determined such that the maximum number of ships that are larger than a design vessel amounts to a maximum of 200 ships or 20% of the total number of passing ships. On the other hand, for waters that are difficult to navigate the number of ships that are larger than a design vessel will amount to a maximum of 50 ships or 10% of the total number of ships. In addition, it is suggested that the design vessel's DWT must not be taken less than 5% of the deadweight tonnage of the largest ship, using the waterway. On the other hand, design of the bridge pier for collision with the largest ship that is expected to transit the bridge during the design period will most certainly result in an overconservative and uneconomical design.

According to Olnhausen (1983), if the required information on vessel traffic and navigational condition is not available, a design vessel can be selected based on a minimum of 20 to 100 expected passages per year, depending on the assumed navigational difficulties in the area.

Prucz and Conway (1987) suggested a simple approach for the determination of the size of the design vessel, which is based on the risk assessment model presented in the preceding section. The suggested procedure accepts a certain predetermined amount of risk of bridge interruption and considers a uniform risk of failure associated with each pier. It proceeds as follows:

1. Information on the expected annual vessel traffic volume and its size distribution should be obtained.
2. The cumulative vessel size distribution should be plotted. An example is shown in Figure 7–1.
3. Select an acceptable value of an annual probability of bridge interruption (P). A value of 10^{-4} has usually been considered. However, this should be related to the consequences of bridge failure.
4. The percentage of vessels equal to or larger than the design vessel (N_{ci}) to be selected for pier i is calculated from formula (7–7):

$$N_{ci} = \frac{100P}{F_c P_a P_{gi} N_v N}. \tag{7–7}$$

   All parameters in this formula are defined in the previous section.
5. Finally the design vessel is selected from the design chart depicted in Figure 7–1 for a given N_{ci}.

More information on subject matter is found in AASHTO (1991).

7.4 SHIP COLLISION IMPACT FORCES

Naturally, collision of a high-energy (large momentum) vessel generates extraordinary impact force. Knott (1987) figuratively equates impact energy of a 30 000 DWT ship colliding at 5 knots to a fully loaded Boeing 747 aircraft crashing at a speed of 120 knots.

The process of determining the impact load on a bridge pier during a ship collision is complex and calculation results depend on shape and structure of a ship bow; the volume of ballast carried in the forepeak; the size of the ship and speed of movement; the geometry, strength, and elastic characteristics of the pier; and the geometry of collision.

It is usually assumed that the forces developed during collision are distributed evenly

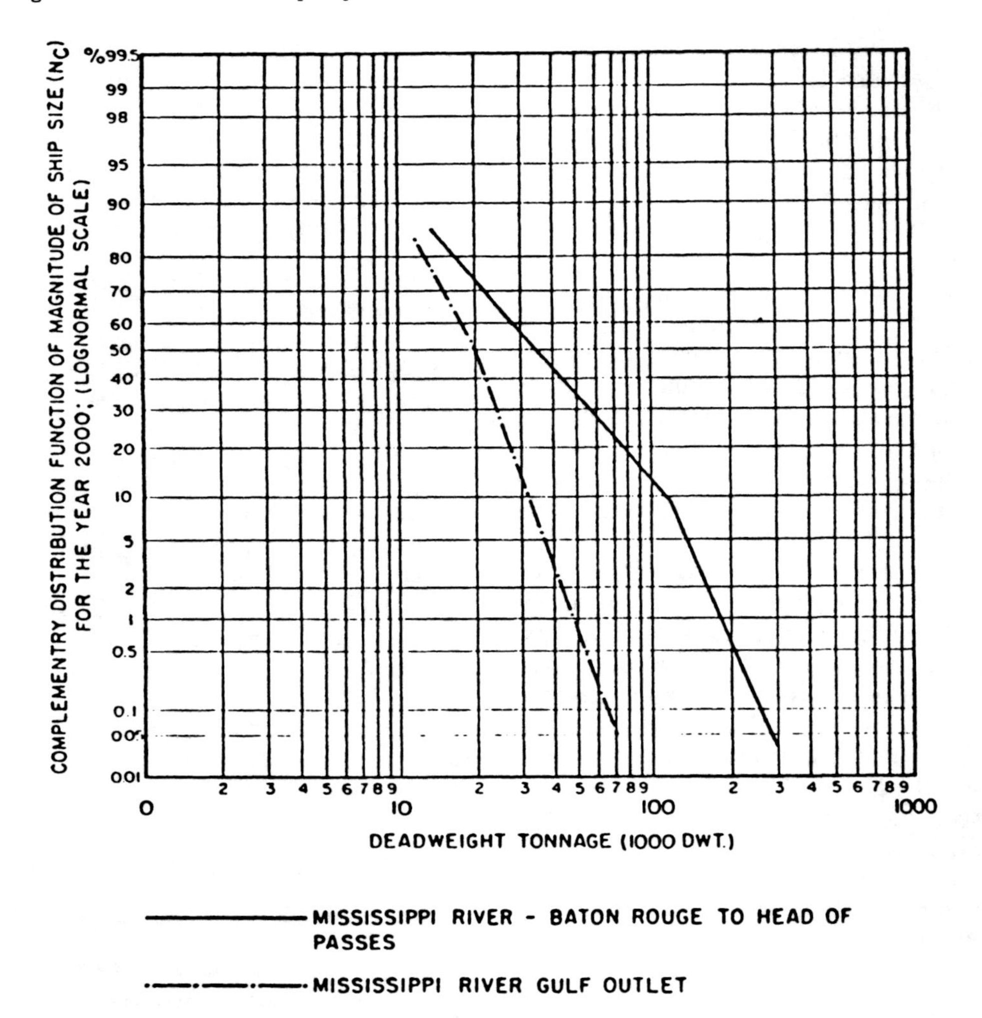

Figure 7–1. Design ship deadweight tonnage versus the complementary distribution of ship size. (From Prucz and Conway, 1987.)

over the cross-section of the vessel's bow and are time dependent. However, in many cases, the width of a pier is less than the width of the ship. In these instances the whole cross-section of the ship will then not take part in the collision. Naturally, this will cause a reduction in impact load. However, this fact is usually ignored and conservatively is not considered in estimating the collision impact force.

Physical model tests and theoretical investigations conducted in Europe and Japan have resulted in the development of a number of empirical relationships for estimating the crushing load of a ship's bow, which is an upper limit of the collision force (Minorsky, 1959; Olnhausen, 1966, 1983; Woisin and Gerlach, 1970; Woisin, 1971, 1976, 1979; Saul and Swensson, 1982; Iwai et al., 1983; Ohinishi et al., 1983).

The correlations between collision loads and deadweight tonnage as recommended for the design by various researchers and authorities are summarized in Figure 7–2.

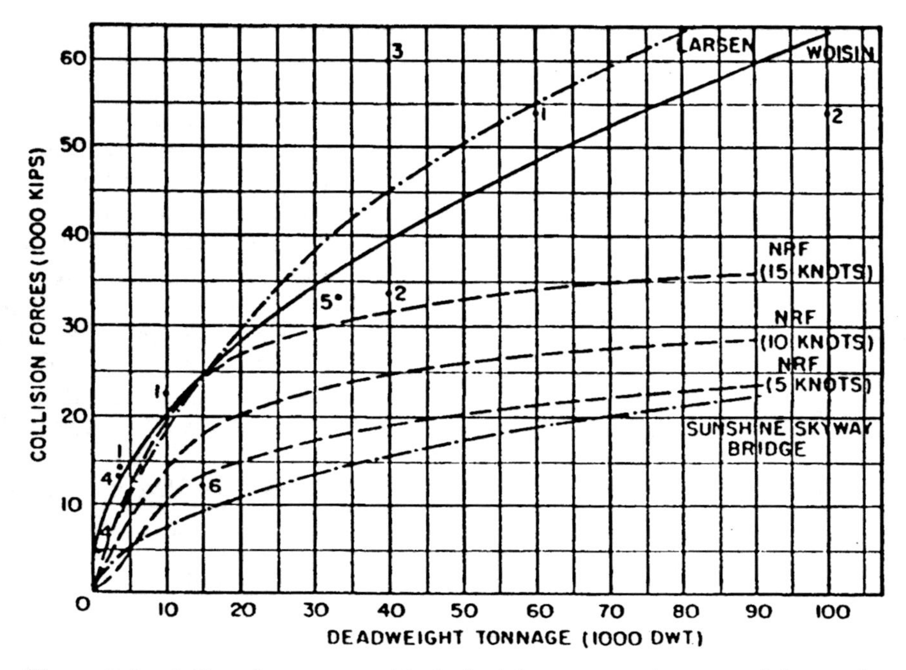

Figure 7–2. Collison forces versus ship deadweight tonnage as recommended or used in various projects. (From Prucz and Conway, 1987.)

As previously noted the collision load is a function of the bow construction and therefore it depends on the type of vessel involved in pier collision (e.g., tanker, container ship, ice breaker, navy vessel, or other).

The loads recommended by Woisin (1976), as indicated in Figure 7–2, are based on measurements made in scale models of direct head-on collisions of various types of ships with a rigid obstruction. According to Prucz and Conway (1987) and Saul and Svensson (1982), the loads (in MN) can be approximated by the formula (7–8):

$$P_{\max} = 0.88(\text{DWT})^{0.5} \pm 50\%. \qquad (7\text{–}8)$$

The variation of ±50% in Woisin's formula accounts for, among other things, the structural type of a ship of the same size, shape of a ship bow, and the degree to which the forepeak of a ship is filled with ballast water.

According to Woisin and Gerlach (1970), the maximum load $P_{\max}$ occurred at the very beginning of the collision and for a short duration (0.1 to 0.2 s) only (Figure 7–3), and then dropped to a mean value of $P_m \approx 0.5P_{\max}$.

Based on studies performed for the Sunshine Skyway Bridge, Greiner Engineering Sciences Inc. (1985) suggested that Woisin's formulation can be modified to reflect reductions in the maximum impact load ($P_{\max}$) in cases where the vessel travels at speeds slower than the maximum (estimated 16 knots) and when the vessel is travelling in a partly loaded, ballasted, or a light (empty) condition. The equation proposed by Greiner Inc., for the ship collision load (in MN) is as follows:

$$P_{\max} = 0.88(\text{DWT})^{0.5}(V/16)^{2/3}(D_{\text{act}}/D_{\max})^{1/3} \qquad (7\text{–}9)$$

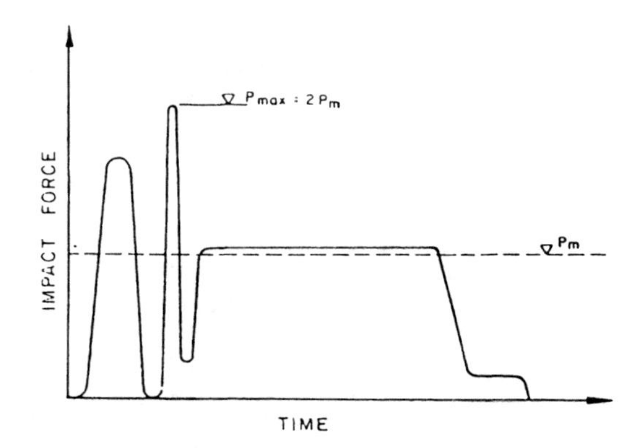

Figure 7–3. Impact force history from a collision test between the bow models of the passenger liner T/W Bremen against the side model of the N/S Otto Han. (From Woisin and Gerlach, 1970.)

where

V = the vessel speed at the time of impact (in knots)

D_{act} = the vessel displacement at the time of impact (tons)

D_{max} = the vessel maximum displacement (tons).

The deformation behavior of a vessel bow has a profound effect on the collision impact load developed. To date various theoretical and experimental studies on ship deformation during collision have been published and there is a large range of different recommendations for both the design collision loads and ship deformations by various authorities.

Saul and Svensson (1982) suggested that the correlation between ship kinetic energy (E_k), median impact collision load (P_m), and damaged length (a) can be expressed as follows:

$$a = E_k / P_m \tag{7–10}$$

where

$E_k = WV^2/2g$

W = ship virtual mass which constitutes the sum of the ship displacement tonnage and the added (hydrodynamic) mass. Depending on type of ship for head-on collision $W = (1.05 \text{ to } 1.1) \times (\text{DWT})$. Low numbers are usually used for the streamlined vessels and larger ones for wide fully loaded ships such as tankers and bulk carriers.

V = speed of ship traveling

g = gravity acceleration.

Minorsky (1959) found that a linear correlation exists between the volume of the ship's steel deformed in the collision and absorbed energy. He proposed a formula that describes ship collision energy in terms of volume of structural steel deformed as a result of the collision. Minorsky's formula, according to Roivainen and Tikkanen (1983), can be presented in SI units as follows:

$$E_k = 32.37 + 47.09R_s \tag{7–11}$$

where

R_s = volume of deformed steel (m^3).

Since initial publication of Minorsky's formula it has been validated continuously through the results of real collisions and model tests.

By simplifying the ship bow into the shape of a triangle (Figure 7–4) Roivainen and Tikkanen recommended the following expression for the energy consumed during collision:

$$E_k(x) = 2.74x^2 + 32.37 \tag{7–12}$$

where

x = penetration of a ship into a bridge structure (in m), or deformed length of a ship's bow (Figure 7–4).

The derivative of Eq. (7–12) gives the following expression for the collision force:

$$P(x) = dE_k(x)/dx = 5.485x \tag{7–13}$$

The following example of the calculation of ship damage due to an actual collision is provided by Saul and Svensson (1982).

A fully laden tanker, "Gerd Maersk" (approximately 38 000 DWT), with 45 000

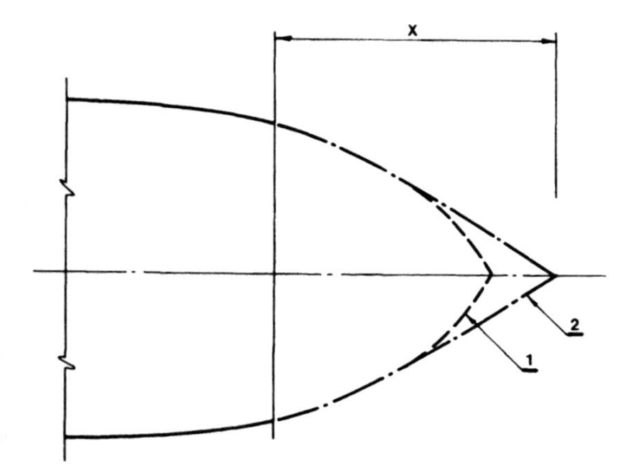

Figure 7–4. Simplified shape of a ship bow for calculation of the amount of energy consumed during collision. (As proposed by Roivainen and Tikkanen, 1983.)

displacement tonnage at the time of accident, collided with one of the main piers of the Newport Bridge. (For details see Kuesel, 1983.) The estimated speed of the ship at the impact event was 6 knots. As a result of collision the bow of the ship was crashed and flattened over a distance of 3.5 m.

Calculations in accordance with formula (7–8) produce the following result:

$$P_{max} = 0.88(38\,000)^{0.5} \pm 50\% = (172 \pm 82)\text{MN}$$

and

$$\text{average impact force } P_m = 0.5P_{max} = (86 \pm 41)\text{MN}$$

Ship kinetic energy

$$E_k = 0.5 \times 45 \times 3^2 \times 1.05 = 213\,\text{MN}$$

and subsequently damage length

$$a = 213/(86 \pm 41) = 4.96 \text{ to } 1.65\,\text{m}.$$

The actual damage depth of 3.5 m indicates that the above calculations can provide results that are in reasonable agreement with observations.

Furthermore, the design charts depicted in Figure 7–5 can also be used for preliminary determination of the ship collision loads.

The vessel impact force is usually applied as a quasi-static load to the bridge structure. The primary reason for this, aside from its simplicity, is that relatively little is known or understood about the dynamic forces associated with the collapse of the vessel's bow structure. To the best of this author's knowledge, the only physical model dynamic tests conducted by Woisin and other researchers have primarily utilized static scale model tests and mathematical models to investigate impact forces during a ship–bridge collision.

Insights into dynamic ship–bridge interaction phenomena are found in works by Petersen and Pedersen (1981), Fauchart (1983a and b), Roivainen and Tikkanen (1983), and Modjeski and Masters Consulting Engineers (1985).

A review of some available data on inland barges operating in a shallow draft waterway indicates that, depending on barge size and speed, the collision load may be on the order of 1150 to 1600 tonnes and may be as much as 3000 tonnes (German Traffic Ministry, 1988). For more information consult AASHTO (1991).

7.5 PIER PROTECTION ALTERNATIVES

In the event of collision the amount of kinetic energy possessed by the ship passing a bridge is in most cases too large to be absorbed by the pier and ship through elastic deformation alone. If the bridge pier is not designed to withstand a ship impact load then an intermediate cushion (protective system) is usually required to

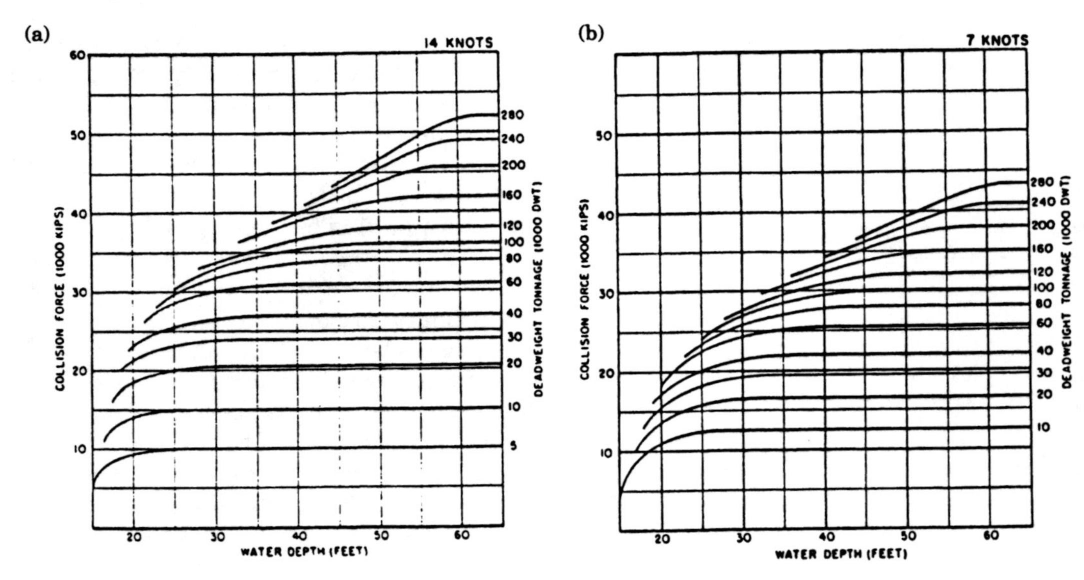

Figure 7–5. Collision design loads as a function of the design ship size selected and the water depth. (a) For ship velocity of 14 knots. (b) For ship velocity of 7 knots. (From Pruez and Conway, 1987)

help the pier–ship system to absorb the required amount of energy.

Kuzmanovič and Sanchez (1992) discuss some practical aspects of direct ship–pier collision and in particular the response of the pile pier to the ship impact.

A number of protective systems have been proposed in recent years for use as bridge pier protection. The solutions include: protective islands placed around the piers, independent structural barriers of miscellaneous constructions, or barriers attached to the pier, dolphins of a variety of designs, protective cells, moored pontoons, moored cable arrays, and different kinds of fenders (timbers, rubber, mechanical, and gravity).

As a protective system commercially available rubber fenders are usually used to absorb the energy of relatively small ships traveling at low-impact speeds and at oblique angles to a pier. In general, fenders of any construction alone have proven to provide inadequate protection for the pier when impacted by large ships. Even the largest commercially available fenders cannot dissipate the energy of a large ship hitting the pier during a head-on collision, and therefore they cannot protect the pier effectively. The simplest type of fender used for pier protection is fabricated out of wood.

A typical timber fender system routinely installed for protection of a bridge pier from minor impact forces is shown in Figure 7–6.

Recently, huge pneumatic rubber fenders (4.5 diameter and 9.5 and 12.0 m long) in combination with a special collapsible multi-cell structure have been installed to protect the bridge piers of the Bisano-Seto Bridge in Japan from head-on collision by relatively small (500 to 4000 gross tonnes) ships (Iwai et al., 1983).

The selection of the system to be used for bridge protection is typically based on the criteria of effectiveness, expected damage to the ship, constructibility, cost-effectiveness, maintenance, safety, and environmental impact.

The two most effective and common types of bridge protection systems used in the past few years in North America are large-diameter sheet-pile cells filled with granular material and protective islands placed around the vulnerable bridge piers.

Figure 7–6. Conventional timber protection system typically installed at a bridge pier for protection from minor impact forces.

7.5.1 Large-Diameter Sheet-Pile Cells

This type of protective structure can be employed at locations where soil condition permits driving of sheet-piles.

The sheet-pile cells that are most often employed for bridge protection are depicted in Figure 7–7.

Timber rubbing strips are sometimes placed on the outer perimeter of the cell to act as an anti-sparking surface in the event of collision with a steel hulled vessel carrying flammable products.

The circular shape of the cell can help to deflect an aberrant ship away from a bridge pier under glancing-blow situations. The cell, however, should be designed for the maximum loading in case of a head-on impact.

Steel sheet-pile protective cells are in general considered to be a relatively "soft" type of structure, as opposed to steel or concrete caissons filled with granular material used for the same purpose. When ruptured by ship collisions, sheet-pile cells tend to rip and burst (Hahn and Rama, 1982). Such a mode of cell collapse helps to stop the ship gradually with minimal damage to the ship. Naturally, after destruction by ship impact, these cells must be completely rebuilt.

If a sheet-pile cell is stronger than the vessel such as is the case when a small vessel hits the cell designed to stop a large vessel, then the vessel will absorb the impact energy through crushing of its bow or side.

Design analyses for sheet-pile cells are usually based on a consideration of the energy changes that take place during the impact. The typical energy/displacement relationship model for a ship collision event is developed for the forces associated with crashing of a vessel bow, lifting of a vessel bow, and friction between ship and cell and a ship keel and a waterway floor. For all cell structure and ship impact particulars, a deformation of a cell/ship system can be developed. Naturally, deformation of the system will stop when all the impact energy is dissipated.

The ability of a sheet-pile cell to absorb more energy during a major ship collision

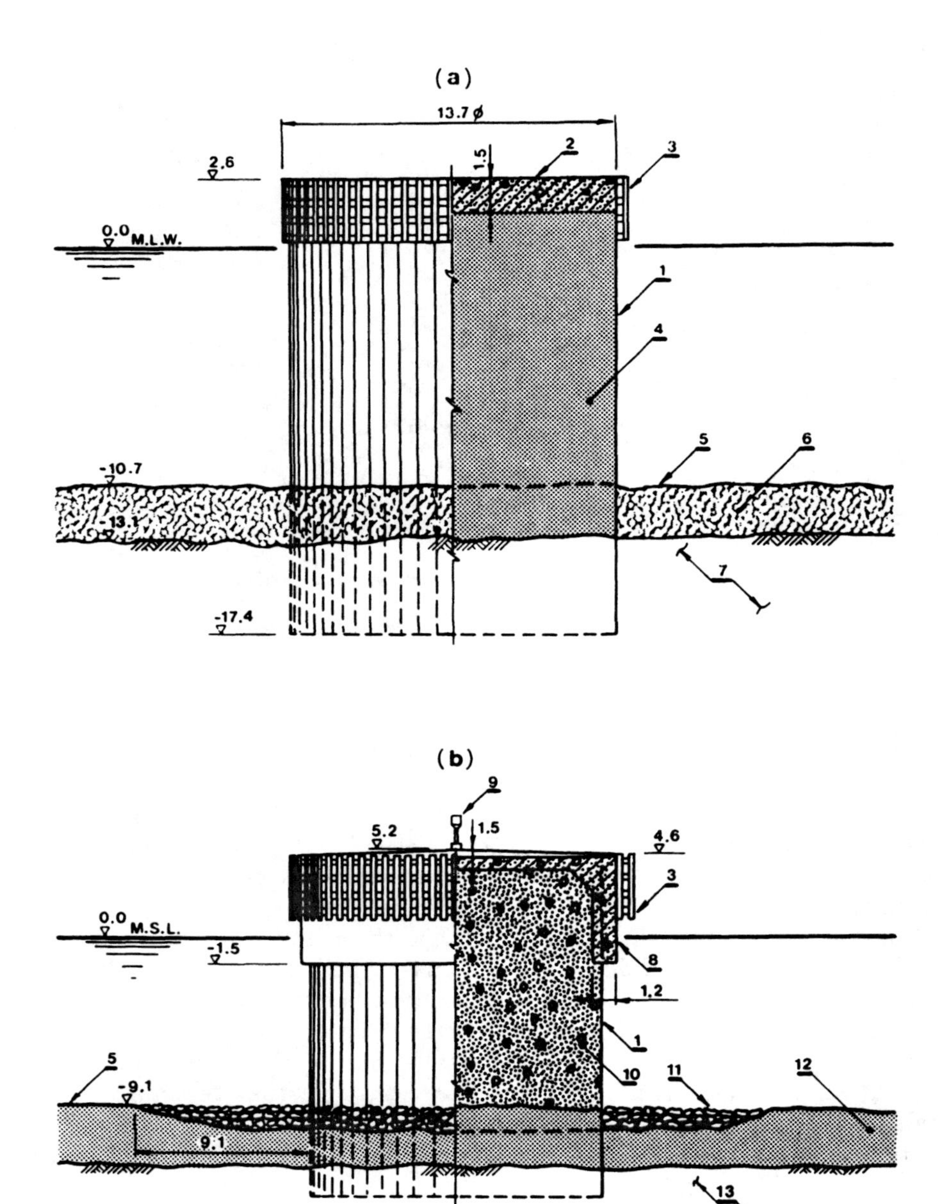

Figure 7–7. Large-diameter steel sheet-pile protective cell. (a) Typical cell installed at the outer bridge crossing, Staten Island, New York. (From Rama, 1981.) (b) Typical cell installed at the Sunshine Skyway Bridge. (From Knott, 1986.) 1—Steel sheet-pile cell, 2—concrete slab, 3—timber fender system, 4—sand fill, 5-mud line, 6—organic silt, 7—consolidated sand and gravel, 8—concrete cap, 9—navigation light, 10—crushed stone fill, 11—rockfill scour protection, 12—overburden, 13—hawthorne formation.

event may be enhanced by keeping the top of the cell rigidly together. This can be accomplished by welding the sheet-pile interlocks near the top of a cell down to water level, and by enclosing the top of the cell with a concrete cap containing reinforcing steel penetrating through holes in the sheet piling. A thick concrete slab that encircles the top of a cell helps to cary the high hoop stresses that occur in a cell structure during collision (Figure 7–7b). According to Knott (1986), this structural element can effectively hold a cell together during the 1- to 3-s collision interval.

It should be noted, however, that although the presence of a very stiff diaphragm at the top of a cell increases cell strength and stability, it may also cause substantial damage to the ship.

The rupture pattern of the steel sheet-pile protective cell with a stiff reinforced concrete cap at a bridge has been reported by Englot (1988). The approximately 14.0-m diameter cell was hit head-on by a 42 000 DWT tanker. As a result the top of the cell was displaced laterally approximately 4.5 m. Englot identified the following three basic deformations observed at the damaged structure:

1. The sheet piles buckled and failed in tension on the side of impact and in compression on the opposite side. The sides of the cell deformed in shear. All deformations were commensurate with the lateral displacement of the concrete cap which served as a rigid diaphragm and remained virtually intact. The cap remained lodged within the cell, although it was pushed down below the water level.
2. On impact, the local inward sheet-pile deformation displaced the sand filling, which caused the heavy hoop tensile forces on the sheet-pile interlocks and pressure on the underside of the concrete cap. As deformation and rupturing of the cell wall occurred, sandfill was lost at the cap perimeter. Investigations conducted by Guillermo and Lawrence (1987) on sheet-pile cell stability brought them to the following conclusion: (i) The maximum sheet-pile cell deflections in the form of a localized bulge occur at elevations of concentrated load applications. (ii) Lateral movements of the cell walls diminish below the level of a concentrated load application, (iii) The stiffer fill may reduce the size of the yielding zones in the foundation soil; however, it produces little impact on the degree of the yielding in the cell fill.
3. The local sheet-pile rupturing occurred at the point of vessel impact. The latter controls the level of damage to the impacting ship.

A description of the construction and operation of several sheet-pile cell bridge protection systems is found in Ostenfeld (1965), Rama (1981), Hahn and Rama (1982), Knott (1986), and Englot (1988).

7.5.1.1 Layout

The number of cells required for bridge protection and their layout basically depends on the geometry of the navigation channel (waterway), hydraulic conditions, vessel traffic and vessel traffic management, and vessel characteristics. In a wide-open waterway where the aberrant ship may approach the bridge pier virtually from any direction the principal pier is usually protected by a cluster of at least six cells placed around a pier (Figure 7–8a). This, for example, was done for the principal pier protection at the new Sunshine Skyway Bridge (Knott and Flanagan, 1983). In the case of less important bridges or for protection of a pier of the approach bridge, the number of protection cells may be reduced to two, one at each side of a pier, subject to careful investigation.

In the case of narrow waterways the protective cells are usually placed in a way so as to protect the pier and/or to guide an aberrant vessel in between closely spaced bridge piers (Figure 7–8b).

In the latter case the minimum number of two protective cells, one at each side of a pier, is used, subject to careful consideration of all parameters involved.

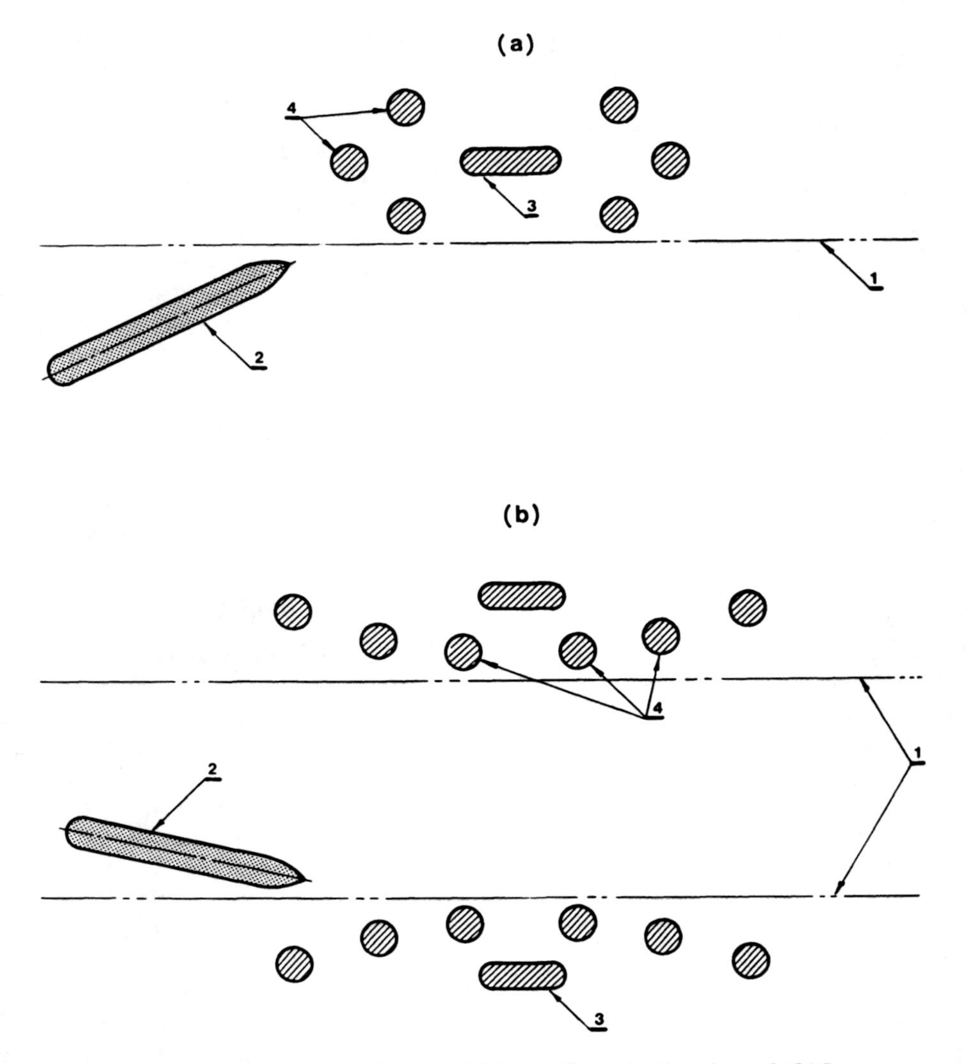

Figure 7–8. Typical protective cell layout. (a) In a wide navigation channel. (b) In a narrow nagivation channel. 1—Boundaries of navigation, 2—aberrant ship, 3—bridge pier, 4—protective cell.

The space between adjacent cells should be given particular attention. For example, too close spacing of the adjacent cells may require installation of an extra cell, and too large spacing between cells cannot serve as a reliable protection, because a vessel may squeeze between cells and hit the bridge. In the former case the solution will be less cost-effective, and in the latter case the bridge may be endangered.

An example of a ship that penetrated a protection system composed of sheet-pile cells and then damaged the bridge is shown in Figure 7–9.

7.5.1.2 Gravity-Type Protective Caissons

This type of protective structure can be built at locations where the sheet-pile cells

Figure 7–9. Collision of general cargo ship "Jalagodavari" (13 455 gross tons) with St. Louis Bridge, Beauharnoi's Canal. (Courtesy of The St. Lawrence Seaway Authority.)

prove to be impractical. Typically this may happen where the seafloor is either "too soft", for example, composed of loose silt deposits of substantial depth, or "too hard", when composed of bedrock, very stiff clay, or similar materials that do not permit sheet-pile driving.

Gravity-type caissons are usually built in dry docks, or at locations where they can be launched, then towed to the designated location, installed upon a previously prepared rock fill mattress, and finally filled with granular material.

Similar to sheet-pile cells, caissons are usually designed in a circular shape in order to redirect the ship away from the piers in glancing blow situations. To absorb the energy of small collisions protective caissons are usually equipped with conventional type fenders.

Gravity-type protective caissons are usually designed to yield to the impact force by sliding and tilting. However, because of the very short time period of ship impact, this type of structure may not have sufficient time to respond, and therefore the impact force is not significantly reduced.

In the latter case, the impact force could be quite damaging to the ship. It must be noted that in ice-affected waterways ice forces, rather than ship impact, may dictate the size of a bridge protective caisson, and therefore the caisson can be much less effective in terms of minimizing damage to the ship.

Reinforced concrete, steel, or a combination of both materials can be used for caisson construction. Typical examples of gravity-type caissons are depicted in Figure 7–10.

7.5.1.3 Sliding Caissons (Mobile Protection System)

Structures such as sheet-pile cells and gravity-type structures of different construction are primarily designed to protect the bridge pier, usually with little consideration for ship protection. The most likely result of ship collision with the above protective sys-

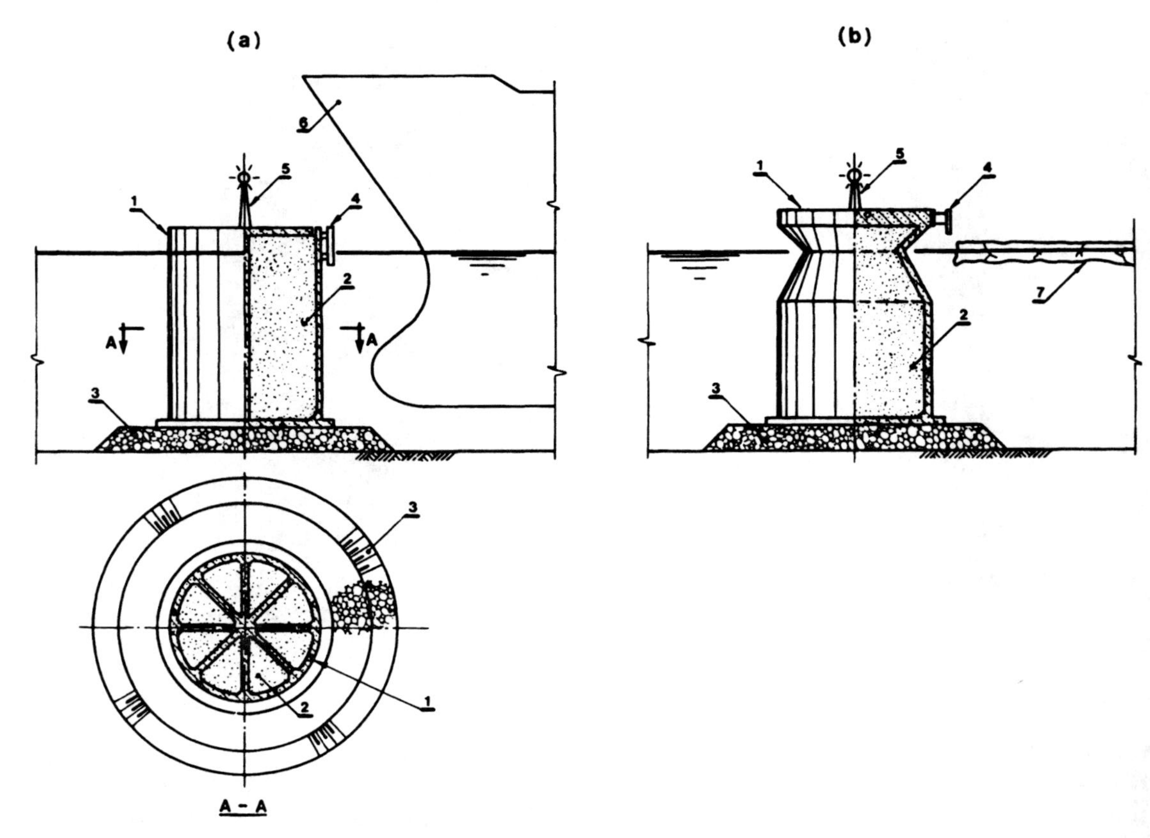

Figure 7–10. Gravity-type concrete caissons. (a) Conventional. (b) For installation in ice-affected waterway. 1—concrete caisson, 2—granular fill, 3—rockfill mattress, 4—fender system, 5—navigation light, 6—ship, 7—ice.

tems is crushing and buckling of the ship bow and/or sides of the hull. Obviously damage to the ship could compromise safety of navigation due to the possibility of blockage of the navigation channel by a sunk vessel or by wreckage of a destroyed protective structure. It may also result in heavy pollution due to oil tank rupture, and spill of miscellaneous pollutants carried by the vessel. In order to mitigate the damages to a ship attributed to a collision event, Saudi-Danish Consultants (1978) proposed the installation of a 23-m diameter concrete sliding caisson in order to protect the Bahrain Causeway Bridge from collision with a 25 000 DWT ship. Located in a 6-m water depth, the caisson was planned to be installed upon a prepared rip-rap mattress to be then filled with sand (Figure 7–11).

The caisson was designed to absorb ship collision energy partly by crushing of the ship's bow and partly by sliding on the foundation mattress. It was designed in such a way as to withstand impact force without damage to the outer shell. After absorbing the ship impact by sliding, the caisson can be emptied, floated back to its original position, and filled with granular material again. However, as in the cases of the sheet-pile protective cells and miscellaneous gravity-type structures, the sliding caisson of proposed construction, owing to previously stated reasons, cannot completely prevent a ship from sustaining substantial damage during collision.

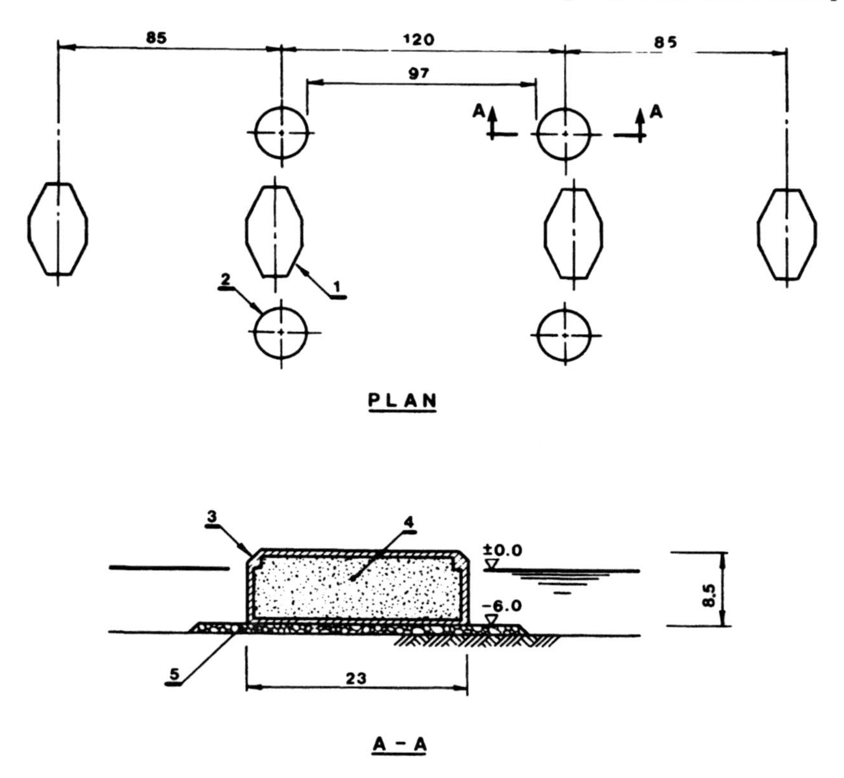

Figure 7–11. Concrete sliding caisson proposed for protection of main pier of the Bahrain Causeway Bridge. (From Saudi-Danish Consultants, 1978.) 1—Bridge pier, 2—protection concrete dolphin, 3—sliding concrete caisson, 4—granular fill, 5—rockfill mattress.

Tsinker et al. (1990) proposed the concept of a movable protection system that is able to protect both a bridge and a ship of any size from being severely damaged during collision (Figures 7–12 and 7–13). The idea behind this concept is to deflect the ship from the collision course, rather than to bring it to a complete stop. The latter, however, is not excluded, and is considered as the most critical, although extraordinary, event.

The system is protected with conventional fenders that are able to absorb small collision energy as generated by a small ship. Conventional fenders, for the first line of energy absorption, are attached to the multi-cell type, or composite type buffer, which is placed around the basic structure (second line of energy absorption). Multi-cell buffers were developed in Japan and were recently used for pier protection by Honshu-Shikoku Bridge Authority (Matsuzaki and Jin, 1983). They are composed of steel box-like cells which, if required, are filled with brittle materials such as hard urethane foam (Figure 7–14). During a head-on collision, when a ship wedges into the multi-cell buffer structure, it crushes cells in a steadily growing number, and therefore penetrates the buffer with a steadily growing resistance force. In order to prevent damage to the ship, the resistance force at any time of penetration shall not exceed the strength of the ship bow. However, even the combined capacity of both conventional fenders and the multi-cell buffer practically cannot absorb the amount of energy of a large ship during collision and stop it from advancing toward

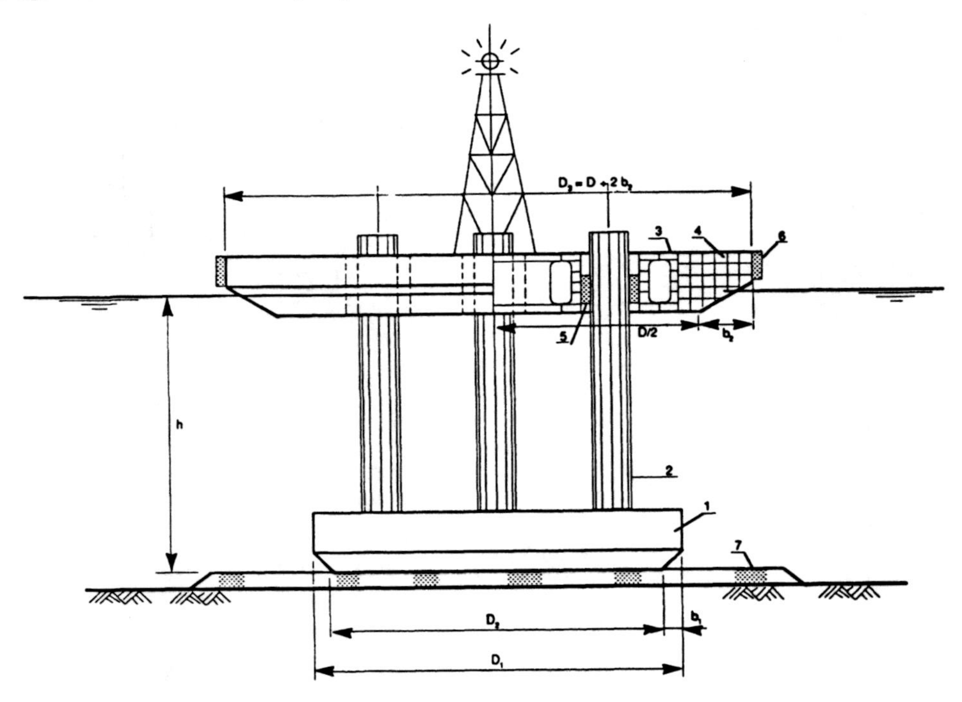

Figure 7-12. Mobile bridge protection system. (From Tsinker et al., 1990.) 1—Floating base (steel or concrete), 2—steel column, 3—guided pontoon (steel or concrete), 4—Multi-cell buffer system, 5—timber or rubber sleeve, 6—conventional fender, 7—rockfill mattress.

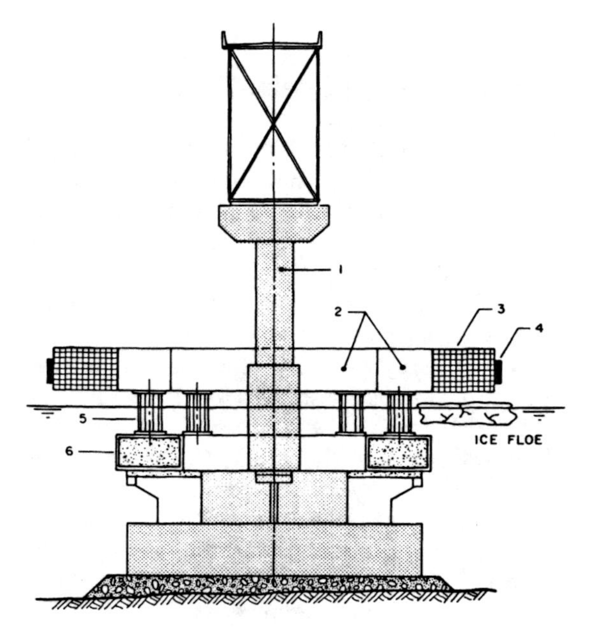

Figure 7-13. Mobile bridge protection system placed around a bridge pier. (From Tsinker et al., 1990.). 1—Bridge pier, 2—superstructure (steel or concrete), 3—multi-cell buffer system, 4—conventional buffer system, 5—column (steel or concrete), 6—base (steel or concrete) filled with ballast as required.

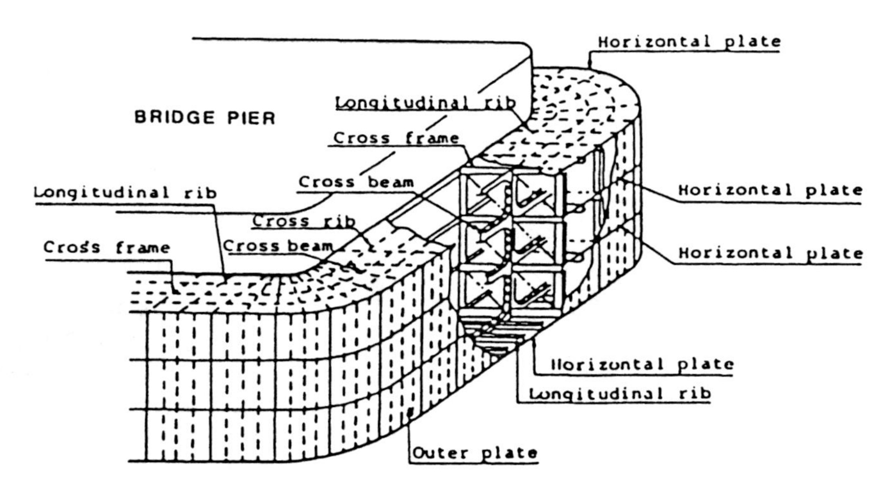

Figure 7–14. Multi-cell buffer system for Bisau-Seto Bridge, Japan. (From Matsuzaki and Jin, 1983.)

the bridge. Therefore the entire system is designed to resist an impact force by absorbing ship energy by the combined fender system and in a sliding mode of failure. Naturally, the ultimate resistance force would be equal to the weight of the structure multiplied by the appropriate friction coefficient (static and dynamic).

Because system resistance to sliding is designed not to exceed the impact force, the entire system will be set in motion during ship penetration into the combined fender/multi-cell buffer structure. Displacement of the system will depend on the remaining kinetic energy of the ship and materials comprising the protection structure and foundation. Furthermore, it should also be noted that some amount of energy would be absorbed by the elastoplastic deformation of the protection structure and ship hull.

Furthermore, the system must be able to sustain environmental forces such as waves, current, wind, and ice with a sufficient factor of safety to resist sliding and overturning. Therefore, a balance between resistance to sliding due to environmental forces and sliding due to ship impact has to be maintained. This is basically achieved by selecting an appropriate stiffness of the buffer structures in relationship to the shape and strength of the design vessel.

In the case of an angular approach of a ship to the protection system, the latter is able to yield some distance, generating a sustained force against the ship hull to cause it to veer off before it can hit a bridge pier.

A mathematical model has been developed to simulate the collision of a ship and predict the response to a movable bridge protection system. The equations of motion for the bodies taking part in the collision (Figure 7–15) are described by Newton's second law as follows:

$$m_1\ddot{x}_1(t) = K(x_2 - x_1) - F_{D(1)} \qquad (7\text{–}14)$$

$$m_2\ddot{x}_2(t) = K(x_1 - x_2) - F_{D(2)} - F_F \qquad (7\text{–}15)$$

where

m_1 and m_2 = mass of the ship and protection system, respectively

$\ddot{x}_1$ and $\ddot{x}_2$ = acceleration of the ship and protection system, respectively

K = variable stiffness of the protection system

and

$F_{D(i)}$ = drag force ($i = 1; 2$)

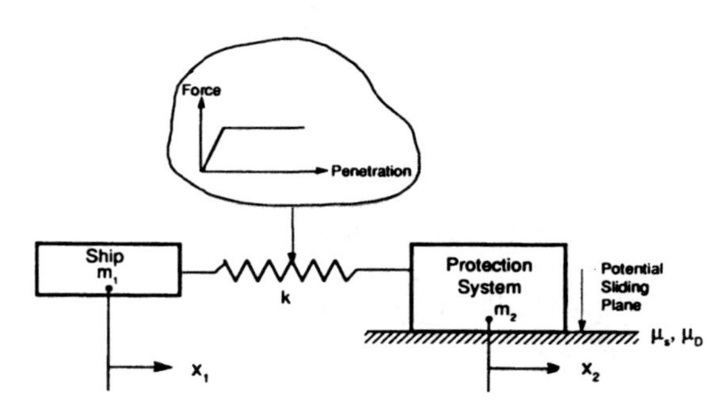

Figure 7–15. Dyanmic model of a ship/movable protection system collision. (From Tsinker et al., 1990.)

F_F = friction force.

The drag forces $F_{D(i)}$ ($i = 1$; 2) are calculated using

$$F_{D(i)} = 0.5\rho_w C_D A_i |\dot{x}_i| \dot{x}_i \qquad (7\text{–}16)$$

where

ρ_w = water density
C_D = drag coefficient $\simeq 1.2$
A_i = area exposed to water pressure
$\dot{x}_i$ = current velocity.

Friction force, F_F, is related to the submerged weight of the protection structure in the water W

$$F_F = \mu_s W = \text{static friction} \qquad (7\text{–}17)$$

$$F_F = \mu_D W = \text{dynamic friction.} \qquad (7\text{–}18)$$

where

μ_s = static friction coefficient
μ_D = dynamic friction coefficient.

The equations of motion are highly nonlinear due to the presence of the friction term, the drag resistance terms, and the nonlinear force–deformation relation for the fender (i.e., stiffness varies with deformation, as shown in Figure 7–16). Thus, the numerical solution has been applied using the standard fourth-order Runge–Kutta–Gill method. The initial conditions for the equations of motion are as follows:

$x_1 = x_2 = 0$ (displacement), $x_1 = V_0$ (initial velocity of the ship) and $x_2 = 0$ (initial velocity of the protection system).

The friction forces acting on the bottom of the protection system are related to the system's velocity. The static friction is applied

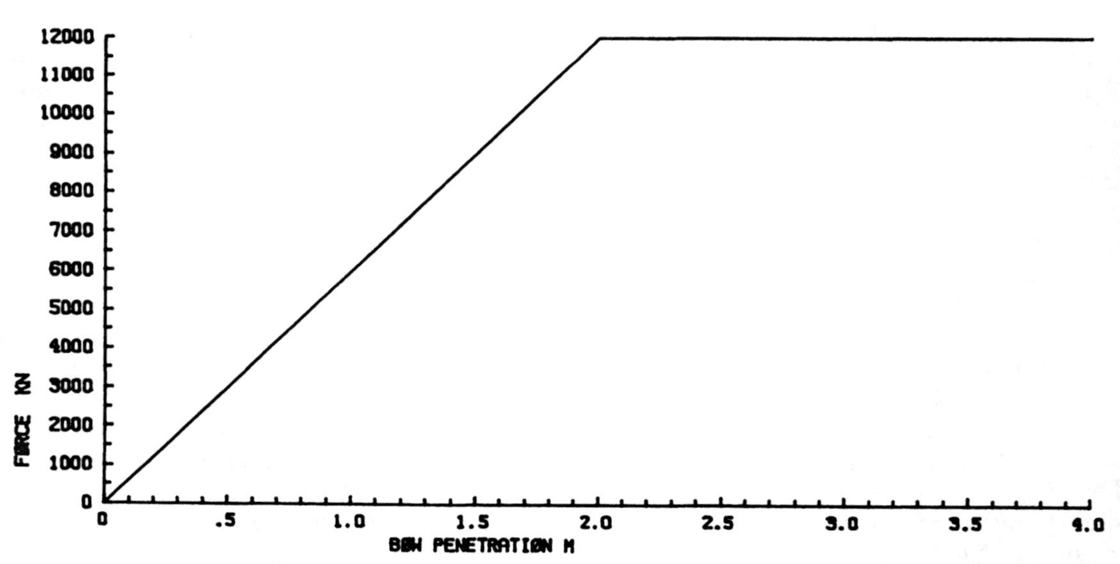

Figure 7–16. Force/bow penetration curve for the bridge protection system. (From Tsinker et al., 1990.)

when the system is at rest, and the dynamic friction when the system is sliding over the bottom. A flow chart of the computer program for simulation of the ship-movable protection system collision is presented in Figure 7–17.

7.5.1.4 Numerical Example

An aberrant ship ($m_1 = 15\,000\,t$) moving at a speed of $V_0 = 1.0$ m/s collides with a mobile bridge protection system ($m_2 = 1240\,t$) similar to that depicted in

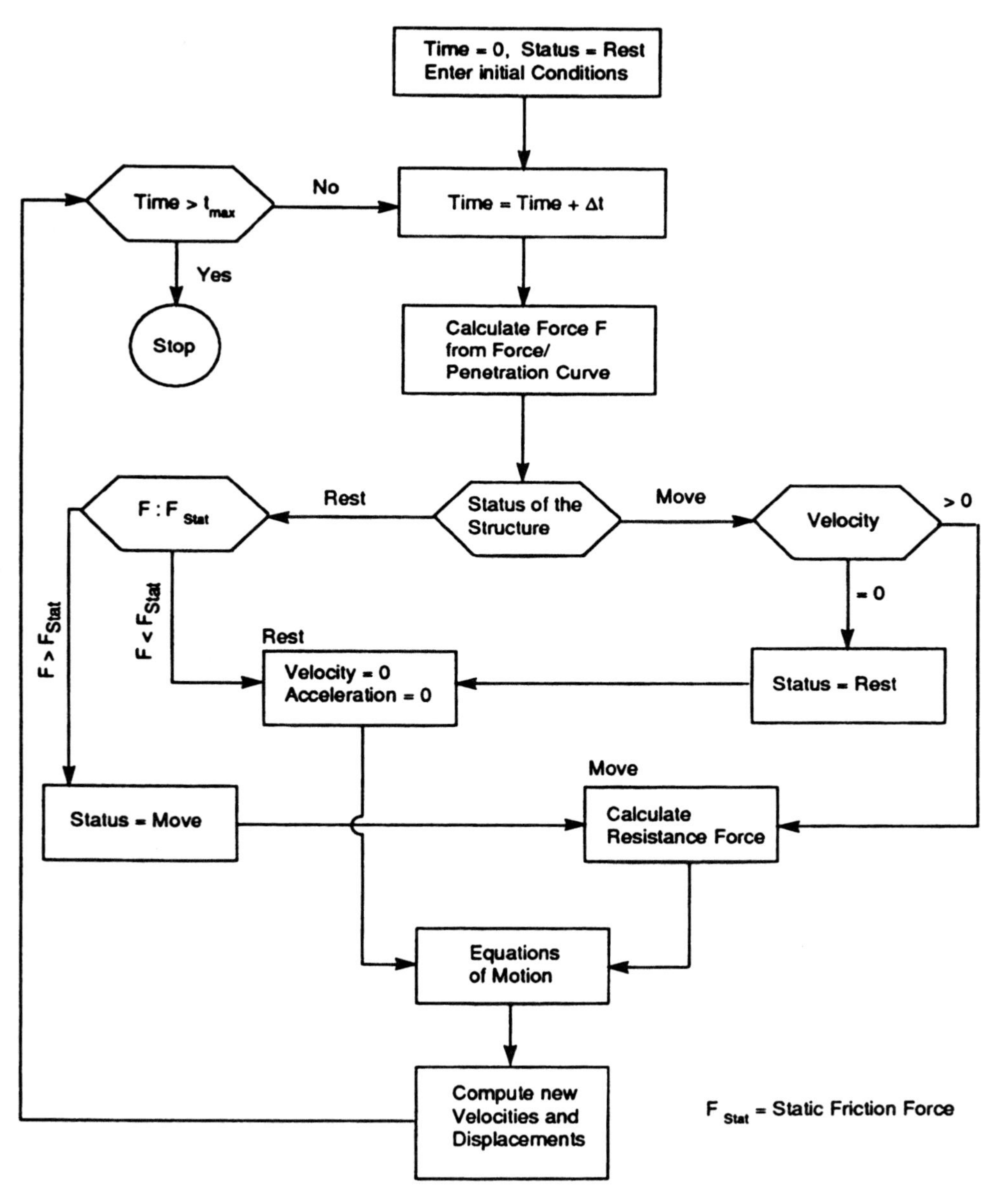

Figure 7–17. Flow chart of the computer program for simulation of the ship/movable protection system collision. (From Tsinker et al., 1990.)

Figure 7–12. The bow penetration curve of the combined fender/multi-cell buffer system is shown in Figure 7–16.

The resulting ship and protection system displacements during collision are shown in Figure 7–18. The resulting dynamic loads acting on the ship bow and the protective system which are presented in Figure 7–19 show that the maximum load has been developed during the initial moments of collision when the inertia of the system contributed the most to the load.

7.5.1.5 Man-Made Islands

Protective islands are often used where a large navigation span permits placing islands around bridge piers adjacent to the navigation channel without significant reduction in width of the channel. Because

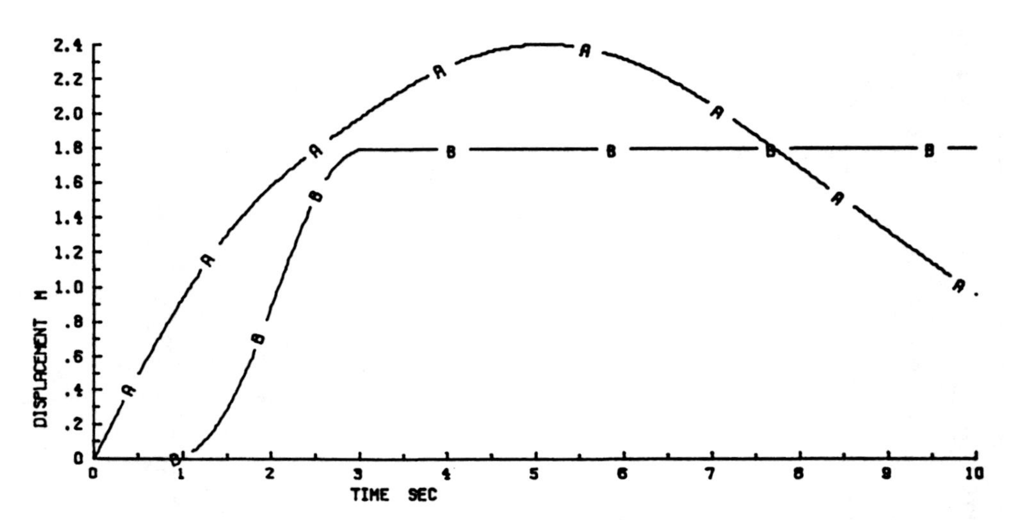

Figure 7–18. Displacements of bodies taking part in collision (A = ship, B = protection system). (From Tsinker et al., 1990.)

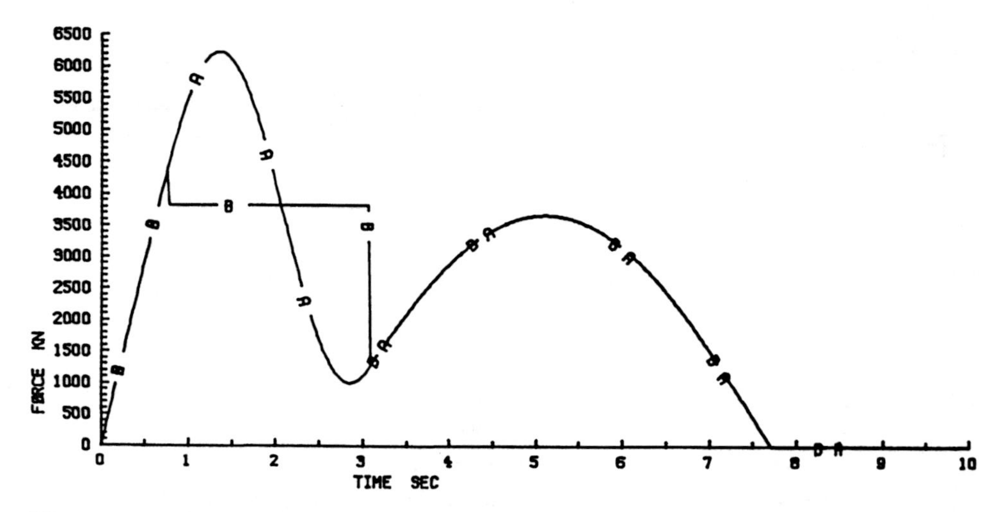

Figure 7–19. Forces during collision. (A = bow, B = protection system). (From Tsinker et al., 1990.)

islands completely encircle the bridge pier, they are generally an effective means of reducing the risk of a ship collision with a bridge pier.

Islands offer a high degree of protection in addition to a long life expectancy and relatively low cost of maintenance. However, as stated earlier the use of islands for pier protection is generally limited to long navigation span bridges and also to the locations where suitable construction materials are readily available. Depending on the location of the project, the cost of materials such as sand, rocks, and rip-rap armor layer can be either relatively inexpensive or very expensive.

It should be noted that environmental concerns and some geotechnical aspects of a navigation channel bed could sometimes prohibit construction of islands for bridge pier protection. The former is usually associated with a potential increase in current velocities, scouring, and adverse effects on flushing characteristics of a waterway. Sometimes, the bottom of the waterway is an environmentally sensitive habitat and objections to the island construction become a major issue.

The best examples of protective island construction for bridge pier protection from around the world are Verranzano Narrows Bridge, New York; The Chesapeake Bay Bridge; Sunshine Skyway Bridge, Tampa Bay, Florida; the Loire Bridge near St. Nazaire, France; Westgate Bridge, Australia; Taranto Bridge, Italy; Laviollette Bridge across the St. Lawrence Seaway; Annacis Island Bridge crossing the Fraser River near Vancouver, Canada; and Orwell Bridge near the Port of Ipswich, Great Britain.

Protective islands may consist of sand, gravel, or rubble embankments, suitably armored against erosion by waves, currents, ice, and propeller-induced scour. Typical examples are depicted in Figure 7–20.

Typically, the colliding ship tends to plow into the island and then ride up on it. Hence, the island must be large enough to bring a ship to rest before the bow hits the pier. Also, the island's geometry should be proportioned in a way to ensure that the ship impact force transmitted through the island's material to the bridge pier does not exceed the lateral design capacity of the pier and pier foundation. In developing the island geometry, consideration should also be given to whether the design vessel travels empty, in ballast, or fully laden. In sizing the island, consideration should also be given to the overhang, or flair, over a vessel bow which should be added to the required stopping distance of the vessel.

Naturally, the ship impact energy is absorbed by deformation of the island material, deformation of the ship hull, and by the friction of the hull (underkeel) sliding up against the island. Furthermore, the increase in virtual weight of a ship while sliding up the island, as well as waves and water turbulence generated by a displaced vessel, may contribute to energy dissipation.

The distance of ship penetration into the island and/or sliding up depend on the island and ship geometry, island material, and speed of the ship.

The inclusion of all the above factors in a mathematical model of a ship–island interaction is very difficult so that usually some simplifications have been adopted. According to Bouvet (1963), in 69% of tanker groundings studied, the ship keel structure was damaged to a depth of less than 0.5 m. The above and similar data sometimes bring the designers to the conclusion that the crushing of the hull of the ship (which, in turn, may heavily depend on the type of ship and protective island construction) can be ignored in the design computation (Fletcher et al., 1983). In a conservative approach to the protection island design two limiting cases are usually considered.

1. It is assumed that the island's material does not permit any substantial penetration by a ship bow. Therefore a ship could be brought

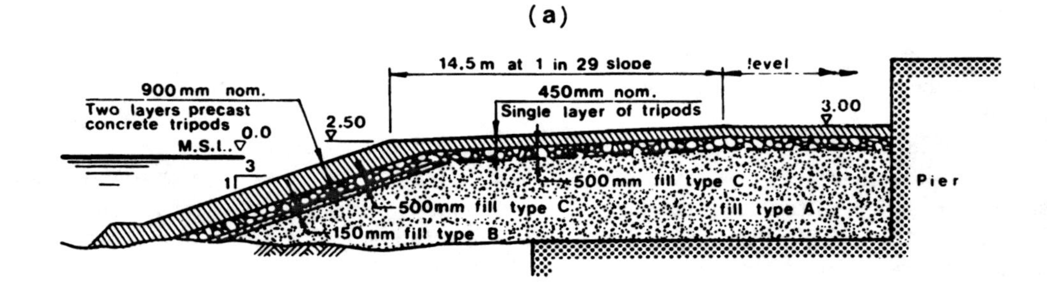

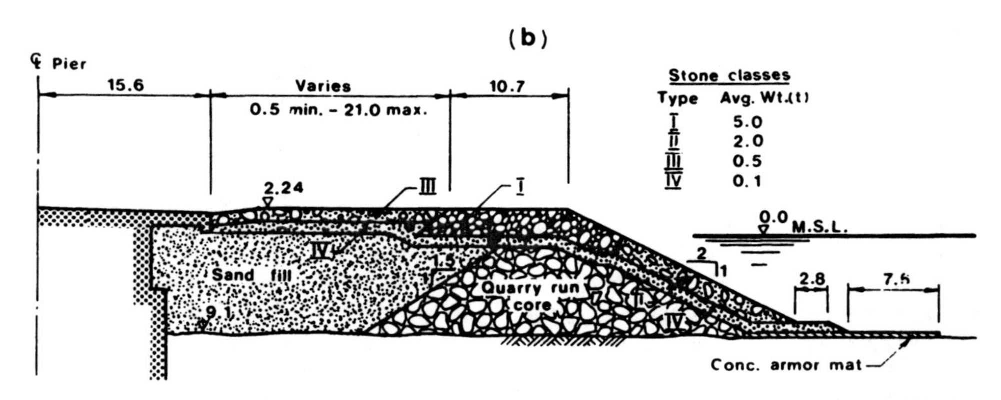

Figure 7–20. Protective islands. (a) Typical cross- section for the Orwell Bridge, U.K. (From Fjeld, 1979.) (b) Sunshine Skway Bridge, Tampa, Florida (Federal Highway Administration, 1983.)

to rest only by rotation about its center of gravity and by friction between the ship hull and the face of an island. Accordingly, the friction coefficient depends on the ship hull and island materials. For a steel hull and granular island materials, the friction coefficient is usually taken as equal to 0.4 to 0.6 (smaller values are used for wet and larger for dry material).

2. Second, it is assumed that no rotation of a ship would take place and that all of the energy would be dissipated by the ship penetrating the island material.

In both design assumptions the presence of the armored layer on the island's face is usually ignored. However, in reality, a combination of both cases usually takes place during ship collision. In general, the simplified but nevertheless realistic design model depicted in Figure 7–21 can be considered in island design.

Similar types of mathematical models have been used for the design of protective islands at the Sunshine Skyway Bridge (Havno and Knott, 1986) and the proposed Northumberland Crossing between New Brunswick and Prince Edward Island, Canada (Acres International Limited, 1988).

A relatively complex mathematical model that takes into account all six degrees of freedom of the vessel movements, as well as three-dimensional geometry of the island, has been developed by the Danish Hydraulic Institute (Briuk-Kjaer et al., 1982). However, the best result in island proportioning can be obtained by performing large-scale physical model tests.

Comprehensive physical model tests on protective islands have been conducted for

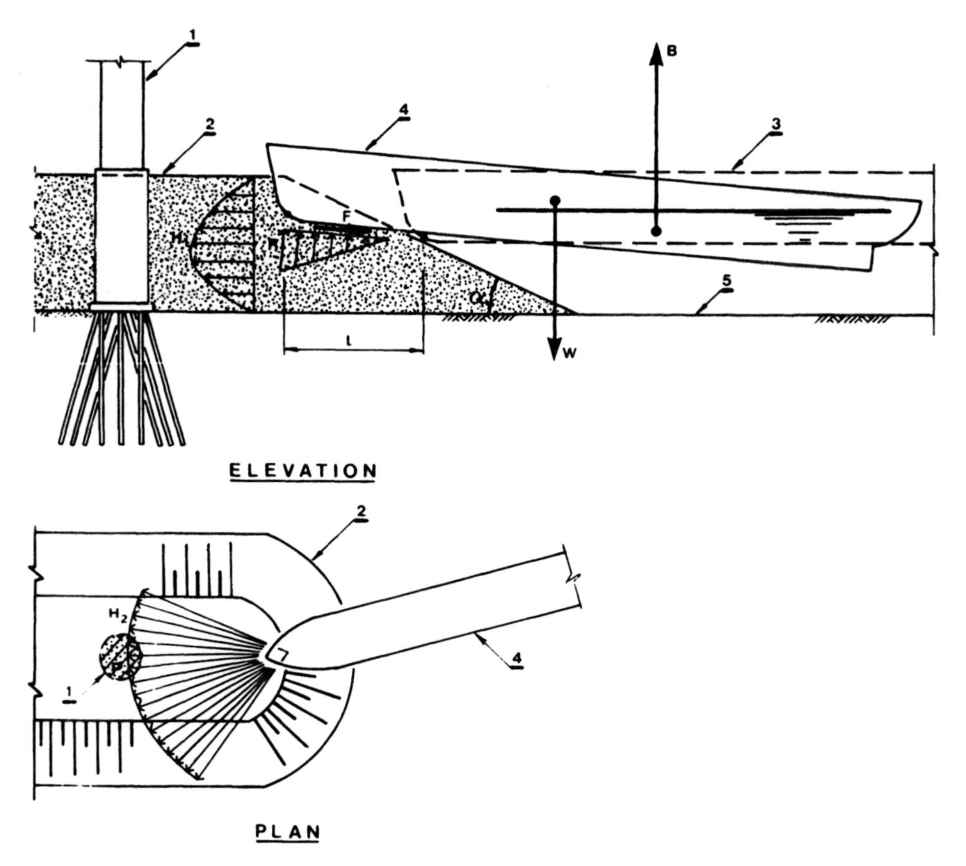

Figure 7–21. Vertical and horizontal distribution of ship impact force through a protective island. (After Havno and Knott, 1986.) 1—Bridge pier, 2—protective island, 4—aberrant ship before and after collision respectively, 5—mudline.

the Great Belt Bridge in Denmark by the Danish Hydraulic Institute (Briuk-Kjaer et al., 1982; Denver, 1983). Data obtained from these tests have been used for calibrating and verifying the mathematical model. One of the major conclusions of both mathematical and physical modeling was that the island's cross-section should be developed to maximize its deflective potential, thus reducing the horizontal forces ultimately transferred to the bridge pier. This is due to the fact that the horizontal collision forces between the vessel and the island decrease rapidly when the vessel is deflected up.

If the ship penetrates the island, then a primary means of collision energy absorption with the island system is the deformation and displacement of the island materials (Denver, 1983; Sexsmith, 1983). Although under these circumstances the vessel's bow will also deform, the effect of this on energy dissipation is expected to be considerably less than would occur during collision with a rigid protective system.

Description of the model set-up and results of other important physical model tests are found in Havno and Knott (1986), Fletcher et al. (1983), and Sexsmith (1983). Because of the relatively gentle nature of an island stopping a ship on its collision course with a pier, it is usually favored by all those concerned with safety of navigation, such as ship masters, pilots, environmental agencies, and others.

7.5.2 Other Protective Systems

In the past, various attempts have been made to design a system that could effectively protect both the bridge pier and a ship. Consideration was also given to the design of floating barriers, moored pontoons, or use of old out-of-service small ships to protect both pier and ship. In some protective schemes, the energy-absorbing mechanism of the ship, such as plastic deformation of certain parts of the ship's hull, was considered.

However, almost all of the above-mentioned solutions, including, as noted previously, use of a commercially available rubber fender system of miscellaneous construction, appeared to have some serious limitations and questionable applicability to the protection of a major bridge pier against high-energy large vessels.

Discussion on some of the above-mentioned systems is provided in the following paragraphs.

7.5.2.1 Cable Systems

These are usually considered for pier protection against ship collisions at locations where there are no substantial ice movements, the water is too deep, and/or geotechnical conditions are not suitable for construction of such types of protective systems as large-diameter cells, dolphins of miscellaneous designs, or man-made islands. The basic principle of the cable protective system is to arrest a ship movement toward a pier by absorbing its kinetic energy with relatively small forces but large deformations of the arrestor system. Examples of the two typical layouts of the cable protective systems are depicted in Figure 7–22.

The protective system depicted in Figure 7–22a has actually been installed to protect the Taranto Bridge which crossed the Mare Piccolo waterway in Italy (Saul and Svensson, 1983). The system was designed to stop ships up to 15 000 DWT traveling at a speed of up to 6 knots.

A similar system has been proposed for protection of the Honshu-Shikokn Bridge crossing the Akashi strait in Japan against impacts from small, 1000 DWT vessels. There, the kinetic energy of a ship was designed to be absorbed by the elastic deformation of the entire system. This included resistance to ship movement by the arrestor chain (cable) and by the anchors dragging on the bay floor.

As shown in Figure 7–22a, this system includes the arrestor chain (cable) spanning between support buoys, which in turn are secured to anchors by anchor chains or cables. In some cases in order to prevent blocking of the waterway the anchor chains can be depressed to the waterway floor by the tension (concrete or metal) sinkers. The cable system depicted in Figure 7–22b was used for protection of a drilling rig in the Akashi channel, Japan (Oda and Nagai, 1976).

The primary disadvantages of a cable protective system include concerns over system corrosion, necessary adjustments of anchor cable lengths due to large tide height fluctuations, and, as indicated by Maunsell et al. (1978), potential ship roll over the arrestor.

In the latter case ships with a bulbous bow have a better chance of being stopped by the arrestor chain. Ships with a raking bow, as well as pontoons and barges, may easily pass over the arrestor (Figure 7–23). Additional drawbacks to the use of cable systems include uncertainty of the anchor resistance to the pull-out forces, and a potential bottom scour which can undermine the anchors. Further details on anchor and cable systems design and anchor–soil interaction are given in Chapter 4 and Tsinker (1986).

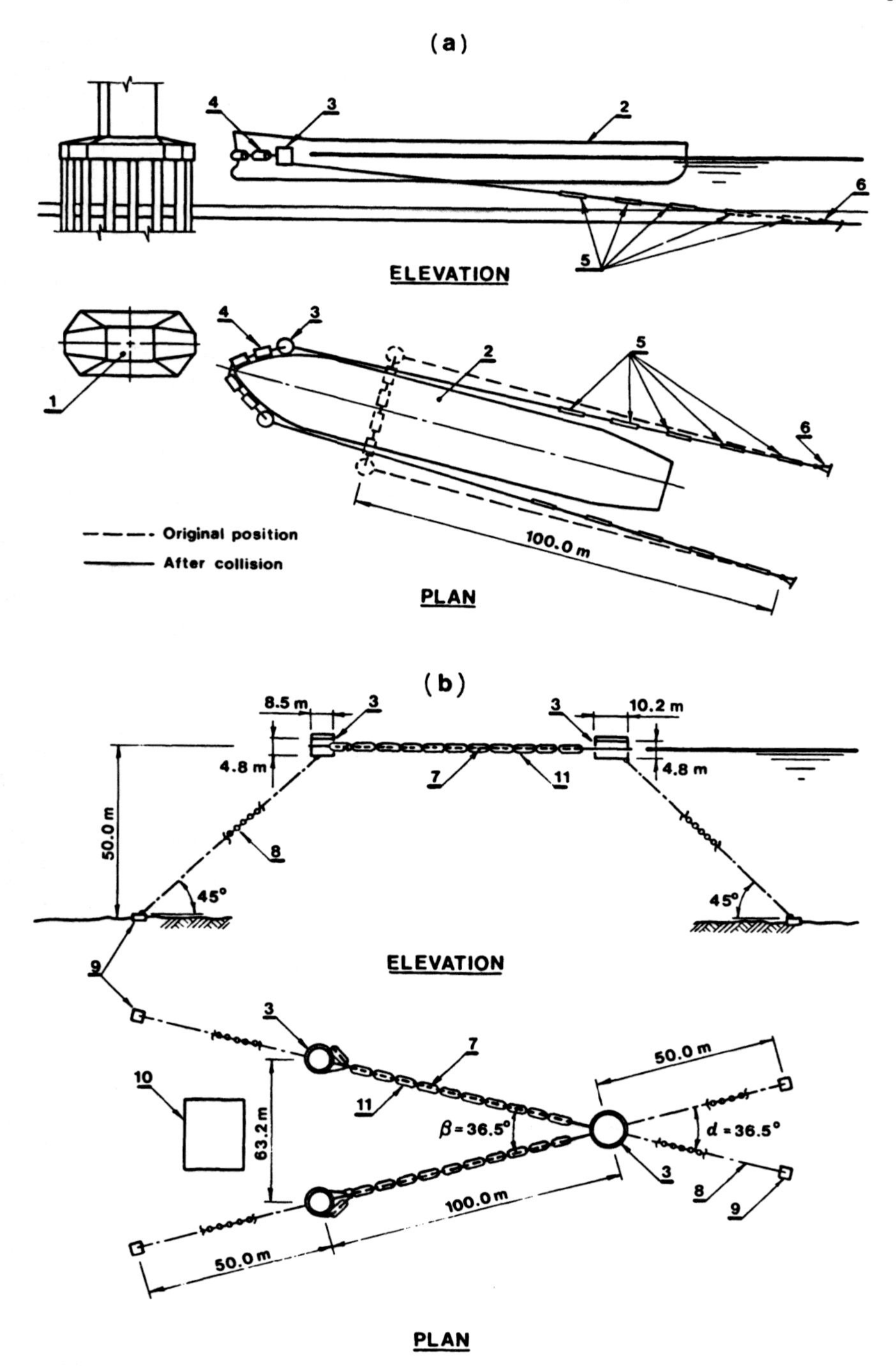

Figure 7-22. Examples of a cable protection system. (a) Protection of bridge piers of the Taranto Bridge across the Mare Piccolo, Italy. (From Rowe Research and Engineering Associates, Inc., 1985.) (b) Protection of the drilling rig in the Akashi channel, Japan. (From Oda and Nagai, 1976.) 1—Bridge pier, 2—aberrant ship, 3—buoy, 4—arrestor cable with buoyancy cylinders, 5—dampers, 6—conventional anchor, 7—chain 95.0 m diameter, 8—anchor chain, 9—gravity-type anchor, 10—drilling rig, 11—pneumatic rubber fender 2.0 m in diameter, 5.0 m long.

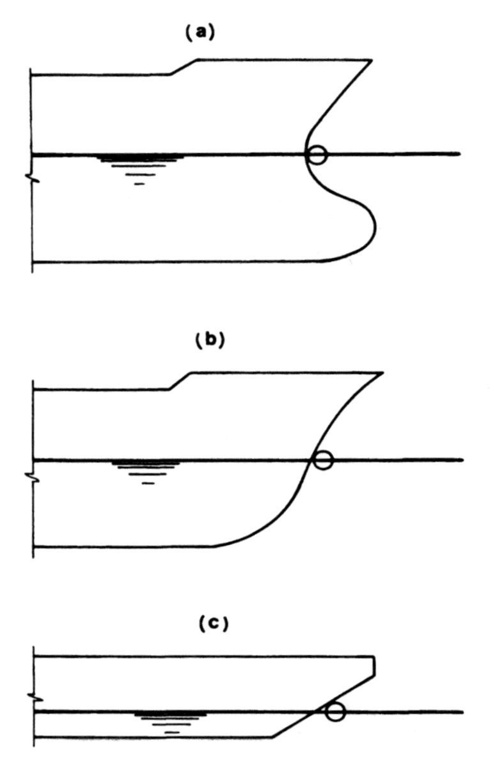

Figure 7–23. Ship/arrestor cable interaction as a function of the shape of the vessel bow. (a) Ship with balboas bow. (b) Ship with raking bow. (c) Pontoon (barge).

7.5.2.2 Anchored Pontoons

Anchored pontoons (or small ships), as used for pier protection, consist of a floating structure that is held in place by anchor cables secured to miscellaneous anchor systems or by post-tensioned tethers (Figure 7–24).

The ship impact is basically absorbed by deformation of the pontoon hull and in part by the pontoon displacement by impact force. The pontoon's effectiveness as a protective device can be significantly enhanced by providing it with a fender system similar to that proposed by Tsinker et al. (1990) for the mobile protective structure as discussed in previous sections.

A special anchored floating protective system has been proposed in Argentina for protection of a major bridge (Mondorf, 1983).

7.5.2.3 Miscellaneous Barriers Erected Around a Pier

Single standing piles and pile clusters (wooden, steel, and concrete) placed around a pier, and pile founded barriers (fenders) of miscellaneous constructions, have long been used for pier protection. These structures are usully able to absorb the energy of a ship collision by crushing or by plastic deformation. Therefore, after the collision, all or part of such a structure has to be replaced.

In several reported cases protection against ship impact was provided by an independent piled fender "ring" placed around a pier. One such system was proposed for protection of the Tasman Bridge, Australia (Maunsell et al., 1978). The proposed system consisted of eight 3-0 m diameter prestressed concrete piles used to support a rigid concrete fender beam (Figure 7–25). It was designed to take an impact from a 35 000 DWT ship traveling at 8 knots. It was assumed that the piles under impact load would form plastic hinges at the top and at the fixity points at the bottom to absorb the impact energy through the system's displacement.

The most recent case of a similar construction, for protection of the Sungai Perak Bridge, Malaysia, was reported by Stanley (1990). In the latter case the heavily reinforced concrete fender "ring" beam supported on sixteen 1800-mm diameter steel pipe piles was placed around the two pile caps built to support bridge pillars (Figure 7–26). The fender "ring" beam was built by using precast two-wall formwork units erected upon piles and then filled with concrete. A very similar structure was built to protect the Santubong Bridge, Sarawak (Buckby and Sim, 1990). In the latter case the 4 m deep by 2.2 m wide fender "ring"

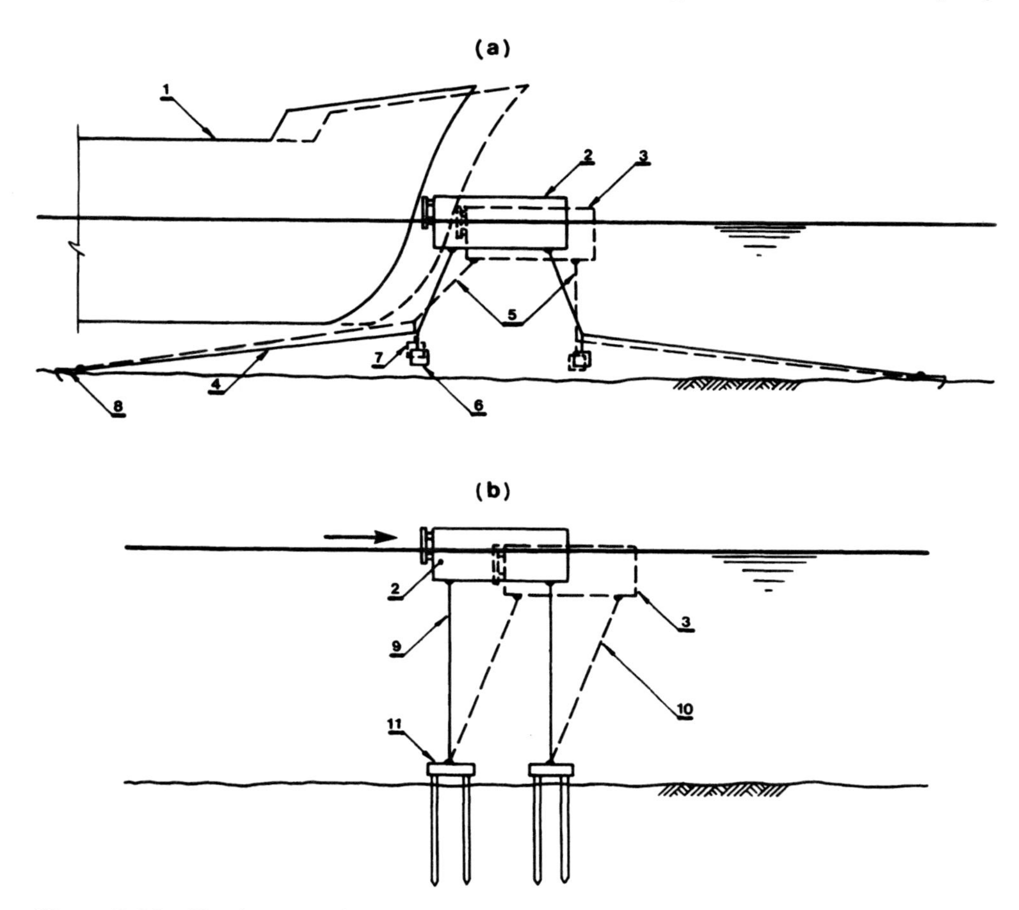

Figure 7–24. Floating protective systems. (a) Pontoon anchored by prestressed tethers. 1—Aberrant ship, 2—and 3—pontoon in original and displaced positions, 4— and 5—anchor cable in original and displaced position, 6— and 7— sinker in original and displaced positions, 8—anchor, 9—and 10—tether in original and displaced positions, 11—bottom-fixed anchor system.

beam was supported on eight 1.4-m diameter bored piles taken down to bedrock.

An unusual barrier around the bridge pier for protection from a relatively small ship impact has been recently constructed for the strengthened and refurbished Severn River Crossing, U.K. (Flint, 1992). There the protective works comprised concrete structures composed of solid concrete bull-nosed ends with reinforced concrete side walls all anchored to the bedrock and brought to a sufficient height to protect both the concrete bridge piers and steel box columns from accidental ship impact (Figure 7-27). There the walls were formed from the precast planks. These walls were installed into precast concrete channels (kickers) anchored to the bedrock backed up by the cast-in-place concrete wall. Above high water level the wall was capped by the cast-in-place heavy concrete slab provided with openings for two concrete columns comprised of the bridge pier structure.

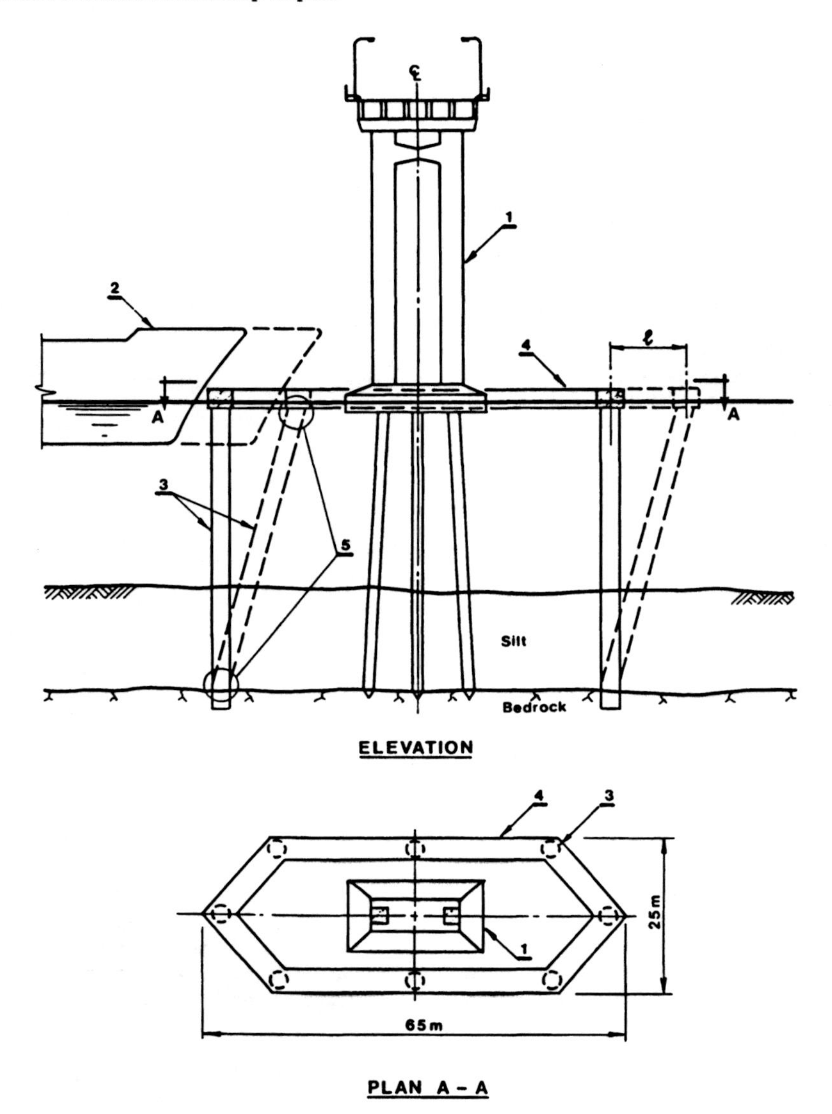

Figure 7-25. Pile-supported concrete barrier proposed for Tasman Bridge, Australia. (From Maunsel and Partners and Brady, 1978.) 1—Bridge pier, 2—aberrant ship, 3—prestressed concrete caissons 3050 mm in diameter in original and displaced positions, 4—concrete cap beam, 5—assumed plastic hinges.

7.6 COST-EFFECTIVENESS CRITERIA

The economics of development of a bridge protection system are usually based on such considerations as the risk of economic consequences of a catastrophic ship collision with a bridge, and cost of the construction of a bridge protection system. As with any project, the economic evaluation of construction of a bridge protective system involves comparison of the cost of bridge protection against the benefits of the reduction of a

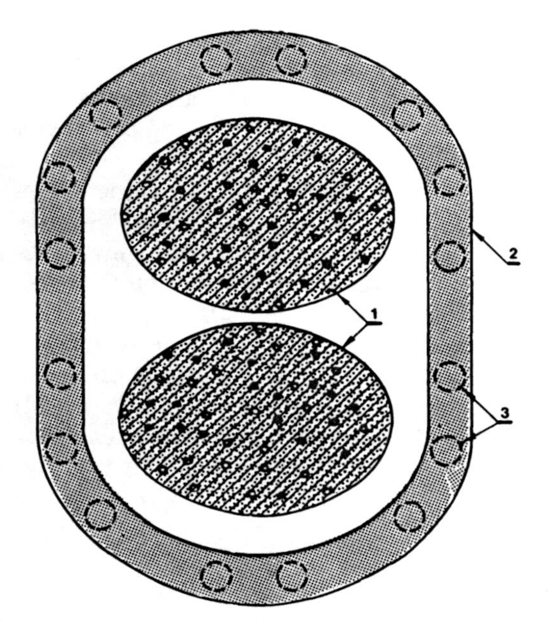

Figure 7-26. Pier protection at Sungai Perak Bridge, Malaysia. (From Stanley, 1990.) 1—Piled bridge pier cap, 2—concrete ring beam, 3—steel caissons 1800 mm in diameter.

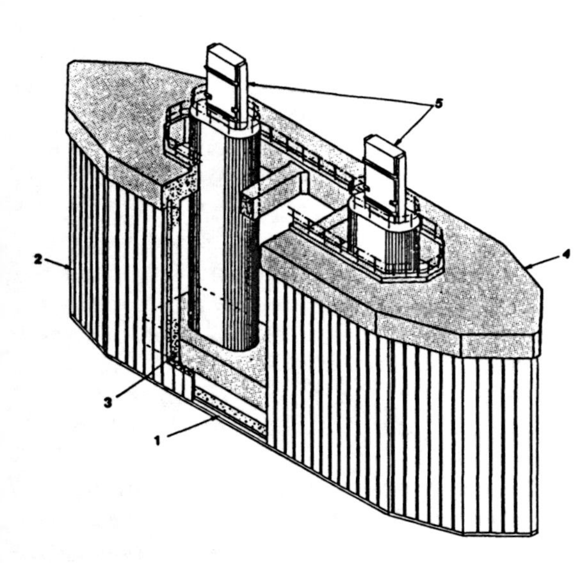

Figure 7-27. Pier protection at Severn River Crossing, U.K. (From Flint, 1992.) 1—Precast concrete channels (kickers), 2—precast concrete planks, 3—cast-in-situ concrete tremie concrete wall, 4—cast-in-situ concrete capping slab, 5—bridge piers.

risk, or complete prevention of a ship hitting the bridge.

In certain situations (e.g., deep water, strong current, heavy ice, or others) the cost of construction of a protective system can be prohibitive, as for example was the case of the collapsed Tasman Bridge, Australia. There, the cost of the bridge protection system exceeded the cost of construction of a completely new bridge (Leslie et al. 1983).

The California Department of Transportation designed and built piled platforms to protect three piers flanking the main and secondary navigation channels at Richmond-San Rafael Bridge across San Pablo Bay (ENR, October 22, 1987). Each platform is a concrete deck of honeycomb construction supported on about 140 366-mm diameter concrete-filled pipe piles ranging in length from 32 to 44 m. Six to one side cylindrical rubber fenders have been wedged between the platform's deck and the pier in order to help absorb the energy of impact. However, despite such a massive construction the protective platform is not capable of absorbing the energy of a direct hit of the design vessel and is basically designed as "sacrificial."

In the cases mentioned, the bridge owner must be prepared to take some risk of economic and social disruption associated with the bridge closure due to a ship collision.

Several approaches for evaluating the costs associated with vessel collision and pier protection have been reported in the literature, for example, Nordic Road Engineering Federation (1980), Sexsmith (1983), Buckland and Taylor (1982), and Leslie et al. (1983). A comprehensive methodology for cost-effectiveness evaluation of bridge pier protection has been developed by Greiner Engineering Sciences (1984, 1985, 1986).

The basic concept of the proposed methodology was developed during investigation of the pier protection system for the Sunshine Skyway, Florida replacement bridge and further refined for the other projects. In the proposed cost-effectiveness analysis (CEA), the costs for implementing and maintaining alternative protection systems are compared against the benefits that would be derived from having that system in place. In the above analysis, all expected costs and benefits over the lifetime of the system are expressed in present constant dollars, and those that occur in future years are "brought back" to present values using standard discounting procedures.

In the Greiner Inc. methodology of bridge protection CEA, the risk of collision is represented by both an annual frequency and by annual exposure expressed in dollars per year. The exposure value represents the annual disruption costs associated with vessel accident and includes noncatastrophic as well as catastrophic collisions. The benefits in CEA are assumed to be reduction in annual exposure which would result from having a particular protection system in place.

The annual exposure is calculated for each event involving a specific vessel impacting a specific pier or span of the bridge. The sum of all collision events would give the exposure of that particular bridge element.

Typically, the collision event exposure is derived by multiplying the cost of event (EC) by its annual frequency (AF). However, the latter approach may not always be satisfactory because an event may have a wide variety of associated costs depending on a great many external circumstances. To account for at least some of these, Greiner Inc. recommends computing the exposure by multiplying the annual frequency of an event of any severity by the expected event cost per event, generated using statistical techniques. Hence the annual exposure (in dollars) is expressed as follows:

$$EX = A_f \sum_s [EC(s)P(AS, s)] \tag{7–19}$$

where

EX = annual exposure

A_f = annual frequency of collision

EC = event cost as a function of severity (s)

P = event frequency as a function of severity (s) and average severity (AS). It is computed using a 10-step Poisson distribution function.

Knott and Bonyun (1982) suggested that the annual frequency of vessel collision (A_f) and subsequent annual frequency of bridge component collapse due to ship impact (A_{fc}) can be estimated for each event category using the following equations:

$$A_f = NP_aP_zP_gP_e \tag{7–20}$$

and

$$A_{fc} = A_fP_c \tag{7–21}$$

where

N, P_a, and P_g are the same as in Section 7.2.

P_z = probability that an aberrant ship is located in a zone in front of a particular pier or pier grouping

P_e = probability that the ship master or pilot has not taken successful evasive action to avoid the collision

P_c = probability of a bridge total collapse.

Values for P_z can be estimated using a normal probability distribution. The median value is usually established at the centerline of the navigation channel. Based on historical worldwide ship collisions, a standard deviation value (S_d) of 457.2 m may be chosen for the distribution (Knott and Bonyun, 1982).

The collision zone for each pier or group of piers is usually defined as the distance between the span centerlines on the adjacent sides of the pier once the boundaries are known; a value of P_z can be computed based on the area under the normal distribution for that particular zone.

As recommended by Knott and Bonyun the value of P_e can be modeled using the following formulation:

$$P_e = e^{-(x/S_d)} \tag{7–22}$$

where

x = the distance from the channel centerline to the centerline of the collision zone

S_d = the standard deviation value used for calculation of P_z

e = base of the natural logarithm.

Equation (7–22) models the observed action of piloting where the closer an aberrant vessel is to the channel, the less probable it is that the pilot is aware that he is aberrant; and similarly, the further away from the channel the aberrant vessel becomes, the probability increases that the pilot is aware that he is out of the channel and will take evasive action to avoid the collision.

As previously discussed, the probability of bridge collapse varies for each event. In addition, it is a function of the bridge pier (or span) strength, or ability to resist an impact force. Naturally, the less the pier (span) ability to resist impact load, the greater the probability of collapse, and vice versa. The following relationship is recommended by Knott and Bonyun to compute (P_c):

$$P_c = A^{H/P_b} \tag{7–23}$$

where

A = a constant expressing the estimated probability of pier (span) collapse when $H = P_b$. A value of $A = 0.1$ was used for the design of the New Sunshine Skyway Bridge in Florida. This value represents one out of every 10 collisions causing bridge collapse, with the remainder causing only slight damage to the bridge.

H = the ultimate impact load that can be sustained by the pier (span)

P_b = the average vessel bow collapse force.

The example of a typical event frequency (P) distribution over severity is provided in detail in Knott and Bonyun (1982).

The severity with which an event impacts an asset can be measured as the proportion of the total asset value affected, the length of time that the asset is denied to the organization, or both. For example, a large ship will cause significantly more damage than a small ship, all else being equal.

The value of severity (s) where the collision event results in catastrophic damage to the bridge is termed the critical severity (CR) and is computed by choosing a value of average severity (AS) such that the sum of the discrete Poisson distribution steps greater than or equal to (CR) is equal to the probability of pier (span) collapse (PC). Because the sum of all the Poisson steps equals unity, the value $[1-(PC)]$ represents the noncatstrophic event. The event cost (EC) as a function of severity (s) is computed as follows:

$$EC = (s)(PRC) \quad \text{for } s < CR \tag{7–24}$$

and

$$EC = (s)(PRC + SRC) + (MIC) + (PIC) + (ENC) + (H)] \quad \text{for } s \geq CR \tag{7–25}$$

where

PRC = pier replacement cost

SRC = span replacement cost

MIC = motorist inconvenience cost

PIC = port interruption cost

ENC = environmental damage cost

H = loss of human life cost.

Critical severity (CR) is defined as the step in the Poisson distribution in which a discontinuity occurs in the event cost formula. For severity less than critical (e.g., 7–24), the ship impact results in relatively minor damage and expenditure of funds for repair of the structure. For severities equal to or greater than critical (e.g., 7–25), the ship impact causes a total collapse of the bridge element and requires expenditure of funds to replace the structure, in addition to the costs associated with loss of life and commerce.

Typically *MIC* represents a significant and growing portion of the total disruption costs. The cost-effectiveness of protection alternatives can be computed using standard procedures, once the construction and maintenance costs and structure life time are estimated. The latter values represent the "cost" side of the recommended analysis while the reduction in bridge exposure represents the "benefit" side.

Typically, for each alternative, the CEA determines the net present value, benefit/cost ratio, payback period, and the internal rate of return.

REFERENCES

AASHTO, 1991. "Guide Specification and Commentary for Vessel Collision Design of Highway Bridges." American Association of State Highway and Transportation Officials, February.

ACRES INTERNATIONAL LIMITED, 1988. "Northumberland Crossing." Preliminary Design. Unpublished report.

AREA, 1986. "Manual for Railway Engineering." Chapter 6, Part 23, Pier Protection Systems at Spans over Navigable Streams.

BLAAUW, H. G. and BUZEK, F. J., 1988. "Collision of my 'Forth bank' with the Van Brienemoored Bridge—a Judicial–Technical Approach." The Dock and Harbour Authority.

BOUVET, D. M., 1963. "Preliminary Analysis of Tanker Collisions and Groundings." U.S. Coast Guard, January.

BRIUK-KJAER, O., BRODERSON, F. P., and NIELSEN, H. A., 1982. "Modeling of Ship Collision Against Protective Structures." IABSE Colloquium, Introductory Report. Copenhagen, Denmark.

BUCKBY, R. J., and SIM, T. K., 1990. "Construction of the Santubong Bridge, Sarawak." Proceedings of the Institution of Civil Engineering. Part 1. Design and Construction, August, Vol. 88. London, U.K.

BUCKLAND, P. and TAYLOR, 1982. "Annasis Island Bridge." Report No. 3 (Final). Ship Collision Risk Analysis. Prepared for the British Columbia Ministry of Transportation and Highways.

CANADIAN COAST GUARD, 1982. "Vulnerability of Bridges in Canadian Waters." Aids and Waterways Branch of the Canadian Coast Guard, January.

CHICKEN, J. C. and HARBISON, S. A., 1986. "Difference Between Industries in the Definition of Acceptable Risk." SRA, Annual Conference, Boston, Massachusetts.

CHICKEN, J. C. and HAYUS, M. R., 1986. "Assessment of the Acceptability of the Proposals for a Fixed Link Across the English Channel." SRA, Annual Conference, Boston, Massachusetts.

COWIconsult, MHAI and PBI, 1987. "Bored and Immersed Railway Tunnel Under the Eastern Channel." Operational Risk Analysis. Unpublished Report, November.

COWIconsult, 1987a. "Study of Protection of Bridge Piers Against Ship Collision and Evaluation of Collision Risk for a Bridge Across the Strait of Gibraltar." Report No. 7, prepared for the Society National d'Etudies du Detroit.

COWIconsult, 1987b. "General Principles of Risk Evaluation of Ship Collisions, Strandings, and Contact Incidents." Unpublished Technical Notes, January.

Criteria for the Design of Bridge Piers with Respect to Vessel Collision in Louisiana Waterways, 1985. Louisiana Department of Transportation. Baton Rouge, Louisiana.

DANISH ROAD DIRECTORATE, 1984. "Load Regulations for Road Bridges." First edition, October.

DENVER, H., 1983. "Geotechnical Model Test for the Design of Protective Islands." IABSE Colloquium, Copenhagen, Denmark.

EDECTECH ASSOCIATION, 1985. "Short Range Aids to Navigation Systems." Design Manual for Restricted Waterways. Final Report CG-D-18-85 Prepared for the U.S. Coast Guard Office of Research and Development, June.

ENGLOT, J. P., 1988. "Collision Protection of Arthur Kill Bridges." New York ASCE Chapter Structures Conference Proceedings, May.

ENGINEERING NEWS RECORD (ENR), 1987. Flexible Fenders Protect Piers.

FAIRLEY, W.B., 1981. "Assessment for Catastrophic Risks." Risk Analysis, Vol. 1, No. 3, Plenum Press, New York.

FAUCHART, J., 1983a. "Impact of a Ship on a Ductile Obstacle." IABSE Colloquium, Copenhagen, Denmark.

FAUCHART, J., 1983b. "Consequences of a Ship Collision with the Verdon Bridge." IABSE Colloquium, Preliminary Report, Copenhagen, Denmark.

FEDERAL HIGHWAY ADMINISTRATION, 1983. Pier Protection and Warning Systems for Bridges Subject to Ship Collision. Technical Advisory T5140.19.February 11. Washington, D.C.

FJELD, S., 1979. "Offshore Platforms – Design Against Accidental Loads. Behaviour of Offshore Structures." August. London, U.K.

FJELD, S., 1982. "Design Assumptions and Influence on Design of Offshore Structures." IABSE Colloquium, Introductory Report, Vol. 41, Copenhagen, Denmark.

FLETCHER, M. S., MAY, R. W., and PERKINS, J. A., 1983. "Pier Protection by Man-Made Islands for Orvel Bridge, UK." IABSE Colloquium, Copenhagen, Denmark.

FLINT, A. R., 1992. "Strengthening and Refurbishment of Severn Crossing." Part 2: Design. Paper 9846. Proceedings of the Institution of Civil Engineers. Structures and Buildings. February.

FRANDSEN, A. G., 1983. "Accidents Involving Bridges." IABSE Colloquium, Vol. 1, Copenhagen, Denmark.

FRENCH MINISTRY OF PUBLIC WORKS, 1971. Cahies des Prescriptions Commus Fascicule No. 61, 28 December.

FUJII, Y., 1983. "Integrated Study on Marine Traffic Accidents." IABSE Colloquium, Preliminary Report, Copenhagen, Denmark.

FUJII, Y. and SHIOBORA, R., 1978. "The Estimation of Losses Resulting from Marine Accidents." Journal of Navigation, Vol. 31, No. 1.

FUJII, Y., YAMANOUCHE, H., and MIZUKI, N., 1974. "The Probability of Stranding." Journal of Navigation, Vol. 27, No. 2.

GERMAN TRAFFIC MINISTRY, 1988. Safeguarding the Piers of Rhine Bridges Against Impact Forces from Barges. (Letter from August 8).

GREINER ENGINEERING SCIENCES, INC., 1984. "Bridge/Vessel Safety Study for the Dams Point Bridge, Jacksonville, Florida." Report Prepared for Svedrup Parcel, Inc./Jacksonville Transportation Authority, July.

GREINER ENGINEERING SCIENCES, INC., 1985. Pier Protection for the Sunshine Skyway Bridge Replacement—Ship Collision Risk Analysis. Report, Prepared for the Florida Department of Transportation, December.

GREINER ENGINEERING SCIENCES, INC., 1986. Ship Collision Risk Analysis for the Centennial Bridges, Chatham, New Brunswick. Report Prepared for the Canadian Coast Guard/Department of Transport Canada, March.

GUILLERMO, E. C. and LAWRENCE, S. A., 1987. "Design of Aida Seaport for the De Long Mountain Regional Transportation System at the Chukchi Sea, Alaska." Proceedings of the Ninth International Conference POAC, August, Fairbanks, Alaska.

HAHN, D. M. and RAMA, H. E., 1982. "Cofferdams Protecting New York Bridges from Ship Collisions." Civil Engineering, ASCE, February.

HAVNO, K. and KNOTT, M., 1986. "Risk Analysis and Protective Island Design for Ship Collisions." IABSE Symposium on Safety and Quality Assurance of Civil Engineering Structures. September, Tokyo, Japan.

HEMING, W. C., 1981. "Fendering Problems in the Third Coast Guard District." Proceedings of Specialty Conferences on Bridge and Pier Protective System and Devices. Stevens Institute of Technology, Hoboken, New Jersey, December.

INTERIM PROPOSAL FOR SHIP IMPACTS, 1988. American Association of the State Highways and Transportation Officials (AASHTO). Washington, D.C.

INTERNATIONAL STANDARD ORGANIZATION, 1987. Accidental Actions Due to Human Activities. Working Group TC 98/SC3/WG4, Working Draft, November.

IWAI, A. et al., 1983. "Safeguard System of the Bisan-Seto Bridge in Japan." IABSE Colloquium, Copenhagen, Denmark.

JENSEN, A. O. and SORENSEN, E. A., 1983. "Ship Collision and the Faro Bridges." IABSE

Colloquium, Preliminary Report, Vol. 42. Copenhagen, Denmark.

KNOTT, M. A., 1986. "Pier Protection System for the Sunshine Skyway Bridge Replacement." Proceedings of Third Annual International Bridge Conference, Pittsburgh, Pennsylvania.

KNOTT, M. A., 1987. "Ship Collision with Bridges." PIANC, Bulletin No. 57.

KNOTT, M. and BONYUN, D., 1982. "Ship Collision Against the Sunshine Skyway Bridge." IABSE Colloquium, Preliminary Report, Vol. 42. Copenhagen, Denmark.

KNOTT, M. and FLANAGAN, M., 1983. "Pier Protection for the Sunshine Skyway Bridge." IABSE Colloquium, Copenhagen, Denmark.

KUESEL, T. R., 1983. "Newport Bridge Collision." IABSE Colloquium, Copenhagen, Denmark.

KURODA, K. and KITA, H., 1983. "Probabilistic Modeling of Ship Collision with Bridge Piers." IABSE Colloquium, Copenhagen, Denmark.

KUZMANOVIĆ, B. O. and SANCHEZ, M. R., 1992. "Design of Bridge Pile Foundation for Ship Impact." ASCE Journal of Structural Engineering, Vol. 118, No. 8, August.

LARSEN, O. D., 1982. "Ship Collision Risk Assessment for Bridges." IABSE Colloquium, Introductory Report, Vol. 41.

LESLIE, J., CLARK, N., and SEGEL, J., 1983. "Ship and Bridge Collisions—The Economics of Risk." IABSE Colloquium, Preliminary Report.

MACDUFF, T., 1974. "The Probability of Vessel Collisions." Ocean Industry, Vol. 9, No. 9, September.

MATSUZAKI, Y. and JIN, H., 1983. "Design Specification of Buffer Structure." IABSE Colloquium, Copenhagen, Denmark.

MAUNSELL AND PARTNERS and BRADY, P. J. E., 1978. "Second Hobart Bridge—Risk of Ship Collision and Methods of Protection." Technical Report Prepared for Department of Main Road, Tasmania, Australia.

MCDONALD, J., 1983. "Bulk Shipping." WWS/World Ports, April/May.

MINORSKY, V. U., 1959. An Analysis of Ship Collision with Reference to Protection of Nuclear Plants. Journal of Ship Research, Vol. 3.

MODJESKI AND MASTERS CONSULTING ENGINEERS, 1985. Criteria for The Design of Bridge Piers with Respect to Vessel Collision in Louisiana Waterways. The Louisiana DOT and Development and the Federal Highway Administration, July.

MONDORF, P. E., 1983. "Floating Pier Projections Anchored by Prestressing Tendons." IABSE Colloquium, Preliminary Report.

NORDIC COMMITTEE FOR DESIGN CODES FOR BUILDINGS (NCDCB), 1978. NKB Report No. 35.

NORDIC ROAD ENGINEERING FEDERATION, 1980. Load Regulations for Road Bridges. NVF Report No. 4 (In Norwegian).

NORWEGIAN PUBLIC ROAD ADMINISTRATION, 1986. Load Regulations for Bridges and Ferry Ramps in the Public Road System. Preliminary Edition.

ODA, K. and NAGAI, S., 1976. "Protection of Maritime Structures Against Ship Collision." ASCE, Proceedings 15th Coastal Engineering Conference, Honolulu, July.

OHINISHI, T., et al., 1983. "Ultimate Strength of Bow Construction." IABSE Colloquium, Copenhagen, Denmark.

OLNHAUSEN, W. VON, 1966. "Ship Collision with Bridge Piers." Tekuish Tidskriff, No. 17. (In Swedish, Summary in English and German.)

OLNHAUSEN, W. VON, 1983. "Ship Collisions with Bridges in Sweden." IABSE Colloquium, Preliminary Report.

OSTENFELD, C., 1965. "Ship Collision Against Bridge Piers." Publication IABSE.

PETERSEN, M. J. and PEDERSON, P. T., 1981. "Collision Between Ships and Offshore Platforms." Proceedings of Offshore Technology Conference, Houston, Texas.

PRUCZ, Z. and CONWAY, W. B., 1987. "Design of Bridge Piers Against Ship Collision." Proceedings of ASCE Structural Congress, 1987. Bridge and Transmission Line Structures. Orlando, Florida, August 17–20.

RAMA, H. E., 1981. "Pier Protection of Staten Island." Proceedings of Specialty Conference on Bridge and Pier Protective Systems and Devices. Stevens Institute of Technology, Hoboken, New Jersey.

RASMUSSEN, B. H., 1982. "Design Assumptions and Influence on Design of Bridges." IABSE Colloquium, Introductory Report, Vol. 41. Copenhagen, Denmark.

ROIVAINEN, T. and TIKKANEN, E., 1983. "Effects of a Ship Collision with a Bridge." IABSE Colloquium, Preliminary Report, Copenhagen, Denmark.

ROW RESEARCH AND ENGINEERING ASSOCIATES, INC., 1985. "Top-Down Risk Analysis of the Alternative Crossings Over Storebaelt." Report to the Ministeriet for Offentlige Arbejder (In Danish), Copenhagen, Denmark.

ROWE, W. D., 1979. "What Is an Acceptable Risk and How Can It Be Determined?" Energy Risk Management, edited by G. T. Goodman and W. D. Row, Academic Press, London.

ROWE, W. D., 1982. "Acceptable Levels of Risk for Technological Undertakings." IABSE Colloquium, Introductory Report. Copenhagen, Denmark.

SAUDI-DANISH CONSULTANTS, 1978. Saudi Arabia—Bahrain Causeway, Design Memorandum No. 4. "Review of Ship Impact on Bridge Pier." Riyadh, Copenhagen, August.

SAUL, R. and SVENSSON, H., 1982. On the Theory of Ship Collision Against Bridge Piers. IABSE Proceedings, pp. 51.

SAUL, R. and SVENSSON, H., 1983. "Means of Reducing the Consequences of Ship Collision with Bridges and Offshore Structures." IABSE Colloquium, Introductory Report, Copenhagen, Denmark.

SEXSMITH, R. G., 1983. "Bridge Risk Assessment and Protective Design for Ship Collision." IABSE Colloquium, Preliminary Report.

STANLEY, 1990. "Sungei Perak Bridge, Malaysia." Proceedings Institution of Civil Engineering, London, U.K.

TSINKER, G. P., 1986. "Floating Ports." Gulf Publishing Company. Houston, Texas.

TSINKER, G. P., WOSNIAK, Z. A., and CURTIS, D. D., 1990. "New Concept for Protection Systems for Bridges Across Navigable Waterways." Proceedings PIANC, Osaka, Japan.

U.S. DEPARTMENT OF COMMERCE, MARITIME ADMINISTRATION, 1978. Merchant Fleet Forecast of Vessels in U.S. Foreign Trade. Prepared by Temple, Barker, and Sloane, Inc. Wellesley Hills, Massachusetts, May.

U.S. DEPARTMENT OF TRANSPORTATION, (DOT), 1986. National Transportation Statistics. Annual Report DOT-TSC-RSPA-86-3, July.

U.S. NATIONAL RESEARCH COUNCIL, 1983. "Ship Collisions with Bridges." The Nature of the Accidents, Their Prevention and Mitigation. National Academy Press, Washington, D.C.

U.S. NATIONAL TRANSPORTATION SAFETY BOARD, 1980. "Ramming of the Sunshine Skyway Bridge by the Liberian Bulk Carrier Summit Venture." Marine Accident Report NTSB-MAR-81-3, May. Tampa Bay, Florida.

VINCK, W. F., GILBY, E. V., and CHICKEN, J. C., 1987. "Quantified Safety Objectives in High Technology: Meaning and Demonstrations." Nuclear Safety, Vol. 28, No. 4, edited by E. G. Silver. National Laboratory, Oak Ridge, Tennessee.

WOISIN, G., 1971. "Ship–Structural Investigation for the Safety of Nuclear Power Merchant Vessels." (In German) Jahrbuch der Schiffbautechnischen Geselshaft, Vol. 65.

WOISIN, G., 1976. "The Collision Tests of the GKSS" (In German). Jahrbuch der Schiffbautechnischen Geselshaft, Vol. 70.

WOISIN, G., 1979. "Structures Against the Effects of Ship Collision" (In German). Shift und Hafen/Kommando-Brucke 31, Vol. 12.

WOISIN, G. and GERLACH, W., 1970. "Valuation of Forces due to Ship Collision on Offshore Lighthouses" (In German). German Technical Reports for the 8th International Seamark Conference. Stockholm, Sweden.

Index

CPSIA information can be obtained at www.ICGtesting.com
Printed in the USA
LVOW09s1618160716

496607LV00004B/5/P